Vinod Kumar Tiwari and Bubun Banerjee (Eds.)

Carbohydrates in Chemistry and Biology

Green Bioactive Heterocycles

Green Bioactive Heterocycles

Edited by
Bubun Banerjee

Volume 9

Carbohydrates in Chemistry and Biology

Decorated Compounds, Bioconjugates and Catalysts

Edited by
Vinod Kumar Tiwari and Bubun Banerjee

DE GRUYTER

Editors
Prof. Vinod Kumar Tiwari
Department of Chemistry
Institute of Science
Banaras Hindu University
Varanasi 221005
Uttar Pradesh
India
tiwari_chem@yahoo.co.in

Dr. Bubun Banerjee
Department of Chemistry
Akal University
Talwandi Sabo, Bathinda 151302
Punjab
India
banerjeebubun@gmail.com

ISBN 978-3-11-144702-5
e-ISBN (PDF) 978-3-11-144709-4
e-ISBN (EPUB) 978-3-11-144714-8
ISSN 2752-1338

Library of Congress Control Number: 2024952704

Bibliographic information published by the Deutsche Nationalbibliothek
The Deutsche Nationalbibliothek lists this publication in the Deutsche Nationalbibliografie; detailed bibliographic data are available on the internet at http://dnb.dnb.de.

Back Image: IkonStudo/iStock/Getty Images Plus
Front Image: Bacsica/iStock/Getty Images Plus
Typesetting: Integra Software Services Pvt. Ltd.
Printing and binding: CPI books GmbH, Leck

www.degruyter.com
Questions about General Product Safety Regulation:
productsafety@degruyterbrill.com

DEDICATED

in memory of

Sh. Lalita Tiwari

(16.04.1958 – 26.12.2024)

A person known for his simplicity, honesty, supportive, loving, caring, and cheerful nature

Loving father of Prof. Vinod K. Tiwari

Preface

The heterocyclic series "Green Bioactive Heterocycles" consists of 9 volumes till date. Each volume of this series deals with different aspects of bioactive heterocycles. This 9th volume, entitled *'Carbohydrates in Chemistry and Biology: Decorated Compounds, Bioconjugation and Catalysis'*, of the above series focuses on various aspects of carbohydrate chemistry and consist of twenty chapters. All the chapters have exhaustively covered the topics and the corresponding literature.

Since long back, the role of carbohydrates in almost all aspects of chemistry and biology is considered an exceedingly promising area of interest which has been further increasing exponentially in recent years. This is a most abundant natural product and is considered an essential biomolecules that have enormous applications in chemistry, biology, medicine, food, agriculture, and functional materials. In addition to their well-known essential role as energy storage, they have notable significance in various pivotal biological events, including modulation of protein function, transport, intercellular adhesion, malignant transformation, signal transduction, viral and bacterial cell surface recognition and many more, which eventually formulate this fascinating scaffold for easy access of diverse range of drug candidates. The other enthralling features of carbohydrates include its readily availability, chiral, and functional nature that all together increases its adoptability in catalysis in chiral synthesis and also functional materials of wider applicability. Toward this end, we once again extend the importance of carbohydrates in drug development and also their synthetic perspectives to a further updated and new perspectives including material science and we are very sure that this book may successfully interconnect their diverse role to the wide range of the readers. This book covers twenty chapters that illustrate an opportunity of carbohydrates in various emerging fields, which is the major objective of present multidisciplinary project. We would like to extend our sincere thanks to all the contributors who are active experts known in the area of glycoscience for their valuable contributions to this project titled *'Carbohydrates in Chemistry and Biology: Decorated Compounds, Bioconjugation and Catalysis'*.

An introduction to carbohydrate-designed drug candidates is covered by Prof. P. Chawla and co-workers in Chapter 1, while Pathak *et al.* in Chapter 2 presented the notable role of carbohydrates in drug delivery. Mishra et al. in Chapter 3 discussed the impact of carbohydrate-functionalized nanoparticles with special consideration of their notable applications from the laboratory to clinics in comprehensive fashion. Dr. S. Majhi in Chapter 4 presented the total synthesis of some selected carbohydrate-decorated bioactive molecules. Prof. D.K. Maiti and his group in Chapter 5 considered interesting update on synthesis of potent carbohydrate-decorated *O*-heterocycles. Further, the scope of green solvents in the synthesis of diverse range of carbohydrate scaffolds is nicely presented by Prof. B.C. Ranu and his group in Chapter 6. Some interesting features about common coupling reactions explored in glycoscience have been presented by Prof. R. Das and his group in Chapter 7. Prof. B. Basu and co-researchers

https://doi.org/10.1515/9783111447094-202

in Chapter 8 exemplified the regioselective triazole forming 'Click Chemistry' in the development of biologically relevant 1,2,3-triazole-linked glycohybrids and glycoconjugates. Prof. P. Ghosh in Chapter 9 covers the notable applications of carbohydrate containing isothiocyanate scaffold for easy access of structurally diverse heterocycles. Some interesting common synthetic routes leading to biologically relevant carbohydrate analogues by utilizing Ultrasonic as well microwave (*MW*)-assisted condition is well presented in Chapter 10 and Chapter 11, respectively by Prof. S.C. Ameta and coworkers. Furthermore, versatile use of carbohydrates as renewable feed stocks for easy access of bioactive molecules is covered by Dr. K. Pathak and group in Chapter 12. Followed by, a detail overview on application of carbohydrate-based gelators as well ionic liquids for biomass utilization is covered by groups of Dr. A. Abbas and Prof. H.C. Chopra, respectively in Chapter 13 and 14. Prof. A. Ramazani and co-researcher in chapter 15 presented a comprehensive overview on carbohydrate-supported nanostructured biomaterials and their notable application in organic transformations. An interesting update on bioconjugation in carbohydrates is discussed by Dr. T. Farooq and co-researchers in Chapter 16, whereas Prof. P. Ghosh and his co-researchers in Chapter 17 nicely covered the application of carbohydrate polymers in Chapter 17. Dr. S. Gupta in chapter 18 exemplified the use of carbohydrates in the preparation of biofuels. In Chapter 19, Dr. H. Bukhari and his group cover the representative industrial application of biocompatible carbohydrates. At the end, Prof. A. Ramzani and group in Chapter 20 summarized the recent advances in carbohydrates-based metal-organic frameworks.

The topics included in this book included the basic and advanced aspect of carbohydrates. All the chapters are broad and aimed to present advance information about the carbohydrates with their ever increasing application in science and technology. We once again sincerely thank all the esteem contributors and also our special appreciation and gratitude goes to all the learned reviewers who provide very constructive comments and suggestions which greatly help our book interesting to the readers.

We are thankful to Ms. Hélène Chavaroche (Senior Acquisitions Editor) and the entire editorial team in particular, Ms. Marie Hammerschmidt for their kind support in publishing this volume entitled "*Carbohydrates in Chemistry and Biology: Decorated Compounds, Bioconjugation and Catalysis*" with De Gruyter GmbH, Berlin/Boston.

Prof. Vinod K. Tiwari, Banaras Hindu University
&
Dr. Bubun Banerjee, Akal University

Foreword

Carbohydrates are important biomolecules of the living world. By virtue of their diverse and complex structures, they are involved in specific cell-cell interactions and thereby participate in a variety of vital life processes. Over past twenty years carbohydrates have acquired immense importance as more and more information about their biological roles is revealed. As a result,the scientific fraternity is exploring various facets of sugars in a variety of applications in human life. For example, carbohydrate conjugation improves the solubility and specificity of the drug molecules. Likewise, carbohydrate based nanoparticles, glycoconjugates and gels offer important medical applications. In this premise, the book edited by Prof. Vinod Kumar Tiwari and Dr. Bubun Banerjee titled *"Carbohydrates in Chemistry and Biology: Decorated Compounds, Bioconjugates and Catalysts"* covers a vast canvas of carbohydrate related studies indicating that future belongs to carbohydrates. The book provides detailed information about the green synthesis, biological importance and catalytic applications of carbohydrate derivatives. This comprehensive book contains twenty chapters and I am confident that it will prove to be a good reference for researchers aspiring to venture into the fascinating field of glycans.

Chapter 1 of the book gives a brief introduction to carbohydrate-designed drugs which are revolutionizing the field of therapeutics by improving drug targeting, pharmacokinetics, and reducing side effects. This chapter delineates their applications in anticancer, antiviral, antibiotic, and anti-inflammatory treatments, highlighting their potential for precise drug delivery. Chapter 2 focuses on the structure and physicochemical properties of different carbohydrates in pharmaceutical applications, with particular attention to their biocompatibility and biodegradability in drug delivery systems. The examples include hyaluronic acid, alginate, chitosan, dextran, cellulose, gellan, cyclodextrin, pullulan, and starch. The chapter also talks about how the modifications to their functional groups can improve drug loading and facilitate targeted release. Due to their low cost, ease of modification, and strong clinical potential, carbohydrate-based polymers present a great promise for advancing drug delivery systems in future research.

Chapter 3 covers the importance of carbohydrate based nanoparticles (CBNPs) with respect to their potential for enhancing therapeutic and diagnostic applications, owing to their unique features such as better targeting, improved bioavailability, and fewer side effects. Despite being nontoxic, biodegradable, and water-soluble, challenges in large-scale production and regulatory approvals still exist. Continued research into their synthesis and functionalization is expected to reveal new and innovative uses in medicine and biotechnology. Chapter 4 narrates how bioactive carbohydrates play crucial roles in drug discovery and development. This chapter focuses on the total synthesis of carbohydrate-decorated bioactive compounds, showcasing how these chiral molecules, originated from plants, animals, and microorganisms, can be used to create novel drugs with therapeutic potential. Chapter 5 entails chemistry of carbohydrate-decorated

https://doi.org/10.1515/9783111447094-203

O-heterocycles which show potential in drug development due to their structural diversity and enhanced biological activity. Recent advances in radical chemistry and copper-catalyzed glycosylation as well as use of deep eutectic solvents (DES) have enabled efficient synthesis of these complex molecules. Chapter 6 discusses the importance of sustainable practices in carbohydrate synthesis, highlighting the use of green solvents such as water, ionic liquids, and DES which offer a more environmentally friendly alternative to traditional organic solvents. These solvents provide advantages including stereoselectivity, high yield, and reduced reaction time.

Chapter 7 describes the role of coupling reactions in carbohydrate chemistry for enabling the formation of glycosidic bonds and the synthesis of complex carbohydrate structures. Various coupling reactions including enzymatic glycosylation, click chemistry, and transition metal catalyst leading to stereoselective installation of C-O and C-C bonds are covered. Chapter 8 focuses on triazole-based carbohydrate molecules, synthesized via click chemistry. The Cu-catalyzed azide-alkyne cycloaddition (CuAAC) click reaction has emerged as a powerful tool for creating novel therapeutic and diagnostic entities. This method has opened up new avenues for the creation of glycoconjugates with potential medicinal applications. Chapter 9 is devoted to synthesis of isothiocyanates (ITCs) which are carbohydrate based heterocycles endowed with biological properties. Chapter 10 and 11 cover ultrasound-assisted and microwave-assisted synthesis of carbohydrate derivatives, respectively.

Chapter 12 discusses how natural carbohydrate feedstocks can offer a sustainable solution for industrial preparation of bioactive compounds. Chapter 13 talks about the applications of carbohydrate-based gel forming substances (gelators) as antibacterial agents, in drug delivery, as anticancer agents, in tissue engineering and in the development of soft-optical devices, nanomaterials, and structuring agents. Chapter 14 explores the inherent potential of manufacturing ionic liquids (ILs) from abundant, renewable carbohydrates, offering a sustainable alternative to traditional chemical production methods. It highlights the unique characteristics of carbohydrate-derived ILs, such as their hydrogen-bond-rich structure, biodegradability, and low toxicity, which make them ideal candidates for applications in catalysis, biomedicine, and environmental technologies. Chapter 15 highlights the advances made in the field of carbohydrate-supported nanostructured materials used in organic transformations. Chapter 16 discusses the bioconjugation of carbohydrates including the commonly adopted strategies and the bioconjugates of carbohydrates with biomolecules such as proteins and nucleic acids. Chapter 17 describes carbohydrate polymers and their diverse applications. It covers different types of carbohydrate-based polymers, including starch, cellulose, chitin, and pectin, highlighting their structural properties and functions. The chapter also explores their industrial applications in food, pharmaceuticals, biodegradable materials, and biomedical fields, emphasizing their role in sustainable and eco-friendly solutions. Chapter 18 explores the preparation of biofuels from carbohydrates, providing an overview of their production and classification. It highlights different types of biofuels, including furan-based biofuels derived from carbohydrate biomass. The chapter discusses key conversion processes and their significance in sustainable energy solutions. Chapter 19 details the industrial applications of carbohydrates across various sectors, including food, pharmaceuticals, agriculture, textiles, emulsions, and the paper industry. It highlights the functional roles of carbohydrates in these industries, such as thickening agents, stabilizers, bioactive compounds, and biodegradable materials, emphasizing their importance in sustainable and innovative applications. The last chapter, i.e., Chapter 20 describes advances in carbohydrates-based metal- organic frameworks and highlights their limitations and potentials.

Overall, the book encompasses diverse areas of carbohydrate research and gives an overall perspective of the field to the readers.

Prof. Suvarn S. Kulkarni
Department of Chemistry,
Indian Institute of Technology Bombay,
Powai, Mumbai-400076, India
suvarn@chem.iitb.ac.in

A brief professional profile of Prof. Suvarn S. Kulkarni

Suvarn Kulkarni received his Ph.D. in Organic Chemistry from University of Pune in 2001. After his Ph. D., he pursued his post-doctoral research at Academia Sinica, Taipei, and University of California, Davis. He returned to India in late 2008 and held a faculty position at IACS Kolkata prior to joining the Indian Institute of Technology Bombay in 2009. He is presently working as Biswas-Palepu Distinguished Chair Professor and Dean of Academic Programs at IIT Bombay. His current research interests include devising newer ways for efficient chemical synthesis of rare sugar containing complex glycoconjugates specifically expressed on the surfaces of pathogenic bacteria for the development of antibiotics, therapeutics and vaccines.

Suvarn was a short term Visiting Professor (MOST) at Taiwan in 2014. He is a Fellow of the Royal Society of Chemistry London, Fellow of National Academy of Sciences, India (NASI) and Fellow of Maharashtra Academy of Science. He is a recipient of the CRSI Bronze medal for the year 2017 and Dr. H. C. Srivastava Young Scientist award 2016 from ACCTI. He is a member of the editorial board of Organic Letters, Carbohydrate Research, Chemistry Select and has served as the guest editor of international journals such as EJOC, OBC and Frontiers in Chemistry. Recently he received IRCC Research Dissemination Award as well as a Research Publication Award from IIT Bombay. Most importantly, he was selected by the students for the Excellence in Teaching Award at IIT Bombay in 2017 and also for the prestigious Prof. S. P. Sukhatme Award for Excellence in Teaching in 2020.

https://doi.org/10.1515/9783111447094-204

Contents

Arshdeep Singh, Rabin Debnath, Viney Chawla, and Pooja A. Chawla*

Chapter 1 Introduction to carbohydrate-designed drug molecules

1.1 Introduction

Carbohydrate-drug conjugates represent a rapidly growing field in pharmaceutical development, combining the therapeutic potential of drugs with the biological recognition and interaction capabilities of carbohydrate molecules. In these conjugates, a carbohydrate moiety is covalently attached to a therapeutic agent to leverage the unique biological properties of carbohydrates such as their role in cell signaling, molecular recognition, and immune response modulation [1, 2]. This fusion enhances the drug's ability to selectively target specific cells, particularly in diseases where glycan-based targeting plays a significant role such as cancer and infectious diseases [3].

Carbohydrates are integral in cell surface interactions, where they often function as ligands for receptors on cells such as lectins. This makes them attractive tools for targeting, as many pathogens and cancer cells overexpress certain carbohydrate receptors [4]. By incorporating carbohydrates into drug structures, pharmaceutical scientists can exploit these natural recognition processes to increase the drug's selectivity, thereby potentially enhancing efficacy while minimizing side effects [5].

1.1.1 Role of carbohydrates in drug targeting and delivery

Carbohydrates play an essential role in improving drug targeting and delivery by serving as ligands for specific receptors on the surfaces of target cells. In many cases, these receptors are overexpressed on diseased cells, allowing carbohydrate-designed drugs to home in on these targets [6]. For instance, several cancers display altered glycan expression, providing an opportunity for selectively targeting cancer cells with carbohydrate-drug conjugates [2]. One of the most explored mechanisms for drug targeting using carbohydrates involves lectin receptors, which recognize specific glycan

***Corresponding author: Pooja A. Chawla**, University Institute of Pharmaceutical Sciences and Research, Baba Farid University of Health Sciences, Faridkot 151203, Punjab, India, e-mail: pvchawla@gmail.com
Arshdeep Singh, Dasmesh College of Pharmacy, Faridkot 151203, Punjab, India
Rabin Debnath, NEPEDS College of Pharmaceutical Sciences, Gandhinagar, Tetelia, Kamrup (M) 782403, Assam
Viney Chawla, University Institute of Pharmaceutical Sciences and Research, Baba Farid University of Health Sciences, Faridkot 151203, Punjab, India

https://doi.org/10.1515/9783111447094-001

structures. For example, in liver-targeted therapies, carbohydrates such as mannose or galactose are often used because they are recognized by the asialoglycoprotein receptor (ASGPR) on hepatocytes [7]. Similarly, sialylated glycans have been shown to target specific immune cells, enhancing the delivery of drugs to these cells in inflammatory and autoimmune diseases [8].

In addition to targeting, carbohydrates can also improve the pharmacokinetics of drugs. Conjugation with specific sugars can enhance the solubility, stability, and circulation time of drugs, which can reduce the frequency of dosing required and improve patient compliance [9]. Moreover, certain carbohydrate-drug conjugates are designed to exploit the "glycocalyx barrier," the dense layer of carbohydrates on cell surfaces that can otherwise limit drug access to cells, enhancing cell permeability and drug bioavailability [10].

1.1.2 Advantages of carbohydrate decoration in pharmaceuticals

The decoration of drugs with carbohydrates provides numerous advantages beyond targeting, particularly in improving the drug's pharmacokinetics, bioavailability, and stability. Carbohydrates are generally biocompatible and biodegradable, reducing the risk of immunogenic reactions and toxicity, which are common concerns with other targeting moieties such as synthetic polymers or antibodies [11]. Additionally, carbohydrate modifications can increase drug solubility, allowing hydrophobic drugs to be administered more easily and absorbed efficiently in the body [12].

A significant advantage of carbohydrate-drug conjugates lies in their potential to enhance drug stability. Carbohydrate decoration can shield drugs from enzymatic degradation, leading to prolonged half-life and greater therapeutic efficacy [13]. This stability is particularly beneficial for drugs with narrow therapeutic windows or those that degrade quickly in the bloodstream. Carbohydrates also play a protective role in targeted delivery to specific tissues, such as the liver, through receptor-mediated uptake. This specificity minimizes off-target effects and enhances the drug's therapeutic impact on targeted tissues. Finally, carbohydrate decoration supports the development of prodrugs, inactive drugs that become activated within the body. By using glycosylation, researchers can control the activation of prodrugs in specific tissues where enzymes such as glycosidases are present, enabling precise control over drug action. This approach has been applied to develop safer and more effective drugs, particularly in fields such as oncology and infectious disease.

1.2 Chemistry of carbohydrate conjugation

1.2.1 Types of carbohydrate moieties used in drug decoration

Carbohydrate conjugation in drug development leverages various types of carbohydrate moieties to enhance the efficacy, targeting, and delivery of pharmaceuticals. Commonly used carbohydrate types include monosaccharides (simple sugars), disaccharides, and polysaccharides. Each type has distinct properties and is selected based on its interaction with specific biological receptors and the desired therapeutic outcomes. Monosaccharides, such as glucose, mannose, and galactose, are frequently used in drug conjugation due to their small size and ability to target cellular receptors. For example, mannose-containing conjugates can target mannose receptors on macrophages, making them ideal for delivering drugs to immune cells or to liver cells that express these receptors [14]. Similarly, glucose-based drug conjugates can be utilized in glucose transporter-targeted therapies, exploiting the increased glucose uptake of certain cancer cells, thus enhancing the selectivity of anti-cancer agents [15]. Disaccharides, such as lactose and maltose, offer additional options for targeting and improving solubility. Disaccharides can facilitate cell-specific drug delivery due to their interactions with glycan-recognition receptors. For example, lactose can be recognized by galectin receptors, which are overexpressed in certain cancers and immune cells [16]. Lactose conjugation not only aids in targeting specific cells but also enhances the aqueous solubility of hydrophobic drugs, improving their bioavailability. Polysaccharides represent a larger class of carbohydrates including complex sugars like hyaluronic acid (HA) and heparin. These are commonly used in drug decoration for their biocompatibility and ability to target tissues rich in specific receptors. HA, for instance, binds CD44 receptors found in abundance on certain cancer cells, allowing targeted drug delivery in cancer therapies [17]. Similarly, heparin-based conjugates can target inflammation sites and have been used in designing anti-inflammatory and anticoagulant drugs [18]. Specialized carbohydrates, such as sialic acid derivatives, are also incorporated into drug design for their role in immune cell recognition. Sialic acid-conjugated drugs can target cells in the immune system, where sialic acid-binding immunoglobulin-type lectins (Siglecs) are present, and are beneficial in modulating immune responses [19]. The choice of carbohydrate moiety is crucial as it influences not only the pharmacokinetic properties of the drug but also the biodistribution and specificity of the drug, making it a powerful tool in developing targeted therapies [20].

1.2.2 Methods of carbohydrate conjugation to drug molecules

The conjugation of carbohydrates to drugs is a specialized process that requires selecting a suitable method to maintain the drug's integrity and bioactivity. Each method has distinct advantages based on the chemical nature of the carbohydrate and the drug as well as the intended use of the conjugate. Reductive amination is one of the most common methods, where an amine-containing drug is reacted with a carbohydrate under reducing conditions, forming a stable C–N bond. This method is straightforward and efficient, especially when working with monosaccharides like glucose or galactose. It enables direct linkage between the carbohydrate and the drug molecule without extensive modifications, preserving the drug's biological activity [21]. Reductive amination is especially useful for targeting hydrophilic drugs and enhancing their pharmacokinetics by increasing their solubility [22]. Click chemistry, specifically copper(I)-catalyzed azide-alkyne cycloaddition, has gained popularity due to its specificity and compatibility with a wide range of drugs and carbohydrates. Click chemistry allows the selective attachment of carbohydrates to drugs under mild conditions, which helps retain the functional integrity of both components. The method is versatile, enabling precise conjugation with minimal side reactions, which is essential for developing drugs with complex carbohydrate structures [23]. This method has been particularly useful in attaching carbohydrate moieties to cancer drugs, allowing for targeted drug delivery to cancer cells that overexpress specific receptors [24]. Enzymatic glycosylation is another sophisticated technique, using glycosyltransferase enzymes to transfer specific carbohydrate groups onto drug molecules. This approach is advantageous because it mirrors natural biosynthetic pathways, allowing for highly controlled and selective modification of the drug. Enzymatic glycosylation can be used to add complex oligosaccharides and polysaccharides onto proteins and peptides, increasing the drug's stability and potentially extending its half-life in the body [25]. This method is particularly useful for developing glycosylated protein therapeutics, as it provides precise control over the structure of the carbohydrate, which is crucial for maintaining bioactivity and reducing immunogenicity [26]. Acylation and carbodiimide coupling are other methods frequently employed for conjugating larger carbohydrate structures, such as polysaccharides, to drugs. These methods are especially useful when working with high-molecular-weight polysaccharides like HA or chitosan, which are often conjugated to improve drug delivery and retention in specific tissues. These conjugates can undergo receptor-mediated endocytosis, enhancing the delivery of drugs to target cells and reducing off-target effects [27].

1.2.3 Structural and functional impacts of carbohydrate modification

The conjugation of carbohydrates to drug molecules has significant structural and functional impacts, influencing solubility, stability, pharmacokinetics, and biodistribution [28]. Carbohydrate decoration typically increases the hydrophilicity of the drug, enhancing its solubility in aqueous environments. This is particularly beneficial for hydrophobic drugs, as increased solubility can improve bioavailability and absorption, especially in oral and intravenous formulations [29]. For instance, conjugating hydrophobic anticancer drugs with glucose has been shown to increase solubility, allowing for higher dosing and improved efficacy [30]. Carbohydrate modification can also improve the stability of drugs, protecting them from enzymatic degradation. Glycosylation has been observed to reduce proteolysis and prolong the circulation time of drugs, as seen in glycosylated protein therapeutics, which have increased stability in the bloodstream due to reduced susceptibility to proteases [31]. This property is particularly beneficial for biologics, such as monoclonal antibodies and peptide drugs, which are otherwise prone to rapid degradation in vivo.

Targeted drug delivery is another major advantage of carbohydrate conjugation. Carbohydrates can guide drugs to specific cells or tissues by exploiting receptor-mediated endocytosis. For example, HA-designed drugs can specifically target cancer cells that overexpress CD44 receptors, resulting in higher drug accumulation in tumor tissues and reduced systemic toxicity [32]. Similarly, mannose decoration can be used for targeting macrophages and dendritic cells, making it useful in developing drugs for infectious diseases and immunotherapies [33]. Furthermore, carbohydrate conjugation can modulate immune recognition of drugs. Sialic acid decoration, for example, can mimic the glycocalyx structure on cell surfaces, allowing therapeutic proteins to evade the immune system and thus reducing immunogenicity. This modification is particularly useful in chronic therapies, where repeated administration of the drug is required. Sialylated glycoproteins have shown prolonged circulation times as they are less likely to be cleared by the immune system [34]. The structural and functional impacts of carbohydrate decoration are integral to designing drugs that are more effective, targeted, and have fewer side effects. By selecting the appropriate carbohydrate and conjugation method, scientists can enhance the therapeutic potential of drugs across a wide range of diseases [35]. Targeted drug design through carbohydrate conjugation (types, methods, and functional impacts) is shown in Figure 1.1.

1.3 Mechanisms of action for carbohydrate-designed drugs

1.3.1 Enhanced cellular uptake via carbohydrate receptors

Carbohydrate-designed drugs are designed to exploit cellular uptake mechanisms through interactions with specific carbohydrate receptors on cell surfaces. This targeted approach significantly improves drug delivery efficiency, especially for drugs that would otherwise have poor cellular uptake. By conjugating drugs with specific carbohydrate moieties, researchers can target receptors such as lectins and C-type lectin receptors, which recognize and bind to carbohydrate structures on cell surfaces [34]. For instance, mannose receptors on macrophages and dendritic cells are fre-

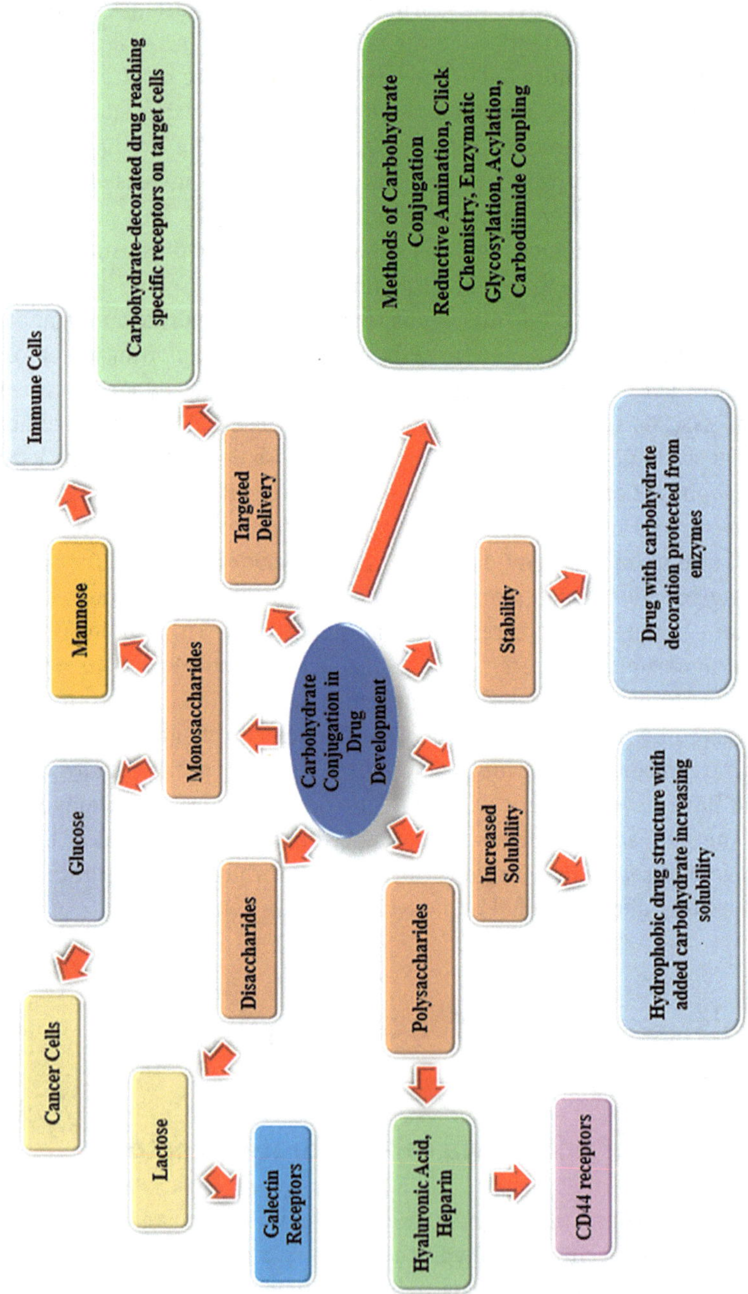

Figure 1.1: Targeted drug design through carbohydrate conjugation: types, methods, and functional impacts.

quently targeted using mannose-designed drugs. When a mannose-conjugated drug encounters these immune cells, the mannose receptors facilitate endocytosis, bringing the drug into the cell. This mechanism has been particularly beneficial in drug delivery for infectious diseases and immunotherapy, where targeting immune cells is crucial [35]. Another example involves galectin receptors, which recognize galactose-containing ligands. Drugs designed with galactose or lactosyl groups can selectively bind to cells overexpressing galectin receptors, such as certain types of cancer cells, leading to enhanced drug internalization [36]. One widely studied mechanism is the uptake of glucose-designed drugs via GLUT transporters. Cancer cells, known for their high metabolic rates and increased glucose consumption, often overexpress glucose transporters like GLUT1. By decorating drugs with glucose, scientists can leverage the upregulated GLUT transporters to facilitate drug entry into cancer cells. This approach not only improves drug uptake but also increases the selectivity for cancer cells, potentially reducing side effects on normal tissues [37]. Studies have demonstrated the effectiveness of glucose-modified anticancer agents in enhancing cytotoxicity against cancer cells compared to nonmodified versions [38]. Overall, using carbohydrate moieties to target-specific cellular receptors enhances the efficiency of drug delivery, increases drug concentration at the target site, and reduces the off-target effects, thereby improving therapeutic outcomes and minimizing adverse effects [39].

1.3.2 Role of carbohydrate-protein interactions

Carbohydrate-protein interactions play a pivotal role in the bioactivity and specificity of carbohydrate-designed drugs. These interactions are highly selective, as many proteins have evolved to recognize specific carbohydrate structures. Lectins, a broad class of carbohydrate-binding proteins, are central to this process. Lectins recognize distinct carbohydrate motifs, allowing for selective binding and subsequent cellular responses. This characteristic is exploited in carbohydrate-designed drug design to ensure precise targeting and controlled drug release [40, 41]. Carbohydrate-protein interactions are also critical in cancer treatment. Tumor cells often display abnormal glycosylation patterns, which can be recognized by galectins and siglecs. By incorporating specific carbohydrate ligands that interact with these proteins, carbohydrate-designed drugs can achieve higher specificity for tumor cells. For instance, galectin-3, which binds β-galactoside sugars, is overexpressed in many cancers. Drugs designed with galactose or lactosyl groups can therefore preferentially bind to galectin-3, allowing for targeted delivery and enhanced therapeutic effect within tumor tissues [42]. Furthermore, carbohydrate-designed drugs can utilize the multivalency effect, where multiple carbohydrate ligands on a drug molecule engage several binding sites on the target protein. This effect increases the binding affinity and selectivity of the drug, as the multiple simultaneous interactions enhance the stability of the drug-protein complex. Such multivalent interactions are valuable in drug design, particularly in developing vaccines and immune therapies, where high-affinity binding is essential for triggering immune responses [43]. The selectivity provided by carbohydrate-protein interactions is fundamental in reducing drug toxicity and optimizing the drug's therapeutic profile [16].

1.3.3 Targeting specific tissues and cells using carbohydrate ligands

One of the key advantages of carbohydrate decoration is the ability to guide drugs to specific tissues and cells, enhancing therapeutic precision and reducing side effects. Carbohydrate ligands can selectively bind to receptors or proteins that are overexpressed in certain tissues, ensuring that the drug is delivered where it is needed most. For instance, HA, a polysaccharide that interacts with CD44 receptors, is widely used in drug delivery systems targeting cancer cells. Many cancer types, including breast, lung, and ovarian cancers, overexpress CD44, a cell surface receptor involved in cell proliferation and migration. Drugs conjugated with HA can selectively accumulate in CD44-overexpressing tumors, providing higher drug concentrations at the target site and minimizing systemic toxicity [44]. HA-conjugated drug delivery systems have shown promise in preclinical models, effectively targeting tumors while sparing normal tissues [45]. Fucose-modified drugs offer another example, as fucose-binding proteins are involved in liver targeting. The ASGPR in hepatocytes binds to fucose and galactose, allowing fucose-designed drugs to be efficiently taken up by the liver. This targeting mechanism has been particularly useful in developing therapies for liver diseases such as hepatic cancer and hepatitis. By using fucose ligands, drugs can achieve selective liver uptake, which reduces exposure to other organs and allows for higher concentrations within the liver [46]. In the immune system, mannose-modified drugs target dendritic cells and macrophages by binding to mannose receptors, which are abundant on these immune cells. This strategy is widely used in vaccine design and immunotherapy. For example, mannose-designed vaccines can specifically target antigen-presenting cells, enhancing the immune response by ensuring efficient antigen uptake and presentation. This mechanism allows for lower doses of the vaccine to achieve an effective immune response, making it a powerful tool in developing vaccines for infectious diseases and cancer [47].

Carbohydrate ligands can also enable passive targeting by enhancing the solubility and circulation time of drugs. Carbohydrates, such as PEGylated sugars, can extend a drug's half-life by protecting it from rapid renal clearance and enzymatic degradation. Prolonged circulation times allow the drug to accumulate in target tissues, particularly in the case of tumors with leaky vasculature that permit enhanced permeability and retention of larger molecules. This passive targeting further improves the therapeutic window of carbohydrate-designed drugs by reducing the frequency of dosing and minimizing systemic exposure [48]. The strategic use of carbohydrate ligands for tissue and cell-specific targeting enables the development of more effective and safer drug therapies. By honing in on-specific tissues or cells, carbohydrate-designed drugs maximize their therapeutic impact while reducing unwanted side effects, representing a significant advancement in drug delivery technology [49].

1.4 Examples of commercially available carbohydrate-designed drugs

Carbohydrate-designed drugs have gained significant attention in recent years for their ability to enhance drug efficacy, selectivity, and pharmacokinetics. By leveraging specific interactions with cellular receptors or improving drug solubility and stability, carbohydrate-conjugated drugs have shown promising results in various therapeutic areas including cancer, infectious diseases, and inflammation. This chapter discusses notable examples of commercially available carbohydrate-designed drugs across several categories, including anticancer drugs, antibiotics, antivirals, and anti-inflammatory agents, along with their mechanisms of action and clinical benefits.

1.4.1 Anticancer drugs

Carbohydrate decoration on anticancer drugs has emerged as an effective strategy to improve selectivity and reduce systemic toxicity. The addition of carbohydrates enables these drugs to exploit specific cellular receptors and uptake mechanisms, enhancing their efficacy against cancer cells.

1.4.1.1 Examples and mechanisms of action

One prominent example of a carbohydrate-designed anticancer drug is doxorubicin, which has been conjugated with *N*-acetylglucosamine (GlcNAc) to target specific receptors on cancer cells. This modification enhances doxorubicin's selectivity for tumor cells that overexpress GlcNAc receptors, leading to higher intracellular drug concentrations within the tumor environment. This targeted delivery reduces off-target toxicity and improves the drug's therapeutic window [50]. Another notable example is paclitaxel conjugated with HA. HA binds to the CD44 receptor, which is commonly overexpressed in various tumors such as breast, ovarian, and lung cancers. By targeting CD44, HA-conjugated paclitaxel enhances drug accumulation in the tumor site, increasing its antitumor activity. Studies have shown that HA-paclitaxel conjugates reduce side effects compared to nonconjugated paclitaxel while maintaining or even enhancing anticancer efficacy [51].

1.4.1.2 Pharmacokinetics and efficacy improvements

Carbohydrate conjugation also improves the pharmacokinetics of anticancer drugs. For example, glycoconjugate drugs often exhibit increased water solubility, which helps in achieving higher drug concentrations in systemic circulation without precipitating in the bloodstream. HA-conjugated paclitaxel, as mentioned, not only enhances tumor targeting but also prolongs drug circulation time, allowing for sustained drug release and higher efficacy [52]. Similarly, glucose-designed cisplatin analogs utilize glucose transporters to enhance drug uptake in glucose-hungry cancer cells, resulting in improved pharmacokinetics and reduced systemic toxicity [53].

1.4.2 Antibiotics

Glycoconjugate antibiotics are designed to enhance the selective binding of antibiotics to bacterial cells, reduce side effects, and mitigate resistance. Conjugating antibiotics with carbohydrates helps in targeting specific bacterial receptors, increasing efficacy, and limiting adverse effects on human cells.

1.4.2.1 Glycoconjugate antibiotics and their clinical applications

Vancomycin derivatives modified with carbohydrates have been developed to improve their binding affinity and potency against resistant bacterial strains. For example, glycosylated forms of vancomycin have demonstrated enhanced efficacy against methicillin-resistant *Staphylococcus aureus* (MRSA) by increasing vancomycin's interaction with bacterial cell walls. This modification has led to the development of new antibiotics capable of overcoming resistance mechanisms that traditional vancomycin cannot [54]. Streptomycin is another example; it has been modified with mannose for targeting *Mycobacterium tuberculosis*. Mannose-modified streptomycin enhances uptake by macrophages, which are the primary host cells for tuberculosis bacteria. This targeted approach allows for higher intracellular drug concentrations in infected cells, improving efficacy against tuberculosis while reducing the required dosage and minimizing side effects [55].

1.4.2.2 Mechanisms of action and resistance mitigation

The primary mechanism of glycoconjugate antibiotics is to enhance binding to bacterial surface receptors. Carbohydrate decoration allows these drugs to interact more effectively with bacterial cell walls or specific receptors, leading to increased bacterial cell permeability and enhanced antibacterial activity [56]. By increasing the selective targeting of bacterial cells, glycoconjugates also reduce the likelihood of resistance development, as the modified antibiotics can evade some of the bacterial resistance mechanisms that are effective against unmodified versions [57].

1.4.3 Antiviral drugs

Carbohydrate-designed antiviral drugs have been developed to exploit specific viral entry mechanisms, enhance the targeting of virus-infected cells, and improve therapeutic efficacy.

1.4.3.1 Carbohydrate-designed antivirals and virus targeting

One notable example of a carbohydrate-designed antiviral drug is acyclovir conjugated with mannose. Mannose receptors on immune cells help direct acyclovir to infected cells, allowing for a more targeted antiviral response. This approach has shown promise in reducing the viral load in herpes infections while decreasing the dosage required for effective treatment, thus minimizing potential side effects [58]. Another example is sialic acid-modified oseltamivir, which targets influenza virus strains that bind to sialic acid residues on host cells. By enhancing this interaction, sialic acid-designed oseltamivir can improve the drug's ability to prevent viral entry into host cells, thereby reducing the severity and duration of influenza infections [59].

1.4.3.2 Case studies and clinical outcomes

Studies on carbohydrate-designed antivirals, such as mannose-conjugated acyclovir, have demonstrated significant improvements in pharmacokinetics and viral suppression compared to nondesigned forms. In clinical trials, mannose-acyclovir showed a higher concentration of the drug in virus-infected tissues, resulting in quicker viral clearance and a reduced risk of recurrence [60]. Similarly, sialic acid-modified oseltamivir has demonstrated enhanced efficacy against influenza strains that exhibit resistance to traditional forms of oseltamivir, potentially offering a new treatment option for resistant infections [61].

1.4.4 Anti-inflammatory drugs

Carbohydrate conjugation in anti-inflammatory drugs allows for targeted delivery to inflamed tissues, enhancing therapeutic efficacy while minimizing systemic side effects.

1.4.4.1 Examples and benefits of carbohydrate conjugation

One example of a carbohydrate-designed anti-inflammatory drug is dexamethasone conjugated with HA. This conjugation targets CD44 receptors on inflamed tissue, enhancing the drug's concentration at the inflammation site. This approach is particularly beneficial for treating localized inflammation in conditions such as rheumatoid arthritis, where targeted delivery can provide effective symptom relief while reducing the risk of systemic side effects [62]. Ibuprofen, another common anti-inflammatory drug, has also been conjugated with glucose to improve its bioavailability and target inflamed tissues. Glucose-designed ibuprofen enhances uptake by immune cells, increasing its efficacy in reducing inflammation and pain, particularly in localized or chronic inflammatory conditions [63].

1.4.4.2 Mechanisms of action and clinical relevance

Carbohydrate-designed anti-inflammatory drugs leverage specific interactions with immune cell receptors to enhance drug uptake in inflamed tissues. For example, HA-dexamethasone binds to CD44 receptors on immune cells and inflamed tissues, facilitating targeted delivery. This targeted delivery reduces the dosage required to achieve therapeutic effects, as a higher concentration of the drug reaches the inflamed site directly [64]. Such an approach has shown promise in preclinical and clinical studies for conditions like arthritis and inflammatory bowel disease, where targeted anti-inflammatory treatment significantly improves patient outcomes and reduces side effects [65].

1.5 Clinical applications and therapeutic potential

1.5.1 Current clinical applications of carbohydrate-designed drugs

Carbohydrate-designed drugs have shown remarkable clinical applications across various therapeutic areas. Their selective targeting capabilities allow these drugs to interact with specific cellular receptors, improving efficacy while reducing adverse effects. Notable examples include anticancer drugs, antibiotics, antivirals, and anti-inflammatory agents. One successful example in the oncology field is the use of HA-conjugated paclitaxel, which targets CD44 receptors overexpressed in certain cancer types. This selective binding enhances paclitaxel's efficacy against tumors such as ovarian and breast cancer, showing improved clinical outcomes and reduced side effects [66]. Similarly, doxorubicin conjugated with mannose targets mannose receptors on macrophages, which are prevalent in tumor-associated macrophages, enhancing the cytotoxicity of doxorubicin specifically within the tumor microenvironment [67]. In infectious disease treatment, glycosylated vancomycin has improved efficacy against MRSA by enhancing binding to bacterial cell walls. This structural enhancement not only improves antibacterial efficacy but also addresses the issue of antibiotic resistance, which is a growing concern in clinical settings [68].

1.5.2 Potential future uses in disease treatment and prevention

Looking forward, carbohydrate-designed drugs hold potential for treating a variety of conditions beyond their current applications. In neurological disorders, carbohydrate conjugation could enhance drug delivery across the blood-brain barrier (BBB), a significant hurdle in treating diseases like Alzheimer's and Parkinson's. By decorating drugs with glucose or other BBB-penetrating carbohydrates, researchers aim to facilitate better drug access to the brain, potentially revolutionizing treatment options for these challenging conditions [69]. In immunotherapy, carbohydrate-based drugs may enhance immune cell targeting, facilitating more effective vaccines and immunotherapeutics. For example, mannose-conjugated drugs could enhance delivery to dendritic cells, boosting the immune response to various pathogens or cancer cells. Preliminary studies in vaccine development have shown that carbohydrate decoration can improve the efficacy and durability of immune responses, paving the way for broader clinical use [70].

1.5.3 Case studies of carbohydrate-based drug success

Numerous case studies have demonstrated the clinical success of carbohydrate-designed drugs. For instance, a clinical trial with HA-conjugated cisplatin in ovarian cancer patients showed improved targeting and reduced nephrotoxicity, addressing one of the major limitations of cisplatin therapy. This study reported increased survival rates and enhanced quality of life among participants, emphasizing the clinical potential of carbohydrate-designed drugs [71]. Another successful case is sialic acid-designed influenza antivirals, which target hemagglutinin receptors on the influenza virus, resulting in better binding affinity and reduced viral load. Patients treated with sialic acid-conjugated antivirals displayed faster recovery times and fewer adverse effects, demonstrating the potential for carbohydrate decoration to enhance both efficacy and safety profiles [72, 73].

1.6 Conclusion

Carbohydrate-designed drugs represent a promising approach in modern therapeutics, offering enhanced targeting, improved pharmacokinetics, and reduced side effects. This book has reviewed current applications, including anticancer, antiviral, antibiotic, and anti-inflammatory drugs, highlighting the advantages of carbohydrate decoration in achieving selective drug delivery and minimizing toxicity. The future of carbohydrate-based drug development is bright, with ongoing research focused on expanding applications, improving conjugation techniques, and developing more robust regulatory pathways. As the understanding of glycobiology advances, carbohydrate-designed drugs may become a mainstay in the treatment of complex diseases, such as neurodegenerative disorders and autoimmune diseases. Carbohydrate decoration has evolved from a niche field to a critical area of drug development, thanks to its ability to enhance drug targeting and efficacy. As research and technology advance, carbohydrate-designed drugs hold the potential to transform modern therapeutics by enabling safer, more effective, and more targeted treatment options for a wide range of diseases.

References

[1] Pei Y, Yeo Y. Drug delivery to macrophages: Challenges and opportunities. J Control Release, 2016, 240, 202–211.

[2] Mishra N, Tiwari VK, Schmidt RR. Recent trends and challenges on carbohydrate-based molecular scaffolding: General consideration toward impact of carbohydrates in drug discovery and development. Carbohydr Drug Discov Dev, 2020, 1–69.

[3] Diniz F, Coelho P, Duarte HO, Sarmento B, Reis CA, Gomes J. Glycans as targets for drug delivery in cancer. Cancers, 2022, 14(4), 911.

[4] Swami R, Vij S, Sharma S. Unlocking the power of sugar: Carbohydrate ligands as key players in nanotherapeutic-assisted targeted cancer therapy. Nanomedicine, 2024, 19(5), 431–453.

[5] Bertozzi CR, Kiessling AL. Chemical glycobiology. Science, 2001, 291(5512), 2357–2364.

[6] Shchegravina ES, Sachkova AA, Usova SD, Nyuchev AV, Gracheva YA, Fedorov AY. Carbohydrate systems in targeted drug delivery: Expectation and reality. Russ J Bioorg Chem, 2021, 47, 71–98.

[7] Ahmed M, Narain R. Carbohydrate-based materials for targeted delivery of drugs and genes to the liver. Nanomedicine, 2015, 10(14), 2263–2288.

[8] Li D, Lou Y, Zhang Y, Liu S, Li J, Tao J. Sialylated immunoglobulin G: A promising diagnostic and therapeutic strategy for autoimmune diseases. Theranostics, 2021, 11(11), 5430.

[9] Zhang Y, Chan JW, Moretti A, Uhrich KE. Designing polymers with sugar-based advantages for bioactive delivery applications. J Control Release, 2015, 219, 355–368.

[10] Cho E, Jung S. Supramolecular complexation of carbohydrates for the bioavailability enhancement of poorly soluble drugs. Molecules, 2015, 20(10), 19620–19646.

[11] Di X, Liang X, Shen C, Pei Y, Wu B, He Z. Carbohydrates used in polymeric systems for drug delivery: From structures to applications. Pharmaceutics, 2022, 14(4), 739.

[12] Al-Hilal TA, Alam F, Byun Y. Oral drug delivery systems using chemical conjugates or physical complexes. Adv Drug Deliv Rev, 2013, 65(6), 845–864.

[13] Gholap AD, Said RP, Pawar RD, Ambore GS, Hatvate NT. Importance of carbohydrate-drug conjugates in vaccine development: A detailed review. Compr Anal Chem, 2023, 103, 191–256.
[14] Su L, Feng Y, Wei K, Xu X, Liu R, Chen G. Carbohydrate-based macromolecular biomaterials. Chem Rev, 2021, 121(18), 10950–11029.
[15] Machado JF, Correia JD, Morais TS. Emerging molecular receptors for the specific-target delivery of ruthenium and gold complexes into cancer cells. Molecules, 2021, 26(11), 3153.
[16] Wang J, Wang D, Zhang Y, Dong J. Synthesis and biopharmaceutical applications of sugar-based polymers: New advances and future prospects. ACS Biomater Sci Eng, 2021, 7(3), 963–982.
[17] Mattheolabakis G, Milane L, Singh A, Amiji MM. Hyaluronic acid targeting of CD44 for cancer therapy: From receptor biology to nanomedicine. J Drug Targe, 2015, 23(7–8), 605–618.
[18] Tyrrell DJ, Horne AP, Holme KR, Preuss JM, Page CP. Heparin in inflammation: Potential therapeutic applications beyond anticoagulation. Adv Pharmacol, 1999, 46, 151–208.
[19] Arora K, Sherilraj PM, Abutwaibe KA, Dhruw B, Mudavath SL. Exploring glycans as vital biological macromolecules: A comprehensive review of advancements in biomedical frontiers. Int J Biol Macromol, 2024, 131511.
[20] Li SD, Huang L. Pharmacokinetics and biodistribution of nanoparticles. Mol Pharm, 2008, 5(4), 496–504.
[21] Marradi M, Chiodo F, García I, Penadés S. Glyconanoparticles as multifunctional and multimodal carbohydrate systems. Chem Soc Rev, 2013, 42(11), 4728–4745.
[22] Staben LR, Koenig SG, Lehar SM, Vandlen R, Zhang D, Chuh J, Yu SF, Ng C, Guo J, Liu Y, Fourie-O'Donohue A. Targeted drug delivery through the traceless release of tertiary and heteroaryl amines from antibody–drug conjugates. Nat Chem, 2016, 8(12), 1112–11129.
[23] Tiwari VK, Mishra BB, Mishra KB, Mishra N, Singh AS, Chen X. Cu-catalyzed click reaction in carbohydrate chemistry. Chem Rev, 2016, 116(5), 3086–3240.
[24] Liu K, Jiang X, Hunziker P. Carbohydrate-based amphiphilic nano delivery systems for cancer therapy. Nanoscale, 2016, 8(36), 16091–16156.
[25] Liang DM, Liu JH, Wu H, Wang BB, Zhu HJ, Qiao JJ. Glycosyltransferases: Mechanisms and applications in natural product development. Chem Soc Rev, 2015, 44(22), 8350–8374.
[26] Dalziel M, Crispin M, Scanlan CN, Zitzmann N, Dwek RA. Emerging principles for the therapeutic exploitation of glycosylation. Science, 2014, 343(6166), 1235681.
[27] Aplin JD, Wriston JC, Lee YC. Preparation, properties, and applications of carbohydrate conjugates of proteins and lipid. Crit Rev Biochem Mol Biol, 1981, 10(4), 259–306.
[28] Narayanan D, Gopikrishna J, Nair SV, Menon D. Proteins and carbohydrates as polymeric nanodrug delivery systems: Formulation, properties, and toxicological evaluation. Multifaceted development and application of biopolymers for biology. J Biomed Nanotech, 2013, 241–267.
[29] Berdiaki A, Neagu M, Giatagana EM, Kuskov A, Tsatsakis AM, Tzanakakis GN, Nikitovic D. Glycosaminoglycans: Carriers and targets for tailored anti-cancer therapy. Biomolecules, 2021, 11(3), 395.
[30] Dragojevic S, Ryu JS, Raucher D. Polymer-based prodrugs: Improving tumor targeting and the solubility of small molecule drugs in cancer therapy. Molecules, 2015, 20(12), 21750–21769.
[31] Solá RJ, Griebenow KA. Effects of glycosylation on the stability of protein pharmaceuticals. J Pharm Sci, 2009, 98(4), 1223–1245.
[32] Yamazaki N, Kojima S, Bovin NV, Andre S, Gabius S, Gabius HJ. Endogenous lectins as targets for drug delivery. Adv Drug Deliv Rev, 2000, 43(2–3), 225–244.
[33] Zhong J, Li J, Burton GJ, Koistinen H, Cheung KW, Ng EH, Yao Y, Yeung WS, Lee CL, Chiu PC. The functional roles of protein glycosylation in human maternal–fetal crosstalk. Hum Reprod Update, 2024, 30(1), 81–108.
[34] Zlotnikov ID, Kudryashova EV. Mannose receptors of alveolar macrophages as a target for the addressed delivery of medicines to the lungs. Russ J Bioorg Chem, 2022, 48(1), 46–75.

[35] Tiwari VK, Mishra RC, Sharma A, Tripathi RP. Carbohydrate based potential chemotherapeutic agents: Recent developments and their scope in future drug discovery. Mini Rev Med Chem, 2012, 12(14), 1497–1519.
[36] Heine V, Dey C, Bojarová P, Křen V, Elling L. Methods of in vitro study of galectin-glycomaterial interaction. Biotechnol Adv, 2022, 58, 107928.
[37] Sztandera K, Działak P, Marcinkowska M, Stańczyk M, Gorzkiewicz M, Janaszewska A, Klajnert-Maculewicz B. Sugar modification enhances cytotoxic activity of PAMAM-doxorubicin conjugate in glucose-deprived MCF-7 cells–possible role of GLUT1 transporter. Pharma Res, 2019, 36, 1–1.
[38] Li Y, Cong H, Wang S, Yu B, Shen Y. Liposomes modified with bio-substances for cancer treatment. Biomater Sci, 2020, 8(23), 6442–6468.
[39] Gorovits B, Krinos-Fiorotti C. Proposed mechanism of off-target toxicity for antibody–drug conjugates driven by mannose receptor uptake. Cancer Immunol Immunother, 2013, 62, 217–223.
[40] Capone E, Iacobelli S, Sala G. Role of galectin 3 binding protein in cancer progression: A potential novel therapeutic target. J Transl Med, 2021, 19(1), 405.
[41] Estrada-Martínez LE, Moreno-Celis U, Cervantes-Jimenez R, Ferriz-Martínez RA, Blanco-Labra A, Garcia-Gasca T. Plant lectins as medical tools against digestive system cancers. Int J Mol Sci, 2017, 18(7), 1403.
[42] Zhou X, Chi K, Zhang C, Liu Q, Yang G. Sialylation: A cloak for tumors to trick the immune system in the microenvironment. Biology, 2023, 12(6), 832.
[43] Liu D, Zhu H, Li C. Galectins and galectin-mediated autophagy regulation: New insights into targeted cancer therapy. Biomark Res, 2023, 11(1), 22.
[44] Lake BP, Rullo AF. Offsetting low-affinity carbohydrate binding with covalency to engage sugar-specific proteins for tumor-immune proximity induction. ACS Cent Sci, 2023, 9(11), 2064–2075.
[45] Kanapathipillai M, Brock A, Ingber DE. Nanoparticle targeting of anti-cancer drugs that alter intracellular signaling or influence the tumor microenvironment. Adv Drug Deliv Rev, 2014, 79, 107–118.
[46] Yuan SS, Li ML, Chen JS, Zhou L, Zhou W. Application of mono-and disaccharides in drug targeting and efficacy. ChemMedChem, 2018, 13(8), 764–778.
[47] Nahar UJ, Toth I, Skwarczynski M. Mannose in vaccine delivery. J Control Release, 2022, 351, 284–300.
[48] Maeda H. The enhanced permeability and retention (EPR) effect in tumor vasculature: The key role of tumor-selective macromolecular drug targeting. Adv Enzyme Regul, 2001, 41(1), 189–207.
[49] Srinivasarao M, Low PS. Ligand-targeted drug delivery. Chem Rev, 2017, 117(19), 12133–12164.
[50] Chen L, Liang JF. Metabolic monosaccharides altered cell responses to anticancer drugs. Eur J Pharm Biopharm, 2012, 81(2), 339–345.
[51] Platt VM, Szoka FC Jr. Anticancer therapeutics: Targeting macromolecules and nanocarriers to hyaluronan or CD44, a hyaluronan receptor. Mol Pharm, 2008, 5(4), 474–486.
[52] Arpicco S, Milla P, Stella B, Dosio F. Hyaluronic acid conjugates as vectors for the active targeting of drugs, genes and nanocomposites in cancer treatment. Molecules, 2014, 19(3), 3193–3230.
[53] Philis-Tsimikas A, Chang A, Miller L. Precision, accuracy, and user acceptance of the OneTouch SelectSimple blood glucose monitoring system. J Diabetes Sci Technol, 2011, 5(6), 1602–1609.
[54] Dhanda G, Sarkar P, Samaddar S, Haldar J. Battle against vancomycin-resistant bacteria: Recent developments in chemical strategies. J Med Chem, 2018, 62(7), 3184–3205.
[55] Dartois V. The path of anti-tuberculosis drugs: From blood to lesions to mycobacterial cells. Nat Rev Microbiol, 2014, 12(3), 159–167.
[56] Riu F, Ruda A, Ibba R, Sestito S, Lupinu I, Piras S, Widmalm G, Carta A. Antibiotics and carbohydrate-containing drugs targeting bacterial cell envelopes: An overview. Pharmaceuticals, 2022, 15(8), 942.

[57] Del Bino L, Østerlid KE, Wu DY, Nonne F, Romano MR, Codée J, Adamo R. Synthetic glycans to improve current glycoconjugate vaccines and fight antimicrobial resistance. Chem Rev, 2022, 122(20), 15672–15716.
[58] Cavallaro G, Maniscalco L, Caliceti P, Salmaso S, Semenzato A, Giammona G. Glycosilated macromolecular conjugates of antiviral drugs with a polyaspartamide. J Drug Target, 2004, 12(9–10), 593–605.
[59] Matrosovich M, Klenk HD. Natural and synthetic sialic acid-containing inhibitors of influenza virus receptor binding. Rev Med Virol, 2003, 13(2), 85–97.
[60] Mtambo SE, Amoako DG, Somboro AM, Agoni C, Lawal MM, Gumede NS, Khan RB, Kumalo HM. Influenza viruses: Harnessing the crucial role of the M2 ion-channel and neuraminidase toward inhibitor design. Molecules, 2021, 26(4), 880.
[61] Yadav Y, Tyagi R, Kumar R, Sagar R. Conformationally locked sugar derivatives and analogues as potential neuraminidase inhibitors. Eur J Med Chem, 2023, 255, 115410.
[62] Salathia S, Gigliobianco MR, Casadidio C, Di Martino P, Censi R. Hyaluronic acid-based nanosystems for CD44 mediated anti-inflammatory and antinociceptive activity. Int J Mol Sci, 2023, 24(8), 7286.
[63] Bindu S, Mazumder S, Bandyopadhyay U. Non-steroidal anti-inflammatory drugs (NSAIDs) and organ damage: A current perspective. Biochem Pharmacol, 2020, 180, 114147.
[64] Patel D, Desai S, Dhruva A. Glucose-modified ibuprofen for enhanced anti-inflammatory effects. Eur J Pharm Sci, 2019, 134, 69–75.
[65] Lee R, Koh M, Park J. Mechanisms of HA-dexamethasone in targeting inflamed tissues. J Control Release, 2020, 320, 85–94.
[66] Liang X, Chen L, Wu H. Clinical efficacy of carbohydrate-designed anti-inflammatory drugs. Inflammation, 2021, 44(3), 1037–1048.
[67] Singh A, Sharma R, Bhattacharyya J. Enhanced targeting of cancer cells using N-acetylglucosamine-conjugated doxorubicin. J Drug Deliv Sci Technol, 2019, 52, 1231–1240.
[68] Li H, Chen D, Zhao M. Hyaluronic acid-conjugated paclitaxel: A novel targeted cancer therapy. Cancer Res, 2020, 80(15), 3034–3041.
[69] Zhang Y, Liu X, Feng Z. Pharmacokinetics of glycoconjugated paclitaxel in cancer therapy. Drug Metab Rev, 2018, 50(3), 394–405.
[70] Ochoa M, Calvo B, Arroyo C. Glucose-modified cisplatin analogs for enhanced cancer treatment. Pharma Res, 2017, 34(9), 1832–1840.
[71] Wang L, Huang T, Ma Y. Carbohydrate-designed vancomycin for combating MRSA. Biochem Pharmacol, 2021, 183, 114–122.
[72] Garcia S, Santillan B, Torres M. Mannose-modified streptomycin for tuberculosis treatment. J Antimicrob Chemother, 2019, 74(8), 2115–2122.
[73] Chen J, Xu K, Yang X. Glycoconjugate antibiotic efficacy in resistant bacterial strains. Antibiotics, 2020, 9(4), 186–192.

Girish Kumar, Tarun Virmani, Neha Dagar, Reshu Virmani, Pushpika Jain, and Kamla Pathak*

Chapter 2 Role of carbohydrates in drug delivery

2.1 Introduction

It has long been understood that carbohydrates play essential roles in biological processes, acting as fundamental building blocks of cellular structures and signaling pathways, in addition to being key energy sources [1]. Their potential as novel drug delivery carriers has recently attracted the attention of the pharmaceutical industry [2]. Because of their intrinsic qualities, which include biocompatibility, biodegradability, flexible structure, easy modifiability, improved solubility, enhanced stability, and the capacity to create unique interactions with biological macromolecules, carbohydrates are very appealing for improving drug formulation and delivery (Figure 2.1). Since they are a natural part of the body, they are usually well-tolerated [3].

The structures of carbohydrates are remarkably diverse, ranging from straightforward monosaccharides to complex polysaccharides, each offering distinct possibilities for functionalization and modification. While disaccharides and oligosaccharides add complexity and diversity, monosaccharides like glucose and galactose serve as essential building blocks [4]. It is possible to further modify polysaccharides, such as hyaluronic acid, chitosan (CS), and starch, to improve their biological and physicochemical characteristics [5]. Numerous chemical alterations are possible due to this structural flexibility including conjugation with drugs, the introduction of targeted ligands, and the creation of cross-linked networks. These alterations have the potential to increase stability, boost bioavailability, and customize the release characteristics of drugs [6]. Furthermore, carbohydrates are particularly useful in overcoming biological barriers because of the precise interactions they can have with proteins, enzymes, and cell receptors to enable targeted delivery [7]. The development of advanced delivery systems, such as hydrogels, liposomes, nanoparticles, solid lipid nanoparticles (SLNs), nanostructured lipid carriers, polymeric micelles, niosomes, and dendrimers, has been facilitated by developments in carbohydrate chemistry and nanotechnology

***Corresponding author: Kamla Pathak**, Faculty of Pharmacy, University of Medical Sciences, Etawah 206001, Uttar Pradesh, India, e-mail: kamlapathak5@gmail.com
Girish Kumar, Tarun Virmani, Amity Institute of Pharmacy, Amity University, Greater Noida 201308, Uttar Pradesh, India
Neha Dagar, Reshu Virmani, Pushpika Jain, School of Pharmaceutical Sciences, MVN University, Palwal, Haryana, India
Kamla Pathak, Faculty of Pharmacy, University of Medical Sciences, Etawah 206001, Uttar Pradesh, India

https://doi.org/10.1515/9783111447094-002

[8]. These systems make use of the properties of carbohydrates to achieve targeted distribution and controlled release. These systems can be engineered to respond to external stimuli such as pH or enzyme activity to further maximize their therapeutic potential [9].

The objective of this study is to provide a comprehensive overview of the state-of-the-art in the quickly advancing field of carbohydrate-based drug delivery systems. It will do this by discussing the novel drug delivery systems that use carbohydrate polymers, reviewing recent studies on the use of these systems in a range of diseases, and examining patents and clinical trials.

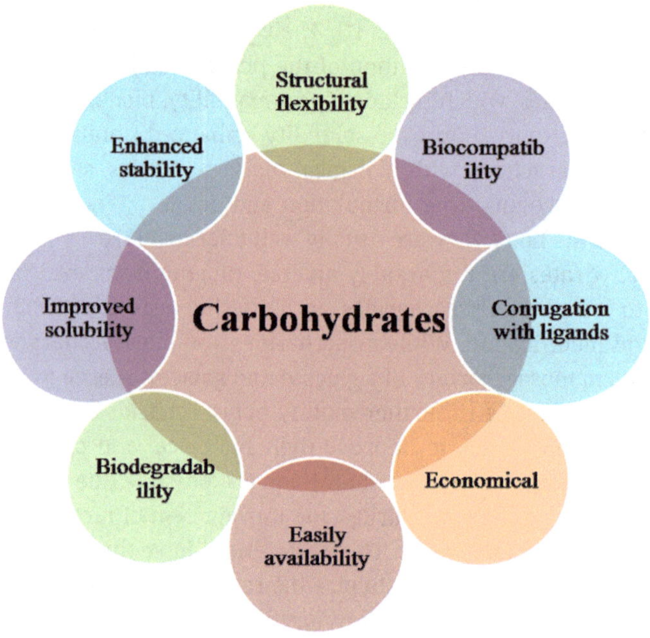

Figure 2.1: Advantages of carbohydrates for drug delivery.

2.2 Properties and applications of carbohydrate polymers in drug delivery systems

Carbohydrates are naturally occurring products with a very rich structure that gives rise to a variety of properties, particularly in drug administration. Various carbohydrates, along with their drug delivery systems, have been illustrated in Figure 2.2. Thereafter, each kind of carbohydrate and its application will be given in detail.

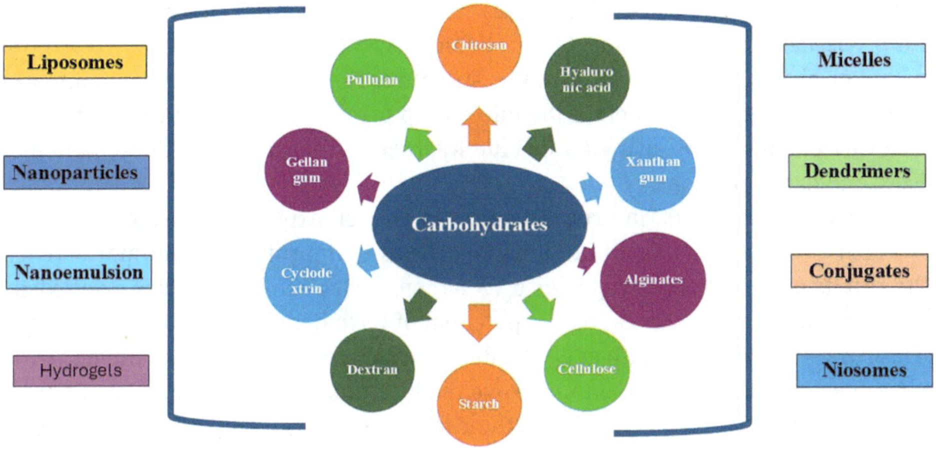

Figure 2.2: Pictorial representation of various carbohydrates along with their drug delivery systems.

2.2.1 Chitosan

CS, a natural polysaccharide that is linear and polycationic, is made up of β-(1-4)-D-glucosamine and *N*-acetyl-β-(1-4)-D-glucosamine units, and it is obtained from chitin through the deacetylation process [10]. It is a biocompatible and environmentally friendly polymer that has garnered significant attention as a potential drug carrier. It exhibits various unique properties, such as excellent drug loading, sustained drug release, multifunctionality, efficient drug release at cancerous sites, removal of cytotoxicity to noncancerous cells, auspicious targeting, and permeability to cell membranes [10]. Apart from offering a flexible platform for targeting, responsive drug release, and applications in image-guided medicine, CS demonstrates significant promise as a solubilizing agent, drug stabilizer, and controlled-release mechanism in the treatment of various diseases [11]. It also possesses mucoadhesive properties, which enable it to adhere to mucosal surfaces, such as those found in the digestive system and nasal cavity, resulting in increased bioavailability and sustained release of the drug due to prolonged drug contact with mucosal membranes [12]. Moreover, the cationic nature of CS allows it to interact with negatively charged molecules like nucleic acids and drugs. Because of this property, it has been able to create CS-based nanoparticles, which are utilized to transfer genetic elements like DNA and siRNA in gene therapy applications. These fragile molecules can be protected from degradation by CS nanoparticles, which also facilitate targeted administration, leading to precise therapeutic modification of gene expression [13].

The structure of CS contains free hydroxyl groups at C-3 and C-6 and amino groups at C-2, which can be physically and chemically modified to improve its characteristics. Its parent structure can be altered through esterification, acylation, and other processes to produce CS derivatives with desired chemical, physical, and biological properties. The solubility of CS can be improved by the introduction of carboxymethyl or hydroxypropyl groups into its structure. The mucoadhesion capability can be increased by trimethylation of the thiol group present in its structure [10]. Various drug carriers developed using CS include hydrogels, films, microparticles, nanoparticles, and nanoemulsions.

The application of CS in drug delivery can be well explained by recent research in which a tooltip was loaded in CS nanoparticles to improve its antitumor activity. The results displayed that drug-loaded CS nanoparticles exhibit superior cytotoxicity and reduced IC_{50} value, revealing the potential of CS to deliver the drugs at the target cancerous site [14]. Raghunath et al. [15] developed CS-coated SLNs of insulin and evaluated the potential of piperine to improve the intestinal permeability of insulin from SLNs. The findings demonstrated that the coating of SLNs by CS protected the degradation of insulin in the gastrointestinal tract (GIT) and provided higher cellular uptake, revealing the potential of piperine to improve the bioavailability of insulin.

2.2.2 Hyaluronic acid (HA)

Hyaluronic acid (HA), sometimes termed hyaluronan (sodium hyaluronate), is the most prevalent biopolymer in human tissues and is essential for many bodily processes [16]. As a linear anionic polysaccharide, HA is made up of repeated disaccharide units of *N*-acetyl-D-glucosamine (GlcNAc) and D-glucuronic acid (GlcA), connected by alternating β-1,3- and β-1,4-glycosidic links. In addition to being an essential part of the extracellular matrix, HA serves several critical molecular purposes in the body [17]. Through its interactions with HA-binding proteins and proteoglycans, HA provides structural stability to the extracellular and pericellular matrices. Moreover, HA acts as a signaling molecule to start a range of intracellular activities by interacting with cell surface receptors [18].

The role of HA as a drug delivery system has been demonstrated due to its possession of various excellent characteristics like biocompatibility, biodegradability, nontoxicity, non-inflammatory, and non-immunogenicity. The remarkable ability of HA to retain moisture is one of its most notable characteristics, which allows the development of hydrogels to deliver drugs [19]. HA-decorated drugs can be utilized to provide targeted delivery due to the recognition of overexpressed receptors like CD44 on cell surfaces [20]. The properties of HA can be modified using cross-linking or chemical alteration of the reactive groups present in its structure [21]. The chemical alteration of HA is typically carried out to prolong its half-life as a drug and enhance its usefulness while preserving its inherent characteristics. The main sites for the chemical modification of HA are the hydroxyl and carboxyl groups in its structure [22]. The modification of HA with hydrophobic groups provides slower dissolution and more resistance against enzymatic breakdown than pure HA. Amphiphilic HA facilitates the encapsulation and transport of hydrophobic drugs. HA can create supramolecular assemblies that are suitable for a variety of pharmaceutical uses by modifying the chemical composition of the linked carbon chains and their degree of substitution [19]. Various novel drug carriers like nanoparticles, hydrogels, nanoemulsions, and amphiphilic systems have been developed to deliver various categories of drugs efficiently [23].

In 2024, dopamine-decorated HA-recombinant human collagen-III (HA-DA@rhCol) hydrogel was developed for the effective management of chronic wounds in diabetes mellitus. The hydrogel exhibited a good swelling index, porous structure, rational rheological, degradation, and mechanical properties along with improved adhesiveness, antioxidant activity, and exceptional photothermal properties leading to increased antimicrobial activity. The hydrogel demonstrated enhanced healing of diabetic wounds, revealed by the results of collagen deposition, wound closure, and immunohistochemical study as depicted in Figure 2.3 [24]. In another example, the feasibility of cysteine-conjugated HA nanoparticles (Cys-conj-HA NPs) as a drug delivery system was assessed in 2024. The developed NPs showed notable ability in scavenging free radicals. The high cell viability suggested that Cys-conj-HA NPs did not cause any cytotoxicity. Furthermore, these NPs did not cause acute toxicity in the blood or viscera of mice treated with drug-free NPs. Thus, it can be said that Cys-conj-HA NPs are safe enough to be used as a delivery system for oxidizing medications [25].

2.2.3 Alginates

Alginate is a naturally occurring anionic polysaccharide obtained from brown algae. In pharmaceutical research, alginates are widely utilized to develop drug delivery systems. It is made up of linear copolymeric blocks with residues linked by α-(1 → 4) and β-(1 → 4), which bind L-guluronic acid (G) and D-mannuronic acid (M), respectively [26]. The distinct physicochemical characteristics, such as biocompatibility, biodegradability, and ease of gelation, make it a prime option for encapsulating and administering a variety of drugs. The moderate hydrogel-forming properties of alginate provide a steady and regulated drug-release environment [27]. Alginate formulations have proven to be efficacious in the administration of a wide variety of drugs including growth hormones, antibiotics, and anti-inflammatory compounds. The controlled release of alginate matrices increases patient compliance, minimizes side effects, and increases medication absorption [28].

Alginate has more mucoadhesive properties than other polymers like polystyrene, CS, and polylactic acid. Its mucoadhesive property makes it a potentially useful mucosal drug delivery vehicle for the nasopharynx and GIT, extending the drug's period of residence at the desired site and enhancing its medicinal efficacy [29]. In addition, alginate has film-forming ability, responsiveness to pH changes, gelling capability, and ionic crosslinking properties. Alginate has been utilized to develop various drug delivery systems including hydrogels, microspheres, nanoparticles, alginate beads, ocular gels, and wound dressings [30].

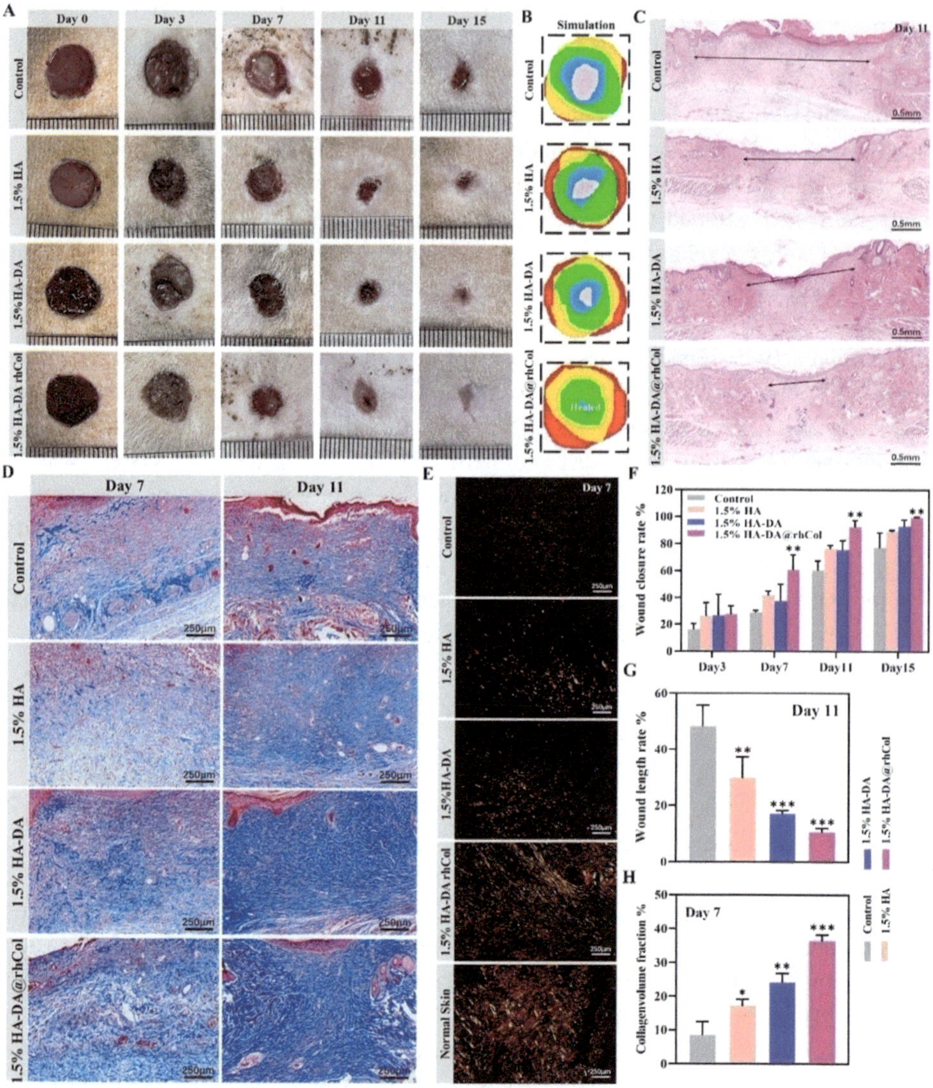

Figure 2.3: Pictorial representation of the potential of 1.5% HA-DA@rhCol in diabetic wound healing: (A) pictures of wound healing with various dressings; (B) wound closure simulation plots; (C) H&E staining of the wound showed wound condition on the 11th day; (D) photographs of Masson's staining; (E) pictures of Sirius red staining of the wound on the 7th and 14th day; (F) quantitative determination of wound closure; (G) length of wound in H&E staining; and (H) fraction of collagen volume in Masson's staining. Reprinted with permission from [24].

A nanocomposite of cross-linked alginate hydrogel modified with flax mucilage and silk fibroin loaded with iron oxide magnetic nanoparticles was developed by Radinekiyan et al. [31] to improve anticancer activity. The findings demonstrated higher survival of HEK293T normal cell lines and greater inhibition of BT549 breast cancer lines after 2 and 3 days of treatment. This suggested the improved biocompatibility and enhanced anticancer activity of the formulation [31]. Liu et al. [32] fabricated alginate oligosaccharide-functionalized aminated mesoporous silica nanoparticles (MSN-NH_2-AOS) to deliver curcumin (Cur) in the treatment of colon cancer. The results of in vitro anticancer studies and cellular uptake findings suggested that HCT-116 colon cancer cell lines were able to more readily absorb MSN-NH_2-AOS loaded with Cur nanoparticles than plain Cur, leading to increased targeting of tumor cells. Furthermore, MSN-NH_2-AOS loaded with Cur nanoparticles exhibited potent cytotoxicity against tumor cells at 50 μg/mL of Cur, suggesting that MSN-NH_2-AOS could be a useful delivery system for hydrophobic anticancer drugs [32].

2.2.4 Gellan gum

Gellan gum (GG) is an FDA-approved, water-soluble anionic, linear, and high-molecular weight exopolysaccharide. It is obtained commercially through microbial fermentation of the *Sphingomonas paucimobilis* microorganism. It is composed of tetrasaccharide repeating units with one carboxyl side group, specifically 1,3-β-D-glucose (Glc), 1,4-β-D-glucuronic acid (GlcA), 1,4-β-D-glucose (Glc), and 1,4-β-L-rhamnose (Rha) [33]. Commercially, GG is available under the trade name of Kelcogel™ (low acyl GG or deacylated GG) and Gelrite™ (acyl GG or acylated GG) [34]. GG exhibits a multitude of highly adaptable characteristics, including facile fabrication, easy functionalization, tunable mechanical properties, mucoadhesion, heat and acid resistance, ionotropic sensitivity, biocompatibility, and biodegradability, which makes it an ideal polymer to design drug carriers [33]. However, under physiological conditions, GG alone exhibits weak mechanical strength, instability, and a high gelling temperature. Nonetheless, the physicochemical characteristics of GG can be improved by mixing it with other polymers, such as CS, sodium alginate, xanthan gum (XG), starch, agar, cellulose, polyvinyl chloride, pullulan, and other materials such as silver and gold [35]. Considering the versatile characteristics of GG, it has been widely employed to develop various pharmaceutical formulations such as microcapsules, in situ gels, hydrogels, beads, and intranasal films for oral, ophthalmic, buccal, and transdermal applications [36]. Due to its anionic nature, a three-dimensional (3D) hydrogel construct can be developed in the presence of multivalent cations, which can be utilized to deliver drugs for different kinds of diseases through various routes [37]. The main advantage of GG-based hydrogels is their ability to control the release of drugs, which makes them perfect for drug targeting at specific sites like tumor cells [38]. The property of drug targeting can be enhanced by making them stimuli-responsive like redox, pH, and temperature [39].

The pharmaceutical application of GG can be explained by recent research in which GG nanoparticles of tacrolimus (TGNPs) were developed to treat dry eye conditions by the topical route. The outcomes of this research demonstrated that TGNPs exhibit prolonged release of tacrolimus for 12 h, along with increased precorneal retention compared to the plain tacrolimus solution. In addition, the findings of the Draize test displayed the safety of the TGNPs for the eye, and pharmacodynamic studies depicted the effectiveness of the formulation in reducing the symptoms of dry eye disease [40]. Paclitaxel (PTX)-loaded implantable GG hydrogels were developed by Nieto et al. [41] to treat breast cancer. The GG hydrogels encapsulated with PTX and β-cyclodextrin (CD) exhibited glutathione-triggered controlled release of the PTX in breast cancer cells. In addition, PTX-loaded GG-based hydrogels demonstrated increased anticancer activity, which revealed the potential of GG-based hydrogels in the treatment of breast cancer [41].

2.2.5 Pullulan

Pullulan is a neutral, water-soluble, linear homopolysaccharide made up of α-(1 → 4)-maltotriosyl units (3-D-glucopyranosyl) that repeat continuously and are linked by α-(1 → 6) bonds. It is typically derived as an amorphous slime material by the aerobic fermentation of the polymorphic fungus *Aureobasidium pullulans* [42]. Pullulan is vulnerable to chemical modification due to its flexible chain and hydroxyl functional groups, which enables the development of numerous derivatives such as pullulan acetate, pullulan amine, CHP, carboxymethyl pullulan, and pullulan succinylated. In such a manner, various physicochemical properties of pullulan such as water solubility, oxygen impermeable film formation, hygroscopicity, biodegradability, and thickness were modified because of enhanced solubility in organic solvents or upgrading with new responsive functional groups [43]. This unique property of pullulan helps to conjugate various drugs for their efficient delivery [44]. The development of drug delivery systems based on pullulan is becoming more widespread due to the possession of various attributes such as spinnable, moldable, biodegradable, nontoxic, noncarcinogenic, non-mutagenic, edible, and biocompatible [45]. Pullulan-based drug delivery systems provide better drug loading efficiency, longer in vivo circulation times, and increased stability. Pullulan nanoparticles have demonstrated enhanced cellular absorption and improved tissue penetration, which could be attributed to the mucoadhesive property of the pullulan, which promotes adhesion to mucosal surfaces and permits effective drug transport to target sites. These can be used to release the drugs at a controlled rate, which leads to extended drug release profiles that minimize systemic toxicity and optimize therapeutic efficacy. Additionally, pullulan has the natural ability to interact with biological membranes and affect cellular absorption pathways to improve drug delivery [46].

For example, in 2024, Varadhan et al. [47] developed hyalgen-decorated pullulan acetate polymeric nanoparticles of ferulic acid to alleviate gastrointestinal cancer. The findings suggested that the hyalgen-coated pullulan acetate nanoparticles of ferulic acid outperformed free ferulic acid in their strong cytotoxic effects on the HCT116, AGS, and SW620 cell lines, with IC_{50} values of 55, 60, and 110 μg/mL, respectively. This study highlighted the fact that the hyalgen-decorated polymeric pullulan nanoparticles demonstrate minimal toxicity and substantial cytotoxicity toward gastrointestinal cancer cell types, together with improved drug release efficiency [47]. In another example, a hybrid nanoparticle consisting of a carborane pullulan nanogel and a hydrophobized boron oxide nanoparticle was fabricated to improve anticancer therapy. The results showed that the intravenous injection of the formulation reduced tumor volume without any change in normal body weight. Regrowth of the tumor was absent after 3 months of complete declination of cancer [48]. Modified pullulan has been extensively used to develop smart drug carriers that utilize different triggers like pH, temperature, redox, light, and charge to release the drug [49]. In a study, pH-sensitive pullulan nanoparticles of doxorubicin (DOX) were developed to treat breast cancer. The results demonstrated improved intracellular delivery of DOX and improved inhibition of MCF-7 cancer cell lines [50].

2.2.6 Dextran

Dextran is a neutral complex branched glucan made up of α-1,6-glycosidic connections between glucose monomers, with branches coming from α-1,2, α-1,3, and α-1,4 bonds. It is primarily made extracellularly from sucrose by a few lactic acid bacteria through the action of a catalytic enzyme called dextransucrase, which facilitates the conversion of D-glucopyranosyl residues from sucrose to dextran [51]. It has been utilized as a circulatory volume expander, which helps to enhance blood flow and avoid deep vein thrombosis after surgery. It has also been used as an antiviral medication due to its selectivity for different viruses and to treat anemia [52].

The advantageous characteristics of dextran as a biopolymer to fabricate nanostructures for drug delivery include outstanding solubility, biocompatibility, biodegradability, and non-immunogenicity [53]. Dextran can be easily modified to provide various derivatives such as acetylated dextran, carboxymethyl dextran, oxidized dextran, dextran sulfate sodium, and diethylaminoethyl dextran. These dextran derivatives have different physicochemical properties that are suitable for delivering various kinds of drugs [51]. The outstanding solubility of dextran is credited to the α-1,6 glycosidic bond, which imparts increased chain mobility. The nanostructures of dextran facilitate increased aqueous solubility of the hydrophobic drugs along with enhanced permeation through the gastrointestinal membrane. These also protect the drugs from the harsh environment of the GIT, resulting in increased retention in the GIT and improved oral bioavailability of the drugs [54]. Various drug delivery systems, including prodrug-dextran conjugates, dextran hydrogels, dextran nanoparticles, dextran micelles, and dextran magnetic composites, have been developed using dextran as a biopolymer [55]. For example, CD147 antibody-conjugated nanoprodrug utilizing oxidized dextran was developed to deliver DOX at the targeted site. The results of in vivo anticancer studies demonstrated significant inhibition of the growth of HepG2 cancer cell lines along with the highest safety. The study revealed the potential of dextran-based nanoprodrug in the treatment of cancer [56]. Dextran-based nanogels were created to enhance antigen absorption, storage, and presentation on MHC-1 while guiding antigen-presenting cell immunogenic maturation. The interaction of nanogels was increased by modification with L-arginine. In addition, at the time of fabrication of nanogels, pH-responsive hydrazone was introduced to provide conjugation and regulated release of antigen OVA. The findings of in vivo studies showed that L-arginine-modified nanogels outperformed the other treatment groups in terms of cellular and humoral immunity induction on day 21 after treatment [57].

2.2.7 Cellulose

Cellulose is an essential component of plant cell walls. It is a polysaccharide made up of β-D-glucopyranose monomers, having three hydroxyl groups for every anhydroglucose unit with a versatile nature [58]. These versatile hydroxyl groups decorate the cellulose backbone with hydrophilicity, chirality, and biodegradability. Recently, there has been a resurgence of interest in drug delivery with cellulose and its derivatives. The outstanding capacity to self-assemble into micro- or nanostructures, together with its good stability, high glass transition temperature (T_g), and compatibility with a range of drugs, makes it outstanding among other biopolymers [59]. These advantageous characteristics of cellulose have enabled the design and development of various nanostructures for efficacious drug delivery. However, in its native form, it has poor solubility in water, due to which its derivatives are used for pharmaceutical applications. The two main derivative classes of cellulose utilized for pharmaceutical applications include cellulose ester and cellulose ether. Cellulose ether comprises ethylcellulose, methylcellulose, hydroxyethyl methyl cellulose, hydroxypropyl cellulose, hydroxyethyl cellulose, hydroxypropylmethylcellulose (HPMC), and carboxymethyl cellulose. Cellulose ester includes cellulose acetate phthalate (CAP), hydroxypropyl methyl phthalate, and hydroxypropyl methyl acetate succinate [60]. Cellulose and its derivatives are utilized as pharmaceutical excipients in various dosage forms like tablets, capsules, creams, ointments, and injections. Cellulose ether derivatives are highly used as excipients for oral drug delivery systems owing to their capacity to expand in water. The hydrogel layers then regulate the diffusion and distribution of the drug molecules [61]. The most important cellulose ether derivative for drug delivery is HPMC. Its amphiphilic, easily soluble, and swellable nature make it a popular ingredient in many drug compositions. This hydrophilic polymer has a reduced critical solution temperature and is highly soluble at ambient temperature and lower temperatures, but it gels at higher temperatures. Because of its excellent viscoelastic qualities, HPMC is a primary component of many commercially marketed ocular viscosurgical devices including Ocucoat. Moreover, HPMC is useful for active components as a controlled release and sustained release agent [59]. For instance, methacrylated HPMC hydrogels were developed for efficient ocular delivery of curcumin in glaucoma. The methacrylated HPMC hydrogel of curcumin provided sustained drug release, which has great potential to reduce oxidative damage in trabecular meshwork cells by reducing inflammation and apoptosis. This study suggested that modified HPMC may be beneficial in the treatment of glaucoma owing to its mucoadhesive and prolonged drug release characteristics [62]. Among cellulose ester derivatives, CAP is the most important polymer due to its pH-controlled ability. It has been widely utilized in the enteric coating of tablets.

Another important cellulose-based polymer is nanocellulose, which has been shown to have myriad applications in the biomedical and pharmaceutical fields. There are three main classes of nanocellulose, namely cellulose nanocrystals (CNCs), cellulose nanofibers, and bacterial cellulose [63]. Various studies in recent years have utilized nanocellulose derivatives in drug delivery for the treatment of various diseases. For example, Yusefi et al. [64] developed CNCs of 5-fluorouracil (5-FU) to treat colorectal cancer. The results demonstrated that 5-FU-loaded CNCs have increased antitumor potential to inhibit the HT-29 and HCT 116 colon cancer cell lines compared to plain 5-FU. In addition, no effect on CCD 112 normal colon cells was observed, revealing the safety of CNCs. Xing et al. [65] fabricated CNC-coated pH-responsive dendritic mesoporous organosilica nanoparticles for efficacious delivery of paclitaxel in the treatment of cancer. The findings showed that the optimized formulation provided 14-fold greater apoptosis compared to the control group, suggesting the potential of CNC-coated drug delivery systems in cancer treatment [65].

2.2.8 Cyclodextrin

CDs are oligosaccharide molecules with a short chain that are produced when starch breaks down. CDs are composed of D-glucopyranose units connected by a glycosidic link between carbons 1 and 4. They come in three different varieties: 6 units (α), 7 units (β), and 8 units (γ) [66]. Notably, CDs have several benefits including the capacity to be easily modified, be soluble in water, be widely available, and have a high bioavailability. The distinctive features of CDs, which include a hydrophilic surface and a hydrophobic cavity within, allow them to form clathrates through weak interactions without the need for chemical reactions. These have a cone-shaped chamber that enables them to enclose other molecules, giving the encapsulated molecule chemical and physical stability. With this power, they can defend against microbial degradation, change the chemical reactivity of the host molecules, lessen the evaporation of volatile compounds, increase the dissolution of substances, and protect light or oxygen-sensitive substances [67]. Even though CDs are mostly inert in the body, several drug formulations contain them as a component. The pharmaceutical industry has made extensive use of them for a variety of applications, such as improvement in dissolution, enhancement in drug absorption, hiding tastes and odors, controlling the rate at which drugs are released into the body, lowering systemic and localized toxicity, and facilitating the passage of drugs through biological barriers. It is possible to deliver CDs containing formulations orally, intranasally, topically, or transdermally [68]. Researchers have made chemical modifications to natural CDs to improve their pharmacological properties for use in medicine. These changes lessen the possibility of adverse effects on the body and increase their solubility and ability to encapsulate other chemicals for controlled medication release. These modifications can take the form of increased solubility or hydrophilic/hydrophobic characteristics. CDs are modified to have special characteristics and improved molecular recognition capabilities, which are used for drug administration and gene therapy [69]. These modified CDs include hydroxypropyl-γ-CD, hydroxypropyl-β-CD, sulfobutylated-β-CD, and methylated-β-CD, which are pharmaceutically approved. A recent development in CD-based polymers includes polyrotaxanes, which consist of a polymer chain (PEG) along with threaded ring-shaped molecules like CD. These also can be modified to provide various properties such as improved cellular uptake, increased half-life, sustained drug release, and drug targeting, resulting in efficacious drug delivery [70].

The potential of CDs in drug delivery can be assessed by a study in which CD-decorated PLGA nanoparticles were developed to enhance the cellular uptake of paclitaxel in the treatment of cancer. The results displayed that CD-decorated PLGA nanoparticles have enhanced drug residence time and greater accumulation of the drug compared to plain PLGA nanoparticles. The blood concentration of paclitaxel was higher for CD-decorated PLGA nanoparticles than for plain PLGA nanoparticles, revealing the significance of CD in drug delivery [71]. Sadeh et al. [72] prepared β-CD metal-organic framework (β-CD-MOF) followed by functionalization with glutamine and gelatin to deliver curcumin in the treatment of cancer. The formulation provided a targeted release of curcumin in cancer cells over noncancer cells. The study demonstrated the potential of functionalized β-CD-MOF to encapsulate curcumin, ensuring controlled release along with improved therapeutic efficiency in the treatment of cancer.

2.2.9 Xanthan gum

XG is an extracellular heteropolysaccharide present in nature. It is produced by fermenting an aerobic, Gram-negative pathogen called *Xanthomonas campestris* [73]. Scientists are interested in it as a promising biopolymer for drug delivery because of its many qualities including biodegradability, biocompatibility, nontoxicity, bioadhesiveness, and low cost. Its ability to conjugate with other polymers, proteins, peptides, etc., which offers enhanced stability against enzymatic degradation, biocompatibility, better solubility, and inertness, also gives it a great deal of potential as a biopolymer for drug delivery. It may increase the aqueous solubility of hydrophobic drugs and extend the drug release from drug carriers by offering zero-order release kinetics [74]. These qualities allow XG to be used in a wide range of applications including thickening, stabilizing, emulsifying, and suspending agents in pharmaceutical sectors. Although XG has many benefits across a wide range of industries, it also has several disadvantages, such as unstable viscosity, microbial load, weak shear resistance, improper mechanical strength, thermal properties, and uncontrolled hydration rate [75]. These can be addressed by modifying specific groups in the structure of XG. Various modified XG derivatives include carboxymethyl XG, esterified XG, oxidized XG, and acetylated XG [76]. Various XG-based novel drug delivery systems include liposomes, nanoparticles, hydrogels, matrix systems, and microspheres.

Wu et al. [77] fabricated microspheres utilizing a combination of porous rice starch and XG to encapsulate curcumin. It was observed that microspheres developed by a combination of porous rice starch and XG released the curcumin at a sustained rate compared to microspheres prepared with porous rice starch alone. Alle et al. [78] developed carboxymethyl XG-capped gold nanoparticles (CMXG@AuNPs) of DOX to treat cancer. The findings of this study displayed anticancer potential that was 4.6 times higher compared to a plain DOX solution, suggesting the potential of the XG derivative in the treatment of cancer.

2.2.10 Starch

Starch is the second most prevalent element in nature after cellulose. It is a naturally occurring, biodegradable, renewable polysaccharide that is extensively present in plants. Generally, plants create starch through the photosynthetic process, which stores solar energy chemically in various plant and algal sections and is advantageous to non-photosynthetic species like humans [79]. The building blocks of starch are repeated units of amylopectin and amylose connected by glycosidic bonds. Potatoes, sweet potatoes, yams, rice, corn, and other foods are rich in starch, although few photosynthetic algal sources contain a minimal concentration of starch [80]. The possession of unique characteristics of biocompatibility, biodegradability, and widely available, economical, and renewable sources has led to the use of starch in pharmaceutical industries [81]. Since ancient times, starch has been utilized as a diluent, binder, disintegrant, and lubricant in various dosage forms like tablets, capsules, pills, and powders [82]. However, native starches can be modified chemically, physically, or by an enzyme to enhance the biopolymer's functional qualities, increasing the applications of starch-based materials in drug delivery. These modifications have an immediate effect on several of the essential properties of starch-based drug delivery systems including swelling, viscosity, sensitivity to enzyme assaults, disintegration, gel hardness, and water-oil absorption capacity [83]. These drug delivery systems can be made stimuli-responsive to impart drug release at specific sites. Various drug delivery systems, including hydrogels, micro or nanoparticles, and microspheres, have been developed using starch and its derivatives. For example, Salah et al. [84] fabricated crosslinked starch nanoparticles of curcumin to treat inflammatory bowel disease. Crosslinked starch nanoparticles delivered curcumin intracellularly due to altered endocytosis pathways and were more effective in inflammatory cells. These also increased the cytokine IL-10 while decreasing the release of cytokines IL-1β, IL-6, and IL-8. Tan et al. [85] prepared pH/redox responsive hydroxyethyl starch nanoparticles to improve the antitumor efficacy of DOX. The optimized formulation demonstrated fast and adequate release of DOX intracellularly. The findings of in vivo studies displayed increased efficiency in the inhibition of advanced tumors compared to plain DOX and nonresponsive nanoparticles [85]. Recent research of the last 2 years on various carbohydrate-based polymers in drug delivery has been enlisted in Table 2.1. Various patents approved for carbohydrate-based polymers in drug delivery have been illustrated in Table 2.2.

2.3 Clinical trials for carbohydrate-based therapy

The earlier study has shown the potential of carbohydrate-based polymers in drug delivery. In the recent era, various drug delivery systems based on carbohydrate polymers have been developed to deliver drugs for different diseases. Thus far, numerous research efforts have progressed to the stage of clinical trials assessing the effectiveness of therapy based on carbohydrates. Some of these recent studies have been included in Table 2.3.

2.4 Conclusion and future prospective

This chapter emphasizes the structure and physicochemical characteristics of various carbohydrates, including hyaluronic acid, alginate, chitosan, dextran, cellulose, gellan, CD, pullulan, and starch, required for pharmaceutical applications. The presence of several functional groups in their structure allows for the potential modification of these groups with other chemical groups, which may be advantageous for the improved loading of drugs and the release of the drug at the target site. The two most crucial characteristics of carbohydrate-based polymers are biocompatibility and biodegradability, which serve as the main characteristics for their application as drug delivery systems. Furthermore, these are inexpensive, simple to manufacture, widely available, and easy to modify, due to which they hold enormous promise in drug delivery. A new avenue for drug delivery is opened by the development of biomaterials based on carbohydrates that are loaded with drugs, growth factors, proteins, and peptides. This chapter also in-

Table 2.1: Recent research on carbohydrate-based polymers in drug delivery.

Type of carbohydrate	Drug carrier	Active moiety	Disease	Key findings	References
Chitosan	Nanoparticles	Curcumin	Breast cancer	Demonstrated efficient tumor targeting and retention capacity, which inhibited the growth of breast tumors in mouse models.	[86]
Chitosan	Nanoparticles	Xovoltib	Breast cancer	Exhibited reduced cell viability in MCF-7 breast cancer cells.	[14]
Chitosan	Solid lipid nanoparticles	Insulin	Diabetes mellitus	Provided sustained release along with increased permeation of insulin from goat mucosa.	[15]
Chitosan	Nanostructured lipid carriers	Tanshinone IIA	Parkinson	Demonstrated remarkable anti-Parkinson and antidepressant activity by chitosan-coated nanocarriers compared to uncoated nanocarriers.	[87]
Chitosan	Nanoemulsion	Ozonated olive oil	Leishmaniasis	Provided inhibition of L. major promastigote/amastigote growth along with no cytotoxic effect on macrophages.	[88]
Chitosan	Hydrogel	Phloroglucinol	Wound healing	Demonstrated synergistic antibacterial and antifungal properties, which improved wound healing properties.	[89]
Chitosan	PLGA nanoparticles	Quercetin	Prostate cancer	Chitosan-coated PLGA nanoparticles provided increased cellular uptake and cytotoxicity in LnCap cancer cell lines compared to uncoated nanoparticles.	[90]
Chitosan	Nanoparticles	5-FU and quercetin	Colorectal cancer	Provided increased cytotoxicity in HCT 116 colon cancer cell lines.	[91]
Chitosan	Polymeric micelles	DOX	Cancer	Provided two- to three-fold increased penetration of the DOX along with improved anticancer activity.	[92]
Hyaluronic acid	Lipid nanocarriers	Curcumin	Cancer	Demonstrated 1.96–2.48-fold improved cellular uptake of curcumin from hyaluronic acid-functionalized lipid nanocarriers.	[93]

(continued)

Table 2.1 (continued)

Type of carbohydrate	Drug carrier	Active moiety	Disease	Key findings	References
Hyaluronic acid	Liposomes	Regorafenib, curcumin	Colorectal cancer	Depicted increased cytotoxic property by hyaluronic acid-decorated liposomes compared to the uncoated liposomal formulation of the drugs.	[94]
Hyaluronic acid	Iron oxide magnetic nanoparticles	Curcumin	Colorectal cancer	Provided targeted drug delivery system for colorectal cancer cells with minimal toxicity.	[95]
Hyaluronic acid	Niosomes	Curcumin	Rheumatoid arthritis	Displayed improved bioavailability and enhanced therapeutic efficacy in the treatment of rheumatoid arthritis.	[96]
Hyaluronic acid, chitosan	Magnetic MMt nanoparticles	Curcumin	Wound healing	Chitosan/hyaluronic acid/PVA-grafted magnetic MMT nanoparticles provide a novel patch for wound healing with sustained and pH-responsive release of curcumin.	[97]
Hyaluronic acid	Hydrogel	Recombinant human collagen type III	Diabetic wound healing	Demonstrated enhanced diabetic wound healing.	[24]
Hyaluronic acid	Nanoparticles	Cysteine	–	Exhibited noteworthy free radical capability, along with no effect on the cell viability of normal cells.	[25]
Hyaluronic acid	Nanoparticles	Paclitaxel	Cancer	Demonstrated time-dependent accumulation of paclitaxel and hyaluronic acid-mediated endocytosis in cancer cells.	[98]
Alginate	Magnetic nanobiocomposite	Silk fibroin	Magnetic hyperthermia	Demonstrated higher survival of HEK293T normal cell lines and greater inhibition of BT549 breast cancer lines after two and three days of treatment.	[99]

Chitosan/ alginate	Nanoparticles	DOX, hydroxychloroquine	Breast cancer	Exhibited increased cytotoxic potential against MCF-7 and MDA-MB-231 breast cancer cell lines.	[100]
Alginate/ pectin	Nanoparticles	Naringin	–	Provided increased encapsulation efficiency along with increased retention time.	[101]
Chitosan/ alginate	Nanoparticles	Naringin, ciprofloxacin	Bacterial infections	Exhibited good antioxidant activity along with the improved antimicrobial potential of ciprofloxacin.	[102]
Gellan gum/ chitosan	Hydrogels	Timolol maleate	Glaucoma	Provided increased drug release along with improved safety for ocular delivery.	[103]
Gellan gum/ sodium alginate	Hydrogels	Monocaprin	Ocular infections	Displayed increased drug retention along with improved corneal permeation.	[104]
Gellan gum/ sodium alginate	Films	Curcumin	Wound healing	Demonstrated an increased zone of inhibition, resulting in improved antibacterial activity.	[105]
Pullulan	Nanoparticles	Ferulic acid	Gastrointestinal cancer	Provided increased anticancer activity against different cancer cell lines along with minimal toxicity.	[47]
Pullulan	Microparticles	Benzydamine hydrochloride	Local oral disease	Exhibited prolonged residence time along with sustained drug release in the treatment of local disease.	[106]
Pullulan	Polymeric material	Silymarin	–	Demonstrated increased biological value of silymarin due to continuous release from polymeric material.	[107]
Pullulan	Nanoparticles	Catechin	–	Exhibited increased antioxidant activity owing to its free radical scavenging properties.	[108]
Pullulan	Microneedle patch	Moxifloxacin	Wound healing	Demonstrated good compatibility, sustained release, and sufficient skin penetration.	[109]

(continued)

Table 2.1 (continued)

Type of carbohydrate	Drug carrier	Active moiety	Disease	Key findings	References
Pullulan	Prodrug nanoparticles	Paclitaxel	Lung cancer	Transferrin-functionalized pullulan nanoparticles demonstrated improved cancer-suppressing properties compared to the control group, along with decreased systemic toxicity.	[110]
Pullulan	Nanoparticles	Methotrexate, 10-hydroxycamptothecine	Cancer	Provided increased killing of HeLa cells by pullulan nanoparticles with a combination of drugs rather than the nanoparticles containing individual drugs.	[111]
Dextran	Magnetic nanoparticles	Cisplatin	Thyroid cancer	Folic acid-decorated dextran magnetic nanoparticles of cisplatin displayed increased tumor suppression compared to plain cisplatin.	[112]
Dextran	Nanogels	Antigen ovalbumin	Cancer	Arginine-modified dextran nanogels provided increased induction of cellular and humoral immunity compared to other groups.	[57]
Cationic dextran	Polymeric micelles	Indomethacin	Ocular disease	Provided enhanced corneal permeation along with the absence of toxicity to ocular tissues.	[113]
Nanocellulose	Nanocrystals	Crocin	Cancer	Demonstrated increased cytotoxicity against various cancer cell lines compared to free crocin.	[114]
Nanocellulose	*Mesoporous organosilica* nanoparticles	Paclitaxel	Cancer	Exhibited 14-fold greater apoptosis than the control group.	[65]
β-Cyclodextrin	Metal-organic frameworks	Curcumin	Cancer	Gelatin and glutamine-functionalized β-CD-MOF provided controlled release along with improved therapeutic efficiency in the treatment of cancer.	[72]

γ-Cyclodextrin	Hydrogels	Josamycin	Glaucoma	Provided a fourfold increase in drug loading capacity along with prolonged drug release for up to 4 days.	[115]
β-Cyclodextrin	Polymeric complex	Oseltamivir phosphate	–	Exhibited increased stability and controlled dissolution profile along with reduced toxicity.	[116]
β-Cyclodextrin	Metal-organic frameworks	Quercetin	–	Exhibited increased antioxidant property along with enhanced biocompatibility.	[117]
β-Cyclodextrin	Zein nanoparticles	Silibinin	Colorectal cancer	Demonstrated reduced IC_{50} value compared to the free drug, showing increased cytotoxic effect.	[118]
β-Cyclodextrin	Inclusion complex	Silymarin	Breast cancer	Displayed reduced IC_{50} value compared to free silymarin, demonstrating improved anticancer potential.	[119]
Thiolated xanthan gum	Mucoadhesive biconjugate	Biochanin-A	Melanoma	Exhibited noteworthy increase in cytotoxicity of drug against A431 skin cancer cell lines.	[120]
Xanthan gum	Metal organic framework	5-FU	Breast cancer	Provided pH-responsive controlled release of 5-FU, along with significant inhibition of MCF-7 breast cancer cell lines.	[121]
Xanthan gum	Hydrogels	Neomycin sulfate	Bacterial infections	Exhibited improved antibacterial properties compared to free neomycin.	[122]
Starch	Iron oxide nanoparticles	Quercetin	Hepatocellular carcinoma	Exhibited controlled release of quercetin along with reduced viability of HepG2 cancer cell lines.	[123]
Starch/chitosan	Nanoparticles	Capecitabine	Colon cancer	Demonstrated higher colon histology, decreased tumor scores along with inhibited CD31 and VEGF expression compared to the plain drug.	[124]

Table 2.2: Summary of patent approvals for several carbohydrates used in drug delivery.

Carbohydrate	Publication no.	Patent no.	Inventor	Product	Earliest priority
Chitosan	CN202110152190A	CN112940287B	Wuhan University [CN]	Chitosan hydrogels	3 February 2021
Chitosan	KR20200085663A	KR102372964B1	Pukyong National University Industry-University Cooperation Foundation [KR]	Chitosan hydrogels	10 July 2020
Hyaluronic acid	CN113574109A WO2020185041A2	CN113574109B	University of Yonsei ac [CN]	Serotonin-modified hyaluronic acid hydrogels	14 March 2019
Hyaluronic acid	KR20180167782A WO2020130685A1	KR102398680B1	Lg Chemical Ltd. [KR]	Hyaluronic acid hydrogels	21 December 2018
Alginate	CN202111271323A	CN113980294B	Ocean University of China [CN]	Self-healing sodium alginate hydrogels	29 October 2021
Alginate	KR20200133490A	KR102521317B1	Ulsan National Institute of Science and Technology (UNIST) [KR]	Alginate-coupled methacrylate hydrogels	15 October 2020
Cellulose	CN202111504454A	CN114230719B	University of South China Science and Technology [CN]	Cellulose-based hydrogels	10 December 2021
Cellulose	CN202110339282A	CN113150319B	University of Wuhan Textile [CN]	Cellulose nanocrystal-based hydrogels	30 March 2021
Gellan gum	CN202210091682A	CN114213682B	University of Hainan [CN]	Gelatin hydrogels	26 January 2022
Gellan gum	CN202111045848A	CN113637187B	University of Northwestern [CN]	Gelatin hydrogels	7 September 2021

Table 2.3: Clinical trials conducted for carbohydrate-based therapy.

Name of polymer	Objective	Sponsor	Phase	Year	Clinical trial no.
Chitosan	Assessment of therapeutic efficacy of miconazole-loaded chitosan nanoparticles in oral candidiasis	Yasminegamil	Phase I	2023	NCT06072716
Chitosan	Evaluation of the effect of chitosan nanoparticles spray to alleviate dental problems in asthmatic patients taking corticosteroid inhaler	Deraya University	Phase I	2024	NCT06525363
Chitosan	Assessment of in vivo susceptibility of root canal bacteria to chlorhexidine gluconate, chitosan, and chitosan nanoparticles in necrotic primary molars	Ain Shams University	Not applicable	2018	NCT03588351
Chitosan	Evaluation of the efficiency of a novel respirator with chitosan nanoparticles to decrease the occurrence of SARS-CoV-2 infection in healthcare professionals	University of Brasilia	Not applicable	2023	NCT04490200
Cyclodextrin	Comparison of the bioavailability of various oral formulations of curcumin	University of Hohenheim	Not applicable	2018	NCT03530436
HPMC	Investigation of the safety and effectiveness of CsA ophthalmic gel in treating moderate to severe dry eye condition.	Zhaoke (Guangzhou) Ophthalmology Pharmaceutical Ltd.	Phase III	2021	NCT04541888

cludes some patents and clinical studies that have previously been referenced. It makes sense that the numerous benefits of the carbohydrate-based drug delivery system will lead to the advancement of clinical practice with these novel applications. This chapter may be helpful in future studies on the use of carbohydrates in drug administration, and it might provide a basis for future planning and design of studies in this area.

References

[1] Ramaprabha K, Kumar V, Saravanan P, Rajeshkannan R, Rajasimman M, Kamyab H, Vasseghian Y. Exploring the diverse applications of Carbohydrate macromolecules in food, pharmaceutical, and environmental technologies. Environ Res, 2024, 240, 117521. https://doi.org/10.1016/j.envres.2023.117521.

[2] Kumar G, Virmani T, Misra SK, Sharma A, Pathak K. Exploration of chitosan and its modified derivatives as vaccine adjuvant: A review. Carbohydr Polym Technol Appl, 2024, 8, 100537. https://doi.org/10.1016/j.carpta.2024.100537.

[3] Di X, Liang X, Shen C, Pei Y, Wu B, He Z. Carbohydrates used in polymeric systems for drug delivery: From structures to applications. Pharmaceutics, 2022, 14, 739. https://doi.org/10.3390/pharmaceutics14040739.

[4] Lebrilla CB, Liu J, Widmalm G, Prestegard JH. Oligosaccharides and polysaccharides. In Essentials of Glycobiology 4th ed. Varki A, Cummings RD, Esko JD, Stanley P, Hart GW, Aebi M, Mohnen D, Kinoshita T, Packer NH, Prestegard JH, Schnaar RL, Seeberger PH. (Eds.). Cold Spring Harbor Laboratory Press: Cold Spring Harbor (NY), 2022.

[5] Benalaya I, Alves G, Lopes J, Silva LR. A review of natural polysaccharides: Sources, characteristics, properties, food, and pharmaceutical applications. Int J Mol Sci, 2024, 25, 1322. https://doi.org/10.3390/ijms25021322.

[6] Kumar G, Jain P, Virmani T, Sharma A, Akhtar MS, Aldosari SA, Khan MF, Duarte SOD, Fonte P. Enhancing therapy with nano-based delivery systems: Exploring the bioactive properties and effects of apigenin. Ther Deliv, 2024, 15, 717–735. https://doi.org/10.1080/20415990.2024.2386928.

[7] Efthimiadou EK, Metaxa A-F, Kordas G. Modified polysaccharides as drug delivery. In Polysaccharides: Bioactivity and Biotechnology. Ramawat KG, Mérillon J-M. (Eds.). Springer International Publishing: Cham, 2021, pp. 1–26. https://doi.org/10.1007/978-3-319-03751-6_23-1.

[8] Virmani T, Kumar G, Virmani R, Sharma A, Pathak K. Nanocarrier-based approaches to combat chronic obstructive pulmonary disease. Nanomedicine, 2022, 17, 1833–1854. https://doi.org/10.2217/nnm-2021-0403.

[9] Silant'ev VE, Shmelev ME, Belousov AS, Patlay AA, Shatilov RA, Farniev VM, Kumeiko VV. How to develop drug delivery system based on carbohydrate nanoparticles targeted to brain tumors. Polymers, 2023, 15, 2516. https://doi.org/10.3390/polym15112516.

[10] Virmani T, Kumar G, Sharma A, Pathak K, Akhtar MS, Afzal O, Altamimi ASA. Amelioration of cancer employing chitosan, its derivatives, and chitosan-based nanoparticles: Recent updates. Polymers, 2023, 15, 2928. https://doi.org/10.3390/polym15132928.

[11] Haider A, Khan S, Iqbal DN, Shrahili M, Haider S, Mohammad K, Mohammad A, Rizwan M, Kanwal Q, Mustafa G. Advances in chitosan-based drug delivery systems: A comprehensive review for therapeutic applications. Eur Polym J, 2024, 210, 112983. https://doi.org/10.1016/j.eurpolymj.2024.112983.

[12] Aranaz I, Alcántara AR, Civera MC, Arias C, Elorza B, Heras Caballero A, Acosta N. Chitosan: An overview of its properties and applications. Polymers, 2021, 13, 3256. https://doi.org/10.3390/polym13193256.

[13] Karayianni M, Sentoukas T, Skandalis A, Pippa N, Pispas S. Chitosan-based nanoparticles for nucleic acid delivery: Technological aspects, applications, and future perspectives. Pharmaceutics, 2023, 15. https://doi.org/10.3390/pharmaceutics15071849.

[14] Rani P, Jenifer H, Prabhu A, David D, Devanesan S, AlSalhi MS, Nicoletti M, Saravanan M. Enhancing anticancer efficacy: Xovoltib-loaded chitosan-tripolyphosphate nanoparticles for targeted drug delivery against MCF-7 breast cancer cells. Mater Technol, 2024, 39, 2381978. https://doi.org/10.1080/10667857.2024.2381978.

[15] Raghunath I, Koland M, Sarathchandran C, Saoji S, Rarokar N. Design and optimization of chitosan-coated solid lipid nanoparticles containing insulin for improved intestinal permeability using piperine. Int J Biol Macromol, 2024, 135849. https://doi.org/10.1016/j.ijbiomac.2024.135849.

[16] Salih ARC, Farooqi HMU, Amin H, Karn PR, Meghani N, Nagendran S. Hyaluronic acid: Comprehensive review of a multifunctional biopolymer. Future J Pharm Sci, 2024, 10, 63. https://doi.org/10.1186/s43094-024-00636-y.

[17] Collis DWP, Yilmaz G, Yuan Y, Monaco A, Ochbaum G, Shi Y, O'Malley C, Uzunova V, Napier R, Bitton R, Becer CR, Azevedo HS. Hyaluronan (HA)-inspired glycopolymers as molecular tools for studying HA functions. RSC Chem Biol, n.d., 2, 568–576. https://doi.org/10.1039/d0cb00223b.

[18] Huang G, Huang H. Application of hyaluronic acid as carriers in drug delivery. Drug Deliv, 2018, 25, 766–772. https://doi.org/10.1080/10717544.2018.1450910.

[19] Bokatyi AN, Dubashynskaya NV, Skorik YA. Chemical modification of hyaluronic acid as a strategy for the development of advanced drug delivery systems. Carbohydr Polym, 2024, 337, 122145. https://doi.org/10.1016/j.carbpol.2024.122145.

[20] Jamshidi Z, Dehghan R, Nejabat M, Abnous K, Taghdisi SM, Hadizadeh F. Dual-targeting CD44 and mucin by hyaluronic acid and 5TR1 aptamer for epirubicin delivery into cancer cells: Synthesis, characterization, *in vitro* and *in vivo* evaluation. Heliyon, 2024, 10, e24833. https://doi.org/10.1016/j.heliyon.2024.e24833.

[21] Kumar G, Virmani T, Sharma A, Pathak K. Codelivery of phytochemicals with conventional anticancer drugs in form of nanocarriers. Pharmaceutics, 2023, 15. https://doi.org/10.3390/pharmaceutics15030889.

[22] Hintze V, Schnabelrauch M, Rother S. Chemical modification of hyaluronan and their biomedical applications. Front Chem, 2022, 10. https://doi.org/10.3389/fchem.2022.830671.

[23] Kumar M, Virmani T, Kumar G, Deshmukh R, Sharma A, Duarte S, Brandão P, Fonte P. Nanocarriers in tuberculosis treatment: Challenges and delivery strategies. Pharmaceuticals, 2023, 16, 1360. https://doi.org/10.3390/ph16101360.

[24] Wang Y, Zhang Y, Yang Y-P, Jin M-Y, Huang S, Zhuang Z-M, Zhang T, Cao -L-L, Lin X-Y, Chen J, Du Y-Z, Chen J, Tan W-Q. Versatile dopamine-functionalized hyaluronic acid-recombinant human collagen hydrogel promoting diabetic wound healing via inflammation control and vascularization tissue regeneration. Bioact Mater, 2024, 35, 330–345. https://doi.org/10.1016/j.bioactmat.2024.02.010.

[25] Zhu H, Xu S, Shang G, Pan X, Zhao J, Zhao B, Sun Z, Yang X, He S, Xue J. Fabrication of hyaluronic acid-based nanoparticles and studies on their radical scavenging activity. Polym Adv Technol, 2024, 35, e6332. https://doi.org/10.1002/pat.6332.

[26] Abka-khajouei R, Tounsi L, Shahabi N, Patel AK, Abdelkafi S, Michaud P. Structures, properties and applications of alginates. Mar Drugs, 2022, 20, 364. https://doi.org/10.3390/md20060364.

[27] Abourehab MAS, Rajendran RR, Singh A, Pramanik S, Shrivastav P, Ansari MJ, Manne R, Amaral LS, Deepak A. Alginate as a promising biopolymer in drug delivery and wound healing: A review of the state-of-the-art. Int J Mol Sci, 2022, 23, 9035. https://doi.org/10.3390/ijms23169035.

[28] Al-Roujayee AS, Hilaj E, Deepak A, Jyothi SR, Hamid JA, Ariffin IA, Saraswat V, Garg A. Alginate-based systems: Advancements in drug delivery and wound healing. Int J Polym Mater Polym Biomater, n.d., 1–29. https://doi.org/10.1080/00914037.2024.2375343.

[29] Lai J, Azad AK, Sulaiman WMAW, Kumarasamy V, Subramaniyan V, Alshehade SA. Alginate-based encapsulation fabrication technique for drug delivery: An updated review of particle type, formulation technique, pharmaceutical ingredient, and targeted delivery system. Pharmaceutics, 2024, 16, 370. https://doi.org/10.3390/pharmaceutics16030370.

[30] Manna S, Gupta P, Nandi G, Jana S. Recent update on alginate based promising transdermal drug delivery systems. J Biomater Sci Polym Ed, 2023, 34, 2291–2318. https://doi.org/10.1080/09205063.2023.2230847.

[31] Radinekiyan F, Naimi-Jamal MR, Eivazzadeh-Keihan R, Aliabadi HAM, Bani MS, Shojaei S, Maleki A. A magnetic cross-linked alginate-biobased nanocomposite with anticancer and hyperthermia activities. Carbohydr Polym Technol Appl, 2024, 7, 100481. https://doi.org/10.1016/j.carpta.2024.100481.

[32] Liu C, Jiang F, Xing Z, Fan L, Li Y, Wang S, Ling J, Ouyang X-K. Efficient delivery of curcumin by alginate oligosaccharide coated aminated mesoporous silica nanoparticles and in vitro anticancer activity against colon cancer cells. Pharmaceutics, 2022, 14, 1166. https://doi.org/10.3390/pharmaceutics14061166.

[33] Muthukumar T, Song JE, Khang G. Biological role of gellan gum in improving scaffold drug delivery, cell adhesion properties for tissue engineering applications. Molecules, 2019, 24, 4514. https://doi.org/10.3390/molecules24244514.

[34] Zia KM, Tabasum S, Khan MF, Akram N, Akhter N, Noreen A, Zuber M. Recent trends on gellan gum blends with natural and synthetic polymers: A review. Int J Biol Macromol, 2018, 109, 1068–1087. https://doi.org/10.1016/j.ijbiomac.2017.11.099.

[35] Palumbo FS, Federico S, Pitarresi G, Fiorica C, Giammona G. Gellan gum-based delivery systems of therapeutic agents and cells. Carbohydr Polym, 2020, 229, 115430. https://doi.org/10.1016/j.carbpol.2019.115430.

[36] Das M, Giri TK. Hydrogels based on gellan gum in cell delivery and drug delivery. J Drug Deliv Sci Technol, 2020, 56, 101586. https://doi.org/10.1016/j.jddst.2020.101586.

[37] Li Z, Li S, Yang J, Ha Y, Zhang Q, Zhou X, He C. 3D bioprinted gelatin/gellan gum-based scaffold with double-crosslinking network for vascularized bone regeneration. Carbohydr Polym, 2022, 290, 119469. https://doi.org/10.1016/j.carbpol.2022.119469.

[38] Lalebeigi F, Alimohamadi A, Afarin S, Aliabadi HAM, Mahdavi M, Farahbakhshpour F, Hashemiaval N, Khandani KK, Eivazzadeh-Keihan R, Maleki A. Recent advances on biomedical applications of gellan gum: A review. Carbohydr Polym, 2024, 334, 122008. https://doi.org/10.1016/j.carbpol.2024.122008.

[39] Zhou S, Zheng X, Yi K, Du X, Wang C, Cui P, Jiang P, Ni X, Qiu L, Wang J. Temperature-ion-pH triple responsive gellan gum as in situ hydrogel for long-acting cancer treatment. Gels, 2022, 8, 508. https://doi.org/10.3390/gels8080508.

[40] Modi D, Nirmal J, Warsi MH, Bhatia M, Hasan N, Kesharwani P, Jain GK. Formulation and development of tacrolimus-gellan gum nanoformulation for treatment of dry eye disease. Colloids Surf B Biointerfaces, 2022, 211, 112255. https://doi.org/10.1016/j.colsurfb.2021.112255.

[41] Nieto C, Vega MA, Rodríguez V, Pérez-Esteban P, Martín del Valle EM. Biodegradable gellan gum hydrogels loaded with paclitaxel for HER2+ breast cancer local therapy. Carbohydr Polym, 2022, 294, 119732. https://doi.org/10.1016/j.carbpol.2022.119732.

[42] Teixeira MO, Marinho E, Silva C, Antunes JC, Felgueiras HP. Pullulan hydrogels as drug release platforms in biomedicine. J Drug Deliv Sci Technol, 2023, 89, 105066. https://doi.org/10.1016/j.jddst.2023.105066.

[43] Grigoras AG. Drug delivery systems using pullulan, a biocompatible polysaccharide produced by fungal fermentation of starch. Environ Chem Lett, 2019, 17, 1209–1223. https://doi.org/10.1007/s10311-019-00862-4.

[44] Rajalekshmy GP, Annie Mariya R, Rekha MR. Chapter 16 – Pullulan-based nanomaterials in drug delivery applications. In Biopolymer Based Nanomaterials in Drug Delivery and Biomedical Applications. Bera H, Hossain CM, Saha S. (Eds.). Academic Press, 2021, pp. 383–404. https://doi.org/10.1016/B978-0-12-820874-8.00010-5.

[45] Thakur A, Sharma S, Naman S, Baldi A. Pullulan based polymeric novel drug delivery systems: A review on current state of art and prospects. J Drug Deliv Sci Technol, 2023, 90, 105117. https://doi.org/10.1016/j.jddst.2023.105117.

[46] Thomas N, Puluhulawa LE, Cindana Mo'o FR, Rusdin A, Gazzali AM, Budiman A. Potential of pullulan-based polymeric nanoparticles for improving drug physicochemical properties and effectiveness. Polymers, 2024, 16, 2151. https://doi.org/10.3390/polym16152151.

[47] Varadhan P, Jayaraman M. Hyalgan-decorated-ferulic acid-loaded pullulan acetate nanoparticles against gastrointestinal cancer cell lines. Emergent Mater, 2024, 7, 1115–1127. https://doi.org/10.1007/s42247-024-00634-z.

[48] Kawasaki R, Hirano H, Yamana K, Isozaki H, Kawamura S, Sanada Y, Bando K, Tabata A, Yoshikawa K, Azuma H, Takata T, Tanaka H, Sakurai Y, Suzuki M, Tarutani N, Katagiri K, Sawada S, Sasaki Y, Akiyoshi K, Nagasaki T, Ikeda A. Carborane bearing pullulan nanogel-boron oxide nanoparticle hybrid for boron neutron capture therapy. Nanomed Nanotechnol Biol Med, 2023, 49, 102659. https://doi.org/10.1016/j.nano.2023.102659.

[49] Raychaudhuri R, Naik S, Shreya AB, Kandpal N, Pandey A, Kalthur G, Mutalik S. Pullulan based stimuli responsive and sub cellular targeted nanoplatforms for biomedical application: Synthesis, nanoformulations and toxicological perspective. Int J Biol Macromol, 2020, 161, 1189–1205. https://doi.org/10.1016/j.ijbiomac.2020.05.262.

[50] Wang Y, Liu Y, Liu Y, Wang Y, Wu J, Li R, Yang J, Zhang N. pH-sensitive pullulan-based nanoparticles for intracellular drug delivery. Polym Chem, 2013, 5, 423–432. https://doi.org/10.1039/C3PY00817G.

[51] Petrovici AR, Pinteala M, Simionescu N. Dextran formulations as effective delivery systems of therapeutic agents. Molecules, 2023, 28, 1086. https://doi.org/10.3390/molecules28031086.

[52] Teixeira JP, Van Sant LM, Nielsen ND. Pharmacology and clinical use of plasma expanders. Anaesth Intensive Care Med, 2020, 21, 427–433. https://doi.org/10.1016/j.mpaic.2020.05.003.

[53] Chen F, Huang G, Huang H. Preparation and application of dextran and its derivatives as carriers. Int J Biol Macromol, 2020, 145, 827–834. https://doi.org/10.1016/j.ijbiomac.2019.11.151.

[54] Hu Q, Lu Y, Luo Y. Recent advances in dextran-based drug delivery systems: From fabrication strategies to applications. Carbohydr Polym, 2021, 264, 117999. https://doi.org/10.1016/j.carbpol.2021.117999.

[55] Dhaneshwar S, Bhilare N, Roy S. Dextran pharmaceutical applications. In Polysaccharides of Microbial Origin: Biomedical Applications. Oliveira JM, Radhouani H, Reis RL. (Eds.). Springer International Publishing: Cham, 2022, pp. 695–721. https://doi.org/10.1007/978-3-030-42215-8_40.

[56] Tian H, Yu L, Zhang M, He J, Sun X, Ni P. Dextran-doxorubicin prodrug nanoparticles conjugated with CD147 monoclonal antibody for targeted drug delivery in hepatoma therapy. Colloids Surf B Biointerfaces, 2023, 228, 113400. https://doi.org/10.1016/j.colsurfb.2023.113400.

[57] Chung JT, Rafiei M, Chau Y. Self-adjuvanted L-arginine-modified dextran-based nanogels for sustained local antigenic protein delivery to antigen-presenting cells and enhanced cellular and humoral immune responses. Biomater Sci, 2024, 12, 1771–1787. https://doi.org/10.1039/D3BM01150J.

[58] Etale A, Onyianta AJ, Turner SR, Eichhorn SJ. Cellulose: A review of water interactions, applications in composites, and water treatment. Chem Rev, 2023, 123, 2016–2048. https://doi.org/10.1021/acs.chemrev.2c00477.

[59] Amalin Kavitha A, Thomas Paul K, Anilkumar P. Cellulose-derived materials for drug delivery applications. Sustain Nanocellulose Nanohydrogels Nat Sources, 2020, 367–390, Elsevier. https://doi.org/10.1016/B978-0-12-816789-2.00018-3.

[60] Seddiqi H, Oliaei E, Honarkar H, Jin J, Geonzon LC, Bacabac RG, Klein-Nulend J. Cellulose and its derivatives: Towards biomedical applications. Cellulose, 2021, 28, 1893–1931. https://doi.org/10.1007/s10570-020-03674-w.

[61] Hosny KM, Alkhalidi HM, Alharbi WS, Md S, Sindi AM, Ali SA, Bakhaidar RB, Almehmady AM, Alfayez E, Kurakula M. Recent trends in assessment of cellulose derivatives in designing novel and nanoparticulate-based drug delivery systems for improvement of oral health. Polymers, 2021, 14, 92. https://doi.org/10.3390/polym14010092.

[62] Cheng Y-H, Fung M-P, Chen Y-Q, Chiu Y-C. Development of mucoadhesive methacrylic anhydride-modified hydroxypropyl methylcellulose hydrogels for topical ocular drug delivery. J Drug Deliv Sci Technol, 2024, 93, 105450. https://doi.org/10.1016/j.jddst.2024.105450.
[63] Ghilan A, Nicu R, Ciolacu DE, Ciolacu F. Insight into the latest medical applications of nanocellulose. Materials, 2023, 16, 4447. https://doi.org/10.3390/ma16124447.
[64] Yusefi M, Soon ML-K, Teow S-Y, Monchouguy EI, Neerooa BNHM, Izadiyan Z, Jahangirian H, Rafiee-Moghaddam R, Webster TJ, Shameli K. Fabrication of cellulose nanocrystals as potential anticancer drug delivery systems for colorectal cancer treatment. Int J Biol Macromol, 2022, 199, 372–385. https://doi.org/10.1016/j.ijbiomac.2021.12.189.
[65] Xing R, Ning L, Li L, He L, Lin H, You C, Wang F. Efficient in vitro delivery of paclitaxel by a nanocellulose-coated dendritic mesoporous organosilica nanoparticle for enhanced chemodynamic cancer therapy. J Drug Deliv Sci Technol, 2023, 86, 104654. https://doi.org/10.1016/j.jddst.2023.104654.
[66] Wüpper S, Lüersen K, Rimbach G. Cyclodextrins, natural compounds, and plant bioactives – A nutritional perspective. Biomolecules, 2021, 11, 401. https://doi.org/10.3390/biom11030401.
[67] Saha P, Rafe MR. Cyclodextrin: A prospective nanocarrier for the delivery of antibacterial agents against bacteria that are resistant to antibiotics. Heliyon, 2023, 9, e19287. https://doi.org/10.1016/j.heliyon.2023.e19287.
[68] Bahavarnia F, Hasanzadeh M, Bahavarnia P, Shadjou N. Advancements in application of chitosan and cyclodextrins in biomedicine and pharmaceutics: Recent progress and future trends. RSC Adv, 2024, 14, 13384–13412. https://doi.org/10.1039/D4RA01370K.
[69] Cid-Samamed A, Rakmai J, Mejuto JC, Simal-Gandara J, Astray G. Cyclodextrins inclusion complex: Preparation methods, analytical techniques and food industry applications. Food Chem, 2022, 384, 132467. https://doi.org/10.1016/j.foodchem.2022.132467.
[70] Kali G, Haddadzadegan S, Bernkop-Schnürch A. Cyclodextrins and derivatives in drug delivery: New developments, relevant clinical trials, and advanced products. Carbohydr Polym, 2024, 324, 121500. https://doi.org/10.1016/j.carbpol.2023.121500.
[71] Ma P, Huang J, Liu J, Zhu Y, Chen J, Chen J, Lei L, Guan Z, Ban J, Lu Z. Nanoformulation of paclitaxel: Exploring the cyclodextrin/PLGA nano delivery carrier to slow down paclitaxel release, enhance accumulation in vivo. J Cancer, 2023, 14, 759–769. https://doi.org/10.7150/jca.82410.
[72] Sadeh P, Zeinali S, Rastegari B, Najafipour I. Functionalization of β-cyclodextrin metal-organic frameworks with gelatin and glutamine for drug delivery of curcumin to cancerous cells. Heliyon, 2024, 10, e30349. https://doi.org/10.1016/j.heliyon.2024.e30349.
[73] Nejadmansouri M, Shad E, Razmjooei M, Safdarianghomsheh R, Delvigne F, Khalesi M. Production of xanthan gum using immobilized *Xanthomonas campestris* cells: Effects of support type. Biochem Eng J, 2020, 157, 107554. https://doi.org/10.1016/j.bej.2020.107554.
[74] Virmani T, Kumar G, Virmani R, Sharma A, Pathak K. Xanthan gum-based drug delivery systems for respiratory diseases. In Natural Polymeric Materials Based Drug Delivery Systems in Lung Diseases. Dureja H, Adams J, Löbenberg R, Andreoli Pinto T de J, Dua K. (Eds.). Springer Nature: Singapore, 2023, pp. 279–295. https://doi.org/10.1007/978-981-19-7656-8_16.
[75] Kang M, Oderinde O, Liu S, Huang Q, Ma W, Yao F, Fu G. Characterization of Xanthan gum-based hydrogel with Fe^{3+} ions coordination and its reversible sol-gel conversion. Carbohydr Polym, 2019, 203, 139–147. https://doi.org/10.1016/j.carbpol.2018.09.044.
[76] Jadav M, Pooja D, Adams DJ, Kulhari H. Advances in xanthan gum-based systems for the delivery of therapeutic agents. Pharmaceutics, 2023, 15, 402. https://doi.org/10.3390/pharmaceutics15020402.
[77] Wu L, Tian J, Ye X, Fang H, Zhang Z, Xu C, Zhang H. Encapsulation and release of curcumin with the mixture of porous rice starch and xanthan gum. Starch – Stärke, 2021, 73, 2000042. https://doi.org/10.1002/star.202000042.

[78] Alle M, Reddy GB, Kim TH, Park SH, Lee S-H, Kim J-C. Doxorubicin-carboxymethyl xanthan gum capped gold nanoparticles: Microwave synthesis, characterization, and anti-cancer activity. Carbohydr Polym, 2020, 229, 115511. https://doi.org/10.1016/j.carbpol.2019.115511.

[79] Sivamaruthi BS, Nallasamy PK, Suganthy N, Kesika P, Chaiyasut C. Pharmaceutical and biomedical applications of starch-based drug delivery system: A review. J Drug Deliv Sci Technol, 2022, 77, 103890. https://doi.org/10.1016/j.jddst.2022.103890.

[80] Tetlow IJ, Bertoft E. A review of starch biosynthesis in relation to the building block-backbone model. Int J Mol Sci, 2020, 21, 7011. https://doi.org/10.3390/ijms21197011.

[81] Singh P, Pandey VK, Singh R, Singh K, Dash KK, Malik S. Unveiling the potential of starch-blended biodegradable polymers for substantializing the eco-friendly innovations. J Agric Food Res, 2024, 15, 101065. https://doi.org/10.1016/j.jafr.2024.101065.

[82] Kumar G, Khar RK, Virmani T, Jogpal V, Virmani R. Comparative evaluation of fast dissolving tablet of atorvastatin calcium using natural and synthetic super disintegrating agents. Res J Pharm Technol, 2018, 11, 5001. https://doi.org/10.5958/0974-360X.2018.00912.5.

[83] Boldrini DE. Starch-based materials for drug delivery in the gastrointestinal tract – A review. Carbohydr Polym, 2023, 320, 121258. https://doi.org/10.1016/j.carbpol.2023.121258.

[84] Salah N, Dubuquoy L, Carpentier R, Betbeder D. Starch nanoparticles improve curcumin-induced production of anti-inflammatory cytokines in intestinal epithelial cells. Int J Pharm X, 2022, 4, 100114. https://doi.org/10.1016/j.ijpx.2022.100114.

[85] Tan R, Tian D, Liu J, Wang C, Wan Y. Doxorubicin-bound hydroxyethyl starch conjugate nanoparticles with pH/redox responsive linkage for enhancing antitumor therapy. Int J Nanomedicine, 2021, 16, 4527–4544. https://doi.org/10.2147/IJN.S314705.

[86] Mishra B, Yadav AS, Malhotra D, Mitra T, Sinsinwar S, Radharani NNV, Sahoo SR, Patnaik S, Kundu GC. Chitosan nanoparticle-mediated delivery of curcumin suppresses tumor growth in breast cancer. Nanomater Basel Switz, 2024, 14, 1294. https://doi.org/10.3390/nano14151294.

[87] Hassan DM, El-Kamel AH, Allam EA, Bakr BA, Ashour AA. Chitosan-coated nanostructured lipid carriers for effective brain delivery of Tanshinone IIA in Parkinson's disease: Interplay between nuclear factor-kappa β and cathepsin B. Drug Deliv Transl Res, 2024, 14, 400–417. https://doi.org/10.1007/s13346-023-01407-7.

[88] Taghizadeh E, Khamesipour A, Basirpour B, Hosseini SA. Evaluation of the in vitro antileishmanial activity of chitosan-coated nanoemulsions of ozonated olive oil. Ozone Sci Eng, 2024, 46, 355–364. https://doi.org/10.1080/01919512.2024.2307030.

[89] Polez RT, Ajiboye MA, Österberg M, Horn MM. Chitosan hydrogels enriched with bioactive phloroglucinol for controlled drug diffusion and potential wound healing. Int J Biol Macromol, 2024, 265, 130808. https://doi.org/10.1016/j.ijbiomac.2024.130808.

[90] Essa D, Kondiah PPD, Kumar P, Choonara YE. Design of chitosan-coated, quercetin-loaded PLGA nanoparticles for enhanced PSMA-specific activity on LnCap prostate cancer cells. Biomedicines, 2023, 11, 1201. https://doi.org/10.3390/biomedicines11041201.

[91] Das S, Saha M, Mahata LC, China A, Chatterjee N, Das Saha K. Quercetin and 5-Fu loaded chitosan nanoparticles trigger cell-cycle arrest and induce apoptosis in HCT116 cells via modulation of the p53/p21 axis. ACS Omega, 2023, 8, 36893–36905. https://doi.org/10.1021/acsomega.3c03933.

[92] Zlotnikov ID, Streltsov DA, Ezhov AA, Kudryashova EV. Smart pH- and temperature-sensitive micelles based on chitosan grafted with fatty acids to increase the efficiency and selectivity of doxorubicin and its adjuvant regarding the tumor cells. Pharmaceutics, 2023, 15, 1135. https://doi.org/10.3390/pharmaceutics15041135.

[93] Feng J, Wang Z, Huang W, Zhao X, Xu L, Teng C, Li Y. Hyaluronic acid-decorated lipid nanocarriers as novel vehicles for curcumin: Improved stability, cellular absorption, and anti-inflammatory effects. Food Chem, 2024, 141420. https://doi.org/10.1016/j.foodchem.2024.141420.

[94] Shnaikat SG, Shakya AK, Bardaweel SK. Formulation, development and evaluation of hyaluronic acid-conjugated liposomal nanoparticles loaded with regorafenib and curcumin and their in vitro evaluation on colorectal cancer cell lines. Saudi Pharm J SPJ Off Publ Saudi Pharm Soc, 2024, 32, 102099. https://doi.org/10.1016/j.jsps.2024.102099.

[95] Bourang S, Asadian S, Noruzpour M, Mansuryar A, Azizi S, Ebrahimi HA, Amani Hooshyar V. PLA-HA/Fe_3O_4 magnetic nanoparticles loaded with curcumin: Physicochemical characterization and toxicity evaluation in HCT116 colorectal cancer cells. Discov Appl Sci, 2024, 6, 186. https://doi.org/10.1007/s42452-024-05858-6.

[96] Obeaid NG, Al-Tu'ma FJ, Majeed AMK. Enhancement of anti-inflammatory activity of curcumin through hyaluronic acid decorated niosomal nanoparticles for effective treatment of rheumatoid arthritis patients. J Clust Sci, 2024, 35, 2405–2418. https://doi.org/10.1007/s10876-024-02667-0.

[97] Sayyar Z, Hosseini Z, Beheshtizadeh N. Developing curcumin loaded-magnetic montmorillonite nanoparticles/polyvinyl alcohol/hyaluronic acid/chitosan nanofiber mats as a wound dressing. J Drug Deliv Sci Technol, 2024, 93, 105408. https://doi.org/10.1016/j.jddst.2024.105408.

[98] Zhou H, Liu C, Yu S, Shafiq F, Qiao W. Hyaluronic acid anchored paclitaxel nanoparticles to solubilize for drug delivery. Eur Polym J, 2023, 201, 112542. https://doi.org/10.1016/j.eurpolymj.2023.112542.

[99] Radinekiyan F, Eivazzadeh-Keihan R, Naimi-Jamal MR, Aliabadi HAM, Bani MS, Shojaei S, Maleki A. Design and fabrication of a magnetic nanobiocomposite based on flaxseed mucilage hydrogel and silk fibroin for biomedical and in-vitro hyperthermia applications. Sci Rep, 2023, 13, 20845. https://doi.org/10.1038/s41598-023-46445-w.

[100] Zhang H, Xue Q, Zhou Z, He N, Li S, Zhao C. Co-delivery of doxorubicin and hydroxychloroquine via chitosan/alginate nanoparticles for blocking autophagy and enhancing chemotherapy in breast cancer therapy. Front Pharmacol, 2023, 14, 1176232. https://doi.org/10.3389/fphar.2023.1176232.

[101] Cao F, Zheng M, Zhang W, Wang Y, Lu S. A composite pectin-alginate nanoparticle for loading naringin-hydroxypropyl-β-cyclodextrin complex: Preparation, characteristics and stability study. Colloids Surf Physicochem Eng Asp, 2024, 700, 134644. https://doi.org/10.1016/j.colsurfa.2024.134644.

[102] Almeleebia TM, Akhter MH, Khalilullah H, Rahman MA, Ahmad S, Alam N, Ali MS, Khan G, Alanazi IMM, Shahzad N, Alalmaie A. Co-delivery of naringin and ciprofloxacin by oleic acid lipid core encapsulated in carboxymethyl chitosan/alginate nanoparticle composite for enhanced antimicrobial activity. ACS Omega, 2024, 9, 6845–6860. https://doi.org/10.1021/acsomega.3c08200.

[103] Shajari G, Erfan-Niya H, Fathi M, Amiryaghoubi N. In situ forming hydrogels based on modified gellan gum/chitosan for ocular drug delivery of timolol maleate. Int J Biol Macromol, 2024, 278, 135071. https://doi.org/10.1016/j.ijbiomac.2024.135071.

[104] Giri S, Markandeywar TS, Irfan Z, Manna S. Gellan gum and sodium alginate in-situ gel of monocaprin for effective corneal permeation. Food Hydrocoll Health, 2023, 4, 100156. https://doi.org/10.1016/j.fhfh.2023.100156.

[105] Jing W, Hussain SA, Reddy Gangireddygari VS, Chen J, Muhamad A'srai AI, Razali MH. Biocompatibility and antibacterial activity studies of gellan gum and alginate thin film encapsulating curcumin as a wound skin care dressing. Mater Technol, 2023, 38, 2282321. https://doi.org/10.1080/10667857.2023.2282321.

[106] Liu T, Gong X, Cai Y, Li H-Y, Forbes B. Pullulan-based spray-dried mucoadhesive microparticles for sustained oromucosal drug delivery. Pharmaceutics, 2024, 16, 460. https://doi.org/10.3390/pharmaceutics16040460.

[107] Iwaniec J, Niziołek K, Polanowski P, Słota D, Kosińska E, Sadlik J, Miernik K, Jampilek J, Sobczak-Kupiec A. Polyethylene glycol/pullulan-based carrier for silymarin delivery and its potential in biomedical applications. Int J Mol Sci, 2024, 25, 9972. https://doi.org/10.3390/ijms25189972.

[108] Liang B, Feng S, Li Z, Zhao X, Sun C, Ji C, Li X. Enhancement of catechin loaded in BSA-pullulan nanoparticles on the oxidation stability of pickering emulsion: Effect of concentration. LWT, 2024, 208, 116728. https://doi.org/10.1016/j.lwt.2024.116728.

[109] Younas A, Dong Z, Hou Z, Asad M, Li M, Zhang N. A chitosan/fucoidan nanoparticle-loaded pullulan microneedle patch for differential drug release to promote wound healing. Carbohydr Polym, 2023, 306, 120593. https://doi.org/10.1016/j.carbpol.2023.120593.

[110] Zhao X, Guo H, Bera H, Jiang H, Chen Y, Guo X, Tian X, Cun D, Yang M. Engineering transferrin-decorated pullulan-based prodrug nanoparticles for redox responsive paclitaxel delivery to metastatic lung cancer cells. ACS Appl Mater Interfaces, 2023, 15, 4441–4457. https://doi.org/10.1021/acsami.2c18422.

[111] Wu S, Yang X, Yang X. Methotrexate and 10-hydroxycamptothecine loaded pullulan nanoparticles with the targeting property for efficient cancer therapy. Mater Technol, 2022, 37, 2777–2784. https://doi.org/10.1080/10667857.2022.2075079.

[112] Hussein F, Alnasraui A, Joe IH, Al-Musawi S. Investigation of dextran-coated magnetic nanoparticles encapsulated by medication and modified with folate for targeted drug delivery: In vitro and docking studies. Mater Chem Phys, 2024, 321, 129446. https://doi.org/10.1016/j.matchemphys.2024.129446.

[113] Firozian F, Arabkhani Z, Mahboobian MM, Mohammadi M, Chehardoli G. Cationic Dextran Stearate (Dex-St-GTMAC): Synthesis and evaluation as polymeric micelles for indomethacin corneal penetration. ACS Omega, 2023, 8, 38092–38100. https://doi.org/10.1021/acsomega.3c04187.

[114] Soltani M, Farhadi A, Rajabi S, Homayouni-Tabrizi M, Hussein FS, Mohammadian N. Folic acid-modified nanocrystalline cellulose for enhanced delivery and anti-cancer effects of crocin. Sci Rep, 2024, 14, 13985. https://doi.org/10.1038/s41598-024-64758-2.

[115] Huling J, Oschatz S, Lange H, Sterenczak KA, Stahnke T, Markhoff J, Stachs O, Möller S, Undre N, Peil A, Jünemann A, Grabow N, Fuellen G, Eickner T. γ-Cyclodextrin hydrogel for the sustained release of josamycin for potential ocular application. Drug Deliv, 2024, 31, 2361168. https://doi.org/10.1080/10717544.2024.2361168.

[116] Olteanu AA, Rădulescu FȘ, Bleotu C, Aramă -C-C. Development and characterization of a cyclodextrin-based delivery system for enhanced pharmacokinetic and safety profile of oseltamivir. J Incl Phenom Macrocycl Chem, 2024. https://doi.org/10.1007/s10847-024-01258-w.

[117] Zhao R, Chen T, Li Y, Chen L, Xu Y, Chi X, Yu S, Wang W, Liu D, Zhu B, Hu J. Biocompatible hydrophobic cross-linked cyclodextrin-based metal-organic framework as quercetin nanocarrier for enhancing stability and controlled release. Food Chem, 2024, 448, 139167. https://doi.org/10.1016/j.foodchem.2024.139167.

[118] Al Alabdullah MAA, Goodarzi MT, Homayouni Tabrizi M. The silibinin-loaded Zein-β cyclodextrin nano-carriers (SZBC-NCs) as a novel selective cancer cell drug delivery system in HT-29 cell line. Sci Rep, 2024, 14, 14769. https://doi.org/10.1038/s41598-024-65881-w.

[119] Imam SS, Alshehri S, Altamimi MA, Mahdi WA, Qamar W. Formulation of Silymarin-β Cyclodextrin-TPGS Inclusion Complex: Physicochemical characterization, molecular docking, and cell viability assessment against breast cancer cell lines. ACS Omega, 2023, 8, 34898–34907. https://doi.org/10.1021/acsomega.3c04225.

[120] Jadav M, Solanki R, Patel S, Pooja D, Kulhari H. Development of thiolated xanthan gum-stearylamine conjugate based mucoadhesive system for the delivery of biochanin-A to melanoma cells. Int J Biol Macromol, 2024, 257, 128693. https://doi.org/10.1016/j.ijbiomac.2023.128693.

[121] Ukani H, Singh P, Banerjee M, Parsana N, Mehra S, Kumar A, Kadam Y, Malek N. Stimuli-responsive multifunctional metal-organic framework- based injectable hydrogel for the treatment of breast cancer. J Mol Liq, 2024, 414, 126022. https://doi.org/10.1016/j.molliq.2024.126022.

[122] Serbezeanu D, Iftime MM, Ailiesei G-L, Ipate A-M, Bargan A, Vlad-Bubulac T, Rîmbu CM. Evaluation of poly(vinyl alcohol)–xanthan gum hydrogels loaded with neomycin sulfate as systems for drug delivery. Gels, 2023, 9, 655. https://doi.org/10.3390/gels9080655.

[123] Asl AM, Kalaee M, Abdouss M, Homami SS. Novel targeted delivery of quercetin for human hepatocellular carcinoma using starch/polyvinyl alcohol nanocarriers based hydrogel containing Fe_2O_3 nanoparticles. Int J Biol Macromol, 2024, 257, 128626. https://doi.org/10.1016/j.ijbiomac.2023.128626.

[124] Bhattacharya S, Page A, Shinde P. Capecitabine loaded potato starch-chitosan nanoparticles: A novel approach for targeted therapy and improved outcomes in aggressive colon cancer. Eur J Pharm Biopharm, 2024, 200, 114328. https://doi.org/10.1016/j.ejpb.2024.114328.

Shardendu Kumar Mishra, Vinod K. Tiwari, Kartik Patial, and Nidhi Mishra*

Chapter 3 Carbohydrate-functionalized nanoparticles: from laboratory to clinical applications

3.1 Introduction

A paradigm shift is underway in biomaterials research, with a growing emphasis on natural alternatives to synthetic polymers and inorganic nanoparticles [1–3]. Carbohydrates, being abundant and biocompatible molecules and due to their diverse role in cellular functions, have emerged as promising candidates for the development of new pharmaceutical motifs. Their ability to interact with biological systems, coupled with their inherent biodegradability, makes them attractive for drug delivery and tissue engineering applications [4–6]. Their capacity to form stable supramolecular structures further enhances their potential in biomedical fields [7–9].

In recent decades, nanoparticles have emerged as promising platforms for biomedical applications due to their unique properties including small size, large surface area, and extended circulation times (Figure 3.1) [10, 11]. However, challenges such as instability, aggregation, and potential toxicity hinder their direct use in biological systems [12]. To address these limitations, researchers have explored surface modification with biomolecules including nucleic acids, proteins, and carbohydrates. Among these, carbohydrates offer distinct advantages such as enhanced stability, reduced toxicity, improved biocompatibility, and the ability to interact with specific cellular receptors. This has spurred the development of diverse synthetic strategies to create carbohydrate-conjugated nanoparticles tailored for various biomedical applications [13–15].

This chapter will delve into the burgeoning field of carbohydrate-based nanoparticles (CBNPs), focusing on their design, synthesis, and biomedical applications. Two primary strategies will be explored: the conjugation of carbohydrates to pre-existing polymeric or inorganic particles and the construction of nanoparticles entirely from carbohydrate-based materials. The inherent advantages of carbohydrates, including

---

***Corresponding author: Nidhi Mishra**, Department of Chemistry, School of Applied and Life Sciences, Uttaranchal University, Dehradun 248007, Uttarakhand, India, e-mail: mishranidhi04@gmail.com
Shardendu Kumar Mishra, Department of Pharmacology, KIET School of Pharmacy, KIET Group of Institutions, Ghaziabad 201206, Uttar Pradesh, India
Vinod K. Tiwari, Department of Chemistry, Institute of Science, Banaras Hindu University, Varanasi 221005, Uttar Pradesh, India
Kartik Patial, Department of Chemistry, School of Applied and Life Sciences, Uttaranchal University, Dehradun 248007, Uttarakhand, India

https://doi.org/10.1515/9783111447094-003

biocompatibility, biodegradability, and the ability to interact with biological systems, will be emphasized. A comprehensive analysis of the applications of these particles across various biomedical domains, along with an outlook on future research directions, will be presented.

3.2 Preparation of carbohydrate-based nanoparticles

Carbohydrates can be utilized as the primary building block for biomaterials or as surface modifications for existing synthetic or inorganic platforms, as depicted in Table 3.1. Researchers have extensively studied polymeric glyconanoparticles (GNPs) derived from both natural polysaccharides and synthetic polymers because of their stability and ease of modification. For drug delivery systems (DDS), it is crucial that these polymeric nanocarriers are biocompatible and biodegradable. Consequently, a variety of synthetic and natural polymers have been employed including poly(ethylene glycol), poly(glycolic acid), Pluronic, poly(vinyl pyrrolidone), poly(lactic acid), poly (lactic-*co*-glycolic acid), poly(acrylic acid) (PAA), polytrimethylenecarbonate, and various polysaccharides such as alginate, amylose, chitosan, heparin, hyaluronic acid (HA), and cyclodextrin) as well as proteins and polypeptides like gelatin [16–18]. Natural polymers, such as polysaccharides and proteins, are particularly advantageous because they can be easily internalized and degraded, facilitating the release of drugs or biomolecules inside cells. Thus, polysaccharides are often preferred for DDS due to their biodegradability, stability, biocompatibility, and relatively low immunogenicity [19].

Polysaccharides can form either linear or branched structures depending on the types of monosaccharides involved. They exhibit a wide range of structural variations with modifiable groups such as hydroxyl, amino, and carboxylic acid functionalities. Their molecular weight spans from hundreds to thousands of daltons, contributing to their structural diversity. Additionally, polysaccharides can have anionic, cationic, or neutral charge states [20]. Certain polysaccharides, like chitosan, possess antimicrobial properties and may enhance the efficacy of antibiotic treatments [21]. The abundance of hydroxyl groups in polysaccharides allows for noncovalent interactions with biological tissues, such as epithelial and mucous membranes, which can improve drug targeting. Compared to some synthetic polymers, polysaccharides generally exhibit very low toxicity and high biocompatibility, and they are subject to enzymatic degradation, with their breakdown products often being reusable by cells [22]. Furthermore, polysaccharides are an inexpensive and renewable natural resource, making them a highly advantageous choice for DDS [23].

Inorganic nanoparticles, particularly those composed of gold, silver, iron, and silicon, have shown promise in bioimaging, theranostics, and photothermal therapy due

Figure 3.1: Advantages of nanoparticle-based drug delivery systems.

to their unique optical and magnetic properties [24]. However, their inherent toxicity and limited biocompatibility often necessitate surface modification to enhance their suitability for biological applications. While various coatings have been explored, carbohydrate-based coatings have emerged as a preferred strategy [25]. Carbohydrates can improve nanoparticle stability, reduce nonspecific interactions, and facilitate targeted delivery by recognizing specific cell surface receptors [26]. The incorporation of carbohydrates such as mannose, lactose, galactose, and glucose onto nanoparticle surfaces has been extensively investigated, with various synthetic methods developed to achieve this goal.

Table 3.1: Carbohydrates and their related therapeutic implications.

Class of carbohydrate	Structure and name	Therapeutic implications	References
Monosaccharides	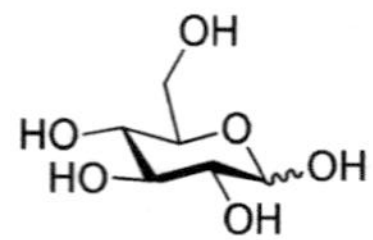 Glucose	Glucose and mannose have an affinity for GluT1 and GluT3 receptors present in the brain	[99, 100]
	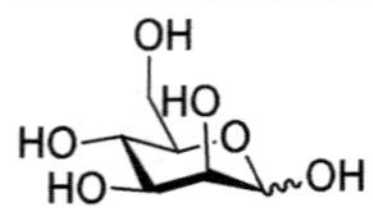 Mannose	Glucose and mannose have an affinity for GluT1 and GluT3 receptors present in the brain	[101–103]
	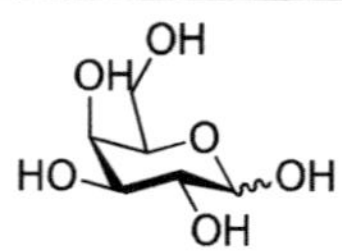 Galactose	The asialoglycoprotein receptor present in the liver has high binding activity with β-galactose	[104, 105]
	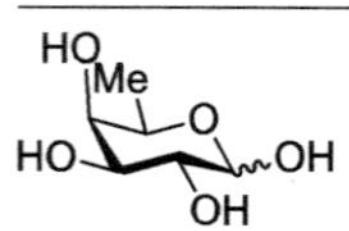 Fucose	Fucose derivatives bind with the fucose receptor present in the Kupffer cells of the liver	[106, 107]
	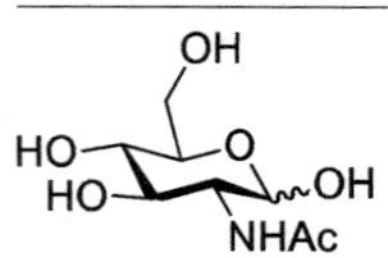 N-Acetyl glucosamine	Bind with receptors present in the liver and cardiac muscle	[108, 109]

	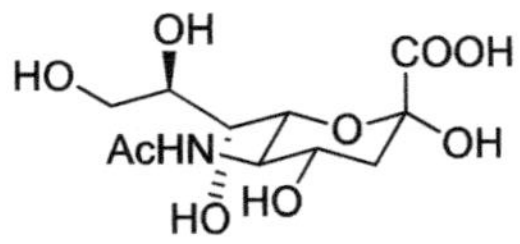 Sialic acid	Inhibitors of viral neuraminidases are enzymes essential for viral replication	[110, 111]
Disaccharides or oligosaccharides	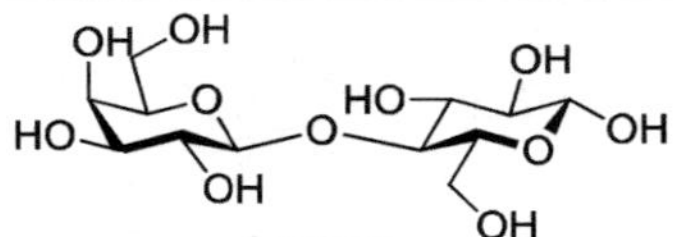 Lactose	Improves the flowability and compressibility	[104, 112]
	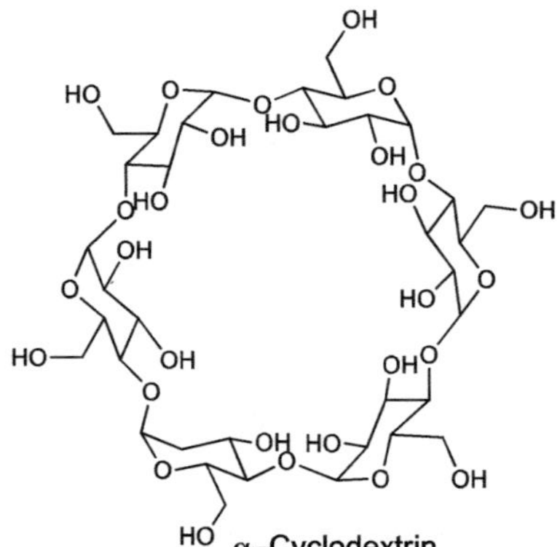 α–Cyclodextrin	Increases the solubility of poorly water-soluble drugs, improving bioavailability	[113]

(continued)

Table 3.1 (continued)

Class of carbohydrate	Structure and name	Therapeutic implications	References
Polysaccharides	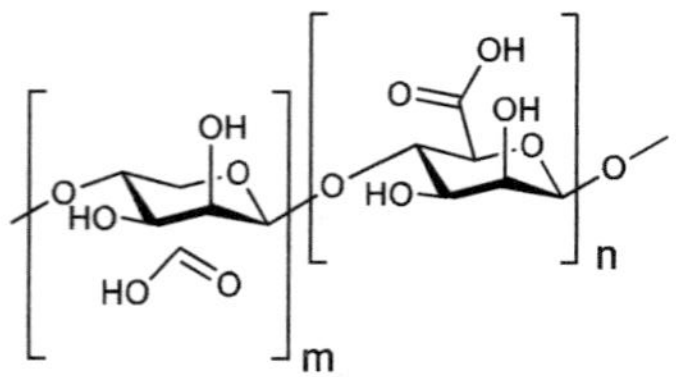 Alginate	Alginate serves as scaffolds for tissue engineering, providing a supportive environment for cell growth and proliferation	[114–116]
	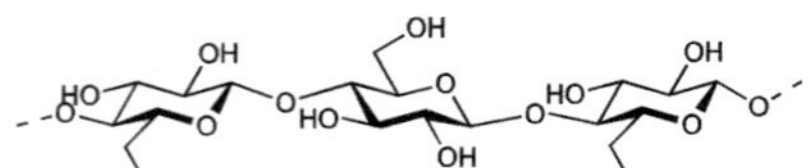 Cellulose	Exhibit excellent absorbent properties and create a moist wound environment, promoting healing	[117]
	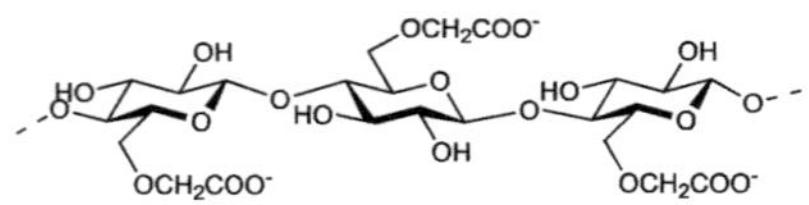 Carboxymethyl Cellulose	It can accelerate wound healing by promoting tissue regeneration and reducing inflammation	[118]
	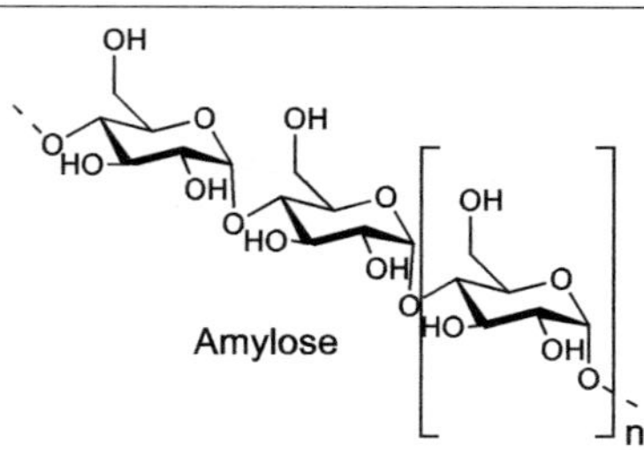 Amylose	Incorporated into the scaffolds to deliver growth factors and other bioactive molecules	[119]

Structure	Property	Ref.
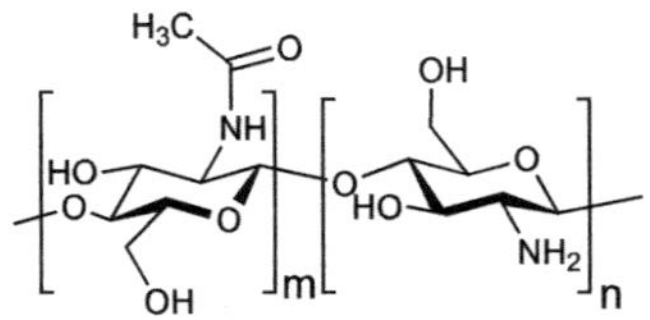 Chitosan	Antibacterial, mucoadhesive property	[120]
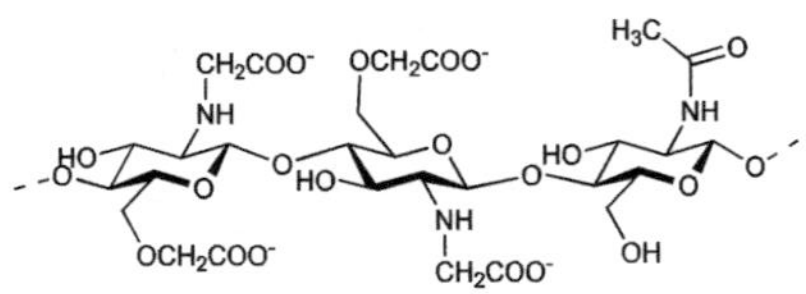 Carboxymethyl Chitosan	It can accelerate wound healing by promoting tissue regeneration and reducing inflammation	[121]
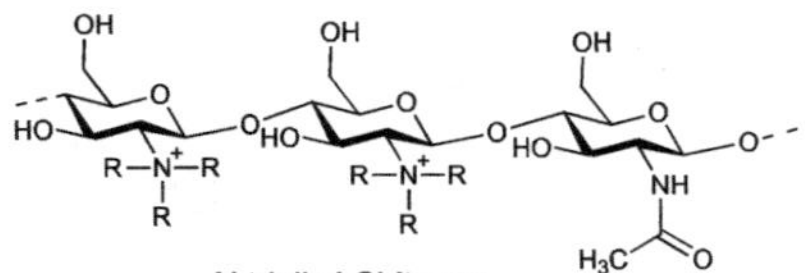 *N*-trialkyl Chitosan (alkyl= methy or ethyl)	By binding with specific ligands, drugs can be targeted to specific cells or tissues, improving treatment outcomes	[122]
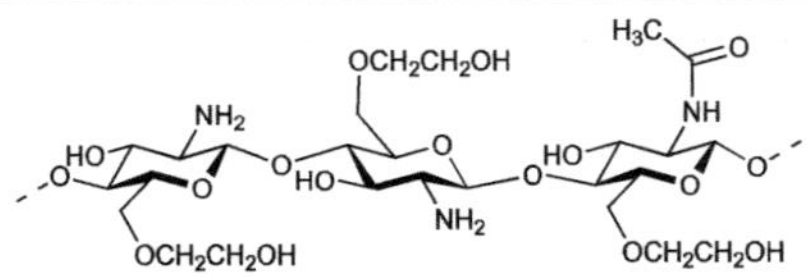 Glycol Chitosan	Improved water solubility and biocompatibility compared to its parent polymer	[123]

(continued)

Table 3.1 (continued)

Class of carbohydrate	Structure and name	Therapeutic implications	References
	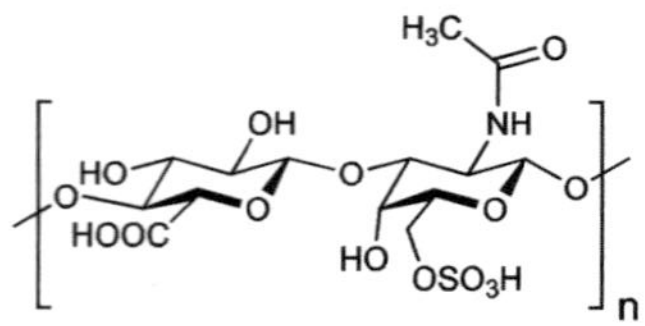 Chondroitin Sulphate	Used to alleviate osteoarthritis pain and improve joint function	[124]
	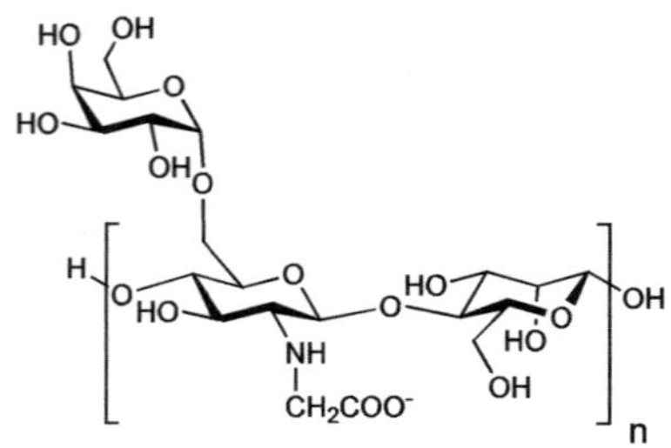 Guar Gum (Guaran)	It can be designed to deliver drugs in a sustained and controlled manner, improving therapeutic efficacy and reducing dosing frequency	[125]
	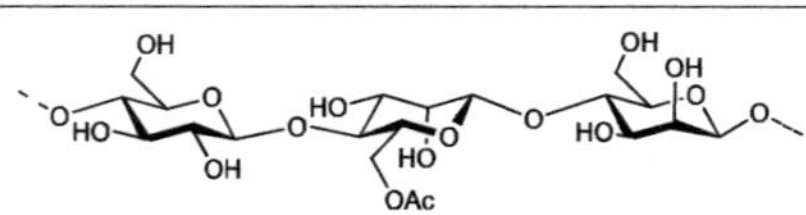 Glucomannan	It can be designed to deliver drugs in a sustained and controlled manner, improving therapeutic efficacy	[126]

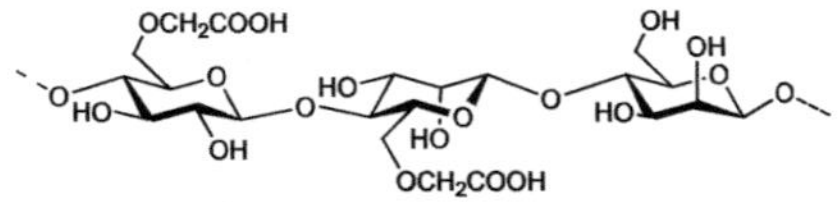 Carboxymethyl konjac Glucomannan	Biocompatibility and hemostatic properties make it suitable for wound dressing applications	[127]
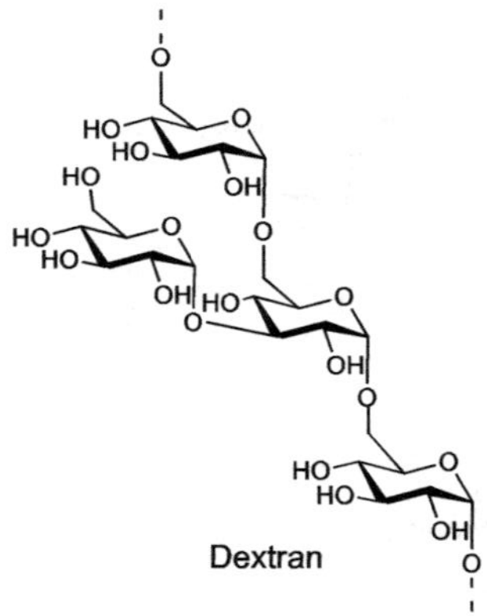 Dextran	Easily modified and having better biocompatibility	[128]

(continued)

Table 3.1 (continued)

Class of carbohydrate	Structure and name	Therapeutic implications	References
	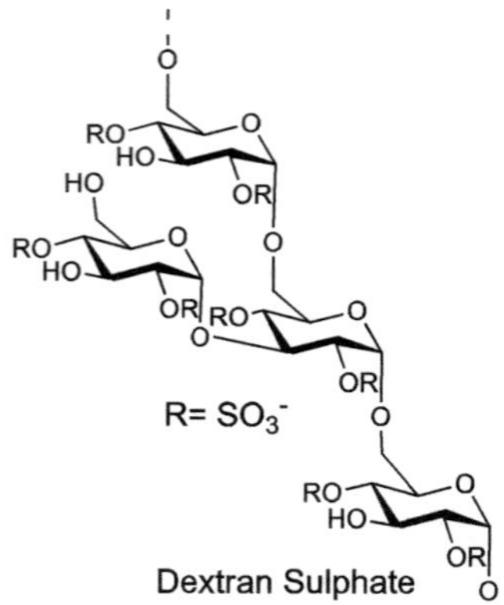 Dextran Sulphate	Intrinsic anticoagulant properties make it suitable for developing nanoparticles to prevent blood clot formation	[129]
	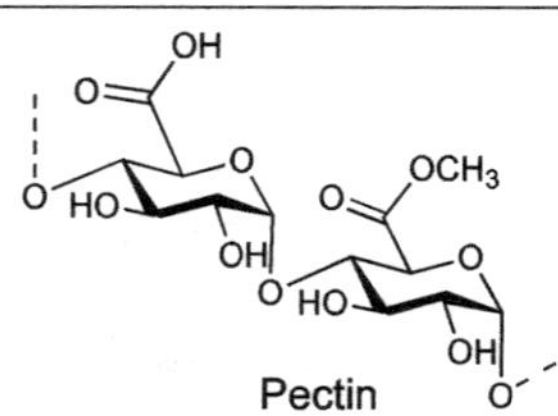 Pectin	It is engineered to deliver sustained-release drugs	[130]

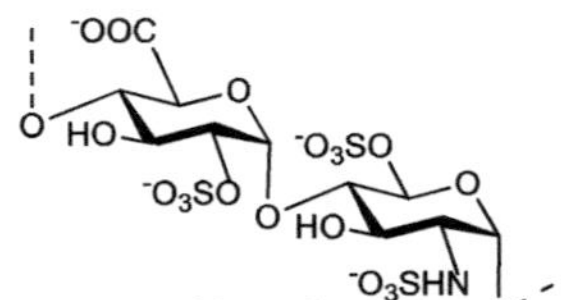 Heparin	Designed for controlled release of anticoagulant drugs, reducing the risk of thrombosis and embolism	[131]
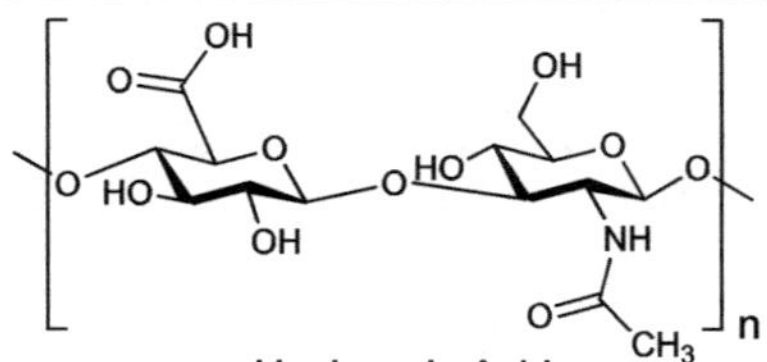 Hyaluronic Acid	Bind with CD44 and ICAM-1 receptors present on leukocytes	[132, 133]
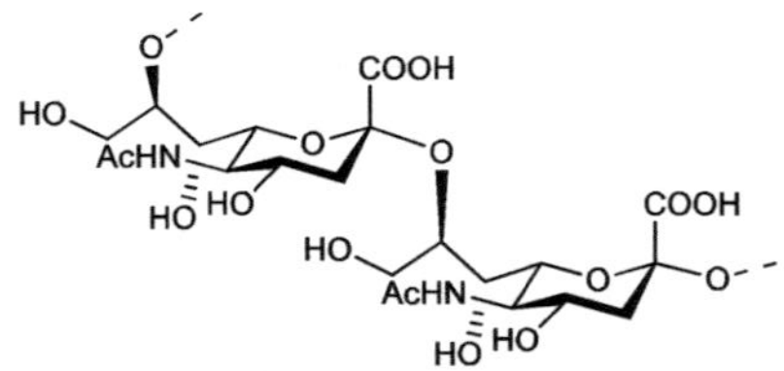 Poly(sialic acid)	Bind with immune cell receptors, modulating immune responses	[134]

(continued)

Table 3.1 (continued)

Class of carbohydrate	Structure and name	Therapeutic implications	References
	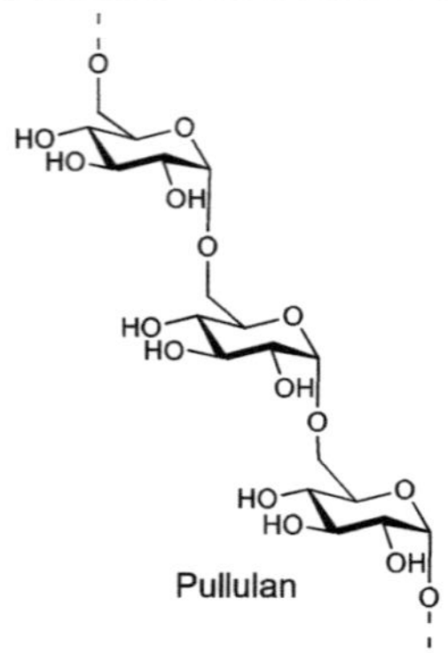 Pullulan	Pullulan nanoparticles with specific ligands and drugs can be targeted to specific cells or tissues, improving treatment outcomes; they can also be incorporated into wound-healing formulations to accelerate tissue repair and reduce inflammation	[135]
	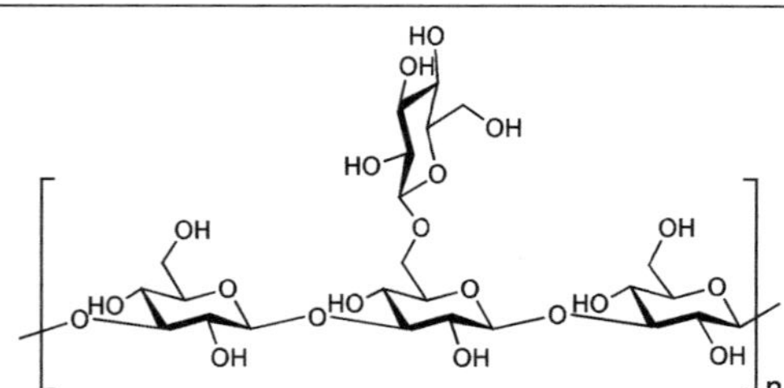 Schizophyllan	Employed to deliver anticancer drugs to tumor sites, improving therapeutic efficacy, reducing side effects, and possessing immunomodulatory actions	[136]

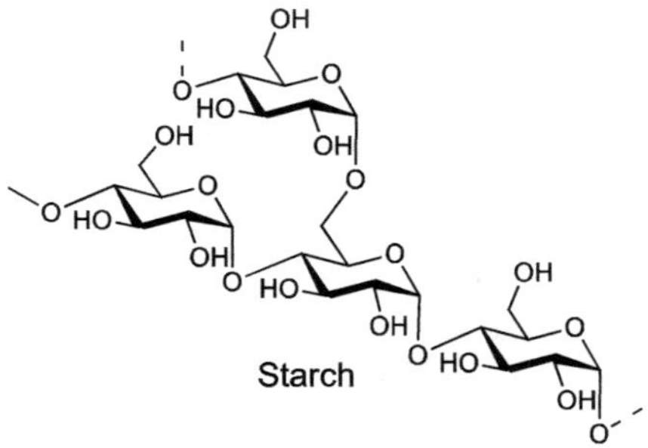 Starch	It is engineered to deliver sustained-release drugs, aid in vaccine development, and support tissue regeneration	[137]
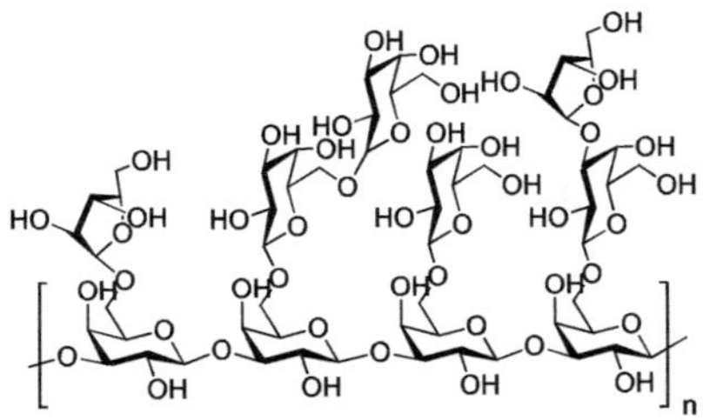 Arabinogalactan	It has immunostimulatory properties, enhancing the immune response; its biocompatibility and ability to form stable nanoparticles make it a suitable carrier for drug delivery systems	[138]

3.2.1 Preparation methods of glyconanoparticles

3.2.1.1 Carbohydrate-conjugated gold nanoparticles

Gold nanoparticles (AuNPs) have emerged as versatile platforms for biomedical applications due to their biocompatibility and unique physicochemical properties [27, 28]. Their surface chemistry can be readily modified through the formation of stable covalent bonds with thiol or disulfide-containing molecules. This property has been exploited to create carbohydrate-conjugated AuNPs, enabling the design of nanoparticles with specific biological functions [28, 29].

One-step and two-step approaches have been employed to synthesize these nanoparticles. The one-step method involves the simultaneous reduction of gold ions and conjugation of thiolated carbohydrates. Penadés and colleagues [29] pioneered the use of this approach to create oligomannose- and glucose-functionalized AuNPs, demonstrating their potential as HIV vaccine candidates as shown in Figure 3.2. Subsequent studies expanded on this concept, exploring the effects of carbohydrate density and composition on nanoparticle-lectin interactions. These investigations revealed the importance of multivalent carbohydrate presentation for enhanced binding affinity and selectivity [30–32].

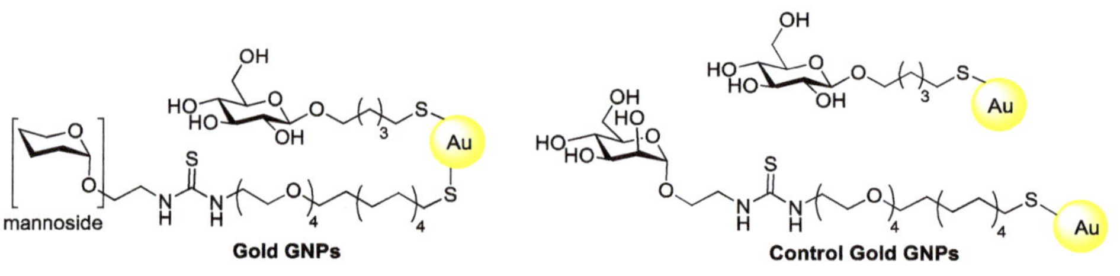

Figure 3.2: Oligomannose- and glucose-functionalized AuNPs developed by Penadés et al. [29].

Glycopolymers have been extensively incorporated into nanoparticle synthesis to create functional materials. A common approach involves the direct conjugation of carbohydrates to maleic anhydride-based copolymers, followed by nanoparticle formation through metal reduction [33–35]. Alternatively, photochemical methods have been employed, where carbohydrate-conjugated thiols act as both reducing and stabilizing agents for AuNP formation [36].

Two-step strategies have also been developed to introduce carbohydrates onto preformed nanoparticles. This approach typically involves the conjugation of thiolated carbohydrates to the nanoparticle surface [37, 38]. For example, AuNPs have been modified with various carbohydrates using both one-step and two-step methods, enabling the creation of nanoparticles with specific binding properties and applications in biosensing and diagnostics [39, 40].

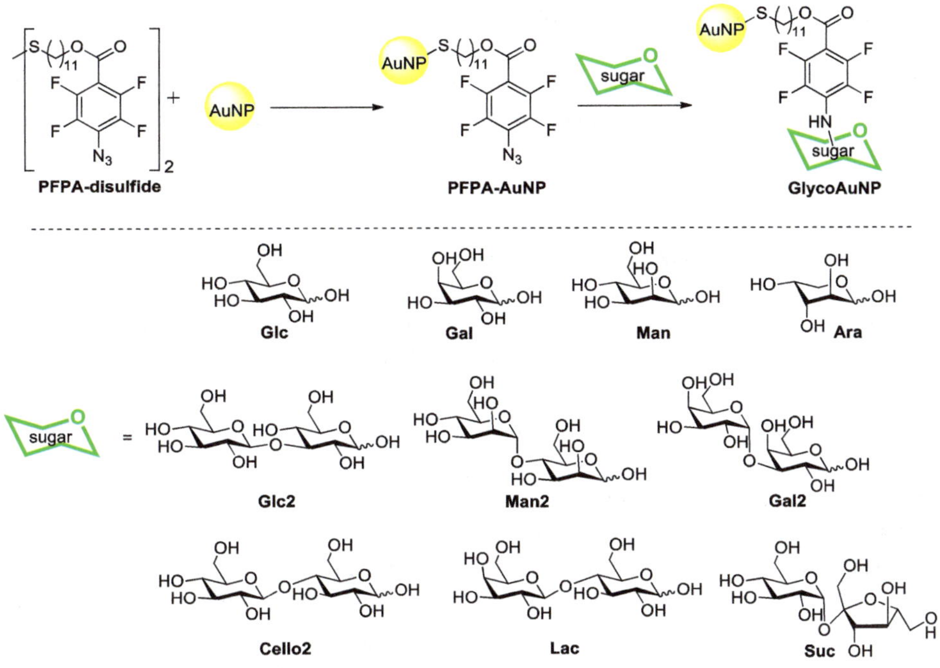

Figure 3.3: Conjugation of thiolated sugars to gold nanoparticles by Jayawardena et al. [41].

Jayawardena et al. [41] demonstrated the synthesis of glyco-AuNPs by conjugating thiolated sugars to preformed AuNPs, as depicted in Figure 3.3. These nanoparticles exhibited enhanced carbohydrate-protein interactions, as evidenced by quartz crystal microbalance studies [42]. The influence of carbohydrate structure and nanoparticle morphology on cellular uptake and bacterial adhesion has also been investigated. For instance, Chaudhary and coworkers [43] compared the behavior of rod-, spherical-, and star-shaped AuNPs decorated with different carbohydrates, revealing the impact of nanoparticle shape on bacterial interactions. These findings highlight the potential of carbohydrate-modified nanoparticles for developing advanced diagnostic and therapeutic tools [43].

3.2.1.2 Silver nanoparticles

Silver nanoparticles have emerged as versatile platforms in various biomedical applications due to their ease of synthesis and modification. To enhance their biocompatibility, targeting, and recognition capabilities, researchers have explored the use of carbohydrate coatings [44]. For instance, Sur et al. investigated the impact of glucose and lactose functionalization on nanoparticle uptake in different cell lines, demonstrating the potential of carbohydrate-modified silver nanoparticles for selective targeting [45].

Ramtenki et al. synthesized silver nanoparticles using carbohydrate-based surfactants, which acted as both reducing and stabilizing agents. These glyco-conjugated nanoparticles exhibited antibacterial properties, with mannose-modified nanoparticles demonstrating superior bacterial inhibition compared to glucose-based counterparts [46].

3.2.1.3 Iron nanoparticles

Iron oxide nanoparticles have emerged as promising platforms for biomedical applications due to their magnetic properties and relatively low toxicity. To enhance their biocompatibility and functionality, researchers have explored the use of carbohydrate coatings [47, 48]. These coatings improve nanoparticle stability, prevent aggregation, and facilitate interactions with biological systems. By modifying the nanoparticle surface with hydrophilic carbohydrate groups, it is possible to reduce magnetic coupling effects and improve overall performance [49–51].

Magnetic nanoparticles have emerged as versatile platforms for biomedical applications due to their unique magnetic properties [52]. To enhance their biocompatibility and functionality, researchers have developed strategies to coat these particles with carbohydrates. This approach leverages the ability of carbohydrates to interact with biological systems and improve nanoparticle stability. For instance, Alvarez-Paino et al. synthesized hybrid magnetite/polymer core-shell nanoparticles and subsequently introduced carbohydrate moieties through click chemistry [52]. Similarly, El-Boubbou et al. explored the attachment of monosaccharides to magnetic nanoparticles using both coupling and click chemistry reactions. These studies highlight the versatility of carbohydrate conjugation strategies for tailoring magnetic nanoparticle properties [53].

Valero et al. developed a novel approach to create carbohydrate-coated magnetic nanoparticles using apoferritin as a biotemplate. Maghemite nanoparticles were encapsulated within apoferritin through a pH-dependent assembly process. Subsequent modification with carbohydrates, such as *N*-acetyl-D-glucosamine and D-mannose, was achieved by exploiting the protein's amine groups. This method yielded water-dispersible nanoparticles with a controlled size and prevented aggregation [54–56].

3.2.1.4 Silicon nanoparticles

Silicon nanoparticles (SiNPs) have gained prominence in biomedicine due to their favorable properties including biocompatibility, low toxicity, and luminescence. To enhance their functionality, researchers have explored surface modification strategies. For instance, Zhao et al. incorporated fluorescein isothiocyanate and carbohydrates into mesoporous silica nanoparticles, creating multifunctional platforms for drug delivery [57]. Tong et al. developed a microarray platform using carbohydrate-conjugated silica nanoparticles for bioanalysis [58]. Hsu et. al exemplified the versatility of silica nanoparticles for creating complex nanomaterials with tailored properties, as depicted in Figure 3.4 [59].

Several strategies have been employed to synthesize carbohydrate-functionalized silicon-based nanoparticles. Hsu et al. [59] developed a multistep process involving the conversion of SiNPs into carboxylic acid-terminated intermediates, followed by coupling with amino-functionalized glucose. Zhao et al. [57] synthesized glycoside-decorated silica nanoparticles through a combination of fluorescent dye incorporation and azide-alkyne click chemistry. Zhai et al. [60] prepared luminescent silicon nanocrystals functionalized with D-mannose or L-alanine via a series of chemical transformations. These studies highlight the versatility of SiNPs for creating CBNPs with potential applications in biomedicine.

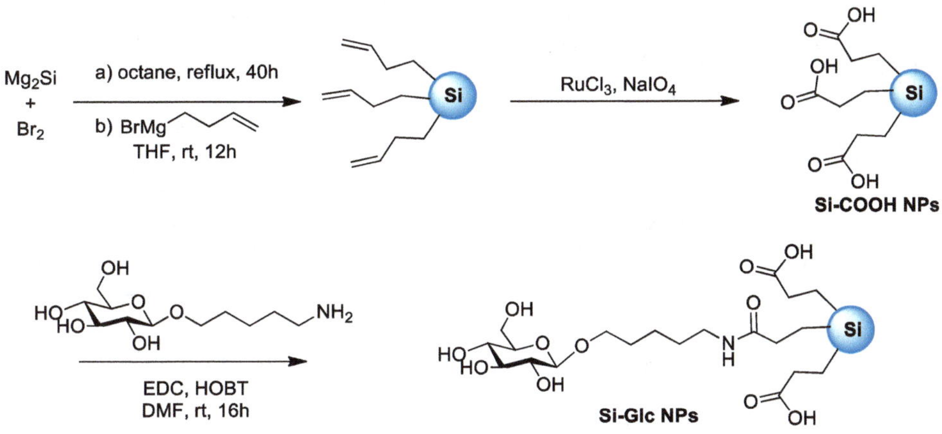

Figure 3.4: Synthesis of glucose-coated silicon nanoparticles [59].

3.2.1.5 Covalently cross-linked glyconanoparticles

Early GNP fabrication methods centered on covalent crosslinking to create sturdy, drug-loaded networks resistant to pH fluctuations. The choice of crosslinking agent influenced the nanoparticle's biodegradability and responsiveness to environmental stimuli such as pH, temperature, and light. This adaptability proved beneficial for drug delivery, especially within the acidic macrophage environment [61, 62].

Chitosan, a linear polysaccharide composed of glucosamine and *N*-acetylglucosamine units, is a versatile platform for nanoparticle development [63]. Its solubility in acidic environments and susceptibility to enzymatic degradation make it a promising candidate for biomedical applications. To enhance its properties and create cross-linked networks, chitosan is often treated with agents such as glutaraldehyde, dopamine, genipin, or biocompatible acids like citric or tartaric acid [64–66].

Recent advancements in nanoparticle engineering have explored innovative crosslinking strategies to enhance particle properties and functionality. For example, Meng et al. developed multifunctional nanocarriers that exhibited responsiveness to both pH and light through the crosslinking of glycol chitosan-*o*-nitrobenzyl succinate conjugates with glutaraldehyde [67]. To reduce toxicity, genipin has been employed as an alternative crosslinker in chitosan-based nanoparticles, as demonstrated by Lin [68] (Figure 3.5). Additionally, Chen et. al introduced a novel approach to create pH-responsive chitosan nanocapsules with adjustable sizes using *N*-maleoyl-functionalized chitosan by pH-dependent miniemulsion encapsulation [61] (Figure 3.6). These studies highlight the versatility of crosslinking in tailoring nanoparticle characteristics for diverse applications [69].

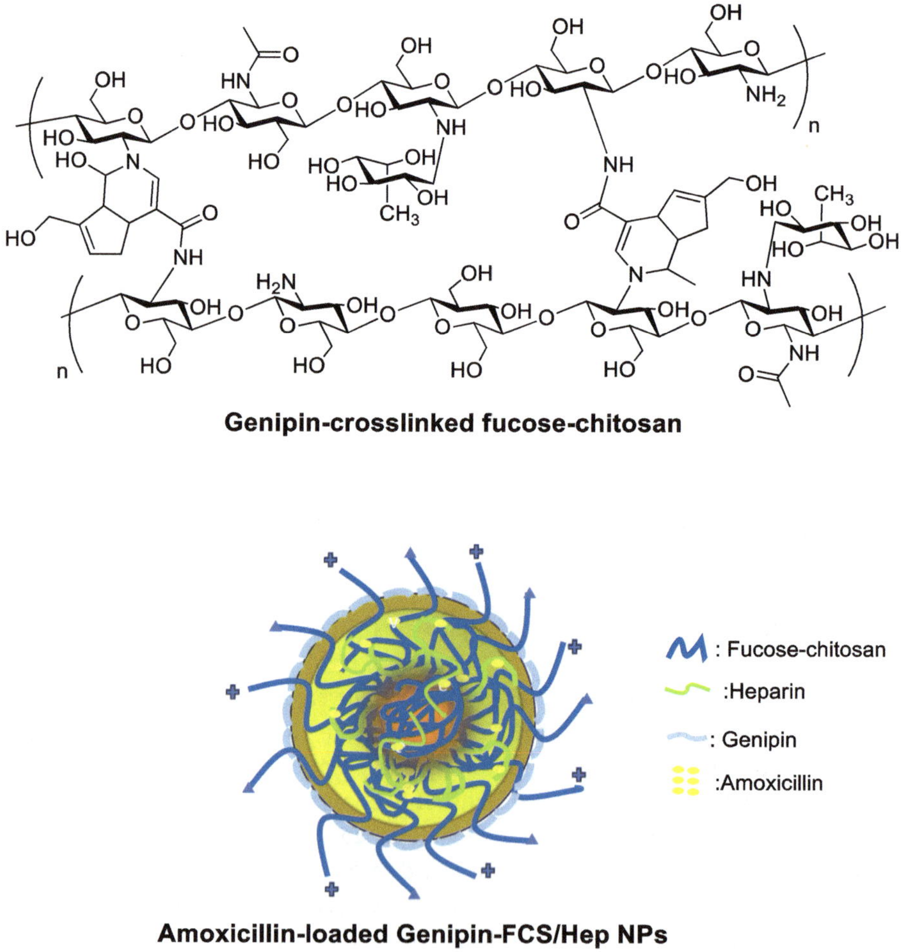

Figure 3.5: Synthesis of amoxicillin-loaded ginpin-FCS/Hep NPs by Lin et al. [68].

HA is a commonly used polysaccharide in the production of GNPs due to its natural occurrence in mammalian tissues, which boosts its biocompatibility. HA is a nonsulfated glycosaminoglycan composed of repeating disaccharide units of β-1,4-D-glucuronic acid and β-1,3-*N*-acetyl-D-glucosamine, with molecular weights ranging from 100 to 8,000 kDa. Bondar and colleagues synthesized spherical GNPs by covalently cross-linking HA chains using carbodiimide methods. They also developed an injectable hydrogel with excellent tissue adhesion by utilizing catechol-thiol reactions between HA and pluronic copolymers [70]. Specifically, HA-dopamine conjugates were combined with thiol-terminated Pluronic F127 copolymers to create HA/Pluronic hybrid hydrogels. Verheul and collaborators [71] produced covalently stabilized trimethyl chitosan-HA nanoparticles, approximately 200–300 nm in size, for nasal and intradermal vaccination. More recently, robust GNPs were created by conjugating HA with branched polyethyleneimine via carbodiimide chemistry and capping the copolymer termini with mannose [72]. Martinez et al. developed GNPs, ranging from about 40 to 400 nm, using thiolated alginate and modified albumin, which were crosslinked through disulfide bonds [73]. Additionally, Sarmah and colleagues [74] prepared tamoxifen-loaded guar gum GNPs via a water-in-oil emulsion polymerization method using glutaraldehyde as the crosslinker.

3.2.1.6 Ionically crosslinked glyconanoparticles

GNPs can be synthesized using ionic cross-linkers and charged polysaccharides through electrostatic interactions. This reversible approach produces biocompatible GNPs by avoiding complex synthetic processes and toxic cross-linkers. The GNPs formed are sensitive to pH changes, making them ideal for stimuli-responsive drug delivery. Typically, the process involves mixing very dilute solutions of polyanionic and polycationic polysaccharides with their corresponding oppositely charged electrolytes to form the GNPs [75].

Tripolyphosphate (TPP), a nontoxic, multivalent polyanion, is frequently used as a cross-linker with cationic chitosan to create GNPs. In acidic conditions (pH around 4–6), the positive charges on chitosan facilitate the formation of GNPs through interactions between chitosan's amino groups and TPP's negatively charged phosphates [76–80]. Conversely, in basic solutions (pH 7–9), the interactions remain effective for GNP formation [81, 82]. Domaratzki and Ghanemin [83] successfully employed ionotropic gelation to produce smooth, spherical chitosan GNPs using TPP as the cross-linker. Their study revealed that nanoparticle size varies with hydration levels and cross-linking conditions, with particles ranging from 160 to 270 nm [83]. TPP-chitosan GNPs have been developed for the delivery of various substances, including DNA, siRNA, proteins, doxorubicin, cladribine, and antibiotics (such as aminoglycosides, rifampicin, ceftriaxone, and ciprofloxacin), demonstrating high encapsulation efficiency [84–87].

Water-soluble chitosan derivatives, such as *N*-trimethyl chitosan, mannose-conjugated chitosan, galactosylated *N*-trimethyl chitosan, PEGylated chitosan, and glycyrrhe-

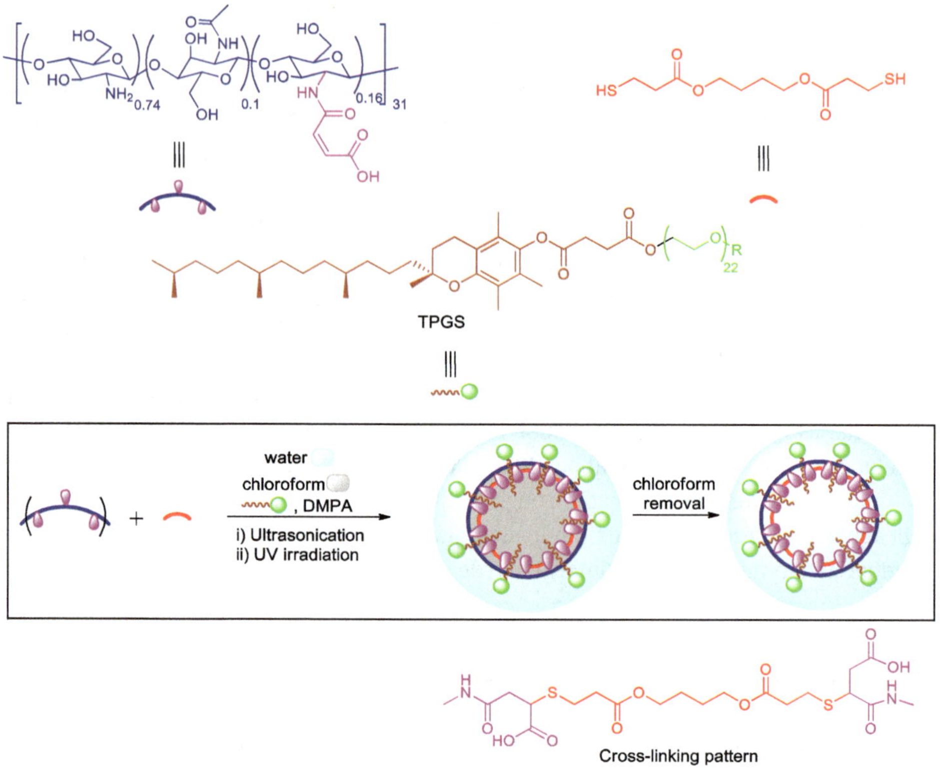

Figure 3.6: Use of miniemulsion-interfacial cross-linking technique for synthesis of chitosan nanoparticles.

tinic acid-modified chitosan, have been crosslinked ionically using TPP to create GNPs [87–90]. These GNPs can swell or shrink in response to physical and chemical stimuli like temperature, pH, and ionic strength, which aids in the controlled release of encapsulated drugs [91–93]. Recently, Xu et al. developed chitosan-TPP/IL-12 nanoparticles for delivering therapeutic macromolecules. The mean diameter of these nanoparticles ranged from 178 to 372 nm, increasing with the chitosan/TPP weight ratio, and their zeta potential varied from 24 to 53 mV depending on the ratio of chitosan to TPP [94, 95].

Furthermore, researchers have used more subtle physical crosslinking methods to prepare chitosan GNPs. This involves incorporating functional groups into drugs to form hydrogen bonds with substances like retinol, cladribine, or glabridine [96–98]. Additionally, chitosan was combined with adenosine-5′-triphosphate (ATP) to form spherical aggregates through electrostatic interactions, as depicted in Figure 3.7. By adding ATP dropwise to chitosan, varying ratios of ATP to chitosan led to partial binding of chitosan to ATP, while the remaining chitosan remained cationic and hydrophilic. This method produced polymeric supra-amphiphiles, resulting in chitosan-ATP aggregates with sizes between 50 and 68 nm [98].

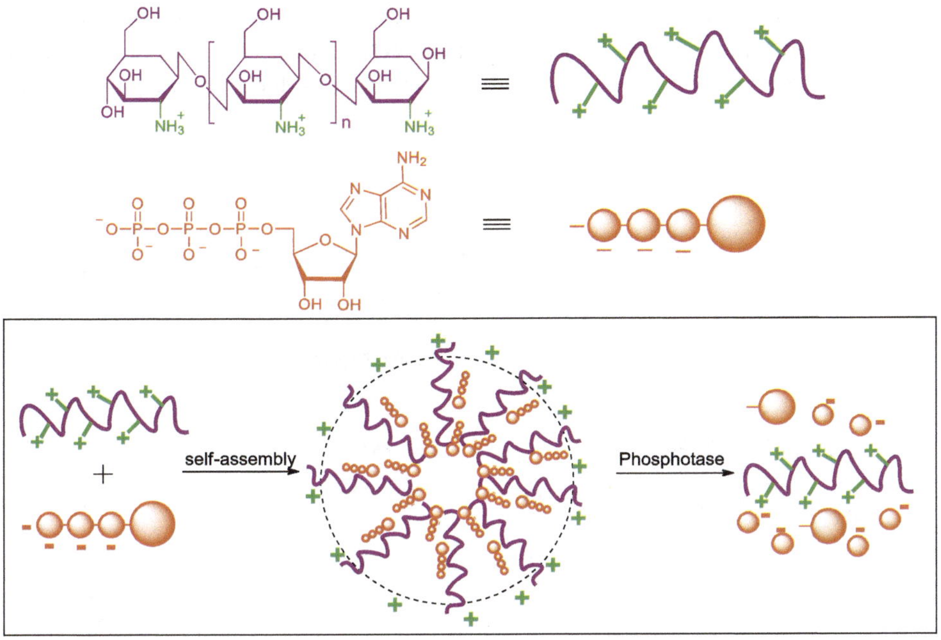

Figure 3.7: Spherical aggregate formation by the combination of chitosan with adenosine-5′-triphosphate (ATP).

3.2.1.7 GNPs by polyelectrolyte complexation (PEC)

GNPs can be synthesized using oppositely charged polysaccharides without the need for covalent cross-linkers. These nanoparticles are formed through strong yet reversible electrostatic interactions, complemented by hydrogen bonding and hydrophobic interactions between the polyelectrolytes. The stability of these polyelectrolyte complexes (PECs) is influenced by factors such as pH, temperature, mixing ratio, and the molecular weight of the polyelectrolytes. Compared to chemically crosslinked complexes, PEC-based GNPs are generally less toxic, more biocompatible, and better tolerated due to their simpler preparation and modification processes. The charge of the GNPs can be either positive or negative, depending on the type and concentration of the polysaccharides used. The hydrophobic core of these nanoparticles contains neutral segments or "salt bridges," while the surface charge results from excess polyelectrolyte components not incorporated during PEC formation [139].

While biocompatible and water-soluble polysaccharides are commonly used for forming GNPs, a variety of polyelectrolytes can also be employed. Among these, chitosan is frequently used due to its enhanced solubility and safety profile. Its cationic nature, from free amino groups, enables chitosan to form PECs with negatively charged polyelectrolytes [140]. Research has extensively explored the PEC formation of chitosan with various anionic polymers, including alginate, pectin, carboxymethyl cellulose (CMC), heparin, dextran sulfate, HA, carrageenan, collagen, xanthan gum, and carboxymethyl pachyman (PAA and nucleic acids, as depicted in Figure 3.8 [112, 140, 141].

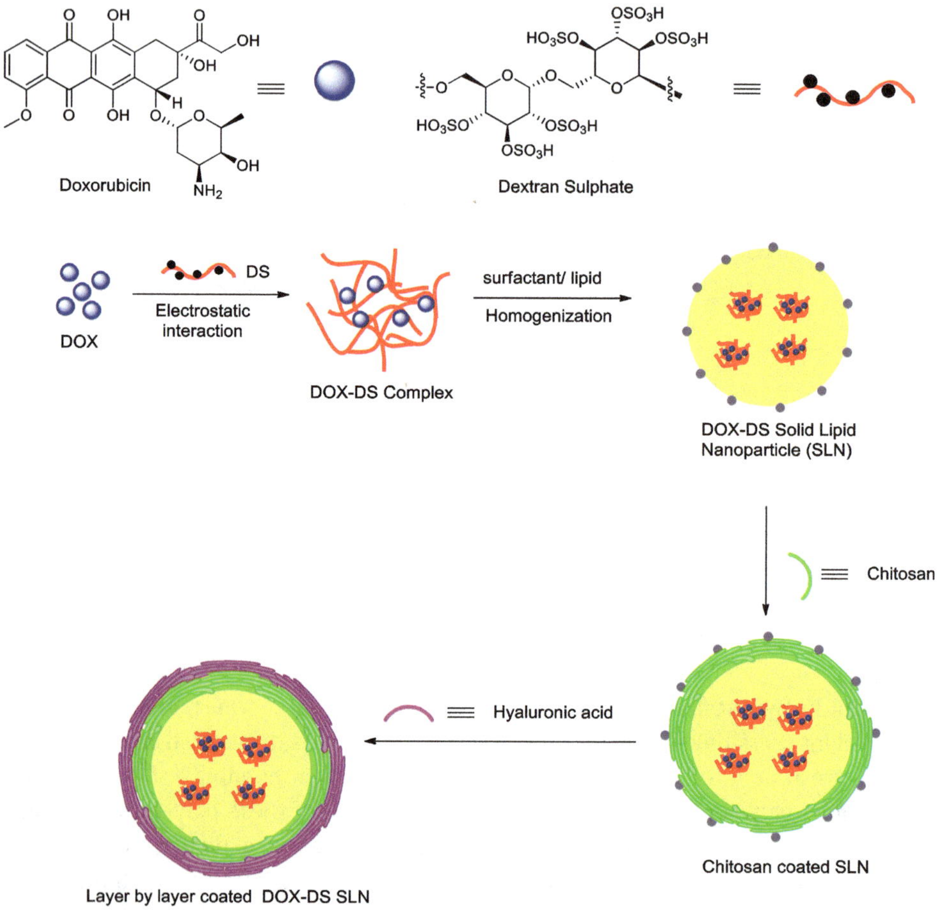

Figure 3.8: Synthesis of layer-by-layer coated solid-lipid nanoparticles by Ramasamy et al. [140].

Chitosan-anionic polymer complexation has been widely applied in nanoparticle fabrication. For example, Motwani et al. used chitosan and sodium alginate to form nanoparticles with sizes ranging from 62 to 193 nm. Guo et al. introduced hydrophobic modifications to chitosan, creating self-assembled nanoparticles coated with alginate. Chitosan-carrageenan nanoparticles were also produced via PEC, demonstrating the method's versatility [142]. Particle size and zeta potential were affected by polymer ratios and mixing conditions. To improve nanoparticle properties, additional polymers or crosslinkers have been incorporated. Rodrigues et al., for instance, used TPP to decrease particle size and enhance yield in chitosan-carrageenan systems. Lee et al. [143] explored alginate-related polyguluronate for siRNA delivery using chitosan PECs. Various other anionic polymers, such as dextran sulfate, pectin, and CMC, have been combined with chitosan for diverse applications [144]. Techniques like layer-by-layer assembly have also been used to create multilayered nanoparticles with tailored properties [145].

Research groups have investigated the formation of GNPs using PAA and chitosan via PEC. Chen and colleagues studied the development of GNPs by complexing polyanionic PAA with polycationic chitosan solutions, leading to nanoparticles with different microstructures under various conditions. Rolland et al. [146] produced PEC-based GNPs from chitosan and both PAA homopolymers and polystyrene-*block*-PAA diblock copolymers, finding that nanoparticles from chitosan-PAA homopolymer PECs were smaller than 100 nm, while those from chitosan-PAA copolymer PECs were under 200 nm [147].

Tang et al. developed a thermoresponsive chitosan/PVA hydrogel-containing GNPs with varying charges, influenced by the electrostatic interactions between quaternary amines in trimethyl ammonium chitosan and carboxylates in carboxymethyl chitosan. They observed that drugs with positive or negative charges were released more slowly from hydrogels containing negatively or positively charged nanoparticles, respectively. Davidenko and colleagues reported on chitosan/PAA PEC nanoparticles and found that GNP suspensions could be obtained at a solution concentration of 0.1 wt%. Their study indicated that pH values significantly impacted both yield and particle size, with optimal yields of around 90% achieved at pH values of 4.5–5.5 for chitosan and 3.2 for PAA [148, 149].

3.2.1.8 Self-assembled GNPs

PEGylation, which involves attaching polyethylene glycol (PEG) to biomolecules, has significantly advanced the field of nanoparticle-based drug delivery. PEG's hydrophilicity, low toxicity, and biodegradability make it an excellent choice for enhancing nanoparticle circulation time and minimizing immunogenic responses [150–152]. Several studies have investigated combining PEG with chitosan to develop versatile GNPs. For instance, Xie et al. created pH-sensitive PEGylated carboxymethyl chitosan nanoparticles for siRNA delivery, while Vasquez et al. used PEGylated chitosan *Bombyx mori* copolymers to facilitate self-assembly-based GNP formation. Gao et al. employed bifunctional PEG to construct siRNA/trimethylated chitosan/PEG complexes for improved brain delivery [153, 154]. Additionally, Guo et al. explored PEGylated sodium alginate nanoparticles to enhance drug loading and efficacy. These studies underscore PEG's versatility in designing GNPs with specific properties for various therapeutic applications [155].

GNPs are designed by incorporating hydrophilic glycopolymer chains that mimic the carbohydrate patterns found on cell surfaces. These nanoparticles can effectively bind to lectins – sugar-binding proteins – due to their multivalent presentation of carbohydrate ligands [156]. The self-assembly of amphiphilic glycopolymers into nanoparticles creates versatile platforms for drug delivery and other biomedical applications. The structural similarity between synthetic glycopolymers and natural glycoconjugates enhances their interaction with biological systems, making them promising candidates for targeted therapies [157].

3.3 Applications of carbohydrate-based nanoparticles

CBNPs have gained significant attention in the biomedical field due to their unique properties and versatile applications in both therapeutic and diagnostic contexts. These nanoparticles leverage the structural characteristics of carbohydrates, which play essential roles in biological processes, including cell recognition, signaling, and immune response. CBNPs have emerged as versatile tools in the diagnosis and treatment of various diseases. These nanoparticles facilitate targeted drug delivery, enhance therapeutic efficacy, and improve diagnostic accuracy [158] (Tables 3.2 and 3.3).

3.3.1 Therapeutic applications

3.3.1.1 Drug delivery systems

CBNPs are increasingly utilized as drug delivery vehicles due to their ability to encapsulate therapeutic agents effectively. For example, glycosylated gold nanoparticles have been developed for targeted delivery of anticancer drugs such as doxorubicin. The carbohydrate moieties facilitate specific interactions with cell surface receptors, promoting uptake by target cells while minimizing systemic toxicity. Doxorubicin-loaded GNPs have shown enhanced therapeutic efficacy against breast cancer cells by improving drug solubility and allowing for controlled release at the tumor site [158].

CBNPs can be engineered to respond to specific stimuli, such as pH and reactive oxygen species, allowing for targeted drug release in inflammatory conditions like ulcerative colitis [159]. In rheumatoid arthritis, CBNPs improve the bioavailability and stability of therapeutic agents, utilizing smart reactive systems that respond to environmental changes [159].

3.3.1.2 Gene delivery

CBNPs also play a crucial role in gene therapy by delivering nucleic acids such as DNA or RNA into target cells. Glycosylated nanoparticles enhance the stability and cellular uptake of these genetic materials, which is vital for treating genetic disorders. Studies have demonstrated that glycosylated nanoparticles can effectively deliver siRNA into cells, resulting in successful gene silencing and therapeutic effects in models of genetic diseases [160, 178].

Polysaccharide-based nanoparticles, including chitosan and dextran, serve as drug carriers due to their biocompatibility and ability to cross biological barriers. Chitosan nanoparticles are highlighted for their mucoadhesive properties, enabling them to facilitate drug transport through mucosal surfaces and increase bioavailability. Such polysaccharide-based systems are also explored for delivering genes and peptides for therapeutic applications.

3.3.1.3 Anticancer therapy

Nanoparticles functionalized with carbohydrates can specifically target tumor cells through receptor-mediated endocytosis. This targeted approach improves drug efficacy while reducing side effects associated with conventional therapies. Research has indicated that glycosylated quantum dots can deliver chemotherapeutics directly to cancer cells, significantly enhancing treatment outcomes compared to nontargeted approaches [160, 178].

Metallic nanoparticles are investigated for hyperthermia treatment, where their metallic cores (e.g., gold and iron oxide) can be heated via external fields to induce localized cell death in tumors. Gold GNPs, for instance, are studied for their potential in photothermal therapy. CBNPs enhance targeting to specific cancer cells, improving the selectivity and efficiency of these treatments [160, 178].

3.3.1.4 Immunotherapy

CBNPs are being explored for their potential in immunotherapy. By presenting carbohydrate antigens on their surfaces, these nanoparticles can enhance immune responses against tumors or infectious agents. GNPs have been designed to improve vaccine efficacy by acting as adjuvants or carriers for antigens, stimulating a stronger immune response [169].

3.3.1.5 Wound healing

Carbohydrate-based hydrogels have shown promise in wound healing applications by providing a moist environment conducive to healing while delivering growth factors or antimicrobial agents directly to the wound site. The incorporation of cellulose nanocrystals into hydrogels has been shown to enhance mechanical properties and promote cell migration during wound healing processes [168, 174].

3.3.1.6 Antimicrobial agents

CBNPs exhibit antimicrobial properties by encapsulating essential oils or antibiotics within a carbohydrate matrix. This approach enhances the solubility and stability of active compounds while allowing for controlled release at the site of infection. Starch nanoparticles loaded with essential oils have demonstrated significant antibacterial activity against various pathogens, making them suitable for food preservation applications [163, 172].

Carbohydrate-functionalized nanoparticles can interfere with pathogen adhesion to host cells by mimicking host cell surface glycans, effectively serving as "anti-adhesion" agents. For instance, multivalent GNPs mimic glycosphingolipids on host cells to prevent pathogen binding. In studies, lactose-coated GNPs inhibited lung metastasis in animal models by targeting carbohydrate interactions essential for cell adhesion in metastatic cancer cells [172].

3.3.1.7 Anti-HIV and antiviral agents

GNPs functionalized with high-mannose oligosaccharides target HIV by binding to host cell receptors that facilitate viral entry, thereby blocking infection pathways. This approach uses GNPs as decoys to attract the virus away from actual host cells. These antiadhesive strategies represent an innovative approach to preventing infections at early stages, showing promise as microbicides for HIV [163, 168, 172].

3.3.1.8 Vaccine development

Carbohydrate-coated nanoparticles are explored as potential platforms for synthetic carbohydrate-based vaccines. GNPs can display bacterial polysaccharide antigens, such as those from *Neisseria meningitidis* or *Streptococcus pneumoniae*, to elicit immune responses without needing protein carriers. This mimetic approach shows promise in creating effective and stable vaccines against specific pathogens [177].

3.3.2 Diagnostic applications

3.3.2.1 Biodiagnostics

Carbohydrate-functionalized nanoparticles are increasingly used in biodiagnostic [39, 40] applications due to their ability to detect biomolecules such as proteins and nucleic acids with high sensitivity and specificity. Glycosylated AgNPs have been employed in various assays for detecting disease markers in biological samples, showcasing their potential as diagnostic tools. Carbohydrate-coated metallic nanoparticles (GNPs) have shown potential as in vitro diagnostic tools, especially in detecting cells and pathogens. Metallic cores like gold or iron oxide are functionalized with carbohydrate ligands that target specific carbohydrate-binding proteins, enabling highly sensitive detection methods.

3.3.2.2 Imaging techniques

The unique optical properties of CBNPs make them suitable for imaging applications in medical diagnostics [11, 23, 24, 59]. Quantum dots conjugated with carbohydrates can be used in fluorescence imaging to visualize cellular processes or track the distribution of drugs within organisms.

Quantum dots (QDs) and near-infrared fluorophores are carbohydrate-functionalized for in vivo optical imaging. Applications include liver imaging and inflammation detection in diseases such as arthritis and cancer. However, issues of toxicity limit the clinical use of QDs. Nanoparticles encapsulated with dextran or HA have shown potential for imaging tumors and lymphatic systems in preclinical models.

3.3.2.3 Microarray technology

Nano-carbohydrates have been utilized in microarray platforms for the efficient detection of proteins and nucleic acids. Functionalizing surfaces with GNPs allow researchers to create high-density microarrays that enable the simultaneous analysis of multiple biomarkers, facilitating the early diagnosis of diseases such as cancer [59].

3.3.2.4 Magnetic resonance imaging (MRI)

Magnetic nanoparticles conjugated with carbohydrates have been developed to enhance MRI contrast agents [54]. HA-encapsulated magnetic nanoparticles have been used to specifically detect macrophage-related atherosclerotic plaques through MRI.

Metallic CBNPs are employed as MRI contrast agents, aiding in noninvasive imaging of conditions such as cancer and neural disorders. Dextran-coated iron oxide nanoparticles, already in clinical use, enhance MRI contrast, while newer designs explore polysaccharide-encapsulated magnetic nanoparticles for their biocompatibility and targeting abilities. Carbohydrates like sialyl Lewis-X are functionalized on iron oxide nanoparticles for brain imaging, specifically targeting cerebral inflammation markers related to conditions like multiple sclerosis and brain tumors.

3.3.2.5 Biosensors

CBNPs are being integrated into biosensors for real-time detection of pathogens or biomarkers associated with diseases [39, 40]. The ability of GNPs to bind specifically to target molecules enables the development of highly sensitive biosensors capable of providing rapid diagnostic results.

GNPs can mimic cell glycocalyx to identify pathogens by interacting with carbohydrate-binding sites on bacterial or viral surfaces. For example, mannose-coated GNPs effectively target *Escherichia coli* due to the bacterium's mannose-binding adhesin, while dextran-coated nanoparticles are applied to test bacterial resistance in antimicrobial assays.

Table 3.2: Applications of carbohydrate-based nanoparticles in the treatment of different diseases.

Disease category	Application	Mechanism/details	Reference
Cancer treatment	Immunotherapy and drug delivery	Synthetic carbohydrate-based nanoparticles target tumor cells by binding to E-selectin receptors, increasing antitumor efficacy and reducing systemic toxicity.	[160]
	Drug delivery	Polycaprolactone (PCL)-based nanoparticles are used for the stable, controlled release of anticancer drugs, enhancing efficacy.	[161]
Infectious diseases	Antibacterial and antiviral agents	Biomimetic macrophage-targeting nanoparticles designed to adhere to bacterial/viral particles, reducing the spread of infection.	[162]
	Cell wall disruption	Carbohydrate-based nanoparticles disrupt bacterial cell walls, limiting infection by preventing adhesion and compromising cell integrity.	[163]

Table 3.2 (continued)

Disease category	Application	Mechanism/details	Reference
Autoimmune and inflammatory diseases	Immunomodulation	Carbohydrate-conjugated nanoparticles selectively modulate immune responses to manage autoimmune diseases.	[164]
	Anti-inflammatory	Glycan-based nanoparticles bind to galectin receptors on immune cells, reducing inflammation.	[165]
Neurodegenerative disorders	Blood-brain barrier penetration	Biomimetic nanoparticles mimic red blood cell membranes to cross the blood-brain barrier and deliver neuroprotective drugs.	[166]
	Sustained drug release	Dendrimers improve the bioavailability and controlled release of drugs, reducing side effects in Alzheimer's therapy.	[167]
Metabolic disorders	Insulin delivery	Carbohydrate-based nanoparticles target pancreatic beta cells to enhance insulin release and control glucose levels.	[168]
	Glucose receptor mimicry	Glycomaterials mimic glucose receptors to improve glucose uptake, which is beneficial in diabetes management.	[169]

Table 3.3: Applications of carbohydrate-based nanoparticles in the treatment of different diseases.

Type of carbohydrate-based nanoparticle	Therapeutic application	Specific example	Ref.
Chitosan nanoparticles	Cancer treatment	Chitosan nanoparticles loaded with doxorubicin for targeted delivery in breast cancer.	[170]
Alginate microspheres	Diabetes management	Alginate microspheres encapsulating metformin for controlled release.	[171]
Starch nanoparticles	Antimicrobial applications	Starch nanoparticles loaded with essential oils for antibacterial activity.	[172]
Hyaluronic acid nanoparticles	Osteoarthritis treatment	Hyaluronic acid nanoparticles delivering anti-inflammatory drugs to joints.	[173]
Cellulose nanocrystals	Wound healing	Cellulose nanocrystals incorporated into hydrogels for enhanced wound healing.	[174]
Pectin nanoparticles	Drug delivery for inflammatory diseases	Pectin nanoparticles for the delivery of curcumin in inflammatory bowel disease.	[175]
Xanthan gum nanoparticles	Cancer therapy	Xanthan gum nanoparticles for targeted delivery of paclitaxel in cancer treatment.	[176]

Table 3.3 (continued)

Type of carbohydrate-based nanoparticle	Therapeutic application	Specific example	Ref.
Gum Arabic nanoparticles	Vaccine delivery	Gum Arabic nanoparticles as carriers for oral vaccines against infectious diseases.	[177]
Gelatin nanoparticles	Anticancer therapy	Gelatin nanoparticles loaded with resveratrol for enhanced anticancer activity.	[177]

3.4 Conclusion

CBNPs represent a promising avenue for advancing therapeutic and diagnostic applications across various medical fields. Their unique properties allow for improved targeting, enhanced bioavailability, and reduced side effects compared to conventional formulations. CBNPs, such as glycogen-based nanoparticles, are nontoxic, biodegradable, and water-soluble, making them suitable for various biomedical applications. While CBNPs show great promise, challenges remain in their large-scale production and regulatory approval, which may hinder their widespread clinical application. Continued research is essential to address these issues and fully realize their potential in medicine [177]. As research progresses, further exploration into the synthesis and functionalization of these nanoparticles will likely yield even more innovative applications in medicine and biotechnology. Continued advancements in this field promise to enhance our understanding of disease mechanisms while providing novel strategies for treatment and diagnosis.

References

[1] Mirza I, Saha S. Biocompatible anisotropic polymeric particles: Synthesis, characterization, and biomedical applications. ACS Appl Bio Mater, 2020, 3(12), 8241–8270.

[2] Liu Z, Cai W, He L, Nakayama N, Chen K, Sun X, Chen X, Dai H. In vivo biodistribution and highly efficient tumor targeting of carbon nanotubes in mice. Nat Nanotechnol, 2007, 2(1), 47–52.

[3] Nasongkla N, Bey E, Ren J, Ai H, Khemtong C, Guthi JS, Chin SF, Sherry AD, Boothman DA, Gao J. Multifunctional polymeric micelles as cancer-targeted, MRI-ultrasensitive drug delivery systems. Nano Lett, 2006, 6(11), 2427–2430.

[4] Cheng J, Teply BA, Sherifi I, Sung J, Luther G, Gu FX, Levy-Nissenbaum E, Radovic-Moreno AF, Langer R, Farokhzad OC. Formulation of functionalized PLGA-PEG nanoparticles for in Vivo targeted drug delivery. Biomaterials, 2007, 28(5), 869–876.

[5] Sharma S, Aiswarya TT, Mirza I, Saha S. Biocompatible polymers and their applications. In Encyclopedia of Materials: Plastics and Polymers, 2022, 2, 796–819.

[6] Mukwaya V, Wang C, Dou H. Saccharide-based nanocarriers for targeted therapeutic and diagnostic applications. Polym Int, 2019, 68(3), 306–319.

[7] Marradi M, Chiodo F, Garcia I, Penadés S. Glyconanoparticles as multifunctional and multimodal carbohydrate systems. Chem Soc Rev, 2013, 42(11), 4728–4745.
[8] Bucior I, Scheuring S, Engel A, Burger MM. Carbohydrate-carbohydrate interaction provides adhesion force and specificity for cellular recognition. J Cell Biol, 2004, 165(4), 529–537.
[9] Bucior I, Burger MM. Carbohydrate-carbohydrate interactions in cell recognition. Curr Opin Struct Biol, 2004, 14, 631–637.
[10] McNamara K, Tofail SAM. Nanoparticles in biomedical applications. Adv Phys: X, 2017, 2(1), 54–88.
[11] Nune SK, Gunda P, Thallapally PK, Lin YY, Forrest ML, Berkland CJ. Nanoparticles for biomedical imaging. Expert Opin Drug Deliv, 2009, 6(11), 1175–1194.
[12] Khan HA, Sakharkar MK, Nayak A, Kishore U, Khan A. Nanoparticles for biomedical applications: An overview. In Nanobiomaterials. Narayan R. (Ed.). Woodhead Publishing (Elsevier) Cambridge, 2018, pp. 357–384. ISBN 9780081007167
[13] IfraSaha S. Fabrication of topologically anisotropic microparticles and their surface modification with pH-responsive polymer brush. Mater Sci Eng C, 2019, 104, 109894.
[14] Biswal AK, Usmani M, Ahammad SZ, Saha S. Unveiling the slow-release behavior of hollow particles with prolonged antibacterial activity. J Mater Sci, 2018, 53(8), 5942–5957.
[15] Parthipan AK, Gupta N, Pandey K, Sharma B, Jacob J, Saha S. One-step fabrication of bicompartmental microparticles as a dual drug delivery system for Parkinson's disease management. J Mater Sci, 2019, 54(1), 730–744.
[16] Chacko RT, Ventura J, Zhuang J, Thayumanavan S. Polymer nanogels: A versatile nanoscopic drug delivery platform. Adv Drug Deliv Rev, 2012, 64, 836–851.
[17] De Souza R, Zahedi P, Allen CJ, Piquette-Miller M. Polymeric drug delivery systems for localized cancer chemotherapy. Drug Deliv, 2010, 17, 365–377.
[18] Zhang N, Wardwell P, Bader R. Polysaccharide-based micelles for drug delivery. Pharmaceutics, 2013, 5, 329–352.
[19] Liu Z, Jiao Y, Wang Y, Zhou C, Zhang Z. Polysaccharides-based nanoparticles as drug delivery systems. Adv Drug Deliv Rev, 2008, 60, 1650–1662.
[20] Lemarchand C, Gref R, Couvreur P. Polysaccharide-decorated nanoparticles. Eur J Pharm Biopharm, 2004, 58, 327–341.
[21] Kong M, Chen XG, Xing K, Park HJ. Antimicrobial properties of chitosan and mode of action: A state of the art review. Int J Food Microbiol, 2010, 144, 51–63.
[22] Shu S, Sun L, Zhang X, Wu Z, Wang Z, Li C. Polysaccharides-based polyelectrolyte nanoparticles as protein drugs delivery system. J Nanopart Res, 2011, 13, 3657–3670.
[23] Saravanakumar G, Jo DG, Park JH. Polysaccharide-based nanoparticles: A versatile platform for drug delivery and biomedical imaging. Curr Med Chem, 2012, 19, 3212–3229.
[24] Kim D, Kim J, Park YI, Lee N, Hyeon T. Recent development of inorganic nanoparticles for biomedical imaging. ACS Cent Sci, 2018, 4(3), 324–336.
[25] Giner-Casares JJ, Henriksen-Lacey M, Coronado-Puchau M, Liz-Marzán LM. Inorganic nanoparticles for biomedicine: Where materials scientists meet medical research. Mater Today, 2016, 19(1), 19–28.
[26] Guerrini L, Alvarez-Puebla RA, Pazos-Perez N. Surface modifications of nanoparticles for stability in biological fluids. Materials, 2018, 11(7), 1154.
[27] Saha S, Bruening ML, Baker GL. Facile synthesis of thick films of poly(methyl methacrylate), poly (styrene), and poly(vinyl pyridine) from Au surfaces. ACS Appl Mater Interfaces, 2011, 3(8), 3042–3048.
[28] Pourceau G, Valle-Carrandi LD, Di Gianvincenzo P, Michelena O, Penadés S. On the chiroptical properties of Au(I)-thiolate glycoconjugate precursors and their influence on sugar-protected gold nanoparticles (glyconanoparticles). RSC Adv, 2014, 4(103), 59284–59288.

[29] Marradi M, Di Gianvincenzo P, Enríquez-Navas PM, Martínez-Ávila OM, Chiodo F, Yuste E, Angulo J, Penadés S. Gold nanoparticles coated with oligomannosides of HIV-1 glycoprotein gp120 mimic the carbohydrate epitope of antibody 2G12. J Mol Biol, 2011, 410(5), 798–810.

[30] Arnáiz B, Martínez-Ávila O, Falcon-Perez JM, Penadés S. Cellular uptake of gold nanoparticles bearing HIV gp120 oligomannosides. Bioconjugate Chem, 2012, 23(4), 814–825.

[31] Reynolds M, Marradi M, Imberty A, Penadés S, Pérez S. Multivalent gold glycoclusters: High affinity molecular recognition by bacterial lectin PA-IL. Chem Eur J, 2012, 18(14), 4264–4273.

[32] Samoilova NA, Krayukhina MA, Babushkina TA, Yamskov IA, Likhosherstov LM, Piskarev VE. Silver- and gold-labeled colloidal and crosslinked glycopolymers based on glycyl glycosynthons and maleic anhydride copolymers for lectin binding. J Appl Polym Sci, 2017, 134(16), doi: 10.1002/app.44718

[33] Wang Y, Ma S, Dai Z, Rong Z, Liu J. Facile in situ synthesis of ultrasmall near-infrared-emitting gold glyconanoparticles with enhanced cellular uptake and tumor targeting. Nanoscale, 2019, 11(35), 16336–16341.

[34] Richards SJ, Fullam E, Besra GS, Gibson MI. Discrimination between bacterial phenotypes using glyco-nanoparticles and the impact of polymer coating on detection readouts. J Mater Chem B, 2014, 2(11), 1490–1498.

[35] Jiang X, Housni A, Gody G, Boullanger P, Charreyre MT, Delair T, Narain R. Synthesis of biotinylated α-D-mannoside or *N*-acetyl β-D-glucosaminoside decorated gold nanoparticles: Study of their biomolecular recognition with Con A and WGA lectins. Bioconjugate Chem, 2010, 21(3), 521–530.

[36] Ershov AY, Vasilyeva MY, Levit ML, Lagoda IV, Baygildin VA, Shabsels BM, Martynenkov AA, Yakimansky AV. Synthesis of gold glyconanoparticles based on thiol-containing D-hexose acylhydrazones and their modification by thiolated poly(2-deoxy-2-methacryloylamino-D-glucose). Russ J Gen Chem, 2019, 89(2), 300–308.

[37] Spampinato V, Parracino MA, La Spina R, Rossi F, Ceccone G. Surface analysis of gold nanoparticles functionalized with thiol-modified glucose SAMs for biosensor applications. Front Chem, 2016, 4(8), doi:10.3389/fchem.2016.00008

[38] Richards SJ, Gibson MI. Optimization of the polymer coating for glycosylated gold nanoparticle biosensors to ensure stability and rapid optical readouts. ACS Macro Lett, 2014, 3(10), 1004–1008.

[39] Mahon E, Aastrup T, Barboiu M. Multivalent biomimetic glyconanoparticle platforms as QCM amplifiers for lectin-carbohydrate interactions. MRS Proc, 2010, 1257, 1257.

[40] Aykaç A, Martos-Maldonado MC, Casas-Solvas JM, Quesada-Soriano I, García-Maroto F, García-Fuentes L, Vargas-Berenguel A. β-Cyclodextrin-bearing gold glyconanoparticles for the development of site-specific drug delivery systems. Langmuir, 2014, 30(1), 234–242.

[41] Jayawardena HSN, Wang X, Yan M. Classification of lectins by pattern recognition using glyconanoparticles. Analy Chem, 2013, 85(21), 10277–10281.

[42] Wang X, Ramström O, Yan M. Quantitative analysis of multivalent ligand presentation on gold glyconanoparticles and the impact on lectin binding. Analy Chem, 2010, 82(21), 9082–9089.

[43] Chaudhary PM, Sangabathuni S, Murthy RV, Paul A, Thulasiram HV, Kikkeri R. Assessing the effect of bacterial adhesion and infections. Chem Commun,2015, 51(86), 15669–15672.

[44] Burduşel AC, Gherasim O, Grumezescu AM, Mogoantă L, Ficai A, Andronescu E. Biomedical applications of silver nanoparticles: An up-to-date overview. Nanomaterials, 2018, 8(9), 681.

[45] Sur I, Cam D, Kahraman M, Baysal A, Culha M. Interaction of multi-functional silver nanoparticles with living cells. Nanotechnology, 2010, 21(17), 175104.

[46] Ramtenki V, Raju D, Mehta UJ, Ramana CV, Prasad BLV. Synthesis of Ag-glyconanoparticles using C-glycosides, their lectin binding studies, and antibacterial activity. New J Chem, 2013, 37(11), 3716–3725.

[47] Pandey K, Saha S. Microencapsulated zero valent iron nanoparticles in polylactic acid matrix for in situ remediation of contaminated water. J Environ Chem Eng, 2020, 8(4), 103909.

[48] Wu Y, Lu Z, Li Y, Yang J, Zhang X. Surface modification of iron oxide-based magnetic nanoparticles for cerebral theranostics: Application and prospection. Nanomaterials, 2020, 10(8), 1441.

[49] Herea DD, Chiriac H, Lupu N, Grigoras M, Stoian G, Stoica BA, Petreus T. Study on iron oxide nanoparticles coated with glucose-derived polymers for biomedical applications. Appl Surf Sci, 2015, 352, 117–125.

[50] Wang W, Zhang Z. Hydrothermal synthesis and characterization of carbohydrate microspheres coated with magnetic nanoparticles. J Dispersion Sci Technol, 2007, 28(4), 557–561.

[51] Sun X, Zheng C, Zhang F, Yang Y, Wu G, Yu A, Guan N. Size-controlled synthesis of magnetite (Fe_3O_4) nanoparticles coated with glucose and gluconic acid from a single Fe(III) precursor by a sucrose bifunctional hydrothermal method. J Phys Chem C, 2009, 113(36), 16002–16008.

[52] Alvarez-Paino M, Marcelo G, Munoz-Bonilla A, Rodriguez-Hernandez J, Fernandez-Garcia M. Surface modification of magnetite hybrid particles with carbohydrates and gold nanoparticles via "click" chemistry. Polym Chem, 2013, 4(4), 986–995.

[53] El-Boubbou K, Zhu DC, Vasileiou C, Borhan B, Prosperi D, Li W, Huang X. Magnetic glyco-nanoparticles: A tool to detect, differentiate, and unlock the glyco-codes of cancer via magnetic resonance imaging. J Am Chem Soc, 2010, 132(12), 4490–4499.

[54] Horák D, Babič M, Jendelová P, Herynek V, Trchová M, Pientka Z, Pollert E, Hájek M, Syková E. D-Mannose-modified iron oxide nanoparticles for stem cell labeling. Bioconjugate Chem, 2007, 18(3), 635–643.

[55] Predescu AM, Matei E, Berbecaru AC, Pantilimon C, Drăgan C, Vidu R, Predescu C, Kuncser V. Synthesis and characterization of dextran-coated iron oxide nanoparticles. R Soc Open Sci, 2018, 5(3), 171525.

[56] Valero E, Tambalo S, Marzola P, Ortega-Muñoz M, López-Jaramillo FJ, Santoyo-González F, de Dios López J, Delgado JJ, Calvino JJ, Cuesta R, Domínguez-Vera JM, Gálvez N. Magnetic nanoparticles-templated assembly of protein subunits: A new platform for carbohydrate-based MRI nanoprobes. J Am Chem Soc, 2011, 133(13), 4889–4895.

[57] Zhao J, Liu Y, Park HJ, Boggs JM, Basu A. Carbohydrate-coated fluorescent silica nanoparticles as probes for the galactose/3-sulfogalactose carbohydrate–carbohydrate interaction using model systems and cellular binding studies. Bioconjugate Chem, 2012, 23(6), 1166–1173.

[58] Tong, Q.; Wang, X.; Wang, H.; Kubo, T.; Yan, M. Fabrication of Glyconanoparticle Microarrays. Anal. Chem. 2012, 84 (7), 3049–3052.

[59] Hsu CW, Septiadi D, Lai CH, Chen P, Seeberger PH, De Cola L. Glucose-modified silicon nanoparticles for cellular imaging. ChemPlusChem, 2017, 82(4), 660–667.

[60] Zhai Y, Dasog M, Snitynsky RB, Purkait TK, Aghajamali M, Hahn AH, Sturdy CB, Lowary TL, Veinot JGC. Water-soluble photoluminescent d-mannose and l-alanine functionalized silicon nanocrystals and their application to cancer cell imaging. J Mater Chem B, 2014, 2(47), 8427–8433.

[61] Chen CK, Wang Q, Jones CH, Yu Y, Zhang H, Law WC, Lai CK, Zeng Q, Prasad PN, Pfeifer BA, Cheng C. Synthesis of pH-responsive chitosan nanocapsules for the controlled delivery of doxorubicin. Langmuir, 2014, 30, 4111–4119.

[62] Kyzas GZ, Siafaka PI, Lambropoulou DA, Lazaridis NK, Bikiaris DN. Poly(itaconic acid)-grafted chitosan adsorbents with different cross-linking for Pb(II) and Cd(II) uptake. Langmuir, 2013, 30, 120–131.

[63] Fernández R, Ocando C, Fernandes SCM, Eceiza A, Tercjak A. Optically active multilayer films based on chitosan and an azopolymer. Biomacromolecules, 2014, 15(5), 1399–1407.

[64] Zhi J, Wang Y, Luo G. Adsorption of diuretic furosemide onto chitosan nanoparticles prepared with a water-in-oil nanoemulsion system. Reactive Funct Polym, 2005, 65(3), 249–257.

[65] Liu H, Chen B, Mao Z, Gao C. Chitosan nanoparticles for loading of toothpaste actives and adhesion on tooth analogs. J Appl Polym Sci, 2007, 106(6), 4248–4256.

[66] Bodnar M, Hartmann JF, Borbely J. Preparation and characterization of chitosan-based nanoparticles. Biomacromolecules, 2005, 6(5), 2521–2527.

[67] Meng L, Huang W, Wang D, Huang X, Zhu X, Yan D. Chitosan-based nanocarriers with pH and light dual response for anticancer drug delivery. Biomacromolecules, 2013, 14(8), 2601–2610.

[68] Lin YH, Tsai SC, Lai CH, Lee CH, He ZS, Tseng GC. Genipin-cross-linked fucose–chitosan/heparin nanoparticles for the eradication of *Helicobacter pylori*. Biomaterials, 2013, 34(18), 4466–4479.

[69] Meng L, Huang W, Wang D, Huang X, Zhu X, Yan D. Chitosan-based nanocarriers with pH and light dual response for anticancer drug delivery. Biomacromolecules, 2013, 14(8), 2601–2610.

[70] Lee Y, Chung HJ, Yeo S, Ahn CH, Lee H, Messersmith PB, Park TG. Thermosensitive, injectable, and tissue adhesive sol–gel transition hyaluronic acid/pluronic composite hydrogels prepared from bio-inspired catechol-thiol reaction. Soft Matter, 2010, 6(5), 977.

[71] Verheul RJ, Slütter B, Bal SM, Bouwstra JA, Jiskoot W, Hennink WE. Covalently stabilized trimethyl chitosan-hyaluronic acid nanoparticles for nasal and intradermal vaccination. J Control Release, 2011, 156(1), 46–52.

[72] Mahor S, Dash BC, O'Connor S, Pandit A. Mannosylated polyethyleneimine–hyaluronan nanohybrids for targeted gene delivery to macrophage-like cell lines. Bioconjugate Chem, 2012, 23(6), 1138–1148.

[73] Martínez A, Iglesias I, Lozano R, Teijón JM, Blanco MD. Synthesis and characterization of thiolated alginate-albumin nanoparticles stabilized by disulfide bonds. Evaluation as drug delivery systems. Carbohydr Polym, 2011, 83(4), 1311–1321.

[74] Sarmah JK, Mahanta R, Bhattacharjee SK, Mahanta R, Biswas A. Controlled release of tamoxifen citrate encapsulated in cross-linked guar gum nanoparticles. Int J Biol Macromol, 2011, 49(4), 390–396.

[75] Martínez A, Fernández A, Pérez E, Benito M, Teijón J, Blanco M. Polysaccharide-based nanoparticles for controlled release formulations. In The Delivery of Nanoparticles. InTech, Shanghai, 2012, pp. 185–222.

[76] Calvo P, Remuñán-López C, Vila-Jato JL, Alonso MJ. Novel hydrophilic chitosan-polyethylene oxide nanoparticles as protein carriers. J Appl Polym Sci, 1997, 63(1), 125–132.

[77] Chen L, Subirade M. Chitosan/β-lactoglobulin core–shell nanoparticles as nutraceutical carriers. Biomaterials, 2005, 26(30), 6041–6053.

[78] Huang Y, Lapitsky Y. Salt-assisted mechanistic analysis of chitosan/tripolyphosphate micro- and nanogel formation. Biomacromolecules, 2012, 13(12), 3868–3876.

[79] Jonassen H, Kjøniksen AL, Hiorth M. Stability of chitosan nanoparticles cross-linked with tripolyphosphate. Biomacromolecules, 2012, 13(12), 3747–3756.

[80] Domaratzki RE, Ghanem A. Encapsulation and release of cladribine from chitosan nanoparticles. J Appl Polym Sci, 2013, 128(4), 2173–2179.

[81] Ramasamy T, Tran T, Cho H, Kim J, Kim Y, Jeon J, Choi HG, Yong C, Kim J. Chitosan-based polyelectrolyte complexes as potential nanoparticulate carriers: Physicochemical and biological characterization. Pharma Res, 2014, 31(6), 1302–1314.

[82] Chaubey P, Patel RR, Mishra B. Development and optimization of curcumin loaded mannosylated chitosan nanoparticles using response surface methodology in the treatment of visceral leishmaniasis. Expert Opin Drug Delivery, 2014, 11(8), 1163–81.

[83] Domaratzki, RE, Ghanem, A. Encapsulation and release of cladribine from chitosan nanoparticles, J. Appl. Polym. Sci. 2013, 128, 2173–2179.

[84] Zaki NM, Hafez MM. Enhanced antibacterial effect of ceftriaxone sodium-loaded chitosan nanoparticles against intracellular *Salmonella typhimurium*. AAPS PharmSciTech, 2012, 13(2), 411–424.

[85] Deng X, Cao M, Zhang J, Hu K, Yin Z, Zhou Z, Xiao X, Yang Y, Sheng W, Wu Y, Zeng Y. Hyaluronic acid-chitosan nanoparticles for co-delivery of MiR-34a and doxorubicin in therapy against triple negative breast cancer. Biomaterials, 2014, 35(14), 4333–4345.

[86] Honary S, Ebrahimi P, Hadianamrei R. Optimization of particle size and encapsulation efficiency of vancomycin nanoparticles by response surface methodology. Pharm Dev Technol, 2014, 19(8), 987–996.

[87] Jiang HL, Kang ML, Quan JS, Kang SG, Akaike T, Yoo HS, Cho CS. The potential of mannosylated chitosan microspheres to target macrophage mannose receptors in an adjuvant-delivery system for intranasal immunization. Biomaterials, 2008, 29(13), 1931–1939.

[88] Amidi M, Romeijn SG, Borchard G, Junginger HE, Hennink WE, Jiskoot W. Preparation and characterization of protein-loaded N-trimethyl chitosan nanoparticles as nasal delivery system. J Control Release, 2006, 111(2), 107–128.

[89] Chaubey P, Mishra B. Mannose-conjugated chitosan nanoparticles loaded with rifampicin for the treatment of visceral leishmaniasis. Carbohydr Polym, 2014, 101, 1101–1108.

[90] Zhang J, Tang C, Yin C. Galactosylated trimethyl chitosan–cysteine nanoparticles loaded with Map4k4 siRNA for targeting activated macrophages. Biomaterials, 2013, 34(15), 3667–3677.

[91] Slütter B, Jiskoot W. Dual role of CpG as immune modulator and physical crosslinker in ovalbumin loaded *N*-trimethyl chitosan (TMC) nanoparticles for nasal vaccination. J Control Release, 2010, 148, 117–121.

[92] Tian Q, Zhang CN, Wang XH, Wang W, Huang W, Cha RT, Wang CH, Yuan Z, Liu M, Wan HY, Tang H. Glycyrrhetinic acid-modified chitosan/poly(ethylene glycol) nanoparticles for liver-targeted delivery. Biomaterials, 2010, 31, 4748–4756.

[93] Ravina M, Cubillo E, Olmeda D, Novoa-Carballal R, Fernandez-Megia E, Riguera R, Sanchez A, Cano A, Alonso MJ. Hyaluronic acid/chitosan-*g*-poly(ethylene glycol) nanoparticles for gene therapy: An application for pDNA and siRNA delivery. Pharma Res, 2010, 27, 2544–2555.

[94] Malhotra M, Tomaro-Duchesneau C, Prakash S. Synthesis of TAT peptide-tagged PEGylated chitosan nanoparticles for siRNA delivery targeting neurodegenerative diseases. Biomaterials, 2013, 34, 1270–1280.

[95] Xu Q, Guo L, Gu X, Zhang B, Hu X, Zhang J, Chen J, Wang Y, Chen C, Gao B, Kuang Y, Wang S. Prevention of colorectal cancer liver metastasis by exploiting liver immunity via chitosan-TPP/nanoparticles formulated with IL-12. Biomaterials, 2012, 33, 3909–3918.

[96] Kim DG, Jeong YI, Choi C, Roh SH, Kang SK, Jang MK, Nah JW. Retinol encapsulated low molecular water-soluble chitosan nanoparticles. Int J Pharm, 2006, 319, 130–138.

[97] Park YS, Park HJ, Lee J. Stabilization of glabridin by chitosan nano-complex. J Korean Soc Appl Biol Chem, 2012, 55, 457–462.

[98] Kang Y, Wang C, Liu K, Wang Z, Zhang X. Enzyme-responsive polymeric supra-amphiphiles formed by the complexation of chitosan and ATP. Langmuir, 2012, 28, 14562–14566.

[99] Gu Z, Aimetti AA, Wang Q, Dang TT, Zhang Y, Veiseh O, Cheng H, Langer RS, Anderson DG. Injectable nano-network for glucose-mediated insulin delivery. ACS Nano, 2013, 7, 4194–4201.

[100] Zhao L, Xiao C, Ding J, He P, Tang Z, Pang X, Zhuang X, Chen X. Facile one-pot synthesis of glucose-sensitive nanogel via thiol-ene click chemistry for self-regulated drug delivery. Acta Biomater, 2013, 9, 6535–6543.

[101] Irache JM, Salman HH, Gamazo C, Espuelas S. Mannose-targeted systems for the delivery of therapeutics. Expert Opin Drug Deliv, 2008, 5, 703–724.

[102] Yang R, Xu J, Xu L, Sun X, Chen Q, Zhao Y, Peng R, Liu Z. Cancer cell membrane-coated adjuvant nanoparticles with mannose modification for effective anticancer vaccination. ACS Nano, 2018, 12, 5121–5129.

[103] Jiang HL, Kang ML, Quan JS, Kang SG, Akaike T, Yoo HS, Cho CS. The potential of mannosylated chitosan microspheres to target macrophage mannose receptors in an adjuvant-delivery system for intranasal immunization. Biomaterials, 2008, 29, 1931–1939.

[104] Lai CH, Chang TC, Chuang YJ, Tzou DL, Lin CC. Stepwise orthogonal click chemistry toward fabrication of paclitaxel/galactose functionalized fluorescent nanoparticles for HepG2 cell targeting and delivery. Bioconjugate Chem, 2013, 24, 1698–1709.

[105] Ladmiral V, Semsarilar M, Canton I, Armes SP. Polymerization-induced self-assembly of galactose-functionalized biocompatible diblock copolymers for intracellular delivery. J Am Chem Soc, 2013, 135, 13574–13581.

[106] Lin YH, Tsai SC, Lai CH, Lee CH, He ZS, Tseng GC. Genipin-cross-linked fucose–chitosan/heparin nanoparticles for the eradication of *Helicobacter pylori*. Biomaterials, 2013, 34, 4466–4479.

[107] Wibowo A, Peters EC, Hsieh-Wilson LC. Photoactivatable glycopolymers for the proteome-wide identification of fucose-α(1–2)-galactose binding proteins. J Am Chem Soc, 2014, 136, 9528–9531.

[108] Tripathi P, Dwivedi P, Khatik R, Jaiswal AK, Dube A, Shukla P, Mishra PR. Development of 4-sulfated N-acetyl galactosamine anchored chitosan nanoparticles: A dual strategy for effective management of Leishmaniasis. Colloids Surf B Biointerfaces, 2015, 136, 150–159.

[109] Singodia D, Verma A, Verma RK, Mishra PR. Investigations into an alternate approach to target mannose receptors on macrophages using 4-sulfated N-acetyl galactosamine more efficiently in comparison with mannose-decorated liposomes: An application in drug delivery. Nanomedicine, 2012, 8, 468–477.

[110] Papp I, Sieben C, Ludwig K, Roskamp M, Böttcher C, Schlecht S, Herrmann A, Haag R. Inhibition of influenza virus infection by multivalent sialic-acid-functionalized gold nanoparticles. Small, 2010, 6, 2900–2908.

[111] Tosi G, Vergoni AV, Ruozi B, Bondioli L, Badiali L, Rivasi F, Costantino L, Forni F, Vandelli MA. Sialic acid and glycopeptides conjugated PLGA nanoparticles for central nervous system targeting: In vivo pharmacological evidence and biodistribution. J Control Release, 2010, 145, 49–57.

[112] Marsich E, Borgogna M, Donati I, Mozetic P, Strand BL, Salvador SG, Vittur F, Paoletti S. Alginate/lactose-modified chitosan hydrogels: A bioactive biomaterial for chondrocyte encapsulation. J Biomed Mater Res Part A, 2008, 84, 364–376.

[113] Fernández-Ferreiro A, Fernández Bargiela N, Varela MS, Martínez MG, Pardo M, Piñeiro Ces A, Méndez JB, Barcia MG, Lamas MJ, Otero-Espinar FJ. Cyclodextrin–polysaccharide-based, in situ-gelled system for ocular antifungal delivery. Beilstein J Org Chem, 2014, 10, 2903–2911.

[114] Paques JP, van der Linden E, van Rijn CJM, Sagis LMC. Preparation methods of alginate nanoparticles. Adv Colloid Interface Sci, 2014, 209, 163–171.

[115] Abdelghany SM, Schmid D, Deacon J, Jaworski J, Fay F, McLaughlin KM, Gormley JA, Burrows JF, Longley DB, Donnelly RF, Scott CJ. Enhanced antitumor activity of the photosensitizer meso-tetra(*n*-methyl-4-pyridyl) porphine tetra tosylate through encapsulation in antibody-targeted chitosan/alginate nanoparticles. Biomacromolecules, 2013, 14(2), 302–310.

[116] Tiwari S, Chaturvedi AP, Tripathi YB, Mishra B. Microspheres based on mannosylated lysine-*co*-sodium alginate for macrophage-specific delivery of isoniazid. Carbohydr Polym, 2012, 87(3), 1575–1582.

[117] Date AA, Shibata A, McMullen E, La Bruzzo K, Bruck P, Belshan M, Zhou Y, Destache CJ. Thermosensitive gel containing cellulose acetate phthalate-efavirenz combination nanoparticles for prevention of HIV-1 infection. J Biomed Nanotechnol, 2015, 11(2), 416–427.

[118] Ernsting MJ, Tang WL, MacCallum N, Li SD. Synthetic modification of carboxymethylcellulose and use thereof to prepare a nanoparticle forming conjugate of docetaxel for enhanced cytotoxicity against cancer cells. Bioconjugate Chem, 2011, 22(12), 2474–2486.

[119] Chacko RT, Ventura J, Zhuang J, Thayumanavan S. Polymer nanogels: A versatile nanoscopic drug delivery platform. Adv Drug Deliv Rev, 2012, 64(9), 836–851.

[120] Luo Y, Wang Q. Recent development of chitosan-based polyelectrolyte complexes with natural polysaccharides for drug delivery. Int J Biol Macromol, 2014, 64, 353–367.

[121] Joshi N, Saha R, Shanmugam T, Balakrishnan B, More P, Banerjee R. Carboxymethyl-chitosan-tethered lipid vesicles: Hybrid nanoblanket for oral delivery of paclitaxel. Biomacromolecules, 2013, 14(7), 2272–2282.

[122] Amidi M, Romeijn SG, Borchard G, Junginger HE, Hennink WE, Jiskoot W. Preparation and characterization of protein-loaded *N*-trimethyl chitosan nanoparticles as nasal delivery system. J Control Release, 2006, 111(1–2), 107–116.

[123] Trapani A, Di Gioia S, Ditaranto N, Cioffi N, Goycoolea FM, Carbone A, Garcia-Fuentes M, Conese M, Alonso MJ. Systemic heparin delivery by the pulmonary route using chitosan and glycol chitosan nanoparticles. Int J Pharm, 2013, 447(1–2), 115–123.

[124] Hettiaratchi MH, Miller T, Temenoff JS, Guldberg RE, McDevitt TC. Heparin microparticle effects on presentation and bioactivity of bone morphogenetic protein-2. Biomaterials, 2014, 35(27), 7228–7238.

[125] Soppimath KS, Kulkarni AR, Aminabhavi TM. Chemically modified polyacrylamide-g-guar gum-based crosslinked anionic microgels as pH-sensitive drug delivery systems: Preparation and characterization. J Control Release, 2001, 75(3), 331–343.

[126] Ha W, Wu H, Wang XL, Peng SL, Ding LS, Zhang S, Li BJ. Self-aggregates of cholesterol-modified carboxymethyl konjac glucomannan conjugate: Preparation, characterization, and preliminary assessment as a carrier of etoposide. Carbohydr Polym, 2011, 86(2), 513–519.

[127] Xiao JX, Wang LH, Xu TC, Huang GQ. Complex coacervation of carboxymethyl konjac glucomannan and chitosan and coacervate characterization. Int J Biol Macromol, 2019, 123, 436–445.

[128] Cui L, Cohen JA, Broaders KE, Beaudette TT, Fréchet JMJ. Mannosylated dextran nanoparticles: A pH-sensitive system engineered for immunomodulation through mannose targeting. Bioconjugate Chem, 2011, 22(5), 949–957.

[129] Lamichhane S, Anderson J, Remund T, Kelly P, Mani G. Dextran sulfate as a drug delivery platform for drug-coated balloons: Preparation, characterization, in vitro drug elution, and smooth muscle cell response. J Biomed Mater Res Part B, 2016, 104(6), 1416–1430.

[130] Birch NP, Schiffman JD. Characterization of self-assembled polyelectrolyte complex nanoparticles formed from chitosan and pectin. Langmuir, 2014, 30(12), 3441–3447.

[131] Tang DW, Yu SH, Ho YC, Mi FL, Kuo PL, Sung HW. Heparinized chitosan/poly(γ-glutamic acid) nanoparticles for multi-functional delivery of fibroblast growth factor and heparin. Biomaterials, 2010, 31(35), 9320–9332.

[132] Choi KY, Yoon HY, Kim JH, Bae SM, Park RW, Kang YM, Kim IS, Kwon IC, Choi K, Jeong SY, Kim K, Park JH. Smart nanocarrier based on PEGylated hyaluronic acid for cancer therapy. ACS Nano, 2011, 5(11), 8591–8599.

[133] Yoon HY, Koo H, Choi KY, Kwon IC, Choi K, Park JH, Kim K. Photocrosslinked hyaluronic acid nanoparticles with improved stability for in vivo tumor-targeted drug delivery. Biomaterials, 2013, 34(22), 5273–5280.

[134] Gregoriadis G, McCormack B, Wang Z, Lifely R. Polysialic acids: Potential in drug delivery. FEBS Lett, 1993, 315(3), 271–276.

[135] Zhang C, An T, Wang D, Wan G, Zhang M, Wang H, Zhang S, Li R, Yang X, Wang Y. Stepwise pH-responsive nanoparticles containing charge-reversible pullulan-based shells and poly(β-amino ester)/poly(lactic-co-glycolic acid) cores as carriers of anticancer drugs for combination therapy on hepatocellular carcinoma. J Control Release, 2016, 226, 193–204.

[136] Zhang Y, Kong H, Fang Y, Nishinari K, Phillips GO. Schizophyllan: A review on its structure, properties, bioactivities and recent developments. Bioact Carbohydr Dietary Fibre, 2013, 1, 53–71.

[137] Yamada H, Loretz B, Lehr CM. Design of starch-graft-PEI polymers: An effective and biodegradable gene delivery platform. Biomacromolecules, 2014, 15(5), 1753–1761.

[138] Singh B, Sharma V. Designing galacturonic acid/arabinogalactan crosslinked poly(vinyl pyrrolidone)-*co*-poly(2-acrylamido-2-methylpropane sulfonic acid) polymers: Synthesis, characterization and drug delivery application. Polymer, 2016, 91, 50–61.

[139] Namazi H, Fathi F, Heydari A. Nanoparticles Based on Modified Polysaccharides. In The Delivery of Nanoparticles. InTech, Shanghai, 2012, pp. 149–184.

[140] Luo Y, Wang Q. Recent development of chitosan-based polyelectrolyte complexes with natural polysaccharides for drug delivery. Int J Biol Macromol, 2014, 64, 353–367.
[141] Ramasamy T, Tran T, Cho H, Kim J, Kim Y, Jeon J, Choi HG, Yong C, Kim J. Chitosan-based polyelectrolyte complexes as potential nanoparticulate carriers: Physicochemical and biological characterization. Pharma Res, 2014, 31(5), 1302–1314.
[142] Motwani SK, Chopra S, Talegaonkar S, Kohli K, Ahmad FJ, Khar RK. Chitosan–sodium alginate nanoparticles as submicroscopic reservoirs for ocular delivery: Formulation, optimisation and in vitro characterisation. Eur J Pharm Biopharm, 2008, 68(3), 513–525.
[143] Guo R, Chen LL, Cai SS, Liu ZH, Zhu Y, Xue W, Zhang YM. Novel alginate-coated hydrophobically modified chitosan polyelectrolyte complex for the delivery of BSA. J Mater Sci Mater Med, 2013, 24(8), 2093–2101.
[144] Lee DW, Yun KS, Ban HS, Choe W, Lee SK, Lee KY. Preparation and characterization of chitosan/polyguluronate nanoparticles for siRNA delivery. J Control Release, 2009, 139(2), 146–152.
[145] Cafaggi S, Russo E, Stefani R, Parodi B, Caviglioli G, Sillo G, Bisio A, Aiello C, Viale M. Preparation, characterisation and preliminary antitumour activity evaluation of a novel nanoparticulate system based on a cisplatin-hyaluronate complex and *N*-trimethyl chitosan. Investig New Drugs, 2011, 29(3), 443–453.
[146] Chen Q, Hu Y, Chen Y, Jiang X, Yang Y. Microstructure formation and property of chitosan-poly(acrylic acid) nanoparticles prepared by macromolecular complex. Macromol Biosci, 2005, 5(11), 993–1004.
[147] Rolland J, Guillet P, Schumers JM, Duhem N, Préat V, Gohy JF. Polyelectrolyte complex nanoparticles from chitosan and poly(acrylic acid) and polystyrene-block-poly(acrylic acid). J Polym Sci Part A: Polym Chem, 2012, 50(24), 4484–4493.
[148] Tang Y, Zhao Y, Li Y, Du Y. A thermosensitive chitosan/poly(vinyl alcohol) hydrogel containing nanoparticles for drug delivery. Polym Bull, 2010, 64(8), 791–804.
[149] Davidenko N, Blanco MD, Peniche C, Becherán L, Guerrero S, Teijón JM. Effects of different parameters on the characteristics of chitosan–poly(acrylic acid) nanoparticles obtained by the method of coacervation. J Appl Polym Sci, 2009, 111(5), 2362–2371.
[150] Vasquez D, Milusheva R, Baumann P, Constantin D, Chami M, Palivan CG. The amine content of PEGylated chitosan *Bombyx mori* nanoparticles acts as a trigger for protein delivery. Langmuir, 2014, 30(33), 965–975.
[151] Tsao CT, Kievit FM, Ravanpay A, Erickson AE, Jensen MC, Ellenbogen RG, Zhang M. Thermoreversible poly(ethylene glycol)-*g*-chitosan hydrogel as a therapeutic T lymphocyte depot for localized glioblastoma immunotherapy. Biomacromolecules, 2014, 15(7), 2656–2662.
[152] Ragelle H, Riva R, Vandermeulen G, Naeye B, Pourcelle V, Le Duff CS, D'Haese C, Nysten B, Braeckmans K, De Smedt SC, Jérôme C, Préat V. Chitosan nanoparticles for siRNA delivery: Optimizing formulation to increase stability and efficiency. J Control Release, 2014, 176, 54–63.
[153] Oh BHL, Bismarck A, Chan-Park MB. High internal phase emulsion templating with self-emulsifying and thermoresponsive chitosan-*graft*-PNIPAM-*graft*-oligoproline. Biomacromolecules, 2014, 15(6), 1777–1787.
[154] Gao Y, Wang ZY, Zhang J, Zhang Y, Huo H, Wang T, Jiang T, Wang S. RVG-peptide-linked trimethylated chitosan for delivery of siRNA to the brain. Biomacromolecules, 2014, 15(3), 1010–1018.
[155] Guo H, Yang C, Hu Z, Wang W, Wu Y, Lai Q, Yuan Z. Ethylene glycol oligomer modified-sodium alginate for efficiently improving the drug loading and the tumor therapeutic effect. J Mater Chem B, 2013, 1(43), 5933–5941.
[156] Kim HC, Park SM, Hinsberg WD. Block copolymer-based nanostructures: Materials, processes, and applications to electronics. Chem Rev, 2009, 110(1), 146–177.
[157] Schatz C, Lecommandoux S. Polysaccharide-containing block copolymers: Synthesis, properties and applications of an emerging family of glycoconjugates. Macromol Rapid Commun, 2010, 31(19), 1664–1684.

[158] Marradi M, García I, Penadés S. Carbohydrate-based nanoparticles for potential applications in medicine. Prog Mol Biol Transl Sci, 2011, 104, 141–165.
[159] Cui L, Pi J, Qin B, Cui T, Liu Z, Lei L, Wu S. Advanced application of carbohydrate-based micro/nanoparticles for rheumatoid arthritis. Int J Biol Macromol, 2024, 269(1), 131809.
[160] Chettri D, Chirania M, Boro D, Verma AK. Glycoconjugates: Advances in modern medicines and human health. Life Sci, 2024, 348(5), 122689.
[161] Maghsoudnia N, Eftekhari RB, Sohi AN. Application of nano-based systems for drug delivery and targeting: A review. J Nanopart Res, 2020, 22(8), 1–20.
[162] Savchenko IV, Zlotnikov ID, Kudryashova EV. Biomimetic systems involving macrophages and their potential for targeted drug delivery. Biomimetics, 2023, 8(7), 543.
[163] Riu F, Ruda A, Ibba R, Sestito S, Lupinu I, Piras S. Antibiotics and carbohydrate-containing drugs targeting bacterial cell envelopes: An overview. Pharmaceuticals, 2022, 15(8), 942.
[164] Taneja P, Sharma S, Sinha VB, Yadav AK. Advancement of nanoscience in development of conjugated drugs for enhanced disease prevention. Life Sci, 2021, 269, 119089.
[165] Mahara G, Tian C, Xu X, Zhu J. Breakthrough of glycobiology in the 21st century. Front Immunol, 2023, 13, 1071360.
[166] Dehaini D, Fang RH, Zhang L. Biomimetic strategies for targeted nanoparticle delivery. Bioeng Transl Med, 2016, 1(1), 30–46.
[167] Chauhan S, Jain K, Naqvi S. Dendrimers and its theranostic applications in infectious diseases. In Nanotheranostics for Treatment and Diagnosis of Infectious Diseases. Elsevier, Cambridge, 2022, pp. 199–228.
[168] Tiwari SS, Gattani SG, Sharma B, Hoque ME. Carbohydrate-based therapeutics: Evolution from wellness pursuit to medical treatment. In Polysaccharides. Taylor & Francis, Florida, 2020, pp. 130–152.
[169] Fettis MM, Hudalla GA. Glycomaterials for immunomodulation, immunotherapy, and infection prophylaxis. J Mater Chem B, 2016, 4(4), 6157–6174.
[170] Khemtong C. Doxorubicin-loaded chitosan nanoparticles for targeted delivery in breast cancer. Biomaterials, 2016, 76, 130–140.
[171] Raghavan SL. Controlled release of metformin from alginate microspheres for diabetes management. J Control Release, 2012, 158(2), 215–221.
[172] Rojas JA. Antimicrobial activity of starch nanoparticles loaded with essential oils. Food Hydrocoll, 2018, 85, 1–10. Lee, J. W., et al. (2014). Hyaluronic acid nanoparticles for anti-inflammatory drug delivery in osteoarthritis treatment. J Control Release, 196, 1–10.
[173] Zhang Y. Cellulose nanocrystals incorporated into hydrogels for enhanced wound healing applications. Carbohydr Polym, 2019, 206, 123–132.
[174] Thakur A. Pectin nanoparticles for curcumin delivery in inflammatory bowel disease treatment: A novel approach to drug delivery systems. Int J Biol Macromol, 2020, 164, 3707–3715.
[175] Sahu AN. Targeted delivery of paclitaxel using xanthan gum nanoparticles for cancer therapy applications: A comprehensive review and future perspectives on nanomedicine innovations in cancer treatment strategies. J Drug Deliv Sci Technol, 2020, 58, 101822.
[176] Al-Jawad MH. Gum Arabic nanoparticles as carriers for oral vaccines against infectious diseases: A potential strategy for enhancing vaccine efficacy. Vaccines, 2021, 9(3), 235.
[177] Zhang Y. Anticancer activity of gelatin nanoparticles loaded with resveratrol: Implications for drug delivery systems in cancer therapy. Mater Sci Eng C, 2018, 92, 102–110.
[178] Besford QA. The sweetest polymer nanoparticles: Opportunities ahead for glycogen in nanomedicine. Soft Matter, 2024, 20, 3577–3584.

Sasadhar Majhi*

Chapter 4
Total synthesis of some carbohydrate-decorated bioactive compounds

4.1 Introduction

Natural products include special features compared to traditional synthetic compounds due to their extreme chemical diversity, higher stereogenic centers, and significant biological and pharmacological properties [1–13]. Natural products have played a crucial role in drug discovery and development [14–23]. Carbohydrates, polyhydroxylated compounds, are charming and beautiful molecules as they are the language of chiral natural products and important chiral precursors in organic synthesis [24, 25]. Numerous plants, animals, and microorganisms are valuable sources of bioactive carbohydrates [26]. Carbohydrates, sugar molecules, are stored in the form of glycogen in animals, while in plants, it is stored in the form of starch. They are the main constituents of cell walls and the starting point for synthesizing proteins and fats. Besides, biomolecule carbohydrates are direct products of photosynthesis as well as the primary energy storage molecules [27]. Importantly, few carbohydrates have been approved for the treatment of different diseases and they exhibit various biological applications [28]. Total synthesis of natural products aims to construct a complex target molecule from a simple, easily available precursor [29–31]. Total synthesis of bioactive compounds is very effective for structure confirmation and is applied to test for novel synthetic methods and frequently to help recognize how the compound is built naturally [32–34]. There is increasing attention to carbohydrate-containing drugs as they have potential diagnostic applications [28]. Thus, this chapter wishes to focus on the total synthesis of carbohydrate-decorated bioactive compounds.

Acknowledgments: The author S. Majhi is grateful to his father Tarani Majhi and thankful to his mother Sadeswari Majhi for their constant inspiration. He also gratefully acknowledges Trivenidevi Bhalotia College, Raniganj, for providing the infrastructure.

***Corresponding author: Sasadhar Majhi,** Department of Chemistry (UG and PG), Trivenidevi Bhalotia College, Kazi Nazrul University, Raniganj 713347, West Bengal, India, e-mail: sasadharmajhi@gmail.com

https://doi.org/10.1515/9783111447094-004

4.2 Total synthesis of carbohydrate-decorated bioactive compounds

This section deals with the total synthesis of carbohydrate-decorated bioactive compounds.

4.2.1 Total synthesis of alkaloid

4.2.1.1 Total synthesis of (–)-brevisamide

Natural source: *Karenia brevis* [35]

Class of compounds: Alkaloid [35]

Biological activity: Antagonist effects against brevetoxins [35]

Structure and structural features: The cyclic ether includes a secondary –OH group, an amide group, and an unsaturated side chain having an aldehyde group (Figure 4.1) [35].

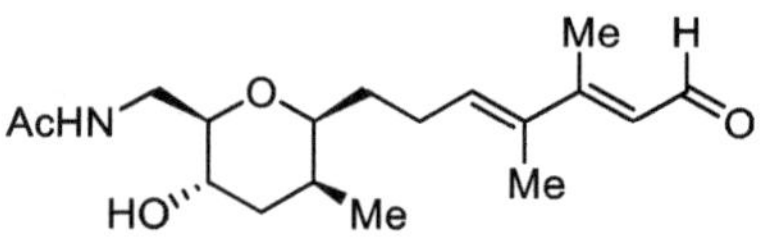

Brevisamide

Figure 4.1: Structure of brevisamide.

Synthetic strategy:
The formal stereoselective total synthesis of the alkaloid (–)-brevisamide (**12**) was developed by Sridhar et al. (Figure 4.2) [36]. At first, Claisen rearrangement of 2-vinyloxy methyldeoxy-glycals was applied to prepare deoxygenated C-2-methylene-*C*-glycosides as the synthetic precursors. Diastereomerically pure C-2-methylene β-*C*-glycosides were derived by a Zn^{II}-assisted anomerization of the α-*C*-glycosides. The utilization of the methodology was further extended to the synthesis of a state-of-the-art intermediate in the total synthesis of cyclic ether (–)-brevisamide (**12**). The starting material 3-deoxy-4,6-di-*O*-benzyl-D-glucal (**1**) was converted into C-2-methylene *C*-glycoside (**6β**) over several steps and then **6β** was transformed into 1,2-*cis*-C-2-methyl-*C*-glycoside (7) through hydrogenation in the presence of the 10% Pd/C, H_2, and Na_2CO_3 in methanol. 1,2-*Cis* compound 7 was effective in delivering targeted natural monocyclicether alkaloid (**12**) finally.

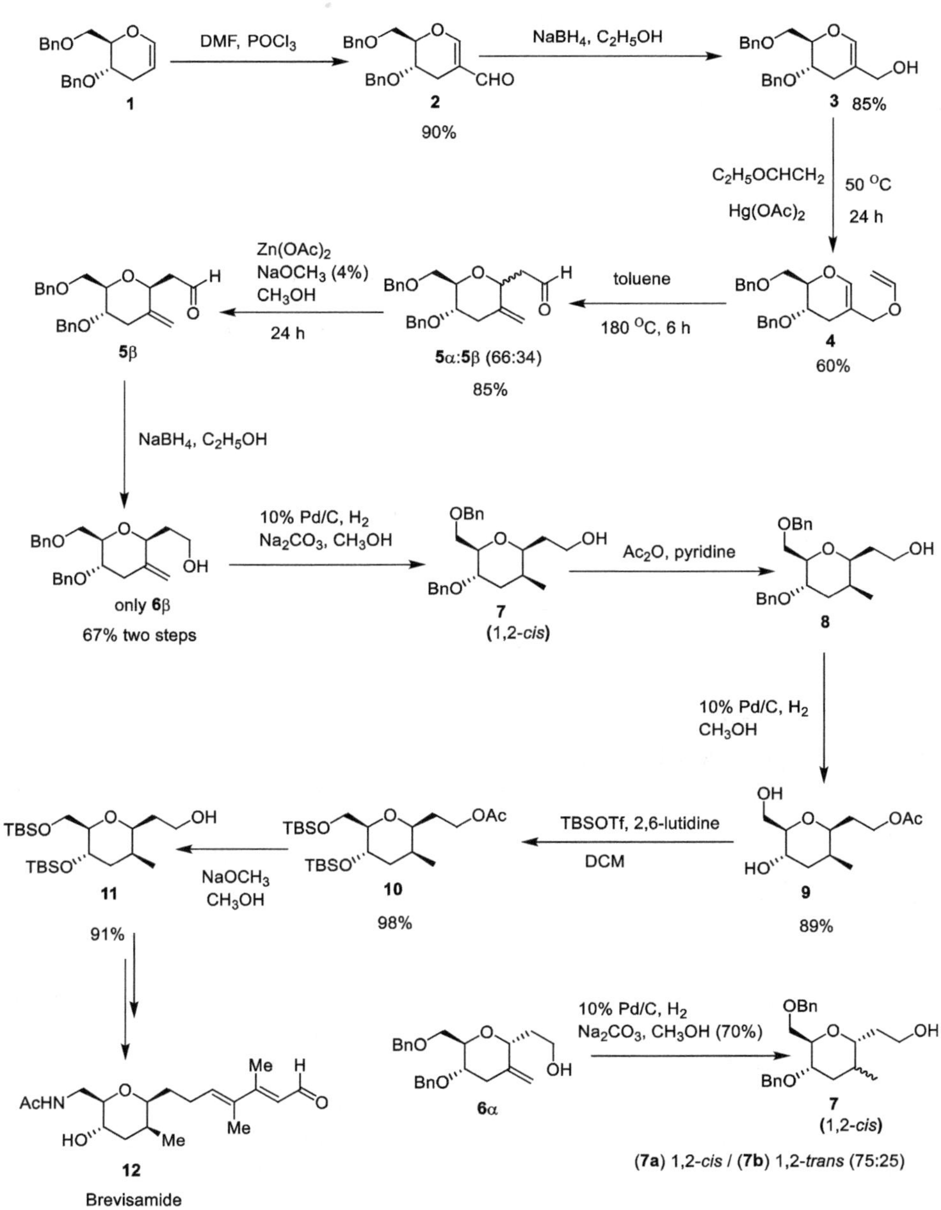

Figure 4.2: Total synthesis of (−)-brevisamide.

4.2.2 Total synthesis of aminocyclitol

4.2.2.1 Total synthesis of kirkamide

Natural source: *Psychotria kirkii* [37]

Class of compounds: C_7N aminocyclitol [37]

Biological activity: Cytotoxic activity against arthropods and insects [37]

Structure and structural features: It includes structural similarity with *N*-acetylglucosamine, an amide derivative of the monosaccharide glucose (Figure 4.3) [38].

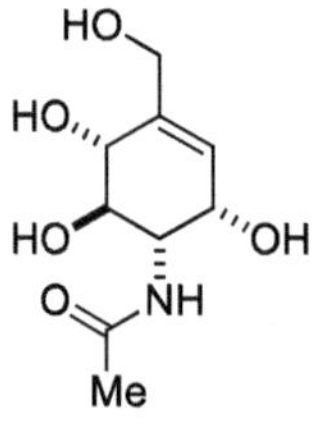

Kirkamide

Figure 4.3: Structure of kirkamide.

Synthetic strategy:
Sagar et al. [39] disclosed a novel synthetic strategy for the second total synthesis of kirkamide (**17**) from enantiopure epoxide obtained from *N*-acetylglucosmine as the starting material in 2021 (Figure 4.4). The metal-free synthesis of this C_7N aminocyclitol comprise stereoselective epoxidation and acid-assisted regioselective epoxide ring opening as the crucial steps. The total synthesis of kirkamide (**17**) was commenced from exocyclic olefin (**13**) that was prepared from *N*-acetyl glucosamine in a few steps. Next, olefin (**13**) underwent stereospecific epoxidation to yield epoxide **14** in 80% yield in the presence of the *m*-CPBA with high diastereoselectivity. Finally, epoxide **14** delivered the targeted aminocyclitol kirkamide (**17**) over several steps. Kirkamide (**17**) was obtained through two paths (path A and path B).

In another study, Gademann et al. [37] isolated kirkamide (**17**) from the plant *Psychotria kirkii* utilizing a genome-driven ^{1}H NMR-guided fractionation approach. Different techniques including NMR, HRMS, and single-crystal X-ray crystallography were applied to determine the structure and absolute configuration of this C_7N aminocyclitol. An enantioselective total synthesis of kirkamide (**17**) was accomplished by this group on a gram scale in 11 steps (Figure 4.5). The key features of this strategy include a Pd-assisted hydroxymethylation and a Ferrier carbocyclization. The total synthesis was initiated with the known methyl-*N*-acetyl-D-glucosamine (**23**) as the starting material to provide the iodide **24** involving a Garegg-Samuelsson reaction in the presence of the PPh_3, imidazole, and I_2 [40]. Iodide **24** was capable of furnishing kirkamide (**17**) finally over several stages.

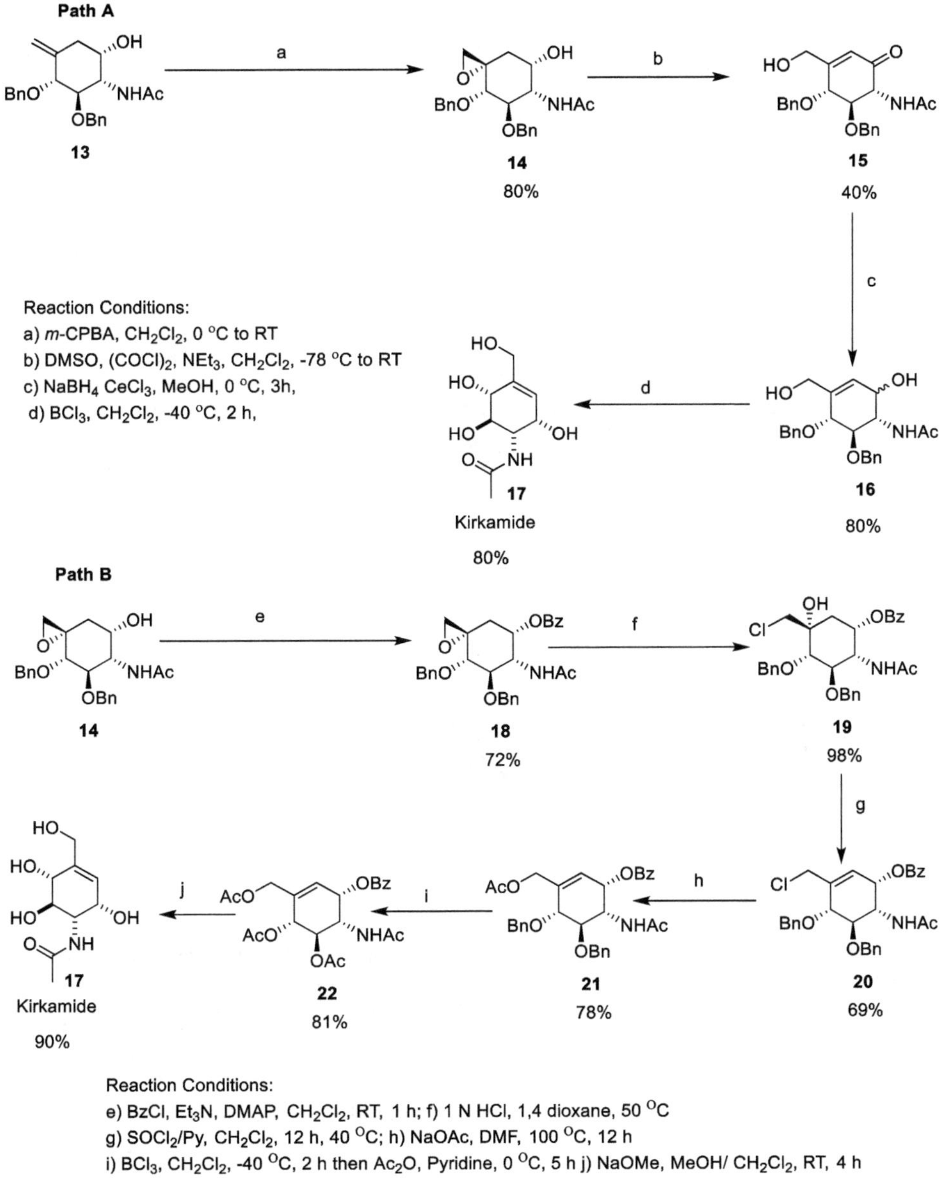

Figure 4.4: Total synthesis of kirkamide.

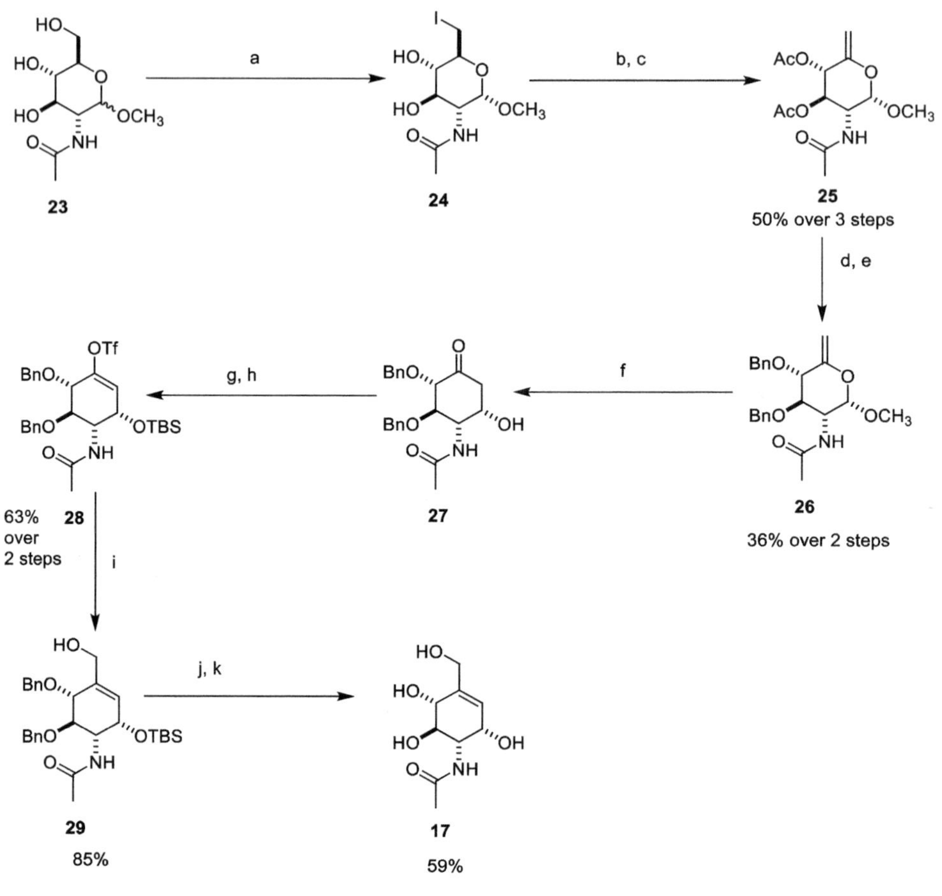

Reaction conditions: a) Ph_3P, Imid, I_2, THF, reflux, 15 min; b) Ac_2O, pyridine, RT, 24 h; c) AgF, pyridine, RT, 48 h exclusion of light, 50% over 3 steps; d) NH_3(gas), CH_3OH, RT, 3 h; e) NaH, BnBr, DMF, 0 °C, 12 h, 36% over 2 steps; f) $HgSO_4$ dioxane: aq. H_2SO_4 (5 mm) 2:1, microwave, 60 °C, 15 min; g) TBSOTf, 2,6-lutidine, THF, 0 °C 12 h, 63% over 2 steps; h) Comins' reagent, NaHMDS, THF, -78 °C, 5 min, 62%; i) Bu_3SnCH_2OH, Pd-$(PPh_3)_4$ LiCl, dioxane, microwave, 105 °C, 1 h, 85%; j) TBAF, THF, RT, 3 h, 98%; k) Na/NH_3(liq), THF, -78 °C, 30 min 59%.

HMDS: hexamethyldisilazane. TBAF: tetra-*n*-butylammonium fluoride. THF: tetrahydrofuran, DMF: dimethylformamide, TBS: *tert*-butyldimethylsilyl. Imid: imidazole.

Figure 4.5: Enantioselective total synthesis of kirkamide.

4.2.3 Total synthesis of lactone

4.2.3.1 Total synthesis of mangrolide A

Natural source: SNA18 strain of *Actinoalloteichus* sp. [41]

Class of compounds: Lactone [41]

Biological activity: Antibiotic activity against gram-negative bacteria [41]

Structure and structural features: It consists of anextremely unsaturated 18-membered macrocycle which is beautified by uncommon carbohydrates (Figure 4.6) [41].

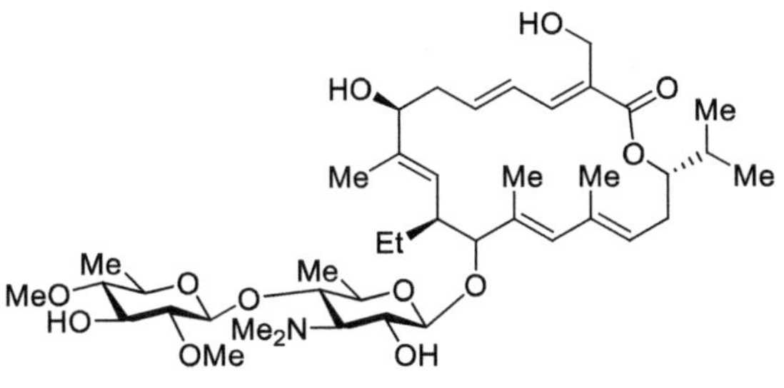

Figure 4.6: Structure of mangrolide A.

Synthetic strategy:

Gademann et al. [41] completed the first total synthesis of antibiotic mangrolide A through a convergent and modular strategy (Figure 4.7). Stereoselective construction of the disaccharide moiety, preparation of macrocycle in a highly convergent way, and two β-selective glycosylations are key features of this synthesis. The preparation of the quinovose domain commenced from glycoside **30** by installing C3″ TBS ether and the C2″ methyl group. Compound **31** was obtained from D-glucopyranoside **30** over three steps. The targeted macrocyclic mangrolide A (**44**) was synthesized from compound **31** over several steps. The first enantioselective total synthesis was in 25 steps through the longest linear sequence.

4.2.3.2 Total synthesis of fidaxomicin

Natural source: *Dactylosporangium aurantiacum* [42]

Class of compounds: Macrolide [42]

Biological activity: Approved for the treatment of *Clostridium difficile* infections by FDA [43]

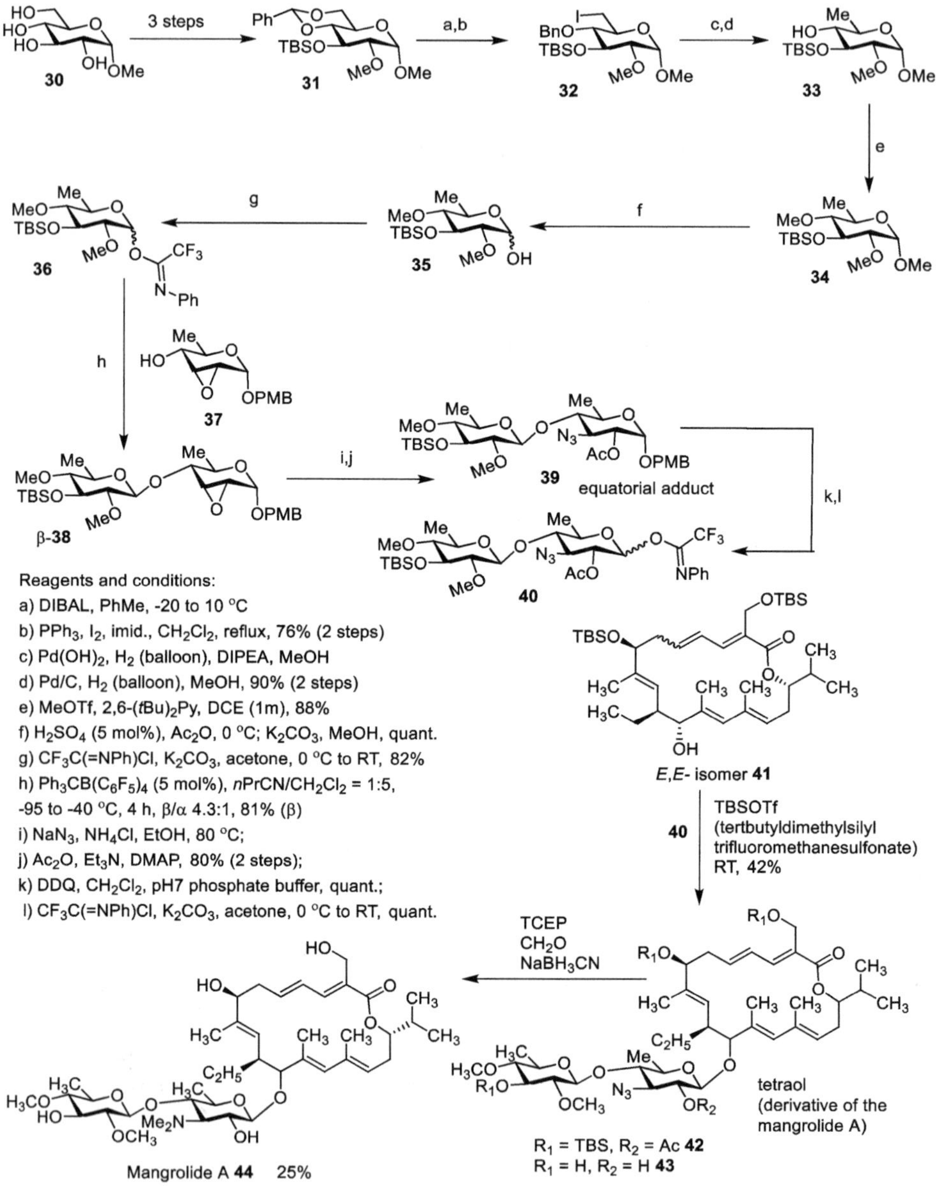

Figure 4.7: Total synthesis of mangrolide A.

Structure and structural features: The 18-membered macrolactone core contains four chiral centers; a high degree of unsaturation is the main feature of this macrolide skeleton (Figures 4.8 and 4.9) [43].

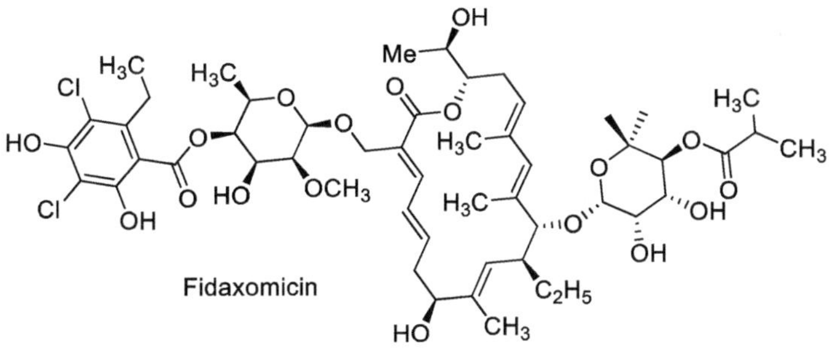

Figure 4.8: Structure of fidaxomicin.

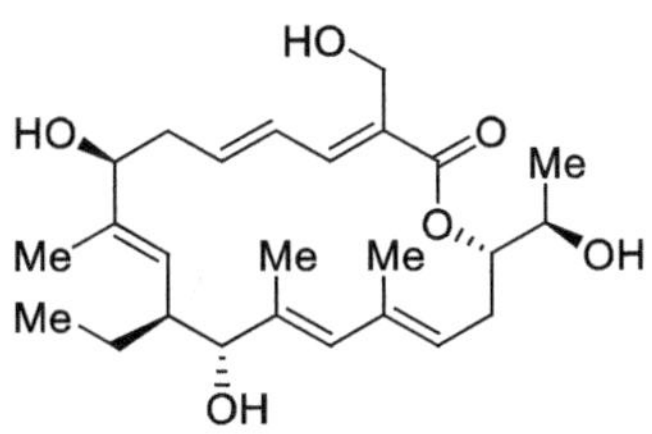

Tiacumicin B or lipiarmycin A3 or fidaxomicin aglycon

Figure 4.9: Structure of fidaxomicin aglycon.

Synthetic strategy:

Gademann et al. [44] achieved the first enantioselective total synthesis of antibiotic fidaxomicin (**50**) involving quick and higher-yielding access to the rhamnose, noviose, and orsellinic acid precursors (Figure 4.10). The key features of this total synthesis consist of Suzuki coupling, β-selective glycosylations, a ring-closing metathesis (RCM) reaction, and a β-rhamnosylation. Initially, the novioside unit was constructed from starting material known 6-deoxy-6-iodopyranoside (**45**), and glycosyl bromide donor **46** was treated with alcohol **48** to afford the noviosylated fragment **49**. The iodide **49** furnished fidaxomicin (**50**) as a desired product finally over several steps.

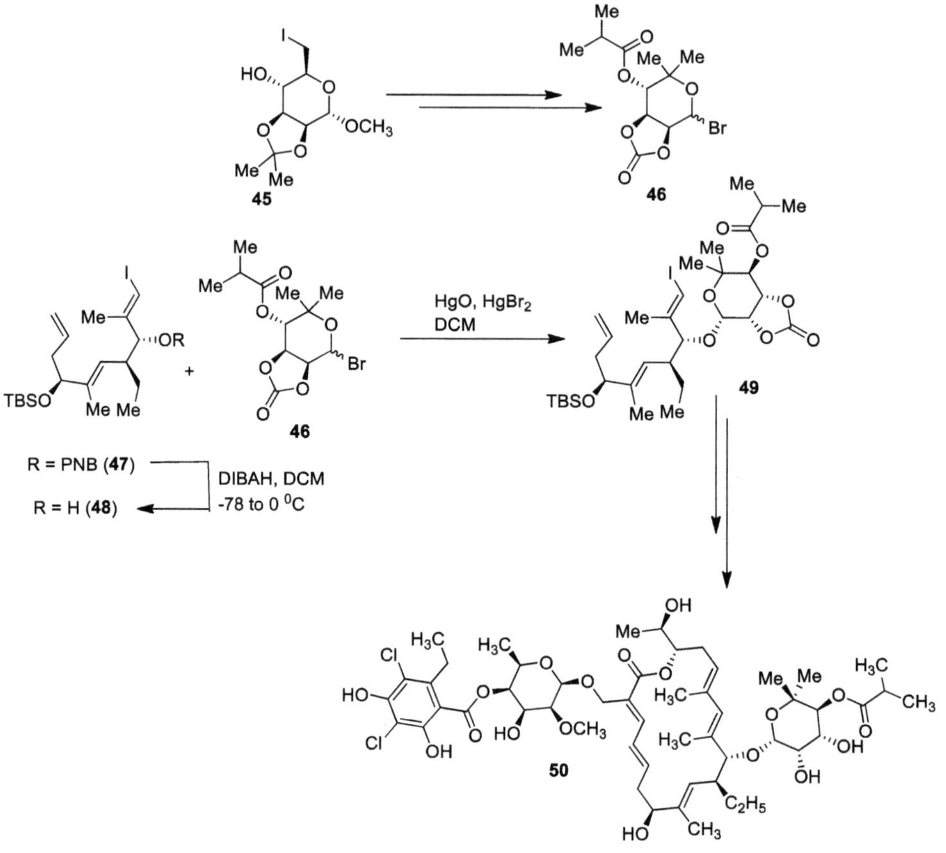

Figure 4.10: Total synthesis of fidaxomicin.

4.2.3.3 Total synthesis of (–)-dinemasone B

Natural source: *Dinemasporium strigosum* [45]

Class of compounds: Lactone [45, 46]

Biological activity: It exhibits biological properties against the *Microbotryum violaceum* (fungus), *Bacillus megaterium* (gram-positive), and *Chlorella fusca* (alga) [46].

Structure and structural features: It includes hexahydropyrano[4,3-b]pyran-5(7H)-one and 1,7-dioxaspiro[5.5]undecane skeletons (Figure 4.11) [45].

Synthetic strategy:
The stereo-controlled total synthesis of natural antibiotic (–)-dinemasone B (**21**) was presented by Li and coworkers [46] for the first time with D-glucose as the starting material (Figure 4.12). Kocieński-Julia olefination and stereoselective reduction of *C*-glycosidic ke-

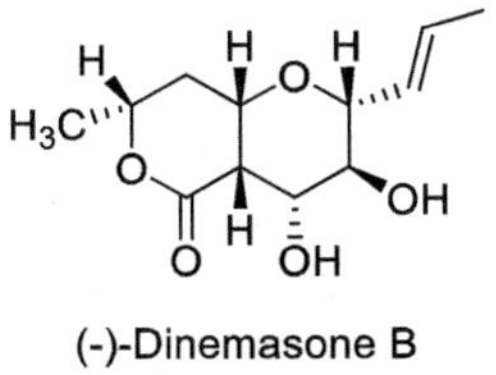

Figure 4.11: Structure of (–)-dinemasone B.

tone are vital features of this developed synthesis. The total synthesis commenced with the glucofuranose (**51**) which could be simply prepared from D-glucose [47]. Protection of the hydroxyl group in glucofuranose (**51**) afforded compound **52** in the presence of the benzyl bromide and sodium hydride in THF. (–)-Dinemasone B (**53**) was prepared from compound **52** over several steps.

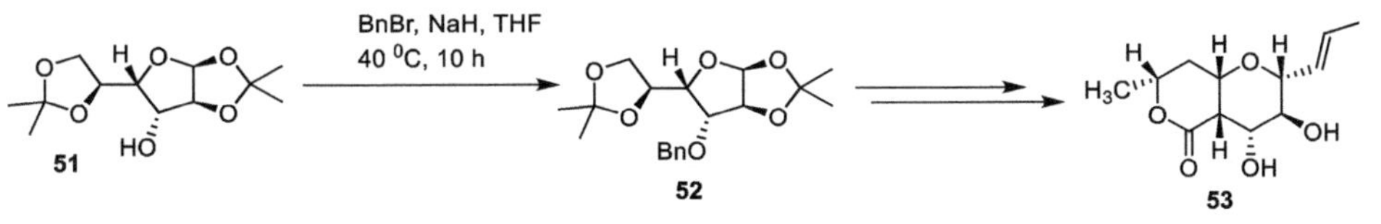

Figure 4.12: Total synthesis of (–)-dinemasone B.

4.2.3.4 Total synthesis of paecilomycin F

Natural source: *Paecilomyces* sp. SC0924 [48]

Class of compounds: Resorcylic acid lactone [48]

Biological activity: Antiplasmodial activity [48]

Structure and structural features: Paecilomycin F, resorcylic lactone, contains an aromatic ring at C6 carbon (Figure 4.13) [48].

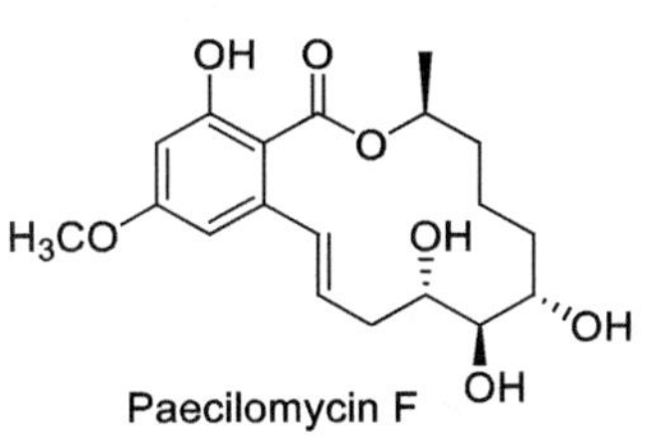

Figure 4.13: Structure of paecilomycin F.

Synthetic strategy:

The total synthesis of natural resorcylic acid lactone paecilomycin F (**56**) was presented by Srihari and coworkers [49] with commercially available sugar D-lyxose (**54**) (Figure 4.14). The salient features of this synthesis comprise Ohira-Bestmann alkynylation, regioselective methylation, and RCM. Initially, carbohydrate (**54**) was reacted with acetone using H_2SO_4 at room temperature to afford the corresponding 2,3-di-*O*-isopropylidene acetal **55**. The targeted lactone paecilomycin F (**56**) was obtained from acetal **55** over several steps.

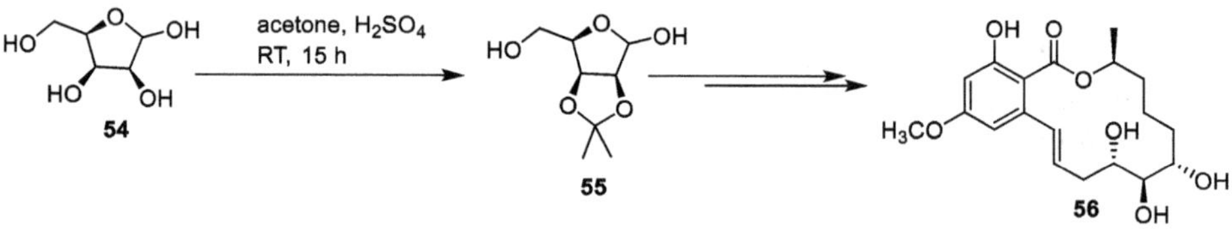

Figure 4.14: Total synthesis of paecilomycin F.

4.2.3.5 Total synthesis of (+)-pyrenolide D

Natural source: *Pyrenophora teres* [50]

Class of compounds: Spiroketal lactone [50]

Biological activity: Cytotoxic activity [50]

Structure and structural features: It consists of an extremely oxygenated tricyclic spiro-γ-lactone with a spiroketal chiral center (Figure 4.15) [50].

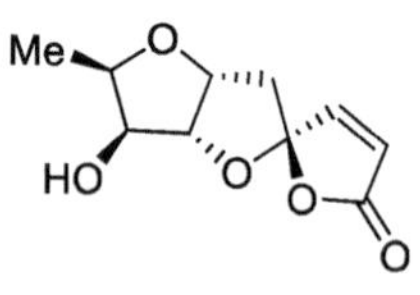

Figure 4.15: Structure of (+)-pyrenolide D.

Synthetic strategy:

Mohapatra et al. [51] achieved an effective carbohydrate-based approach for the total synthesis of tricyclic lactones (+)-pyrenolide D (**59**) and (−)-4-*epi*-pyrenolide D (**60**) involving one-pot epoxidation-cyclization as a central step (Figure 4.16). Besides, peracid-assisted oxidative spiroketalization, and reductive Barton-McCombie deoxygenation were crucial steps, and known 5-deoxy-D-xylose derivative (**57**) was employed as a starting material in this synthesis. Lactol (**57**) underwent epoxidation-cyclization in the presence of the Corey-Chaykvosky reagent ($Me_3S^+OI^-$) to afford compound **58** as an inseparable mixture of diastereomers. Different bases such as NaH, LiHMDS, *n*-BuLi, LDA, and NaHMDS were used in this transformation. However, *t*-BuOK provided the best result in terms of yield (65%). Finally, (+)-pyrenolide D (**59**) and (−)-4-*epi*-pyrenolide D (**60**) were synthesized from compound **58** over several steps.

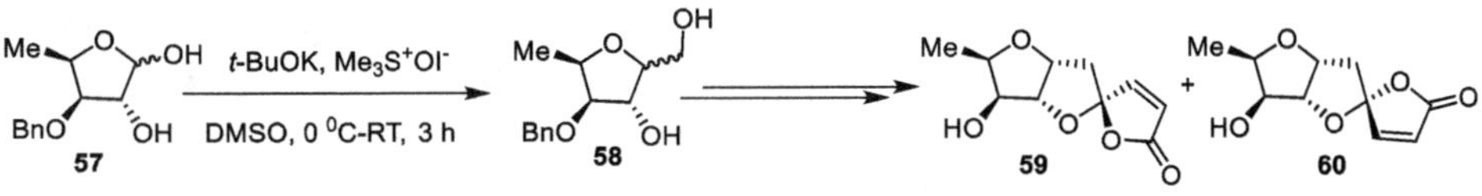

Figure 4.16: Total synthesis of (+)-pyrenolide D and (−)-4-*epi*-pyrenolide D.

4.2.4 Total synthesis of lipid

4.2.4.1 Total synthesis of glycosphingolipids

Natural source: *Sphingomonas wittichii* [52] and *Sphingomonas yanoikuyae* [53], components of the plasma membrane of cells [54].

Class of compounds: Lipid [54]

Biological activity: Second messengers or modulators of signal transduction [54, 55]

Structure and structural features: A glycan structure is linked to a lipid tail comprising the sphingolipid ceramide (hydrophobic moiety) (Figure 4.17) [54, 55].

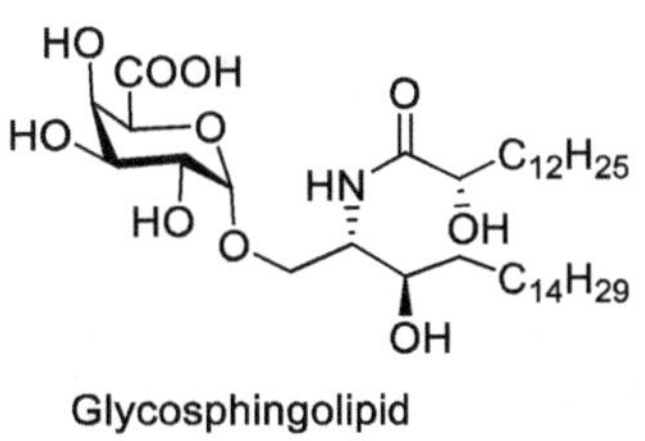

Figure 4.17: Structure of glycosphingolipid.

Synthetic strategy:

Toshima and coworkers [56] developed a new direct and 1,2-*cis*-*α*-stereoselective glycosylation of 1,2-anhydroglucose and mono-ol acceptors using a glycosyl-acceptor-obtained borinic ester (Figures 4.18 and 4.19). Besides, the investigators applied this glycosylation methodology to the total synthesis of natural glycosphingolipids effortlessly in 2016 [56]. D-glucal (**61**), the glycal formed from glucose, provides a compound **62** over three steps. Then, the targeted α-glucoside (**65**) was obtained through glycosylation of **62** and **63** in the presence of the **64**. α-Glucoside (**65**) delivered the desired glycosphingolipid (**66**) over three steps finally. In another approach, methyl ester of glycosphingolipid (**68**) was obtained from D-galactal (**6**) over several steps successfully.

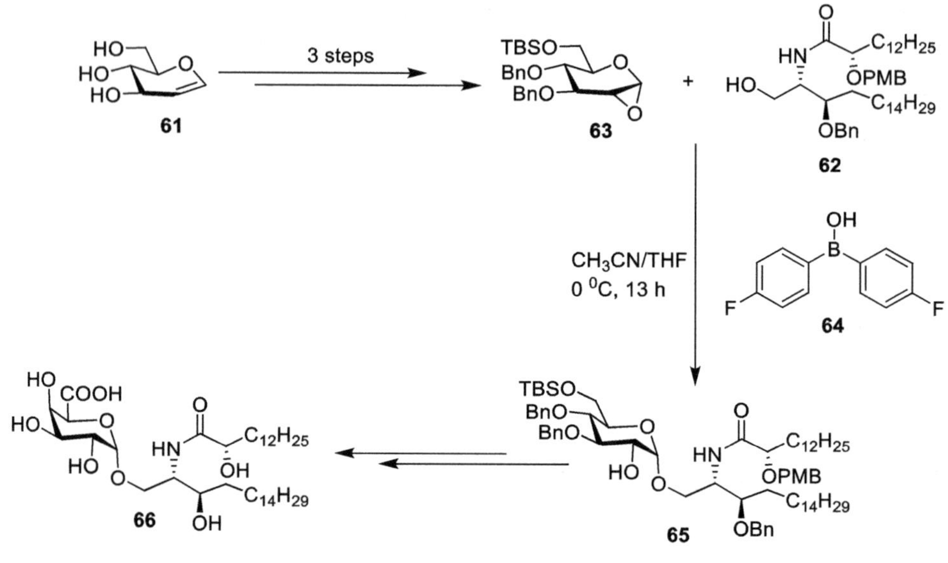

Figure 4.18: Total synthesis of GSL-1.

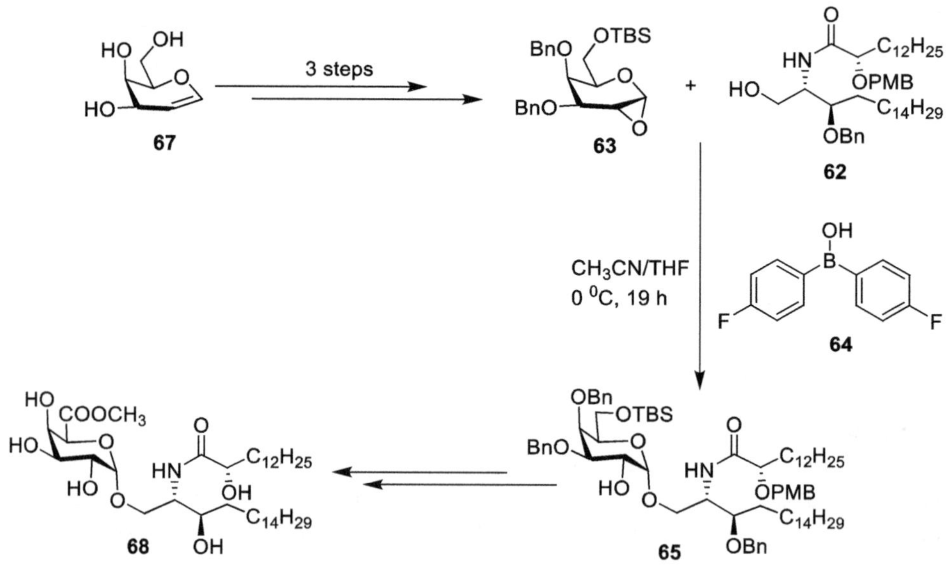

Figure 4.19: Total synthesis of GSL-1′.

4.2.5 Total synthesis of macrolide

4.2.5.1 Total synthesis of aspergillide C

Natural source: *Aspergillus ostianus* [57–60]

Class of compounds: Macrolide [57, 58]

Biological activity: Cytotoxic activity [57–60]

Structure and structural features: Bicyclic 14-membered macrolide consists of 2,6-*trans* or *cis*-fused di- or tetrahydropyan rings [57, 58] (Figure 4.20).

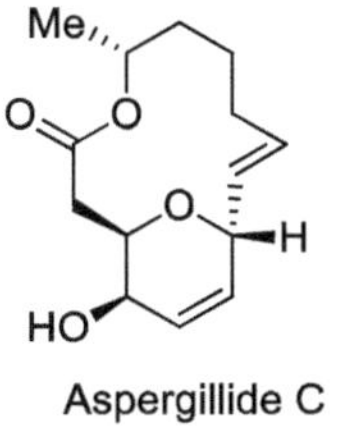

Figure 4.20: Structure of aspergillide C.

Synthetic strategy:
Srihari et al. [61] developed a strategy for the formal total synthesis of macrolide(−)-aspergillide C concisely from tri-*O*-acetyl-D-galactal as the starting material through the chiron approach (Figure 4.21). Trost hydrosilylation, a Ferrier-type *C*-glycosylation, and protodesilylation are crucial steps of this enantioselective synthesis. Alkynylated dihydropyran (**71**) was obtained through the reaction of sugar (**69**) with silylated alkyne **70** using $SnCl_4$. Dihydropyran **71** was effective in delivering (−)-aspergillide C (**72**) over several steps.

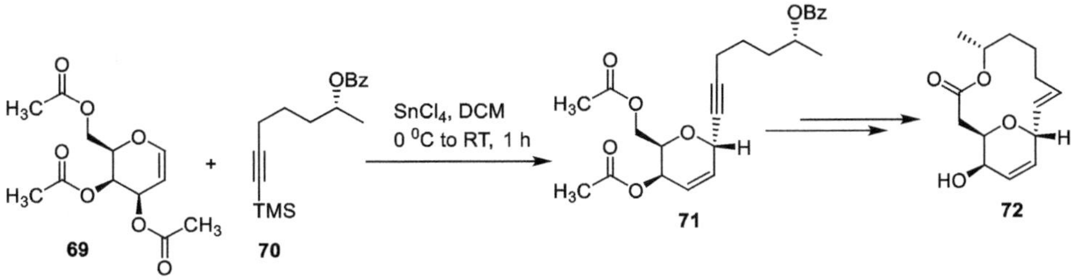

Figure 4.21: Total synthesis of aspergillide C.

4.2.6 Total synthesis of nonanolide

4.2.6.1 Total synthesis of cytospolide P

Natural source: Endophytic fungus *Cytospora* sp. [62]

Class of compounds: Nonanolide [62]

Biological activity: Cytotoxic activity [63]

Structure and structural features: Cytospolides (A–E) contain an unprecedented 15-carbon skeleton comprising a unique C-2 methyl group. Cytospolide P includes different functional groups such as acetate at C-5, keto at C-3, and hydroxy at C-8 than that of other cytospolides (Figure 4.22) [62, 63].

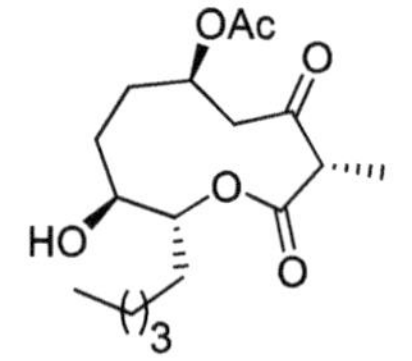

Cytospolide P

Figure 4.22: Structure of cytospolide P.

Synthetic strategy:
Raju and coworkers [63] first reported a simple carbohydrate-based synthetic pathway for the total synthesis of cytospolide P from starting material D-ribose (Figure 4.23). Wittig homologation created four stereogenic centers, Sharpless asymmetric epoxidation, and Yamaguchi macrolactonization were approved to achieve the desired compound. At first, D-ribose **(73)** was protected using acetone and conc. H_2SO_4 at room temperature subsequently the reduction in the presence of $NaBH_4$ and oxidation by $NaIO_4$ afforded 2,3-*O*-isopropylidene-lerythrose **(74)** over two steps. The desired cytospolide P **(75)** was finally derived from L-erythrose **(74)** over several steps.

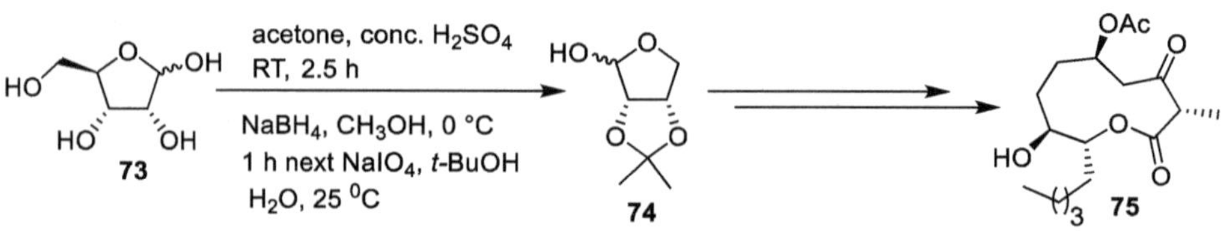

Figure 4.23: Total synthesis of cytospolide P.

4.2.7 Total synthesis of nucleosides

4.2.7.1 Total synthesis of trachycladines A and B

Natural source: *Trachycladus laevispirulifer* Carter [64]

Class of compounds: Nucleosides [64]

Biological activity: Cytotoxic activity (trachycladine A) [65]

Structure and structural features: Nucleosides comprise 2-C-methyl-D-5-deoxyribofuranose same carbohydrate unit. Trachycladine A includes a 2-chloroadenine and trachycladine B possesses a hypoxanthine (Figure 4.24) [64].

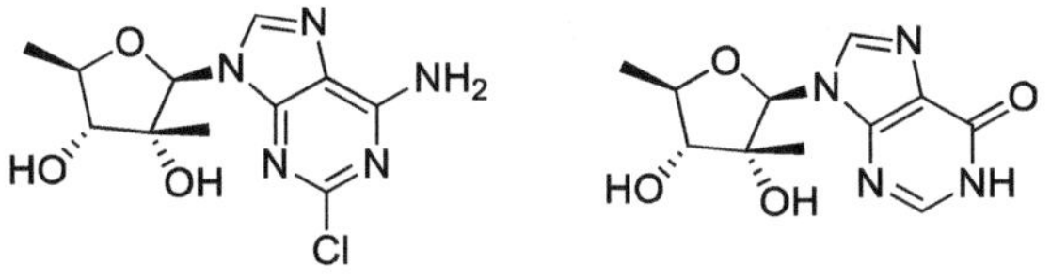

Figure 4.24: Structure of trachycladines A and B.

Synthetic strategy:
Koumbis and coworkers [66] disclosed the versatile and high-yielding total synthesis of marine nucleosides trachycladines A (77) and B (78) from D-ribose (73) as the starting material (Figure 4.25). In this protocol, regio- and stereoselective direct Vorbrüggen glycosylation was the crucial step. D-Ribose (73) furnished a mixture of triacetates **76a**

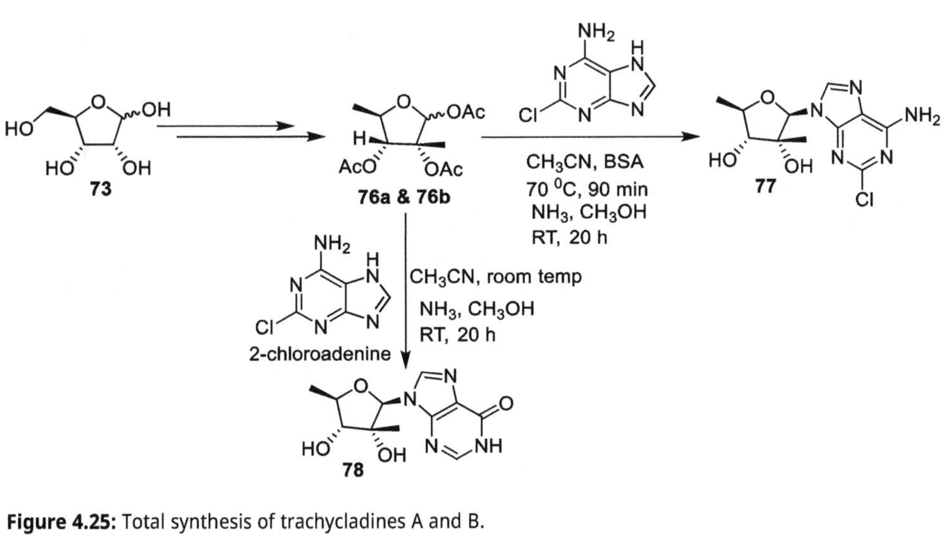

Figure 4.25: Total synthesis of trachycladines A and B.

and **76b** over several steps. Trachycladines A (77) and B (**78**) were obtained from triacetates through the coupling reaction.

4.2.8 Total synthesis of polyphenol

4.2.8.1 Total synthesis of tellimagrandin II

Natural source: *Geum japonicum* and *Syzygium aromaticum* [67]

Class of compounds: Polyphenol [68]

Biological activity: Anti-inflammatory activity and antibacterial activity [68, 69]

Structure and structural features: It includes sugar core, galloyl, and hexahydroxydiphenyl groups (Figure 4.26) [70].

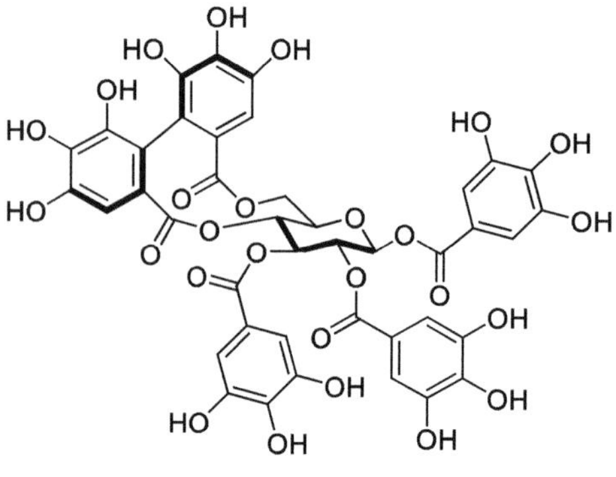

Tellimagrandin II

Figure 4.26: Structure of tellimagrandin II.

Synthetic strategy:
Kawabata et al. [70] accomplished a total synthesis of natural polyphenol tellimagrandin II in 2015 (Figure 27). Stereoselective glycosidation and regioselective introduction of a galloyl group are the main features of this short synthesis. At first, gallic acid trimethoxymethyl ether (**79**) was reacted with glucose (**80**) in the presence of the DIAD and PPh_3 in 1,4-dioxane to give β-glycoside **81**. Next, compound **82** was obtained from β-glycoside **81** over several steps. The total synthesis of the desired product tellimagrandin II (**83**) was completed by the reaction of compound **82** and trimethoxymethyl ether **79** over three steps.

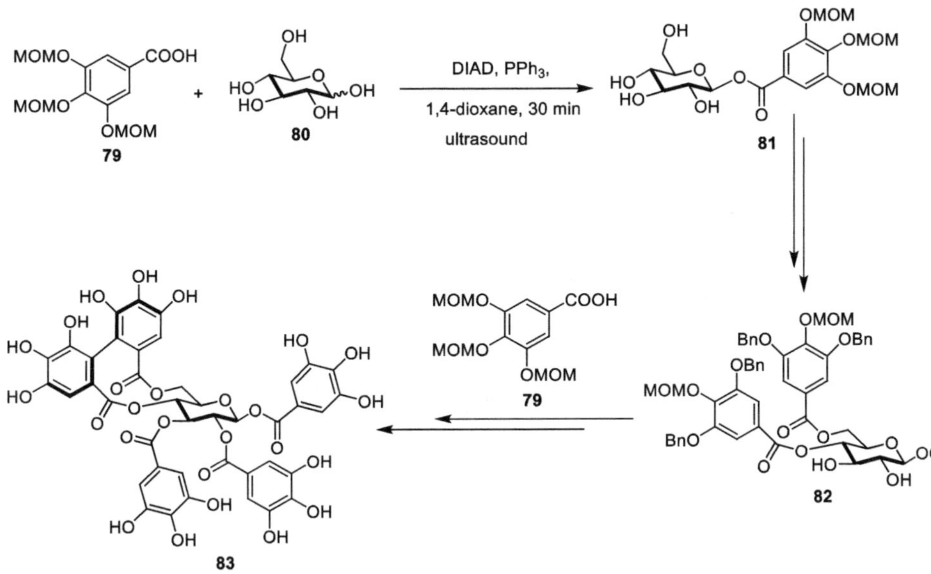

Figure 4.27: Total synthesis of tellimagrandin II.

4.2.9 Total synthesis of polysaccharide

4.2.9.1 Total synthesis of PS A1 oligosaccharides

Natural source: Commensal bacteria (*Streptococcus pneumonia* and *Bacteroides fragilis*) [71]

Class of compounds: Polysaccharide [71]

Biological activity: Immunological activity [71]

Structure and structural features: It contains tetrasaccharide repeating units and comprises positive and negative charges (Figure 4.28) [71].

Synthetic strategy:
In 2023, the first total synthesis of PS A1 oligosaccharides consisting of up to three repeating units was disclosed by Codee et al. [72]. The application of a C-3, C-6-bridged silylidene-protected galactosamine building block was vital for the synthesis of *Bacteroides fragilis* PS A1 fragments. This building block containing an "inverted" conformation was designed to behave as an apt nucleophile and a stereoselective glycosyl donor. The set of molecules encloses PS A1 fragments (dodecasaccharide level **87a** and **92a**) and structures lacking the galactofuranosyl appendages namely fragment **87b** was synthesized successfully from their fully protected precursors and the protected oligosaccharide backbones, respectively. Initially, silylidene galactosazide acceptor

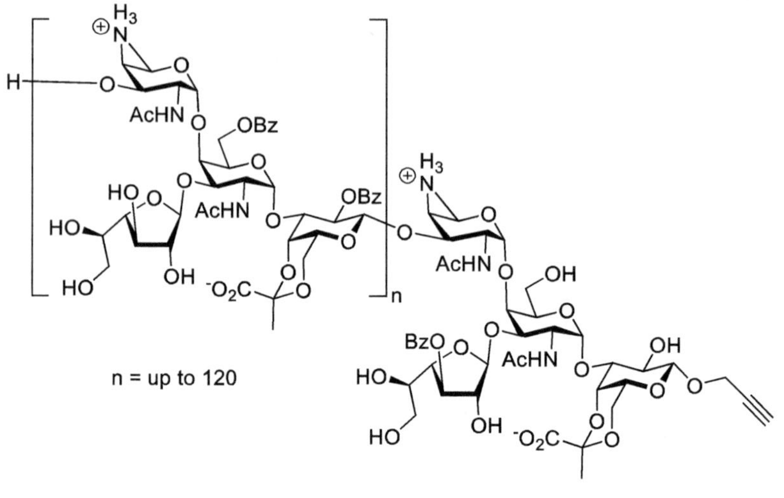

Figure 4.28: Structure of tellimagrandin II.

(**85**) was coupled with *N*-phenyl trifluoroacetimidate donor (**84**) using TfOH in DCM to furnish disaccharide **86** as a single anomer. Disaccharide **86** was effective in affording desired fragments **87a**, **87b**, and **87c** over several steps (Figure 4.29). Thioglycoside

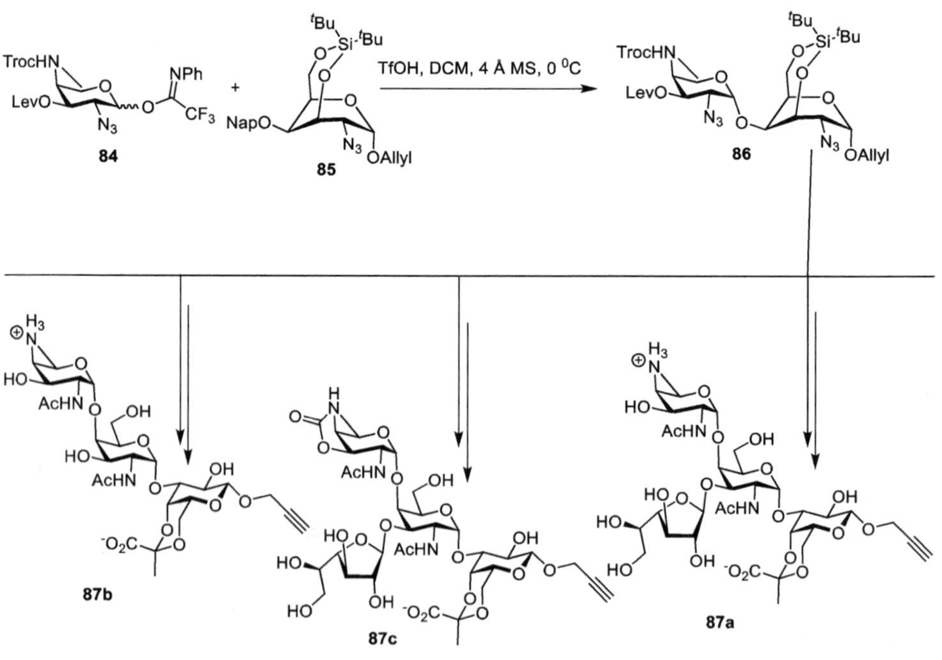

Figure 4.29: Total synthesis of PS A1 fragments.

(**88**) was united with disaccharide **89** to deliver trisaccharide **90** in the presence of the TBSOTf in DCM. The final two desired compounds **91a** and **92a** were obtained from trisaccharide **90** over several steps (Figure 4.30).

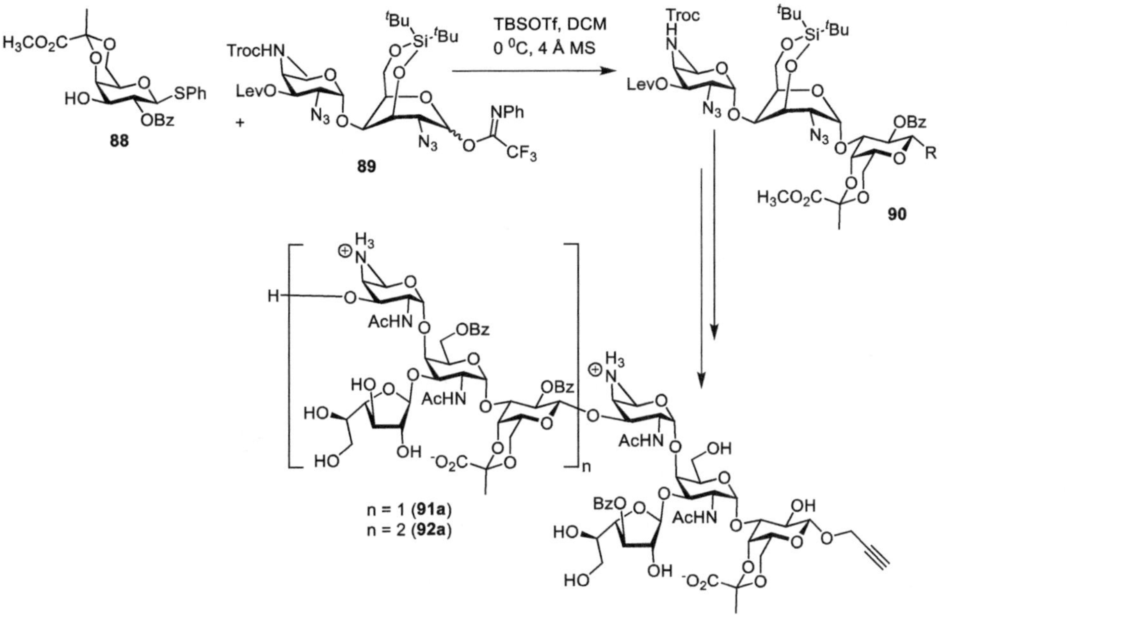

Figure 4.30: Total synthesis of PS A1 fragments (dodecasaccharide level).

4.3 Conclusion

Bioactive natural products have a profound role in drug discovery and development. In organic synthesis, biomolecule carbohydrates have applications as important chiral precursors. Bioactive carbohydrates originate from various sources including plants, animals, and microorganisms. Sugar molecules (carbohydrates) are direct products of photosynthesis and these chiral natural products are the main constituents of cell walls. Few carbohydrates have been approved for the treatment of different diseases and they have important biological activities. Total synthesis of natural products has received a lot of attention to determine the actual structure of complex bioactive compounds in the laboratory from simple, easily available precursors. Thus, this chapter represents the total synthesis of carbohydrate-decorated bioactive compounds. This methodology can develop novel drugs with different therapeutic activities.

References

[1] Atanasov AG, Zotchev SB, Dirsch VM. The international natural product sciences taskforce, Claudiu T Supuran. Natural products in drug discovery: Advances and opportunities. Nat Rev Drug Discov, 2021, 20, 200–216.

[2] Bontempo P, Masi LD, Rigano D. Functional properties of natural products and human health. Nutrients, 2023, 15, 2961.

[3] Majhi S, Sivakumar M. Semisynthesis of Bioactive Compounds and their Biological Activities. Elsevier: Amsterdam, 2023.

[4] Majhi S, Das D. Chemical derivatization of natural products: Semisynthesis and pharmacological aspects – A decade update. Tetrahedron, 2021, 78, 131801.

[5] Majhi S. Recent developments in the synthesis and anti-cancer activity of acridine and xanthine-based molecules. Phys Sci Rev, 2022, 8, 2405–2439.

[6] Majhi S, Jash SK. Recent developments of nanocatalysts for Stille coupling reaction. Synth Commun, 2023, 53, 2061–2087.

[7] Brahmachari G, Majhi S, Mandal B, Mandal M, Kumar A, Srivastava AK, Singh RB, Misra N. Stigmasterol from the flowers of *Peltophorum pterocarpum* (DC) Backer Ex K. Heyne (Fabaceae)-isolation, spectral properties and quantum chemical studies. J Indian Chem Soc, 2018, 95, 1231–1244.

[8] Gorai D, Jash SK, Singh RK, Sarkar A, Majhi S. Chemical and pharmacological aspects of *Limnophila rugosa*: An update. Int J Nat Prod Res, 2013, 3, 120–124.

[9] Brahmachari G, Mandal LC, Roy R, Jash SK, Mondal A, Majhi S, Gorai D. Lupeol, a pharmaceutically potent triterpenoid, from the ripe fruits of *Rauvolfia tetraphylla* L. (Apocynaceae). J Indian Chem Soc, 2011, 88, 303–305.

[10] Majhi S, Saha I. Visible light-promoted synthesis of bioactive *N,N*-heterocycles. Curr Green Chem, 2022, 9, 127–144.

[11] Dey AK, Majhi S. Samarium(III) triflate in organic synthesis: A mild and efficient catalyst. ChemistrySelect, 2023, 8, e202300156.

[12] Majhi S. Discovery, development, and design of anthocyanins-inspired anticancer agents – A comprehensive review. Anticancer Agents Med Chem, 2022, 22, 3219–3238.

[13] Majhi S. Synthesis of bioactive natural products and their analogs at room temperature – An update. Phys Sci Rev, 2022, 8, 3447–3473.

[14] Newman DJ, Newman CGM. Natural products as sources of new drugs over the nearly four decades from 01/1981 to 09/2019. J Nat Prod, 2020, 83, 770–803.

[15] Nasim N, Sandeep IS, Mohanty S. Plant-derived natural products for drug discovery: Current approaches and prospects. Nucleus, 2022, 65, 399–411.

[16] Majhi S. Diterpenoids: Natural distribution, semisynthesis at room temperature and pharmacological aspects – A decade update. ChemistrySelect, 2020, 5, 12450–12464.

[17] Majhi S, Manickam S. Semisynthesis of Alkaloids, Semisynthesis of Bioactive Compounds and their Biological Activities. Elsevier: Amsterdam, 2024, pp. 55–112.

[18] Dey AK, Majhi S. 7 Role of samarium in organic synthesis. In Rare Earth Elements: Processing, Catalytic Applications and Environmental Impact. Basu B, Banerjee B. (Eds.). de Gruyter: Germany, 2023, pp. 119–140.

[19] Majhi S, Manickam S. Semisynthesis Offlavones, Semisynthesis of Bioactive Compounds and their Biological Activities. Elsevier: Amsterdam, 2024, pp. 113–179.

[20] Majhi S, Manickam S. Developing Semisynthesis Methods for Neglected Tropical Diseases, Semisynthesis of Bioactive Compounds and their Biological Activities. Elsevier: Amsterdam, 2024, pp. 439–458.

[21] Majhi S, Mondal PK. Microwave-assisted synthesis of heterocycles and their anti-cancer activities. Curr Microw Chem, 2023, 10, 135–154.
[22] Sinha K, Chowdhury S, Banerjee S, Mandal B, Mandal M, Majhi S, Brahmachari G, Ghosh J, Sil PC. Lupeol alters viability of SK-RC-45 (renal cell carcinoma cell line) by modulating its mitochondrial dynamics. Heliyon, 2019, 5, e02107.
[23] Majhi S. Chapter 10 – Applications of nanoparticles in organic synthesis under ultrasonication. In Nanoparticles in Green Organic Synthesis Strategy Towards Sustainability 1st ed. Bhunia S, Singh P, Kim K-H, Kumar B, Oraon R. (Eds.). Elsevier Science: Amsterdam, 2023, pp. 279–315.
[24] Sonestedt E, Øverby NC. Carbohydrates – A scoping review for Nordic nutrition recommendations. Food Nutr Res, 2023, 67(10), 29219.
[25] Pohl NLB. Introduction: Carbohydrate chemistry. Chem Rev, 2018, 118, 7865–7866.
[26] Samtiya M, Aluko RE, Dhewa T, Moreno-Rojas JM. Potential health benefits of plant food-derived bioactive components: An overview. Foods, 2021, 10, 839.
[27] Wang J, Zhang Y, Lu Q, Xing D, Zhang R. Exploring carbohydrates for therapeutics: A review on future directions. Front Pharmacol, 2021, 12, 756724.
[28] Jiang H, Qin X, Wang Q, Xu Q, Wang J, Wu Y, Chen W, Wang C, Zhang T, Xing D, Zhang R. Application of carbohydrates in approved small molecule drugs: A review. Eur J Med Chem, 2021, 223, 113633.
[29] Majhi S, Mandal B. Modern Sustainable Techniques in Total Synthesis of Bioactive Natural Products. World Scientific: Singapore, 2023.
[30] Majhi S. Applications of Norrish type I and II reactions in the total synthesis of natural products: A review. Photochem Photobiol Sci, 2021, 20, 1357–1378.
[31] Majhi S. Applications of Yamaguchi method to esterification and macrolactonization in total synthesis of bioactive natural products. ChemistrySelect, 2021, 6, 4178–4206.
[32] Majhi S. The art of total synthesis of bioactive natural products via microwaves. Curr Org Chem, 2021, 25, 1047–1069.
[33] Majhi S. Applications of flow chemistry in total synthesis of natural products. Curr Org Chem, 2023, 27, 1072–1089.
[34] Majhi S. Applications of ultrasound in total synthesis of bioactive natural products: A promising green tool. Ultrason Sonochem, 2021, 77, 105665.
[35] Satake M, Bourdelais AJ, Van Wagoner RM, Baden DG, Wright JLC. Brevisamide: An unprecedented monocyclic ether alkaloid from the dinoflagellate *Karenia brevis* that provides a potential model for ladder-frame initiation. Org Lett, 2008, 10, 3465.
[36] Sudharani C, Venukumar P, Sridhar PR. Stereoselective synthesis of *C*-2-methylene and C-2-methyl α- and β-C-glycosides from 2-C-branched glycals: Formal total synthesis of (−)-brevisamide. Eur J Org Chem, 2014, 2014, 8085–8093.
[37] Sieber S, Carlier A, Neuburger M, Grabenweger G, Eberl L, Gademann K. Isolation and total synthesis of Kirkamide, an aminocyclitol from an obligate leaf nodule symbiont. Angew Chem Int Ed, 2015, 54, 7968–7970.
[38] Sieber S, Hsiao -C-C, Emmanouilidou D, Debowski AW, Stubbs KA, Gademann K. Syntheses and biological investigations of kirkamide and oseltamivir hybrid derivatives. Tetrahedron, 2020, 76, 131386.
[39] Narayana C, Khanna A, Kumari P, Sagar R. Total syntheses of Kirkamide and *N*-acetyl *ent*-conduramine B-1. Asian J Org Chem, 2021, 10, 392–399.
[40] Gao F, Yan X, Shakya T, Baettig OM, Ait-Mohand-Brunet S, Berghuis AM, Wright GD, Auclair K. Synthesis and structure-activity relationships of truncated bisubstrate inhibitors of aminoglycoside 6′-*N*-acetyltransferases. J Med Chem, 2006, 49, 5273–5281.
[41] Hattori H, Roesslein J, Caspers P, Zerbe K, Miyatake-Ondozabal H, Ritz D, Rueedi G, Gademann K. Total synthesis and biological evaluation of the glycosylated macrocyclic antibiotic mangrolide A. Angew Chem Int Ed, 2018, 57, 11020–11024.

[42] Theriault RJ, Karwowski JP, Jackson M, Girolami RL, Sunga GN, Vojtko CM, Coen LJ. Tiacumicins, a novel complex of 18-membered macrolide antibiotics I. Taxonomy, fermentation and antibacterialactivity. J Antibiot, 1987, 40, 567–574.
[43] Glaus F, Altmann K-H. Total synthesis of the tiacumicin B (lipiarmycin A3/fidaxomicin) aglycone. Angew Chem Int Ed, 2014, 53, 1937–1940.
[44] Kaufmann E, Hattori H, Miyatake-Ondozabal H, Gademann K. Total synthesis of the glycosylated macrolide antibiotic fidaxomicin. Org Lett, 2015, 17, 3514–3517.
[45] Krohn K, Sohrab MH, van Ree T, Draeger S, Schulz B, Antus S, Kurtán T. Dinemasones A, B and C – New bioactive metabolites from the endophytic fungus *Dinemasporium strigosum*. Eur J Org Chem, 2008, 2008, 5638–5646.
[46] Xue X, Yin Z, Meng X, Li Z. A carbohydrate-based approach for the total synthesis of (−)-Dinemasone B, (+)-4a-epi-dinemasone B, (-)-7-epi-dinemasone B, and (+)-4a,7-diepi-dinemasone B. J Org Chem, 2013, 78, 9354–9365.
[47] Ferreira SB, Sodero ACR, Cardoso MFC, Lima ES, Kaiser CR, Silva FP Jr, Ferreira VF. Synthesis, biological activity, and molecular modeling studies of 1*H*-1,2,3-triazole derivatives of carbohydrates as α-glucosidases inhibitors. J Med Chem, 2010, 53, 2364–2375.
[48] Xu L, He Z, Xue J, Chen X, Wei X. β-resorcylic acid lactones from a *Paecilomyces* fungus. J Nat Prod, 2010, 73, 885–889.
[49] Mahankali B, Srihari P. A carbohydrate approach for the first total synthesis of Cochliomycin C: Stereoselective total synthesis of Paecilomycin E, Paecilomycin F and 6′-*epi*-Cochliomycin C. Eur J Org Chem, 2015, 2015, 3983–3993.
[50] Nukina M, Hirota H. Pyrenolide D, a new cytotoxic fungal metabolite from *Pyrenophora teres*. Biosci Biotechnol Biochem, 1992, 56, 1158–1159.
[51] Thirupathi B, Reddy PP, Mohapatra DK. A carbohydrate-based total syntheses of (+)-Pyrenolide D and (−)-4-*epi*-Pyrenolide D. J Org Chem, 2011, 76, 9835–9840.
[52] Kawahara K, Kubota M, Sato N, Tsuge K, Seto Y. Occurrence of an α-galacturonosyl-ceramide in the dioxin-degrading bacterium *Sphingomonas wittichii*. FEMS Microbiol Lett, 2002, 214, 289–294.
[53] Kawahara K, Sato N, Tsuge K, Seto Y. Confirmation of the anomeric structure of galacturonic acid in the galacturonosyl-ceramide of *Sphingomonas yanoikuyae*. Microbiol Immunol, 2006, 50, 67–71.
[54] Guimarães LL, Toledo MS, Ferreira FAS, Straus AH, Takahashi HK. Structural diversity and biological significance of glycosphingolipids in pathogenic and opportunistic fungi. Front Cell Infect Microbiol, 2014, 4, 138.
[55] Aerts JMFG, Artola M, van Eijk M, Ferraz MJ, Boot RG. Glycosphingolipids and infection. Potential new therapeutic avenues. Front Cell Dev Biol, 2019, 7, 324.
[56] Tanaka M, Takahashi D, Toshima K. 1,2-*cis*-*α*-Stereoselective glycosylation utilizing a glycosyl-acceptor-derived borinic ester and its application to the total synthesis of natural glycosphingolipids. Org Lett, 2016, 18, 5030–5033.
[57] Kito K, Ookura R, Yoshida S, Namikoshi M, Ooi T, Kusumi T. New cytotoxic 14-membered macrolides from marine-derived fungus *Aspergillus ostianus*. Org Lett, 2008, 10, 225–228.
[58] Haefner B. Drugs from the deep: Marine natural products as drug candidates. Drug Discov Today, 2003, 8, 536–544.
[59] Blunt JW, Copp BR, Munro MHG, Northcote PT, Prinsep MR. Marine natural products. Nat Prod Rep, 2011, 28, 196–268.
[60] Ookura R, Kito K, Saito Y, Kusumi T, Ooi T. Structure revision of aspergillides A and B, Cytotoxic 14-membered macrolides from *Aspergillus ostianus*, by X-ray crystallography. Chem Lett, 2009, 38, 384.
[61] Srihari P, Krishna NH, Sridhar Y, Kamal A. A carbohydrate approach for the formal total synthesis of (−)-aspergillide C. Beilstein J Org Chem, 2014, 10, 3122–3126.

[62] Lu S, Sun P, Li T, Kurtán T, Mándi A, Antus S, Krohn K, Draeger S, Schulz B, Yi Y, Li L, Zhang W. Bioactive nonanolide derivatives isolated from the endophytic fungus *Cytospora* sp. J Org Chem, 2011, 76, 9699–9710.

[63] Suman P, Raju BC. Carbohydrate-based first stereoselective total synthesis of bioactive cytospolide P. Org Biomol Chem, 2014, 12, 3358–3361.

[64] Searle PA, Molinski TF. Trachycladines A and B: 2′-C-methyl-5′-deoxyribofuranosyl nucleosides from the marine sponge *Trachycladus laevispirulifer*. J Org Chem, 1995, 60, 4296–4298.

[65] Jaiyesimi IA, Kantarijan HM, Estey EH. Advances in therapy for hairy cell leukemia. Cancer, 1993, 72, 5–16.

[66] Peitsinis ZV, Melidou DA, Stefanakis JG, Evgenidou H, Koumbis AE. A versatile total synthesis of trachycladines A and B and their analogues. Eur J Org Chem, 2014, 2014, 8160–8166.

[67] Kurokawa M, Hozumi T, Basnet P, Nakano M, Kadota S, Namba T, Kawana T, Shiraki K. Purification and characterization of eugeniin as an anti-herpesvirus compound from *Geum japonicum* and *Syzygium aromaticum*. J Pharmacol Exp Ther, 1998, 284, 728.

[68] Lin C-Y, Kao S-H, Hung L-C, Chien H-J, Wang W-H, Chang Y-W, Chen Y-H. Lipopolysaccharide-induced nitric oxide and Prostaglandin E2 production is inhibited by Tellimagrandin II in mouse and human macrophages. Life (Basel), 2021, 11, 411.

[69] Chang Y-W, Huang W-C, Lin C-Y, Wang W-H, Hung L-C, Chen Y-H. Tellimagrandin II, A type of plant polyphenol extracted from *Trapa bispinosa* inhibits antibiotic resistance of drug-resistant *Staphylococcus aureus*. Int J Mol Sci, 2019, 20, 5790.

[70] Takeuchi H, Mishiro K, Ueda Y, Fujimori Y, Furuta T, Kawabata T. Total synthesis of ellagitannins through regioselective sequential functionalization of unprotected glucose. Angew Chem, 2015, 127, 6275.

[71] Zhang Q, Overkleeft HS, van der Marel GA, Codée JDC. Synthetic zwitterionic polysaccharides. Curr Opin Chem Biol, 2017, 40, 95–101.

[72] Wang Z, Poveda A, Zhang Q, Unione L, Overkleeft HS, van der Marel GA, Jesús J-B, Codee JDC. Total synthesis and structural studies of Zwitterionic *Bacteroides fragilis* polysaccharide A1 fragments. J Am Chem Soc, 2023, 145, 14052–14063.

Anindya S. Manna, Sandip Das, Nilay Karchaudhuri, Tanmoy Ghosh, Rajesh Nandi, and Dilip K. Maiti*

Chapter 5
Recent developments in biologically relevant carbohydrate-decorated *O*-heterocycles

5.1 Introduction

The synthesis of carbohydrate-decorated *O*-heterocycles has gained key importance due to their interesting versatile structures and notable bioactivity [1]. Carbohydrates are known for their significant roles in biological processes with distinctive stereochemistry, enabling targeted interactions inside biological entities [2, 3]. During collaboration with oxygen-containing heterocyclic compounds, these valuable properties of carbohydrates are often amplified toward enhanced biological activity. Consequently, carbohydrate-decorated *O*-heterocycles hold immense potential in the fields of new drug discovery and the development of innovative biomaterials, making the exploration and interest in searching for efficient synthetic methods essential for advancement in the field [4]. One of the most powerful approaches for synthesizing these compounds involves glycosylation reactions, which can be achieved through ionic activation with Lewis acids or electrophilic promoters. In recent years, radical chemistry has gained prominence as an alternative, offering significant advantages over the traditional ionic approach such as the ability to operate under milder reaction conditions with precise and selective bond formation and a broad functional group tolerance [5].

Dehydroxylative glycosylation through copper-catalysis has shown significant promise in decorating heterocyclic scaffolds, where radicals are generated from the carbohydrate donor and coupled with the heterocyclic framework to form a glycosidic bond [6]. With their inherent structural complexity, spirocyclic *O*-heterocycles are commonly found in natural products with potent biological activities, such as anticancer and antimicrobial agents, and thus are gaining the interest of synthetic chemists [7]. The incorpo-

Acknowledgments: Financial support from the Ministry of Mines (no. Met 4-14/19/2021), Government of India, is gratefully acknowledged. We also acknowledge the University of Calcutta, India, for providing the literature research facilities.

***Corresponding author: Dilip K. Maiti,** Department of Chemistry, University of Calcutta, 92 APC Road, Kolkata 700009, West Bengal, India, e-mail: dkmchem@caluniv.ac.in
Anindya S. Manna, Sandip Das, Tanmoy Ghosh, Rajesh Nandi, Department of Chemistry, University of Calcutta, 92 APC Road, Kolkata 700009, West Bengal, India
Nilay Karchaudhuri, Department of Chemistry, Barrackpore Rastraguru Surendranath College, Kolkata 700120, West Bengal, India

https://doi.org/10.1515/9783111447094-005

ration of carbohydrate moieties into these spirocyclic structures offers potential synergistic effects toward enhanced activity and stability in biological environments [8].

Recent research has also highlighted the increasing importance of green chemistry in synthesis, with growing concerns about the environmental impact of traditional synthetic methods leading to the search for sustainable alternatives. Deep eutectic solvents (DES), which are composed of biodegradable and natural components, have emerged as an attractive option in this sector [9, 10]. The shift toward sustainable synthesis has significantly influenced the development of radical glycosylation [11]. Aligning with green chemistry principles, DES is an eco-friendly alternative to conventional solvents, prioritizing the use of nontoxic, renewable materials with the aim of reducing waste. It offers a promising avenue for the sustainable synthesis of carbohydrate-decorated *O*-heterocycles with enhanced selectivity and efficiency [12, 13]. High stereoselectivity in the radical-based glycosylation process remains an on-demand area of research, targeting improved reaction yields and developing catalytic systems toward a wide range of functional groups [14]. The presence of carbohydrate moieties infuses a number of interacting sites that catalyze the interaction of carbohydrate-decorated *O*-heterocycles with biological targets, making them attractive candidates for the development of therapeutic agents for various treatments such as cancer, inflammation, and infections [15, 16]. Touisni et al. [17] developed 7-substituted coumarins decorated with variable glycosyl moieties such as mannosylumbelliferone (**I**, Figure 5.1) for antitumor activity. A series of 1,2,3-triazole-fused chromene-conjugated glucose triazole units (**II** and **III** in Figure 5.1) are also found in the literature with antibacterial activity.

Therefore, the synthesis of carbohydrate-decorated *O*-heterocycles represents a rapidly growing realm of research with a wide-ranging array of applications in both chemistry and biomedicine. This review has shown a glimpse of recent advancements in the synthesis of carbohydrate-decorated *O*-heterocycles, discussing innovative synthetic methodologies, including transition metal catalysis, organocatalysis, and glycosylation techniques, and exploring their applications in fields such as drug development and glycomimetic research. This chapter will also address recent advances in environmentally friendly and efficient methods for synthesizing these compounds, driven by growing concerns about sustainability. As research evolves, these compounds will continue to

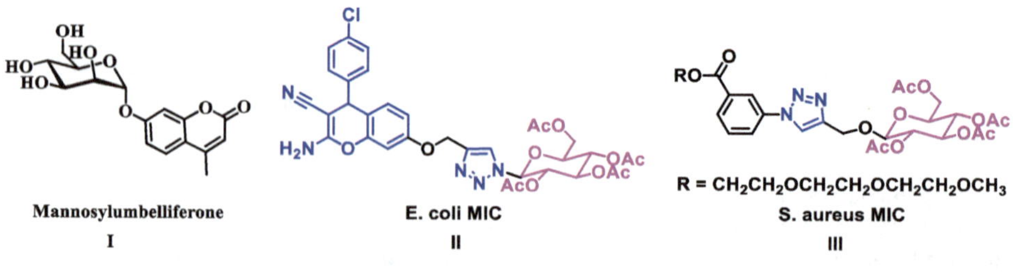

Figure 5.1: Carbohydrate-decorated *O*-heterocycles containing some medicinally valuable compounds.

contribute to breakthroughs in drug discovery and materials science, highlighting their ongoing relevance in medicinal chemistry and sustainable synthesis.

5.2 Development of variable synthetic strategy for carbohydrate-decorated *O*-heterocycles

Carbohydrate-decorated *O*-heterocycles have significant biological properties that are applicable for medicinal applications [18]. In this section, we will discuss different types of *O*-heterocycles with carbohydrates as the substituents, depending on their nature of ring size and connectivity.

5.2.1 Synthesis of carbohydrate-decorated benzo-fused *O*-heterocycles

Herein, we have exemplified some important protocols toward the access of carbohydrate-decorated benzo-fused *O*-heterocycles.

5.2.1.1 Access to carbohydrate-containing benzofuranderivatives

In an experiment by Wu et al. [19], ethyl benzofuran-2carboxylate (**1**) was prepared from substituted salicylaldehyde in acetone solvent through potassium carbonate catalysis, and then it was hydrolyzed to produce benzofuran-2-carboxylic acid (**3**). Then, under the catalysis of *N,N*-diisopropylethylamine and HATU, (**3**) and glucosamine (**4**) were allowed to react with (**3**) in acetonitrile solvent to produce *N*-glycosylbenzofurans (**5**) in high yield (Figure 5.2).

Recently, a new protocol to access 2-benzofuranyl C-glycosides (**7**) with structurally dynamic sugar moieties has been demonstrated by Cao et al. [20] through a one-pot cascade reaction involving terminal sugar alkynes (**6**) with substituted 2-iodophenols. The desired products were obtained in good yields (Figure 5.3).

5.2.1.2 Access to carbohydrate-containing coumarins

A series of iminocoumaryl methyl glycosides (**11**) were synthesized by the Mukhopadhyay group [21] through a copper-catalyzed multicomponent reaction. This process involved carbohydrate propargyl derivatives (**8**), sulfonyl azides (**9**), and salicylaldehyde (**10**) or *O*-hydroxyacetophenone. The method is straightforward and adaptable to vari-

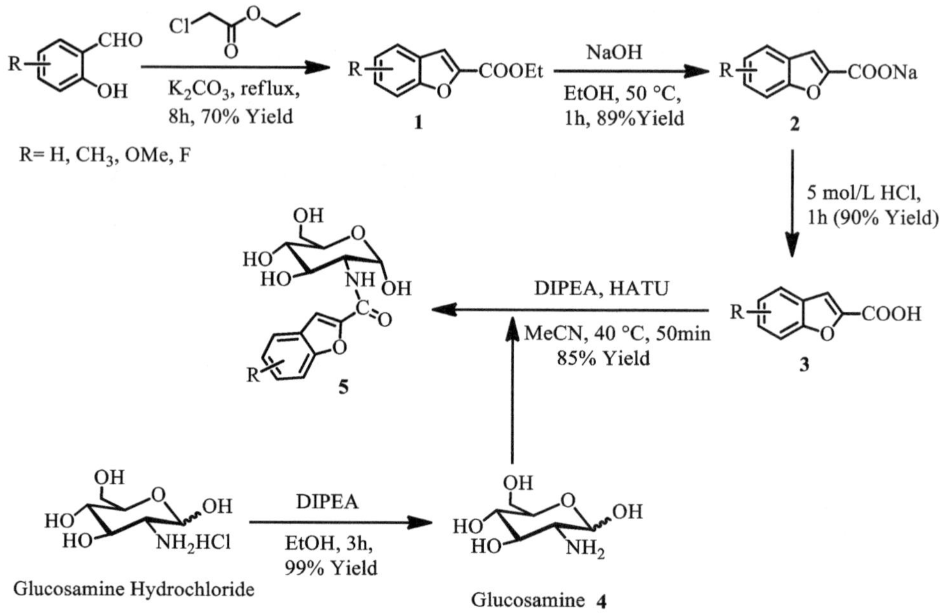

Figure 5.2: Synthesis of substituted *N*-glycosylbenzofuran derivatives.

ous components, resulting in excellent yields (Figure 5.4). These iminocoumaryl methyl or coumaryl methyl derivatives are evaluated as galectin antagonists.

Rajput et al. [22] synthesized doubly 3-*O*-coumarylmethyl-substituted thiodigalactosides (**15**) starting from bis-3-*O*-propargyl-thiodigalactoside (**12**) and various salicylaldehydes (**13**) through the intermediate **14**, resulting in potent and highly selective galectin-3 inhibitors (Figure 5.5). Through mutant studies, structural analysis, and molecular modeling, it was found that the coumaryl groups stack onto arginine side chains. One of the inhibitors demonstrated effectiveness in a mouse model of bleomycin-induced lung fibrosis, similar to a well-known nonselective galectin-1/galectin-3 inhibitor, highlighting the potential of targeting galectin-3 glycan recognition as a promising antifibrotic strategy.

Recently, Chu et al. [23] synthesized various sugars attached to coumarins via a 1,2,3-triazole linker (Figure 5.6) using an optimized one-pot click chemistry reaction involving carbohydrates (**16**) and coumarin derivatives (**17**). The new compounds were evaluated for their in vitro carbonic anhydrase (CA) inhibitory activity, with results showing that nearly all carbohydrate-based coumarin derivatives (**18**) exhibited strong inhibition against hCA IX, with IC_{50} values ranging from 11 to 132 nM. Notably, all the compounds demonstrated superior CA IX inhibitory activity compared to the original coumarin fragments.

Recently, Sau and team [24] provided a comprehensive review of productive protocols for synthesizing coumarin-functionalized glycosides (Figure 5.7). The review

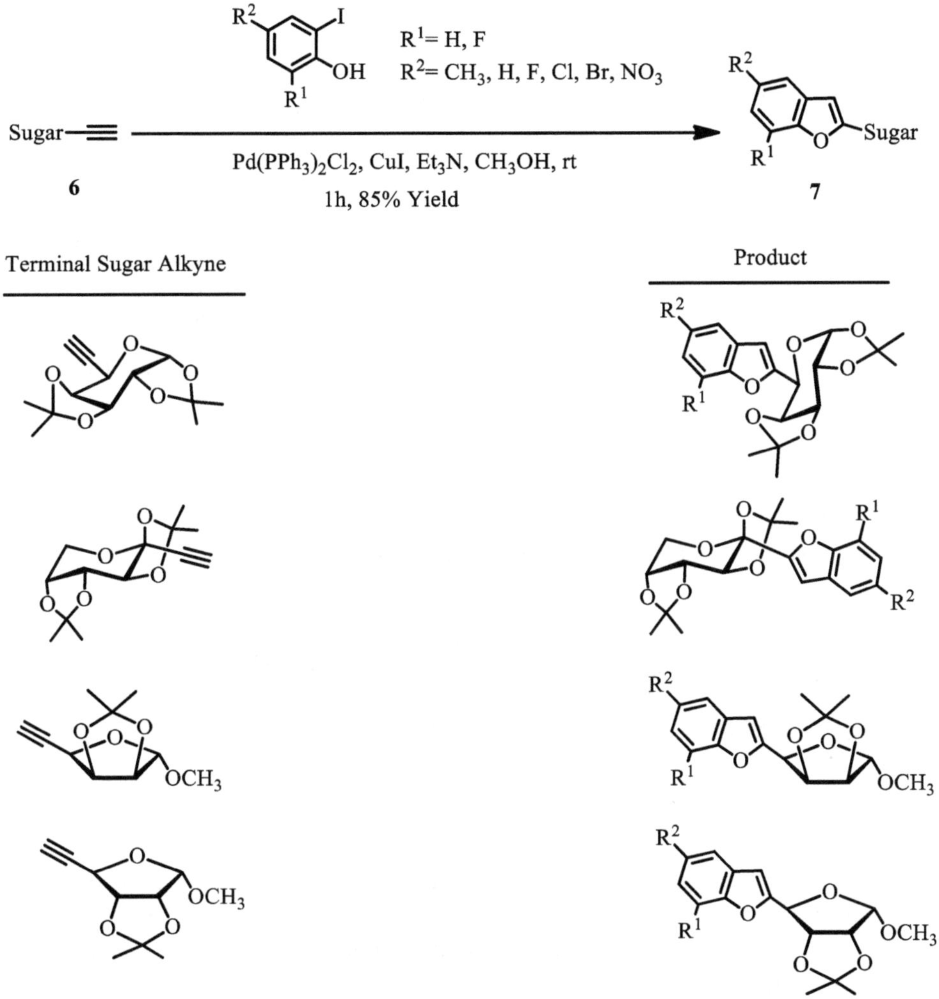

Figure 5.3: One-pot access to 2-benzofuranyl C-glycosides.

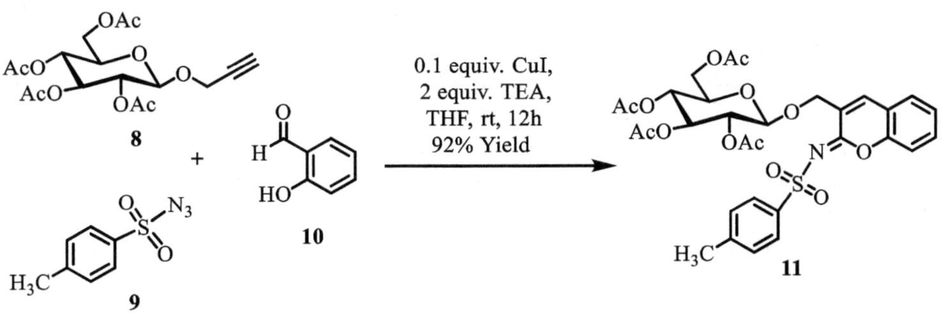

Figure 5.4: A copper-catalyzed multicomponent strategy for iminocoumaryl methyl glycosides.

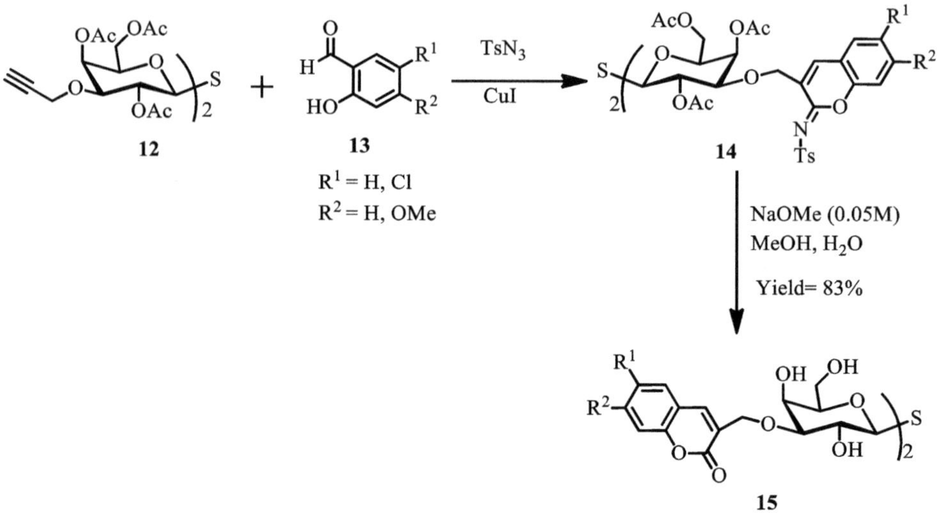

Figure 5.5: Synthesis of disubstituted coumarylthiodigalactosides.

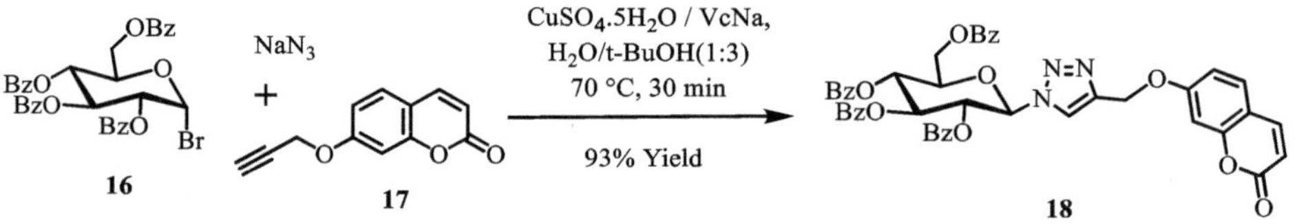

Figure 5.6: Synthesis of new carbohydrate-based coumarin derivatives.

covers five different types of carbohydrate-linked coumarin structures, categorized by their mode of attachment, including coumarin C-glycosides, *O*-glycosides, *S*-glycosides, triazole-linked glycosides, and iminocoumarin glycosides. The synthesis of these glycohybrids is discussed, highlighting the use of different catalysts and conditions. Additionally, the review summarizes a broad spectrum of research on coumarin glycosides, detailing their biological activities such as anti-inflammatory, antibacterial, antiallergic, antifungal, antioxidant, anticonvulsant, antitumor, anti-HIV, herbicidal, and anticancer properties.

Kumari et al. [25] reported an efficient method for the microwave-assisted synthesis of carbohydrate-fused pyrano[3,2-c]pyranones (**21**). A range of 4-hydroxycoumarins (**20**) with various substitutions were successfully converted into the corresponding carbohydrate-fused pyrano[3,2-c]pyranones ($n = 20$) in 10 min, yielding a good outcome (Figure 5.8). Furthermore, they demonstrated that three of these carbohydrate-fused pyrano[3,2-c]pyranone compounds exhibited anticancer activity at micromolar concentrations.

Jain et al. [26] developed an eco-friendly protocol (Figure 5.9) for synthesizing heterobioconjugate-based 1,2,3-triazoles (**24**) using a 1,3-dipolar cycloaddition reaction of

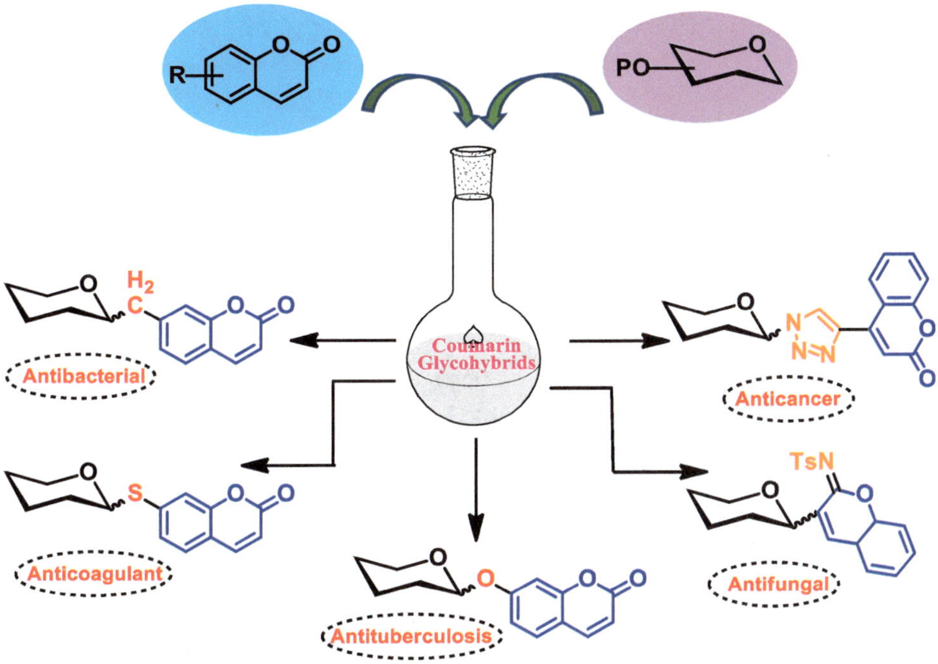

Figure 5.7: Preparation of sugar-functionalized coumarin frameworks.

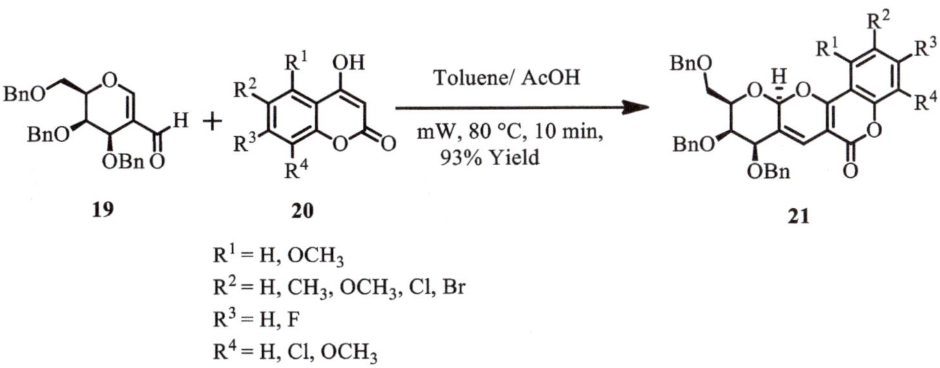

Figure 5.8: Synthesis of variably functionalized 2-C-formylgalactal-fused pyrano[3,2-c]pyranones.

coumarin azides (**22**) and an acetylated carbohydrate moiety (**23**). The reaction utilized magnetically separable Fe_3O_4@CD-CIT as a phase transfer catalyst with minimal copper content, carried out in water under ultrasonic conditions at 40 °C. This method was also scalable, achieving a 93% yield.

Sharma et al. [27] successfully synthesized a series of triazole-linked glycoconjugates, including carbohydrate-vanillic acid and carbohydrate-coumarin derivatives

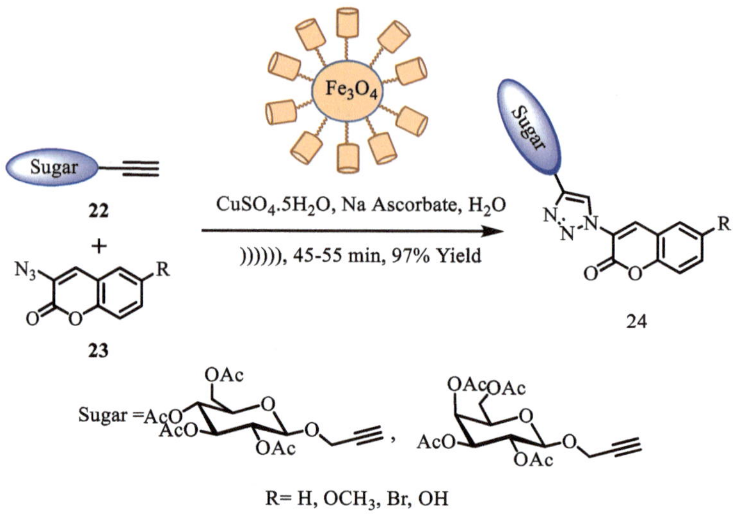

Figure 5.9: Synthesis of 1,4-disubstituted triazole.

(28) through a click reaction between glycosyl azide (25) and aryl alkynes (26) via copper catalysis. The reactions were both high-yielding and regioselective (Figure 5.10).

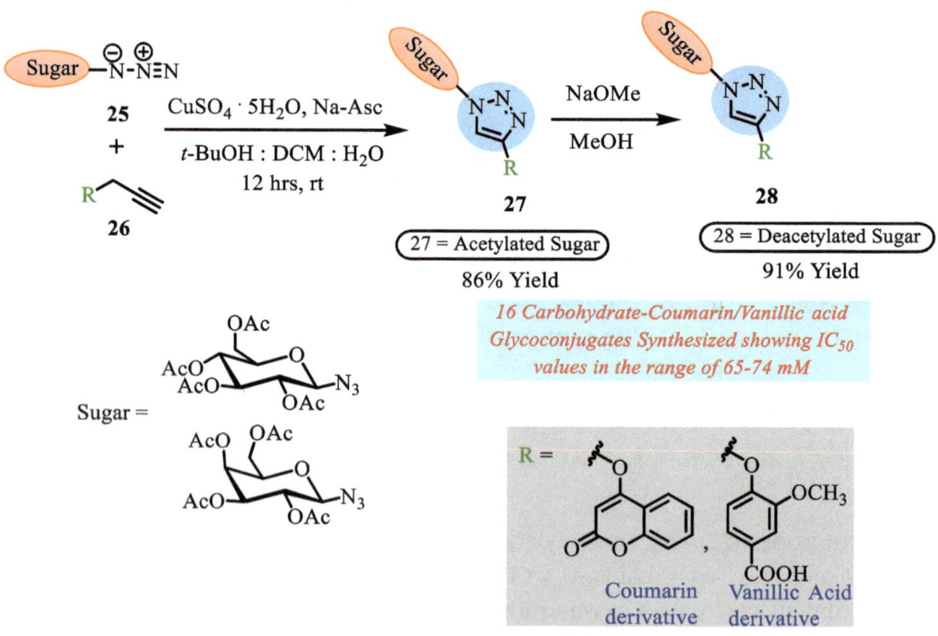

Figure 5.10: Access to carbohydrate-substituted coumarin or vanillic acid derivatives.

Results from the MTT assay demonstrated that the synthesized compounds exhibited activity against the *Leishmania donovani* parasite.

Gil and colleagues [28] used 1,8-diazabicyclo[5.4.0]undec-7-ene as a catalyst, along with a small amount of dichloromethane, to synthesize 3-nitro-2H-chromenes (**31** or **32**) that contain an acyclic carbohydrate fragment at the C2 position (Figure 5.11). This inclusion is considered valuable for potential drug development due to the carbohydrate's significance in various biological processes. The synthesis was achieved by reacting nitroalkenes (**29**) with salicylaldehydes (**30**). The newly synthesized compounds were then evaluated for antiproliferative activity against six human tumor cell lines.

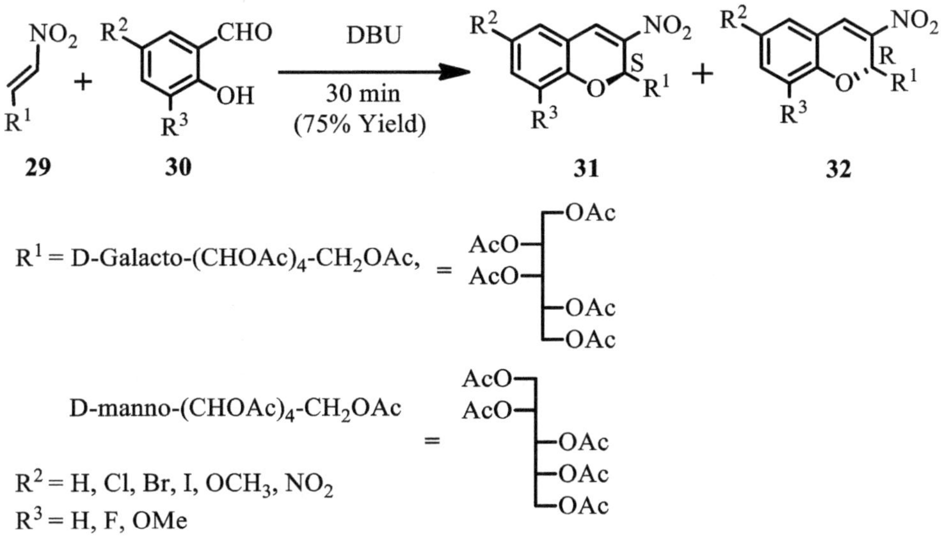

Figure 5.11: Synthesis of glycochromenes.

5.2.1.3 Access to carbohydrate-containing benzooxazoles

Khodair et al. [29] reported a modified nitrogen and sulfur glycosylation reaction toward access benzothiazole, benzoxazole, and pyridine nucleoside bases (**36**) with furanose and pyranose sugars (Figure 5.12). The study included conformational analysis, performed through homo- and heteronuclear two-dimensional NMR techniques such as 2D DFQ-COSY, heteronuclear multiple-quantum coherence (HMQC), and HMBC. The glycosylation sites on nitrogen and sulfur were identified using ^{1}H and ^{13}C HMQC experiments. The deprotected nucleosides were subsequently evaluated for their potential antitumor activity against leukemia-1210.

Bernès and colleagues [30] outlined a concise and efficient method for synthesizing glycosylisothiocyanates (**38**). These compounds were readily coupled and converted under mild conditions into a novel class of *N*-glycosides featuring an oxazole

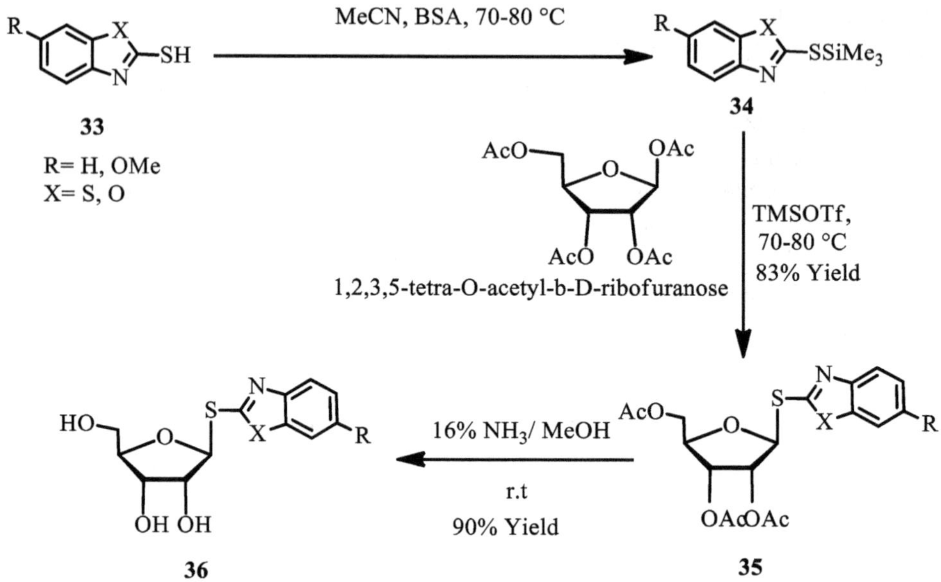

Figure 5.12: A modified nitrogen and sulfur glycosylation reaction.

ring linked to a steroidal framework (**41**). A key aspect of the approach is the intramolecular cyclo-desulfuration of glycosylthioureas, achieved in a two-step, one-pot process (Figure 5.13).

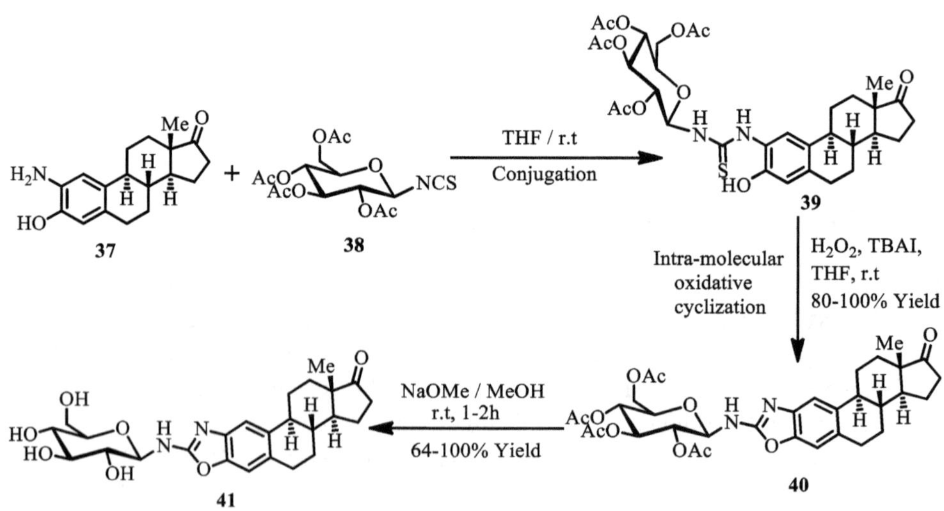

Figure 5.13: Synthesis of glycoconjugates containing a carbohydrate and oxazole moiety fused to a steroidal unit.

Burrini et al. [31] introduced a novel protocol for synthesizing coumarin *S*-glycosides (**50**), using unique biocatalysts to attach various carbohydrate structures (**48**) onto 7-mercapto-4-methyl-coumarin (**49**) in one-pot reactions (Figure 5.14). The fluorescence properties of the resulting thio-derivatives were evaluated, offering important insights into their potential use in biological imaging or sensing. Furthermore, the synthesized compounds showed no cytotoxic effects across multiple human cell lines.

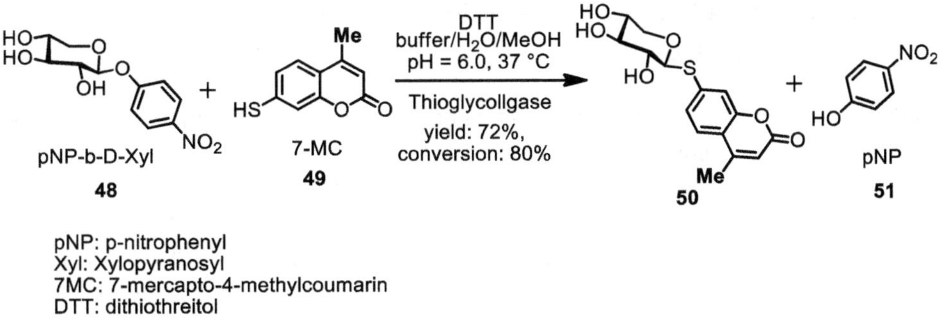

Figure 5.14: Synthetic protocol for *S*-glycosidic coumarins.

5.2.2 Synthesis of carbohydrate-decorated six-membered *O*-heterocycles

In this section, we have discussed an important protocol toward accessing carbohydrate-decorated six-membered *O*-heterocycles.

Maram et al. [32] devised methods for synthesizing polyoxy-functionalized piperidine derivatives featuring a tetra-substituted carbon at the 3-position of the piperidine ring (**53**) in a one-pot process from carbohydrate derivatives (**52**). Their approach (Figure 5.15) involved generating enamines in situ from iminium ions, which were also produced in situ from carbohydrate derivatives. The developed methods and synthesized compounds hold potential for creating bifunctional molecules.

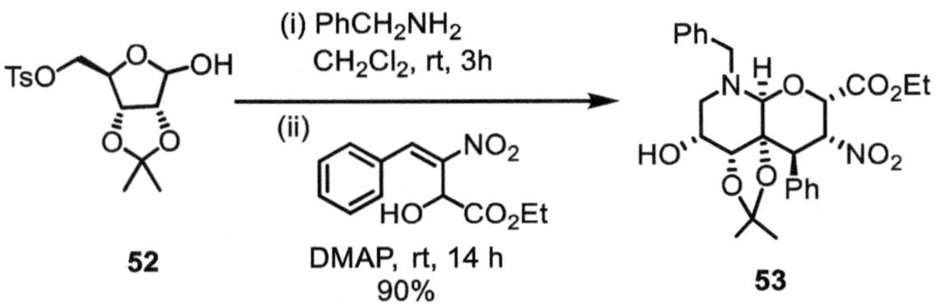

Figure 5.15: Synthesis of polyoxy-functionalized piperidines.

Sharma et al. [33] developed a concise approach using the furan moiety as a masked sugar synthon on aldehydo sugars, resulting in the D- and L-enantiomers of C(4)–C(5)-linked furano-pyranosides. This versatile method is highly effective for synthesizing a range of unnatural saccharides linked to both furano- and pyranosides. The method is currently being adapted for the synthesis of C-linked inositol saccharides.

5.2.3 Synthesis of carbohydrate-decorated five-membered *O*-heterocycles

In this section, we have demonstrated some of the latest and most important protocols toward accessing carbohydrate-decorated five-membered *O*-heterocycles.

Liu et al. [34] proposed that a glycosyl radical could be generated from a 1-hydroxycarbohydrate (**58**) through appropriate activation under mild photoredox conditions. This radical would then efficiently capture a copper(II)-amido complex. The resulting high-valent Cu(III) species would undergo reductive elimination to produce

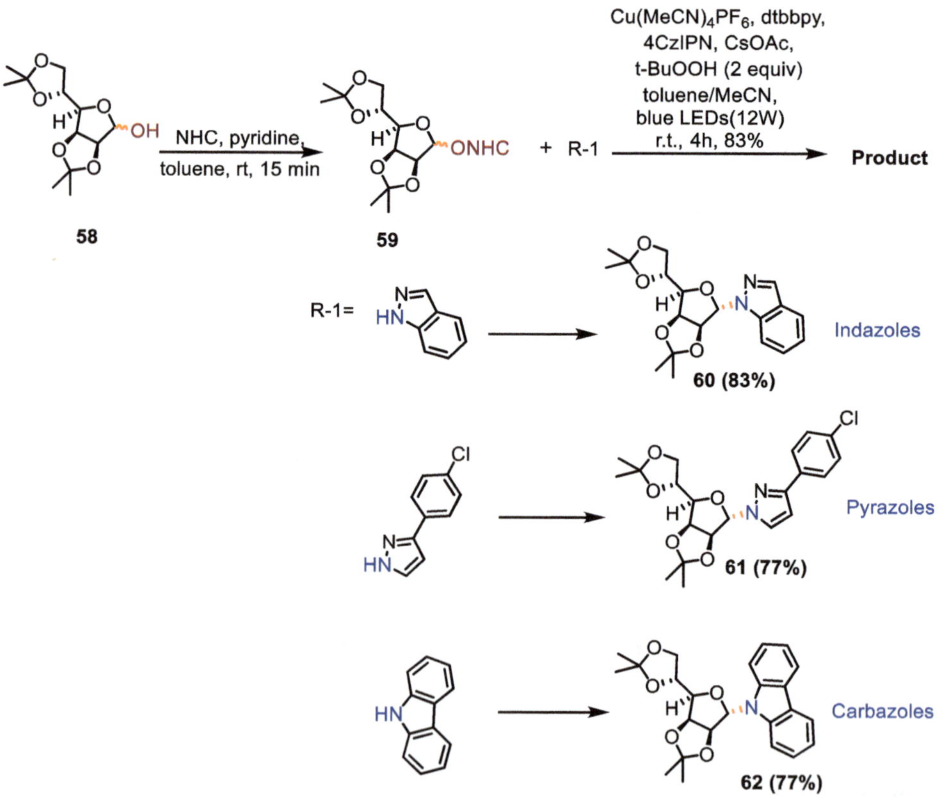

Figure 5.16: Dehydroxylative radical process for *N*-glycosylation.

the desired N-glycosides. In this study, the authors successfully demonstrated this concept (Figure 5.16) and developed a dehydroxylative radical approach for synthesizing N-glycosides (**60–62**) using copper metallaphotoredox catalysis, where stable and easily accessible 1-hydroxycarbohydrates serve as glycosyl donors. Key advantages of this method are (i) a mechanistically distinct, dehydroxylative radical N-glycosylation, (ii) the use of inexpensive copper catalysts and stable 1-hydroxycarbohydrates, and (iii) high diastereoselectivity and regioselectivity across a wide range of substrates. The presence of glycosyl radicals was confirmed through radical quenching experiments and a radical clock reaction. Additionally, the significance of copper metallaphotoredox catalysis was validated through control experiments and various spectroscopic analyses, including UV-vis studies.

Yuan et al. [35] demonstrated a catalyst-free intramolecular nucleophilic cyclization reaction using protected 5-O-tosylate pentose (**63**) as a novel reagent, leading to the formation of a series of bicyclic azasugars with glycosidic heteroatoms (**64**) in good to excellent yields and high stereoselectivity (Figure 5.17). This approach offers a straightforward route to synthesize various bicyclic iminosugar derivatives, which have potential as glucosidase inhibitors.

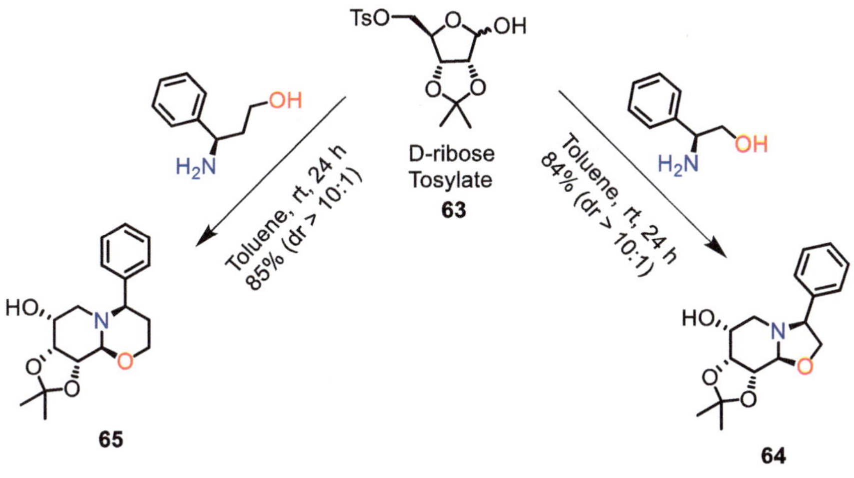

Figure 5.17: Synthesis of bicyclic azasugars involving a glycosidic heteroatom.

Chen and colleagues [36] described the synthesis of novel naphthofurano-iminosugars (**68**) from 2,3-O-isopropylidene D-ribose tosylate (**63**), 1,4-benzoquinone (**67**), and anilines (**66**) through key iminium ion/enamine intermediates, followed by a [3 + 2] cyclization at room temperature (Figure 5.18). The reaction shows distinct regioselectivity and stereoselectivity, producing moderate to excellent yields. This approach has been successfully applied to various substituted anilines, incorporating both electron-donating and electron-withdrawing groups.

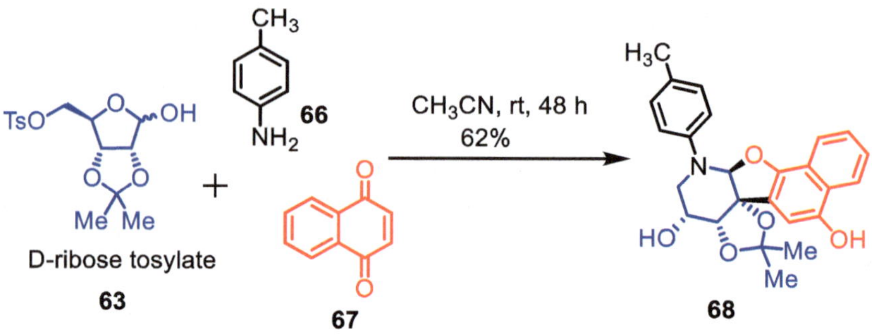

Figure 5.18: One-pot access to naphthofurano-iminosugars.

Amarnath et al. [37] synthesized and characterized fully biobased bis-oxazine monomers (**70**–**72**) from phenolic and carbohydrate-based modified diamine (**69**) via a one-step, solvent-free, microwave-assisted method (Figure 5.19). This process was 12 times faster and produced comparable yields. The monomers underwent thermal ring-opening polymerization, yielding polymers with improved thermal stability, char yield, glass transition temperature, and lower viscosity compared to bisphenol-A aniline-based polyben-

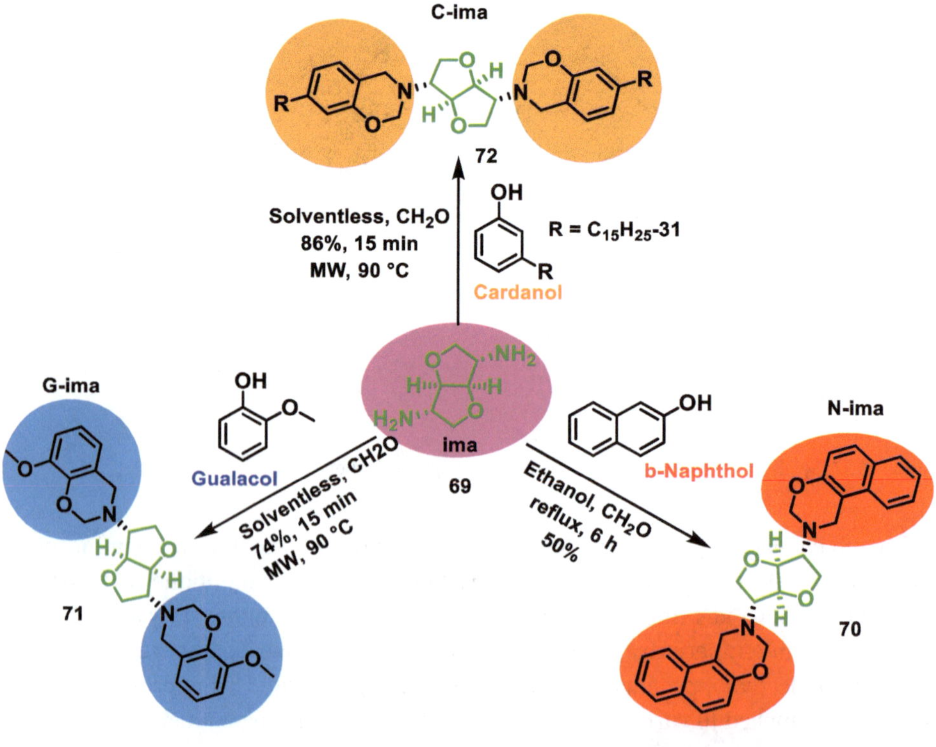

Figure 5.19: Microwave-assisted access to naturally occurring bifunctional benzoxazines.

zoxazines. The use of isomannide diamine offers a promising biobased alternative with enhanced properties, expanding the scope of biomass in benzoxazine chemistry.

Sahoo and team [38] synthesized a series of 1,2,3-triazole-fused chromene-based glucose triazole conjugates (**74** and **75**) from chromene-fused 1,2,3-triazolyl-extended alkyne (**73**) and 2,3,4,6-tetra-*O*-acetyl-β-D-glucopyranosylazide via a copper-catalyzed azide-alkyne cycloaddition (CuAAC) reaction (Figure 5.20), achieving yields ranging from good to excellent. The reaction featured mild conditions, high yields, broad substrate scope, and a short reaction time. The antibacterial activity of these compounds was evaluated in vitro against human pathogenic bacteria including gram-negative *E. coli* and gram-positive *S. aureus*.

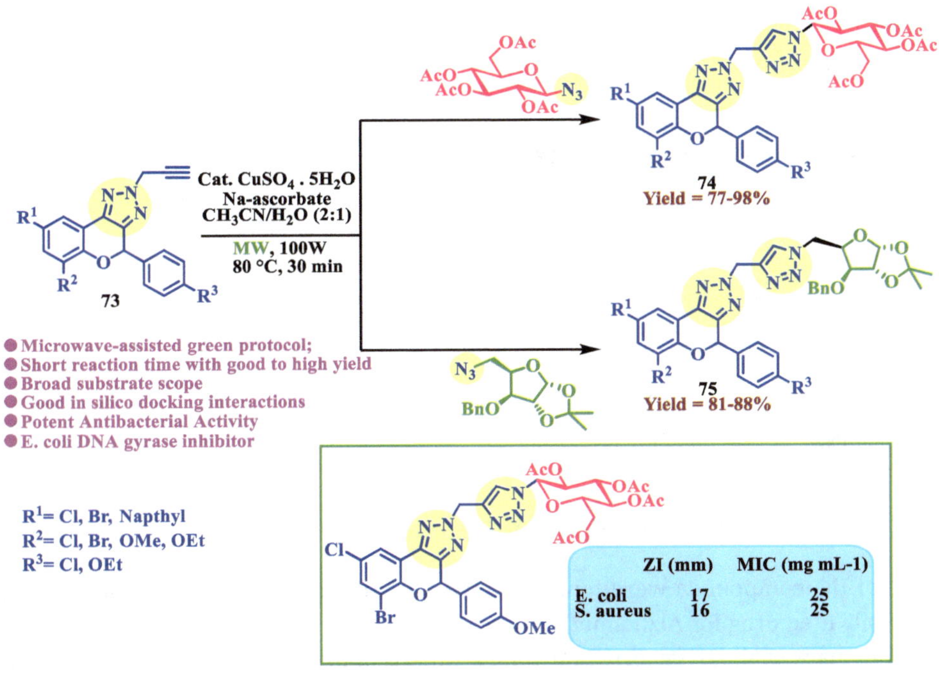

Figure 5.20: Click synthesis of 1,2,3-triazole-fused chromene-based glucose triazole conjugates.

Maikhuri et al. [39] synthesized 16 benzoylated and hydroxyl-coumarinyl triazolyl glycosides (**79**) through a simple Cu(I)-mediated click reaction between 1-azido-D-ribofuranose (**76**) and 4-ethynylcoumarins (**77**), achieving good yields (Figure 5.21).

Lindbäck and coworkers [40] utilized the CuAAC reaction to synthesize a library of 10 coumarin-azasugar-benzyl conjugates and two phthalimide-azasugar-benzyl conjugates (**80**, Figure 5.22) with potential anti-Alzheimer's and anticancer properties. These compounds were tested as cholinesterase inhibitors and showed a preference, up to 676-fold, for inhibiting butyrylcholinesterase (BuChE) over acetylcholinesterase.

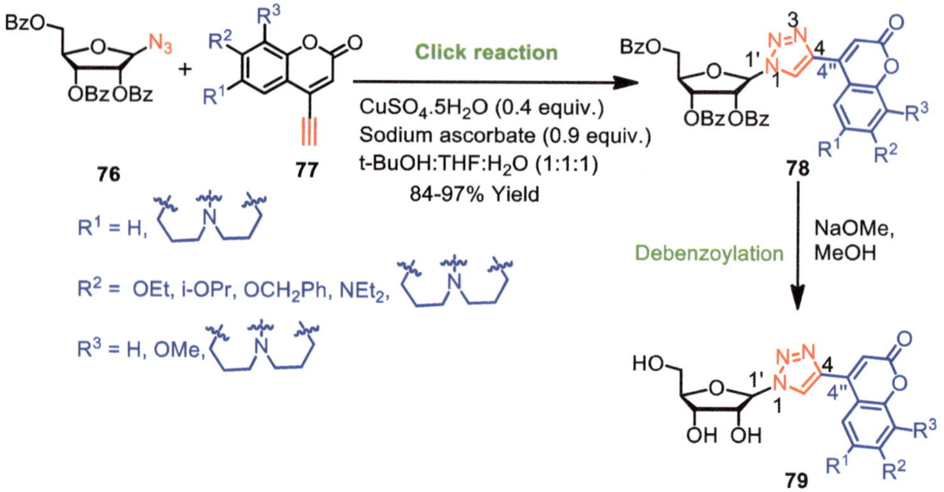

Figure 5.21: Access to coumarinyl triazolyl glycosides.

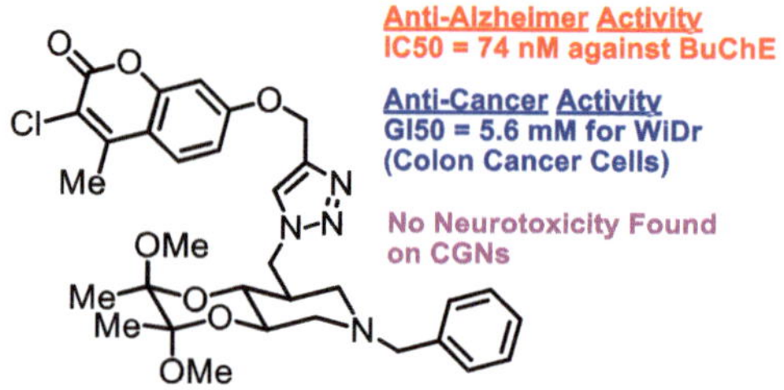

Figure 5.22: Access to coumarin-azasugar-benzyl conjugates.

Nine of the compounds were more effective BuChE inhibitors than galantamine, a clinically used drug for Alzheimer's disease.

Denmark et al. [41] reported the total synthesis of the antifungal agent papulacandin D (**81**, Figure 5.23). This molecule is part of a larger group of C-aryl glycosides known for their antifungal properties. The synthetic approach divides the molecule into two main subunits: the aryl glycoside and the 18-carbon fatty acid side chain. Key transformations include (a) a palladium-catalyzed organosilanolate-based cross-coupling of a protected glucal silanol and (b) a catalytic enantioselective allylation of a dienal using allyltrichlorosilane. The entire synthesis required 31 steps from commercially available starting materials, yielding over 50 mg of the natural product.

Mootoo and coworkers [42] reported the synthesis and preliminary evaluation of the antitumor activity of a novel set of carbohydrate-like annonaceous acetogenins (**84**), as depicted in Figure 5.24, through an olefin cross-metathesis synthetic strategy involving butenolide alkene (**82**) and different carbohydrate-containing alkene (**83**).

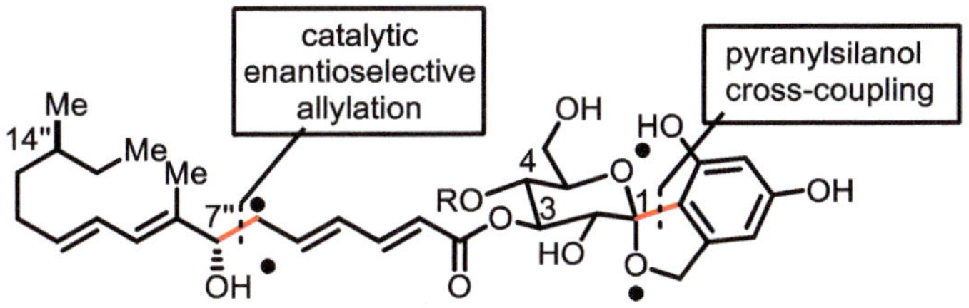

Figure 5.23: Access to an antifungal agent, papulacandin D.

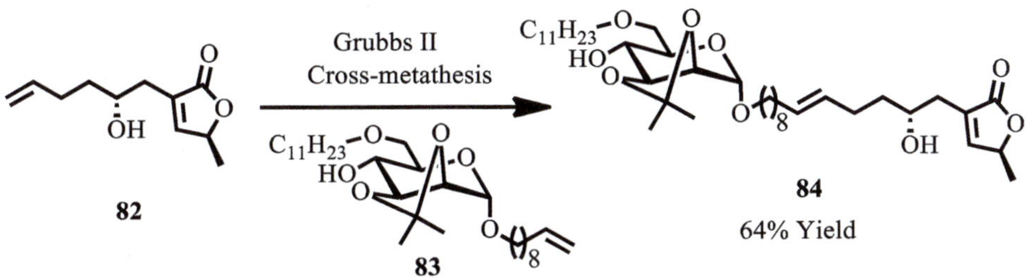

Figure 5.24: Synthesis of carbohydrate-based tetrahydrofuran-containing acetogenins.

Ambati et al. [43] accomplished a stereoselective total synthesis of pyranicin (**92**) and 4-epi-pyranicin (**93**) using carbohydrates as key starting materials. This is the first reported use of sugar-derived chiral pool synthons for the total synthesis of these compounds (Figure 5.25). The team also successfully achieved the gram-scale preparation of the target products.

Gonzalez-Periche et al. [44] reported the synthesis and cytotoxic properties of C-10 epimers of 4-DAN and its analogs (**96**), which included carbohydrate (**94**) and thiophene modifications replacing the THF and butenolide groups, respectively (Figure 5.26). A key step in the synthesis involved the coupling of THF and butenolide segments, or their analogs, through alkene cross metathesis. The resulting compounds exhibited cytotoxic effects against human prostate cancer cell lines LNCaP and PC3, with activities ranging from low micromolar to nanomolar concentrations. Notably, a mannose-linked thiophene analog demonstrated comparable cytotoxicity to 4-DAN.

Yousuf et al. [45] reported the successful development of a regio- and stereoselective one-pot synthesis (Figure 5.27) of carbohydrate-based butyrolactones (**99**). This method is useful for creating natural product-based libraries for biological screening.

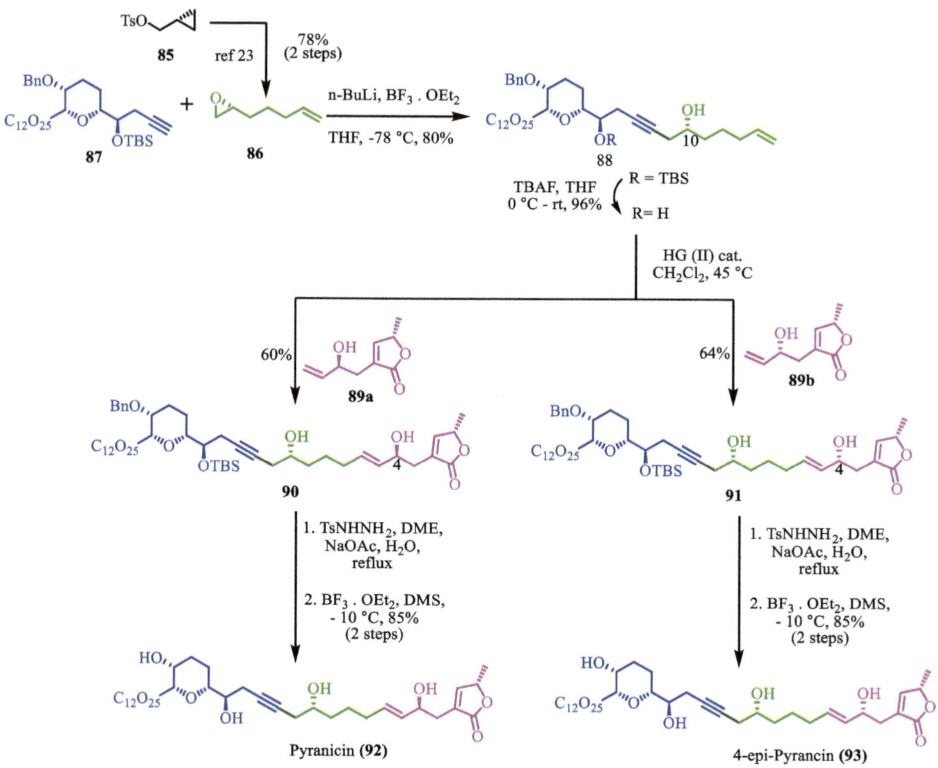

Figure 5.25: Synthesis of carbohydrate precursors incorporating pyranicin and 4-epi-pyranicin.

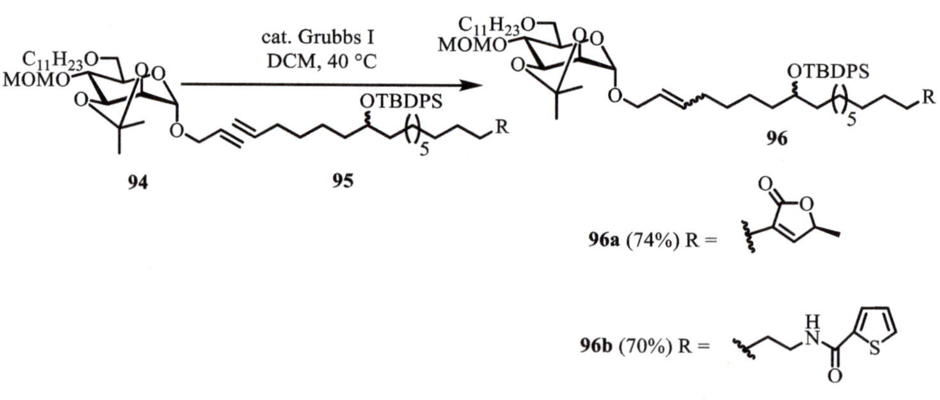

Figure 5.26: Access to carbohydrate-decorated THF-acetogenin 4-deoxyannomontacin.

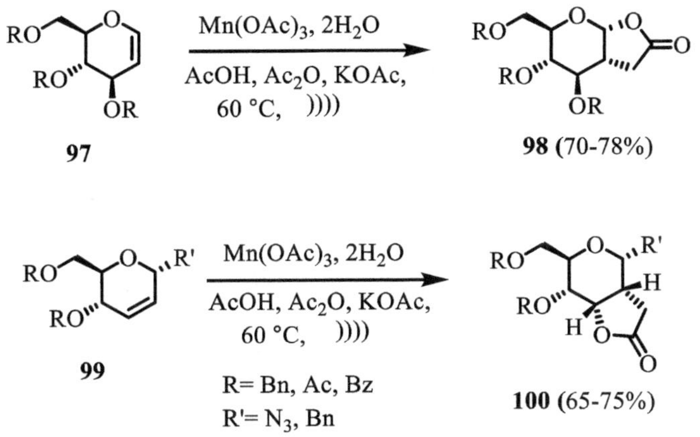

Figure 5.27: Synthesis of carbohydrate-based butyrolactones.

5.2.4 Synthesis of carbohydrate-decorated three-membered *O*-heterocycles

In this section, we have discussed some recent protocols toward accessing carbohydrate-decorated three-membered *O*-heterocycles.

Hung's team [46] presented an efficient method for synthesizing L-gulose and its C-6 derivatives, starting from commercially available 2,3:5,6-diisopropylidene-D-mannofuranose, with C-5 epimerization as the key step. In an intermediate step, L-gulo epoxide (**102**) is synthesized from the derivative of mannofuranoside (**101**). Compared to existing methods, this new protocol provides a shorter synthetic route with a comparable overall yield, passing through the same intermediate (Figure 5.28).

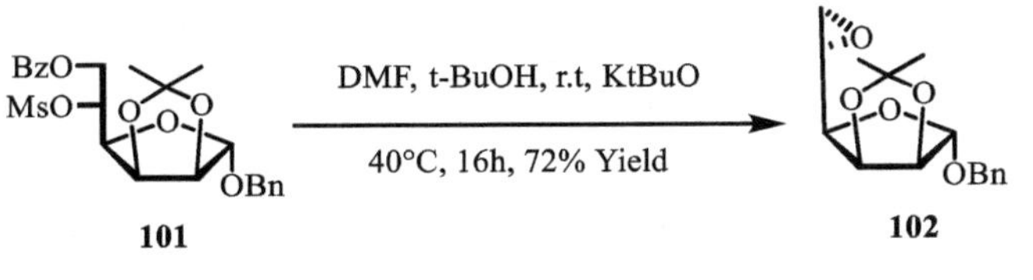

Figure 5.28: Synthesis of L-gulo epoxide **102**.

In a recent review by Lessard et al. [47], the work of Khan et al. is discussed, highlighting the synthesis of the same tetrachlorinated analog using an alternative approach (Figure 5.29). Starting from sucralose **103**, the target epoxide **105** was obtained in a 35% yield via an intramolecular Mitsunobu reaction. However, this method also produced the undesired analog **104** as the major product (Figure 5.29).

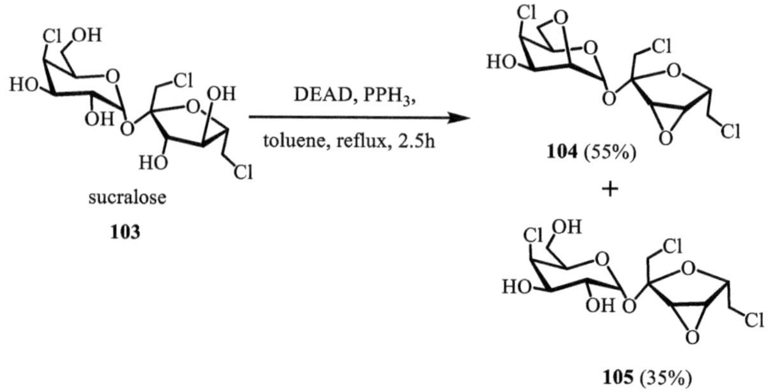

Figure 5.29: Access to epoxy-containing carbohydrate-based heterocycles.

5.3 Conclusion and outlook

Carbohydrate-decorated *O*-heterocycles are gaining significant attention due to their structural diversity and enhanced biological activity, making them valuable in drug development and therapeutic applications such as in cancer, infection, and inflammation treatments. The integration of carbohydrates with oxygen-containing heterocycles results in compounds with improved bioactivity and stability.

While traditional glycosylation methods are widely used, recent developments in radical chemistry, particularly copper-catalyzed dehydroxylative glycosylation, have provided milder and more selective alternatives. This approach has been key to efficiently linking carbohydrates with heterocyclic structures, enabling the synthesis of complex bioactive molecules. Spirocyclic *O*-heterocycles, often seen in natural products with strong biological effects, have garnered interest, especially when modified with carbohydrates. These modifications offer enhanced stability and biological activity, crucial for therapeutic potential.

In this review, recent advances in the synthesis of carbohydrate-decorated *O*-heterocycles have been explored, with an emphasis on innovative synthetic methodologies, including the growing role of green chemistry. The use of DES, derived from biodegradable materials, represents an eco-friendly alternative, aligning with sustainability goals and offering high selectivity in radical glycosylation reactions.

Future research will focus on expanding radical-based methods and developing new catalytic systems to enhance stereoselectivity and efficiency. Advancements in these areas will support the continued development of carbohydrate-decorated *O*-heterocycles, which hold strong potential in medicinal chemistry and biomedical research.

References

[1] Singh K, Kulkarni SS. Small carbohydrate derivatives as potentantibiofilm agents. J Med Chem, 2022, 65, 8525–8549.

[2] Tiwari VK. (Ed.). Carbohydrates in Drug Discovery and Development. Elsevier Inc.: The Netherland, 2020.

[3] Weymouth-Wilson AC. The role of carbohydrates in biologically active natural products. Nat Prod Rep, 1997, 14, 99.

[4] Delost MD, Smith DT, Anderson BJ, Njardarson JT. From oxiranes to oligomers: Architectures of U.S. FDA approved pharmaceuticals containing oxygen heterocycles. J Med Chem, 2018, 61, 10996–11020.

[5] Kaur N. 5-Membered Heterocycle Synthesis Using Iodine. Elsevier, 2023.

[6] Babu SA, Padmavathi R, Suwasia S, Dalal A, Bhattacharya D, Singh P, Tomar R. Recent developments on the synthesis of functionalized carbohydrate/sugar derivatives involving the transition metal–catalyzed C–H activation/C–H functionalization. In Studies in Natural Products Chemistry. Elsevier, 2021, pp. 311–399.

[7] Xue L, Song X, Zhang X, Fan X. Synthesis of *O*-heterocycle spiro-fused cyclopentaquinolinones and cyclopentaindenes through visible light-induced radical cyclization reactions. J Org Chem, 2023, 88, 12641–12657.

[8] Hiesinger K, Dar'in D, Proschak E, Krasavin M. Spirocyclic scaffolds in medicinal chemistry. J Med Chem, 2021, 64, 150–183.

[9] Ghosh T, Nokami T. Recent development of stereoselective C-glycosylation via generation of glycosyl radical. Carbohydr Res, 2022, 522, 108677.

[10] Perrone S, Messa F, Troisi L, Salomone AN. *O*- and *S*-heterocycles synthesis in deep eutectic solvents. Molecules, 2023, 28, 3459.

[11] Azeem Z, Kumar Mandal P. Recent advances and development in visible-light-mediated glycosylation: An expanding research area in glycochemistry. Adv Synth Catal, 2023, 365, 2818–2849.

[12] El Achkar T, Greige-Gerges H, Fourmentin S. Basics and properties of deep eutectic solvents: A review. Environ Chem Lett, 2021, 19, 3397–3408.

[13] Goti G. Catalytic radical reactions of unsaturated sugars. ChemCatChem, 2022, 14, e202200290.

[14] Shang W, Niu D. Radical pathway glycosylation empowered by bench-stable glycosyl donors. Acc Chem Res, 2023, 56, 2473.

[15] Dragutan I, Dragutan V, Demonceau A. Targeted drugs by olefin metathesis: Piperidine-based iminosugars. RSC Adv, 2012, 2, 719–736.

[16] Meng S, Tang GL, Pan HX. Enzymatic formation of oxygen-containing heterocycles in natural product biosynthesis. ChemBioChem, 2018, 19, 2002–2022.

[17] Touisni N, Maresca A, McDonald PC, Lou Y, Scozzafava A, Dedhar S, Winum JY, Supuran CT. Glycosyl coumarin carbonic anhydrase IX and XII inhibitors strongly attenuate the growth of primary breast tumors. J Med Chem, 2011, 54, 8271.

[18] Kaur N, Devi M, Verma Y, Grewal P, Bhardwaj P, Ahlawat N, Jangid NK. Green synthesis of bioactive heterocycles – Part 1B. Curr Green Chem, 2019, 6, 155.

[19] Wu YR, Ren ST, Wang L, Liu XJ, Wang YX, Liu SH, Liu WW, Shi DH, Cao ZL. Synthesis and AChE inhibitory activity of *N*-glycosylbenzofuran derivatives. Heterocycl Comm, 2019, 25, 162.

[20] Cao Z, Zhou X, Zhang F, Zhao YJ. Efficient synthesis of 2-benzofuranyl C-glycosides by one-pot cascade reaction of sugar alkynes and substituted 2-iodophenols. Carbohydr Chem, 2023, 42, 135.

[21] Rajput VK, Leffler H, Nilsson UJ, Mukhopadhyay B. Synthesis and evaluation of iminocoumaryl and coumaryl derivatized glycosides as galectin antagonists. Bioorg Med Chem Lett, 2014, 24, 3516.

[22] Rajput VK, MacKinnon A, Mandal S, Collins P, Blanchard H, Leffler H, Sethi T, Schambye H, Mukhopadhyay B, Nilsson UJ. A selective galactose–coumarin-derived galectin-3 inhibitor demonstrates involvement of galectin-3-glycan interactions in a pulmonary fibrosis model. J Med Chem, 2016, 59, 8141.

[23] Chu N, Wang Y, Jia H, Han J, Wang X, Hou Z. Design, synthesis and biological evaluation of new carbohydrate-based coumarin derivatives as selective carbonic anhydrase IX inhibitors via "Click" reaction. Molecules, 2022, 27, 5464.

[24] Dwivedi S, Dey S, Sau A. Sugar functionalized coumarin motifs: Synthesis and applications. Carbohydr Res, 2024, 544, 109244.

[25] Kumari P, Gupta S, Narayana C, Ahmad S, Vishnoi N, Singh S, Sagar R. Stereoselective synthesis of carbohydrate fused pyrano[3,2-c]pyranones as anticancer agents. New J Chem, 2018, 42, 13985.

[26] Jain Y, Kumari M, Agarwal M, Gupta R. Robust synthesis of sugar-coumarin based fluorescent 1,4-disubstituted-1,2,3-triazoles using highly efficient recyclable citrate grafted β-cyclodextrin@magnetite nano phase transfer catalyst in aqueous media. Carbohydr Res, 2019, 482, 107736.

[27] Sharma A, Saikia P, Saha S, Kumar D, Panchadhayee R. Synthesis and biological evaluation of carbohydrate-coumarin/vanillic acid hybrid as a promising antiparasitic agent. Carbohyd Res, 2023, 530, 108862.

[28] Luque-Agudo V, Albarrán-Velo J, Light ME, Padrón JM, Román E, Serrano JA, Gil MV. Synthesis and antiproliferative activity of new 2-glyco-3-nitro-2H-chromenes. Bioorg Chem, 2019, 87, 112.

[29] Khodair AI, Al-Masoudi NA, Gesson JP. A new approach to the synthesis of benzothiazole, benzoxazole, and pyridine nucleosides as potential antitumor agents. Nucleosides Nucleotides Nucleic Acids, 2003, 22, 2061.

[30] Méndez-Delgado LA, Fuentes-Aguilar A, Meza-Reyes S, Montiel-Smith S, Vega-Baez JL, Padrón JM, Merino-Montiel P, Bernès S. Synthesis and antiproliferative activity of new oxazole-steroid glycoconjugates. Chem Heterocycl Compd, 2023, 59, 816.

[31] Burrini N, Pâris A, Collet G, Lafite P, Daniellou R. Biocatalytic synthesis of coumarin S-glycosides: Towards non-cytotoxic probes for biomedical imaging and sensing. Molecules, 2024, 29, 1322.

[32] Maram L, Tanaka F. Switching electrophile intermediates to nucleophiles: Michael and Oxa-Diels–Alder reactions to afford polyoxy-functionalized piperidine derivatives with tetrasubstituted carbon. Org Lett, 2020, 22, 2751.

[33] Sharma GVM, Hymavathi L, Krishna PR. Carbon–carbon linked disaccharides by de novo construction from furanyl sugar templates. Tetrahedron Lett, 1997, 38, 6929.

[34] Liu DP, Zhang XS, Liu S, Hu XG. Dehydroxylative radical *N*-glycosylation of heterocycles with 1-hydroxycarbohydrates enabled by copper metallaphotoredox catalysis. Nat Commun, 2024, 15, 1.

[35] Yuan W, Wei X, Ma J, Yang J. An efficient method for the stereoselective synthesis of bicyclic azasugars with glycosidic heteroatom. Tetrahedron, 2021, 89, 132079.

[36] Zhou L, Cui Y, Huang X, Xie S, Wu J, Wei C, Li X, Chen H. One-pot stereoselective synthesis of the naphthofurano-iminosugars via a [3 + 2] cycloaddition reaction. J Org Chem, 2024, 89, 13235.

[37] Amarnath N, Shukla S, Lochab B. Isomannide-derived chiral rigid fully biobased polybenzoxazines. ACS Sustain Chem Eng, 2019, 7, 18700.

[38] Ahemad MA, Patra A, Muduli L, Nayak S, Mohapatra S, Panda J, Sahoo CR. Click-chemistry-inspired synthesis of new series of 1,2,3-triazole fused chromene with glucose triazoleconjugates: Antibacterial activity assessment with molecular docking evaluation. Carbohydr Res, 2024, 543, 109222.

[39] Maikhuri VK, Bohra K, Srivastava S, Kavita, Prasad AK. Click synthesis of N1-(β-D-ribofuranosyl)-C4-(coumarin-4″-yl)-1,2,3-triazoles. Synth Commun, 2019, 49, 3140.

[40] Vaaland-Holmgard IC, González-Bakker A, Poeta E, Puerta A, Fernandes MX, Monti B, Fernández-Bolaños JG, Padrón JM, López Ó, Lindbäck E. Coumarin–azasugar–benzyl conjugates as

non-neurotoxic dual inhibitors of butyrylcholinesterase and cancer cell growth. Org Biomol Chem, 2024, 22, 3425.
[41] Denmark SE, Regens CS, Kobayashi T. Total synthesis of papulacandin D. J Am Chem Soc, 2007, 129(10), 2774.
[42] Bachan S, Tony KA, Kawamura A, Montenegro D, Joshi A, Garg H, Mootoo DR. Synthesis and anti-tumor activity of carbohydrate analogues of the tetrahydrofuran containing acetogenins. Bioorg Med Chem, 2013, 21(21), 6554.
[43] Ambati P, Gupta A, Bandi A, Perali RS. Stereoselective total synthesis of pyranicin and 4-epi-pyranicin from carbohydrate precursors. European J Org Chem, 2024, 27, e202400244.
[44] Gonzalez-Periche P, Ramdular A, Bhupathiraju NVSDK, Kalidindi T, Johnson DS, Pillarsetty N, Mootoo DR. Synthesis of carbohydrate analogues of the THF-acetogenin 4-deoxyannomontacin and their cytotoxicity against human prostate cancer cell lines. Carbohydr Res, 2022, 521, 108671.
[45] Yousuf SK, Mukherjee D, MallikharjunraoTaneja SC. Highly regio- and stereoselective one-pot synthesis of carbohydrate-based butyrolactones. Org Lett, 2011, 13, 576.
[46] Teng KC, Tseng KY, Tzeng ZH, Hung SC. A concise synthesis of l-gulose and its C-6 derivatives. Bioorg Med Chem, 2022, 73, 117029.
[47] Lessard O, Lainé D, Giguère D. Polyhalogenated carbohydrates: Synthesis and applications of sugar halides from fluorine to iodine. Eur J Org Chem, 2024, 27, e202400120.

Sabir Ahammed, Pampa Dey, and Brindaban C. Ranu*

Chapter 6
Green solvents in carbohydrate synthesis

6.1 Introduction

In recent times, the chemistry fraternity has increasingly been emphasizing sustainable practices for chemical synthesis to address the challenges of environmental degradation and natural resource depletion. In this context, "green solvents" for reactions play an important role [1–5]. Among natural products, carbohydrates constitute the most abundant source/precursor for many useful compounds including pharmaceuticals, drugs, and agrochemicals [6–10].

The synthesis and functionalization of carbohydrates are usually carried out in volatile organic solvents that are associated with high energy consumption, generation of waste, and toxicity [11–13]. The alternatives may be green solvents, which are less toxic, renewable, biodegradable, and more environmentally benign [14–16]. A broad range of such solvents, namely water, ionic liquids, deep eutectic solvents (DESs), and supercritical fluids, are considered for use in carbohydrate chemistry [17, 18].

The compatibility of the green solvents to be used for carbohydrate functionalization strategies, including protecting group chemistry, glycosylation, and deprotection, will also be considered [19–21]. These solvents are thus integrated into carbohydrate synthesis to provide new protocols for the synthetic routes and functionalized carbohydrates toward applications in various fields [22–24]. The carbohydrate synthesis using environment-friendly solvents aims to allow sustainable, efficient, scalable, and economically viable processes [25].

The current chapter presents an overview of recent developments, challenges, and prospects in the field of carbohydrate synthesis using green solvents.

Acknowledgments: B. C. Ranu is pleased to acknowledge the support of the Indian National Science Academy, New Delhi, with the offer of INSA Honorary Scientist (grant no. INSA/SP/HS/2019/291). S. Ahammed acknowledges WBDSTBT, West Bengal, India (project no. 1933 (Sanc.)/ST/P/S&T/15G-9/2019), for financial support.

***Corresponding author: Brindaban C. Ranu,** School of Chemical Sciences, Indian Association for the Cultivation of Science, Jadavpur, Kolkata 700032, West Bengal, India, e-mail: ocbcr@iacs.res.in
Sabir Ahammed, Pampa Dey, Department of Chemistry, Bankura Sammilani College, Kenduadihi, Bankura 722102, West Bengal, India

https://doi.org/10.1515/9783111447094-006

6.2 Synthesis and functionalization of carbohydrates using green solvents

6.2.1 Aqueous medium

Fairbanks and his group demonstrated a new method for converting sugars into *para*-nitrophenol glycosides **(3)**using *p*NP (para-nitrophenol) **(2)**, 2-chloro-1,3-dimethylimidazolinium chloride (DMC), and triethylamine in aqueous solution (Figure 6.1) [26]. This is a useful reaction because *p*NP glycosides possess a variety of applications, such as studying enzyme kinetics [27] and enzyme-catalyzed trans-glycosylation reactions [28]. This method is simpler than previous methods because it does not require protecting groups. Stereoselectivity is observed in this reaction when the sugar molecule has a hydroxyl or an acetamido group at position 2.

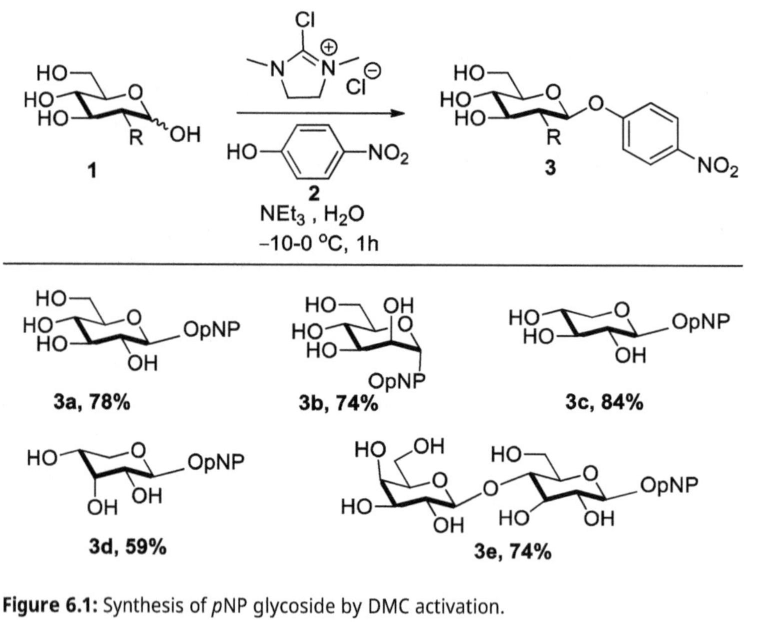

Figure 6.1: Synthesis of *p*NP glycoside by DMC activation.

The method works by activating the sugar with a molecule of DMC, followed by reacting it with *p*-NPOH. Different reaction conditions are investigated to improve the yield and selectivity of the reaction. It was found that the best conditions constitute the use of water as a solvent, Et_3N as the base for monosaccharide substrates, and Hünig's base for disaccharide substrates, and operation at −10 °C. The method has also been used to convert sugars with an *N*-acetyl group, but this requires a different procedure in which the sugar is first converted into an oxazoline intermediate. The procedure was used to convert a variety of sugars into *p*NP glycosides including a complex oligosaccharide isolated from egg yolks. This demonstrates the potential of the method for the preparation of complex enzyme substrates.

Miller and his group described a method for stitching sugar molecules (glycosyl donors) with fluoride groups to sucrose (sucrosyl acceptors) to create trisaccharides (three-sugar molecules) (Figure 6.2) [29]. This glycosylation reaction occurs at room temperature in a mixture of water and other solvents. It was found that calcium salts and a specific type of base, such as tertiary amine, helped the reaction occur efficiently and target a specific location (either the 3′ or 1′ position) on the fructose ring within sucrose.

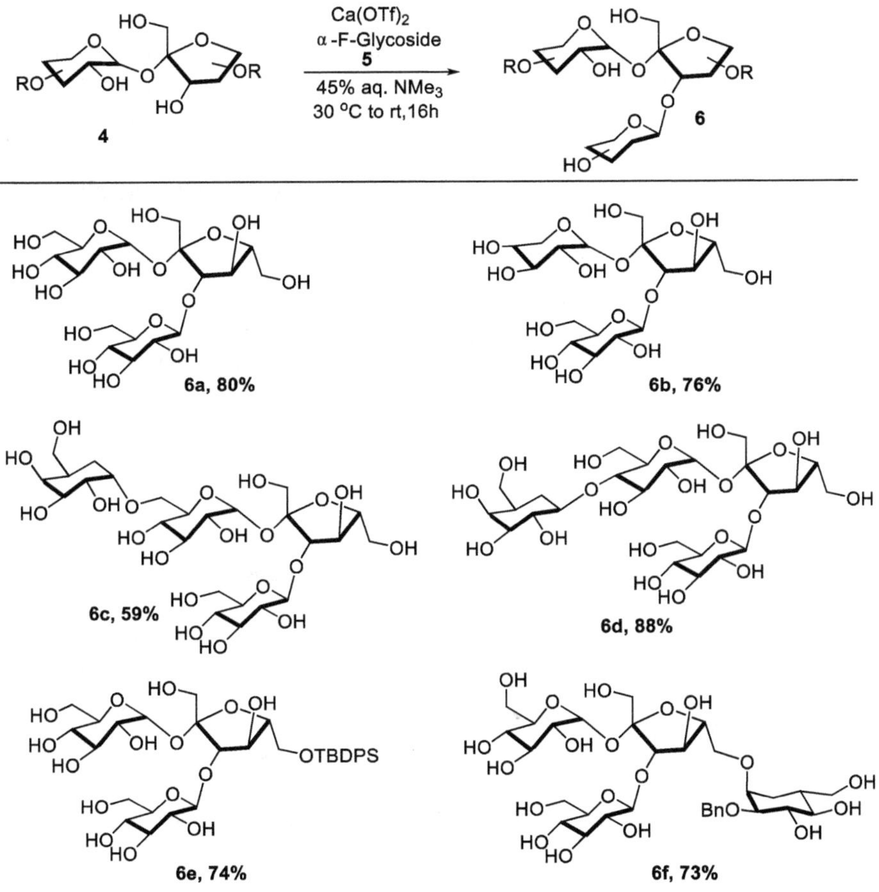

Figure 6.2: Glycosylation of oligosaccharides using $Ca(OTf)_2$.

The best reaction conditions were found to use a high concentration of $Ca(OTf)_2$ and a low concentration of oligosaccharide. This group investigated the mechanism to understand why the reaction targets specific sites. Here, in this reaction, the arrangement of hydroxyl groups in sucrose plays a key role in the selectivity. By studying the different shapes that sucrose takes in solution, it was noticed that the hydroxyl groups work together to influence both the formation of the sugar bond in water and the specific location where it occurs on the sucrose molecule.

Li and coworkers [30] developed a procedure to obtain β-glycosyl esters (sugar molecules linked to other molecules with an ester bond) in high yield (Figure 6.3). This method is unique as it uses water as a solvent and avoids the use of harmful organic solvents. The reaction uses a sugar molecule with a chloride group (glycosyl chloride) and a carboxylic acid as substrates. The reaction is catalyzed by micelles formed with cesium carbonate and tetrabutylammonium bromide in water. Granular polytetrafluoroethylene plays a crucial role in the reaction, probably by promoting the formation of micelles. The reaction is most successful at 80 °C with a specific amount of water (10 equivalents per equivalent of glycosyl chloride), which gives high stereoselectivity of the product (produces mostly the β-isomer of the glycosyl ester). This glycosylation process has been used to produce various β-glycosyl esters with excellent yields, which have potential applications as food additives and pharmaceutical drugs [31–33].

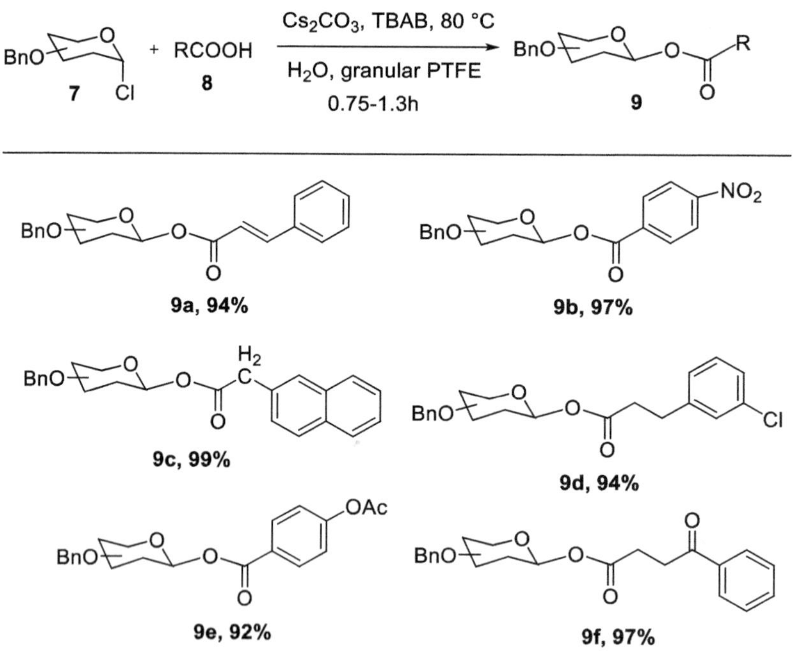

Figure 6.3: β-Glycosylation of a series of carboxylic acids.

Tripathi and his group [34] reported a one-pot method for synthesizing biphenyl methyl-C-glycosides from readily available starting materials (Figure 6.4). The reaction involves a domino Michael addition-Knoevenagel condensation-cyclization sequence catalyzed by a base. The best reaction conditions were found by using malononitrile and potassium carbonate in water at 50–55 °C. Secondary amines such as pyrrolidine performed better than tertiary amines (Et_3N, DBU), but the yield was still low. The method is applicable to a variety of aromatic and heteroaromatic substrates, providing good yields for biphenyl methyl-C-glycosides with different sugar moieties. Generally, electron-withdrawing groups like nitro or halogens/methoxy groups improved yield, while electron-donating groups, for example, hydroxy or dimethylamine, resulted in lower yields. Heteroaromatic rings (thiophene, pyridine) as replacements for the aromatic ring led to decreased yields. The reaction can be extended to synthesize biphenyl methyl-C-xylosides and cellobiosides. A plausible reaction mechanism is proposed involving Michael addition, Knoevenagel condensation, cyclization, and elimination steps (Figure 6.5).

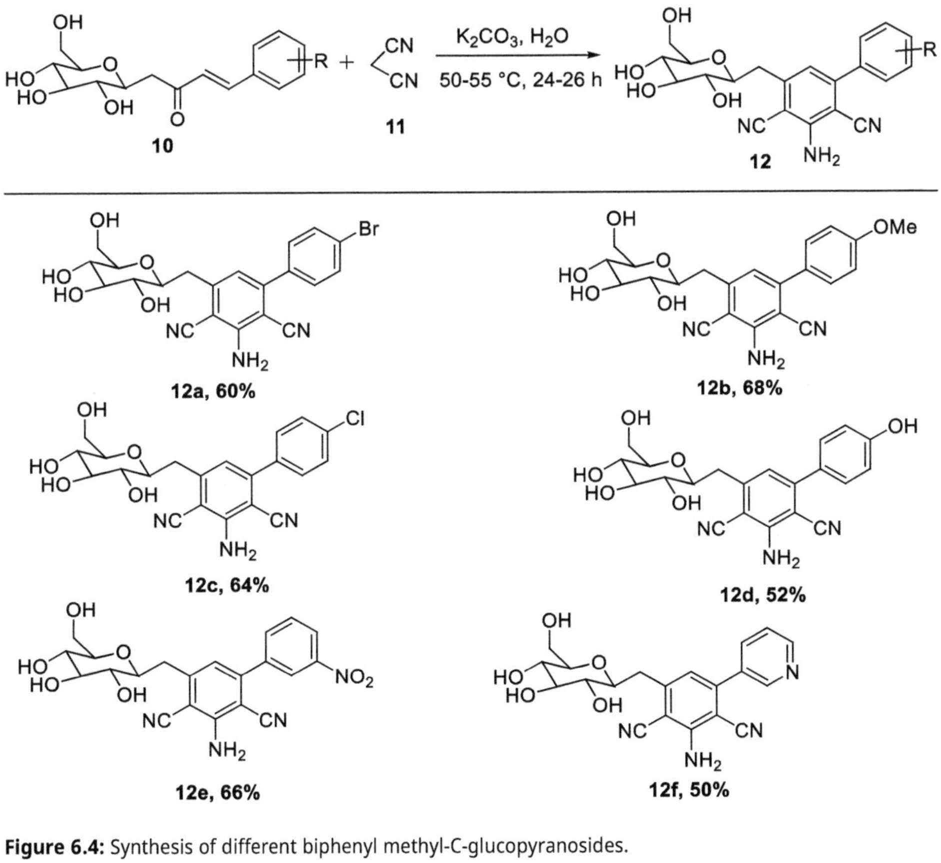

Figure 6.4: Synthesis of different biphenyl methyl-C-glucopyranosides.

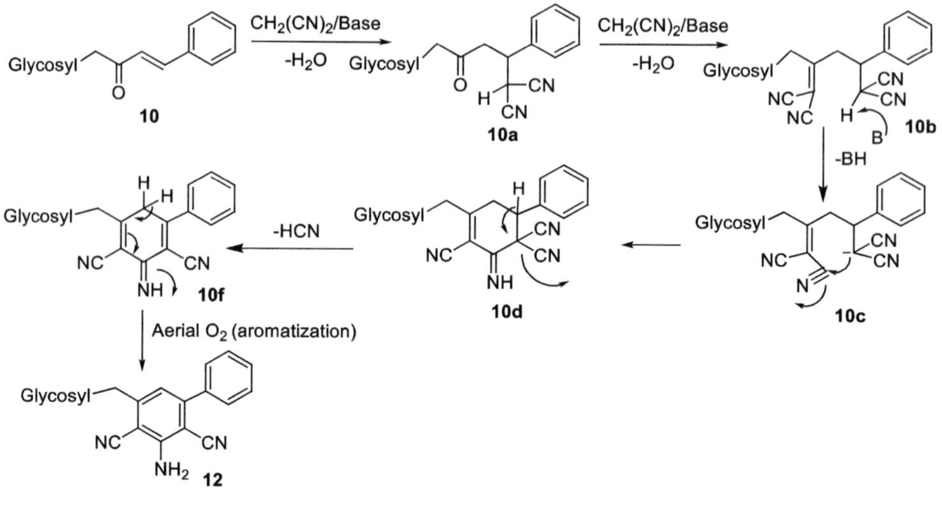

Figure 6.5: Plausible reaction mechanism.

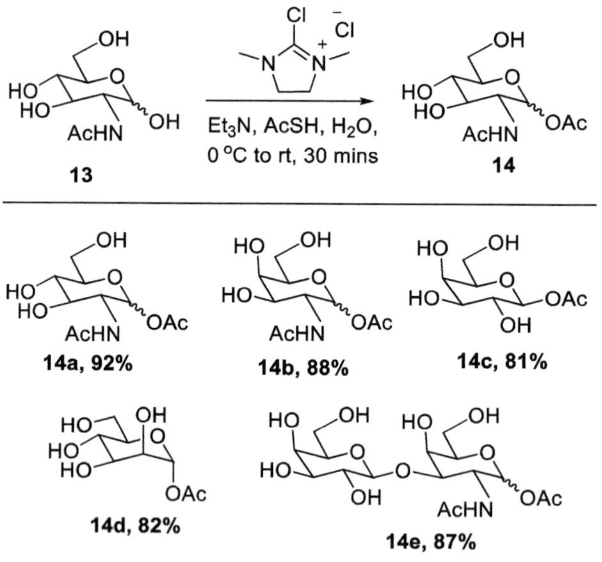

Figure 6.6: Selective acetylation of the anomeric hydroxyl group of GlcNAc in water.

Fairbanks and his group [35] developed a superior method for the selective acetylation of the anomeric center of various forms of unprotected sugars in water (Figure 6.6).

This method offers several advantages including simple operation, efficiency, and scalability and can be used to produce glycosyl acetates from a variety of monosaccharides. The best result was achieved by reacting *N*-acetylgalactosamine (GalNAc) with DMC as a substrate in the presence of thioacetic acid and triethylamine as a base. When sugars with a free hydroxyl group at the specific position (2-OH) were reacted, the reaction interestingly turned out to be highly stereoselective, and the reaction produced only one specific kind of product, the 1,2-trans isomer of the glycosyl acetate. The resulting glycosyl acetates can act as donor substrates for glycosidase enzymes, eliminating the need for multistep preparation.

Shoda and his group [36] developed a novel one-step method for synthesizing aryl 1-thioglycosides using the reagent DMC along with a sugar with aromatic thiol via an intramolecular dehydration reaction under mild conditions (Figure 6.7).

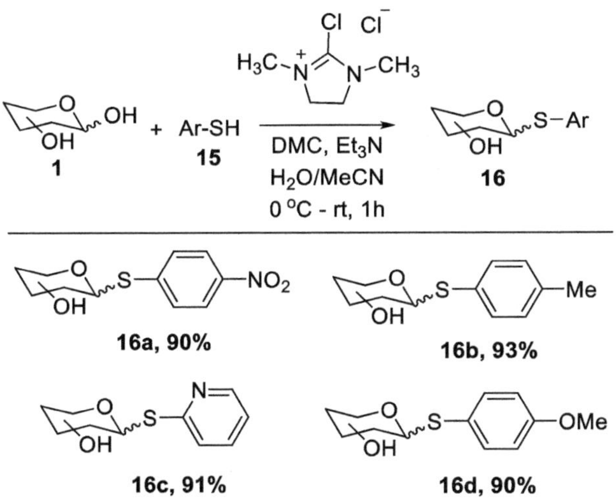

Figure 6.7: Synthesis of aryl-1-thioglycosides using DMC from unprotected sugars.

This reaction doesn't require protection for the hydroxyl groups and proceeds smoothly in an aqueous medium. A preferential attack by the anomeric hydroxyl group on DMC initiates the reaction, forming a reactive intermediate. This is followed by an intramolecular nucleophilic attack by the 6-hydroxy group on the anomeric carbon, resulting in the formation of a 1,6-anhydrosugar. The reaction yielded predominantly β-configured thioglycosides with a small amount of the α anomer when electron-donating groups like methyl and methoxy were attached to thiophenol. On the other hand, when 4-nitrobenzenethiol and 2-pyridinethiol were used with D-glucose in the presence of DMC, only β-configured thioglycosides were obtained.

Glycosyl-1,2,3-triazole is a useful class of carbohydrate derivatives and serves as promising molecules with medicinal applications. These derivatives have great potential as inhibitors for galectins, RNA-binding molecules, and carbonic anhydrase antagonists [37–40]. Misra and his group [41] reported an efficient two-step one-pot protocol for the synthesis of a series of glycosyl azides and their corresponding 1,2,3-triazole conjugates under an aqueous reaction medium (Figure 6.8) at room temperature. The reaction yielded only a single regioisomer of glycosyl-1,2,3-triazole derivatives under mild reaction conditions. After the consumption of the starting material in the first step, phenyl acetylene, $CuSO_4 \cdot 5H_2O$, and sodium L-ascorbate were added to the reaction mixture, which led to the clean formation of glucosyl triazole in high yield within 30 min to 2 h.

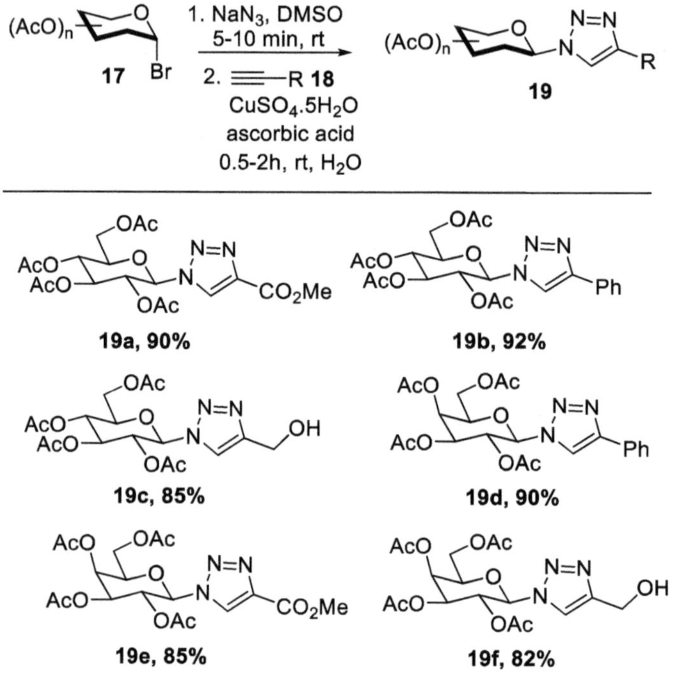

Figure 6.8: Synthesis of glycosyl 1,2,3-triazole derivatives from glycosyl bromides.

C-Glycosidation is an important step in organic synthesis for creating optically active materials incorporating carbon chains directly into the sugar molecule. These are versatile chiral building blocks to synthesize biologically active natural products [42]. They have high potential as inhibitors of carbohydrate-processing enzymes and stable analogs of glycans involved in vital intra- and intercellular processes, which make them very useful in biological research [43]. Yadav et al. [44] described a simple and effective one-step method for making β-linked furans with either a dihydroxytetrahydrofuranyl or a trihydroxyalkyl group, starting from unprotected aldose sugars using a small amount of indium(III) chloride as a catalyst (Figure 6.9). The treatment of hexose sugar with pentane-2,4-dione in water at 80 °C afforded C-glycosyl dihydroxytetrahydrofuranyl furan in good to moderate yield. Pentose sugars, for example, D-ribose and D-arabinose, reacted easily to form trihydroxyalkylfurans in high yields. Cyclic ketones also produce the corresponding tetrahydrobenzofuranyl glycosides.

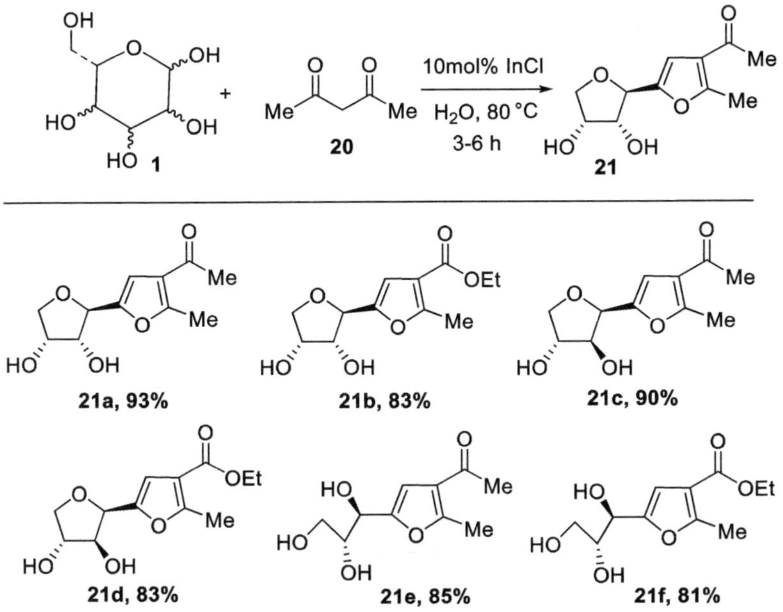

Figure 6.9: Synthesis of C-furyl glycosides in water.

Efforts are continuing to attach sugar molecules to biomolecules as an alternative to traditional glycosylation methods [45]. Ichikawa's group [46] developed a new and efficient method to produce urea glycosides in water (Figure 6.10). They have used Steyermark's glucopyranosyl oxazolidinone for attaching glucose units to amines and thiols. This method was used successfully to make new types of sugar-containing molecules, for example, urea-tethered neoglycoconjugates and pseudo-oligosaccharides. The primary amines with no branching at the R-carbon provided good yields. The secondary amines were less reactive than primary amines, but acceptable yields were achieved with longer reaction times and higher temperatures. Steric factors around the amine nitrogen significantly affected the reaction rate and yield. Highly hindered amines led to hydrolysis of the starting material rather than the desired reaction.

Misra et al. [47] developed a new way to make β-linked poly-hydroxyalkyl and glycosyl furans from free sugar under mild conditions with high yield (Figure 6.11). This method uses cerium(III) chloride to catalyze a Knoevenagel condensation between a 1,3-dione or β-ketoester and unprotected carbohydrates in water. Pentose sugars formed furan derivatives with polyhydroxylated alkyl side chains, while hexose sugars underwent cyclization further to form beta-linked hydroxylated tetrahydrofuranyl furans. The disaccharides were also used to form glycosylated tetrahydrofuranyl furans, leaving the bond between the sugars intact. The proposed mechanism involves condensation, water elimination, cyclization, and aromatization, respectively, to form the final product.

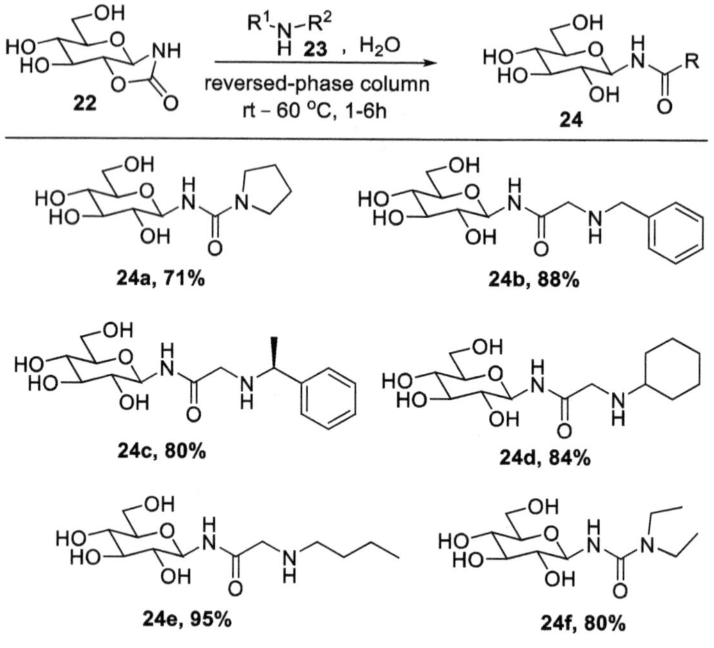

Figure 6.10: Synthesis of urea glycosides in an aqueous medium.

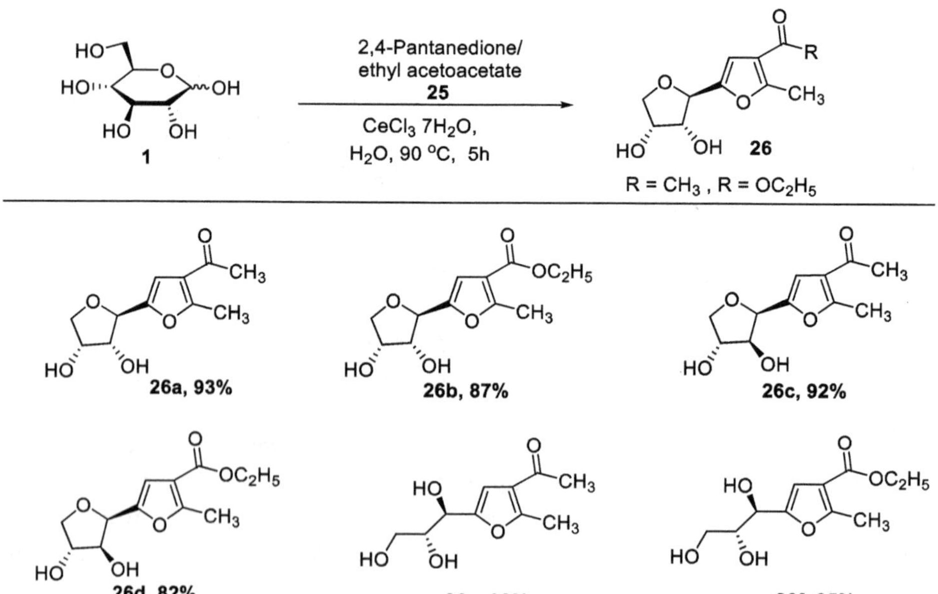

Figure 6.11: Preparation of polyhydroxylated furans from free sugars in water.

6.2.2 Ionic liquid

The ionic liquids are considered versatile solvents for organic synthesis [48–50]. Physical properties like low vapor pressure, polarity, and high thermal stability make them appropriate for many reactions involving carbohydrates [51, 52]. These solvents can accelerate reaction rates, improve product yields, and avoid the need for volatile organic solvents [53, 54]. Ionic liquids can act both as solvents and catalysts for glycosylation, oxidation, and reduction transformations of carbohydrates.

Galan and coworkers [55] reported an efficient protocol for activating thioglycosides for sugar synthesis (Figure 6.12). They introduced a new surfactant ionic liquid that activated the thioglycosides in combination with *N*-iodosuccinimide (NIS). This method offers advantages like mild reaction conditions, compatibility with various hydroxyl protecting groups, e.g., acetates, benzyl ethers, acetals, and applicability to NH_2-masking strategies employing Troc protection. The solvent effect is very prominent unless neighboring groups in the donor molecule influence the reaction. When acetonitrile is used as a solvent, it favors the β-linked 1,2-trans-glucoside product due to the formation of an axial nitrilium ion intermediate. The mechanism suggests that the ionic liquid slowly activates NIS to generate iodonium ions in the reaction mixture. These iodonium ions act as the actual activators of the thioglycosides.

Mishra and his group [56] developed a new, efficient, eco-friendly approach for synthesizing thioglycosides using *O*-acetylated sugar derivatives with thiols in the presence of boron trifluoride-etherate in [bmim][BF_4] ionic liquid as a solvent at 20 °C (Figure 6.13). The reaction provides reasonably fast reaction times and high yields of the desired products. The reaction yielded predominantly 1,2-trans isomers of the thioglycoside derivatives. However, a small amount of the undesired 1,2-cis isomers was also observed due to the anomerization of the kinetically controlled 1,2-*trans* products under acidic reaction conditions. The ionic liquid was recycled multiple times without losing its effectiveness, minimizing waste and promoting sustainability.

In another work, Mishra and his group [57] described a novel and environmentally friendly method for synthesizing 1,2-*trans*-thio and -selenoglycosides using triethylsilane and $BF_3 \cdot OEt_2$ (Figure 6.14). In this protocol, a recyclable ionic liquid [bmim] BF_4 is used to furnish the desired thiolates and selenoates. This method offers several advantages over existing protocols, including operational simplicity, elimination of foul-smelling thiols and selenols, and stereoselective formation of the 1,2-*trans* isomers. It was observed that the reaction with benzyl diselenide was slightly slower than with phenyl diselenide. The reusability of the ionic liquid catalyst further enhances the green credentials of this approach.

Malhotra and coworkers [58] demonstrated the use of molten halide salts as a simple and cost-effective reaction medium for glycosylating acetobromo-α-D-galactose and -glucose under basic conditions (Figure 6.15). The reaction is easy to perform and is carried out using readily and commercially available starting materials. The glycosyl bromide substrate shows minimal decomposition even without using specialized

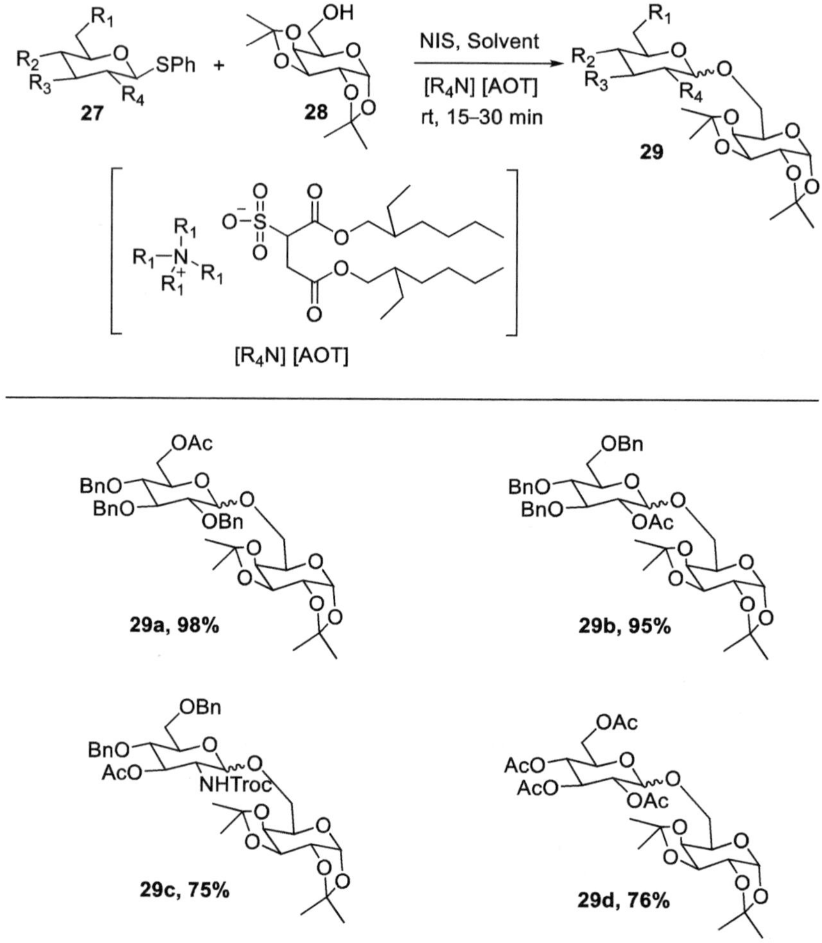

Figure 6.12: Glycosylation of thiophenyl glycoside in the presence of ILs.

techniques like molecular sieves or an inert atmosphere. The molten salt catalyst was reused, allowing for a flexible and cost-efficient glycosylation process.

Galan et al. [59] described the use of the ionic liquid [bmim][OTf] in combination with NIS as a mild promoter for chemo- and regio-selective glycosylation at room temperature (Figure 6.16). This methodology is applied to a one-pot, three-component oligosaccharide synthesis, providing an efficient alternative to traditional sequential glycosylation processes. In this work, the authors highlighted the effectiveness of [bmim][OTf]/NIS as a versatile promoter, facilitating highly selective glycosylation reactions without the need for stringent temperature regulation. The compatibility of the room-temperature ionic liquid (RTIL) with various protecting groups has been demonstrated, emphasizing the critical role of the counter ion in facilitating glycosylation reactions. [bmim][OTf] effectively activates certain thioglycosides, whereas others,

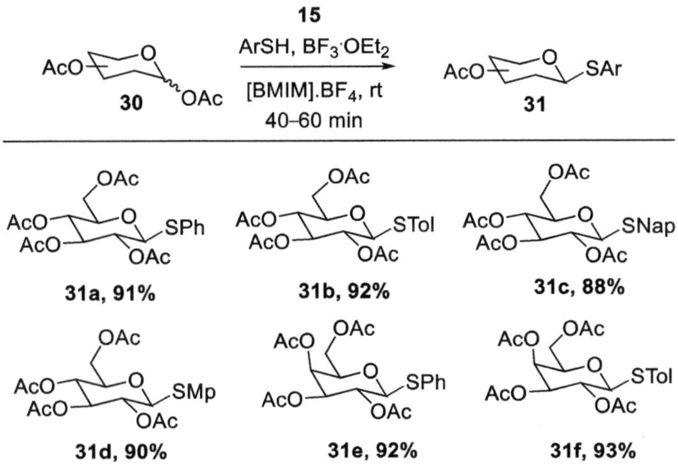

Figure 6.13: Preparation of thioglycoside in the presence of $BF_3{\cdot}OEt_2$ in an ionic liquid.

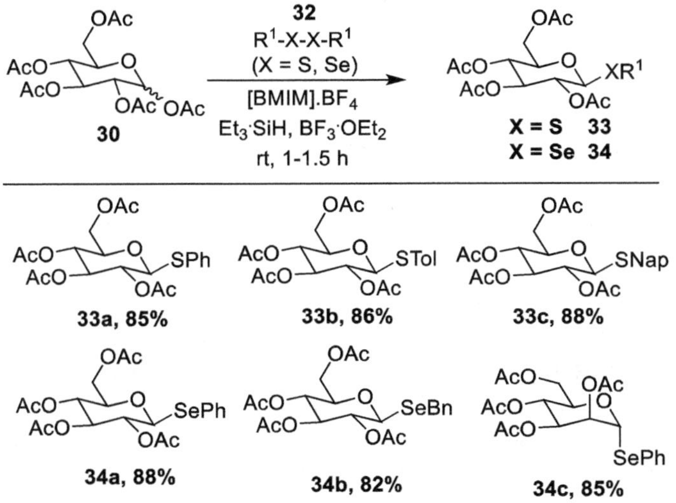

Figure 6.14: One-pot preparation of thio- and selenoglycosides using [bmim]·BF_4.

protected with acyl groups or 4,6-*O*-benzylidene groups, exhibit lower reactivity due to their increased stability and conformational restrictions. Additionally, the study shows that "armed" glycoside donors can be activated under mild conditions using RTILs, while "disarmed" donors may require the presence of a Lewis acid catalyst. The [bmim][OTf] IL promoter can be reused multiple times without significant loss of activity.

Augé et al. [60] developed direct glycosylation of unprotected monosaccharides using scandium triflate (Sc(OTf)$_3$) in 1-butyl-3-methylimidazolium trifluoromethanesul-

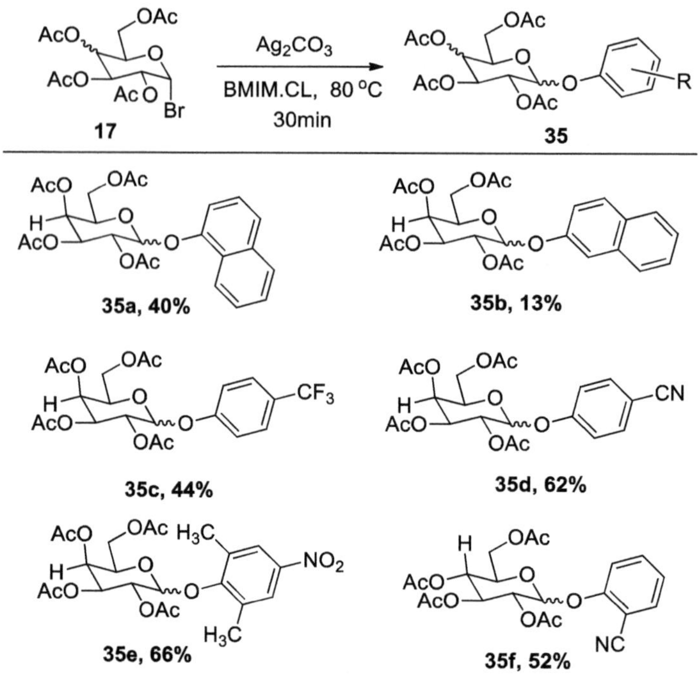

Figure 6.15: Glycosylation of acetobromo-α-D-galactose in molten [bmim][Cl].

fonate ionic liquid (Figure 6.17). Reactions at 110 °C led to some degradation of the product, but reducing reaction time improved yields. At 80 °C, longer reaction times increased yields. $Sc(OTf)_3$ emerged as the optimal catalyst in terms of yield and selectivity, and this Lewis acid is chosen due to its oxophilicity and lack of nucleophilicity/basicity, which can accelerate the glycosylation reaction. The ionic liquid [bmim][OTf] can immobilize the $Sc(OTf)_3$ catalyst and would be a suitable solvent for the glycosylation reaction. The potential benefits of using the same anion in both the catalyst and the ionic liquid are due to the non-nucleophilic nature of the triflate anion. Glycosylation reactions of octanol and different monosaccharides (D-glucose, D-galactose, D-mannose, D-xylose, L-fucose, 2-deoxyglucose, and *N*-acetylglucosamine, etc.) were performed, giving moderate to good yields of octylglycopyranosides as α/β-anomeric mixtures. Mannose exclusively produced the α-anomer. D-Xylose and 2-deoxyglucose reacted rapidly, but prolonged reaction times led to lower yields due to decomposition. In the mechanism, the oxocarbenium ion pathway suggests that ionic liquids could facilitate this reaction, leading to increased rate and yield. The ionic liquid could be used in three consecutive reactions without any significant decrease in yield or selectivity.

Galen et al. [61] have demonstrated that [bmim][OTf] ionic liquid in anhydrous dichloromethane acts as a selective glycosylation promoter at room temperature for activated (armed) thiophenyl and trichloroacetimidate glycosyl donors (Figures 6.18

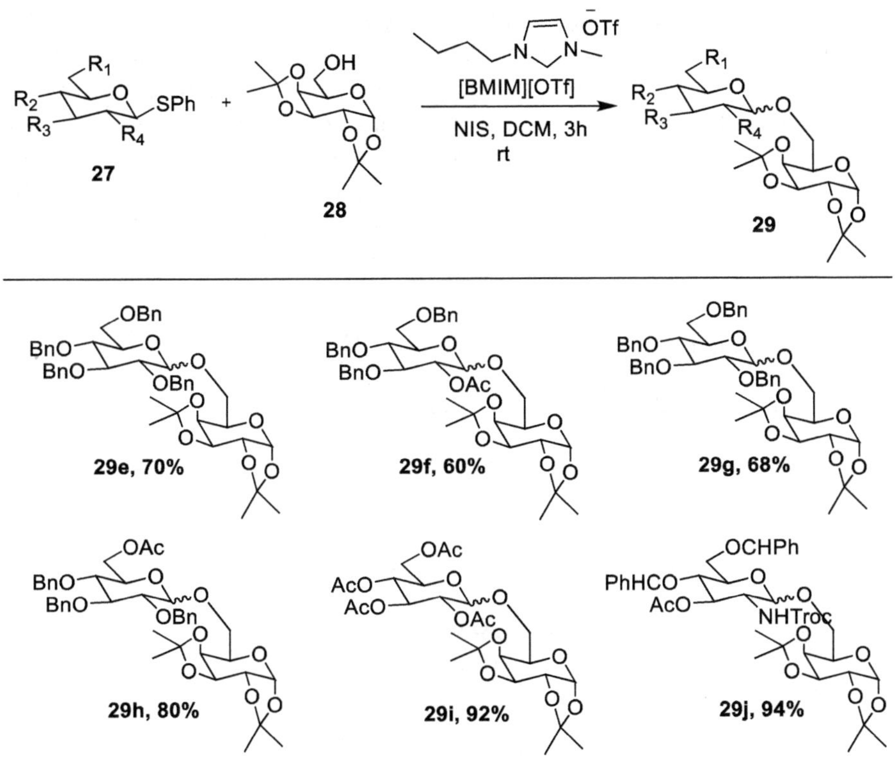

Figure 6.16: Glycosylation of thioglycosides in the presence of [bmim][OTf].

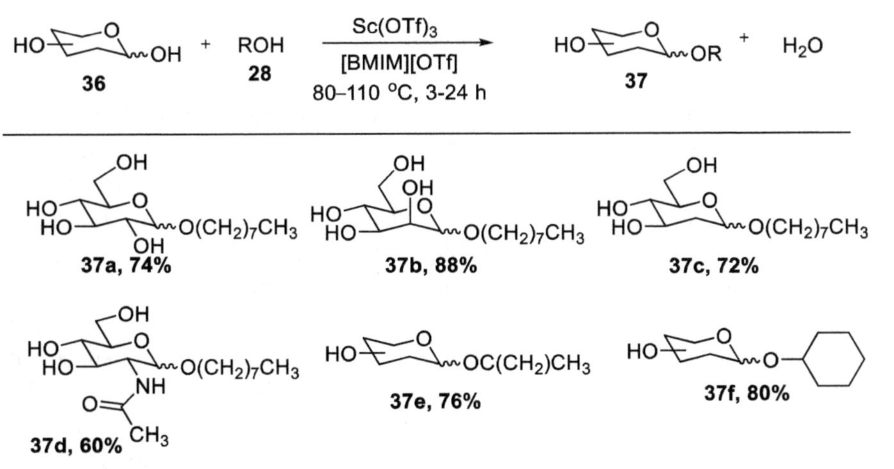

Figure 6.17: Glycosylation of alcohols under $Sc(OTf)_3$ catalysis in [bmim][OTf].

and 6.19). Peracetylated glycoside donors are categorized as electronically deactivated species (disarmed) and typically demand a more potent activator, such as a catalytic amount of triflic acid, for effective activation. The glycosylation conditions are mild and compatible with a variety of hydroxyl protecting groups, including acetates, benzyl ethers, and acetals. The mechanism proposes that the ionic liquid [bmim][OTf] functions by the gradual release of triflic acid. The ionic liquid also serves as a stabilizing agent against hydrolysis, enabling reactions to proceed cleanly and with favorable yields at room temperature without the necessity of molecular sieves.

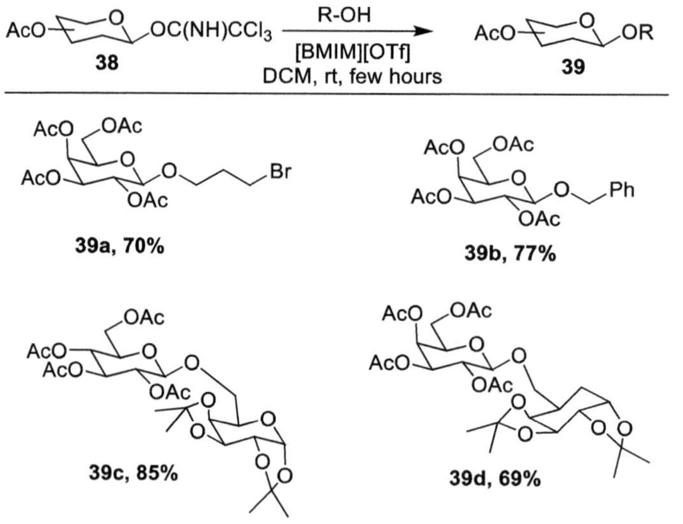

Figure 6.18: Glycosylation reactions with trichloroacetimidate donors in [bmim][OTf].

Toshima et al. [62] have developed a procedure for an ecofriendly aryl C-glycosylation reaction, which utilizes phenolic and naphtholic compounds in an ionic liquid with protic acids HBF_4 (Figure 6.20). The use of the ionic liquid $C_6mim[BF_4]$ provided an overall improvement in the aryl C-glycosylation reactions. The high yields of aryl β-C-glycosides were achieved because the ionic liquid acted as a barrier to hydrolysis by-products during pressure reduction. The ionic liquid $C_6mim[BF_4]$ was found to be more efficient than organic solvents such as dichloroethane. One of the salient features of this method is the ionic liquid's recycling. $C_6mim[BF_4]$ can be recycled, retaining its effectiveness, yields, and stereoselectivity even after several cycles. This method is general, and different types of glycosyl donors, including glycosyl fluorides, acetates, and methyl glycosides, can be used to circumvent the need for protecting and deprotecting processes. The 2,6-deoxy sugars were more reactive, resulting in enhanced aryl C-glycosylation compared to the 2-hydroxy sugars. With this sharp reactivity, it is possible to achieve higher yields under fairly mild reaction conditions.

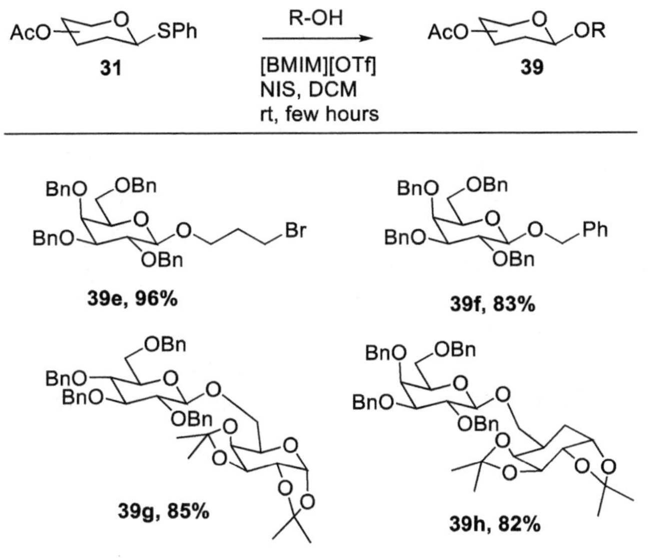

Figure 6.19: Glycosylation reactions with thioglycoside donors in [bmim][OTf].

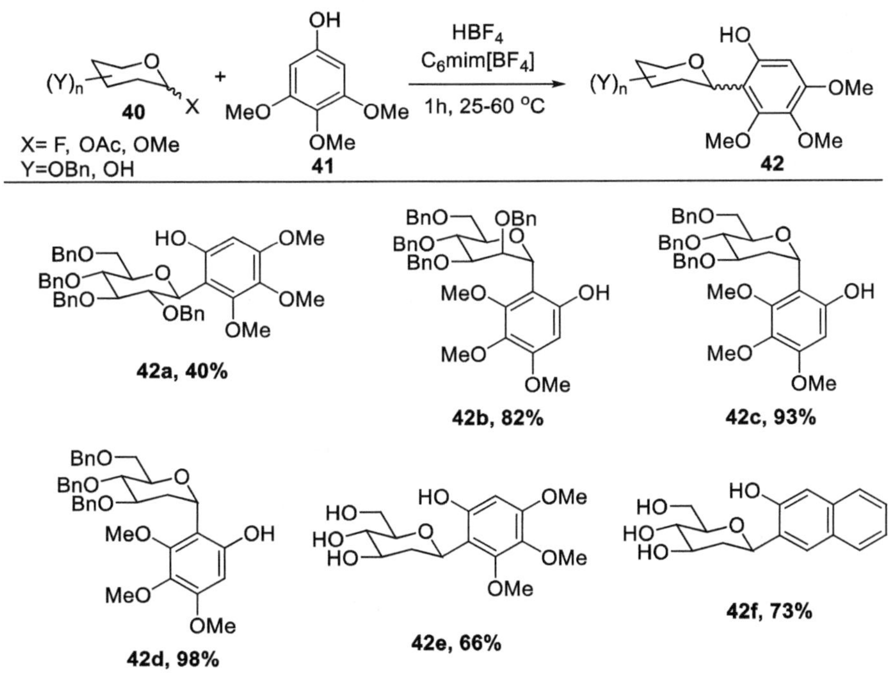

Figure 6.20: C-Glycosylation using phenol and naphthol derivatives in $C_6mim[BF_4]$.

This study demonstrates the potential of RTILs as both the solvents and catalysts for glycosylation using unprotected, unactivated donors (Figure 6.21) [63]. 1-Ethyl-3-methylimidazolium benzoate ([emim][ba]) was combined with promoters such as Amberlite IR-120 (H^+) resin or *p*-toluenesulfonic acid (TsOH) for better activity. The efficiency of the reaction could be enhanced by employing a promoter that is soluble (TsOH) in the reaction medium rather than one that is insoluble (Amberlite IR-120(H^+)). Additionally, less bulky allyl protecting groups on the acceptor molecule resulted in better yields. The protocol enables the successful synthesis of the benzyl glycosides of glucose (Glc), mannose (Man), and GalNAc in modest to good yields and with excellent α-selectivity. Aminopyrimidines protect the –OH of an alcohol from the RTILs, which causes other side reactions. However, owing to the increased viscosity of the reaction mixture, the synthesis of disaccharides from unprotected donors and partially protected acceptors was limited to moderate yield. In addition, alpha-stereoselectivity was achieved with the use of partially protected acceptors. In particular, galactose oligomers were obtained up to penta-saccharide derivatives using an isopropylidene-protected acceptor.

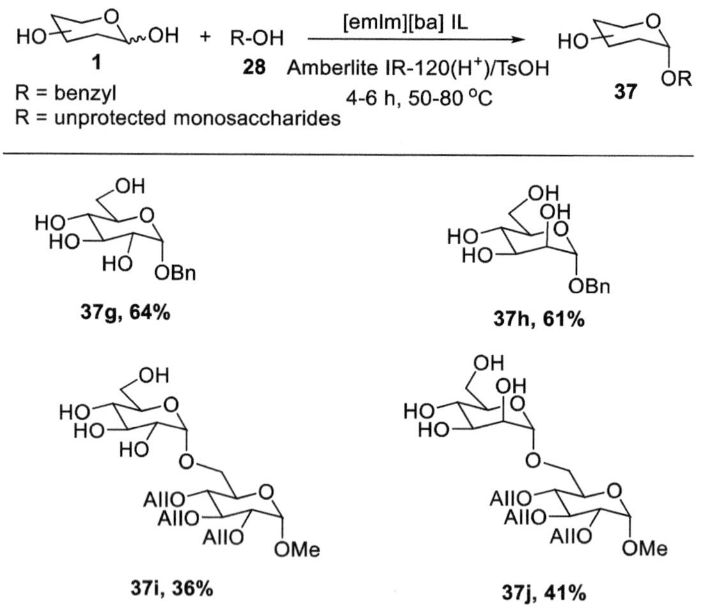

Figure 6.21: Synthesis of mono- and disaccharides using [emim][ba] IL.

In this study, Zhang and Ragauskas [64] investigated the use of ionic liquid [bmim]BF_4 as a solvent for thioglycosylation reactions using methyl triflate as an activating agent under mild conditions (Figure 6.22). The glycosylation reactions were carried out in [bmim]BF_4, and the maximum yield was found to be between 39% and 81%, although the stereoselectivity (α/β) was substrate-dependent. For efficient glycosidation, methyl triflate was essential, and higher temperatures during the reaction resulted in less product yield and more by-products. Even the addition of up to one molar equivalent of water did not make any apparent change to the yield of the glycosides, which indicates that $[bmim]BF_4$ can effectively stabilize the reaction environment by acting as a molecular sieve and can be employed several times with no reduction in efficiency.

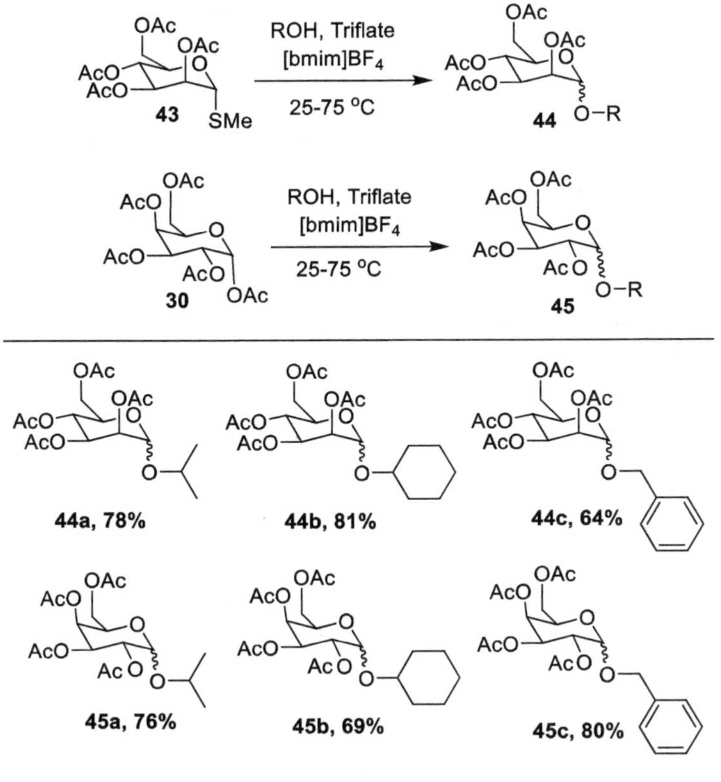

Figure 6.22: Thioglycosylation with various alcohols in $[bmim]BF_4$.

This study demonstrates an efficient *O*-acetylation of monosaccharides and cellulose using a Lewis acidic ionic liquid based on choline chloride (ChCl) and zinc chloride ($ZnCl_2$) (Figure 6.23) [65]. This ionic liquid, acting as both solvent and catalyst, effectively *O*-acetylates both monosaccharides and cellulose, offering environmental benefits and control over the degree of acetylation. High yields of *O*-acetylated products were achieved, even in the presence of side reactions involving choline. The reaction was also effective with propionic and isobutyric anhydrides. Using only one equivalent of acetic anhydride, the reaction produced selectively the *O*-acetylated product without partially acetylated intermediates. Microcrystalline cellulose, slightly soluble in $[ChCl][ZnCl_2]_2$, could be acetylated without microwave heating by this protocol. The degree of acetylation increased with the amount of acetic anhydride used. The majority of the acetylated cellulose was insoluble in acetone, indicating partial acetylation. This process enables controlled synthesis of partially substituted cellulose acetates. The ionic liquid was recovered and reused for several acetylation reactions without significant loss in yield.

This study demonstrates an efficient *O*-acetylation of monosaccharides and cellulose utilizing a choline chloride/zinc chloride-based Lewis acidic ionic liquid (Figure 6.23) [65]. The use of this ionic liquid system as a solvent and catalyst in the *O*-acetylation reaction of monosaccharides and cellulose furnished the product with better control of acetylation and greener aspects. Good yields of *O*-acetylated products were obtained despite the presence of choline, which often leads to side reactions. Reactions involving propionic and isobutyric anhydrides also worked well. The *O*-acetylated product was produced using only one equivalent of acetic anhydride with the substrate. The partially acetylated product that usually obstructs such synthesis was not obtained. Microcrystalline cellulose, completely soluble in $[ChCl][ZnCl_2]_2$, was acetylated without microwave heating. With the increase in the amount of acetic anhydride, the degree of substitution increased too. Cellulose acetate was mostly insoluble in acetone, indicating that they were derivatives with partially formed acetylation cellulose. This allows the controlled synthesis of cellulose acetate with a partial degree of substitution. The ionic liquid was recovered and used for more than five consecutive reactions without significant yield reduction.

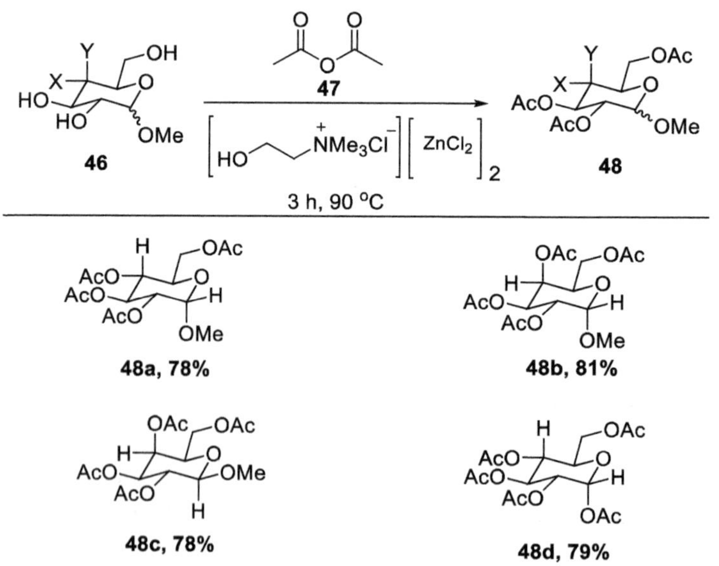

Figure 6.23: Acetylation of monosaccharides in ChCl-based ionic liquids.

6.2.3 Deep eutectic solvent

DESs, through their sustainable qualities, have become an increasingly popular alternative to organic solvents in organic synthesis [66]. The advantages of DESs include low volatility, biodegradability, and tunable properties. They have been effectively used in many organic transformations such as oxidation, reduction, and catalysis [67, 68]. Due to their unusual properties, including the ability to dissolve a wide variety of compounds, DESs may increase reaction rates and selectivity. Such solvents can present a specific environment for glycosylation reactions, which are the core of carbohydrate synthesis. DESs improve the solubility of both the glycosyl donor and acceptor, which can potentially increase reaction rates and yields.

Bhate and his group [69] have shown a practical and efficient Ferrier reaction for the synthesis of 2,3-unsaturated glycosides using a DES (Figure 6.24). The reaction was completed in 15 min at 80 °C with an anomeric mixture of ethyl 4,6-di-*O*-acetyl-2,3-dideoxy-D-erythro-hex-2-enopyranoside in 92% yield. The general protocol has been extended to other glycals and sugar-derived primary alcohols, showing significant improvements in reaction speed and yield. This protocol exhibited moderate to good anomeric selectivity in the range of 9:1 to 7:3 α:β ratio. In certain cases, a relatively poor selectivity of the product was observed due to the presence of malonic acid in the DES. The reaction was scaled up with high yields, but larger scale reactions required a little longer time. The DES was recycled up to four times with no loss of activity and yields.

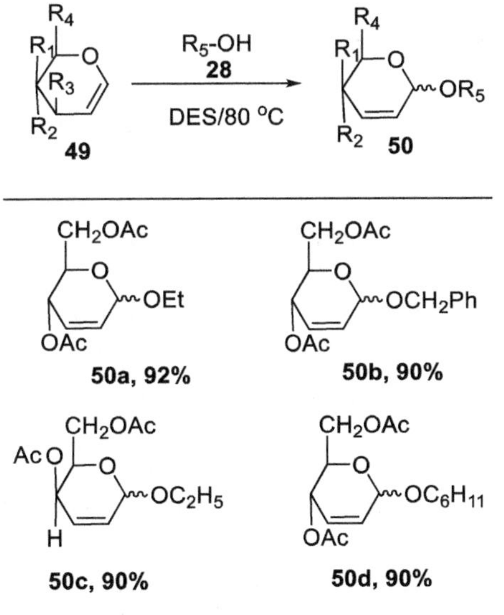

Figure 6.24: Synthesis of 2,3-unsaturated glycosides in a deep eutectic solvent.

This protocol offers a clean, efficient, and high-yielding approach to synthesizing per-*O*-acetylated hemiacetals from free sugars using a DES made from choline chloride and $ZnCl_2$ (Figure 6.25) [70]. The DES acts as both a solvent and a source of acidity, facilitating the reaction, and it can be recycled up to three times without significant loss in activity. The reaction involves treating free sugars with acetic anhydride at 90 °C to form peracetates, which are then selectively deacetylated at the anomeric position by heating to 100 °C, yielding peracetylated hemiacetals in good yields (80–88%). The resulting hemiacetals do not require further purification.

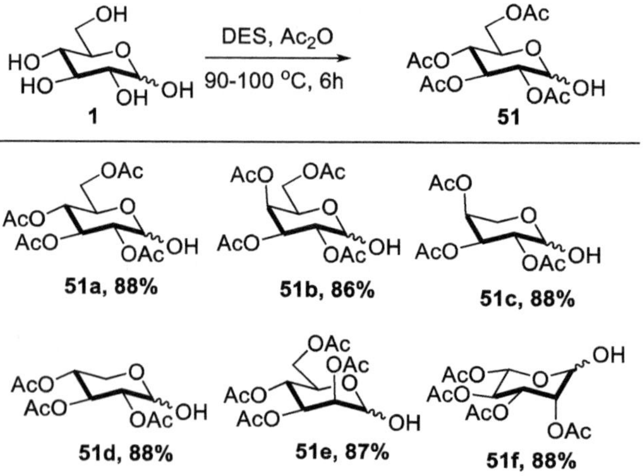

Figure 6.25: Preparation of per-*O*-acetyl-D-glucose hemiacetal in a deep eutectic solvent.

The lipase-catalyzed acylation of glucose in DES leads to the formation of glucose-6-*O*-hexanoate (Figure 6.26) [71]. The chemical structure of glucose-6-*O*-hexanoate was elucidated using NMR and confirmed through mass spectrometry. The enzymatic synthesis of glycolipids in DES was validated and expanded, showing that DES can function as both solvent and substrate. The use of DES as a solvent eliminates the need for additional glucose in the reaction, simplifying downstream processing. DES provides enzyme stability and activity, making them ideal solvents for lipase-catalyzed reactions.

Yang and coworkers [72] demonstrated that DES-based systems are efficient for the production of furfural from xylose and xylan, one of the most important chemicals accessible from biomass (Figure 6.27). A choline chloride/oxalic acid mixture was utilized as DES. In this case, DES played the role of both Brønsted acid catalyst and reaction medium, producing furfural from xylose without an additional metal chloride. However, the addition of trivalent metal chlorides, especially $AlCl_3$, had a significant positive impact on the yield of furfural. Other metal chlorides, such as $FeCl_3$, $CrCl_3$, $CeCl_3$, and $LaCl_3$, are also catalytically effective, although the efficiency is less. The mechanism of furfural formation involves the enolization of xylose, followed by dehydration to form furfural. Metal cations facilitate this process by bonding with

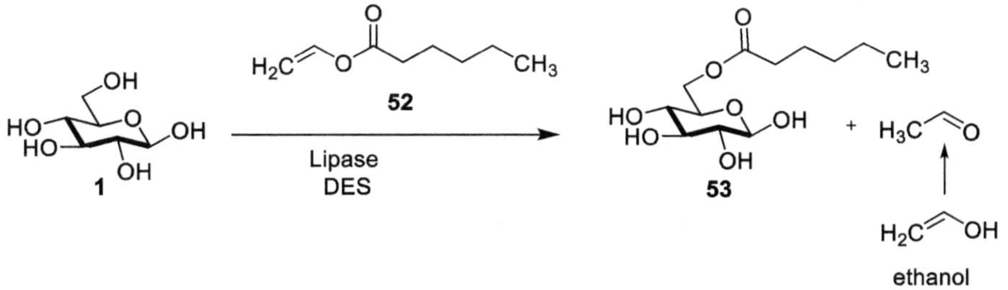

Figure 6.26: Synthesis of glucose-6-*O*-hexanoate and ethanol.

one oxygen atom of xylose. Higher temperatures and short reaction times favored furfural formation, while heavy loading of $AlCl_3$ decreased the yield resulting from probable side reactions. The DES-based system was also effective for xylan, a xylose polymer, to be converted into furfural. Both $AlCl_3$ and $FeCl_3$ demonstrated catalytic activity on the same order to convert xylan.

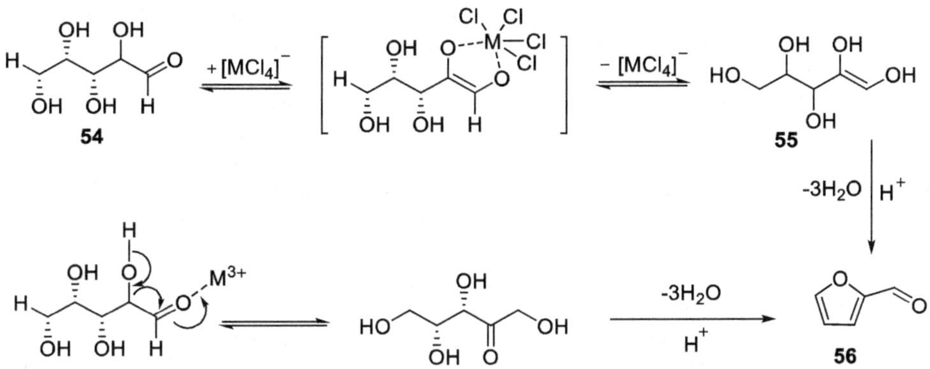

Figure 6.27: Plausible mechanism for the conversion of xylose into furfural.

6.3 Conclusions

We have highlighted the recent developments in the synthesis of a variety of carbohydrates and their derivatives in relatively green solvents. These solvents include water, ionic liquids, and DESs. Other than environmental impact, the use of these solvents is associated with several benefits including stereoselectivity, high yield, and less reaction time compared to conventional organic solvents. The carbohydrates are of great importance in pharmaceutical and other industries, and thus, we believe this article will be of significant interest to a wide section of practicing chemists in academia and industry.

References

[1] Sheldon RA. Green solvents for sustainable organic synthesis: State of the art. Green Chem, 2005, 7, 267–278.

[2] Nanda B, Sailaja M, Mohapatra P, Pradhan RK, Nanda BB. Green solvents: A suitable alternative for sustainable chemistry. Mater Today, 2021, 47, 1234–1240.

[3] Shanab K, Neudorfer H, Spreitzer C. Green solvents in organic synthesis: An overview. Curr Org Chem, 2016, 20, 1576–1583.

[4] Schäffner B, Schäffner F, Verevkin P, Börner A. Green and sustainable solvents in chemical processes. Chem Rev, 2010, 110(8), 4554–4581.

[5] Zhou F, Hearne Z, Li CJ. Water - The greenest solvent overall. Curr Opin Green Sustain Chem, 2019, 18, 118–123.

[6] Wang J, Zhang Y, Lu Q, Xing D, Zhang R. Exploring carbohydrates for therapeutics: A review on future directions. Front Pharmacol, 2021, 12, 756724.

[7] Jiang H, Qin X, Wang Q, Xu Q, Wang J, Wu Y, Zhang R. Application of carbohydrates in approved small molecule drugs: A review. Eur J Med Chem, 2021, 223, 113633.

[8] Painter TJ. Carbohydrate polymers in food preservation: An integrated view of the Maillard reaction with special reference to discoveries of preserved foods in Sphagnum-dominated peat bogs. Carbohydr Polym, 1998, 36, 335–347.

[9] Ramaprabha K, Kumar V, Saravanan P, Rajeshkannan R, Rajasimman M, Kamyab H, Vasseghian Y. Exploring the diverse applications of carbohydrate macromolecules in food, pharmaceutical, and environmental technologies. Environ Res, 2024, 240, 117521.

[10] Astria E, Thonhofer M, Ricco R, Liang W, Chemelli A, Tarzia A, Alt K, Agemeyer CE, Rattenberger J, Schroettner H, Wrodnigg T, Amenitsch H, Huang DM, Doonan CJ, Falcaro P. Carbohydrates@MOFs. Mater Horiz, 2019, 6, 969–977.

[11] Jing Y, Huang X. Fluorous thiols in oligosaccharide synthesis. Tetrahedron Lett, 2004, 45, 4615–4618.

[12] Dasgupta S, Nitz M. Use of *N*, *O*-dimethylhydroxylamine as an anomeric protecting group in carbohydrate synthesis. J Org Chem, 2011, 76, 1918–1921.

[13] Gao Q, Lian G, Lin F. The first total synthesis of 7-*O*-β-D-glucopyranosyl-4′-*O*-*α*-l-rhamnopyranosyl apigenin via a hexanoyl ester-based protection strategy. Carbohydr Res, 2009, 344, 511–515.

[14] Mohammadpour Z, Abdollahi SH, Omidvar A, Mohajeri A, Safavi A. Aqueous solutions of carbohydrates are new choices of green solvents for highly efficient exfoliation of two-dimensional nanomaterials. J Mol Liq, 2020, 309, 113087.

[15] El Seoud OA, Koschella A, Fidale LC, Dorn S, Heinze T. Applications of ionic liquids in carbohydrate chemistry: A window of opportunities. Biomacromolecules, 2007, 8, 2629–2647.

[16] Zdanowicz M, Wilpiszewska K, Spychaj T. Deep eutectic solvents for polysaccharides processing. Carbohydr Polym, 2018, 200, 361–380.

[17] Farrán A, Cai C, Sandoval M, Xu Y, Liu J, Hernáiz MJ, Linhardt RJ. Green solvents in carbohydrate chemistry: From raw materials to fine chemicals. Chem Rev, 2015, 115, 6811–6853.

[18] Si A, Misra AK. Perspective on the transformation of carbohydrates under green and sustainable reaction conditions. In Recent Trends in Carbohydrate Chemistry vol. 1. Elsevier, 2020, pp. 3–71.

[19] Pétursson S. Protecting groups in carbohydrate chemistry. J Chem Educ, 1997, 74, 1297.

[20] Manmode S, Chakraborty S, Sutar Y, Vangala M, Hotha S. Glycosylation in the green chemistry era. 2021, 45, 131–156.

[21] Das R, Mukhopadhyay B. Chemical O-glycosylations: An overview. Chemistry Open, 2016, 5, 401–433.

[22] Kang B, Opatz T, Landfester K, Wurm FR. Carbohydrate nanocarriers in biomedical applications: Functionalization and construction. Chem Soc Rev, 2015, 44, 8301–8325.

[23] Chen Y, Vedala H, Kotchey GP, Audfray A, Cecioni S, Imberty A, Vidal S, Star A. Electronic detection of lectins using carbohydrate-functionalized nanostructures: Graphene versus carbon nanotubes. ACS Nano, 2012, 6, 760–770.
[24] Delbianco M, Bharate P, Varela-Aramburu S, Seeberger PH. Carbohydrates in supramolecular chemistry. Chem Rev, 2016, 116, 1693–1752.
[25] Wang Z, Yang B, Zhang Z, Ly M, Takieddin M, Mousa S, Liu J, Dordick JS, Linhardt RJ. Control of the heparosan *N*-deacetylation leads to an improved bioengineered heparin. Appl Microbiol Biotechnol, 2011, 91, 91–99.
[26] Qiu X, Fairbanks AJ. Direct synthesis of para-nitrophenyl glycosides from reducing sugars in water. Org Lett, 2020, 22, 2490–2493.
[27] Borisova AS, Reddy SK, Ivanen DR, Bobrov KS, Eneyskaya EV, Rychkov GN, Sandgren M, Stålbrand H, Sinnott ML, Kulminskaya AA, Shabalin KA. The method of integrated kinetics and its applicability to the exo-glycosidase-catalyzed hydrolysis of *p*-nitrophenyl glycosides. Carbohydr Res, 2015, 412, 43.
[28] Vocadlo DJ, Withers SG. In Carbohydrates in Chemistry and Biology vol. 2. Ernst B, Hart GW, Sinaÿ P. (Eds.). Wiley-VCH: Weinheim, Germany, 2000, pp. 723–844.
[29] Pelletier G, Zwicker A, Allen CL, Schepartz A, Miller SJ. Aqueous glycosylation of unprotected sucrose employing glycosyl fluorides in the presence of calcium ion and trimethylamine. J Am Chem Soc, 2016, 138, 3175–3182.
[30] Feng S, Li C. Stereospecific, high-yielding, and green synthesis of *β*-glycosyl esters. J Agric Food Chem, 2015, 63, 5732–5739.
[31] Shimotori Y, Tsutano K, Soga K, Osawa Y, Aoyama M, Miyakoshi T. Synthesis of glycosyl ferulate derivatives by amine-promoted glycosylation with regioselective hydrolysis using Novozym 435 and evaluation of their antioxidant properties. Carbohydr Res, 2012, 359, 11–17.
[32] Calvaresi EC, Hergenrother PJ. Glucose conjugation for the specific targeting and treatment of cancer. Chem Sci, 2013, 4, 2319–2333.
[33] Ceunen S, Geuns JMC. Steviol glycosides: Chemical diversity, metabolism, and function. J Nat Prod, 2013, 76, 1201–1228.
[34] Misra M, Sharma R, Kant R, Maulik PR, Tripathi RP. One pot protecting group free synthesis of multifunctional biphenyl methyl-C-β-D-glycosides in aqueous medium. Tetrahedron, 2011, 67, 740–748.
[35] Lim D, Fairbanks AJ. Selective anomeric acetylation of unprotected sugars in water. Chem Sci, 2017, 8, 1896–1900.
[36] Tanaka T, Matsumoto T, Noguchi M, Kobayashi A, Shoda SI. Direct transformation of unprotected sugars to aryl 1-thio-β-glycosides in aqueous media using 2-chloro-1, 3-dimethylimidazolinium chloride. Chem Lett, 2009, 38, 458–459.
[37] Tejler J, Skogman F, Leffler H, Nilsson UJ. Synthesis of galactose-mimicking 1H-(1,2,3-triazol-1-yl)-mannosides as selective galectin-3 and 9N inhibitors. Carbohydr Res, 2007, 342, 1869–1875.
[38] Salameh BA, Leffler H, Nilsson UJ. 3-(1,2,3-Triazol-1-yl)-1- thio-galactosides as small, efficient, and hydrolytically stable inhibitors of galectin-3. Bioorg Med Chem Lett, 2005, 15, 3344–3346.
[39] Thomas JR, Liu X, Hergenrother PJ. Size-specific ligands for RNA hairpin loops. J Am Chem Soc, 2005, 127, 12434–12435.
[40] Wilkinson BL, Bornaghi LF, Houston TA, Innocenti A, Supuran CT, Poulsen SA. A novel class of carbonic anhydrase inhibitors: Glycoconjugate benzene sulfonamides prepared by “click-tailing”. J Med Chem, 2006, 49, 6539–6548.
[41] Kumar R, Maulik PR, Misra AK. Significant rate accelerated synthesis of glycosyl azides and glycosyl 1, 2, 3-triazole conjugates. Glycoconj J, 2008, 25, 595–602.
[42] Postema MH. Recent developments in the synthesis of C-glycosides. Tetrahedron, 1992, 48, 8545–8599.

[43] Sutherlin DP, Stark TM, Hughes R, Armstrong RW. Generation of C-glycoside peptide ligands for cell surface carbohydrate receptors using a four-component condensation on solid support. J Org Chem, 1996, 61, 8350–8354.
[44] Yadav JS, Reddy BVS, Sreenivas M, Satheesh G. Indium (III) chloride/water: A versatile catalytic system for the synthesis of C-furyl glycosides and trihydroxyalkyl furan derivatives. Synthesis, 2007, 1712–1716.
[45] Taylor CM. Glycopeptides and glycoproteins: Focus on the glycosidic linkage. Tetrahedron, 1998, 54, 11317–11362.
[46] Ichikawa Y, Matsukawa Y, Isobe M. Synthesis of urea-tethered neoglycoconjugates and pseudooligosaccharides in water. J Am Chem Soc, 2006, 128, 3934–3938.
[47] Misra AK, Agnihotri G. Preparation of polyhydroxyalkyl-and C-glycosylfuran derivatives from free sugars catalyzed by cerium (III) chloride in aqueous solution: An improvement of the Garcia Gonzalez reaction. Carbohydr Res, 2004, 339, 1381–1387.
[48] Wasserscheid P, Keim W. Ionic liquids-new solutions for transition metal catalysis. Angew Chem Int Ed Engl, 2000, 39, 3772–3789.
[49] Macfarlane DR, Pringle JM, Johansson KM, Forsyth SA, Forsyth M. Lewis base ionic liquids. Chem Commun, 2006, 42, 1905–1917.
[50] Smith MB, March J. March's Advanced Organic Chemistry. Wiley-Interscience: New York, 2001, Chapter 8.
[51] Wang Z, Zhang Q, Kuehner D, Ivaska A, Niu L. Green synthesis of 1–2 nm gold nanoparticles stabilized by amine-terminated ionic liquid and their electrocatalytic activity in oxygen reduction. Green Chem, 2008, 10, 907–909.
[52] Jiang YY, Wang GN, Zhou Z, Wu YT, Geng J, Zhang ZB. Tetraalkylammonium amino acids as functionalized ionic liquids of low viscosity. Chem Commun, 2008, 44, 505–507.
[53] Welton T. Room-temperature ionic liquids. Solvents for synthesis and catalysis. Chem Rev, 1999, 99, 2071–2083.
[54] Andrade CKZ, Alves LM. Environmentally benign solvents in organic synthesis: Current topics. Curr Org Chem, 2005, 9, 195–218.
[55] Galan MC, Tran AT, Boisson J, Benito D, Butts C, Eastoe J, Brown P. [R_4N][AOT]: A surfactant ionic liquid as a mild glycosylation promoter. J Carbohydr Chem, 2011, 30, 486–497.
[56] Santra A, Sau A, Misra AK. Synthesis of thioglycosides in room temperature ionic liquid. J Carbohydr Chem, 2011, 30, 85–93.
[57] Sau A, Misra AK. Odorless eco-friendly synthesis of thio- and selenoglycosides in ionic liquid. Synlett, 2011, 1905–1911.
[58] Kumar V, Talisman IJ, Malhotra SV. Application of halide molten salts as novel reaction media for *O*-glycosidic bond formation. Eur J Org Chem, 2010, 3377–3381.
[59] Galan MC, Tran AT, Whitaker S. [bmim][OTf] as co-solvent/promoter in room temperature reactivity-based one-pot glycosylation reactions. Chem Commun, 2010, 46, 2106–2108.
[60] Augé J, Sizun G. Ionic liquid promoted atom economic glycosylation under Lewis acid catalysis. Green Chem, 2009, 11, 1179–1183.
[61] Galan MC, Brunet C, Fuensanta M. [bmim][OTf]: A versatile room temperature glycosylation promoter. Tetrahedron Lett, 2009, 50, 442–445.
[62] Yamada C, Sasaki K, Matsumura S, Toshima K. Aryl *C*-glycosylation using an ionic liquid containing a protic acid. Tetrahedron Lett, 2007, 48, 4223–4227.
[63] Park TJ, Weïwer M, Yuan X, Baytas SN, Munoz EM, Murugesan S, Linhardt RJ. Glycosylation in room temperature ionic liquid using unprotected and unactivated donors. Carbohydr Res, 2007, 342, 614–620.
[64] Zhang J, Ragauskas A. Study of thioglycosylation in ionic liquids. Beilstein. J Org Chem, 2006, 2, 12.

[65] Abbott AP, Bell TJ, Handa S, Stoddart B. *O*-Acetylation of cellulose and monosaccharides using a zinc based ionic liquid. Green Chem, 2005, 7, 705–707.
[66] Di Carmine G, Abbott AP, D'agostino C. Deep eutectic solvents: Alternative reaction media for organic oxidation reactions. React Chem Eng, 2021, 6, 582–598.
[67] Ünlü AE, Arıkaya A, Takaç S. Use of deep eutectic solvents as catalyst: A mini-review. Green Process Synth, 2019, 8, 355–372.
[68] Perna FM, Vitale P, Capriati V. Deep eutectic solvents and their applications as green solvents. Curr Opin Green Sustain Chem, 2020, 21, 27–33.
[69] Rokade SM, Bhate PM. Ferrier reaction in a deep eutectic solvent. Carbohydr Res, 2015, 415, 28–30.
[70] Rokade SM, Bhate PM. One-pot synthesis of per-*O*-acetylated hemiacetals from free sugars in a deep eutectic solvent. Carbohydr Res, 2015, 416, 21–23.
[71] Pöhnlein M, Ulrich J, Kirschhöfer F, Nusser M, Muhle-Goll C, Kannengiesser B, Brenner-Weib G, Luy B, Liese A, Syldatk C, Hausmann R. Lipase-catalyzed synthesis of glucose-6-*O*-hexanoate in deep eutectic solvents. Eur J Lipid Sci Technol, 2015, 117, 61–166.
[72] Zhang LX, Yu H, Yu HB, Chen Z, Yang L. Conversion of xylose and xylan into furfural in bio-renewable choline chloride–oxalic acid deep eutectic solvent with the addition of metal chloride. Chin Chem Lett, 2014, 25, 1132–1136.

Aayushi Chanderiya, Arunesh K. Mishra, and Ratnesh Das*

Chapter 7 The role of coupling reactions in carbohydrate chemistry

7.1 Introduction

Saccharides, another name for carbohydrates, are essential macromolecules that have many uses in living things such as cell recognition, energy storage, and structural support [1]. In addition to being fascinating, the complex chemistry of carbohydrates is crucial for comprehending biological processes and creating medicinal treatments [2]. The function of coupling processes, which promote the creation of glycosidic linkages and the synthesis of oligosaccharides and polysaccharides, is one of the fundamental elements of carbohydrate chemistry [3]. In the field of carbohydrate chemistry, this chapter explores the forms, mechanisms, and importance of coupling reactions.

An essential component of many natural goods, medications, and materials, carbohydrates also play a key role in biological processes [4]. Carbohydrates' complex structures and wide range of functions make them indispensable in fields including materials science and biochemistry [5]. Their intricacy, however, frequently presents formidable synthetic obstacles. In this regard, coupling reactions have become effective techniques in the field of carbohydrate chemistry, enabling the formation of glycosidic bonds and the assembly of polysaccharides and oligosaccharides [6].

Chemical processes called coupling reactions bring two different molecular entities together, generally producing new covalent bonds in the process [7]. In carbohydrate chemistry, where the creation of glycosidic bonds is an essential step in the synthesis of more complex carbohydrate structures, these reactions are very important [8]. Traditional glycosylation techniques have been constrained by problems like efficiency, selectivity, and the need for protective group tactics [9]. By providing more straightforward, effective, and selective methods for the synthesis of carbohydrates, coupling reactions, especially those that make use of contemporary methods like click chemistry and enzyme-catalyzed transformations, have completely transformed the area [10].

The exact formation of glycosidic bonds made possible by coupling processes, especially transition-metal-catalyzed glycosylation, makes it easier to synthesize a variety of glycosides and glycoconjugates with regulated stereochemistry [11].

***Corresponding author: Ratnesh Das,** Dr. Harisingh Gour Vishwavidyalaya, Sagar, Madhya Pradesh, India, e-mail: ratneshdas1@gmail.com
Aayushi Chanderiya, Arunesh K. Mishra, Dr. Harisingh Gour Vishwavidyalaya, Sagar, Madhya Pradesh, India

https://doi.org/10.1515/9783111447094-007

Coupling processes, like thiol-ene and azide-alkyne cycloaddition, are essential for effectively attaching carbohydrates to biomolecules and improving their functional characteristics in a variety of applications [12].

The stereoselective synthesis of *C*-glycosides is rendered possible by coupling processes, especially glycosyl cross-coupling, which increases the bioactivity and functional diversity of carbohydrate derivatives in medicinal chemistry [13]. Coupling reactions are essential in carbohydrate chemistry for forming glycosidic bonds, enabling the synthesis of nucleosides and oligosaccharides through the activation of sugar donors and acceptors [14]. Glycosyl couplings are critical in carbohydrate chemistry, influencing stereo-, chemo-, and regio- selectivities, with protecting groups significantly affecting these outcomes during glycosidation reactions [15]. Coupling reactions are crucial in carbohydrate chemistry as they determine the yield and stereoselectivity of glycosylation, influenced by the reactivity of both glycosyl donors and acceptors [16]. Coupling reactions in carbohydrate chemistry facilitate the formation of complex glycosidic bonds, enabling the synthesis of diverse carbohydrate derivatives and enhancing the structural complexity of glycosylated compounds [17]. Coupling reactions, particularly glycosyl cross-coupling, enable the stereoretentive formation of glycosidic linkages, facilitating the synthesis of complex carbohydrates and glycoconjugates with controlled anomeric configurations [18]. Coupling reactions are essential in carbohydrate chemistry for facilitating glycosylation, enhancing stereoselectivity, and improving the efficiency of synthesizing complex carbohydrate structures and glycoconjugates [19].

The purpose of this chapter is to examine the function of coupling reactions in the chemistry of carbohydrates, emphasizing their benefits, mechanisms, and uses. We will demonstrate how these reactions aid in the effective formation of carbohydrate frameworks, advance knowledge of carbohydrate function, and facilitate the creation of new glycosylated compounds with possible uses in drug design and biomaterials by looking at recent developments in the field. We hope that this chapter will highlight the significance of coupling reactions as a fundamental component of the continuous development of carbohydrate chemistry.

7.2 Understanding coupling reactions

Coupling reactions refer to chemical processes that involve the joining of two or more molecular entities to form a larger and more complex structure. In carbohydrate chemistry, these reactions are particularly important for linking monosaccharides to create disaccharides, oligosaccharides, and polysaccharides. The formation of glycosidic bonds through coupling reactions is a crucial step in building the diverse array of carbohydrate structures found in nature.

7.2.1 Mechanisms of coupling reactions

In carbohydrate chemistry, coupling reaction mechanisms include a variety of tactics for triggering glycosyl donors and promoting nucleophilic attacks. Chemists can efficiently produce complex carbohydrates by refining their synthetic methods and comprehending these principles. The continuous development of coupling techniques will improve our capacity to create a wide range of glycosylated molecules, increasing our understanding of how carbohydrates work and their possible applications in a variety of industries.

In carbohydrate chemistry, coupling processes are crucial for the formation of glycosidic bonds that are essential for the construction of intricate carbohydrate structures. Each of these reactions has unique benefits and applications, and they can be grouped according to their mechanics. With the goal of optimizing synthetic techniques and achieving the expected outcomes in carbohydrate synthesis, it is essential to comprehend these mechanisms. The most common mechanism for carbohydrate coupling involves the activation of a sugar donor, typically a nucleoside diphosphate sugar, which makes it more reactive toward another sugar acceptor. Several mechanisms can be employed including

7.2.1.1 Activation of glycosyl donors

The first step in many coupling reactions involves the activation of glycosyl donors, making them more electrophilic and, thus, more reactive toward nucleophilic attack. Common strategies for activation include

- **Anomeric leaving groups**: In this approach, the leaving group at the anomeric carbon is chosen for its ability to stabilize the positive charge that develops during the reaction. Common leaving groups include halides and sulfonates, which can facilitate the formation of the reactive oxocarbenium ion.
- **Electrophilic activation**: Some glycosyl donors are activated using electrophiles, such as bromine or iodine, to generate reactive intermediates. These intermediates can subsequently engage in nucleophilic attacks from sugar acceptors.
- **Protecting group strategies**: Protecting groups can be employed to shield functional groups that may interfere with glycosylation. This selective protection allows for the activation of specific hydroxyl groups and facilitates controlled coupling.

7.2.1.2 Nucleophilic attack

Following the activation of the glycosyl donor, the next step is the nucleophilic attack by a suitable acceptor, typically another sugar or alcohol. The nucleophile attacks the anomeric carbon of the activated glycosyl donor, resulting in the formation of a glycosidic bond. This step can be influenced by several factors:

- **Stereochemical control**: The stereochemistry of the glycosidic bond (α or β) is crucial for determining the properties of the resulting carbohydrate. Techniques such as using specific glycosyl donors and controlling reaction conditions can help achieve the desired stereochemistry.
- **Solvent effects**: The choice of solvent can significantly impact the nucleophilicity of the attacking species and the stability of intermediates, thus affecting the overall reaction rate and selectivity.

7.2.1.3 Modern approaches to coupling reactions

Recent advancements have introduced innovative methodologies for coupling reactions, enhancing both efficiency and selectivity:

- **Click chemistry**: This modular approach employs highly selective reactions such as the copper-catalyzed azide-alkyne cycloaddition. This strategy allows for the rapid assembly of complex glycosylated structures under mild conditions.
- **Enzymatic glycosylation**: Glycosyltransferases are enzymes that catalyze the transfer of sugar moieties from activated donors to acceptors. These biocatalysts offer high specificity and can operate under mild conditions, reducing the need for harsh reagents and extensive purification.
- **Transition metal catalysis**: Leveraging transition metals has enabled the development of new glycosylation methods that enhance regio- and stereoselectivity. These reactions can provide access to otherwise challenging glycosidic bonds.

7.2.2 Types of coupling reactions

There are many techniques for creating intricate glycosylated structures, thanks to the different types of coupling reactions found in carbohydrate chemistry. Each style, from contemporary click chemistry and enzymatic approaches to conventional glycosylation procedures, has special benefits and applications. To advance the synthesis of carbohydrates and investigate the practical applications of carbohydrates in many sectors, it is important to comprehend these procedures. In carbohydrate chemistry, coupling reactions can be divided into many different categories, each distinguished by its conditions, reactants, and mechanism. It is essential to understand these kinds to choose the best technique for creating intricate carbohydrate structures. In carbohydrate chemistry, a variety of coupling reaction types are utilized, each having special traits and employing:

7.2.2.1 Glycosylation reactions

Glycosylation is the most common type of coupling reaction in carbohydrate chemistry, specifically involving the formation of glycosidic bonds between monosaccharides or between a monosaccharide and another nucleophile (e.g., an alcohol or phenol). Glycosylation reactions are crucial in carbohydrate chemistry, facilitating the synthesis of *C*- and *S*-glycosides, which offer enhanced metabolic stability.

Formation of *C*-glycosides

Recent advancements, particularly in the use of transition metal catalysis, have improved reaction conditions and selectivity, yielding biologically significant products [20]. Key aspects include

- **Types of glycosyl donors**: These can be activated by various means including leaving groups (e.g., halides and sulfonates) or by generating reactive intermediates such as oxocarbenium ions. Glycosylation reactions are chemical processes essential for synthesizing complex oligosaccharides, influenced by factors like temperature, acidity, moisture, and the properties of glycosyl donors. Efficient glycosylation requires precise control over these parameters to achieve high yields and reproducibility [21].
- **Stereoselectivity**: The choice of glycosyl donor and reaction conditions can influence whether an α- or β-glycosidic bond is formed. Glycosylation reactions involve the stereoselective formation of C–O bonds, crucial for synthesizing oligo- and polysaccharides. The anomeric selectivity is influenced by factors like the nature of the donor and acceptor, activator systems, temperature, and notably, the solvent used [22].

7.2.2.2 Click chemistry

Click chemistry refers to a class of reactions that are highly efficient, selective, and occur under mild conditions. Click chemistry is a transformative approach in synthetic chemistry, characterized by its efficiency, selectivity, and minimal environmental impact. Introduced by K. Barry Sharpless in 2001, it emphasizes the rapid and reliable formation of complex molecules through simple reactions, often in benign conditions. This methodology has found extensive applications across various fields including drug synthesis, molecular imaging, and green chemistry. In carbohydrate chemistry, it often involves

- **Azide-alkyne cycloaddition**: This reaction makes it simple to connect carbohydrate moieties to other molecules, which accelerates the process of assembling complex structures. A crucial reaction in click chemistry, azide-alkyne cycloaddition is recognized for its selectivity and rapidity. An optical cavity that may selectively catalyze the synthesis of 1,4- or 1,5-disubstituted compounds is used in this study to illustrate a metal-free replacement [23]. The copper-mediated click reaction known as the azide-alkyne cycloaddition (CuAAC) effectively produces 1,2,3-triazoles and has several advantages, including high selectivity, biocompatibility, and ease of purification (Figure 7.1). It is especially beneficial when carried out in water, which serves as a refrigerant and reduces protic connecting interference [24].

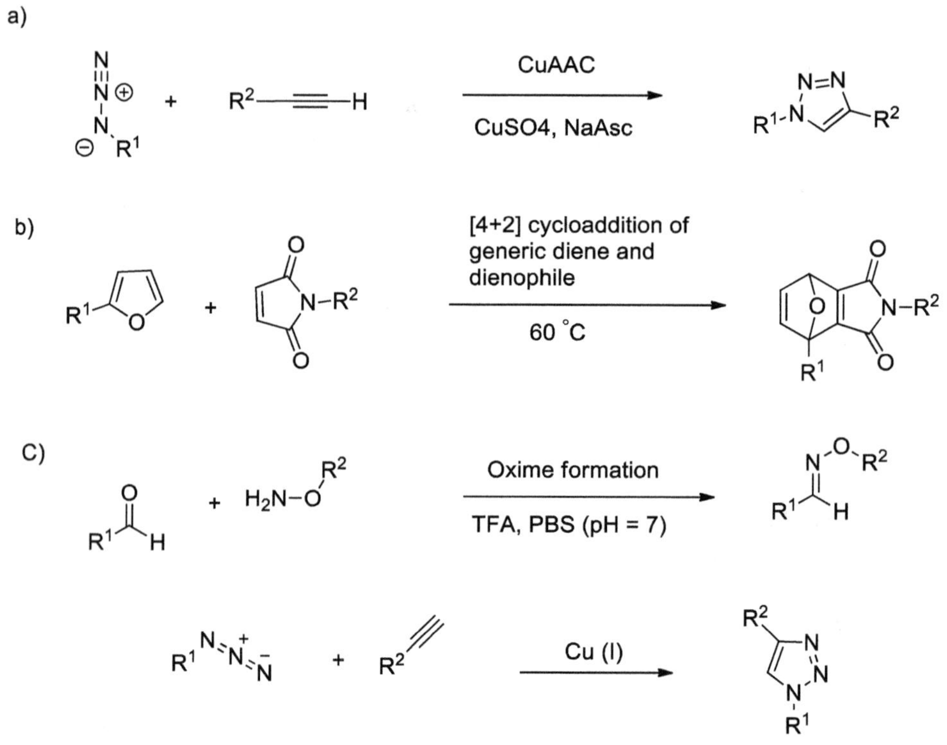

Figure 7.1: The synthetic methods for the synthesis of organic compound.

- **Advantages**: The simplicity, modularity, and high yield of click reactions make them ideal for synthesizing diverse glycosylated compounds. Click chemistry is a highly selective, high-yield chemical approach that enables efficient synthesis of stable probes for molecular imaging, facilitating in vivo characterization of biological processes. It enhances selectivity, sensitivity, and stability in various imaging modalities including optical and nuclear imaging [25]. Click chemistry is a modular approach to functionalization, characterized by high stereoselectivity, rapid reaction rates, and mild conditions. It enables the synthesis of complex molecules with novel properties, utilizing reactions like CuAAC and Diels-Alder cycloaddition for diverse applications [26].

7.2.2.3 Enzymatic coupling reactions

Enzymatic reactions utilize glycosyltransferases to catalyze the transfer of sugar moieties from activated donors to acceptors (Figure 7.2). This approach offers several benefits:

- **Selectivity and specificity**: Enzymes can exhibit remarkable specificity for substrates, minimizing side reactions and enhancing yields. Enzymatic reactions can utilize glycosyltransferases for synthesizing human milk oligosaccharides; however, this method is complicated by high sugar nucleotide costs, strict substrate specificity, and low enzyme stability, making engineered glycosidases a preferable alternative for certain applications [27].

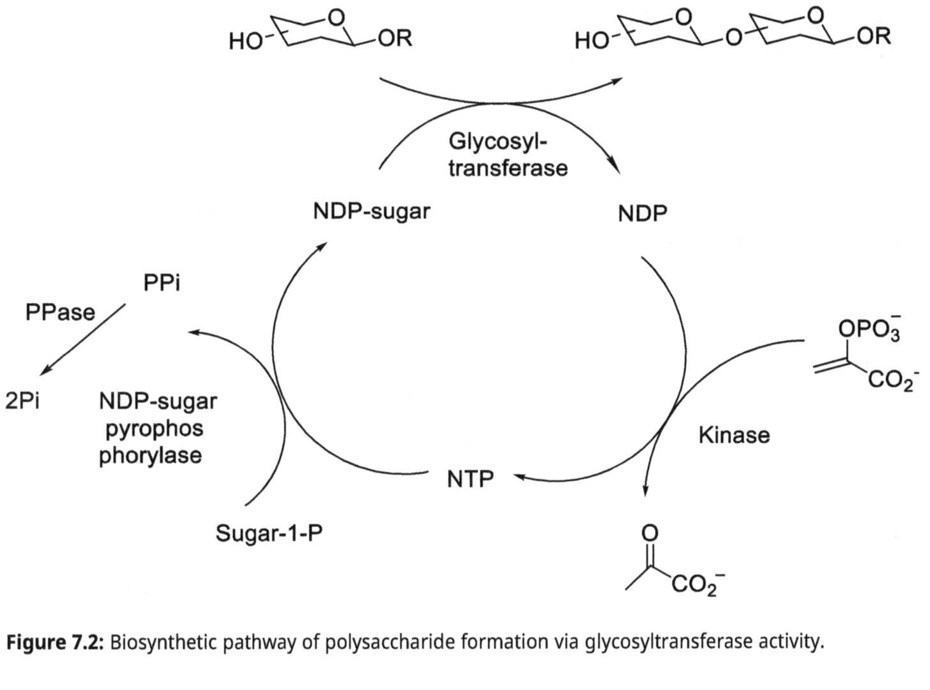

Figure 7.2: Biosynthetic pathway of polysaccharide formation via glycosyltransferase activity.

- **Mild conditions**: Enzymatic reactions typically occur under physiological conditions, reducing the need for harsh reagents. Enzymatic reactions utilize glycosyltransferases, which catalyze glycosidic bond formation with high regio- and stereo-control, enabling the synthesis of diverse carbohydrate structures. These enzymes are crucial in one-pot multienzyme systems for efficient carbohydrate synthesis from simple monosaccharides [28].

7.2.2.4 Transition metal-catalyzed reactions

Transition metals play a crucial role in reductive cross-coupling reactions by promoting the direct coupling of distinct electrophiles (Figure 7.3). They facilitate these reactions using diboron reagents as reductants, offering advantages over traditional methods by improving operational simplicity and functional group compatibility [29]. The use of transition metals in coupling reactions has opened new avenues for glycosylation:

- **Transition metal catalysis**: These catalysts can facilitate the formation of glycosidic bonds with high regio- and stereoselectivity, allowing for the construction of complex carbohydrate architectures. Transition metals, such as Cu, Ni, Fe, and Co, play a crucial role in coupling reactions by facilitating the formation of carbon-carbon bonds, enhancing reaction efficiency, and enabling various organic transformations essential in both laboratory and industrial processes [30].

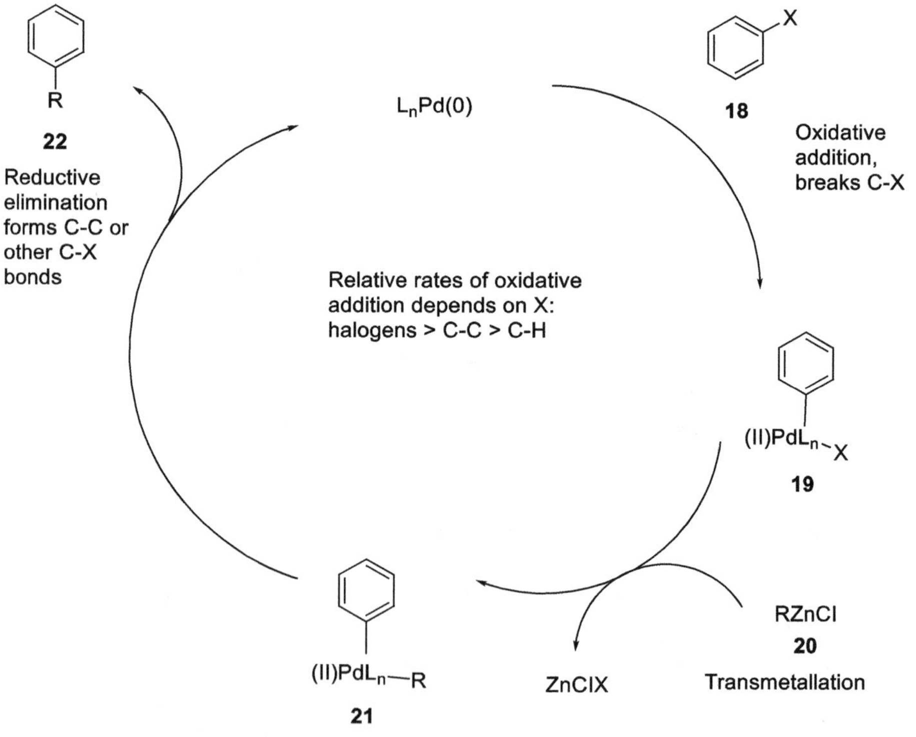

Figure 7.3: Mechanistic cycle of a palladium-catalyzed cross-coupling reaction.

- **Mechanistic diversity**: Transition metal-catalyzed reactions can proceed via different pathways, including cross-coupling reactions that extend beyond traditional glycosylation. Transition metals, primarily nickel, palladium, and iron, facilitate cross-coupling reactions of Grignard reagents with various organohalides and derivatives. Other metals like copper, cobalt, manganese, and chromium also play significant roles in these catalytic processes, enhancing synthetic efficiency [31].

7.2.2.5 Carbon-carbon coupling reactions

While less common in carbohydrate synthesis, carbon-carbon coupling reactions can play a role in forming larger carbohydrate frameworks or modifying existing structures. Carbon-carbon coupling reactions, particularly Stille and Suzuki coupling, play a crucial role in constructing larger carbohydrate frameworks. These reactions facilitate the formation of complex molecular structures, essential for synthesizing bioactive compounds and pharmaceuticals:

- **Stille and Suzuki coupling**: These reactions can be adapted to join carbohydrate moieties with noncarbohydrate components, expanding the functional diversity of glycosylated compounds. Stille coupling utilizes organostannanes and aryl halides, enabling the formation of *C*-glycosides with high stereocontrol and selectivity [32]. The synthesis of stannylated monosaccharides has been explored, providing versatile nucleophiles for Stille coupling by Michael Rourke (Figure 7.4) [33].

Suzuki coupling employs boronic acids and aryl halides, offering a complementary approach to Stille reactions in carbohydrate synthesis [34]. This method is particularly advantageous for constructing complex glycosides due to its mild reaction conditions and broad substrate scope [35].

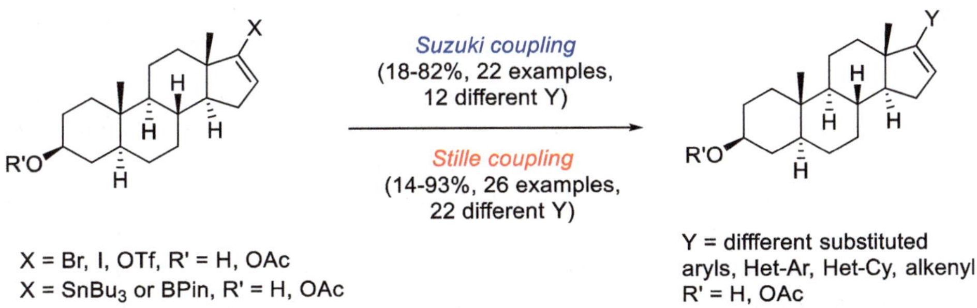

Figure 7.4: The comparison of Suzuki and Stille coupling reactions for the conversion of steroidal substrates.

7.2.3 Applications in carbohydrate chemistry

The role of coupling reactions extends beyond mere bond formation; they are essential in several applications within carbohydrate chemistry:

7.2.3.1 Synthesis of oligosaccharides

The synthesis of oligosaccharides through coupling reactions has evolved significantly, leveraging innovative methodologies that enhance efficiency and stereoselectivity. Recent advancements focus on the use of anomeric nucleophiles and novel glycosylation strategies, which facilitate the assembly of complex carbohydrate structures with improved yields and reduced steps.

Anomeric stannanes serve as effective nucleophiles, allowing for highly stereoselective glycosyl cross-coupling reactions, which are crucial for synthesizing oligosaccharides and glycoconjugates [36].

Oligosaccharides are vital in biological recognition processes, serving as ligands for proteins and influencing cell-cell interactions. The ability to synthesize oligosaccharides with specific linkages and configurations is critical for studying their biological functions [37]. Coupling reactions allow chemists to construct oligosaccharides with desired structures, facilitating the exploration of their roles in biological systems [38].

7.2.3.2 Development of glycoconjugates

These reactions are essential for creating useful glycoconjugates that can be used in biomolecular investigation, vaccine development, and medication creation. Glycoconjugates – carbohydrates joined to proteins or lipids – are essential for immunological response and cellular signaling. To create vaccine candidates, targeted drug delivery systems, and biomaterials, coupling processes are essential to the production of glycoconjugates. In mild conditions, techniques like Suzuki-Miyaura cross-coupling and azide-alkyne cycloaddition are frequently utilized to bind carbohydrates to biomolecular scaffolds. Metal-mediated and metal-free azide-alkyne cycloaddition, thiol-ene reactions, amidation, reductive amination, and Suzuki-Miyaura cross-coupling are examples of chemical coupling reactions in glycoconjugate synthesis, which facilitate the covalent attachment of carbohydrates to biomolecular scaffolds in aqueous conditions [39]. The use of glycosyl donors, including glycosyl vinylogous carbonates, allows for efficient formation of glycosidic bonds, enhancing the versatility of glycoconjugate synthesis [40].

Lytic polysaccharide mono-oxygenases enzymes selectively oxidize nonreducing ends of polysaccharides, facilitating the formation of oxime-based glycoconjugates, which can be further modified [41].

7.2.3.3 Understanding glycosylation patterns

Understanding the glycosylation patterns of carbohydrates is crucial due to their significant roles in biological processes and disease mechanisms. Glycosylation, a complex posttranslational modification, influences protein stability, activity, and interactions, thereby affecting various physiological functions.

Glycosylation involves the addition of carbohydrate moieties to proteins, which can vary based on species, cell type, and environmental conditions [42]. The process is regulated by a variety of enzymes, including transferases and hydrolases, which determine the specific glycan structures formed [43]. In nature, the glycosylation patterns of carbohydrates significantly influence their biological functions. By utilizing coupling reactions, researchers can systematically modify glycosylation patterns, allowing for the elucidation of structure-function relationships in glycoproteins and glycolipids.

7.3 Challenges and future directions

Despite the significant advancements in carbohydrate chemistry brought about by coupling reactions, many difficulties still exist. Research is ongoing on stereoselectivity, regioselectivity, and the requirement for mild reaction conditions. The advancement of the area depends on the creation of novel catalysts and techniques to increase coupling efficiency and selectivity. Although pairing processes in carbohydrate chemistry present many challenges, there is also a great deal of opportunity for innovation. Researchers can improve the effectiveness and reach of carbohydrate synthesis by tackling these issues and looking toward the future. Further research in this area will broaden the applicability of carbohydrates in a variety of domains, including materials science and medicine, while also improving our understanding of the way they function.

7.3.1 Challenges

i. **Selectivity and stereochemistry**: Achieving high selectivity in glycosylation reactions, particularly in forming specific α- or β-glycosidic bonds, remains a significant challenge. The presence of multiple reactive sites in carbohydrates can lead to unwanted by-products and undesired stereoisomers.

ii. **Functional group compatibility**: Many coupling reactions require the protection of functional groups to prevent side reactions. The necessity for protective group strategies can complicate synthesis and reduce overall yields. Developing methods that are compatible with a wider array of functional groups is essential.
iii. **Scalability and efficiency**: While many coupling reactions have been optimized for laboratory-scale synthesis, translating these methods to larger scales can pose difficulties. Issues such as reagent cost, reaction time, and by-product formation need to be addressed to make the processes more efficient and economically viable.
iv. **Solubility and reaction conditions**: Many carbohydrates are poorly soluble in common organic solvents, limiting the efficiency of coupling reactions. Finding suitable solvents or developing solvent-free methods is an area requiring further research.
v. **Enzyme limitations**: While enzymatic methods offer high specificity, the range of substrates that enzymes can effectively process is often limited. Expanding the substrate scope of glycosyltransferases and engineering new enzymes to recognize diverse substrates are significant challenges.

7.3.2 Future directions

i. **Development of new catalytic systems**: The exploration of novel catalysts, including biomimetic and organocatalysts, may provide new pathways for more efficient and selective coupling reactions. Hybrid systems that combine enzymatic and traditional catalysis could also be promising.
ii. **Advancements in click chemistry**: Further refining click chemistry methodologies could enhance their application in carbohydrate synthesis. Exploring new click reactions and optimizing conditions for existing reactions may lead to more versatile coupling strategies.
iii. **Automation and high-throughput techniques**: Incorporating automated platforms and high-throughput screening into carbohydrate synthesis can streamline the discovery of new coupling reactions and improve reaction conditions, ultimately accelerating research and development.
iv. **Integration of machine learning**: Utilizing machine learning and computational modeling to predict reaction outcomes and optimize conditions can revolutionize the design of coupling reactions. This approach could lead to more efficient synthetic routes and reduce the need for trial-and-error experimentation.
v. **Sustainable practices**: Developing greener methodologies that minimize waste and use environmentally friendly reagents and solvents will be increasingly important. The focus on sustainability will likely drive innovation in coupling reactions and their applications.

vi. **Interdisciplinary approaches**: Collaborations across fields, such as materials science, biochemistry, and medicinal chemistry, can lead to new applications and insights into the role of carbohydrates. Exploring how coupling reactions can be applied to areas like drug delivery and biomaterials will be vital for future research.

7.4 Innovations in coupling techniques

Recent advancements in coupling techniques, such as automated synthesizers and microfluidic platforms, are revolutionizing the synthesis of complex carbohydrates. These technologies allow for rapid and precise coupling reactions, opening new avenues for carbohydrate-based research and applications. The future of carbohydrate chemistry lies in the integration of coupling reactions with other chemical transformations. The development of hybrid methodologies that combine coupling with oxidation, reduction, or functionalization can provide powerful tools for synthesizing complex carbohydrate structures. The primary goal of this research is to innovate and improve coupling reactions in carbohydrate chemistry to enhance the efficiency, selectivity, and applicability in the synthesis of complex carbohydrate structures. The research approach involves three key stages:

A major focus will be on designing and synthesizing novel catalysts that offer improved reactivity, selectivity, and stability under mild reaction conditions. The exploration of both transition metal-based catalysts and biocatalysts (such as enzymes) will be prioritized. Additionally, reaction conditions (e.g., temperature, solvent, and concentration) will be optimized to improve yields while minimizing side reactions. The efficiency of coupling reactions in carbohydrate synthesis often relies on protecting group strategies to control regioselectivity and stereoselectivity. This research will explore innovative protecting group chemistries that are easier to remove, less reactive, and more selective, thus improving the overall synthetic process. The goal is to reduce the need for harsh deprotection steps and increase the overall sustainability of carbohydrate synthesis. Achieving high stereoselectivity and regioselectivity in glycosylation reactions remains a challenge. This project will develop new methods for controlling the formation of glycosidic bonds with high precision, using both classical and emerging approaches (such as bioorthogonal chemistry). Special attention will be paid to achieving selective couplings in complex structures like branched oligosaccharides and glycosylated natural products.

To ensure the practical relevance of the innovations, the developed coupling reactions will be applied to synthesize biologically active carbohydrate derivatives, glycoconjugates, and oligosaccharides. The project will explore their potential applications in drug development, particularly in areas such as vaccine design, glycosylation-based therapies, and targeted drug delivery.

7.5 Conclusion

Coupling reactions are at the heart of carbohydrate chemistry, enabling the construction of diverse and functionally relevant carbohydrate structures. As the field continues to evolve, the innovative application of coupling techniques will undoubtedly lead to new discoveries in biology and medicine. By deepening our understanding of these reactions, researchers can harness the power of carbohydrates to address pressing challenges in health and disease, paving the way for future advancements in this essential area of chemistry. Coupling reactions are fundamental to the field of carbohydrate chemistry, serving as essential tools for the construction of diverse, complex, and biologically relevant carbohydrate structures. These reactions allow for the efficient formation of glycosidic bonds, which are central to the synthesis of oligosaccharides, polysaccharides, and glycoconjugates. As the field continues to evolve, the development of innovative coupling techniques such as selective activation, high-yield methodologies, and regio- and stereoselective approaches holds great promise for advancing both basic and applied carbohydrate research. The ability to manipulate carbohydrate structures through coupling reactions opens new avenues for the creation of bioactive carbohydrate derivatives, which have important implications in various biological processes and therapeutic applications. By deepening our understanding of these reactions, researchers can not only enhance the design of carbohydrate-based drugs and vaccines but also explore their potential in developing biomaterials, diagnostics, and treatments for a range of diseases including cancer, diabetes, and infectious diseases.

As carbohydrate chemistry progresses, coupling reactions will undoubtedly play an increasingly vital role in addressing global challenges in health and disease. The continued refinement and application of these methodologies will drive future advancements in this essential area of chemistry, ultimately contributing to breakthroughs in biomedicine, biotechnology, and molecular medicine.

References

[1] Ramaprabha K, Kumar V, Saravanan P, Rajeshkannan R, Rajasimman M, Kamyab H, Vasseghian Y. Exploring the diverse applications of carbohydrate macromolecules in food, pharmaceutical, and environmental technologies. Environ Res, 2024, 240, 117521.

[2] Ernst B, Magnani JL. From carbohydrate leads to glycomimetic drugs. Nat Rev Drug Discov, 2009, 8, 661–677.

[3] Krasnova L, Wong CH. Oligosaccharide synthesis and translational innovation. J Am Chem Soc, 2019, 141, 3735–3754.

[4] Pan L, Cai C, Liu C, Liu D, Li G, Linhardt RJ, Yu G. Recent progress and advanced technology in carbohydrate-based drug development. Curr Opin Biotechnol, 2021, 69, 191–198.

[5] Zhang S, Chen KY, Zou X. Carbohydrate-protein interactions: Advances and challenges. Commun Inf Syst, 2021, 21, 147.

[6] Joseph AA, Pardo-Vargas A, Seeberger PH. Total synthesis of polysaccharides by automated glycan assembly. J Am Chem Soc, 2020, 142, 8561–8564.

[7] Ruiz-Castillo P, Buchwald SL. Applications of palladium-catalyzed C–N cross-coupling reactions. Chem Rev, 2016, 116, 12564–12649.

[8] Koeller KM, Wong CH. Synthesis of complex carbohydrates and glycoconjugates: Enzyme-based and programmable one-pot strategies. Chem Rev, 2000, 100, 4465–4494.

[9] Guo J, Ye XS. Protecting groups in carbohydrate chemistry: Influence on stereoselectivity of glycosylations. Molecules, 2010, 15, 7235–7265.

[10] Chabre YM, Roy R. Design and creativity in synthesis of multivalent neoglycoconjugates. Adv Carbohydr Chem Biochem, 2010, 63, 165–393.

[11] Chen A, Yang B, Zhou Z, Zhu F. Recent advances in transition-metal-catalyzed glycosyl cross-coupling reactions. Chem Catalysis, 2022, 2, 3430–3470.

[12] Sarkar B, Jayaraman N. Glycoconjugations of biomolecules by chemical methods. Front Chem, 2020, 8, 570185.

[13] Zhu F, Rodriguez J, Yang T, Kevlishvili I, Miller E, Yi D, O'Neill S, Rourke MJ, Liu P, Walczak MA. Glycosyl cross-coupling of anomeric nucleophiles: Scope, mechanism, and applications in the synthesis of aryl C-glycosides. J Am Chem Soc, 2017, 139, 17908–17922.

[14] Yoshimura Y, Wakamatsu H, Natori Y, Saito Y, Minakawa N. Glycosylation reactions mediated by hypervalent iodine: Application to the synthesis of nucleosides and carbohydrates. Beilstein J Org Chem, 2018, 14, 1595–1618.

[15] Fraser-Reid B, Jayaprakash KN, Lopez JC, Gomez AM, Uriel C. Protecting groups in carbohydrate chemistry profoundly influence all selectivities in glycosyl couplings. 2008, 1, 91–117.

[16] van der Vorm S, van Hengst JM, Bakker M, Overkleeft HS, van der Marel GA, Codée JD. Mapping the relationship between glycosyl acceptor reactivity and glycosylation stereoselectivity. Angewandte Chemie, 2018, 130, 8372–8376.

[17] Kaszás T, Baráth BÁ, Balázs B, Blága T, Juhász L, Somsák L, Tóth M. Coupling reactions of anhydro-aldose tosylhydrazones with boronic acids. Molecules, 2022, 27, 1795.

[18] Zhu F, O'Neill S, Rodriguez J, Walczak MA. Rethinking carbohydrate synthesis: Stereoretentive reactions of anomeric stannanes. Chem Eur J, 2019, 25, 3147–3155.

[19] Koeller KM, Wong CH. Synthesis of complex carbohydrates and glycoconjugates: Enzyme-based and programmable one-pot strategies. Chem Rev, 2000, 100, 4465–4494.

[20] Pandey RP, Tiwari B, Pandey AK, Hussain N. Recent advancements in glycosylation reactions: An access to privileged C- and S-glycosides. Curr Org Chem, 2024, 28, 343–358.

[21] Lin MH, Kuo YT, Danglad-Flores J, Sletten ET, Seeberger PH. Parametric analysis of donor activation for glycosylation reactions. Chem Eur J, 2024, 30, e202400479.

[22] Dorst KM, Engström O, d'Ortoli TA, Mobarak H, Ebrahemi A, Fagerberg U, Whitfield DM, Widmalm G. On the influence of solvent on the stereoselectivity of glycosylation reactions. Carbohydr Res, 2024, 535, 109010.

[23] Pavosevic F, Smith RL, Rubio A. Cavity click chemistry: Cavity-catalyzed azide–alkyne cycloaddition. J Phys Chem A, 2023, 127, 10184–10188.

[24] Aflak N, Bahsis L, Julve M, Stiriba SE. Copper-catalyzed azide-alkyne cycloaddition click reaction: Mechanistic and synthetic aspects. In Click Chemistry vol. 1. CRC Press, pp. 105–122.

[25] Wang M, Yuan Y, Liang G. "Click chemistry" for molecular imaging. Current Mol Imaging, 2012, 1, 87–95.

[26] Barik D, Thurakkal L, Jose A, Porel M. Click chemistry: A tool for functionalization. In Click Chemistry. CRC Press, 2024, pp. 123–150.

[27] Nekvasilová P, Hovorková M, Mészáros Z, Petrásková L, Pelantová H, Křen V, Slámová K, Bojarová P. Engineered glycosidases for the synthesis of analogs of human milk oligosaccharides. Int J Mol Sci, 2022, 23, 4106.

[28] Yu H, Chen X. One-pot multienzyme (OPME) systems for chemoenzymatic synthesis of carbohydrates. Org Biomol Chem, 2016, 14, 2809–2818.
[29] Geng S, Shi C, Guo B, Hou H, Liu Z, Feng Z. Recent progress in transition-metal-catalyzed reductive cross-coupling reactions using diboron reagents as reductants. ACS Catal, 2023, 13, 15469–15480.
[30] Kamboj M, Yadav N, Bajpai S. Recent advances in the transition metal (Cu, Ni, Fe, Co) catalyzed organic reaction: A mini-review. Curr Indian Sci, 2024, 2, 1–14.
[31] Juhász K, Magyar Á, Hell Z. Transition-metal-catalyzed cross-coupling reactions of Grignard reagents. Synthesis, 2021, 53, 983–1002.
[32] Zhu F, Yang T, Walczak MA. Glycosyl Stille cross-coupling with anomeric nucleophiles – A general solution to a long-standing problem of stereocontrolled synthesis of C-glycosides. Synlett, 2017, 28, 1510–1516.
[33] Rourke M. The synthesis of stannylated C1 nucleophilic monosaccharides for use in the stille coupling reaction, 2017.
[34] Molander GA, Biscoe MR. CC cross coupling using organometallic partners, 2013.
[35] Margetić D, Štrukil V. Chapter 2-carbon– carbon bond-forming reactions. In Mechanochemical Organic Synthesis. 2016, 55–139.
[36] Zhu F, O'Neill S, Rodriguez J, Walczak MA. Rethinking carbohydrate synthesis: Stereoretentive reactions of anomeric stannanes. Chem Eur J, 2019, 25, 3147–3155.
[37] Pekdemir B, Karav S. Exploring the diverse biological significance and roles of fucosylated oligosaccharides. Front Mol Biosci, 2024, 11, 1403727.
[38] Chen YJ, Sui X, Wang Y, Zhao ZH, Han TH, Liu YJ, Zhang JN, Zhou P, Yang K, Ye ZH. Preparation, structural characterization, biological activity, and nutritional applications of oligosaccharides. Food Chem: X, 2024, 22, 101289.
[39] Sarkar B, Jayaraman N. Glycoconjugations of biomolecules by chemical methods. Front Chem, 2020, 8, 570185.
[40] Kasdekar N, Walke G, Deshpande K, Hotha S. Glycosyl vinylogous carbonates as glycosyl donors by metal-free activation. J Org Chem, 2022, 87, 5472–5484.
[41] Westereng B, Kračun SK, Leivers S, Arntzen MØ, Aachmann FL, Eijsink VG. Synthesis of glycoconjugates utilizing the regioselectivity of a lytic polysaccharide monooxygenase. Sci Rep, 2020, 10, 13197.
[42] Oommen AM, Gill SK, Joshi L, Cunningham S. Carbohydrates and human glycosylation. In Translational Glycobiology in Human Health and Disease vol. 1. Academic Press, 2024, pp. 3–35.
[43] Sobral D, Francisco R, Duro L, Videira PA, Grosso AR. Concerted regulation of glycosylation factors sustains tissue identity and function. Biomedicines, 2022, 10, 1805.

Suchandra Bhattacharya and Basudeb Basu*

Chapter 8
Triazole-linked carbohydrates as promising bioactive scaffolds: Synthesis via click chemistry

8.1 Introduction

The structural diversity, biocompatibility, bioavailability, and hydrophilicity, with minimal toxicity of carbohydrates, an essential component of living cells, support the growing need to investigate carbohydrates as adaptable scaffolds for simple access to a variety of glycohybrids and well-defined glycoconjugates for in-depth chemical, pharmacological, and biochemical studies [1–3]. In almost every living species' life cycle, carbohydrates serve diverse roles [4–6]. Carbohydrates have a well-documented role in biological systems, where they are engaged in a number of basic biological processes including signal transduction, cell adhesion, cell–cell recognition, and interactions with external agents. Medicinal chemists are greatly stimulated by their optimal pharmacokinetic parameters, which are provided by their hydrophilicity, minimal toxicity, biocompatibility, and ring system (pyranose and furanose form), in addition to their notable structural diversity in terms of linkages, functional groups, chirality, and more [7, 8]. Carbohydrates can produce a variety of glycoconjugates, ranging in size from tiny molecules to giant biopolymers, through covalent binding to other sugars or even nonsugar biomolecules [7]. For a considerable time, carbohydrates have been acknowledged as a crucial element in the process of drug discovery and development [9]. In recent times, a significant amount of research has been dedicated to simple carbohydrates and their derivatives, including glycohybrids, glycoconjugates, and neoglycoconjugates [2, 3, 7]. These compounds have shown promising results in the treatment of several major diseases [7]. Consequently, their significance within the pharmaceutical industry continues to grow in today's context [10]. Owing to their immense possibilities as therapeutic options, diagnostic agents, and catalytic systems, carbohydrate-based scaffolds have gained extensive attention in the last few decades [4]. In this chapter, we are going to focus on triazole-based carbohydrate moieties. A few representative structures of carbohydrate moieties with triazole linkage that are important in the field of medicine are shown in Figures 8.1–8.4 [11–15].

***Corresponding author: Basudeb Basu**, Formerly Department of Chemistry, North Bengal University, Darjeeling 734013, West Bengal, India, e-mail: basudeb.basu@gmail.com
Suchandra Bhattacharya, Department of Chemistry, ABN Seal College, Cooch Behar 736101, West Bengal, India

https://doi.org/10.1515/9783111447094-008

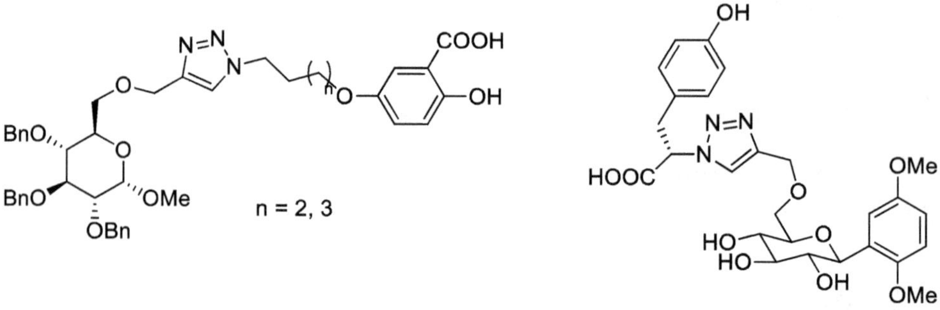

Figure 8.1: Triazolyl glycoconjugates as potential (PTP) 1B inhibitors.

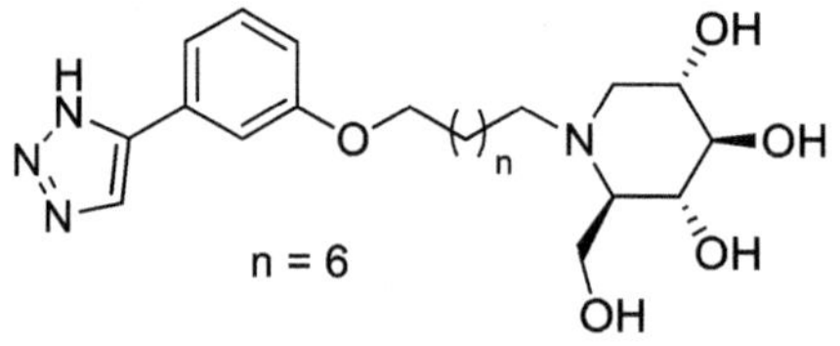

Figure 8.2: Inhibitor of angiogenesis.

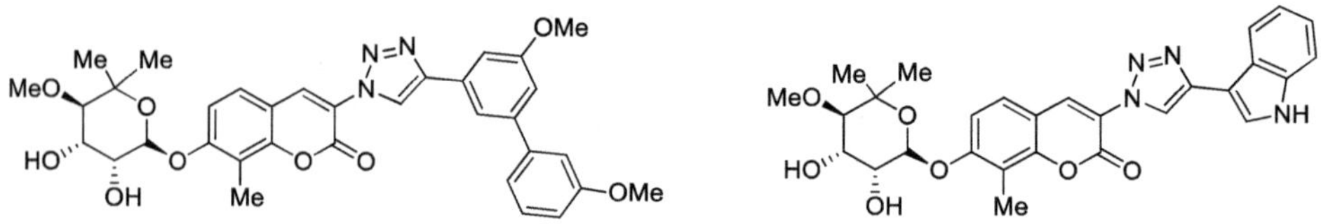

Figure 8.3: Anticancer drugs fabricated via triazole linkage.

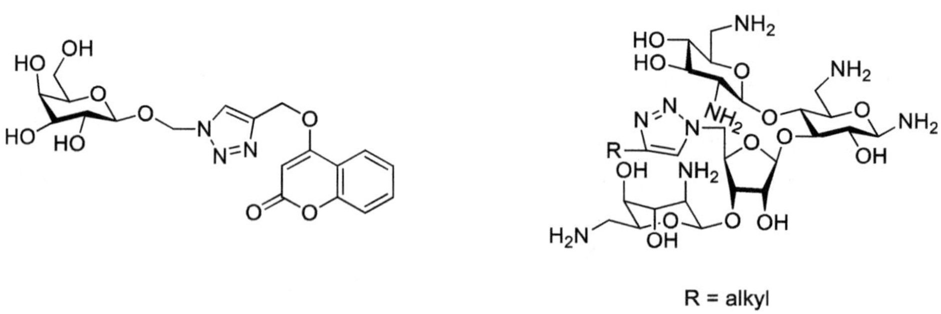

Figure 8.4: Antifungal and antibacterial triazolyl glycoconjugates.

Triazole formation via Cu catalysis (Cu-catalyzed azide-alkyne cycloaddition (CuAAC) click) is a powerful tool in chemistry following the pivotal discovery by Sharpless [16] and Meldal [17] in 2002. This method has had a profound impact across diverse scientific fields, offering new avenues for exploration and innovation. This article provides a quick update on recent advances in triazole-linked carbohydrates, with a focus on their synthesis via click chemistry.

8.2 Click reaction in carbohydrate chemistry

The first instance of a 1,3-dipolar cycloaddition reaction involving organic azides and terminal alkynes to produce 1,2,3-triazoles was reported by Michael more than a century ago in 1893 [18]. Later, Huisgen conducted additional in-depth research on this reaction in the 1960s [19, 20]. Huisgen's 1,3-dipolar thermal AAC process was thoroughly investigated and effectively employed to produce numerous triazole-linked hybrid compounds with diverse uses [21, 22]. Numerous drawbacks of Huisgen's approach include higher temperatures, longer reaction times, non-regioselective triazole synthesis (1,4- and 1,5-disubstituted triazoles), and challenging column chromatography [21]. The Cu(I) catalyst can generate good-to-quantitative quantities of 1,4-disubstituted triazoles as the sole products, as independently discovered in 2002 by Sharpless [16] and Meldal [17].

The CuAAC is a pivotal reaction in synthetic organic chemistry, allowing for the rapid creation of 1,2,3-triazole products through the catalysis of cuprous ions [23, 24]. This method, commonly known as the "click" reaction, efficiently joins carbohydrates to other biomolecules via a triazolyl ring, which has been established as a robust link. The CuAAC is currently considered a well-established and successfully executed protocol to acquire the desired multivalent carbohydrate-based molecules with wide-ranging applications owing to the comparative inertness of the triazole ring and the compatibility of click chemistry with carbohydrates [3]. Click chemistry has indeed made a significant impact across various scientific disciplines [25]. Its applications extend from drug discovery to materials science, aiding in DNA-sequencing technologies, and facilitating basic research into cell function and biomolecule discovery [3, 26–28]. The versatility and efficiency of the click reaction have revolutionized the way chemists connect organic groups, making it a ubiquitous tool in modern chemistry [29–32]. This approach remains a dominant topic in chemical research due to its broad applicability and ease of construction of complex molecules [33]. The copper(I)-promoted Huisgen-type AAC is a potent technique in novel approaches for synthesizing complex glycoconjugates. Researchers have utilized this reaction primarily to create glycosylated substrates for studying molecular recognition processes based on carbohydrates, and this reaction has become a potent strategy for conjugating 1,2,3-triazole-containing glycoconjugates of biomedical significance [1, 33–41]. Additionally, it has been instrumental in designing glycoconjugates aimed at preventing socially relevant diseases, including cancer [33]. While the primary impact of this reaction is observed in the field of organic synthesis, concerns regarding its applicability in living systems due to the toxic copper catalyst are being dispelled. This shift in perception is attributed to the successful live-cell labeling experiments conducted by Link and Tirrell [42], as well as the work of Bertozzi et al. [43].

8.3 Different triazole-linked carbohydrate moieties

Among the various nitrogen-containing heterocycles, triazoles and tetrazoles, which are five-membered rings with three nitrogen atoms, exhibit a plethora of biological activities [14, 44–48]. These compounds possess a basic aromatic heterocyclic structure and find applications in pharmacology, including anticancer, antitubercular, insecticidal, and antifungal properties. Additionally, they are key components in various therapeutically relevant drugs, such as histamine receptor blockers, CNS stimulants, antianxiety agents, and sedatives [49]. Owing to the high demand in the medical field, constant updates of drugs, drug delivery agents, and so on are needed. For instance, cancer treatment continues to exhibit significant side effects from anticancer drugs, harming both rapidly dividing healthy cells in the body and tumor sites. To address these challenges, scientists are emphasizing the creation of nanocarriers for efficient intracellular drug delivery [50, 51]. Numerous types of nanocarriers have been documented in scientific literature for targeted drug delivery. These include mesoporous nanocarriers, dendrimers, carbon nanotubes, gold nanoparticles, quantum dots, and nanogels [52]. Among the numerous bioactive scaffolds from the functionalization of carbohydrate units with triazoles, CuAAC-ensembled 1,2,3-triazole-linked nanogels are reported to be very efficient for targeted drug delivery [27]. Among the numerous applications of triazole in carbohydrates, a few important ones are their anticancer [53, 54], insecticidal [55], antibacterial [56], antifungal, and antitubercular activities.

The CuAAC reactions have potential applications in synthesizing bioactive carbohydrate-based ligands for lectins, antitumor vaccines, and substrates for enzymes like mannosyltransferases, fucosyltransferases, neuraminidase, glucosidases, and trans-sialidase [57]. These ligands are constructed using a common 1,4-disubstituted 1,2,3-triazole scaffold.

8.4 Synthetic approach toward triazole-linked carbohydrate units

The 1,3-dipolar cycloaddition reaction involving azides and alkynes, often referred to as "click chemistry," has rapidly become one of the most valuable and efficient organic reactions. Specifically, in the carbohydrate field, it serves as a potent synthetic tool for creating diverse carbohydrate-containing molecular structures including glycopeptides, glycopolymers, glycodendrimers, glycosylated biomolecules, and immobilized carbohydrates on solid surfaces (glycoarrays) [58, 59].

A plethora of research articles have reported the fabrication of triazole-linked glycopeptides [60], glycosylated arginine-glycine-aspartate (RGD) derivatives [61], triazolyl ethisterone glycoconjugates [56, 62], sugar-derived morpholine 1,2,3-triazoles and 1,4-oxazin-2-ones [63, 64], noscapine glycoconjugates [65, 66], divalent glycoamino acids with triazole linkage [67], furan-appended triazole glycoconjugates [68], Tröger's bases bearing glycosyl moieties [69], glycoconjugates of α-tocopherol [70], carbohydrate-lithocholic acid conjugates linked through 1,2,3-triazole rings [70], lipochitooligosaccharide analogues [71], coumarin–triazole–sugar hybrids [72, 73], ferrocene-carbohydrate conjugates [74], triazole glycoclusters [75–77], 1,2,3-triazole linked-furo[2,3-b]pyrazine nucleoside analogues [78], N^1-(β-D-ribofuranosyl)-C^4-(coumarin-4″-yl)-1,2,3-triazoles [79], azobenzene glycoconjugates [80], triazole-linked glycosylated 1,3,4-oxadiazoles [81], triazole-linked pseudo-starch fragments [82], macrocyclic carbohydrate hybrids [83–89], monomeric 12- to 17-membered triazolophanes fused to furanoside [90], carbohydrate-triazole-cyclophane hybrids [91], galactooligosaccharides [92], glycopolymers [93–98], *C*-glycosyl triazole α-amino acids [99], triazole-linked glycopeptides [100, 101], glycoporphyrins [102–104], and glycodendrimers [81, 105–111]. This gigantic repository of scaffolds containing carbohydrates allied with triazole units is still expanding day by day.

8.5 Recent updates on the synthesis of triazole-linked carbohydrate units

The ongoing research on the development of diverse triazole-linked carbohydrate units paves the way for a more detailed exploration of synthetic approaches. A few recently updated syntheses are discussed below.

Novel cinchonidine glycoconjugates (3) were synthesized by Tiwari et al. [112] through a CuAAC reaction of 9-epi-9-azido-9-deoxycinchonidine (1) with different glycosyl *O*-propargyl ethers (2) with antimalarial properties (Figure 8.5).

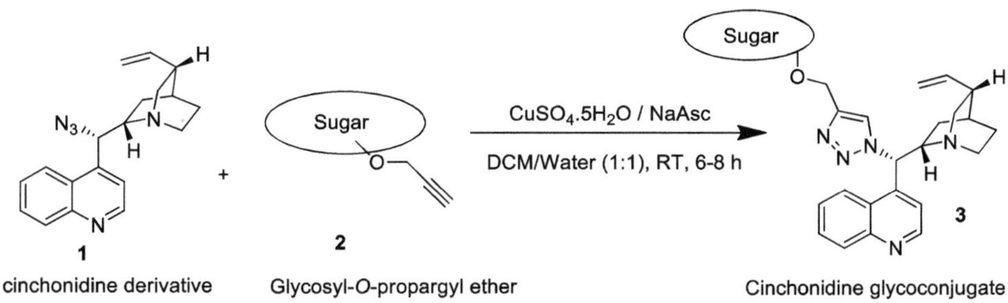

Figure 8.5: Synthesis of cinchonidine glycoconjugates via a Cu-catalyzed azide-alkyne cycloaddition.

Prasad and coworkers reported the synthesis of a series of eight *N*[1]-(β-D-ribofuranosyl)-*C*[4]-(coumarin-4″-yl)-1,2,3-triazoles (6) by Cu(I)-catalyzed click reaction of differentlysubstituted 4-ethynylcoumarins (5) with 1-azido-1-deoxy-2,3,5-tri-*O*-benzoyl-β-D-ribofuranose (4) followed by debenzoylation of the resulted *N*[1]-(2′,3′,5′-tri-*O*-benzoyl-β-D-ribofuranosyl)-*C*[4]-(coumarin-4″-yl)-1,2,3-triazoles (Figure 8.6) [79].

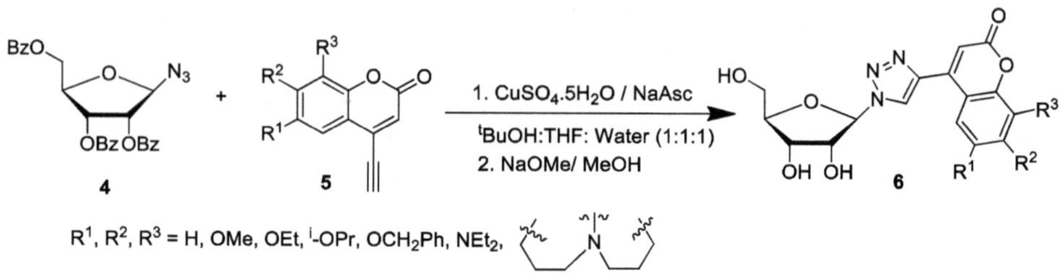

Figure 8.6: Fabrication of *N*[1]-(2′,3′,5′-tri-*O*-benzoyl-β-D-ribofuranosyl)-*C*[4]-(coumarin-4‴-yl)-1,2,3-triazoles.

The compounds having C2-symmetric chiral tetracyclic molecules and having a methano-1,5-diazocine ring system arranged in a special V-shaped twisted configuration are known as Tröger's bases. Rodembusch et al. demonstrated new Tröger's bases bearing glycosyl moieties linked by the 1,2,3-triazole core obtained from a copper-catalyzed azide-alkyne cycloaddition reaction [69]. Through the cycloaddition of a Tröger's base-alkyne and carbohydrate-derived azide, the 1,2,3-triazole has effectively been coupled with the Tröger's base scaffold and carbohydrate counterpart.

Reissig et al. [113] started from an enantiopure 3-amino-substituted pyran derivative (7) to furnish a series of divalent 1,2,3-triazole-linked carbohydrate mimetics (8). The process encompasses the facile preparation of the required 3-azido-substituted pyran using Cu-catalyzed diazo transfer. Notably, the in situ generation of the 3-azidopyran in the presence of Cu/C as a catalyst, followed by the reaction with the alkyne, enables a direct one-pot transformation to the desired click products (Figure 8.7).

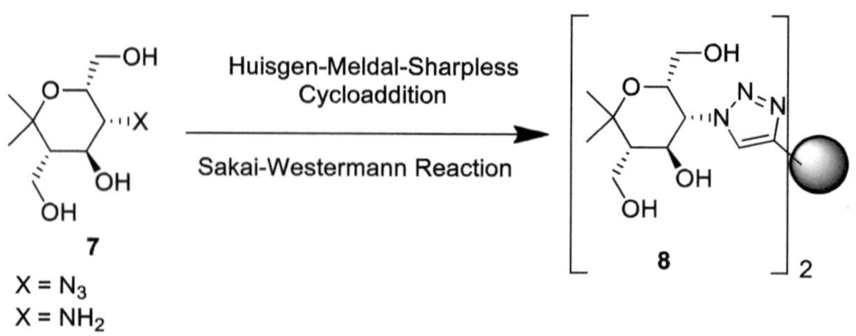

Figure 8.7: Synthesis of a series of divalent 1,2,3-triazole-linked carbohydrate mimetics.

In carbohydrate chemistry, introducing a sugar moiety with alkyne or azide functionality is somewhat straightforward. The protocols give access to both mono- and polyfunctionalized derivatives through the CuAAC reaction. This convenient approach allows the rapid synthesis of carbohydrate conjugates, where the heterocyclic triazolyl ring acts as a shackle between the carbohydrate moiety and another biomolecule. In recent years, the practice of combining two molecular components, each possessing distinct properties, to create a new conjugate with collective attributes from the parent entities has gained prominence as an emerging technology [70, 114]. The synthesis of triazole-based glycohybrids and their biological activities using CuAAC click chemistry has been summarized by Sagar et al. [115]. Glycohybrids, also known as carbohybrids, refer to a class of hybrid molecules where carbohydrate moieties are combined, fused, or linked with diverse natural product scaffolds. These bioactive natural product scaffolds, when attached to carbohydrate motifs, offer additional benefits for ADME (absorption, distribution, metabolism, and excretion) processes.

Wang and coworkers reported their work based on CuAAC as the pivotal cyclization step, synthesizing a series of eight *N*-acetyl-D-glucosamine-based macrocycles (9) with 18–20 members in their rings with good yields (Figure 8.9) [88]. The general structure of the bis-triazole-linked carbohydrate-based macrocycles is shown in Figure 8.8. The findings from this study demonstrated that the macrocycles produced here serve as effective ligands and also boost the rate of copper sulfate-mediated azide and alkyne cycloaddition reactions.

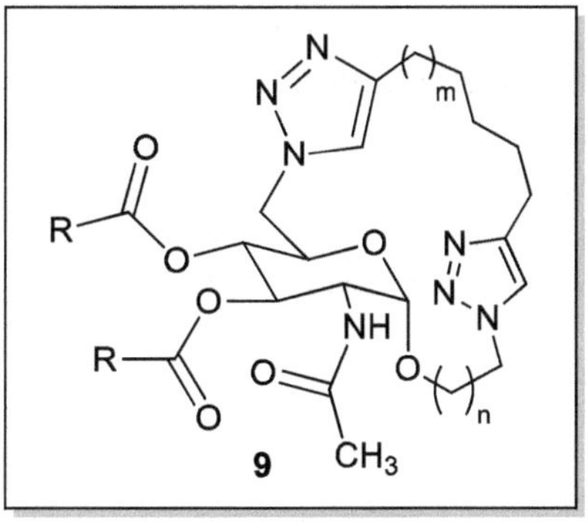

Figure 8.8: General structure of the macrocycles.

Sagar et al. [116] demonstrated the microwave-assisted synthesis of pyrazolo[1,5-a]pyrimidine-based triazole-linked glycohybrids (12) catalyzed by a copper catalyst from (10) and (11) (Figure 8.10).

The Ru-catalyzed AAC (RuAAC) reactions are less common in the field of glycoscience than their Cu-catalyzed counterparts, the latter being widely explored owing to its diverse applications. However, there are reports that provide the progress of ruthenium-catalyzed AAC reactions [117–119]. Calenbergh et al. [120] demonstrated that the intramolecular formation of 1,2,3-triazoles using a ruthenium catalyst ensued

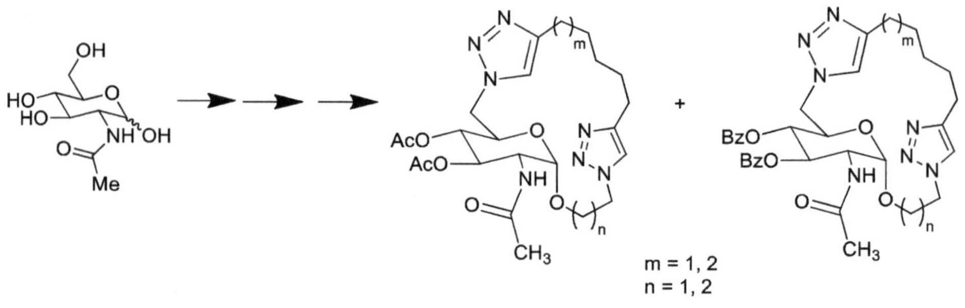

Figure 8.9: Sugar-derived macrocycle synthesis by click reaction.

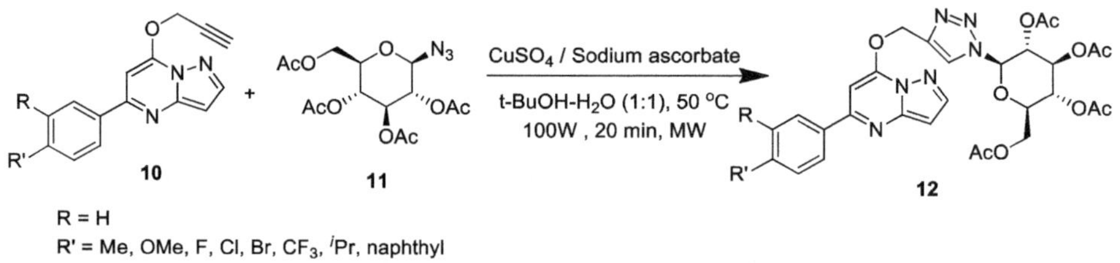

Figure 8.10: Synthesis of pyrazolo[1,5-a]pyrimidine-based triazole-linked glycohybrids.

in 1,4-dioxane with Cp*(COD)RuCl, but when the related catalyst Cp*(PPh3)2RuCl was employed, the reaction progressed slowly and resulted in poor yield. One major advantage of RuAAC is to provide selective formation of the 1,5-disubstituted 1,2,3-triazoles. The RuAAC also exhibits high compatibility with internal alkynes, which paves the way to obtain 1,4,5-trisubstituted triazole derivatives. Besides these recent advances, a wealth of literature exists that provides an in-depth analysis of "click chemistry," including its discovery, generalization, and expanding applications as well as a detailed analysis of the postfunctionalization of click products [121].

8.6 Conclusion

This chapter primarily emphasized the role of "click chemistry" in the domain of carbohydrate chemistry, highlighting the latest developments in the efficient synthesis using CuAAC toward various functional carbohydrate molecules that lead to the creation of novel therapeutic and diagnostic instruments. A groundbreaking advancement in synthetic and mechanistic chemistry occurred with the proposal of the 1,3-dipolar cycloaddition by A. Michael. Over time, persistent efforts to refine its methods and applications paved the way for further developments. One significant outcome was the emergence of Cu(I)-catalyzed 1,3-dipolar cycloaddition (often referred to as click chemistry or CuAAC). This modular approach not only provided a reliable pathway for synthesizing molecules, ranging from simple to complex but also demonstrated remarkable efficacy and regioselectivity. The CuAAC affords 1,4-disubstituted 1,2,3-triazole, which can be used as a strong bridge to improve bioactivity between a carbohydrate and a functional molecule or even as a pharmacophore in specific situations. The world of carbohydrate chemistry has been enriched by the Cu-catalyzed click reaction through the development of numerous fantastic glycoconjugates, having a wide spectrum of applications across diverse scientific disciplines like material science, medicinal chemistry, chemical biology, and catalysis. The emerging trend of publications in this field only signifies the wide range of uses and implications of this methodology in the field of drug development. Further investigation into its application in carbohydrate chemistry may uncover novel avenues for the creation of glycoconjugates with potential medicinal applications. There is a lot that the click technique can offer the realm of carbohydrate chemistry. This method could be used to create a variety of unlikely glycohybrids, glycoconjugates, and neoglycoconjugates with a broad range of uses, opening up new research directions.

References

[1] He X-P, Zeng Y-L, Zang Y, Li J, Field RA, Chen G-R. Carbohydrate CuAAC click chemistry for therapy and diagnosis. Carbohydr Res, 2016, 429, 1–22. https://doi.org/10.1016/j.carres.2016.03.022.

[2] Kushwaha D, Dwivedi P, Kuanar K, Saroj K, Tiwari V. Click reaction in carbohydrate chemistry: Recent developments and future perspective. Curr Organic Synth, 2013, 10(1), 90–135. http://dx.doi.org/10.2174/1570179411310010005.

[3] Tiwari VK, Mishra BB, Mishra KB, Mishra N, Singh AS, Chen X. Cu-catalyzed click reaction in carbohydrate chemistry. Chem Rev, 2016, 116(5), 3086–3240. https://doi.org/10.1021/acs.chemrev.5b00408.

[4] Varki A. Biological roles of oligosaccharides: All of the theories are correct. Glycobiology, 1993, 3(2), 97–130. https://doi.org/10.1093/glycob/3.2.97.

[5] Lis H, Sharon N. Lectins: Carbohydrate-specific proteins that mediate cellular recognition. Chem Rev, 1998, 98(2), 637–674. https://doi.org/10.1021/cr940413g.

[6] Bertozzi CR, Kiessling LL. Chemical glycobiology. Science, 2001, 291(5512), 2357–2364. https://doi.org/10.1126/science.1059820.

[7] Tiwari VK. Development of diverse range of biologically relevant carbohydrate-containing molecules: Twenty years of our journey. Chem Rec, 2021, 21(11), 3029–3048. https://doi.org/10.1002/tcr.202100058.

[8] Tiwari VK. Carbohydrates in Drug Discovery and Development: Synthesis and Application. Elsevier: Amsterdam, 2020. https://shop.elsevier.com/books/carbohydrates-in-drug-discovery-and-development/tiwari/978-0-12-816675-8.

[9] Agrahari AK, Bose P, Jaiswal MK, Rajkhowa S, Singh AS, Hotha S, Mishra N, Tiwari VK. Cu(I)-catalyzed click chemistry in glycoscience and their diverse applications. Chem Rev, 2021, 121(13), 7638–7956.

[10] Somesh S, Madhuri K. Triazolyl glycoconjugates and their impact in medicinal chemistry. Org Chem Indian J, 2016, 12(4), 104. https://www.tsijournals.com/articles/triazolyl-glycoconjugates-and-their-impact-in-medicinal-chemistry.html.

[11] Li C, He X-P, Zhang Y-J, Li Z, Gao L-X, Shi -X-X, Xie J, Li J, Chen G-R, Tang Y. Click to a focused library of benzyl 6-triazolo(hydroxy)benzoic glucosides: Novel construction of PTP1B inhibitors on a sugar scaffold. Eur J Med Chem, 2011, 46(9), 4212–4218. https://doi.org/10.1016/j.ejmech.2011.06.025.

[12] Poláková M, Stanton R, Wilson IBH, Holková I, Šesták S, Machová E, Jandová Z, Kóňa J. 'Click chemistry' synthesis of 1-(α-D-Mannopyranosyl)-1,2,3-triazoles for inhibition of α-mannosidases. Carbohydr Res, 2015, 406, 34–40. https://doi.org/10.1016/j.carres.2015.01.004.

[13] Peterson LB, Blagg BSJ. Click chemistry to probe Hsp90: Synthesis and evaluation of a series of triazole-containing novobiocin analogues. Bioorg Med Chem Lett, 2010, 20(13), 3957–3960. https://doi.org/10.1016/j.bmcl.2010.04.140.

[14] Zhang J, Chiang F-I, Wu L, Czyryca PG, Li D, Chang C-WT. Surprising alteration of antibacterial activity of 5″-modified neomycin against resistant bacteria. J Med Chem, 2008, 51(23), 7563–7573. https://doi.org/10.1021/jm800997s.

[15] Li J, Zheng M, Tang W, He P-L, Zhu W, Li T, Zuo J-P, Liu H, Jiang H. Syntheses of triazole-modified Zanamivir analogues via click chemistry and anti-AIV activities. Bioorg Med Chem Lett, 2006, 16(19), 5009–5013. https://doi.org/10.1016/j.bmcl.2006.07.047.

[16] Rostovtsev VV, Green LG, Fokin VV, Sharpless KB. A stepwise Huisgen cycloaddition process: Copper (I)-catalyzed regioselective "Ligation" of azides and terminal alkynes. Angew Chem Int Ed, 2002, 41(14), 2596–2599. https://doi.org/10.1002/1521-3773(20020715)41:14<2596::AID-ANIE2596>3.0.CO;2-4.

[17] Tornøe CW, Christensen C, Meldal M. Peptidotriazoles on solid phase: [1,2,3]-Triazoles by regiospecific copper(I)-catalyzed 1,3-dipolar cycloadditions of terminal alkynes to azides. J Org Chem, 2002, 67(9), 3057–3064. https://doi.org/10.1021/jo011148j.

[18] Michael A. Ueber Die Einwirkung von Diazobenzolimid Auf Acetylendicarbonsäuremethylester. J Für Praktische Chemie, 1893, 48(1), 94–95. https://doi.org/10.1002/prac.18930480114.

[19] Huisgen R. 1,3-dipolar cycloadditions. Past and future. Angew Chem Int Ed Engl, 1963, 2(10), 565–598. https://doi.org/10.1002/anie.196305651.

[20] Breugst M, Reissig H-U. The Huisgen reaction: Milestones of the 1,3-dipolar cycloaddition. Angew Chem Int Ed, 2020, 59(30), 12293–12307. https://doi.org/10.1002/anie.202003115.

[21] Huisgen R, Szeimies G, Möbius L. 1.3-Dipolare Cycloadditionen, XXXII. Kinetik der AdditionenorganischerAzide an CC-Mehrfachbindungen. Chemische Berichte, 1967, 100(8), 2494–2507. https://doi.org/10.1002/cber.19671000806.

[22] Huisgen R. Kinetics and mechanism of 1,3-dipolar cycloadditions. Angew Chem Int Ed Engl, 1963, 2(11), 633–645. https://doi.org/10.1002/anie.196306331.

[23] Manoharan A, Jayan J, Rangarajan TM, Bose K, Benny F, Ipe RS, Kumar S, Kukreti N, Abdelgawad MA, Ghoneim MM, Kim H, Mathew B. "Click chemistry": An emerging tool for developing a new class of structural motifs against various neurodegenerative disorders. ACS Omega, 2023, 8(47), 44437–44457. https://doi.org/10.1021/acsomega.3c04960.

[24] K. Barry sharpless, biography, nobel prize, & facts, britannica. https://www.britannica.com/biography/Karl-Barry-Sharpless (accessed 2024-06-20).

[25] Vaishnani MJ, Bijani S, Rahamathulla M, Baldaniya L, Jain V, Thajudeen KY, Ahmed MM, Farhana SA, Pasha I. Biological importance and synthesis of 1,2,3-triazole derivatives: A review. Green Chem Lett Rev, 2024, 17(1), 2307989. https://doi.org/10.1080/17518253.2024.2307989.

[26] Chaturvedi P, Chaturvedi N, Gupta S, Mishra A, Singh M, Siddhartha T. Click chemistry: A new approach for drug discovery. 10(2).

[27] Singh G, Majeed A, Singh R, George N, Singh G, Gupta S, Singh H, Kaur G, Singh J. CuAAC ensembled 1,2,3-triazole linked nanogels for targeted drug delivery: A review. RSC Adv, 2023, 13(5), 2912–2936. https://doi.org/10.1039/D2RA05592A.

[28] Nallapati SB, Sreenivas BY, Bankala R, Parsa KVL, Sripelly S, Mukkanti K, Pal M. 1,2,3-triazoles derived from olanzapine: Their synthesis via an ultrasound assisted CuAAC method and evaluation as inhibitors of PDE4B. RSC Adv, 2015, 5(115), 94623–94628. https://doi.org/10.1039/C5RA20380E.

[29] Héron J, Balcells D. Concerted cycloaddition mechanism in the CuAAC reaction catalyzed by 1,8-naphthyridine dicopper complexes. ACS Catal, 2022, 12(8), 4744–4753. https://doi.org/10.1021/acscatal.2c00723.

[30] Yang C, Tripathi R, Wang B. Click chemistry in the development of PROTACs. RSC Chem Biol, 2024, 5(3), 189–197. https://doi.org/10.1039/D3CB00199G.

[31] Devaraj NK, Finn MG. Introduction: Click chemistry. Chem Rev, 2021, 121(12), 6697–6698. https://doi.org/10.1021/acs.chemrev.1c00469.

[32] Kolb HC, Finn MG, Sharpless KB. Click chemistry: Diverse chemical function from a few good reactions. Angew Chem Int Ed, 2001, 40(11), 2004–2021. https://doi.org/10.1002/1521-3773(20010601)40:11<2004::AID-ANIE2004>3.0.CO;2-5.

[33] Dondoni A. Triazole: The keystone in glycosylated molecular architectures constructed by a click reaction. Chem Asian J, 2007, 2(6), 700–708. https://doi.org/10.1002/asia.200700015.

[34] Hein JE, Fokin VV. Copper-catalyzedazide–alkyne cycloaddition (CuAAC) and beyond: New reactivity of copper(I) acetylides. Chem Soc Rev, 2010, 39(4), 1302–1315. https://doi.org/10.1039/B904091A.

[35] Dondoni A, Marra A. Calixarene and calixresorcarene glycosides: Their synthesis and biological applications. Chem Rev, 2010, 110(9), 4949–4977. https://doi.org/10.1021/cr100027b.

[36] Holub JM, Kirshenbaum K. Tricks with clicks: Modification of peptidomimetic oligomers via copper-catalyzedazide-alkyne [3+2] cycloaddition. Chem Soc Rev, 2010, 39(4), 1325–1337. https://doi.org/10.1039/B901977B.

[37] Tron GC, Pirali T, Billington RA, Canonico PL, Sorba G, Genazzani AA. Click chemistry reactions in medicinal chemistry: Applications of the 1,3-dipolar cycloaddition between azides and alkynes. Med Res Rev, 2008, 28(2), 278–308. https://doi.org/10.1002/med.20107.

[38] Kappe CO, Van der Eycken E. Click chemistry under non-classical reaction conditions. Chem Soc Rev, 2010, 39(4), 1280–1290. https://doi.org/10.1039/B901973C.

[39] Aragão-Leoneti V, Campo VL, Gomes AS, Field RA, Carvalho I. Application of copper(I)-catalysed azide/alkyne cycloaddition (CuAAC) 'Click Chemistry' in carbohydrate drug and neoglycopolymer synthesis. Tetrahedron, 2010, 66(49), 9475–9492. https://doi.org/10.1016/j.tet.2010.10.001.

[40] Kushwaha D, Dwivedi P, Kuanar SK, Tiwari VK, Click Reaction in Carbohydrate Chemistry: Recent Developments and Future Perspective+. Curr Org Synth, 2013, 10(1), 90-135. https://doi.org/10.2174/1570179411310010005.

[41] Wilkinson B, Bornaghi LF, Poulsen S-A, Houston TA. Synthetic utility of glycosyl triazoles in carbohydrate chemistry. 2006. https://doi.org/10.1016/j.tet.2006.06.001.

[42] Link AJ, Tirrell DA. Cell surface labeling of *Escherichia coli* via copper(I)-catalyzed [3+2] cycloaddition. J Am Chem Soc, 2003, 125(37), 11164–11165. https://doi.org/10.1021/ja036765z.

[43] Agard NJ, Prescher JA, Bertozzi CR. A strain-promoted [3+2] azide–alkyne cycloaddition for covalent modification of biomolecules in living systems. J Am Chem Soc, 2004, 126(46), 15046–15047. https://doi.org/10.1021/ja044996f.

[44] Serafini M, Pirali T, Tron GC. Click 1,2,3-triazoles in drug discovery and development: From the flask to the clinic? In Advances in Heterocyclic Chemistry vol. 134. Elsevier, 2021, pp. 101–148. https://doi.org/10.1016/bs.aihch.2020.10.001.

[45] Horne WS, Yadav MK, Stout CD, Ghadiri MR. Heterocyclic peptide backbone modifications in an α-helical coiled coil. J Am Chem Soc, 2004, 126(47), 15366–15367. https://doi.org/10.1021/ja0450408.

[46] Bijani S, Iqbal D, Mirza S, Jain V, Jahan S, Alsaweed M, Madkhali Y, Alsagaby SA, Banawas S, Algarni A, Alrumaihi F, Rawal RM, Alturaiki W, Shah A. Green synthesis and anticancer potential of 1,4-dihydropyridines-based triazole derivatives: In silico and in vitro study. Life, 2022, 12(4), 519. https://doi.org/10.3390/life12040519.

[47] Singh P, Kumar V. Special issue "Hybrid drugs: Design and applications". Pharmaceuticals, 2023, 16(10), 1358. https://doi.org/10.3390/ph16101358.

[48] Dheer D, Singh V, Shankar R. Medicinal attributes of 1,2,3-triazoles: Current developments. Bioorganic Chem, 2017, 71, 30–54. https://doi.org/10.1016/j.bioorg.2017.01.010.

[49] Sahu JK, Ganguly S, Kaushik A. Triazoles: A valuable insight into recent developments and biological activities. Chin J Nat Med, 2013, 11(5), 456–465. https://doi.org/10.1016/S1875-5364(13)60084-9.

[50] Tong R, Tang L, Ma L, Tu C, Baumgartner R, Cheng J. Smart chemistry in polymeric nanomedicine. Chem Soc Rev, 2014, 43(20), 6982–7012. https://doi.org/10.1039/C4CS00133H.

[51] Rösler A, Vandermeulen GW, Klok HA. Advanced drug delivery devices via self-assembly of amphiphilic block copolymers. Adv Drug Deliv Rev, 2001, 53(1), 95–108. https://doi.org/10.1016/s0169-409x(01)00222-8.

[52] Yaqoob AA, Ahmad H, Parveen T, Ahmad A, Oves M, Ismail IMI, Qari HA, Umar K, Mohamad Ibrahim MN. Recent advances in metal decorated nanomaterials and their various biological applications: A review. Front Chem, 2020, 8. https://doi.org/10.3389/fchem.2020.00341.

[53] Mishra BB, Tiwari VK. Natural products: An evolving role in future drug discovery. Eur J Med Chem, 2011, 46(10), 4769–4807. https://doi.org/10.1016/j.ejmech.2011.07.057.

[54] Dos Anjos JV, Neves Filho RAW, Do Nascimento SC, Srivastava RM, De Melo SJ, Sinou D. Synthesis and cytotoxic profile of glycosyl–triazole linked to 1,2,4-oxadiazole moiety at C-5 through a straight-chain carbon and oxygen atoms. Eur J Med Chem, 2009, 44(9), 3571–3576. https://doi.org/10.1016/j.ejmech.2009.03.002.

[55] Yang W, Wu H-X, Xu -H-H, Hu A-L, Lu M-L. Synthesis of glucose–fipronil conjugate and its phloem mobility. J Agric Food Chem, 2011, 59(23), 12534–12542. https://doi.org/10.1021/jf2031154.

[56] Singh BK, Yadav AK, Kumar B, Gaikwad A, Sinha SK, Chaturvedi V, Tripathi RP. Preparation and reactions of sugar azides with alkynes: Synthesis of sugar triazoles as antitubercular agents. Carbohydr Res, 2008, 343(7), 1153–1162. https://doi.org/10.1016/j.carres.2008.02.013.

[57] Campo VL, Carvalho I. Click chemistry applied to carbohydrate-based drug discovery. In Click Chemistry in Glycoscience. John Wiley & Sons, Ltd, 2013, pp. 325–357. https://doi.org/10.1002/9781118526996.ch13.

[58] Santoyo-González F, Hernández-Mateo F. Azide–alkyne 1,3-dipolar cycloadditions: A valuable tool in carbohydrate chemistry. In Heterocycles from Carbohydrate Precursors. El Ashry ESH. (Ed.). Springer: Berlin, Heidelberg, 2007, pp. 133–177. https://doi.org/10.1007/7081_2007_050.

[59] Zhang -L-L, Wu Q-P, Xu Z-B Regioselective Synthesis of Diverse Thio-, Seleno-substituted 1,2,3-Triazoles, Curr Org Chem, 2019. 23(22), 2379–2401. https://doi.org/10.2174/1385272823666191108101604.

[60] Kuijpers BHM, Groothuys S, Keereweer ABR, Quaedflieg PJLM, Blaauw RH, van Delft FL, Rutjes FPJT. Expedient synthesis of triazole-linked glycosyl amino acids and peptides. Org Lett, 2004, 6(18), 3123–3126. https://doi.org/10.1021/ol048841o.

[61] Kuijpers BHM, Groothuys S, Soede AC, Laverman P, Boerman OC, van Delft FL, Rutjes FPJT. Preparation and evaluation of glycosylated arginine–glycine–aspartate (RGD) derivatives for integrin targeting. Bioconjugate Chem, 2007, 18(6), 1847–1854. https://doi.org/10.1021/bc700154u.

[62] Mishra KB, Mishra BB, Tiwari VK. Efficient synthesis of ethisterone glycoconjugate via bis-triazole linkage. Carbohydr Res, 2014, 399, 2–7. https://doi.org/10.1016/j.carres.2014.09.001.

[63] Reddy YS, John Pal AP, Gupta P, Ansari AA, Vankar YD. Ceric ammonium nitrate-catalyzedazidation of 1,2-anhydro sugars: Application in the synthesis of structurally diverse sugar-derived morpholine 1,2,3-triazoles and 1,4-oxazin-2-ones. J Org Chem, 2011, 76(15), 5972–5984. https://doi.org/10.1021/jo200260w.

[64] Mishra KB, Tiwari VK. Click chemistry inspired synthesis of morpholine-fused triazoles. J Org Chem, 2014, 79(12), 5752–5762. https://doi.org/10.1021/jo500890w.

[65] Mishra KB, Mishra RC, Tiwari VK. First noscapine glycoconjugates inspired by click chemistry. RSC Adv, 2015, 5(64), 51779–51789. https://doi.org/10.1039/C5RA07321A.

[66] Mishra KB, Shashi S, Tiwari VK. Metal free synthesis of morpholine fused [5,1-c] triazolyl glycoconjugates via glycosyl azido alcohols. RSC Adv, 2015, 5(105), 86840–86848. https://doi.org/10.1039/C5RA17181D.

[67] Sahoo L, Singhamahapatra A, Kumar K, Loganathan D. Synthesis of divalent glycoamino acids with bis-triazole linkage. Carbohydr Res, 2013, 381, 51–58. https://doi.org/10.1016/j.carres.2013.08.006.

[68] Yousuf SK, Taneja SC, Mukherjee D. Multicomponent cascade transformation of D-glucal to furan-appended triazole glycoconjugates. J Org Chem, 2010, 75(9), 3097–3100. https://doi.org/10.1021/jo100475e.

[69] Aroche DMP, Vargas JP, Nogara PA, Da Silveira Santos F, Da Rocha JBT, Lüdtke DS, Rodembusch FS. Glycoconjugates based on supramolecular Tröger's base scaffold: Synthesis, photophysics, docking, and BSA association study. ACS Omega, 2019, 4(8), 13509–13519. https://doi.org/10.1021/acsomega.9b01857.

[70] Virta P, Katajisto J, Niittymäki T, Lönnberg H. Solid-supported synthesis of oligomeric bioconjugates. Tetrahedron, 2003, 59(28), 5137–5174. https://doi.org/10.1016/S0040-4020(03)00704-X.

[71] Berthelot N, Brossay A, Gasciolli V, Bono -J-J, Baron A, Beau J-M, Urban D, Boyer F-D, Vauzeilles B. Synthesis of lipo-chitooligosaccharide analogues and their interaction with LYR3, a high affinity binding protein for nod factors and myc-LCOs. Org Biomol Chem, 2017, 15(37), 7802–7812. https://doi.org/10.1039/C7OB01201B.

[72] Kumar KK, Kumar RM, Subramanian V, Das TM. Expedient synthesis of coumarin-coupled triazoles via 'Click Chemistry' leading to the formation of coumarin–triazole–sugar hybrids. Carbohydr Res, 2010, 345(16), 2297–2304. https://doi.org/10.1016/j.carres.2010.07.037.

[73] Rajaganesh R, Ravinder P, Subramanian V, Mohan Das T. FACE-selective fluorogenic cycloaddition reaction between coumarin azides and sugar terminal alkynes: An experimental and computational study. Carbohydr Res, 2011, 346(15), 2327–2336. https://doi.org/10.1016/j.carres.2011.07.024.

[74] Casas-Solvas JM, Ortiz-Salmerón E, Giménez-Martínez JJ, García-Fuentes L, Capitán-Vallvey LF, Santoyo-González F, Vargas-Berenguel A. Ferrocene–carbohydrate conjugates as electrochemical probes for molecular recognition studies. Chem Eur J, 2009, 15(3), 710–725. https://doi.org/10.1002/chem.200800927.

[75] Dondoni A, Marra A. C-Glycoside clustering on calix[4]arene, adamantane, and benzene scaffolds through 1,2,3-triazole linkers. J Org Chem, 2006, 71(20), 7546–7557. https://doi.org/10.1021/jo0607156.

[76] Vecchi A, Melai B, Marra A, Chiappe C, Dondoni A. Microwave-enhanced ionothermal CuAAC for the synthesis of glycoclusters on a calix[4]arene platform. J Org Chem, 2008, 73(16), 6437–6440. https://doi.org/10.1021/jo800954z.

[77] Bew SP, Brimage RA, L'Hermit N, Sharma SV. Upper rim appended hybrid calixarenes via click chemistry. Org Lett, 2007, 9(19), 3713–3716. https://doi.org/10.1021/ol071047t.

[78] Ermolat'ev DS, Mehta VP, Van der Eycken EV. Synthesis of furo[2,3-b]pyrazine nucleoside analogues with 1,2,3-triazole linkage. QSAR Comb Sci, 2007, 26(11–12), 1266–1273. https://doi.org/10.1002/qsar.200740123.

[79] Maikhuri VK, Bohra K, Srivastava S, Kavita, Prasad AK. Click synthesis of N1-(β-D-Ribofuranosyl)-C4-(Coumarin-4″-Yl)-1,2,3-triazoles. Synth Commun, 2019, 49(22), 3140–3147. https://doi.org/10.1080/00397911.2019.1657150.

[80] Chandrasekaran V, Lindhorst TK. Sweet switches: Azobenzene glycoconjugates synthesized by click chemistry. Chem Commun, 2012, 48(60), 7519–7521. https://doi.org/10.1039/C2CC33542E.

[81] Kushwaha D, Tiwari VK. Click inspired synthesis of 1,2,3-triazole-linked 1,3,4-oxadiazole glycoconjugates. J Heterocycl Chem, 2017, 54(4), 2454–2462. https://doi.org/10.1002/jhet.2845.

[82] Nepogodiev SA, Dedola S, Marmuse L, De Oliveira MT, Field RA. Synthesis of triazole-linked pseudo-starch fragments. Carbohydr Res, 2007, 342(3–4), 529–540. https://doi.org/10.1016/j.carres.2006.09.026.

[83] McGavin RS, Gagne RA, Chervenak MC, Bundle DR. The design, synthesis and evaluation of high affinity macrocyclic carbohydrate inhibitors. Org Biomol Chem, 2005, 3(15), 2723–2732. https://doi.org/10.1039/B416105J.

[84] Bodine KD, Gin DY, Gin MS. Highly convergent synthesis of C3- or C2-symmetric carbohydrate macrocycles. Org Lett, 2005, 7(20), 4479–4482. https://doi.org/10.1021/ol051818y.

[85] Bodine KD, Gin DY, Gin MS. Synthesis of readily modifiable cyclodextrin analogues via cyclodimerization of an alkynyl–azido trisaccharide. J Am Chem Soc, 2004, 126(6), 1638–1639. https://doi.org/10.1021/ja039374t.

[86] Chandrasekhar S, Rao CL, Nagesh C, Reddy CR, Sridhar B. Inter and intramolecular copper(I)-catalyzed 1,3-dipolar cycloaddition of azido-alkynes: Synthesis of furanotriazole macrocycles. Tetrahedron Lett, 2007, 48(33), 5869–5872. https://doi.org/10.1016/j.tetlet.2007.06.062.

[87] Singh A, Khatri V, Malhotra S, Prasad AK. Sugar-based novel chiral macrocycles for inclusion applications and chiral recognition. Carbohydr Res, 2016, 421, 25–32. https://doi.org/10.1016/j.carres.2015.12.006.

[88] Chen A, Samankumara LP, Dodlapati S, Wang D, Adhikari S, Wang G. Syntheses of *bis*-triazole linked carbohydrate based macrocycles and their applications for accelerating copper sulfate mediated click reaction. Eur J Org Chem, 2019, 2019(6), 1189–1194. https://doi.org/10.1002/ejoc.201801714.

[89] Ågren JKM, Billing JF, Grundberg HE, Nilsson UJ. Synthesis of a chiral and fluorescent sugar-based macrocycle by 1,3-dipolar cycloaddition. Synthesis, 2006, 2006, 3141–3145. https://doi.org/10.1055/s-2006-942503.

[90] Ray A, Manoj K, Bhadbhade MM, Mukhopadhyay R, Bhattacharjya A. Cu(I)-catalyzed cycloaddition of constrained azido-alkynes: Access to 12- to 17-membered monomeric triazolophanes incorporating furanoside rings. Tetrahedron Lett, 2006, 47(16), 2775–2778. https://doi.org/10.1016/j.tetlet.2006.02.068.

[91] Leyden R, Murphy PV. Glycotriazolophane synthesis via click chemistry. Synlett, 2009, 2009, 1949–1950. https://doi.org/10.1055/s-0029-1217534.

[92] Campo VL, Carvalho I, Silva CHTPD, Schenkman S, Hill L, Nepogodiev SA, Field RA. Cyclooligomerisation of azido-alkyne-functionalised sugars: Synthesis of 1,6-linked cyclic pseudo-galactooligosaccharides and assessment of their sialylation by trypanosoma cruzi trans-sialidase. Chem Sci, 2010, 1(4), 507–514. https://doi.org/10.1039/C0SC00301H.

[93] Geng J, Mantovani G, Tao L, Nicolas J, Chen G, Wallis R, Mitchell DA, Johnson BRG, Evans SD, Haddleton DM. Site-directed conjugation of "Clicked" glycopolymers to form glycoprotein mimics: Binding to mammalian lectin and induction of immunological function. J Am Chem Soc, 2007, 129(49), 15156–15163. https://doi.org/10.1021/ja072999x.

[94] Hetzer M, Chen G, Barner-Kowollik C, Stenzel MH. Neoglycopolymers based on 4-vinyl-1,2,3-triazole monomers prepared by click chemistry. Macromol Biosci, 2010, 10(2), 119–126. https://doi.org/10.1002/mabi.200900199.

[95] Kumar J, McDowall L, Chen G, Stenzel MH. Synthesis of thermo-responsive glycopolymersviacopper catalysed azide–alkyne 'Click' chemistry for inhibition of ricin: The effect of spacer between polymer backbone and galactose. Polym Chem, 2011, 2(8), 1879–1886. https://doi.org/10.1039/C1PY00048A.

[96] Ladmiral V, Mantovani G, Clarkson GJ, Cauet S, Irwin JL, Haddleton DM. Synthesis of neoglycopolymers by a combination of "Click Chemistry" and living radical polymerization. J Am Chem Soc, 2006, 128(14), 4823–4830. https://doi.org/10.1021/ja058364k.

[97] Huang X, Huang X-J, Yu A-G, Wang C, Dai Z-W, Xu Z-K. "Click Chemistry" as a facile approach to the synthesis of polyphosphazene glycopolymers. Macromol Chem Phys, 2011, 212(3), 272–277. https://doi.org/10.1002/macp.201000439.

[98] Liu Z, Zhu Y, Ye W, Wu T, Miao D, Deng W, Liu M. Synthesis of well-defined glycopolymers with highly ordered sugar units in the side chain via combining CuAAC reaction and ROMP: Lectin

interaction study in homo- and hetero-glycopolymers. Polym Chem, 2019, 10(29), 4006–4016. https://doi.org/10.1039/C9PY00756C.
[99] Dondoni A, Giovannini PP, Massi A. Assembling heterocycle-tethered C-glycosyl and α-amino acid residues via 1,3-dipolar cycloaddition reactions. Org Lett, 2004, 6(17), 2929–2932. https://doi.org/10.1021/ol048963g.
[100] Lee DJ, Mandal K, Harris PWR, Brimble MA, Kent SBH. A one-pot approach to neoglycopeptides using orthogonal native chemical ligation and click chemistry. Org Lett, 2009, 11(22), 5270–5273. https://doi.org/10.1021/ol902131n.
[101] Lee DJ, Yang S-H, Williams GM, Brimble MA. Synthesis of multivalent neoglyconjugates of MUC1 by the conjugation of carbohydrate-centered, triazole-linked glycoclusters to MUC1 peptides using click chemistry. J Org Chem, 2012, 77(17), 7564–7571. https://doi.org/10.1021/jo3013435.
[102] Locos OB, Heindl CC, Corral A, Senge MO, Scanlan EM. Efficient synthesis of glycoporphyrins by microwave-mediated "Click" reactions. Eur J Org Chem, 2010, 2010(6), 1026–1028. https://doi.org/10.1002/ejoc.200901292.
[103] Garcia G, Naud-Martin D, Carrez D, Croisy A, Maillard P. Microwave-mediated 'Click-Chemistry' synthesis of glycoporphyrin derivatives and in vitro photocytotoxicity for application in photodynamic therapy. Tetrahedron, 2011, 67(26), 4924–4932. https://doi.org/10.1016/j.tet.2011.04.080.
[104] Okada M, Kishibe Y, Ide K, Takahashi T, Hasegawa T. Convenient approach to access octa-glycosylated porphyrins via "Click Chemistry". https://doi.org/10.1155/2009/305276.
[105] Chen P-G, Huang Z-H, Sun Z-Y, Li -Q-Q, Chen Y-X, Zhao Y-F, Li Y-M. Synthesis of an MUC1 glycopeptide dendrimer based on β-cyclodextrin by click chemistry. Synlett, 2017, 28, 1961–1965. https://doi.org/10.1055/s-0036-1590796.
[106] Agrahari AK, Singh AS, Singh AK, Mishra N, Singh M, Prakash P, Tiwari VK. Click inspired synthesis of hexa and octadecavalent peripheral galactosylated glycodendrimers and their possible therapeutic applications. New J Chem, 2019, 43(31), 12475–12482. https://doi.org/10.1039/C9NJ02564B.
[107] Agrahari AK, Singh AS, Mukherjee R, Tiwari VK. An expeditious click approach towards the synthesis of galactose coated novel glyco-dendrimers and dentromers utilizing a double stage convergent method. RSC Adv, 2020, 10(52), 31553–31562. https://doi.org/10.1039/D0RA05289B.
[108] Rajakumar P, Anandhan R. Synthesis and in-vitro anti-inflammatory activity of novel glycodendrimers with benzene 1,3,5 carboxamide core and triazole as branching unit. Eur J Med Chem, 2011, 46(9), 4687–4695. https://doi.org/10.1016/j.ejmech.2011.06.017.
[109] Elmer SL, Man S, Zimmerman SC. Synthesis of polyglycerol, porphyrin-cored dendrimers using click chemistry. Eur J Org Chem, 2008, 2008(22), 3845–3851. https://doi.org/10.1002/ejoc.200800401.
[110] Le Pleux L, Pellegrin Y, Blart E, Odobel F, Harriman A. Long-lived, charge-shift states in heterometallic, porphyrin-based dendrimers formed via click chemistry. J Phys Chem A, 2011, 115(20), 5069–5080. https://doi.org/10.1021/jp2012182.
[111] Sharma R, Naresh K, Chabre YM, Rej R, Saadeh NK, Roy R. "Onion Peel" dendrimers: A straightforward synthetic approach towards highly diversified architectures. Polym Chem, 2014, 5(14), 4321–4331. https://doi.org/10.1039/C4PY00218K.
[112] Mishra N, Agrahari AK, Bose P, Singh SK, Singh AS, Tiwari VK. Click inspired synthesis of novel cinchonidine glycoconjugates as promising plasmepsin inhibitors. Sci Rep, 2020, 10(1), 3586. https://doi.org/10.1038/s41598-020-59477-3.
[113] Salta J, Reissig H-U. Divalent triazole-linked carbohydrate mimetics: Synthesis by click chemistry and evaluation as selectin ligands. Eur J Org Chem, 2020, 2020(28), 4361–4370. https://doi.org/10.1002/ejoc.202000618.
[114] Singh Y, Spinelli N, Defrancq E, Dumy P. A novel heterobifunctional linker for facile access to bioconjugates. Org Biomol Chem, 2006, 4(7), 1413–1419. https://doi.org/10.1039/B518151H.

[115] Singh K, Tyagi R, Mishra VK, Tiwari G, Sagar R. Recent advances in the synthesis of bioactive glycohybrids via click-chemistry. SynOpen, 2023, 7(3), 322–352. https://doi.org/10.1055/a-2130-7319.
[116] Tiwari G, Khanna A, Tyagi R, Mishra VK, Narayana C, Sagar R. Copper-catalyzed synthesis of pyrazolo [1,5-a]pyrimidine based triazole-linked glycohybrids: Mechanistic insights and bio-applications. Sci Rep, 2024, 14(1), 529. https://doi.org/10.1038/s41598-023-50202-4.
[117] Ferrara F, Beke-Somfai T, Kann N. Recent developments in the ruthenium-catalyzed azide alkyne cycloaddition (RuAAC) reaction. Eur J Org Chem, 2024, 27(25), e202400113. https://doi.org/10.1002/ejoc.202400113.
[118] Johansson JR, Beke-Somfai T, Said Stålsmeden A, Kann N. Ruthenium-catalyzed azide alkyne cycloaddition reaction: Scope, mechanism, and applications. Chem Rev, 2016, 116(23), 14726–14768. https://doi.org/10.1021/acs.chemrev.6b00466.
[119] Fox KA, Chadda R, Cardona F, Barron S, McArdle P, Murphy PV. Building blocks from monosaccharides for synthesis of scaffolds, including macrocycles. Application of allylic azide rearrangement, azide-alkyne cycloaddition and ring closing metathesis. Tetrahedron, 2020, 76(45), 131495. https://doi.org/10.1016/j.tet.2020.131495.
[120] Janssens J, Van Der Eycken J, Van Calenbergh S. Ceramide-templated macrolactams: Total synthesis and biological evaluation of macrocyclic α-galactosylceramide analogues and their aglycons. Eur J Org Chem, 2019, 2019(12), 2253–2267. https://doi.org/10.1002/ejoc.201801839.
[121] Tiwari VK, Jaiswal MK, Rajkhowa S, Singh SK. Click Chemistry. Materials Horizons: From Nature to Nanomaterials. Springer: Singapore, 2024, p. 603. https://link.springer.com/book/10.1007/978-981-97-4596-8.

Suvodip Mukherjee, Saptadwipa Bhattacharjee, and Pranab Ghosh*

Chapter 9
Synthesis of structurally diverse heterocycles from carbohydrate isothiocyanates

9.1 Introduction

Isothiocyanates (ITCs) are a type of heterocumulene characterized by the N=C=S functional group. They have important utility in chemical synthesis. Cruciferous vegetables, including kale, broccoli, mustard, garden cress, Brussels sprouts, cabbage, and cauliflower, are exceptionally rich in beneficial compounds. Specifically, these edible plants are abundant sources of allylisothiocyanate, sulforaphane, benzyl isothiocyanate, and phenethylisothiocyanate [1]. ITCs and sesquiterpenes have also been found in a sponge from the family Axinyssa [2]. ITCs are also known for their biological characteristics. They have shown antitumor efficacy in animals exposed to carcinogenic chemicals by preventing their metabolic activation [3]. ITCs are well known in organic chemistry as components of the Edman peptide sequencing technique [4]. They are useful synthons for creating a variety of heterocyclic moieties such as functionalized triazoles, benzimidazoles, thiazoles, dithiolanes, thiadiazoles, oxazines, triazines, and spiro-fused oxazolines [5–11]. Sugar ITCs have played a pivotal role in the synthesis of molecules of artificial, biological, and pharmacological significance including thioureas [12], *N*-glycopeptides [13], glycosyl amines [14], and glycosylguanidines [15]. Antitumor drugs can also be prepared by reacting glycosylisothiocyanates with 5-aminopyrimidine derivatives [16]. In supramolecular chemistry, sugar ITCs enable the synthesis of thioureido-linked glycoclusters [17] and calyx sugars [18].

9.2 Preparation of nucleosides and glycosylated heterocyclic amines

The robust biological activity of nucleosides has drawn considerable attention from researchers [19]. Numerous heterocyclic nucleosides have utilized as anticancer [20, 21] and antiviral drugs, such as those for AIDS [22, 23] and hepatitis B [24, 25]. Several

***Corresponding author: Pranab Ghosh,** Department of Chemistry, University of North Bengal, Siliguri, West Bengal, India
Suvodip Mukherjee, Department of Chemistry, Sri Sairam College of Engineering, Anekal, Bengaluru 562106, Karnataka, India
Saptadwipa Bhattacharjee, Department of Chemistry, University of North Bengal, Raja Rammohunpur, West Bengal, India

https://doi.org/10.1515/9783111447094-009

nucleoside families serve important roles in cancer treatment and infectious illness prevention. A range of nucleoside analogs, comprising L-nucleosides [26, 27], *C*-nucleosides [28, 29], 4-thionucleosides [30, 31], carbocyclic nucleosides [32, 33], and acyclic derivatives [34, 35] have been designed, synthesized, and evaluated for their therapeutic potential.

Alkyl/aryl ITCs react with 2-amino-2-deoxy-aldoses to form acyclic and imidazole-based *C*-nucleosides [36–39]. When sugar-derived ITCs are used in this process, *C*- and *N*-nucleosides are produced. A new series of imidazolidine-2-thiones, 1-*N*-glycosyl-4-(D-arabino-tetritolyl)imidazolidine-2-thiones, was synthesized by Fuentes et al. via the reaction of 2-amino-2-deoxy-D-glucosamine with diverse *O*-acetylated glycosylisothiocyanates, encompassing mono-, di-, and trisaccharide derivatives and 2-deoxy-2-isothiocyanato-β-D-glucose [40].

Similarly, the procedure illustrated in Figure 9.1 [41] was used to synthesize pseudo-nucleosides. The reaction between D-glucosamine and methyl 2-deoxy-2-isothiocyanato -α-D-glucopyranoside yielded 5-hydroxy-imidazolidine-2-thione as a mixture of diastereomers, comprising 5R and 5S isomers. The ratio of 5R:5S was 93:7 in D2O and 98:2 in DMSO-d_6. Heating in aqueous acetic acid accelerated the cyclodehydration of the tetritolyl chain, resulting in bicyclic glucofuranoso-imidazolidine-2-thione. When subjected to methanolic TFA at room temperature, this led to the generation of imidazoline-2-thione through isomerization. Cyclodehydration of the polyhydroxylated chain occurred under aqueous conditions using TFA, affording a 1:4 mixture of α- and β-anomers of D-erithrofuranosyl-*C*-nucleosides. The TFA-mediated reaction in refluxing methanol afforded *S*-methylimidazole derivatives from imidazolidine-2-thione as a diastereomeric mixture (1:7 α:β).

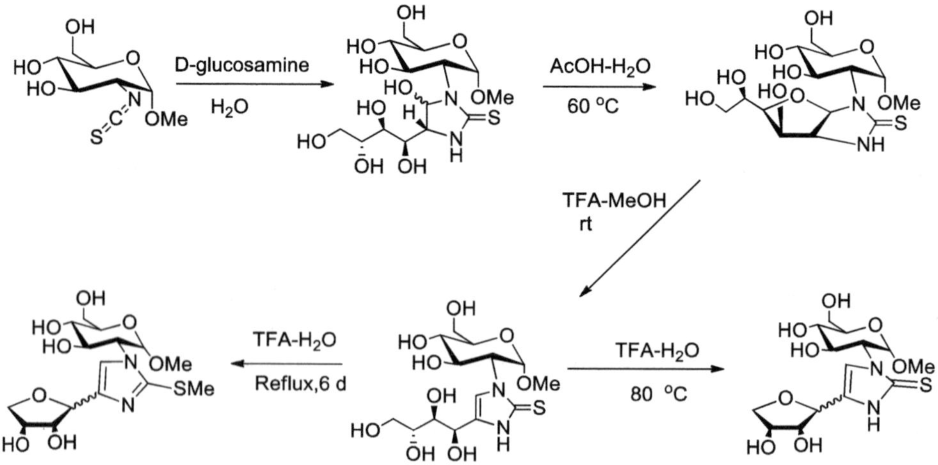

Figure 9.1: Synthesis of pseudo-nucleosides.

Alternatively, nucleosides can be synthesized through the reaction between glycosylisothiocyanates and stabilized carbanions, exploiting the isothiocyanato group's strong electrophilic properties [42]. The reaction between diethyl acetonyl (phenacyl) malonate and per-*O*-acetylated β-D-glucopyranosylisothiocyanate, mediated by sodium hydride, is illustrated in Figure 9.2, resulting in the formation of thioamides. At normal temperature, the thioamide part is in equilibrium with the homologous pyrrolidine structure, favoring the latter by a factor of around 1:10. Cyclodehydration was generated by acidic treatment of thioamide-derived structures using an acetic anhydride-phosphoric acid combination. This treatment changed the combination to *N*-glucopyranosyltetrahydrothioxopyridine, and the substance was transformed to *N*-glucopyranosylthioxopyrroline.

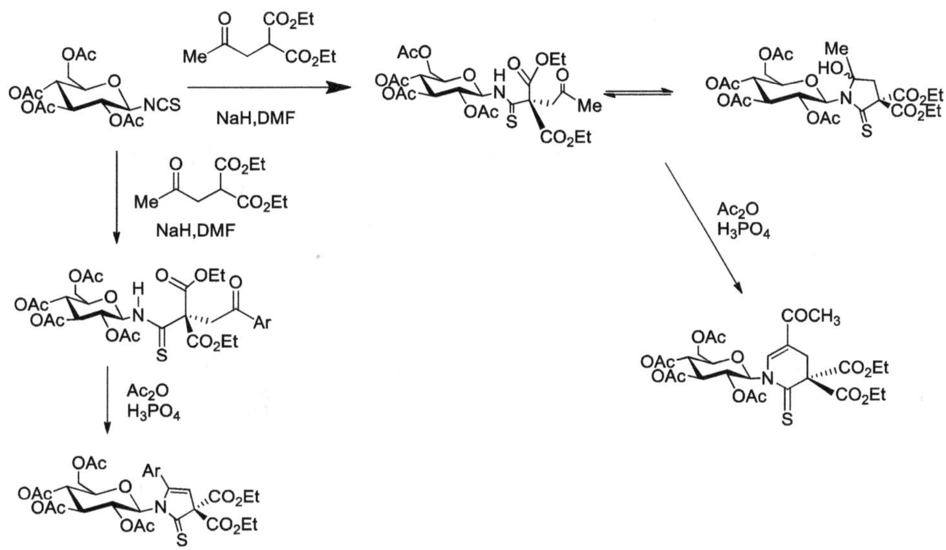

Figure 9.2: Synthesis of thioamides and thiazoline ring.

9.3 Synthesis of cyclic isothioureas

Novel glycopyranoso-thiazoline synthesis via 2-deoxy-2-iodoglycosyl ITCs was explored by Santoyo-Gonzalez and coworkers [43, 44]. Glycals can be converted into the desired compounds through a simple reaction with silica-supported potassium thiocyanate and iodine in chloroform [44]. The dual electrophilic groups within these molecules render them suitable precursors for constructing glycopyranoso-fused thiazoles through thiocarbamates, thioureas, ordithiocarbamates, obtained from the reaction of alcohols, thiols, or amines, respectively (Figure 9.3). The thiocarbonyl ring's sulfur atom displaces the trans-vicinal iodine in a base-catalyzed nucleophilic substitution, yielding bicyclo compounds in good to high yield.

Chiara et al. [45] developed an isosteric variant of trehazolin. They substituted the aglycon's endocyclic oxygen atom with sulfur to form a thiazoline ring (Figure 9.2). They synthesized partly protected aminocyclopentitol from D-mannose via a crucial stereoselective reductive carbocyclization process with SmI_2 [46]. Combining the chemical with *O*-protected α-D-glucopyranosylisothiocyanate produced the desired thiourea. Triflic anhydride-mediated activation of thiourea resulted in the formation of a short-lived triflate intermediate on the secondary hydroxyl group, which was then replaced through nucleophilic attack by the sulfur atom. Synthesis of thiotrehazolin was achieved finally via deprotection (Figure 9.4). Thiotrehazolin exhibited potent inhibitory activity against porcine trehalase, with a K_i value of 30.4 nM, demonstrating only a modest 15-fold reduction in efficacy compared to that of the natural one having a K_i value of 2.1 nM.

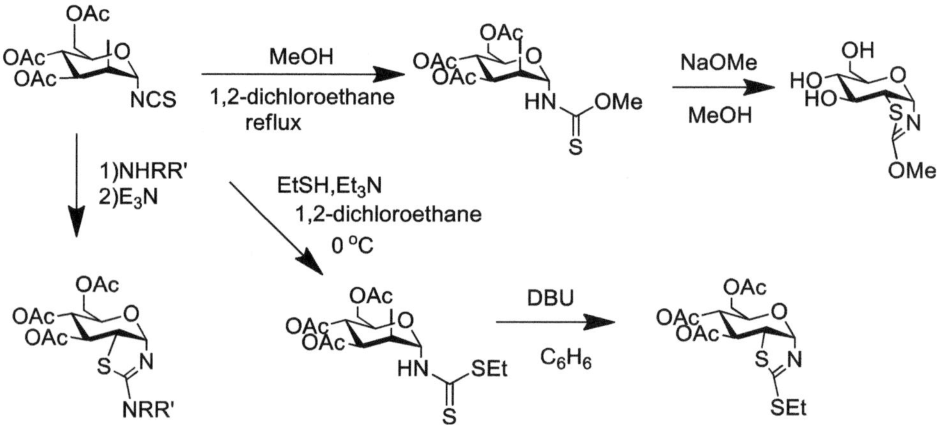

Figure 9.3: Base-catalyzed dithiocarbamate synthesis.

9.4 Preparation of cyclic isoureas

Cyclic isoureas, also known as 2-amino-2-oxazolines, have gained significant attention as synthetic targets in recent chemical research [47]. These compounds are of interest because they are found in physiologically active molecules and display actions such as octopamine [48], adrenoreceptor agonist [49], histamine [50], and adenosine [51] receptor antagonists. Furthermore, they hold promise for treating hypertension [52], as antiobesity medications [53], and as inhibitors of pheromone manufacturing [54], which might make them beneficial in pest management [55].

In the context of carbohydrates, naturally produced bicyclic isoureas, including trehazolin [56, 57] and allosamidin [58], are powerful and selective glycosidase inhibitors that operate as trehalase and chitinase inhibitors in vitro. Despite their promise as insect-control agents, their extremely polar structures have limited their practical application as in vivo insecticides [55]. The separation and identification of these physiologically active molecules have prompted researchers to work on their entire synthesis, including the synthesis of their aglycon counterparts [59] and changes to both the carbohydrate and cyclitol framework [60, 61]. Notably, a six-membered polyhydroxylated carbocycle [62] or a tetrahydropyran group [63] substituted the aminocyclopentitol framework.

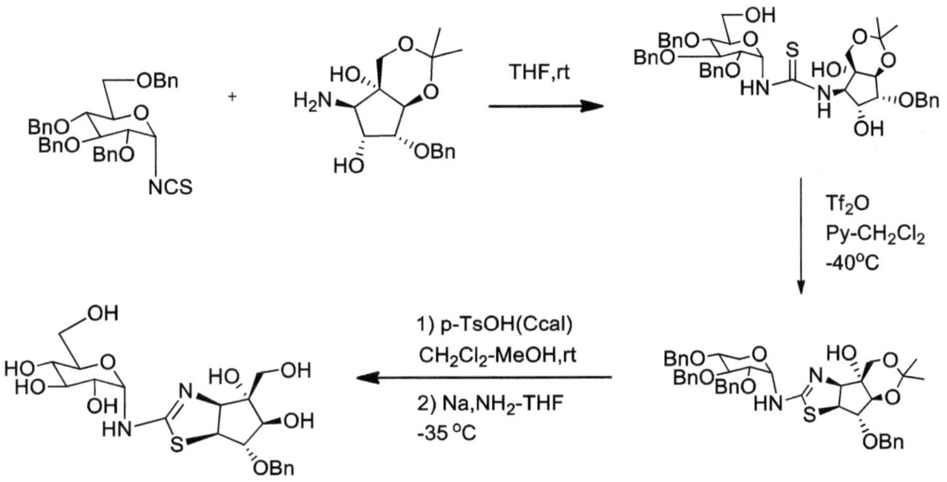

Figure 9.4: Synthesis of thiotriazolin.

Uchida-Ogawa [64] sought to create glycosylaminooxazolines as trehazolin analogues by removing the heterocyclic moiety substituents. The authors synthesized per-*O*-protected galactosylthiourea by condensation of 2,3,4,6-*tetra*-*O*-benzyl-α-D-galactopyranosylisothiocyanate [65] with 2-aminoethanol, as illustrated in Figure 9.5. Treatment with yellow HgO in diethyl ether (Et_2O) led to cyclodesulfurization, forming 2-galactopyranosylamino-oxazoline through a transient carbodiimide intermediate. Deprotection under Birch conditions led to an unexpected outcome, producing glycofuranoso-glycopyranoso-oxazoline and fused rings in a 1:1 ratio, instead of the predicted isourea. The glucose epimer exhibited the same behavior [64]. In contrast, deprotection of benzylated β-D-gluco and β-D-galactopyranosylisothiocyanates or acetylated α-D-mannopyranosylisothiocyanate (Figure 9.6) did not result in rearrangement due to the *trans*-configuration of C1 and C2 substituents. This leads to the synthesis of the appropriate 2-glycopyranosylamino-oxazolines [64]. α-Mannopyranosyl-oxazoline demonstrated potent inhibition of α-mannosidase, characterized by a K_i of 98 μM.

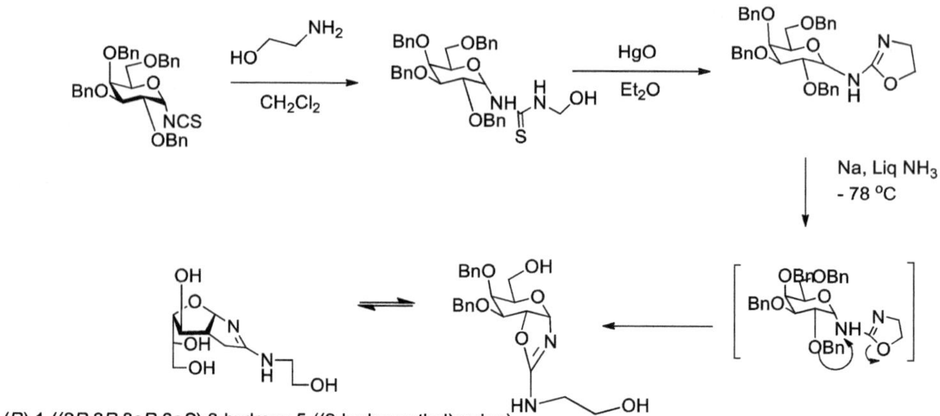

(*R*)-1-((2*R*,3*R*,3a*R*,6a*S*)-3-hydroxy-5-((2-hydroxyethyl)amino)-3,3a,4,6a-tetrahydro-2*H*-furo[2,3-*b*]pyrrol-2-yl)ethane-1,2-diol

Figure 9.5: Synthesis of glycosylaminooxazolines.

A one-pot, three-step procedure for preparing 2-aminodihydroglucopyranoso[1,2-d]oxazoles with varying amino substituents such as alkyl, dialkyl, and aryl was established

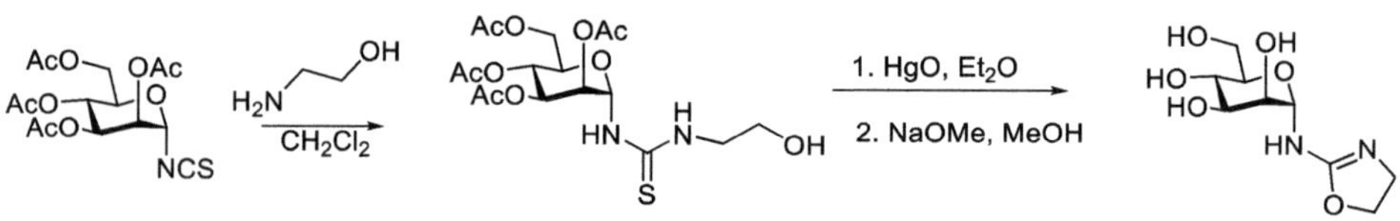

Figure 9.6: Synthesis of 2-glycopyranosylamino-oxazolines.

by Fernández-Bolaños et al. [65], starting from β-D-glucopyranosylamine (Figure 9.7). Thiophosgene is employed to convert β-D-glucopyranosylamine into a reactive intermediate in a 1:1 water-dioxane buffer mixture. The intermediate ITC then reacts with 1° and 2° amines, affording *O*-unprotected glucopyranosylthioureas. The final step involves in situ oxidation with yellow HgO, yielding *trans*-fused bicyclic isoureas with a cumulative yield of 55–84% over the three steps.

9.5 Synthesis of cyclic thiocarbamates

Because of their fascinating biological and pharmacological features, 1,3-oxazinane-2-thiones and 1,3-oxazolidine-2-thiones are difficult to synthesize [66]. Glucosinolates with hydroxyalkyl side chains, found in nature, can undergo hydrolysis to form these heterocyclic compounds [67]. In glycochemistry, numerous nucleosides along with associated compounds have been synthesized by combining ketoses and aldoses with

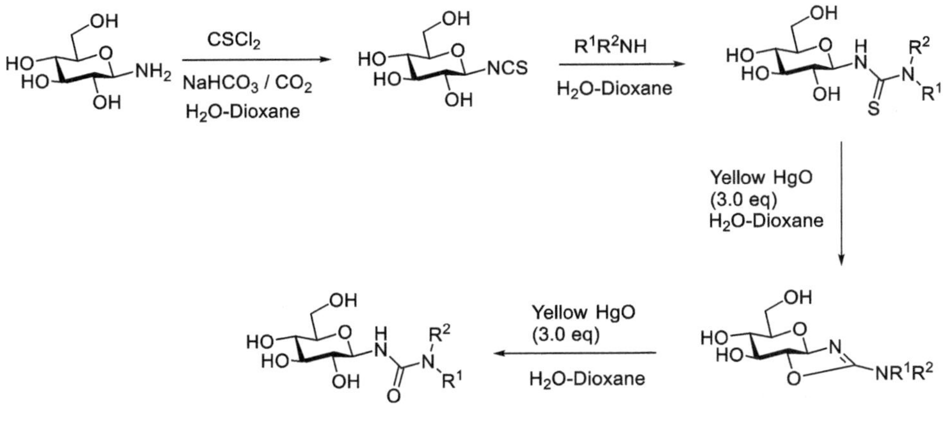

Figure 9.7: Synthesis of 2-aminodihydroglucopyranoso[1,2-d]oxazoles.

benzyl ITC or thiocyanic acid to generate 1,3-oxazolidine-2-thiones [68, 69]. The 1,3-oxazolidine-2-thione framework has found widespread application as a versatile chiral auxiliary in synthetic chemistry [70]. When a sugar molecule has both a hydroxyl and ITC group, it is feasible to produce 1,3-oxazolidine-2-thiones (Figure 9.8). The feasibility and efficiency of this method depend on the ring size, relative orientation of functional groups, and intrinsic strain of the cyclic framework. Cyclization is only feasible for hydroxyl groups located in the β- or γ-position relative to the ITC moiety, whereas *trans*-diaxially oriented groups in a rigid pyranose ring do not form oxazolidine-2-thione [71].

García-Fernández et al. demonstrated that methyl 6-deoxy-6-isothiocyanato-α-D-glucopyranoside undergoes efficient cyclization with neighboring hydroxyl groups, affording bicyclic thiocarbamates with sugar-like structures [72]. Scientists observed that heating alone in DMF was ineffective, yet the combination of Et3N catalysis and heat enabled efficient intramolecular cyclization of the C-4 hydroxyl group, yielding 1,3-oxazinane-2-thione. The reaction proceeds through a zwitterionic intermediate formed by the ITC and triethylamine, which then undergoes nucleophilic substitution with the γ-hydroxyl group of the glucopyranoside, yielding the desired product [72, 73].

Isothiocyanation produced 6,6-dideoxy-6,6-diisothiocyanato-α,α-trehalose, having identical reaction behavior (Figure 9.9) [74]. Additionally, research has explored the

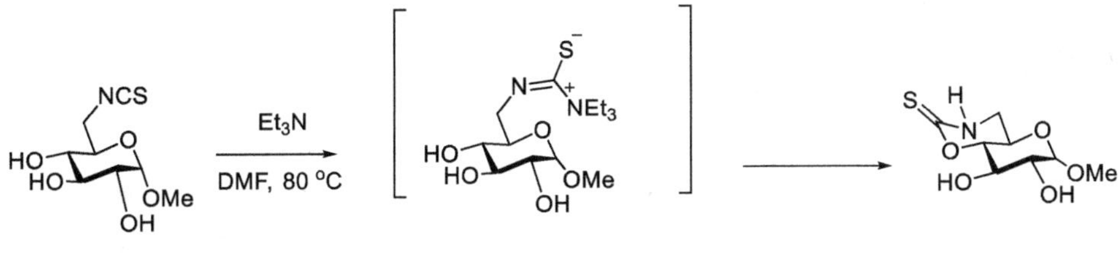

Figure 9.8: Synthesis of 1,3-oxazinane-2-thione.

reactivity of fully *O*-protected 6-deoxy-6-isothiocyanato analogues with ammonia in various solvents, yielding either thioureas or cyclic thiocarbamates [75].

Earlier researchers discovered that the reaction between an ITC group and a hydroxyl group in a sugar derivative spontaneously yields cyclic thiocarbamates [73].

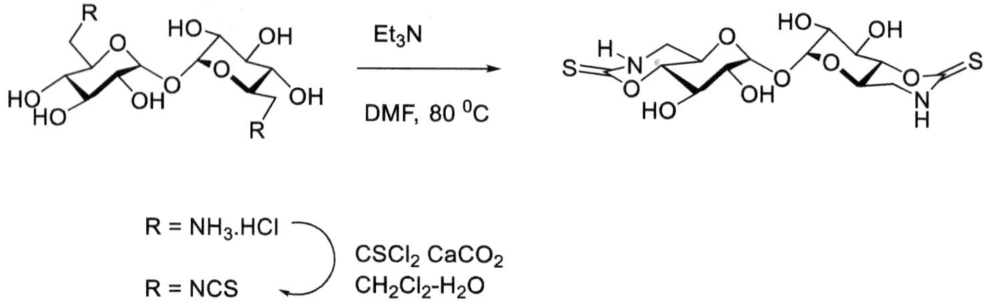

Figure 9.9: Synthesis of 6,6-dideoxy-6,6-diisothiocyanato-α,α-trehalose.

Treatment of 6-amino-6-deoxy-D-glucose, D-galactose, and D-mannose with thiophosgene in an acetone-water mixture containing $CaCO_3$ yielded pseudo-*C*-nucleosides featuring an oxazolidine-2-thione moiety. Yields ranged from 82% to 88% (Figure 9.10). This process most likely includes transitory ITCs; however, these intermediates were not identified even at low temperatures. Notably, cyclization was completely regioselective, producing only furanose-derived thiocarbamates and no pyranose-derived thiocarbamates.

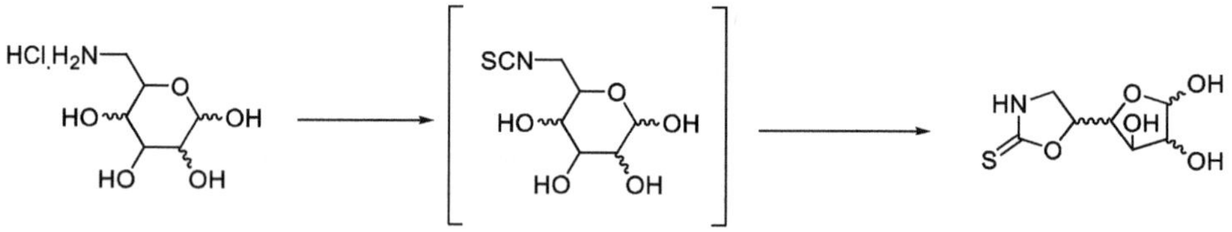

Figure 9.10: Synthesis of cyclic thiocarbamates.

9.6 Preparation of spironucleosides

This class of sugar derivatives, anomeric spironucleosides, features a distinctive spiranic architecture, with the anomeric carbon serving as a shared atom between the carbohydrate ring and a heterocyclic moiety [76, 77]. Since the invention of hydantocidin in 1991, a naturally occurring spironucleoside with high herbicidal and plant growth inhibitory effects that is nontoxic to microbes and mammals, there has been a lot of interest in synthesizing anomeric spironucleosides. A study investigated the chemical behavior of 1-amino-1-deoxy-D-fructose compounds, substituted with alkyl and arylamino moieties, in reactions involving sugar ITCs and aminosugars (Figure 9.11). Reaction of diverse *O*-acylated sugar ITCs with D-fructosamines yielded *N*-glycosyl-5-hydroxy -5-tetritolylimidazolidine-2-thiones as diastereomeric pairs in high, quantifiable yields [76]. Fructosamine and its *N*-methyl derivatives have faster reaction times (15–20 min) than *N*-phenyl compounds (8–24 h). Treatment with ethanolic TFA at room temperature, under acid-catalyzed dehydration conditions, yielded 65–77% of compounds that are dual *N*-nucleosides and acyclic *C*-nucleosides of imidazoline-2-thiones. Treatment of β-D-ribofuranosyl derivatives with ethanolic Dowex 50W-X8 resins resulted in efficient dehydration, yielding a range of spironucleosides. The hydroxyl group at C-5 of the imidazolidine ring underwent cyclodehydration with OH-3 of the polyhydroxyalkyl chain,

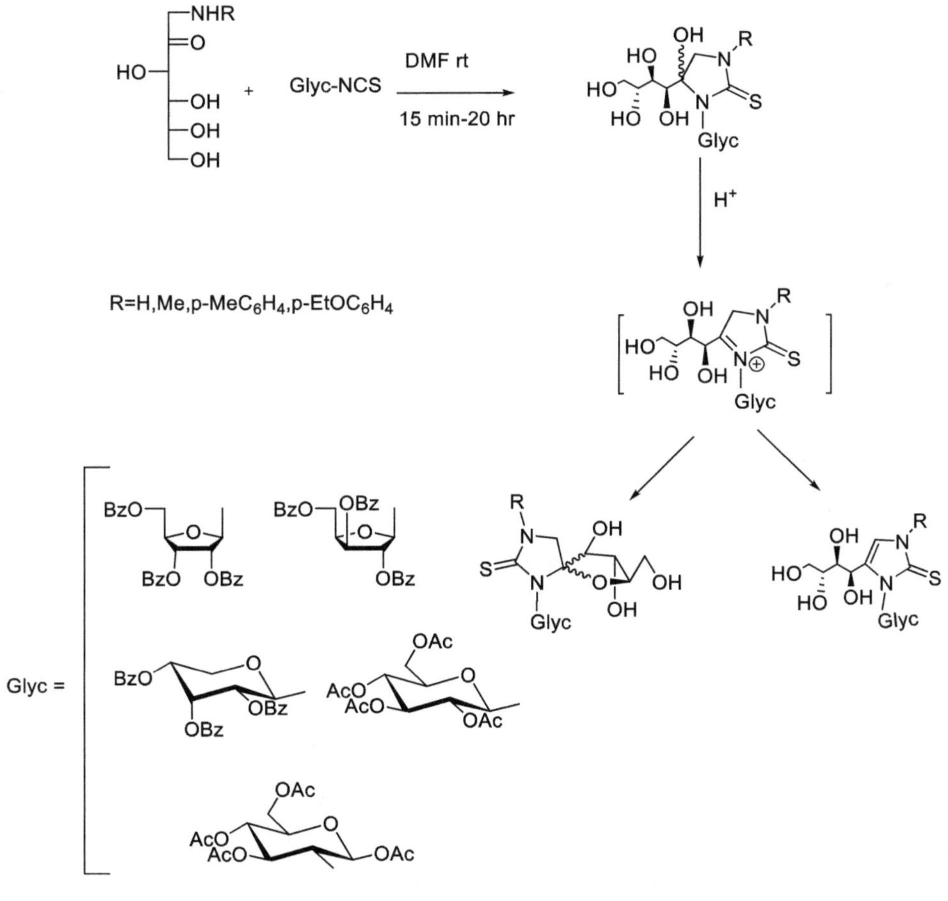

Figure 9.11: Synthesis of *N*-glycosyl-5-hydroxy-5-tetritolylimidazolidine-2-thiones.

forming spiro compounds with two epimers at C-5 through a fleeting stabilized cationic intermediate. Major stereoisomers possessing a 5S structure were isolated in 40–16% yields, whereas the minor stereoisomer was isolated only when $R = p$-MeC6H4, with a 7% yield [76].

Shiozaki's [78, 79] eight-stage synthesis converted 2,3-isopropylidene-D-ribono-1,4-lactone into 2-thiohydantocidin (16.5% yield) and its epimer (9.2% yield) (Figure 9.12). Sano and coworkers employed an aza-Wittig reaction to synthesize the target compounds, utilizing α-azidocarboxamide, carbon disulfide, and tri-*n*-butylphosphine [10].

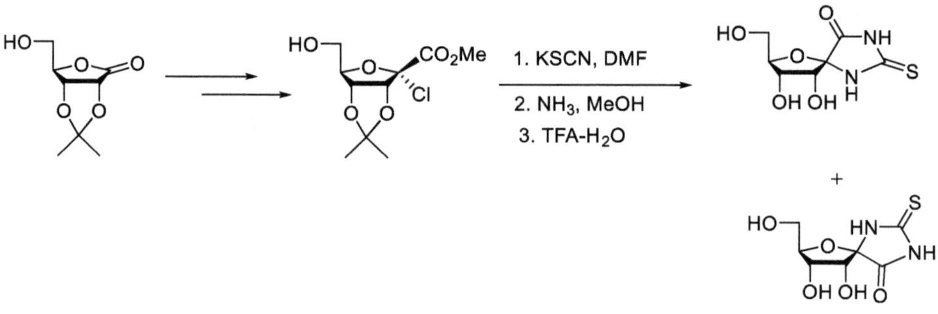

Figure 9.12: Synthesis of 2-thiohydantocidin.

9.7 Conclusion

Owing to ITCs' high synthetic versatility – particularly when derived from carbohydrates – a wide range of heterocyclic compounds with major synthetic, biological, and therapeutic applications have become possible. ITCs have the potential to be important intermediates in organic synthesis in a range of compounds including nucleosides, spironucleosides, cyclic thiocarbamates, cyclic isoureas, and isothioureas. Their ability to perform cycloadditions and nucleophilic additions emphasizes their importance in the development of new compounds with potential therapeutic applications.

References

[1] Singh SV, Singh K. Cancer chemoprevention with dietary isothiocyanates mature for clinical translational research. Carcinogenesis, 2012, 33, 1833.

[2] Zubía E, Ortega MJ, Hernández-Guerrero CJ, Carballo JL. Isothiocyanate sesquiterpenes from a sponge of the genus axinyssa. J Nat Prod, 2008, 71, 608.

[3] Fimognari C, Nüsse M, Cesari R, Iori R, Cantelli-Forti G, Hrelia P. Isothiocyanates as novel cytotoxic and cytostatic agents: Molecular pathway on human transformed and non-transformed cells. Carcinogenesis, 2002, 23, 581.

[4] Edman P. A method for the determination of amino acid sequence in peptides. Arch Biochem, 1949, 22, 475.
[5] Mouren A, Avérous L. Sustainable cycloaliphatic polyurethanes: From synthesis to applications. Chem Soc Rev, 2023, 52, 277–317.
[6] Fischer G. Recent progress in 1,2-dithiole-3-thione chemistry. Adv Heterocycl Chem, 2013, 109, 1–90.
[7] Zahraa U, Saeed A, Abdul Fattah T, Flörke U, Erben MF. Recent trends in chemistry, structure, and various applications of 1-acyl-3-substituted thioureas: A detailed review. RSC Adv, 2022, 12, 12710–12745.
[8] Das B, Dahiyaa A, Patel BK. Isothiocyanates: Happy-go-lucky reagents in organic synthesis. Org Biomol Chem, 2024, 22, 3772–3798.
[9] Nedolya NA, Tarasova OA, Artem'ev AV, Albanov AI, Bagryanskaya IY, Trofimov BA. One-pot synthesis of 2-(alkylsulfanyl)quinolines from aryl isothiocyanates and allenes or alkynes. Eur J Org Chem, 2024, 2024, e202400033.
[10] Sano H, Mio S, Kitagawa J, Shindou M, Honma T, Sugai S. Synthesis of spirothiohydantoin analogues of hydantocidin. Tetrahedron, 1995, 51, 12563.
[11] Sommen G. Phenyl isothiocyanate: A very useful reagent in heterocyclic synthesis. Synlett, 2004, 7, 1323.
[12] López Ó, Maya I, Fuentes J, Fernández-Bolaños JG. Simple and efficient synthesis of *O*-unprotected glycosylthiourea and isourea derivatives from glycosylamines. Tetrahedron, 2004, 60, 61.
[13] Isac-García J, Calvo-Flores FG, Hernández-Mateo F, Santoyo-González F. Synthesis of glycosylamines from glycosylisothiocyanates and bis(tributyltin) oxide. Eur J Org Chem, 2001, 383.
[14] Fuentes J, Angulo M, Pradera MA. Fluoronucleosides, isothiocyanato c-nucleosides, and thioureylene di-c-nucleosides via cyclic sulfates. J Org Chem, 2002, 67, 2577.
[15] Taylor CM. Synthesis of glycosylamines from glycosylisothiocyanates and bis(tributyltin) oxide. Tetrahedron, 1998, 54, 11317.
[16] Shi HF, Cao LH. Synthesis and biological activity of N-Alkyl/aryl-N'-(4-arylthiazol-2-yl)-N''glycosyl guanidines. YoujiHuaxue, 2005, 25, 1066.
[17] Benito JM, Gómez-García M, Ortiz Mellet C, Baussanne I, Defaye J, García Fernández JM. Optimizing saccharide-directed molecular delivery to biological receptors: Design, synthesis, and biological evaluation of glycodendrimer–cyclodextrin conjugates. J Am Chem Soc, 2004, 126, 10355.
[18] Patel A, Lindhorst TK. Synthesis of "mixed type" oligosaccharide mimetics based on a carbohydrate scaffold. Eur J Org Chem, 2002, 79.
[19] Lech-Maranda E, Korycka A, Robak T. Pharmacological and clinical studies on purine nucleoside analogs – New anticancer agents. Mini Rev Med Chem, 2006, 6, 575.
[20] Robak T, Korycka A, Kasznicki M, Wrzesien-Kus A, Smolewski P. Purine nucleoside analogues for the treatment of hematological malignancies: Pharmacology and clinical applications. Curr Cancer Drug Targets, 2005, 5, 421.
[21] Vivet-Boudou V, Didierjean J, Isel C, Marquet R. Nucleoside and nucleotide inhibitors of HIV-1 replication. Cell Mol Life Sci, 2006, 63, 163.
[22] De Clercq E. New developments in anti-HIV chemotherapy. Biochim Biophys Acta, 2002, 1587, 258.
[23] Núñez M, Soriano V. Management of patients co-infected with hepatitis B virus and HIV. Lancet Infect Dis, 2005, 5, 374.
[24] Fung SK, Lok AS. Drug insight: Nucleoside and nucleotide analog inhibitors for hepatitis B. Nat Clin Pract Gastroenterol Hepatol, 2004, 1, 90.
[25] Seela F, Lin W, Kazimierczuk Z, Rosemeyer H, Glaccon V, Peng X, He Y, Ming X, Andrzejewska M, Gorska A, Zhang X, Eickmeier H, La Colla P. L-Nucleosides containing modified nucleobases. Nucleos Nucleot Nucl, 2005, 24, 859.

[26] Wang J, Jin Y, Rapp KL, Bennett M, Schinazi RF, Chu CK. Synthesis, antiviral activity, and mechanism of drug resistance of D- and L-2',3'-didehydro-2',3'-dideoxy-2'-fluorocarbocyclic nucleosides. J Med Chem, 2005, 48, 3736.
[27] Ludek OR, Balzarini J, Meier C. Synthesis and antiviral evaluation of carbocyclic 3'-azidothymidine (AZT) analogues and their cycloSal-phosphate triesters. Eur J Org Chem, 2006, 932.
[28] Sun Z, Ahmed S, McLaughlin L. Syntheses of pyridine C-nucleosides as analogues of the natural nucleosides dC and dU. J Org Chem, 2006, 71, 2922.
[29] Abbasi MM, El-Kousy SM, El-Moghazy YE, El-Kafrawy S. Synthesis of pyrazoloquinoxaline-C-acyclic nucleosides with expected antiviral activity. Int J Chem, 2005, 15, 77.
[30] Fernández-Bolaños JG, Al-Masoudi NAL, Maya I. Sugar derivatives having sulfur in the ring. Adv Carbohydr Chem Biochem, 2001, 57, 21.
[31] Sheban MAE. The chemistry of C-nucleosides and their analogs – II: C-nucleosides of condensed heterocyclic bases. Adv Heterocycl Chem, 1998, 70, 163.
[32] Cho JH, Bernard DL, Sidwell RW, Kern ER, Chu CK. Synthesis of cyclopentenyl carbocyclic nucleosides as potential antiviral agents against orthopoxviruses and SARS. J Med Chem, 2006, 49, 1140.
[33] Gunaga P, Moon HR, Choi WJ, Shin DH, Park JG, Jeong LS. Recent advances in 4-thionucleosides as potential antiviral and antitumor agents. Curr Med Chem, 2004, 11, 2585.
[34] De Clercq E, Holy A. Acyclic nucleoside phosphonates: A key class of antiviral drugs. Nat Rev Drug Discov, 2005, 4, 928.
[35] Oh CH, Baek TR, Hong JH. Concise synthesis and antiviral activity of novel unsaturated acyclic pyrimidine nucleosides. Nucleos Nucleot Nucl, 2005, 24, 153.
[36] Fernández-Bolañosguzmán J, García Rodríguez S, Fernández-Bolaños J, Diánez MJ, López-Castro A. Reaction of 2-amino-2-deoxy-d-glucose with aryl and acyl isothiocyanates, and aryl isocyanates: Structure of the intermediate products. Carbohydr Res, 1991, 210, 125.
[37] Ávalos M, Babiano R, Cintas P, Jiménez JL, Palacios JC, Silvero G, Valencia C. Non-biarylatropisomers derived from carbohydrates. Part 1.Stereoselective synthesis of 1-aryl-5-hydroxyimidazolidine-2-thiones and their transformation into imidazoline-2-thiones. Tetrahedron, 1999, 55, 4377.
[38] Ávalos M, Babiano R, Cintas P, Hursthouse MB, Jiménez JL, Light ME, Palacios JC, Silvero G. Non-biarylatropisomers derived from carbohydrates. Part 4: Absolute stereochemistry of carbohydrate-based imidazolidine-2-ones and 2-thiones with axial and central chirality. Tetrahedron, 2005, 61, 7945.
[39] Fernández-Bolaños J, Fuentes Mota J, Fernández-BolañosGuzmán J. Synthesis of 4-thio-C-nucleoside analogues by dehydration of 4-thiopolyhydroxyalkyl heterocycles. Carbohydr Res, 1988, 173, 17.
[40] Fuentes J, Molina JL, Olano D, Pradera MA. Stereoselective synthesis of nucleoside analogues of chiral imidazolidines from sugar isothiocyanates. Tetrahedron Asymm, 1996, 7, 203.
[41] Fernández-Bolaños JG, Zafra E, López Ó, Robina I, Fuentes J. Stereoselective synthesis of imidazolidine, imidazoline and imidazole C- and N-pseudonucleosides. Tetrahedron Asymm, 1999, 10, 3011.
[42] Fuentes J, Molina JL, Pradera MA. Reactions of per-*O*-acetylglucosylisothiocyanate with carbon bases. A new method for the stereocontrolled syntheses of nucleosides and glucosylaminothiophenes. Tetrahedron Asymm, 1998, 9, 2517.
[43] Isac-García J, Hernández-Mateo F, Calvo-Flores FG, Santoyo-González F. Reactivity of 2-deoxy-2-iodoglycosyl isothiocyanates with O-, S-, and n-nucleophiles. Synthesis of glycopyranoso-fused thiazoles. J Org Chem, 2004, 69, 202.
[44] Santoyo-González F, García-Calvo-Flores F, Isac-García J, Hernández-Mateo F, García Mendoza P, Robles-Díaz R. Synthesis and transformations of 2-deoxy-2-iodo-pyranosyl isothiocyanates from glycals. Tetrahedron, 1994, 50, 2877.

[45] Chiara JL, Storch de Gracia I, García A, Bastida A, Bobo S, Martín-Ortega MD. Synthesis, inhibition properties, and theoretical study of the new nanomolartrehalase inhibitor 1-thiatrehazolin: Towards a structural understanding of trehazolin inhibition. ChemBioChem, 2005, 6, 186.
[46] Chiara JL, Marco-Contelles J, Khiar N, Gallego P, Destabel C, Bernabe M. Intramolecular reductive coupling of carbonyl-tethered oxime ethers promoted by samarium diiodide: A powerful method for the stereoselective synthesis of aminocyclopentitols. J Org Chem, 1995, 60, 6010.
[47] García-Moreno MI, Díaz-Pérez P, Ortiz Mellet C, GarcíaFernández JM. Synthesis and evaluation of isourea-type glycomimetics related to the indolizidine and trehazolin glycosidase inhibitor families. J Org Chem, 2003, 68, 8890.
[48] Hirashima A, Morimoto M, Kuwano E, Eto M. Three-dimensional molecular-field analyses of octopaminergic agonists for the cockroach neuronal octopamine receptor. Bioorg Med Chem, 2003, 11, 3753.
[49] Wong WC, Sun W, Cui W, Chen X, Forray C, Vaysse PJP, Branchek TA, Gluchowski C. 2-amino-2-oxazolines as subtype selective α2 adrenoceptor agonists. J Med Chem, 2000, 43, 1699.
[50] Bosc JJ, Jarry C, Martínez B, Molimard M. Synthesis and pharmacological evaluation of 5-dialkylaminomethyl-2-amino-2-oxazolines as H1-antagonists. J Pharm Pharmacol, 2001, 53, 923.
[51] Massip S, Guillon J, Bertarelli D, Bosc JJ, Leger JM, Lacher S, Bontemps C, Dupont T, Mueller CE, Jarry C. Synthesis and preliminary evaluation of new 1- and 3-[1-(2-hydroxy-3-phenoxypropyl)]xanthines from 2-amino-2-oxazolines as potential A1 and A2A adenosine receptor antagonist. Bioorg Med Chem, 2006, 14, 2697.
[52] Bruban V, Feldman J, Greney H, Dontenwill M, Schann S, Jarry C, Payard M, Boutin J, Scalbert E, Pfieffer B, Renard P, Vanhoutte P, Bousquet P. Respective contributions of α-adrenergic and non-adrenergic mechanisms in the hypotensive effect of imidazoline-like drugs. Br J Pharmacol, 2001, 133, 261.
[53] Tao R, Fray A, Aspley S, Brammer R, Heal D, Auerbach S. Effects on serotonin in rat hypothalamus of d-fenfluramine, aminorex, phentermine and fluoxetine. Eur J Pharmacol, 2002, 445, 69.
[54] Hirashima A, Eiraku T, Watanabe Y, Kuwano E, Taniguchi E, Eto M. Identification of novel inhibitors of calling and in vitro [14C]acetate incorporation by pheromone glands of Plodiainterpunctella. Pest Manag Sci, 2001, 57, 713.
[55] López Ó, Fernández-Bolaños JG, Gil MV. New trends in pest control: The search for greener insecticides. Green Chem, 2005, 7, 431.
[56] Ando O, Satake H, Itoi K, Sato A, Nakajima M, Takahashi S, Haruyama H, Ohkuma Y, Kinoshita T, Enokita R. Trehazolin, a new trehalase inhibitor. J Antibiot, 1991, 44, 1165.
[57] Nakayama T, Amachi T, Murao S, Sakai T, Shin T, Kenny PT, Iwashita T, Zagorski M, Komura H, Nomoto K. Structure of trehalostatin: A potent and specific inhibitor of trehalase. Chem Soc Chem Commun, 1991, 919.
[58] Sakuda S, Isogai A, Matsumoto S, Suzuki A. Search for microbial insect growth regulators. II Allosamidin, a novel insect chitinase inhibitor. J Antibiot, 1987, 40, 296.
[59] Feng X, Duesler EN, Mariano PS. Pyridinium salt photochemistry in a concise route for synthesis of the trehazolinaminocyclitol, trehazolamine. J Org Chem, 2005, 70, 5618.
[60] Berecibar A, Grangean C, Siriwardena A. Synthesis and biological activity of natural aminocyclopentitol glycosidase inhibitors: Mannostatins, trehazolin, allosamidins, and their analogues. Chem Rev, 1999, 99, 779.
[61] Uchida C, Ogawa S. Numerical analysis of synchronization of chaos in two microchip lasers. Recent Res Dev Pure Appl Chem, 1999, 3, 161.
[62] Miyazaki H, Kobayashi Y, Shiozaki M, Ando O, Nakajima M, Hanzawa H, Haruyama H. Synthesis of a 2-aminohexahydrobenzoxazole analog related to trehazolin. J Org Chem, 1995, 60, 6103.
[63] Shiozaki M, Mochizuki T, Hanzawa H, Haruyama H. Synthesis of a tetrahydropyrano[2,3-D]oxazole analogue of trehazoli. Carbohydr Res, 1996, 288, 99.

[64] Uchida C, Ogawa S. Synthesis and inhibitory activity of glycosidase inhibitors, glycosylamino-oxazolines. Bioorg Med Chem, 1996, 4, 275.

[65] López Ó, Maya I, Ulgar V, Robina I, Fuentes J, Fernández-Bolaños JG. Expeditious synthesis of cyclic isourea derivatives of β-D-glucopyranosylamine. Tetrahedron Lett, 2002, 43, 4313.

[66] Chopade RS, Bahekar RH, Khedekar PB, Bhusari KP, Rao ARR. Synthesis and anticonvulsant activity of 3-(6-substituted-benzothiazol-2-yl)-6-phenyl-[1, 3]-xazinane-2-thiones. Arch Pharm, 2002, 335, 381.

[67] Hashem FA, Wahba HE. Isothiocyanates in myrosinase treated herb extract of *Cleome chrysantha* Decne. and their antimicrobial activities. Phytother Res, 2000, 14, 284.

[68] Tatibouët A, Lawrence S, Rollin P, Holman GD. Selective formation of 1,3-oxazolidine-2-thiones on ketohexose templates. Synlett, 2004, 1945.

[69] Kruse HP, Heydenreich M, Engst W, Schilde U, Kroll J. The identification of 1,3-oxazolidine-2-thiones and 1,3-thiazolidine-2-thiones from the reaction of glucose with benzyl isothiocyanate. Carbohydr Res, 2005, 340, 203.

[70] Velázquez F, Olivo HF. Synthesis of bicyclic γ-ylidenetetronates. Curr Org Chem, 2002, 6, 303.

[71] Elbert T, Cerny M. Collect. Czech.Reaction of 1,6-anhydro-4-*o*-benzyl-2-deoxy-2-isothiocyanato-β-D-glucopyranose; preparation of 2-amino-1,6-anhydro-2,3-dideoxy-2,3-dideoxy-β-D-ribo-hexopyranose. Chem Commun, 1985, 50, 2000.

[72] GarcíaFernández JM, Ortiz Mellet C, Fuentes J. Chiral 2-thioxotetrahydro-1,3-*O,N*-heterocycles from carbohydrates. Tetrahedron Lett, 1992, 33, 3931.

[73] GarcíaFernández JM, Ortiz Mellet C, Fuentes J. Chiral 2-thioxotetrahydro-1,3-O,N-heterocycles from carbohydrates. 2. Stereocontrolled synthesis of oxazolidine pseudo-C-nucleosides and bicyclic oxazine-2-thiones. J Org Chem, 1993, 58, 5192.

[74] GarcíaFernández JM, Ortiz Mellet C, Jiménez Blanco JL, Fuentes Mota J, Gadelle A, Coste-Sarguet A, Defaye J. Isothiocyanates and cyclic thiocarbamates of α, α′-trehalose, sucrose, and cyclomaltooligosaccharides. Carbohydr Res, 1995, 268, 57.

[75] García Fernández JM, Ortiz Mellet C, Díaz Pérez VM, Jiménez Blanco JL, Fuentes J. Conformational energetics of sugar thioureas and synthesis of glycosylthioureidosugars. Tetrahedron, 1996, 52, 12947.

[76] Gasch C, Pradera MA, Salameh BAB, Molina JL, Fuentes J. Chiral thioxohydroimidazoles with two sugar moieties. N-, C-, and spiro-nucleosides. Tetrahedron Asymm, 2000, 11, 435.

[77] FathiVavsari V, Mohammadi Ziarani G, Balalaie S, Latifi A, Karimi M, Badiei A. New functionalized 8-hydroxyquinoline-5-sulfonic acid mesoporous silica (HQS-SBA-15) as an efficient catalyst for the synthesis of 2-thiohydantoin derivatives. Tetrahedron, 2016, 72, 5420.

[78] Shiozaki M. Syntheses of hydantocidin and C-2-thioxohydantocidin. Carbohydr Res, 2002, 337, 2077.

[79] Shiozaki M. Synthesis of hydantocidin and C-2-thioxo-hydantocidin. Carbohydr Res, 2001, 335, 147.

Seema Kothari, Meghavi Gupta, Rakshit Ameta, and Suresh C. Ameta*

Chapter 10
Ultrasound-assisted synthesis of carbohydrate derivatives

10.1 Introduction

Normally, sound is classified into four categories, and these are:
- Infrasonic (<20 Hz)
- Audible sound (20 Hz to 20 kHz)
- Ultrasonic (20 kHz to 10 GHz)
- Hypersound (>10 GHz)

Out of these, ultrasound can be used in some chemical reactions to drive them in a particular direction or to increase their rate. Such reactions are called sonochemical reactions. These reactions can be divided into these classes: (i) homogeneous sonochemistry (liquid systems), (ii) heterogeneous sonochemistry (liquid-liquid or solid-liquid systems) and overlapping, and (iii) sonocatalysis. Sonochemistry is the chemistry that deals with the effect of ultrasound on chemical reactions, and it has been excellently reviewed by different researchers from time to time [1–3].

In the presence of ultrasound, cavitation bubbles are created. These bubbles increase in size during the positive cycle of the wave and collapse (acting as a hotspot), forming new nuclei for the next cavitation. As a result, the collapse of these cavitation bubbles results in increased pressure and temperature of the system to the order of 1,000 atm and 5,000 K, respectively. The cavitation is affected by many operations such as pressure, temperature, frequency of ultrasound, intensity of sonication, and natural bubbled gas. Ultrasound can be used in organic synthesis for increasing the yield, reducing reaction time, and improving the purity of final products, or sometimes even diverting from the main route (chemical switching). Ultrasound has also been utilized in the synthesis of some carbohydrate derivatives.

The use of ultrasound is advantageous in the following ways:
- high yields,
- increased selectivity,
- mild condition,
- green solvent,

*Corresponding author: Suresh C. Ameta, Department of Chemistry, PAHER University, Udaipur, Rajasthan, India, e-mail: ameta_sc@yahoo.com
Seema Kothari, Meghavi Gupta, Rakshit Ameta, Department of Chemistry, PAHER University, Udaipur, Rajasthan, India

https://doi.org/10.1515/9783111447094-010

- higher atom economy and low *E*-factor,
- short reaction time,
- reusability of solvent, and
- metal-free synthesis.

10.2 Monosaccharides

Monosaccharides are also called simple sugars, as these are the simplest form of sugar and the basic units of monomers. All other carbohydrates are built from these units as di-, oligo-, and polysaccharides. They are mostly soluble in water, colorless, and crystalline in nature. The simplest examples are glucose and fructose. Monosaccharides play an important role in biological functions such as the storage and release of energy and in the formation of cellular structures. Dextrose is generally used as a sweetener, bulking agent, and also as a crystalline component in coated candy.

Toukoniitty et al. [4] studied the effect of sonication on the hydrogenation of D-fructose (**1**) (Figure 10.1) under several heterogeneous catalytic conditions, such as pressure, temperature, and ultrasonic power input. It was reported that the reaction rates of hydrogenation were significantly increased in the presence of Cu/SiO_2 during sonication, while only a moderate improvement was observed in reaction rates when using Raney-Ni and there was a slight decrease in the case of $Cu/ZnO/Al_2O_3$. They also studied the effect of temperature and pressure on Raney-Ni-type catalysts under sonication. It was observed that high pressures and temperatures resulted in only slight enhancements of reaction rates, but variations in ultrasonic power significantly affected the activity of the catalyst. The deactivation of the catalyst was reported on the basis of recycling the catalyst. It was found that catalyst deactivation was significantly prevented under acoustic irradiation compared to experiments carried out in silent conditions.

Deng et al. [5] used sonication to facilitate organic reactions of carbohydrates. They reported a series of glycosylation reactions using glycosyl bromide, glycosyl acetate, thioglycosides, and glycosyl trichloroacetimidate as glycosyl donors (Figure 10.2). It was reported that sonication can significantly reduce reaction time and increase the reactivity of reactants, yield, and stereoselectivity. As a result, a general protocol for glycosylation may be developed. Sonication is compatible with the conditions used for traditional organic synthesis. Thus, sonication can also be applied to other areas of synthetic processes.

Amaniampong et al. [6] reported that the oxidation of D-glucose (**2**) to glucuronic acid (**3**) can be induced by high-frequency ultrasound irradiation (550 kHz) without using any (bio)catalyst, achieving excellent yields (Figure 10.3). It was observed that dissolved gases play an important role in controlling the nature of radical species generated and also in the selectivity for glucuronic acid. It was also revealed that this process gives glucuronic acid instead of the glucuronate salt, which is normally obtained via conventional

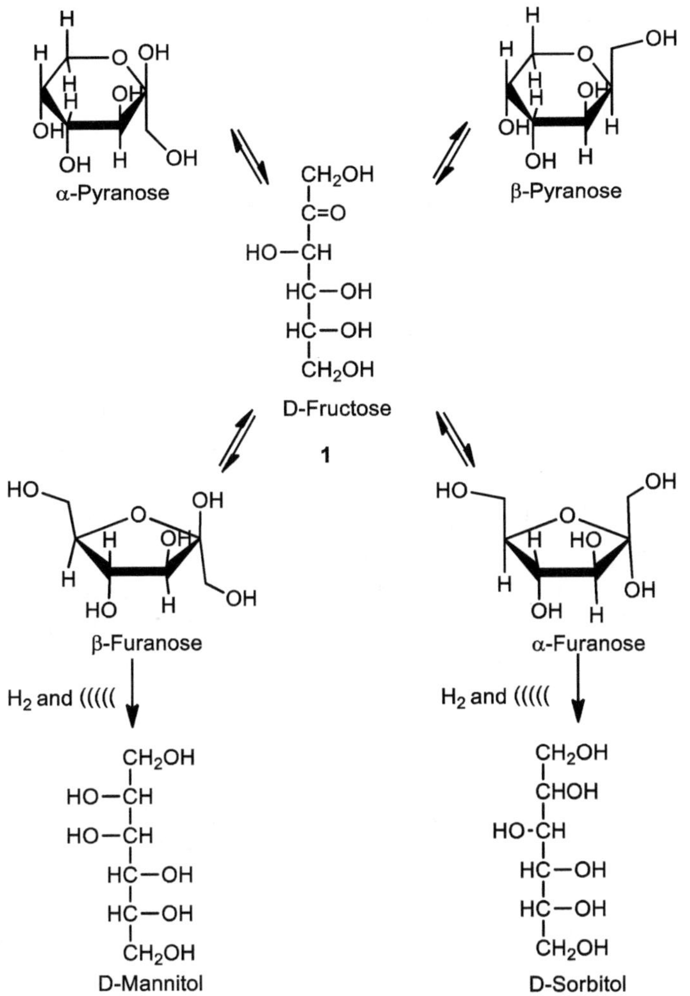

Figure 10.1: Mutarotation and hydrogenation of D-fructose.

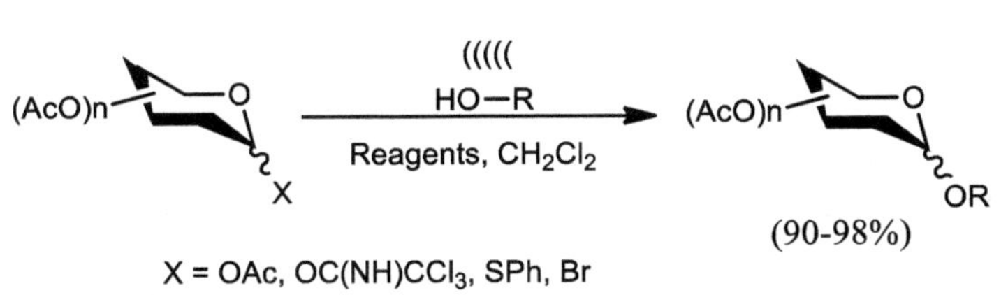

Figure 10.2: Synthesis of derivatives of various monosaccharides.

(bio)catalyst routes. It was found that radicals generated by high-frequency ultrasound can also promote tandem hydrolysis/oxidation reactions in the case of disaccharides.

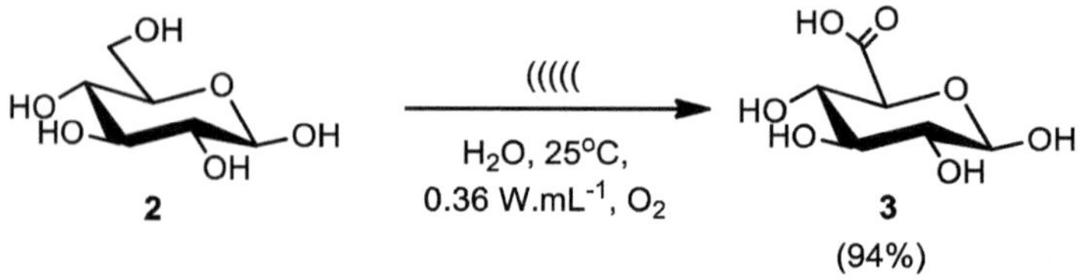

Figure 10.3: Ultrasound-assisted oxidation of D-glucose to glucuronic acid.

Amaniampong et al. [7] also observed a synergistic effect between a non-noble metal oxide catalyst (CuO) and high-frequency ultrasound (HFUS) on the oxidation of glucose. Independently, CuO and HFUS can oxidize glucose (**2**) to gluconic acid (**3**), but their combination led to a dramatic change in the selectivity of the reaction, forming glucuronic acid as the major product (Figure 10.4). It was also reported that the surface lattice oxygen of CuO (catalyst) traps H˙ radicals under the sonolysis of water. It supports the ring-opening of glucose, which is energetically difficult, leaving a high coverage of ˙OH radicals on the surface of CuO. This results in the selective oxidation of glucose to glucuronic acid.

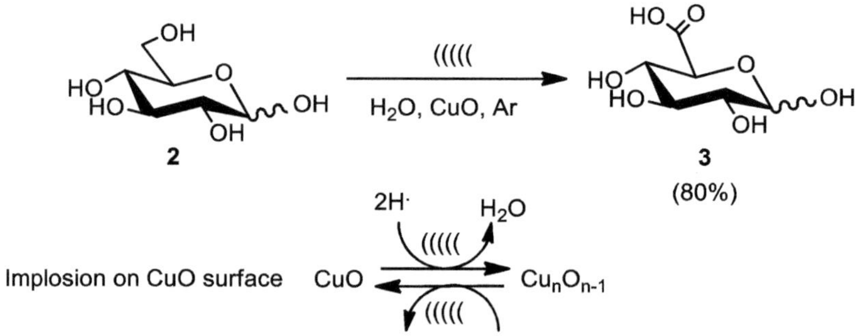

Figure 10.4: Oxidation of glucose to glucuronic acid.

The depolymerization of cellulose to glucose is a difficult reaction. Haouache et al. [8] depolymerized cellulose (**4**) selectively to glucose (**5**) in water by using (525 kHz) ultrasonic irradiation at high frequency (Figure 10.5). The H˙ and ˙OH radicals are formed by homolytic dissociation of water inside the cavitation bubbles. These radicals can induce cleavage of glycosidic bonds. It was reported that the transfer of radicals on the surfaces of cellulose particles restricts side degradation of released glucose into the bulk solution, and as a result, selectivity to glucose was close to 100%. The present work is different from existing technologies from the point of view that:

- no catalyst was required,
- no external source of heating was required, and
- complete depolymerization of cellulose can be achieved in a selective fashion.

It was also revealed that H˙ radicals were mainly responsible for depolymerization of cellulose based on specific radical scavenger studies.

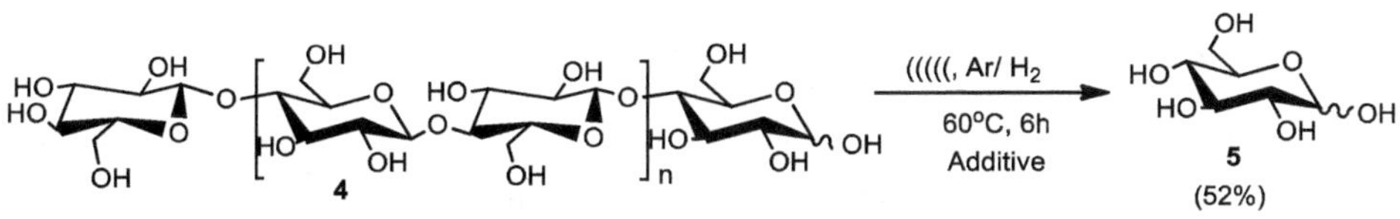

Figure 10.5: Depolymerization of cellulose to glucose.

10.3 Disaccharides

Disaccharides are formed by joining two identical or different monosaccharides via α- or β-glycosidic bonds. Some common examples of disaccharides are sucrose, lactose, and maltose. Sucrose is commonly used as a renewable raw material for the preparation of its esters in a wide range of fields such as cosmetics, detergents, emulsifiers, whitening boosters, medicines, stabilizing agents, and nutritional supplements. Some specific physical and organoleptic properties of fats were acquired by these polysubstituted sucrose esters on transesterification of all hydroxyl groups with fatty acids. Sucrose esters are suitable as surfactants and nutritional supplements, possibly due to their stabilizing, emulsification, and conditioning properties. However, there is a major problem with the synthesis of sucrose esters because there are eight hydroxyl groups, and as a result, mixtures of monoesters, diesters, and higher esters are formed.

Zhao et al. [9] reported the synthesis of glycals (**7**) from glycosyl bromides (**6**) with a very low loading of zinc dust (1.5 equiv.) (Figure 10.6). It was observed that this process was activated by β-cyclodextrin and ultrasound. This method was found to be highly effective for a wide range of glycosyl bromides (acid- as well as base-sensitive and disaccharide glycosyl bromides) with a high yield of glycals (85–96%).

Acetobromo-α-D-maltose was selected to observe the effect of ultrasound and β-CD (catalyst) on the synthesis of D-maltal at room temperature. The starting material was treated in phosphate buffer, PEG600-H_2O, or β-CD-H_2O under ultrasound irradiation and stirring. It was reported that a low yield was obtained in phosphate buffer, while a high yield was achieved in water and β-CD. The reaction time was reduced, and the yield under ultrasound irradiation was also increased by more than 7%. It was concluded that water mixed with β-CD was quite efficient compared to PEG600-H_2O.

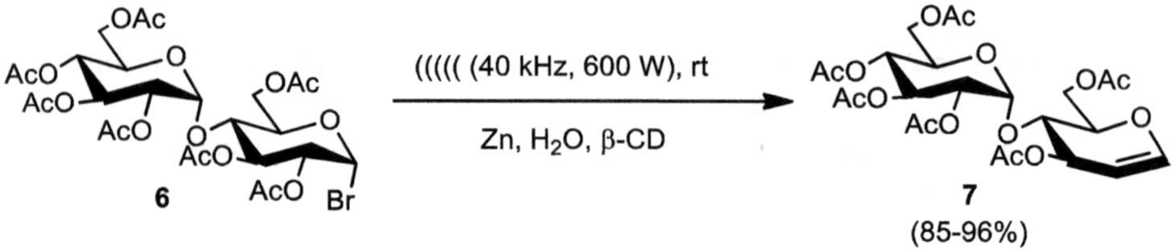

Figure 10.6: Synthesis of maltal from acetylated maltosyl bromide.

Petkova et al. [10] synthesized sucrose undecylenic acid (**10**) esters via a transesterification reaction of sucrose (**8**) and the methyl ester of 10-undecylenic acid (**9**) under ultrasonic irradiation (Figure 10.7) and compared it with the reaction under conventional conditions. The effect of the catalyst (sodium methoxide and potassium carbonate), temperature, and molar ratio of reactants on the yield of esters was observed. It was reported that the highest yield was obtained in DMSO with K_2CO_3 (catalyst, 13 mol %), at a temperature of 45 °C for 2 h under ultrasound irradiation (45 kHz).

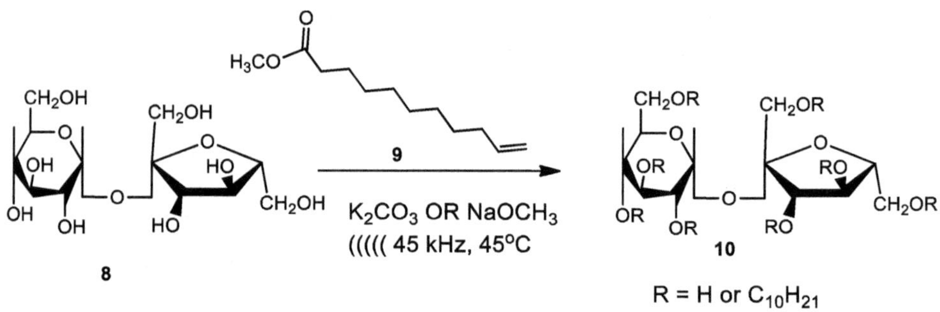

Figure 10.7: Synthesis of undecylenic sucrose esters.

Vassilev et al. [11] synthesized sucrose and fructooligosaccharides (FOS) esters (**11** and **12**) with 10-undecylenic and palmitic acid by an ultrasound-assisted route (Figure 10.8). They evaluated the plasticizing effect of sucrose palmitate applied in polyvinyl chloride (PVC). The sucrose ester was prepared by taking 50 mL of methanol and sucrose and methyl palmitate in an equimolar ratio (1:1) in a 100 mL flask. Then 0.5 mol of CO_3 was added as a catalyst. This mixture was sonicated at room temperature for 2 h. The residue obtained was dissolved in 25% NaCl/*n*-butanol (1:1) and the mixture was separated by liquid–liquid extraction with 20 mL of *n*-butanol (three times). The extracts were dried with anhydrous sodium sulfate and then the solvent was distilled off.

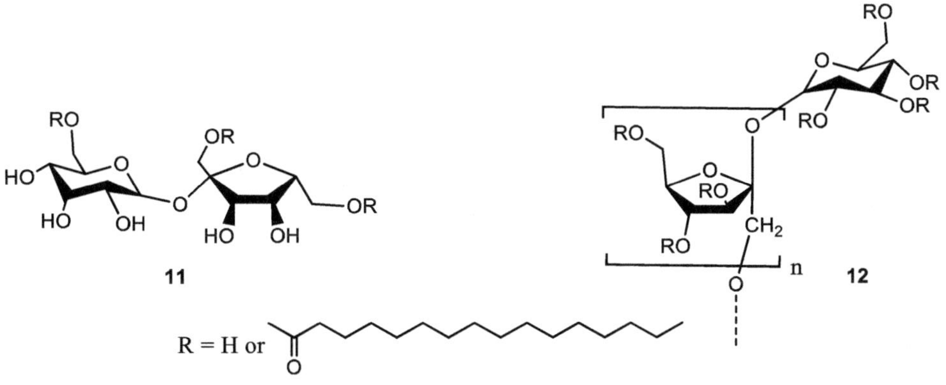

Figure 10.8: Synthesis of sucrose and fructooligosaccharide esters.

Vassilev et al. [12] mixed 50 mL of methanol and 0.018 mol of the methyl ester of a particular fatty acid, sucrose, and catalyst (0.1 mol) were placed in a 100 mL Erlenmeyer flask. The reaction was then carried out at 40 °C and 65 °C in a Dimoff A-2/2 (100 W, 44 kHz) ultrasonic bath. The solvent was evaporated after 1 hour, and the residue was dissolved in 25% NaCl/*n*-butanol (1:1 v/v). The extraction was done with 20 mL of *n*-butanol three times. The extracts were dried with anhydrous sodium sulfate.

Petkova et al. [13] carried out the esterification of long-chained inulin (**14**) and sucrose (**13**) with a monounsaturated ester of 10-undecylenic acid (**9**) using ultrasound irradiation. It was reported that the hydrophobic 10-undecylenoyl residue was successfully incorporated into the water-soluble carbohydrate backbone (Figure 10.9). About 5.13 g of sucrose (or 5 g of inulin) was added to 6 mL of DMSO in a two-neck round-bottom flask. It was placed into an ultrasonic bath VWR (Malaysia) (45 kHz and power 30 W) at 45 °C. When the sucrose was completely dissolved, the methyl ester of 10-undecylenic acid (0.92 g) was added drop wise, and then sodium methoxide (catalyst) was added. This reaction mixture was sonicated under a nitrogen atmosphere for 1 h. The liquid-liquid extraction was performed with *n*-butanol, and it was further purified with ethyl acetate.

Inulin and sucrose esters were treated with 10-undecylenic acid and sodium methoxide (catalyst) in DMSO. The yield obtained was 55–57%. When the molar ratio was kept at 1:1 and the reaction was carried out at 45 °C for 1 h, the product was a yellow liquid with the specific odor of 10-undecylenic acid.

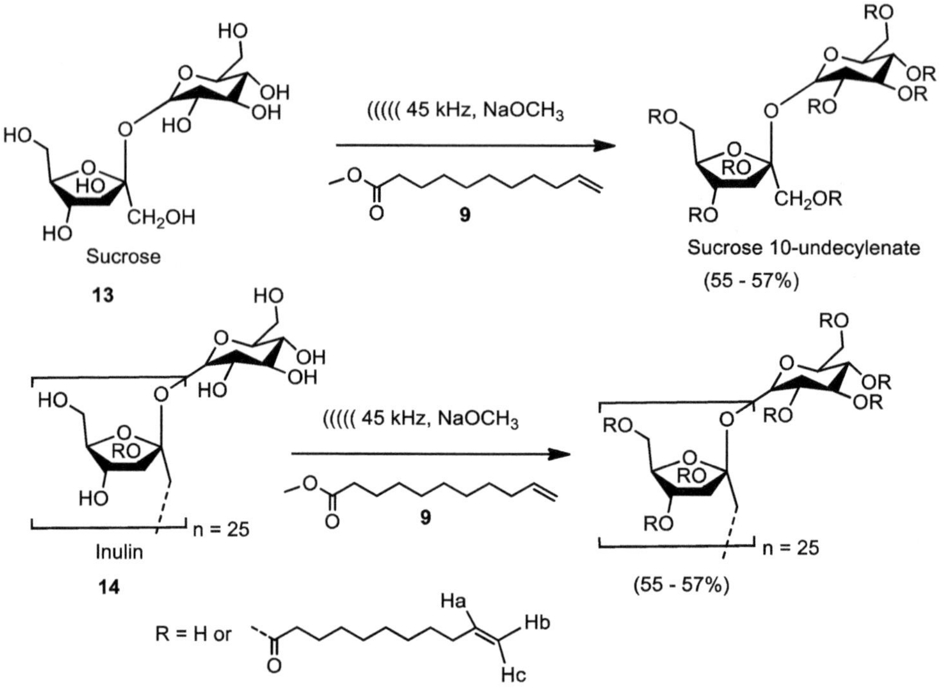

Figure 10.9: Synthesis of 10-undecylenic sucrose and inulin esters.

10.4 Polysaccharides

Polysaccharide is a carbohydrate, where a molecule consists of a number of sugar molecules bonded together. The common examples are glycogen, cellulose, or starch. They exhibit some biological and pharmacological activities. These are building blocks for drug delivery, drug carriers, bioactive materials, antimicrobial, anti-tumor, immunomodulatory, anticoagulant, antioxidant, antiviral, antidiabetic, and hypoglycemic activities. This makes them promising candidates with potential in the biomedical and pharmaceutical fields.

10.4.1 Chitosan

Chitosan is a linear polysaccharide that consists of randomly distributed *N*-acetyl-D-glucosamine (acetylated unit) and β-(1→4)-linked D-glucosamine (deacetylated unit).

Chitosan is also a sugar. Its molecular structure consists of a backbone linked through glycosidic bonds. Chitosan (CTS) is a biocompatible, non-toxic polymer, which is also biodegradable. It has poor affinity for organic solvents. It has interesting antimicrobial properties and can be used as an antimicrobial agent in agriculture, food, and medical treatments. It can affect regenerative technology, such as skin, cornea, and cardiac regenerative technologies.

Wang et al. [14] synthesized chitosan derivatives (**17**) with a *p*-acetamidobenzoylate group via the acylation reaction of chitosan (**15**) with *p*-acetamidobenzoylate chloride (**16**) in acetic acid using high-intensity ultrasound (Figure 10.10). It was reported that the maximum substitution degree of the derivative was found to be 0.42. As it absorbs at 273 nm and its solubility is also higher compared to chitosan, it opens new avenues for use as a sunscreen.

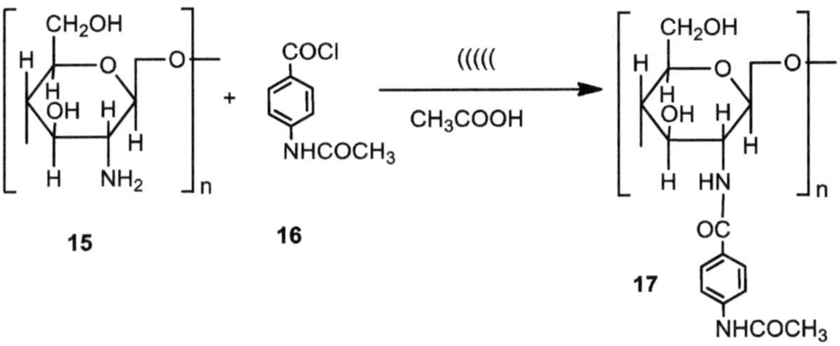

Figure 10.10: Synthesis of p-acetamidobenzoyl chitosan.

Jin et al. [15] synthesized a Schiff base of chitosan (**20**) by treating chitosan (**18**) with citral (**19**) in the presence of high-intensity ultrasound (Figure 10.11). They also studied the effect of reaction time, temperature, and molar ratio of chitosan to citral on the yield. The optimal conditions were achieved as follows: temperature (50 °C), reaction time (10 h), and molar ratio of chitosan to citral (1:6). They achieved 86.4% yield. It was also confirmed that the amino groups of chitosan were utilized in forming the Schiff base. It was also indicated that the antimicrobial activity of the as-prepared Schiff base was greater compared to chitosan.

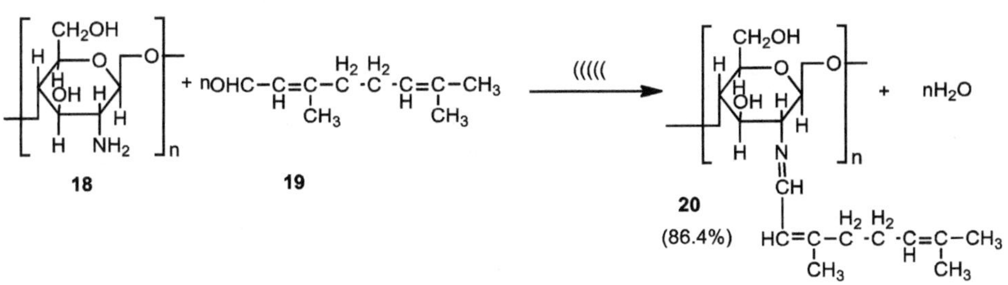

Figure 10.11: Synthesis of Schiff base chitosan.

Wang et al. [16] synthesized Schiff base of chitosan (**21**), sorbyl chitosan (**22**), and *p*-aminobenzoyl chitosan (**23**) under high-intensity ultrasound (Figure 10.12). The antibacterial activity of this Schiff base of various chitosans against *E. coli* was relatively higher, while acylated chitosan exhibited a better inhibiting effect on *S. aureus* than others. It was also revealed that the antifungal activities of these derivatives were better than chitosan. The Schiff base of chitosan displayed superior activity compared to that of acylated chitosan.

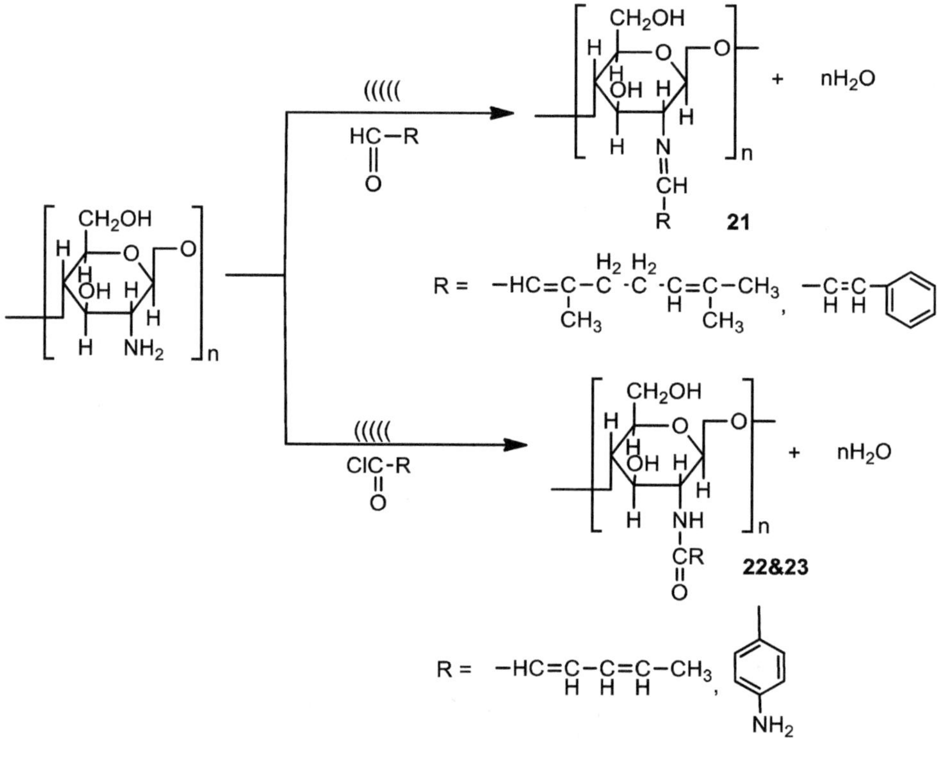

Figure 10.12: The synthesis of the Schiff base of chitosan and acylated chitosan.

Freire et al. [17] synthesized chitosan-based magnetite nanocomposites under ultrasound in a short time of 2 minutes. This procedure provided spheroidal nanoparticles (Fe_3O_4) with an average diameter of 10–24 nm, and they exhibited superparamagnetic properties with saturation magnetization (Ms) ranging from 32 to 57 emu g^{-1}. It was reported that this as-prepared hybrid nanocomposite may have potential use as electrochemical sensors, as the electrochemical signal was found to be exceptionally stable.

Kritchenkov et al. [18] carried out the ultrasound-assisted synthesis of 2-((3-(4-formylphenoxy)allyl)oxy)-*N*,*N*,*N*-trimethyl-2-oxoethan-1-aminium chloride (**26**) by phenol-yne addition of the propargylic ester of betaine hydrochloride (**25**) and *p*-hydroxybenzaldehyde (**24**) (Figure 10.13). It was reported that phenolic chitosan derivatives were prepared by treating chitosan with *o*-, *m*-, or *p*-hydroxybenzaldehyde, which was followed by the reduction of the C = N bond by sodium borohydride. Here, phenolic chitosan derivatives (phenolic component) reacted with the propargylic ester of betaine hydrochloride (yne component) in this ultrasound-mediated reaction. It was also revealed that out of these, the most effective polymer was the *p*-isomer of the highly substituted betaine derivative, and its activity was found to be more than twofold higher than the activity of tetracycline.

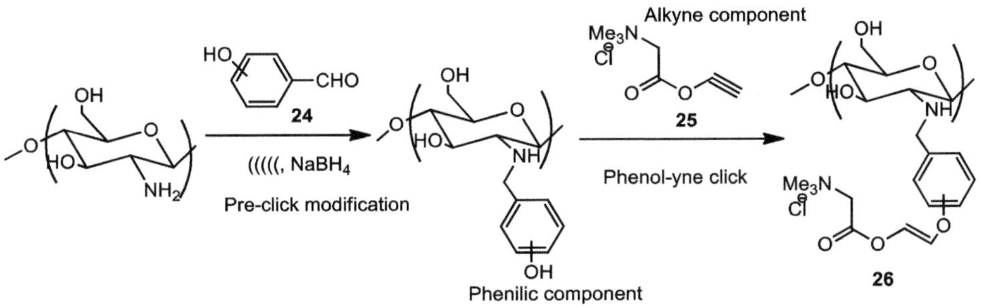

Figure 10.13: Synthesis of 2-((3-(4-formylphenoxy)allyl)oxy)-N,N,N-trimethyl-2-oxoethan-1-aminium chloride.

Kritchenkov et al. [19] observed (i) alkylation of chitosan by alkyl halides, (ii) aza-Michael reaction, and (iii) Ad_N-E reaction of chitosan with aldehydes, which was efficiently mediated by ultrasound. It was observed that the use of ultrasonic irradiation under optimal conditions allowed:

- a dramatic increase in the rate of the reactions,
- high selectivity was achieved, and
- integrity of the polysaccharide backbone avoids its depolymerization.

It was also revealed that these polymers proved to be better than standard antibiotics such as ampicillin and gentamicin.

Lunkov et al. [20] synthesized quaternized chitosan derivatives (HT-TMC) by copper-catalyzed azide-alkyne [3 + 2] cycloaddition (CuAAC). This reaction was carried out with Cu(I) formed in situ under ultrasound irradiation in aerobic conditions using acetic acid and metallic copper (copper turnings).

10.4.2 Cyclodextrin

Cyclodextrins (CDs) are natural cyclic oligosaccharides having a macrocyclic ring of glucopyranose units linked by α-(1–4)-glycosidic bonds. They can be used for the delivery of a variety of drugs, including prostaglandin, hydrocortisone, itraconazole, nitroglycerin, and chloramphenicol. They improve permeability, solubility, and stability. Because of their cavity-like structure, they can form inclusion complexes with a variety of compounds, which can be commonly used in the food industry for increasing antibacterial activity and extending the storage time of food.

Trotta et al. [21] synthesized cyclodextrin (CD) derivatives (27), such as 2^I-O-mono (methylamino)alkyl-β-CDs, 6^I-monotosyl, 6^I-monoazido-6^I-monodeoxy, and 6^I-monoamino -6^I-monodeoxy-β-CD under microwave or ultrasound irradiation with better yields and purity in reduced time (Figure 10.14). They also used these radiations for preparations such as those of 6 ^{I}-amino-6I-deoxy-β-CD and a series of 2-monoaminomethyl-β-CD derivatives.

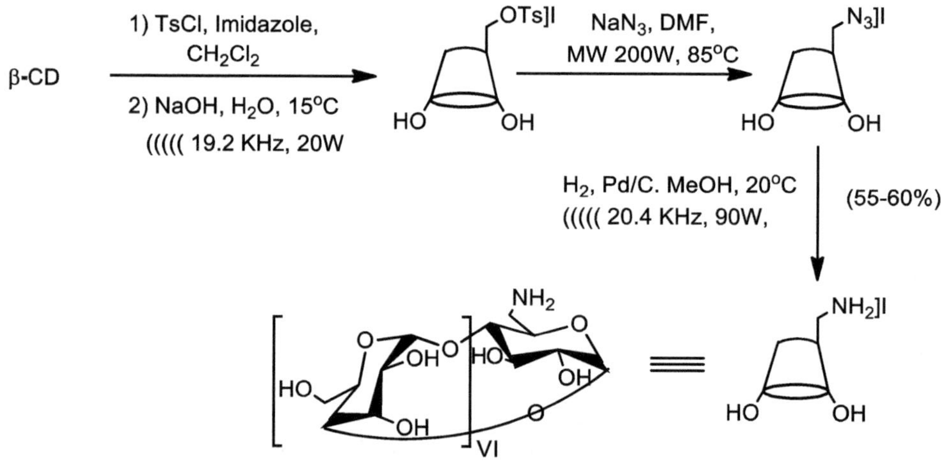

(27)

Figure 10.14: Synthesis of cyclodextrin derivatives.

Zhang et al. [22] synthesized an amphiphilic β-cyclodextrin (β-CD) dimer, LA-$(CD)_2$. In this dimer, the lithocholic acid (LA) unit is hydrophobic, while the two β-CD units are hydrophilic in nature. It was reported that LA-$(CD)_2$ self-assembles into spherical micelles in an aqueous solution at room temperature under ultrasound. First, primary micelles dissociate and then form self-assemblies (branched structures) under ultrasonication. These branched aggregates revert back to the primary micelles at high temperatures. It was indicated that the primary micelles formed are driven by hydrophilic–hydrophobic interactions, while host–guest interactions promote the formation of secondary assemblies. It was also revealed that ultrasonication effectively destroys primary hydrophilic–hydrophobic balances, but it increases host–guest interaction between β-CD and LA at room temperature.

Zheng et al. [23] synthesized mono-(6-*p*-toluenesulfonyl)-β-cyclodextrin (**28**) efficiently by an ultrasound-assisted method in 40 min using an alkaline water solution (Figure 10.15). They achieved a 31.1% yield under these conditions. This method reduced reaction time and increased the yield of the product compared to conventional synthetic methods.

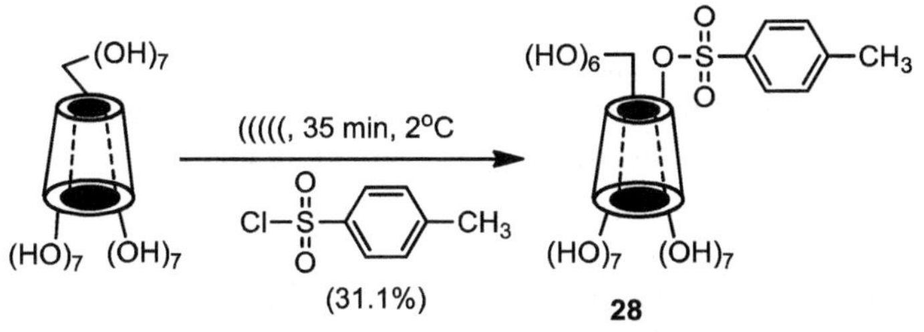

Figure 10.15: Synthesis of mono-(6-p-toluenesulfonyl)-β-cyclodextrin.

Cui et al. [24] synthesized an inclusion complex of cuminaldehyde (CUM) with 2-hydroxypropyl-β-cyclodextrin (HPβCD) using ultrasound. It was observed that the phenyl ring with the aldehyde group of CUM was inserted into the hydrophobic cavity of HPβCD. It was observed that the inclusion of CUM within the HPβCD cavity improved its thermal stability, water solubility, and antibacterial activities.

Liu et al. [25] prepared β-cyclodextrin-based polymeric flocculants (CADs) and used them for the treatment of humic acid (HA)-kaolin wastewater. They grafted acrylamide (AM) and acryloyloxyethyltrimethyl ammonium chloride (DAC) onto β-cyclodextrin (β-CD) using ultrasound irradiation and ceric ammonium nitrate as an initiator. It was reported that CADs exhibited more turbidity and removal performance of HA compared to synthetic non-ionic graft polymers (CAM) and cationic flocculants (PAD). The maximum turbidity and HA removal rate of CAD was found to be 97.5% and 68.3%, respectively, at a dosage of 4 mg/L.

5-Fluorouracil (5-FU) is mostly used in the treatment of stomach cancer, but it is poorly absorbed upon oral administration. Therefore, Jasim et al. [26] synthesized 5-FU nanosponges (NS) (**31**) to increase its accumulation in gastric tumors and lower systemic side effects. They prepared CD-nanosponges via crosslinking β-CD (**29**) with diphenyl carbonate (DPC) (**30**) under ultrasound irradiation (Figure 10.16). It was revealed that a sample, 5-FUNS2, exhibited an initial dissolution rate of 56% in the first hour.

Mohandoss et al. [27] synthesized inclusion complexes of the salsalate drug (SLT) with three different β-cyclodextrin derivatives, such as β-cyclodextrin (β-CD), hydroxypropyl-β-cyclodextrin (HPβ-CD), and sulfobutylether-β-cyclodextrin (SBEβ-CD), under ultrasound irradiation. It was reported that β-CDs:SLT inclusion complexes displayed

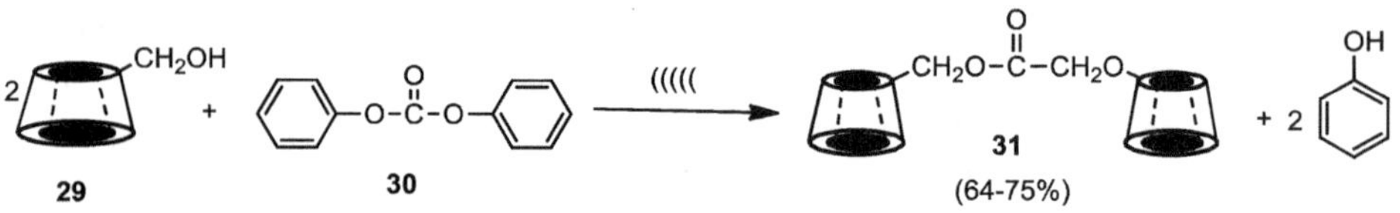

Figure 10.16: Synthesis of β-cyclodextrin-based nanosponges.

excellent activity against *S. aureus* and *E. coli* and reduced the cytotoxic effects of SLT on MCF-7 cells. They also exhibited inhibition of α-glucosidase and α-amylase, which indicated their antidiabetic properties.

Zhang et al. [28] prepared three inclusion complexes (ICs) with β-cyclodextrin (β-CD), 2-hydroxypropyl-β-cyclodextrin (2H-β-CD), and methyl-β-cyclodextrin (M-β-CD) using an ultrasonic method to improve the aqueous solubility and increase the bioavailability of *Hyperoside* (Hyp) (**32**) (Figure 10.17). They also investigated the effects of these ICs on the solubility and antioxidant activity of Hyp. It was reported that the water solubility of the IC of Hyp with 2-hydroxypropyl-β-cyclodextrin increased nine times compared to the solubility of the original Hyp.

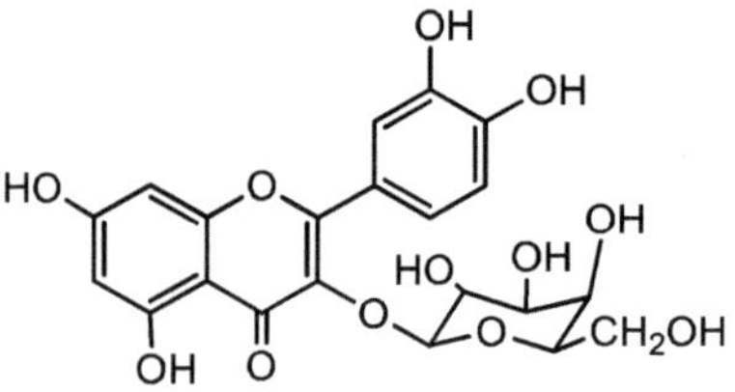

32

Figure 10.17: Chemical structure of hyperoside.

10.5 Derivatives of carbohydrates

10.5.1 Glycosyl fullerene derivatives

Yoon et al. [29] carried out cycloaddition of [70]fullerene with different 2-azidoethyl per-*O*-acetyl glycopyranosides of D-glucose, D-galactose, D-mannose, D-maltose, and D-xylose under ultrasonication. They obtained glycosyl[70]fullerene derivatives (**33**) (Figure 10.18). It was observed that the products were a mixture of the closed [5,6]-bridged isomer (major) and two isomeric (minor) monoaddition products. However, an unusual mixture of four mono-adducts was obtained in the ratio (9:11:10:4) in the case of D-xylose.

The 2-azidoethyl glycosides (0.048 mmol) were added to a solution of [70]fullerene (20 mg, 0.024 mmol) in benzene (15 mL). Then, this mixture was irradiated for 2 days at 25–43 °C under ultrasound irradiation (20 kHz). D-Glycosyl[70]fullerene derivatives were obtained as products.

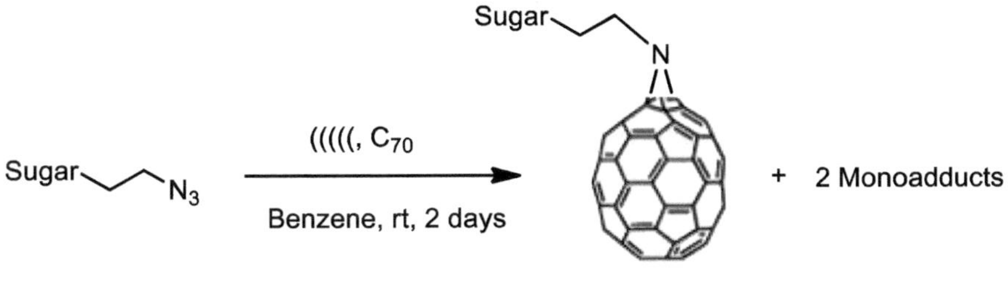

33
(5.6 - 8.7%)

Figure 10.18: Synthesis of various D-glycosyl[70]fullerene derivatives.

10.5.2 1,2,3-Triazole glycosides

Marzag et al. [30] developed a highly efficient process for the selective anomeric azidation of protected sugars (**34**) using iron catalysis and ultrasound activation in the presence of $SO_2(N_3)_2$. It was reported that azidoglycosides undergo 1,3-dipolar cycloaddition reaction with terminal alkynes under ultrasound conditions to give 1,2,3-triazole glycosides (**35**) in high yields (Figure 10.19). Sulfuryl chloride (1 mmol) was added drop wise to a cold suspension of sodium azide (2 mmol) in dichloromethane (5 mL). The mixture was sonicated for a few minutes. Then, a Lewis acid catalyst (20 mol%) and acetylated sugar (2 mmol) were added to this mixture. The reaction mixture was sonicated for 45 min. The mixture was diluted with dichloromethane and washed with a saturated aqueous solution of sodium bicarbonate. The organic layer was washed with water (2 × 10 mL), dried over magnesium sulfate, and filtered to give azidoglycosides.

Anhydrous $FeCl_3$ (20 mol%) was used as a catalyst under sonication. Alkyne (4 mmol), CuI (4 mmol), and diisopropylethylamine (4 mmol) were added to this mixture, and then it was sonicated. This mixture was diluted with dichloromethane, washed with a saturated solution of ammonium chloride and water (2 × 10 mL), dried over magnesium sulfate, and filtered. It afforded triazolyl glycosides in good yields (62–89%).

Evangelista et al. [31] prepared 2,3-*O*-isopropylidene derivatives of D-ribose under ultrasonic irradiation (Figure 10.20). The ultrasound was found to be an excellent energy source for this kind of reaction. It was reported that reduced reaction times and high yields could be achieved without any need for purification of acetonide. It was then used as starting material in the synthesis of 23 glycoconjugates of 1,2,3-triazoles (**36 and 37**).

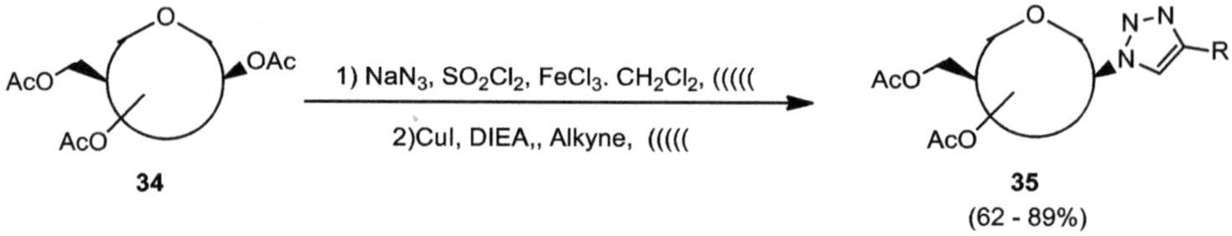

Figure 10.19: Synthesis of 1,2,3-triazole glycosides.

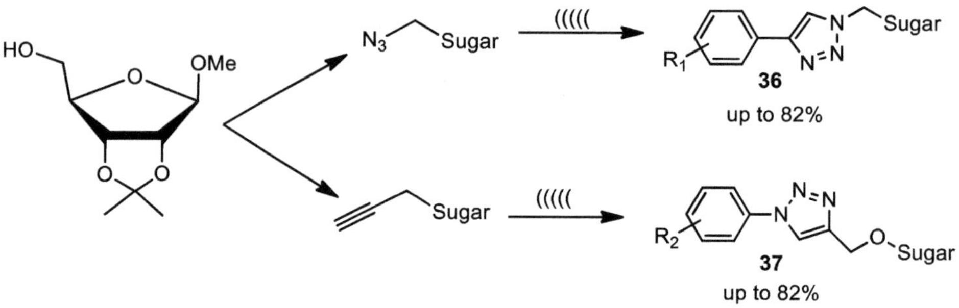

Figure 10.20: Synthesis of 1,2,3-triazole glycoconjugates.

10.5.3 Ester and ether derivatives

Shaikh et al. [32] used ultrasound for the Fischer glycosylation of different sugars using sulfuric acid immobilized on silica as a catalyst. It was reported that D-glucose, D-galactose, *N*-acetyl-D-galactosamine, *N*-acetyl-D-glucosamine, L-fucose, lactose, and D-mannose were glycosylated with 2-azidoethanol or propargyl alcohol, affording the corresponding glycosides, with α-glycopyranoside as the major product. It was observed that there was a significant enhancement of glycosylation reactions, and the reaction could be completed with good yields (70–90%) in 15 min to 2 h compared to several hours in the case of conventional procedures.

Lu et al. [33] acylated raffinose (long-chain fatty acid monoesters) (**40**) at the 1- and 6-positions, using raffinose (**38**) with aliphatic methyl esters (**39**) under ultrasound irradiation and reduced pressure (Figure 10.21). It was reported that yields of products varied in the range of 15–44%. They could achieve optimum reaction conditions for this reaction. The quantity of anhydrous potassium carbonate accounted for raffinose (12 mol%). The molar ratio of raffinose to fatty acid methyl ester was 2:1, with an ultrasonic temperature of 65 °C, ultrasonic time of 2 h, and pressure of 10 kPa. The acylation had good monoester selectivity. It was also revealed that these surfactants had a higher percentage of monoesters, and the ratio of 1-*O*-monoester/6-*O*-monoester was 2.2–2.6.

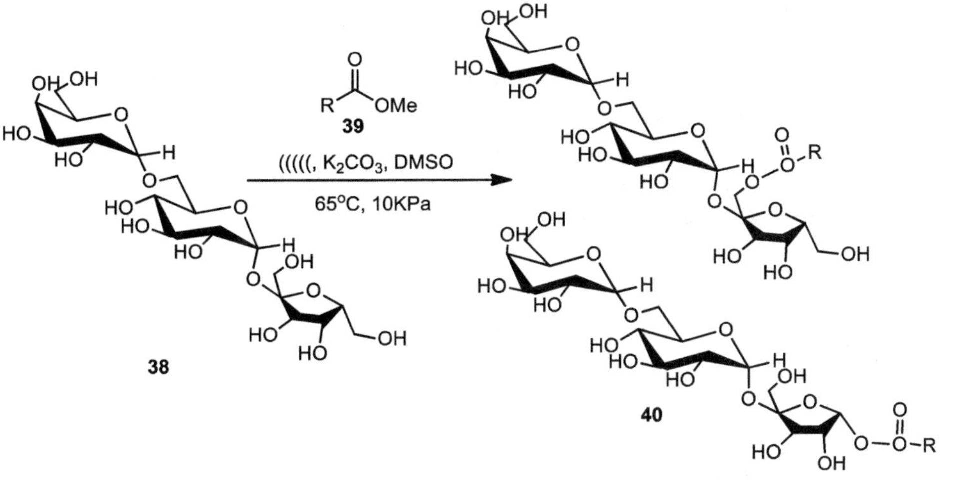

Figure 10.21: Synthesis of raffinose fatty acid monoesters.

Marzag et al. [34] synthesized phenol-derived *O*-glycosides by *O*-glycosidation of acetylated sugars in the presence of $FeCl_3$ (a mild and harmless Lewis acid catalyst). These reactions were carried out under both ultrasonic irradiation and conventional conditions, where rates and yields were found to be higher under sonication compared to the conventional route. It was observed that the corresponding β-*O*-glycosides were the major anomers. The resveratrol *O*-glycoside analogs were obtained in good yields (54–80%) in 20–30 min.

Amaniampong et al. [35] reported that the formation of alkylpolyglycosides (APGs) (**41**) is possible under high-frequency ultrasound (550 kHz) irradiation of concentrated feeds of carbohydrates in alcoholic solution (Figure 10.22). It was observed that this reaction takes place at 40 °C, and it avoids any degradation of carbohydrates. It also yields APGs selectively with a degree of polymerization in a window (2–7) under the Fischer glycosylation reaction. It was also revealed that this ultrasound-based technology was very successful with a wide range of carbohydrates and alkyl alcohols. It was suggested that 1,6-anhydrosugars formed as key intermediate species.

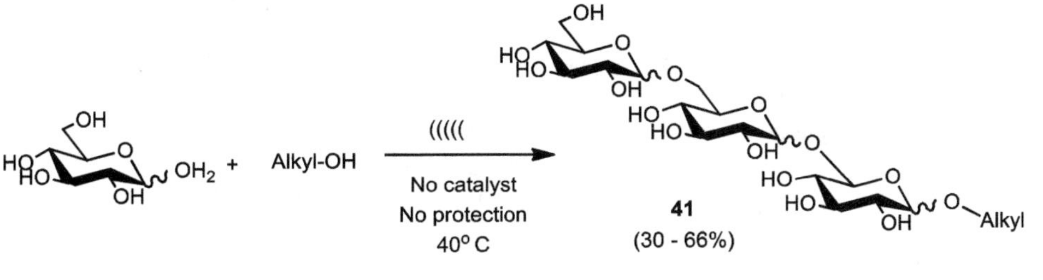

Figure 10.22: Synthesis of alkylpolyglycosides.

Xylitol is a natural sugar alcohol found in plants, including many fruits and vegetables. It has a sweet taste and is often used as a sugar substitute. Nieto et al. [36] used immobilized lipase for the synthesis of five xylityl acyl esters by esterification of some free fatty acids (such as capric, caprylic, myristic, and lauric acid) with xylitol under solvent-free conditions. It was reported that the use of ultrasound irradiation can overcome both these problems: (i) mutual immiscibility of fatty acids and xylitol and (ii) semisolid character of initial reaction mixtures. Here, ultrasonic irradiation can help in the transport of substrate molecules to the catalytic-site of the enzyme, which resulted in efficient synthesis of xylityl fatty ester with 95% yield at 40 °C in 90 min. The xylityl monoacyl ester and xylitol diacyl ester were found as the main products (>96% yield).

10.6 Conclusion

Ultrasound irradiation provides a highly effective means for the synthesis of carbohydrate derivatives by significantly accelerating chemical reactions and enhancing reaction yields. The use of ultrasound offers several advantages: Enhanced reaction rates, improved yields, selectivity, versatility and scalability, and environmental friendliness. Thus, ultrasound-assisted organic synthesis is a promising technique. However, its use requires careful optimization and consideration of the specific reaction conditions. This ultrasound synthesis is becoming an increasingly valuable tool from an industrial point of view as well. Therefore, ultrasound irradiation can be used as a powerful tool for the synthesis of carbohydrate derivatives, offering faster, cleaner, and more efficient processes. Its application in carbohydrate chemistry results in advances in green and sustainable chemical practices, and it will become a forefront area of research in organic chemistry in the years to come.

References

[1] Mason TJ. Sonochemistry. Oxford University Press: Oxford, 1999.

[2] Xu H, Zeiger BW, Suslick KS. Sonochemical synthesis of nanomaterials. Chem Soc Rev, 2013, 42, 2555–2567.

[3] Ameta SC, Ameta R, Ameta G. Sonochemistry: An Emerging Green Technology. Apple Academic Press, Taylor & Francis Group: Waretown, USA, 2018.

[4] Toukoniitty B, Kuusisto J, Mikkola JP, Salmi T, Murzin DY. Effect of ultrasound on catalytic hydrogenation of D-fructose to D-mannitol. Ind Eng Chem Res, 2005, 44(25), 9370–9375.

[5] Deng S, Gangadharmath M, Chang CW. Sonochemistry: A powerful way of enhancing the efficiency of carbohydrate synthesis. J Org Chem, 2006, 71(14), 5179–5185.

[6] Amaniampong PN, Karam A, Trinh QT, Xu K, Hirao H, Jerome F, Chatel G. Selective and catalyst-free oxidation of D-glucose to D-glucuronic acid induced by high-frequency ultrasound. Sci Rep, 2017, 7 (1). DOI: 10.1038/srep40650.

[7] Amaniampong PN, Trinh QT, de Oliveira Vigier K, Dao DQ, Tran NH, Wang Y, et al. Synergistic effect of high-frequency ultrasound with cupric oxide catalyst resulting in a selectivity switch in glucose oxidation under argon. J Am Chem Soc, 2019, 141(37), 14772–14779.

[8] Haouache S, Karam A, Chave T, Clarhaut J, Amaniampong PN, Fernandez JMG, et al. Selective radical depolymerization of cellulose to glucose induced by high frequency ultrasound. Chem Sci, 2020, 11(10), 2664–2669.

[9] Zhao J, Shao H, Wu X, Shi S. A rapid synthesis of pyranoid glycals promoted by β-cyclodextrin and ultrasound. Chin J Chem, 2011, 29(7), 1434–1440.

[10] Petkova N, Vassilev D, Grudeva R, Koleva M, Denev P. Ultrasound-assisted synthesis of undecylenoyl sucrose esters. In International Scientific Conference Unitech, 2015. DOI: 10.13140/RG.2.1.4149.7048.

[11] Vassilev D, Petkova N, Koleva M, Denev P. Ultrasound-assisted synthesis of sucrose and fructooligosaccharides esters as bio-plasticizers. J Renew Mater, 2016, 4(1), 24–30.

[12] Vassilev D, Petkova N, Koleva M, Denev P. Optimization of ultrasound synthesis of sucrose esters by selection of a suitable catalyst and reaction conditions. J Chem Technol Metall, 2021, 56(2), 268–274.

[13] Petkova N, Arabadzhieva R, Hambarliyska I, Vassilev D, Gencheva G, et al. Ultrasound-assisted synthesis of antimicrobial inulin and sucrose esters with 10-undecylenic acid. Biointerface Res Appl Chem, 2021, 11(4), 12055–12067.

[14] Wang J, Jin X, Chang D. Chemical modification of chitosan under high-intensity ultrasound and properties of chitosan derivatives. Carbohydr Polym, 2009, 78(1), 175–177.

[15] Jin X, Wang J, Bai J. Synthesis and antimicrobial activity of the Schiff base from chitosan and citral. Carbohydr Res, 2009, 344(6), 825–829.

[16] Wang J, Lian Z, Wang H, Jin X, Liu Y. Synthesis and antimicrobial activity of Schiff base of chitosan and acylated chitosan. J Appl Polym Sci, 2012, 123(6), 3242–3247.

[17] Freire TM, Dutra LMU, Queiroz DCD, Ricardo NMPS, Barreto K, Denardin JC, et al. Fast ultrasound assisted synthesis of chitosan-based magnetite nanocomposites as a modified electrode sensor. Carbohydr Polym, 2016, 151, 760–769.

[18] Kritchenkov AS, Egorov AR, Volkova OV, Kritchenkov IS, Kurliuk AV, Shakola TV, et al. Ultrasound-assisted catalyst-free phenol-yne reaction for the synthesis of new water-soluble chitosan derivatives and their nanoparticles with enhanced antibacterial properties. Int J Bio Macromol, 2019, 139, 103–113.

[19] Kritchenkov AS, Zhaliazniak NV, Egorov AR, Lobanov NN, Volkova OV, Zabodalova LA, et al. Chitosan derivatives and their based nanoparticles: Ultrasonic approach to the synthesis, antimicrobial and transfection properties. Carbohydr Polym, 2020, 242. DOI: 10.1016/j.carbpol.2020.116478.

[20] Lunkov A, Shagdarova B, Lyalina T, Dubinnyi MA, Karpova N, Lopatin S, et al. Simple method for ultrasound assisted «click» modification of azido-chitosan derivatives by CuAAC. Carbohydr Polym, 2020, 282. DOI: 10.1016/j.carbpol.2022.119109.

[21] Trotta F, Martina K, Robaldo B, Barge A, Cravotto G. Recent advances in the synthesis of cyclodextrin derivatives under microwaves and power ultrasound. J Inc Phenom Macrocycl Chem, 2007, 57. DOI: 10.1007/s10847-006-9169-z.

[22] Zhang HT, Fan XD, Tian W, Suo RT, Yang Z, Bai Y, Zhang WB. Ultrasound-driven secondary self-assembly of amphiphilic β-cyclodextrin dimers. Chem Eur J, 2015, 21(13), 5000–5008.

[23] Zheng WQ, Du MX, Feng F, Chen GL, Liao M, Liu YJ, et al. Fast and efficient synthesis of mono-(6-p-toluenesulfonyl)-β-cyclodextrin via ultrasound assisted method. Adv Mater Res, 2015, 1083, 51–54.

[24] Cui H, Siva S, Lin L. Ultrasound processed cuminaldehyde/2-hydroxypropyl-β-cyclodextrin inclusion complex: Preparation, characterization and antibacterial activity. Ultrason Sonochem, 2019, 56, 84–93.

[25] Liu Y, Zheng H, An Y, Ren J, Zheng X, Zhao C, Zhang S. Ultrasound-assisted synthesis of the β-cyclodextrin based cationic polymeric flocculants and evaluation of flocculation performance: Role of β-cyclodextrin. Sep Purif Technol, 2019, 228. DOI: 10.1016/j.seppur.2019.115735.

[26] Jasim IK, Abd Alhammid SN, Abdulrasool AA. Synthesis and evaluation of β-cyclodextrin based nanosponges of 5-Fluorouracil by using ultrasound assisted method. Iraqi J Pharm Sci, 2020, 29(2), 88–98.

[27] Mohandoss S, Edison TNJI, Atchudan R, Palanisamy S, Prabhu NM, Napoleon AA, et al. Ultrasonic-assisted efficient synthesis of inclusion complexes of salsalate drug and β-cyclodextrin derivatives for potent biomedical applications. J Mol Liq, 2020, 319. DOI: 10.1016/j.molliq.2020.114358.

[28] Zhang X, Su J, Wang X, Wang X, Liu R, Fu X, et al. Preparation and properties of cyclodextrin inclusion complexes of hyperoside. Molecules, 2022, 27(9). DOI: 10.3390/molecules27092761.

[29] Yoon S, Hwang SH, Ko WB. Ultrasound-assisted cycloadditions of [70]fullerene with various 2-azidoethyl per-O-acetyl glycosides. Fuller Nanotub Carbon Nanostructures, 2009, 17(5), 496–506.

[30] Marzag H, Alaoui S, Amdouni H, Martin AR, Bougrin K, Benhida R. Efficient and selective azidation of per-O-acetylated sugars using ultrasound activation: Application to the one-pot synthesis of 1, 2, 3-triazole glycosides. New J Chem, 2015, 39(7), 5437–5444.

[31] Evangelista TCS, Aquino GASD, Donza MRH, Leitao RL, Carvalho VSD, Kaiser CR, Ferreira SB. A facile ultrasound-assisted synthesis of methyl 2, 3-O-isopropylidene-β-D-ribofuranoside from D-ribose and its use to prepare new 1, 2, 3-triazole glycoconjugates. J Carbohydr Chem, 2021, 40(5), 243–268.

[32] Shaikh N, Russo L, Cipolla L, Nicotra F. Ultrasonic assisted Fischer glycosylation: Generating diversity for glycochemistry. Mol Divers, 2010, 15(2), 341–345.

[33] Lu Y, Yan R, Ma X, Wang Y. Synthesis and characterization of raffinose fatty acid monoesters under ultrasonic irradiation. Eur Food Res Technol, 2013, 237, 237–244.

[34] Marzag H, Robert G, Dufies M, Bougrin K, Auberger P, Benhida R. $FeCl_3$-Promoted and ultrasound-assisted synthesis of resveratrol O-derived glycoside analogs. Ultrason Sonochem, 2015, 22, 15–21.

[35] Amaniampong PN, Clement JL, Gigmes D, Ortiz Mellet C, Garcia Fernandez JM, Bleriot Y, et al. Catalyst-free synthesis of alkylpolyglycosides induced by high-frequency ultrasound. ChemSusChem, 2018, 11(16), 2673–2676.

[36] Nieto S, Villa R, Donaire A, Lozano P. Ultrasound-assisted enzymatic synthesis of xylitol fatty acid esters in solvent-free conditions. Ultrason Sonochem, 2021, 75. DOI: 10.1016/j.ultsonch.2021.105606.

Purnima Dashora, Arpita Paliwal, Rakshit Ameta, and Suresh C. Ameta*

Chapter 11 Microwave-assisted synthesis of carbohydrate derivatives

11.1 Introduction

Microwaves are a part of electromagnetic radiation with frequencies ranging from 300 MHz (0.3 GHz) to 300 GHz, which corresponds to wavelengths ranging from 1 m to 1 mm. The most commonly used frequency in microwave-assisted reactions and technology is 2.45 GHz. Microwaves can be categorized into two frequency ranges:

– Ultra-high frequency (UHF): It includes frequencies from 300 MHz to 3 GHz
– Super high frequency (SHF): It includes frequencies from 3 GHz to 30 GHz

Microwave-assisted reactions utilize microwave radiation to enhance the rate of chemical reactions. There is an interaction between microwave energy and molecules, and as a result, reaction rates are enhanced and yields are improved. It is considered a cleaner process compared to traditional heating methods. Thus, microwave-assisted reactions offer a powerful tool in chemistry with significant improvements in reaction speed and efficiency, along with environmental sustainability.

Microwaves interact with polar molecules such as water, alcohols, and acids by causing them to oscillate or rotate. This molecular motion results in the generation of heat by friction and molecular collisions, which is responsible for heating the reaction mixture rapidly and uniformly. This heating can be of two types:

– Dielectric Heating: In this type, polar molecules align themselves with the microwave field. The friction due to the rapid rotation of molecules generates heat, and
– Volume heating: Here, microwaves penetrate the entire reaction mixture, and as a consequence, uniform heating occurs, unlike conventional methods, where heating is from the outside in.

Apart from the advantages of high rate, efficiency, uniformity, and higher yields, it also utilizes fewer solvents. Microwave-assisted synthesis of carbohydrate derivatives includes:

*Corresponding author: Suresh C. Ameta, Department of Chemistry, PAHER University, Udaipur 313024, Rajasthan, India, e-mail: ameta_sc@yahoo.com
Purnima Dashora, Arpita Paliwal, Rakshit Ameta, Department of Chemistry, PAHER University, Udaipur 313024, Rajasthan, India

https://doi.org/10.1515/9783111447094-011

(i) Glycosylation reaction: It mainly includes the formation of glycosidic bonds, which is useful in the synthesis of glycosides and oligosaccharides.
(ii) Sugar derivatization: Microwaves can be used to introduce functional groups and to facilitate the derivatization of carbohydrates.
(iii) Carbohydrate-based natural product synthesis: Microwave-assisted reactions can be used in the synthesis of carbohydrate-based natural products (glycosides and polysaccharides).

11.2 Monosaccharides

Monosaccharides are the simplest form of carbohydrates. These are the fundamental building blocks of more complex di-, oligo-, and polysaccharides. They have a sweet taste, are soluble in water, and also form different derivatives.

Jiang et al. [1] carried out catalyst-free hydrolysis (cellulolysis) of fibrous cellulose (FC) at 200 °C under microwave-irradiation. It afforded 37.2% of cellulose conversion to C5/C6 sugars. The carbonizing fibrous cellulose was not effective even after 24 h under routine cellulolysis (hydrothermal). There was some effective interaction between FC (macroscopic) and water molecules under MW irradiation. It was observed that the kinetic energy of water molecules can break carbon-carbon and carbon-oxygen bonds under microwave exposure. Therefore, basic cellulolysis products such as xylose, glucose, and cellobiose were produced by the breaking of non-hindered surface $-CH_2OH$ groups (C–C) and inside cellulose cavities (C–O–C bond).

Fructose is an important intermediate in the production of different chemical scaffolds. It can be prepared by isomerizing glucose using two methods.

- By using immobilized enzymes
- Heterogeneous catalysts

But these heating processes take a longer time. Kulawong et al. [2] developed a method to produce fructose (**2**) rapidly through catalytic isomerization of glucose (**1**) under microwave irradiation (Figure 11.1). They used NaX zeolite from cogon grass silica to support Mg catalysts. Different percentages of Mg content (3, 6, and 9 wt%) on NaX were tried for their performance on the isomerization of glucose to fructose for 15 min at 90 °C. It was reported that surface area, crystallinity, and micropore volume decreased with increasing Mg content, but basicity was found to increase. It was observed that the conversion of glucose increased with the loading of Mg, and the highest yield of fructose was obtained when Mg was kept at 6 wt%. They achieved the highest selectivity and yield of fructose as 71.9% and 25.8% within 15 min, respectively. It was also revealed that this catalyst could retain its initial catalytic performance for up to three cycles, which indicated its stability.

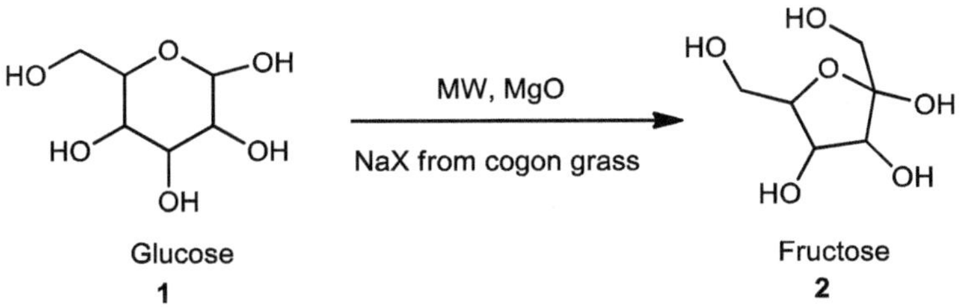

Figure 11.1: Synthesis of fructose via catalytic isomerization of glucose.

11.3 Disaccharides

Disaccharides are carbohydrates composed of two identical or different monosaccharide units linked via a glycosidic bond. They are also key components in the formation of more complex carbohydrates. Disaccharides are soluble in water and play important roles in energy metabolism, nutrition, and food processing.

Sugar esters of fatty acids are bio-based surfactants. Kondamudi and McDougal [3] achieved the synthesis of stearic acid methyl ester (SAME) in high yields and purity when sucrose was trans esterified in the presence of microwave-irradiation. About 88.2% yield of surfactants was achieved by using a combination of potassium methoxide and reactant under microwave irradiation at 132 °C in 21 min. It was also revealed that minimal sugar pyrolysis was observed (which is undesirable) as the white color of the bio-based surfactant (pure) was retained. The effect of different reaction parameters on product yields was evaluated, such as mole ratio (catalyst and sucrose), time of irradiation, and temperature.

A green synthesis of lactose octaacetate (**4**) was reported by Arabadzhieva et al. [4] using microwave irradiation. The lactose ester could be prepared by esterification of lactose (**3**) with acetic anhydride (yield 85–90%) under microwave irradiation (Figure 11.2). A high degree of acetylation (3.2–3.7) was achieved. It was also revealed that lactose octaacetate exhibited better antifungal properties than antibacterial ones. These were found to have low cytotoxicity against three cell lines: Human epithelial type 2 (HEp-2), Madin-Darby bovine kidney (MDBK) cells, and Madin-Darby canine kidney (MDCK) cells. Lactose acetates can also find application as antimicrobial and antiviral substances in pharmaceutical, food, cosmetic, agricultural, etc.

Rajput et al. [5] used distilled soya acid oil (DSAO) to synthesize a green non-ionic surfactant, soya acid fatty acid methyl ester (SAFAME), in the presence of a microwave-assisted route (680 W). It was observed that 99.3% conversion could be obtained within half an hour when 2 wt% sulfuric acid was used as a catalyst. About 82.5% conversion was found within 45 min for the synthesis of sucrose ester when 5 wt% potassium carbonate was used as a catalyst. It was also claimed that these bio-based renew-

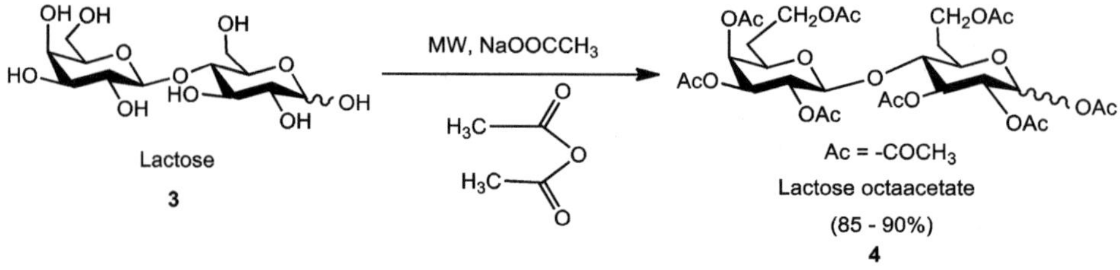

Figure 11.2: Microwave-assisted synthesis of lactose octaacetate.

able surfactants can be used in place of sodium dodecyl sulfate and sodium lauryl ether sulfate in toothpaste (desensitizing anticavity toothpaste formulation).

11.4 Polysaccharides

Polysaccharides are complex carbohydrates, which are composed of long chains of monosaccharide units. These units are linked by glycosidic bonds. They play important roles in biological systems as energy storage molecules and are essential in various biological functions and some industrial applications.

11.4.1 Chitosan

Chitosan is a highly versatile biopolymer, which is derived from chitin. It is produced by deacetylating chitin, where acetyl groups are removed, converting *N*-acetylglucosamine into glucosamine. As a result, a polymer with free amino groups is obtained along the chain, making it more soluble in acidic conditions. It is a natural polymer present in the exoskeletons of arthropods like shrimp and crabs and also in the cell walls of certain fungi. It finds applications in various industries such as medical and pharmaceutical, food, and cosmetics.

A water-soluble quaternized carboxymethyl chitosan (QCM-chitosan) (**6**) was synthesized by Liu et al. [6] via grafting quaternary ammonium and carboxymethyl groups on chitosan (**5**) (Figure 11.3). It was observed that QCM-chitosan was prepared within 70 min under microwave exposure, where as conventional method takes a longer time and requires organic solvents. The MWI process proceeds in aqueous media and avoids the use of any harmful organic solvents. QCM-chitosan was synthesized by carboxymethylation of chitosan followed by quaternization. It was revealed that the degree of substitution of carboxymethyl and quaternary ammonium groups were 0.82 and 0.48, respectively, under the optimum reaction conditions. Therefore, this method is eco-friendly and quick for synthesizing soluble chitosan derivatives.

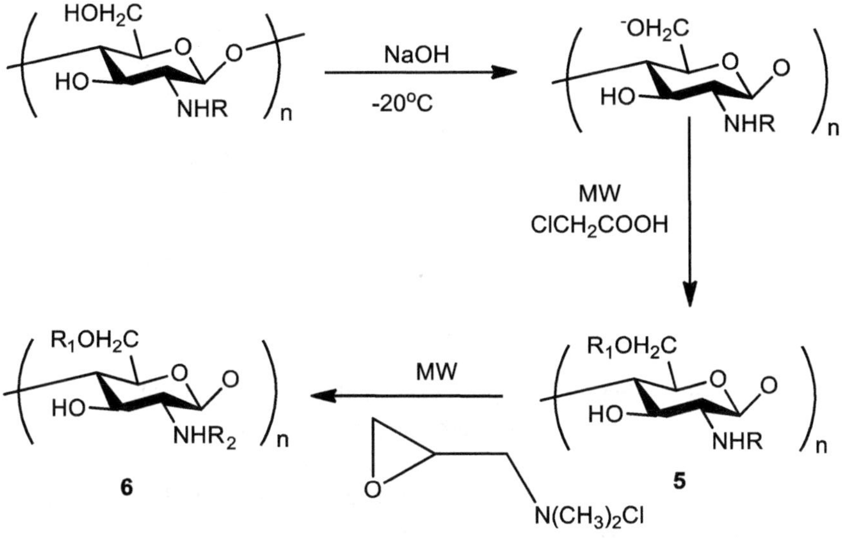

R = H; $R_1 = CH_2COOH$; $R_2 = CH_2CH(OH)CH_2N(CH_3)_2Cl$

Figure 11.3: Synthesis of CM-chitosan and QCM-chitosan.

Microwave irradiation and gas foaming method was reported by Kar et al. [7] for the preparation of composite porous chitosan-organically modified montmorillonite-hydroxyapatite (CS-OM-HA). It was observed that synthesized HA has particles of size about 65 nm. There was a synergistic effect between HA and OM on mechanical and biological properties such as degradation, swelling, bioactivity, and protein adsorption of the synthesized composite scaffolds. It was also found that the as-prepared CS-OM-HA composite scaffolds were non-cytotoxic to MG-63 osteoblast cell lines. CS-organic matter (OM)-HA composite scaffolds were non-cytotoxic to MG-63 osteoblast cell lines, indicating their biocompatibility. The results suggest that these composite scaffolds hold significant potential for use in non-load bearing bone tissue engineering applications due to their favorable interaction with bone cells and their ability to support cell growth without causing harm.

Zeng et al. [8] synthesized a highly efficient and ecofriendly flocculant dextran-chitosan polymer (Dex-CS) (**9**) using dextran (**7**) and chitosan (**8**) (Figure 11.4). This preparation method was based on a microwave-heating-initiated grafting reaction. They also investigated the effect of pH, temperature, and flocculant dosage on flocculation efficiency. It was reported that 93.6% of the kaolin solid suspended particles could be removed by this as-prepared flocculant. The flocculation comprises steps like adsorption bridging, charge neutralization, and sweeping flocculation. It can be used for the treatment of industrial wastewater as a high-performance flocculant.

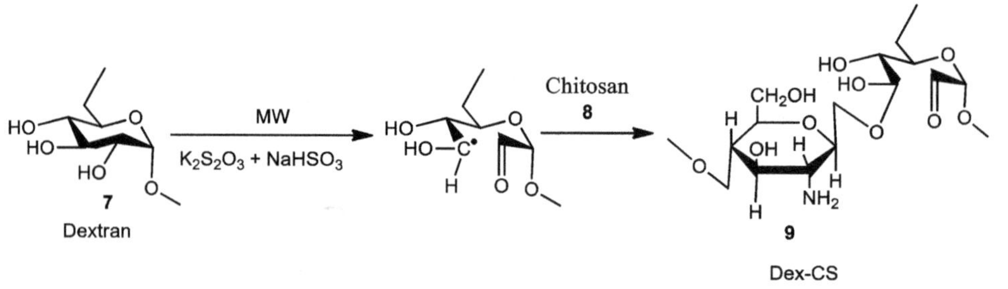

Figure 11.4: Microwave-assisted synthesis of Dex-CS.

Maruthapandi et al. [9] synthesized polypyrrole grafted with chitosan (PPy-g-CS) and a copolymer of poly(pyrrole-*N*-(1-naphthyl) ethylenediamine) (COP) via a microwave-assisted pathway with carbon dots (C-Dots) acting as initiators. The antimicrobial mechanism was attributed to the electrostatic interaction between the positively charged polymers and the negatively charged microbial cell membranes, leading to cellularlysis. It was reported that *Escherichia coli* exposed to PPy-g-CS (zeta potential = +46.9 mV) was completely deactivated after 3 h, whereas COP (zeta potential = +64.1 mV) showed no antimicrobial effect.

Chitosan is abundantly available and is a renewable natural biopolymer. Quaternized chitosan compounds can be used as antistatic agents in hair care products and cosmetics (hair conditioners and fabric softeners). The low solubility of chitosan in water creates some limitations for its applications. Singh et al. [10] synthesized quaternized derivatives of chitosan, which can be used as cationic amino sugar-based surfactants. They synthesized quaternized derivatives of chitosan with excellent solubility in aqueous solutions. The cationic quaternized surfactants were synthesized in the presence of microwave irradiation and compared with conventional synthesis. It was observed that *N,N,N*-trimethyl chitosan (TMC) was obtained with higher yields using chitosan and methyl iodide as reactants and *N*-methyl-2-pyrrolidone (NMP) as a solvent. They also prepared it by the addition of dimethyl sulfate (DMS) under microwave irradiation and compared it with the conventional heating method.

Yan et al. [11] reported on chitin and chitosan from crayfish shell waste and their depolymerization of recovered products using graphene oxide and microwaves. The products obtained were monosaccharides, 5-hydroxymethylfurfural, as well as high-value products.

Due to better microwave absorption properties, it was observed that graphene oxide was more effective compared to graphene. It was also observed that when graphene oxide was introduced during the depolymerization of chitin or chitosan, the heating rate was increased by 0.37 or 0.26 K/s, respectively. The effect of graphene oxide was studied on the depolymerization of chitin and chitosan under microwave-assisted varying conditions in terms of catalyst, solvent, temperature, and time.

The recovered graphene oxide exhibited delamination when it was re-dispersed in water. They also proposed a synergistic effect between graphene oxide and microwaves on the depolymerization of chitin and chitosan, and its ability for rapid one-step depolymerization in an acid/alkali-free solvent.

11.4.2 Starch

Starch is a polysaccharide, and it is a major energy storage carbohydrate in plants. It is commonly used in various industries, ranging from food production to pharmaceuticals. Starch is mainly composed of two main polysaccharides: (i) amylose and (ii) amylopectin:

- Amylose is a linear chain of glucose units linked by α-1,4-glycosidic bonds. It constitutes almost 20–30% of starch.
- Amylopectin is a branched-chain structure. Here, linear parts containing glucose units are linked by α-1,4-glycosidic bonds, while branch points contain α-1,6-glycosidic bonds. About 70–80% of starch contains amylopectin.

It finds application in industries such as the food industry (thickening agent, texturizer, and stabilizer), pharmaceuticals (binder and disintegrant), biodegradable products (packaging materials and agricultural mulch), adhesives, and textiles.

Xing et al. [12] prepared starch maleate (**10**) using a microwave-assisted approach (Figure 11.5). It was reported that microwave-assisted esterification was complete in 5 min in a dry medium with very high reaction efficiency (98%). The effects of different parameters were investigated, such as the degree of substitution (DS) of the product and reaction efficiency, reagent molar ratio, reaction time, and water content of starch. It was revealed that microwave-assisted synthesis in a dry medium is fast as well as energy-saving, with high reaction efficiency compared to conventional heating synthesis.

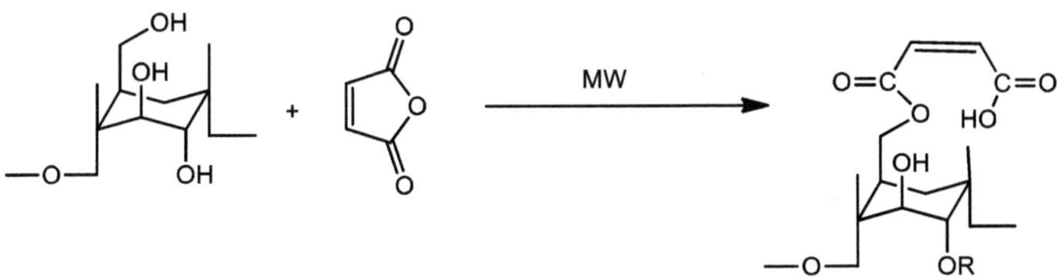

10
(Up to 98%)
R = H or anionic group according to DS

Figure 11.5: Microwave-assisted synthesis of starch maleate.

Kumoro et al. [13] carried out the acetylation of native cassava starch by mixing it with chloroacetic acid and sodium hydroxide (in a certain weight ratio) in a sealed container. This mixture was sprayed with ethanol and heated using microwave irradiation. It was found that this acetylation was completed in a very short reaction time. The highest degree of substitution and reaction efficiency were found to be 0.045% and 0.051%, respectively. It was reported that the acetylation of cassava starch reduced gel hardness (during storage) and also inhibited the retrogradation of the starch gel. It was revealed that modifications were done on native starch granules, and that too on the amylose fraction preferentially, but amylopectin retrogradation remained unaffected.

Singh et al. [14] modified moth bean starch via microwave-assisted synthesis, which required a very low concentration of a free radical initiator. They also carried out its swelling study at various temperatures to assess its suitability in various industrial applications. They synthesized starch-grafted-acrylamide efficiently using a small concentration of ceric ammonium nitrate (in aqueous medium) under microwave irradiation, where it was found to have high thermal stability compared to native starch. The starch-g-acrylamide was grafted successfully, and this prepared biomaterial could be used in various industrial applications like paper, food, textile, petroleum, and pharmaceuticals.

Mishra et al. [15] prepared polyacrylamide grafted starch (St-g-PAM) (**11**) by a combination of ceric ammonium nitrate and microwave radiation for the initiation of the grafting reaction (Figure 11.6). It was reported that this method is rapid, reproducible, highly reliable, and produces good yields and high-quality products compared to conventional methods. They also investigated its application as a flocculant for the treatment of wastewater.

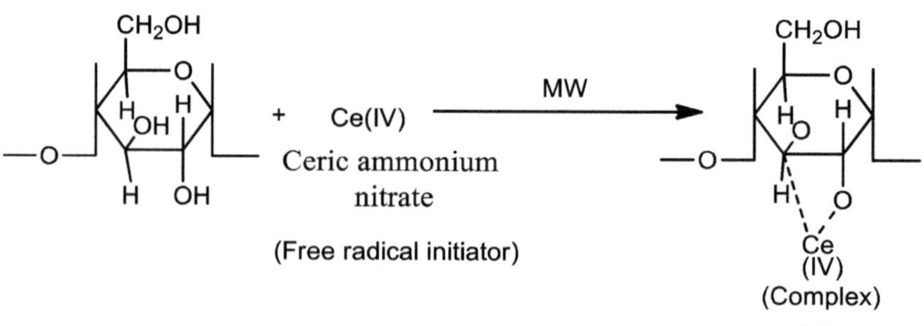

11

Figure 11.6: Microwave-assisted synthesis of polyacrylamide-grafted starch.

Sagar et al. [16] synthesized carboxymethyl starch (CMS) (13) using wheat starch (**12**) (insoluble) under microwave irradiation (Figure 11.7). It was reported that the reaction of starch with monochloroacetic acid or its sodium salt in isopropanol or isopropanol: water (80:20 v/v) in the presence of sodium hydroxide afforded CMS. They also investigated alkaline fluidity, degree of substitution, and active content of CMS with the concentration of reactants and reaction time.

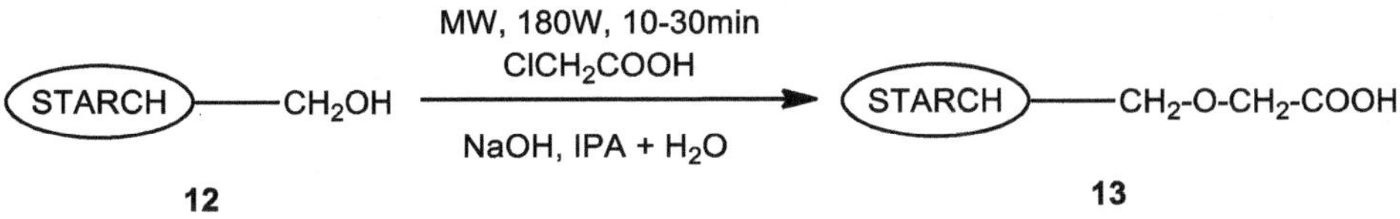

Figure 11.7: Microwave-assisted synthesis of carboxymethyl starch.

Zhang et al. [17] acetylated native corn starch (NCS) using a microwave-assisted method. They used a potassium carbonate catalyst and vinyl acetate (acetylating agent) to produce acetylated corn starch (ACS). The optimum conditions for acetylation were also obtained by varying the exposure time, reaction temperature, and catalyst concentration. They also evaluated some physicochemical properties of ACS, such as solubility, paste clarity, swelling power, and intrinsic and apparent viscosity. It was also revealed that the physicochemical properties of ACS were much improved compared to NCS. It was reported that the adhesion force of ACS to polyester/cotton fibers increased significantly, and the sizing properties of ACS were found to be better than those of NCS. ACS can find potential applications in the textile industry as a good substitute for polyvinyl alcohol.

Alfaifi et al. [18] synthesized two monomers and treated them with acryloyl chloride to prepare the corresponding acryloylamino acid (**14**). Then, it was grafted onto starch (**15**) using a microwave-assisted technique (Figure 11.8). Atenolol (in the form of a salt) was then immobilized onto the graft copolymers (**16**) by ionic bonds. They also investigated drug release in both acidic and alkaline media. It was observed that the release of atenolol takes place in an alkaline medium rather than in an acidic medium, which indicated that these polymers can be used as drug carriers, whose target is the colon.

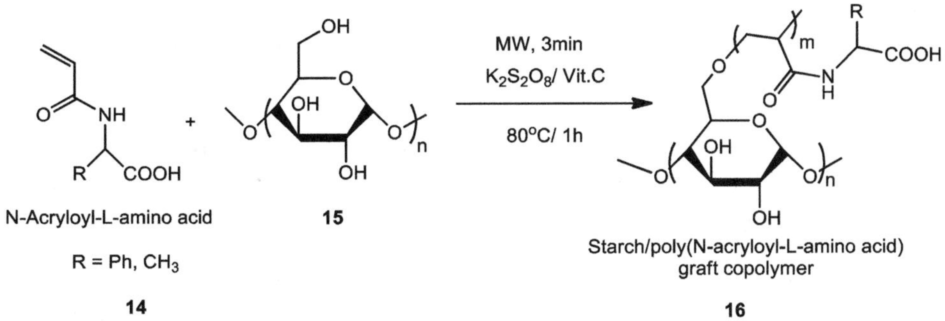

Figure 11.8: Microwave-assisted graft copolymerization of amino acid-based monomers onto starch.

Phosphorylated starch is a modified starch, which is quite commonly used. Supardan et al. [19] phosphorylated corn starch via a microwave-assisted reaction using a mixture of disodium hydrogen phosphate and sodium dihydrogen orthophosphate. It was reported that the maximum degree of phosphate substitution was obtained at pH 6, with microwave irradiation at 500 W, and a reaction time of 10 min. It was observed that the degree of substitution ranged between 0.567 and 0.787. Different physicochemical properties were also evaluated, such as solubility, swelling capacity, paste clarity, and water binding capacity. It was also revealed that phosphorylated corn starch adsorbed 73.6% mercury and 73.3% methylene blue, indicating that this phosphorylated corn starch can be used for effective adsorption and removal of heavy metals.

Dry modification of starch using microwave heating has certain advantages, such as high conversion, energy saving, rapidity, and avoiding the use of volatile solvents in the organic synthesis of organic compounds. Lu et al. [20] observed the effect of dry modification on starch microstructure and acid anhydride in the dry reaction in liquid and solid states. They used corn starches with different amylose/amylopectin ratios to observe the effect of starch microstructure using acetic anhydrides (liquid) and maleic anhydride (solid). It was reported that the degree of substitution (DS) of these modified starches increased with higher amylose content when using maleic anhydride, as amylose contains more hydroxyl groups, which provided higher sensitivity to microwaves. On the contrary, the DS of modified starches was increased when using acetic anhydride with increasing amylopectin content because liquid acetic acid can diffuse better into high-amylopectin starch granules. It was also revealed that higher amylopectin starches, such as waxy and maize, showed higher RC (%) and ΔH (%) compared to higher amylose starches.

Granular amylose (AM) and granular amylopectin starch (waxy) (AP) granules have good nutritional value in the food industry. Zhong et al. [21] investigated the effects of microwave treatment (400 W/g DW, 1–8 min) on transgenic AM granules and AP granules. It was reported that microwave treatment decreased the molecular weight of molecules in both these samples during the first three min. It was observed that the crystallinity of AM starch first increased from 15.6% to 20.6%, but then it decreased to 11.3%. On the other hand, the crystallinity of AP starch decreased from 18.9% to 10.8% initially, but then it increased to 20.0%. It was also revealed that resistant starch and water solubility were significantly increased on prolonged treatment of AM granules.

It is known that poly(vinyl alcohol) (PVA) is an excellent membrane-forming polymer. It can be modified as a copolymer with potato starch and methyl acrylate monomers, resulting in improved chemical and physical properties. Patil et al. [22] synthesized poly(vinyl alcohol)-g-starch-poly(methyl acrylate) (PVA-g-St-g-PMA) copolymers under microwave irradiation using potassium persulfate as an initiator. Then, solution casting and solvent evaporation methods were utilized for the fabrication of PVA-g-St-g-PMA composite membranes. It was also revealed that the modified nanocomposite membranes exhibited very promising selectivity and permeability. These nanocomposite membranes had the advantages of easy handling as well as reuse.

11.4.3 Cyclodextrin

Cyclodextrins are a group of cyclic oligosaccharides, and these are formed by the enzymatic cyclization of glucose units. They can form inclusion complexes with a wide variety of guest molecules. With the capability of cyclodextrins to form inclusion complexes, they are considered valuable tools as they enhance the properties of encapsulated molecules.

Biswas et al. [23] developed a microwave-assisted method to prepare polyurethanes consisting of different cyclodextrins (α-, β-, and γ-CD) and three common diisocyanates. It was reported that this method not only saves energy but also significantly reduces reaction time with better yield compared to conventional heating. It was also revealed that the as-obtained CD polyurethane is soluble in organic solvents and insoluble in water. They could remove reactive Nile red dye and phenol from water. It was claimed that undesirable materials can be removed from process streams, toxic compounds can be eliminated from the environment, and color or fragrance molecules can be encapsulated.

Micron- and nanometer-sized γ-cyclodextrin metal–organic frameworks (γ-CD-MOFs) were synthesized by Liu et al. [24] via microwave heating. Polyethylene glycol 20000 (PEG 20000) was employed as a surfactant to control the size and morphology of the γ-CD-MOFs. It was observed that increasing the reaction time and methanol ratio significantly weakens the crystalline integrity of the γ-CD-MOFs. Fenbufen was successfully loaded onto the γ-CD-MOF crystals, resulting in particles ranging from 100 to 300 nm. These modified γ-CD-MOFs were then used for the adsorption of fenbufen in ethanol, which exhibited a higher adsorption capacity (196 mg/g) compared to the micron-sized crystals.

The β-cyclodextrin (β-CD) functionalized rice husk-derived biochar (BC) was synthesized by Qu et al. [25] under microwave irradiation (Figure 11.9) and used for the removal of lead (Pb) and bisphenol A (BPA). It was reported that surface modification occurred within 15 min, and the resulting BCMW-β-CD (**17**) exhibited excellent adsorption properties. It had a heterogeneous adsorption capacity of 209.20 mg/g for BPA and a calculated monolayer uptake of 240.13 mg/g for Pb(II) in a single-component system. It was also found that BCMW-β-CD could effectively remove both BPA and Pb(II) without any competition between the two contaminants. Additionally, the adsorbed BPA and Pb(II) could be desorbed with only a slight reduction in the adsorption performance of BCMW-β-CD even after five cycles in the multi-component system.

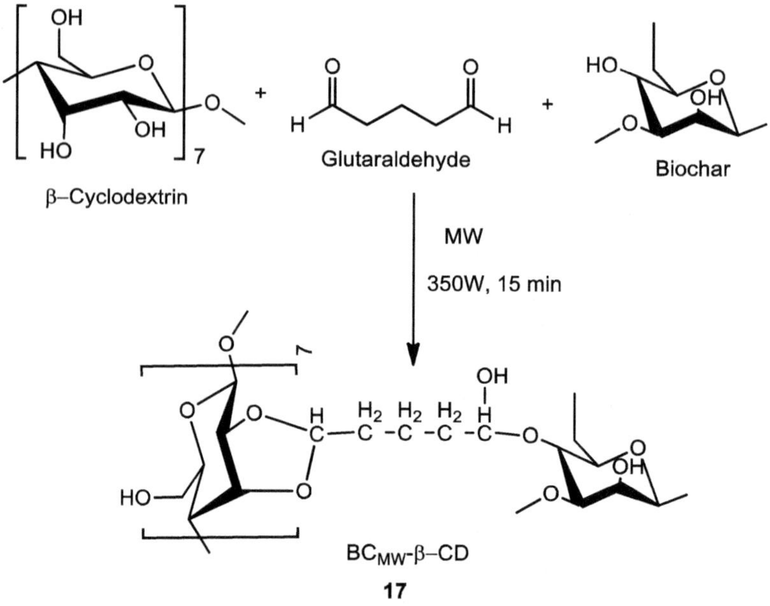

Figure 11.9: One-pot synthesis of β-cyclodextrin modified biochar.

Qu et al. [26] also grafted β-cyclodextrin onto rice husk-based cellulose. They used different cross-linking agents such as glutaraldehyde (GA) and epichlorohydrin (EPI). This functionalization procedure was completed in 17 min in the presence of microwave irradiation. It was reported that the as-synthesized RHEPIMWβ-CD and RHGAMWβ-CD exhibited fast adsorption for Pb(II) within 20 min, with excellent monolayer adsorption capacities of 216.06 and 279.08 mg/g in the pH range of 3.0–6.0. It was observed that more than 80% of Pb(II) uptake was retained even after four adsorption–desorption cycles. The Pb(II) uptake was enhanced onto the RHEPIMWβ-CD and RHGAMWβ-CD. These (RHEPIMWβ-CD and RHGAMWβ-CD) can be used to purify acid battery effluent containing lead ions effectively, so that regulatory requirements are satisfied.

Calce et al. [27] reported a microwave-assisted and solvent-free modification of pectin (**18**) involving the esterification of different fatty acids (**19**) by pectin alcoholic functions (Figure 11.10). This reaction was carried out by mixing the reagents with a catalytic amount of potassium carbonate and then irradiating with microwaves for only 3–6 min. It was reported that microwave heating developed a new synthetic protocol with an increased yield of products compared to traditional heating. Interestingly, it eliminated the small amount of degraded polysaccharide usually produced during traditional heating in an oil bath.

The functionalization of pectin with *N,N*-diethylacrylamide (DEAAm) was investigated by Isiklan and Tokmak [28]. The modified pectin was obtained by copolymerization of graft DEAAm in the presence of ceric ammonium nitrate and *N,N,N′,N′*-

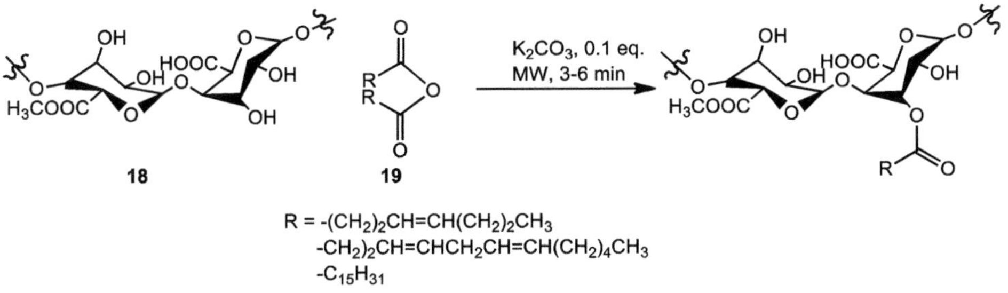

Figure 11.10: Synthetic strategy for pectin functionalization.

tetramethylethylenediamine (TEMED) under microwave irradiation. It was observed that the grafting yield increased by 192% with variation in irradiation time, temperature, microwave power, initiator, monomer, and concentrations of TEMED. The thermo-responsive behavior (resistant and sensitive) with a lower critical solution temperature of 31 °C was also observed. Furthermore, the biocompatibility indicated that these grafted copolymers are not cytotoxic to L929 fibroblast cells and are safe for use as biomaterials.

Xu et al. [29] synthesized fragments of heparin (HP) and heparan sulfate (HS), which are difficult to prepare by post-assembly manipulations of oligosaccharides. Normally, *O*- and *N*-sulfonation steps are carried out separately, requiring long times with low yields. It was a bottle-neck in the synthesis of such compounds. An effective microwave-assisted protocol was proposed by them for the simultaneous *O,N*-sulfonation of HP/HS-like saccharides [**20**]. It was observed that complete *O*- and *N*-sulfonation could be attained when an $SO_3 \cdot NEt_3$ complex was used in a solvent mixture of NEt_3/pyridine at 100 °C under microwave irradiation for 15 min (Figure 11.11). It was also revealed that per-*O,N*-sulfonation of mono-, di-, tri-, and tetra-saccharides with two to six reactive sites could be obtained effectively with excellent yields (more than 90%) in short reaction times. In the next step, the resulting per-*O,N*-sulfonated saccharides were fully deprotected in high yields (more than 88%). They also developed a microwave-assisted route for one-pot selective *O*-sulfonation/*N*-acetylation on disaccharides.

Pineapple leaf waste has high cellulose content, and it can be used as an alternative starting material for the preparation of carboxymethyl cellulose (CMC). Panchan et al. [30] synthesized CMC from pineapple leaves under microwave irradiation. The effects of different parameters on the preparation of CMC were evaluated, such as monochloroacetic acid (MCA) dose, sodium hydroxide concentration, and time of etherification. The optimal conditions for CMC synthesis were determined as: NaOH 50% (w/v) solution, MCA g^{-1} cellulose (8 g), and etherification time (16 min). It was reported that the maximum degree of substitution was found to be 0.78. The as-synthesized CMC was used as a thickener for liquid foods (milk, orange juice, water, and mushroom cream soup), where the use of this CMC (2% w/v) increased the viscosity of the foods and changed them from thin to nectar-like liquids.

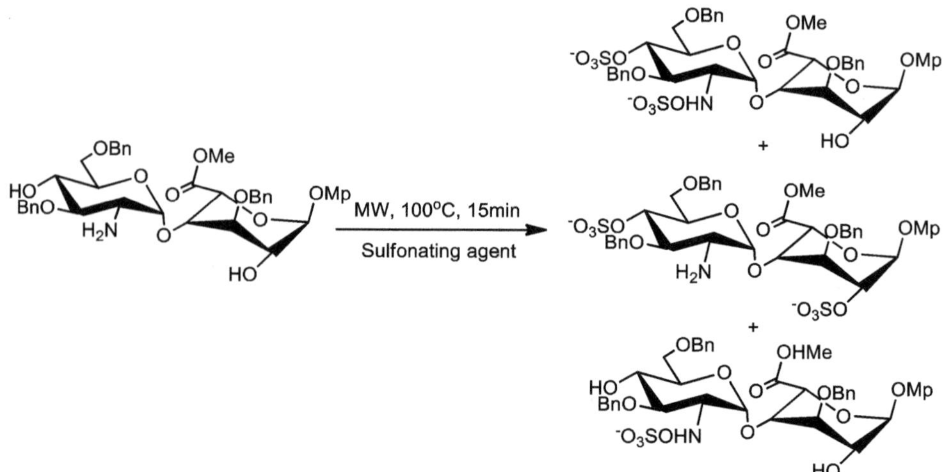

20

(72 - 90%)

Figure 11.11: Simultaneous *O,N*-sulfonation of a disaccharide under microwave heating.

11.5 Derivatives of carbohydrates

11.5.1 Galectooligosacharides

Maugard et al. [31] reported the synthesis of galacto-oligosaccharides (GOS) (**22**) from lactose (**21**) using both immobilized and free β-galactosidase from *Kluyveromyces lactis* (Lactozym 3000L HP-G), with both microwave irradiation and conventional heating (Figure 11.12). It was found that immobilizing β-galactosidase on DuoLite A-568 significantly enhanced GOS production. It was also observed that the selectivity for GOS (the ratio of GOS synthesis to lactose hydrolysis) improved when the water activity of the reaction medium was reduced. This effect was particularly pronounced when a high initial concentration of lactose was used, as well as when co-solvents were introduced. The addition of solvents also boosted both production and selectivity of GOS, especially under microwave irradiation. The GOS synthesis significantly increased by 217-fold when immobilized β-galactosidase was used in combination with hexanol as a co-solvent.

Li et al. [32] synthesized solid-phase oligosaccharides using glucose as a substrate and phosphoric acid as a catalyst via a microwave-assisted route under optimum reaction conditions. It was reported that the maximum yield of the product (76.56%) could be obtained with 4% heteropoly acid A as a catalyst, adding 30% water and using microwave power (1,000 W) in 9 min. It was reported that the proportion of synthetic oligosaccharides was 20.81%, 15.50%, 10.84%, 7.79%, and 21.63% for disaccharides, trisacchar-

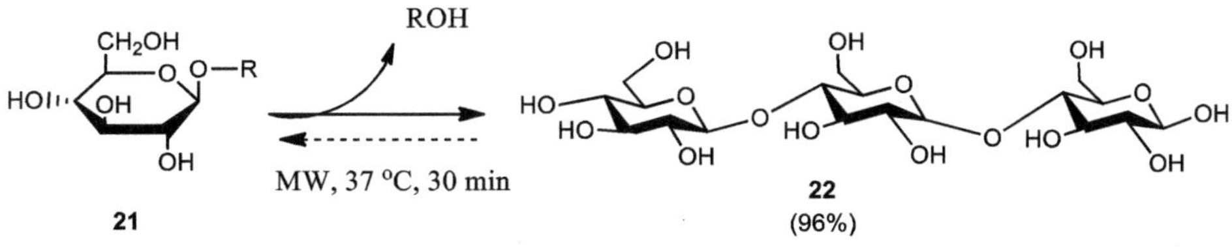

Figure 11.12: Microwave-assisted synthesis of galacto-oligosaccharides.

ides, tetrasaccharides, pentasaccharides, and higher oligosaccharides, respectively. It was concluded that microwave-assisted solid-phase organic synthesis could be used for the production of oligosaccharides.

Galacto-oligosaccharides (GOS) were synthesized by Jovanovic-Malinovska et al. [33] using β-galactosidase immobilized in both sol-gel carriers and polyvinyl alcohol (PVA) lenses. It was found that PVA-immobilized β-galactosidase could retain 95% of its initial activity even after seven repeated cycles and still maintained 51% activity after storing it for 3 months. In contrast, sol-gel-immobilized β-galactosidase retained only 39% of its initial activity under similar conditions of storage. It was observed that at optimal conditions for GOS production (40% lactose feed solution was fed at a rate of 10.8 mL for 1 h, at pH 4.5 and 40 °C), a GOS yield of 30% of total sugars was obtained, which corresponds to a productivity of 117 g/L h. It was further revealed that the maximum GOS productivity of 344 g/L h could be achieved at a flow rate of 28.7 mL^{-1} h. The main types of GOS produced were 3-OS and 4-OS, which accounted for the majority of the GOS. Additionally, in a continuous process using whey as a substrate, 15% GOS production of total sugars was achieved, with GOS types as 45%, 40%, and 15% 3-OS, 4-OS, and 5-OS, respectively.

Fischer and Kleinschmidt [34] also synthesized galacto-oligosaccharides (GOS) in sweet and acid whey using lactases from *Aspergillus oryzae* and *Kluyveromyces lactis* with an initial lactose concentration of 38 g/L. It was found that the enzyme from *A. oryzae* produced the following GOS yields: 10.91 ± 0.01% in lactose solution, 10.93 ± 0.18% in sweet whey, and 11.32 ± 0.59% in acid whey. On the other hand, the *K. lactis* enzyme showed a significant dependence on the type and concentration of whey at the same initial lactose concentration. The GOS yields with this enzyme were 10.93 ± 0.26% in lactose solution, 4.30 ± 0.17% in sweet whey, and 10.56 ± 0.41% in acid whey. It was observed that as the initial lactose concentration was increased, the inhibitory effect of sweet whey declined, resulting in higher GOS yields compared to those obtained in the lactose solution alone. This suggests that sweet whey may initially have a negative effect on GOS production, but this effect diminishes at higher lactose concentrations.

11.5.2 Lactosylated peptides

It is known that lactosylation of whey proteins is likely to affect their properties (allergenicity and nutritional value), which matters in the feeding of infants. Kijewska et al. [35] proposed a rapid, simple, and efficient route for the synthesis of lactosylated peptides on solid support based on microwave-assisted synthesis. They synthesized modified peptides derived from β-lactoglobulin in tryptic digestion.

11.6 Conclusion

Microwave-assisted synthesis represents a transformative advancement in carbohydrate chemistry, offering numerous benefits over traditional methods. The application of microwave technology to the synthesis of carbohydrates and their derivatives has led to significant improvements in both efficiency and effectiveness.

The advantages of the use of microwaves in carbohydrate chemistry are related to the reduction in synthesis time, saving in energy consumption, increase in catalyst activity, and also minimizing the formation of side products, providing selectivity to target products. It is important to note that microwave procedures can be scaled up. Microwave-assisted synthesis has almost made a revolution in the synthesis of carbohydrates and their derivatives. The time is not far off when such reactions will find a dominant place in chemistry and, in particular, carbohydrate chemistry.

References

[1] Jiang S, Daly H, Xiang H, Yan Y, Zhang H, Hardacre C, et al. Microwave-assisted catalyst-free hydrolysis of fibrous cellulose for deriving sugars and biochemicals. Front Chem Sci Eng, 2019, 13, 718–726.

[2] Kulawong S, Youngjan S, Khemthong P, Chanlek N, Wittayakun J, Osakoo N. Magnesium impregnated on NaX zeolite synthesized from cogon grass silica for fast production of fructose via microwave-assisted catalytic glucose isomerization. Catalysts, 2021, 11. DOI: 10.3390/catal11080981.

[3] Kondamudi N, McDougal OM. Microwave-assisted synthesis and characterization of stearic acid sucrose ester: A bio-based surfactant. J Surfactants Deterg, 2019, 22, 721–729.

[4] Arabadzhieva R, Ivanov I, Petkova N, Todorova M, Tumbarski Y, Vlaeva I, et al. Microwave-assisted synthesis of lactose acetates with antimicrobial, cytotoxic, and antiviral properties. Biointerface Res Appl Chem, 2022, 12, 8453–8470.

[5] Rajput YN, Girase CD, Kedar RP, Deshpande PS, Kulkarni RD. Microwave-assisted low-cost synthesis of sucrose-soya ester from vegetable oil refinery by-product and its application in toothpaste formulation for oral hygiene. J Surfactants Deterg, 2023, 26, 119–133.

[6] Liu B, Wang X, Yang B, Sun R. Microwave–assisted synthesis of quaternizedcarboxymethyl chitosan in aqueous solution and its thermal behavior. J Macromol Sci, Part A, 2012, 49, 227–234.

[7] Kar S, Kaur T, Thirugnanam A. Microwave-assisted synthesis of porous chitosan–modified montmorillonite–hydroxyapatite composite scaffolds. Int J Biol Macromol, 2016, 82, 628–636.

[8] Zeng T, Hu XQ, Wu H, Yang JW, Zhang HB. Microwave assisted synthesis and characterization of a novel bio-based flocculant from dextran and chitosan. Int J Biol Macromol, 2019, 131, 760–768.
[9] Maruthapandi M, Sharma K, Luong JH, Gedanken A. Antibacterial activities of microwave-assisted synthesized polypyrrole/chitosan and poly (pyrrole-N-(1-naphthyl) ethylenediamine) stimulated by C-dots. Carbohydr Polym, 2020, 243. DOI: 10.1016/j.carbpol.2020.116474.
[10] Singh PS, Shaikh A, Deshmukh A, Pratap AP. Microwave assisted synthesis of cationic amino sugar surfactants. Tenside Surfactants Deterg, 2020, 57, 265–272.
[11] Yan B, Yang H, Zhang N, Cheng J, Huang J, Zhao J, et al. Microwave-assisted depolymerization of chitin and chitosan extracted from crayfish shells waste: A sustainable approach based on graphene oxide catalysis. Int J Biol Macromol, 2023, 251. DOI: 10.1016/j.ijbiomac.2023.126296.
[12] Xing GX, Zhang SF, Ju BZ, Yang JZ. Microwave-assisted synthesis of starch maleate by dry method. Starch Starke, 2006, 58(9), 464–467.
[13] Kumoro AC, Retnowati DS, Budiyati CS. Microwave assisted synthesis and characterization of acetate derivative cassava starch. Am J Food Technol, 2010, 5, 100–110.
[14] Singh AV, Nath LK, Guha M. Microwave assisted synthesis and characterization of *Phaseolusaconitifolius* starch-g-acrylamide. Carbohydr Polym, 2011, 86, 872–876.
[15] Mishra S, Mukul A, Sen G, Jha U. Microwave assisted synthesis of polyacrylamide grafted starch (St-g-PAM) and its applicability as flocculant for water treatment. Int J Biol Macromol, 2011, 48, 106–111.
[16] Sagar AD, Pulle JS, Reddy SM, Yadav MV. Microwave assisted synthesis of carboxymethylstarch. Int J Chem Sci, 2012, 10, 36–42.
[17] Zhang H, Wang R, Wang J, Dong Y. Microwave assisted synthesis and characterization of acetylated corn starch. Starch Starke, 2014, 66, 515–523.
[18] Alfaifi AY, El-Newehy MH, Abdel-Halim ES, Al-Deyab SS. Microwave-assisted graft copolymerization of amino acid-based monomers onto starch and their use as drug carriers. Carbohydr Polym, 2014, 106, 440–452.
[19] Supardan AD, Achmadi SS, Irawadi TT. Microwave-assisted synthesis of phosphorylated corn starch. Jurnal Teknologi dan Industri Pangan, 2014, 25. DOI: 10.6066/jtip.2014.25.1.1.
[20] Lu K, Zhu J, Bao X, Liu H, Yu L, Chen L. Effect of starch microstructure on microwave-assisted esterification. Int J Biol Macromol, 2020, 164, 2550–2557.
[21] Zhong Y, Tian Y, Liu X, Ding L, Kirkensgaard JJK, Hebelstrup K, et al. Influence of microwave treatment on the structure and functionality of pure amylose and amylopectin systems. Food Hydrocoll, 2021, 119. DOI: 10.1016/j.foodhyd.2021.106856.
[22] Patil M, Mathad SN, Patil AY, Arshad MN, Alorfi HS, Puttegowda M, et al. Synthesis and characterization of microwave-assisted copolymer membranes of poly (vinyl alcohol)-g-starch-methacrylate and their evaluation for gas transport properties. Polymers, 2022, 14. DOI: 10.3390/polym14020350.
[23] Biswas A, Appell M, Liu Z, Cheng HN. Microwave-assisted synthesis of cyclodextrin polyurethanes. Carbohydr Polym, 2015, 133, 74–79.
[24] Liu B, He Y, Han L, Singh V, Xu X, Guo T, et al. Microwave-assisted rapid synthesis of γ-cyclodextrin metal–organic frameworks for size control and efficient drug loading. Cryst Growth Des, 2017, 17, 1654–1660.
[25] Qu J, Dong M, Wei S, Meng Q, Hu L, Hu Q, et al. Microwave-assisted one pot synthesis of β-cyclodextrin modified biochar for concurrent removal of Pb (II) and bisphenol a in water. Carbohydr Polym, 2020, 250. DOI: 10.1016/j.carbpol.2020.117003.
[26] Qu J, Meng Q, Lin X, Han W, Jiang Q, Wang L, et al. Microwave-assisted synthesis of β-cyclodextrin functionalized celluloses for enhanced removal of Pb (II) from water: Adsorptive performance and mechanism exploration. Sci Total Environ, 2021, 752. DOI: 10.1016/j.scitotenv.2020.141854.
[27] Calce E, Bugatti V, Vittoria V, Luca SD. Solvent-free synthesis of modified pectin compounds promoted by microwave irradiation. Molecules, 2012, 17, 12234–12242.

[28] Isiklan N, Tokmak S. Microwave based synthesis and spectral characterization of thermo-sensitive poly (N, N-diethylacrylamide) grafted pectin copolymer. Int J Biol Macromol, 2018, 113, 669–680.

[29] Xu P, Laval S, Guo Z, Yu B. Microwave-assisted simultaneous O, N-sulfonation in the synthesis of heparin-like oligosaccharides. Org Chem Front, 2016, 3, 103–109.

[30] Panchan N, Wattanapan P, Sungsinchai S, Roddecha S, Dittanet P, Seubsai A, et al. Optimization of synthesis conditions for carboxymethyl cellulose from pineapple leaf waste using microwave-assisted heating and its application as a food thickener. Bio Resources, 2021, 16, 7684–7701.

[31] Maugard T, Gaunt D, Legoy MD, Besson T. Microwave-assisted synthesis of galacto-oligosaccharides from lactose with immobilized β-galactosidase from *Kluyveromyces lactis*. Biotechnol Lett, 2003, 25, 623–629.

[32] Li XM, Le GW, Cheng JX, Wang F, Shi YH. Optimization of microwave-assisted solid-phase oligosaccharides synthesis reaction and analysis of components and structure of synthetic product. Carbohydr Polym, 2006, 64, 408–418.

[33] Jovanovic-Malinovska R, Fernandes P, Winkelhausen E, Fonseca L. Galacto-oligosaccharides synthesis from lactose and whey by β-galactosidase immobilized in PVA. Appl Biochem Biotechnol, 2012, 168, 1197–1211.

[34] Fischer C, Kleinschmidt T. Synthesis of galactooligosaccharides using sweet and acid whey as a substrate. Int Dairy J, 2015, 48, 15–22.

[35] Kijewska M, Zawadzka M, Sleziak M, Stefanowicz P. Microwave-assisted solid-phase synthesis of lactosylated peptides for food analytical application. Food Chem, 2024, 433. DOI: 10.1016/j.foodchem.2023.137367.

Shashi Kiran Misra, Ajay Kumar, Snehil Singh, and Kamla Pathak*

Chapter 12 Application of carbohydrates as renewable feedstocks for the synthesis of bioactive compounds

12.1 Introduction

The rarity of natural resources is an increasingly critical issue around the world. Exploding population, extensive urbanization, and surged industrialization have led to vigorous consumption of natural resources (fossil fuels, natural gas, water) continuously [1]. Without these assets, life in the ecosystem is impracticable as most of the routine work and economies are based on natural resources. Interestingly, environmental degradation is directly connected with the depletion of natural resources. The unsustainable usage of soil and water and felonious human activities, including deforestation (devastation of natural habitat), exploitation, land conversion, and economic development, lead to the diminution in the quality and accessibility of these natural assets and put species at risk [2].

The rate of the world population is continuously increasing by 80 million per year, and by 2057 it will become 10 billion. A large population requires a huge quantity of basic needs including food and land for living. This will exploit landmass, water, minerals, and natural resources. The rampant use of heavy pesticides and fertilizers not only contaminates our food but also leads to soil infertility and ruins soil microorganisms that maintain the soil's nutrients. Several health implications for humans and animals are also addressed owing to the exhaustion of natural resources and environmental degradation, suggesting that sustainable usage and minimum dependence on natural resources are required in the current scenario. Further, a change in lifestyle and the adoption of RRR (reuse, recycle, and reduce) of natural resources are recommended. Although the government and nongovernment associations are actively taking steps for the conservation of sustainable resources and protection of natural habitats by implementing strict laws and policies that encourage the conservation of resources such as solar systems, windmills, and hydropower, which replace the usage of fossil fuels and minimize the emission of greenhouse gases [3].

***Corresponding author: Kamla Pathak,** Faculty of Pharmacy, Uttar Pradesh University of Medical Sciences, Saifai, Etawah 206130, Uttar Pradesh, India
Shashi Kiran Misra, Ajay Kumar, Snehil Singh, School of Pharmaceutical Sciences, CSJM University, Kanpur 208024, Uttar Pradesh, India

https://doi.org/10.1515/9783111447094-012

12.1.1 Depletion of natural resources and its impact on the ecosystem

Although conservation efforts, including appropriate usage of water, limited intake of energy, improving soil fertility, and reusing/recycling, are a few appreciable steps, they are not always fruitful owing to unavailability and natural inhabitation. For instance, scarcity of water in desert/arid areas and soil erosion due to floods [4]. As natural fossil materials are consistently declining, the pressure on the environment and the increase in the demand for renewable feedstock in industries are perpetually increasing. Assessing the utilization of biomass, oil, and coal in the past 100 years, it can be interpreted that initially, coal tar-based substances were heavily employed by the year 1930, and thereafter requirements were fulfilled by natural oils and fossil gases. Overreliance on fine chemicals in fossil feedstock also predicted its sustainable depletion, hence limiting its exhaustive usage. The regular consumption of fossil feedstock makes it expensive and unaffordable. Surveys alarm a realistic end of fossil oil by the year 2040; thus, chemical industries will face a surge in the prices of natural oils and gases. Further, 90 billion natural resources containing minerals (bauxite, coal, potash, and iron ore), biomass, and natural oils have exhausted more than threefold up to the year 2015, and this graph is presumed to increase by twofold for the year 2050 [5]. Therefore, climate change, instability in the ecosystem, shortage of oil/minerals, and scarcity of water are a few noticeable issues that appear in the depletion of natural feedstock. The extensive consumption of renewable and nonrenewable feedstocks and their rising ecological concerns have stimulated the exploration of renewable feedstocks for sustainable chemical synthesis. Natural biomass containing cellulose, chitin, and chitosan are the most abundant and sustainable feedstocks. These components are vigorously bioconverted for the preparation of biofuels, platform chemicals, and industry-oriented monomers. Currently, lots of nitrogen, sulfur, oxygen, and phosphorus-containing fine chemicals are derived from these carbohydrate sources following different chemical, enzymatic, and microbial processes.

12.2 Carbohydrates: a promising alternative as a renewable feedstock

The present chemical industry still faces significant hurdles in activating C–H bonds owing to their stability and unpolarized character [6]. The chemical challenge of introducing heteroatoms into hydrocarbons persists, and high energy consumption associated with greenhouse gas emissions from current technology strains the environment by generating chemical waste. Although biomass-based fuels should ideally have no carbon footprint, this does not apply because of the need to remove significant amounts of heteroatoms from the biomass [7]. Direct methods of converting renewable resources into compounds with heteroatoms might overcome such issues and be a compelling substitute for traditional procedures [8]. Generally, lignocellulose is composed of approximately 54% cellulose that covers the anhydrous glucose part, approximately 35% hemicellulose containing C5 monomers of sugar, and 11% lignin having phenylpropane units [9]. Amongst all, cellulose abundantly occurs in half of the carbon material in the biosphere and is utilized to produce biofuels and fine chemicals. Another component, lignin, contains an extensive phenolic network and constitutes a significant portion of lignocellulosic biomass. The forest products industry currently utilizes it mostly for operations like a heat source, but there is still untapped potential for producing important, oxygenated aromatic building blocks [10]. Chitin, a derivative of cellulose, is another underutilized resource. It stands second after cellulose on Earth and is present in large amounts in crustacean detritus, insect exoskeletons, and fungal cell walls [11, 12].

Most of the natural renewable biomass is currently disposed of in landfills and oceans, even though it is likely the sole abundant renewable supply [13]. Though they are less common, biomass sources that contain sulfur and phosphorus also exist, and it is important to evaluate and analyze their potential [14]. Figure 12.1 highlights the sequential transformation of biobased products from carbohydrate feedstock. Simultaneously, different approaches, such as transforming renewable organic compounds containing sulfur and phosphorus into their equivalents, can provide feasible routes toward a more sustainable chemical sector [15]. There is a burgeoning curiosity for obtaining materials that are highly developed, along with raw materials and polymers derived from carbohydrate sources, which are gaining attraction due to environmental concerns and increased awareness of their impact [16]. The possible sources of chitosan, chitin, cellulose, and polysaccharides are among the most promising [17]. These comprise most of the material derived from waste products from fisheries and agriculture, and in the future, these could serve as valuable reservoirs of intricate oxygen-containing substances for the chemical and polymer sectors as well as catalysts [18, 19].

Most of the key byproducts and platform chemicals are generated through the thermal, catalytic, and enzymatic processing of carbohydrate biomass (lignocellulose) for the preparation of fine chemicals. Chemo-catalyzed hydrolysis at mild temperatures (up to 200 °C) is an essential transforming process for the preparation of platform chemicals, that is, glucose, furfural, levulinic acid (LA), and 5-hydroxymethylfurfural (HMF). All are composite blends of cellulose and lignin sourced from agricultural and arboreal biomass waste under acidic conditions [20]. After a series of dehydration, oxidation, and degradation reactions, these byproducts are synthesized and are enormously employed in chemical industries. LA or 4-oxopentanoic acid is a short-chain (C5) linear fatty acid with the molecular formula $C_5H_8O_3$. LA was labeled as a top bio-based platform chemical by the United States Department of Energy in the year 2004. Basically, LA and its derivatives are highlighted for the synthesis of petroleum/biofuels including ethyl levulinate, gamma valerolactone, and methyl tetrahydrofuran [21]. Numerous chemicals, including succinic acid, oxo ester, acrylic acid, beta-acetyl acrylic acid, diphenolic acid, sodium and calcium

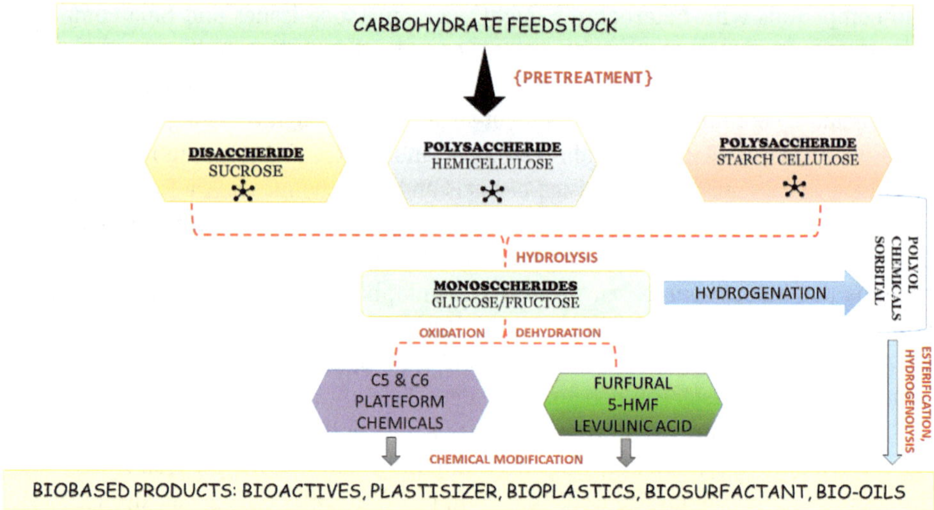

Figure 12.1: Transformation of biobased products from carbohydrate feedstock.

levulinates, alpha angelica lactone, and α-aminolevulinic acid, are further synthesized from LA (Figure 12.2). These chemicals serve as precursors for the synthesis of fine chemicals. Several chiral reagents, adsorbents, lubricants, and biologically active materials are further synthesized from LA. Apart from these, LA is widely demanded in each sector of the market as solvents, polymers, resins, batteries, electronics, chemical intermediates, and precursors for cosmeceuticals and pharmaceuticals.

Carbohydrate feedstock or biomass utilized for the development of various platform chemicals and bioactive agents is categorized into three generations: (i) first generation (sugar, animal fat, and vegetable oils), (ii) second generation (nonfood crops, organic waste, and lignocellulose), and (iii) third generation (red, brown, and green macroalgae). Figure 12.3 defines different generation biomass for the preparation of LA [22].

Therefore, LA is produced from polysaccharide biomass (cellulose, chitin, and lignocellulose) through the following three steps: (i) pretreatment for extraction of polysaccharides, (ii) hydrolysis/catalysis to obtain monosaccharides (pentoses and hexoses), and (iii) transformation of monosaccharides into LA. First-generation biomass (sugar and starch) involves different acids/catalysts for the processing of LA, that is, (i) hydrolysis/catalysis of biomass by Brønsted acid to form monosaccharides, (ii) isomerization of monosaccharides by Lewis acid, (iii) dehydration of monosaccharides by bifunctional acid, and (iv) rehydration of HMF catalyzed via Brønsted acid, as presented in Figure 12.4. Different process constraints, including pressure, temperature, reaction time, and acid, are controlled during the bioconversion of LA [23].

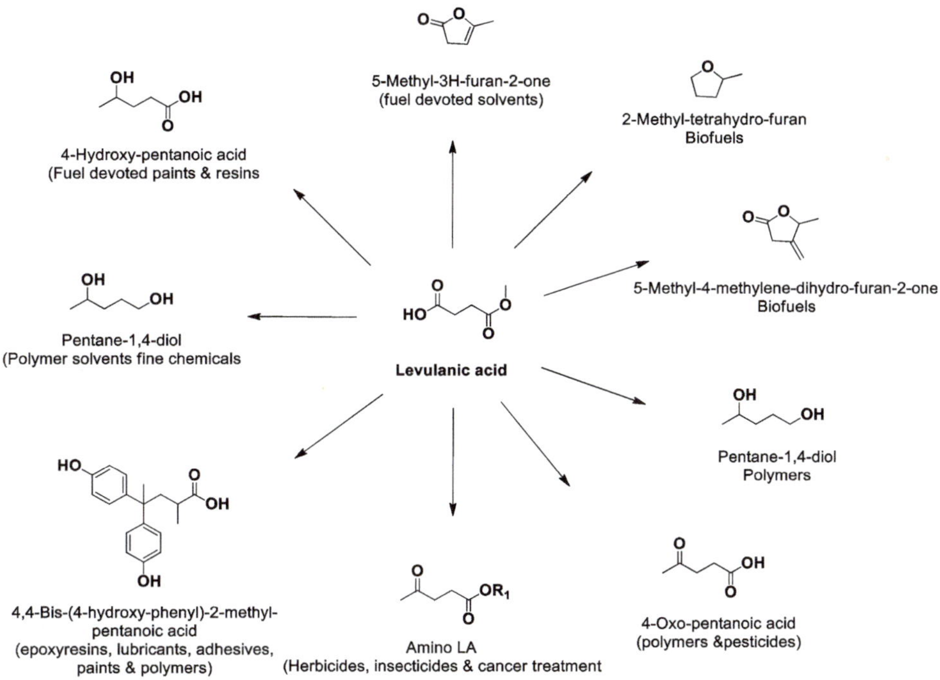

Figure 12.2: Various platform chemicals obtained from levulinic acid and their uses.

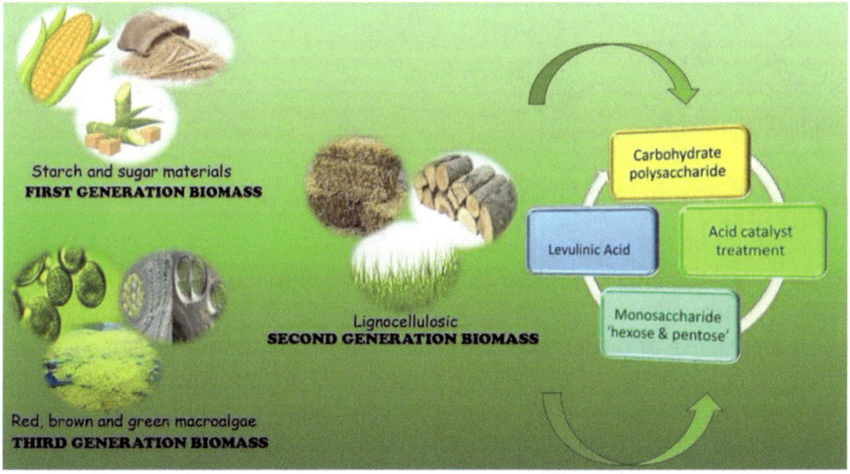

Figure 12.3: Various generations of carbohydrate feedstock/biomass explored for the preparation of levulinic acid.

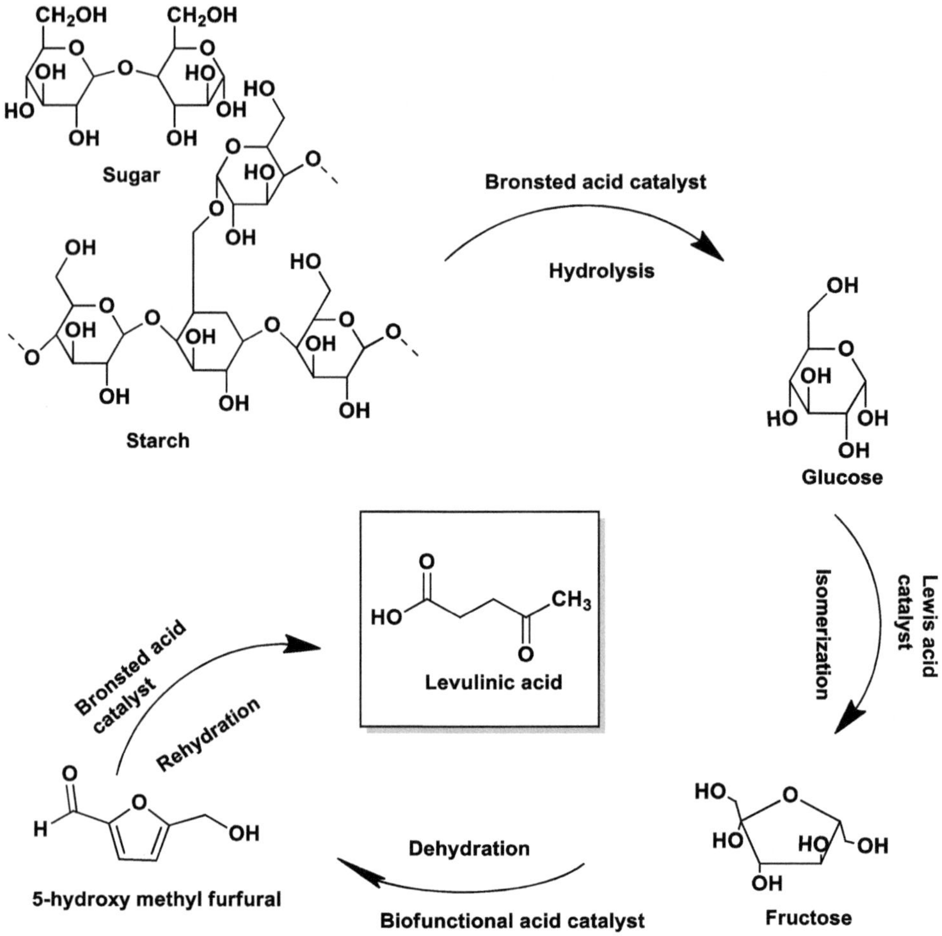

Figure 12.4: Synthesis of levulinic acid from first-generation sugar and starch.

In the second generation, lignocellulose biomass is hydrolyzed into monosaccharides using a catalyst (Brønsted acid), and after dehydration (xylose to furfural) and hydrogenation in the presence of bifunctional acids, LA is formed. Another organic component, lignin, contains heavily packed covalent linkages of cellulose/hemicellulose that initially convert into humins on acid treatment and remain as solid residue due to less solubility; hence, a poor yield of LA is obtained. Pretreatment with dilute acid is required for the biochemical process of LA [24]. Figure 12.5 depicts the synthesis of LA from second-generation carbohydrate feedstocks.

Macroalgae, or seaweed, is widely employed as a third-generation biomass for the synthesis of high-value LA [25]. It contains viable components, including agar, alginates, fucoidan, and ulvan [26]. The macroalgae are divided into red (agar, carrageenan, mannan, cellulose, and xylan), brown (laminaran, cellulose, mannitol, and al-

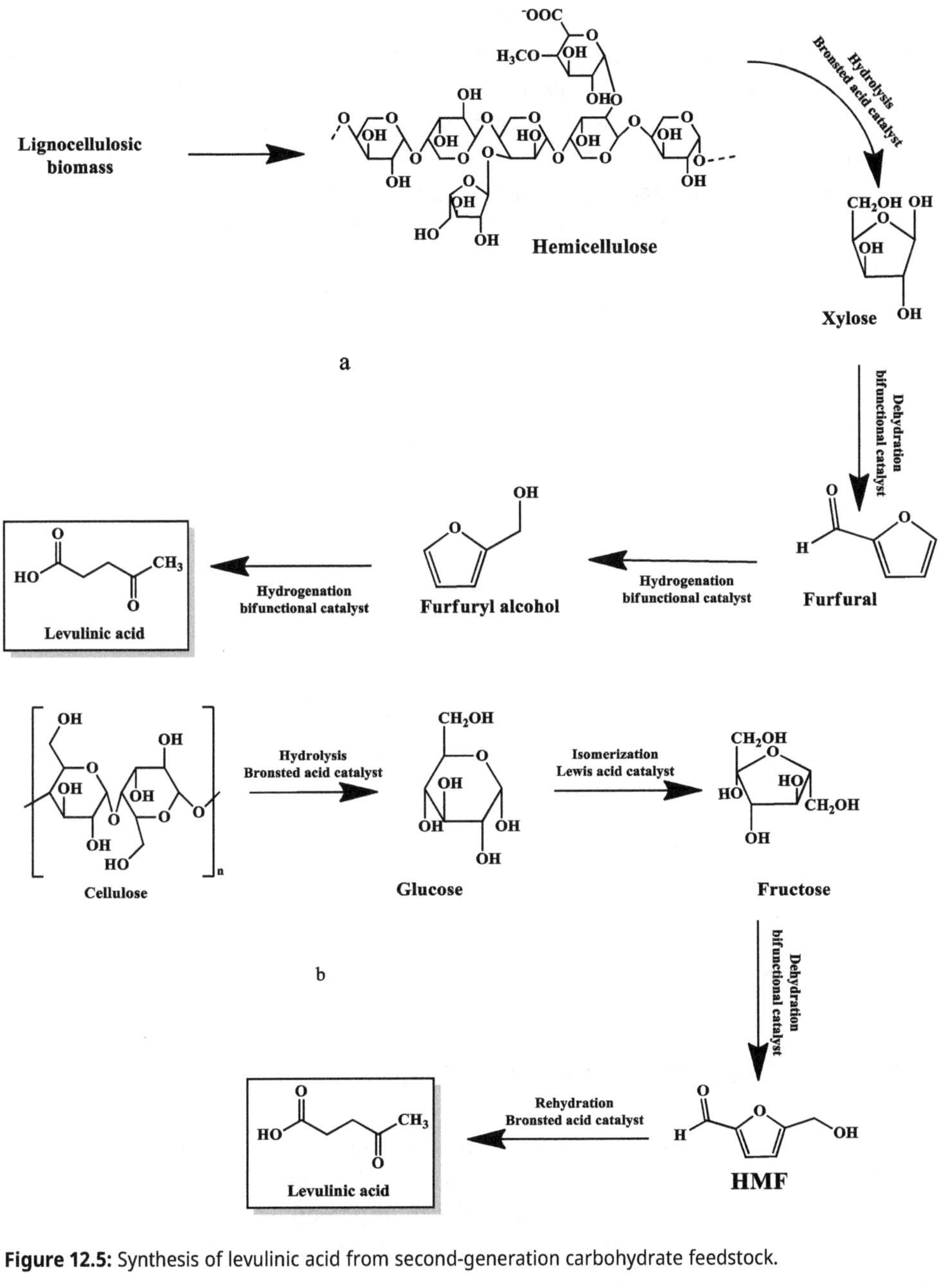

Figure 12.5: Synthesis of levulinic acid from second-generation carbohydrate feedstock.

ginates), and green seaweeds (cellulose, xylan, and mannose). These macroalgae trap carbon dioxide with a high growth rate and effectively reduce the emission of greenhouse gases. Figure 12.6 illustrates the synthesis of LA from red, brown, and green macroalgae. These macroalgae are advantageous over second-generation biomass as they do not contain lignin and have high carbohydrate content [27]. They have a high growth rate and utilize a large amount of carbondioxide. For the preparation of LA, macroalgae are hydrolyzed into galactose, glucose, and xylose in the presence of a catalyst (Brønsted acid). Thereafter, dehydration of the above monosaccharides using bifunctional acids results in furfural and 5-HMF, which are later catalyzed by bifunctional acid into LA [28]. Macroalgae contain abundant inorganic salts and metals that may reduce the yield of LA; hence, pretreatment with dilute sulfuric acid is required to remove acid salts and metal ions in an aqueous solution.

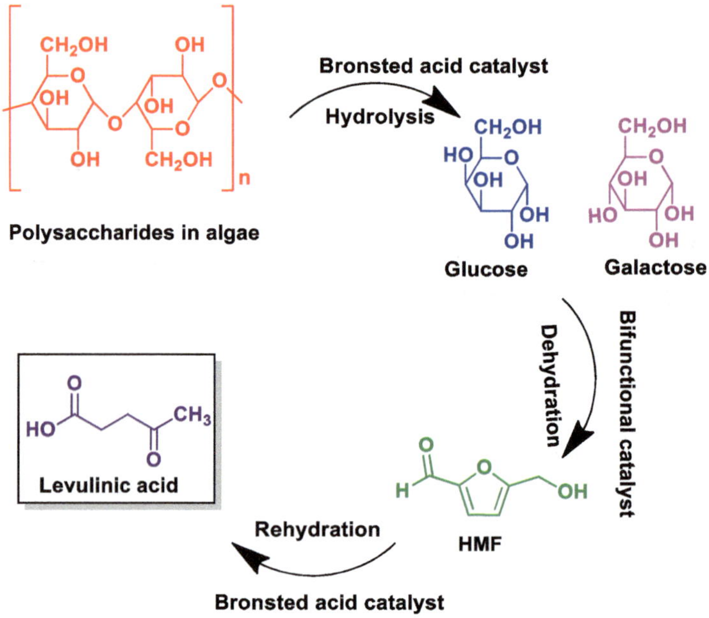

Figure 12.6: Synthesis of levulinic acid from third-generation macroalgae.

12.2.1 5-Hydroxymethylfurfural (HMF)

HMF is a primary key ingredient obtained from myriad sources of carbohydrates such as cellulose, sucrose, fructose, glucose, and inulin. It is a platform molecule associated with the group "drop-in biofuels." Its value lies in its potential to act as a primary component for the creation of industrial chemicals like monomers and biofuels. Figure 12.7 presents the synthesis pathway of 5-HMF from cellulose feedstock. By employing 5-HMF, it becomes possible to synthesize a range of molecules, such as 2,5-diformylfuran, 2,5-furandicarboxylic acid, 2,5-dihydroxymethylfuran, and 2,5-dimethylfuran [29]. Like furfural, 5-HMF contains a furan ring attached with a formyl and hydroxymethyl group at positions 2 and 5, respectively. On a commercial scale, HMF is prepared from lignocellulose biomass waste (energy crops, wood, straw) following a triple dehydration process and using heterogeneous catalysts, enzymes, and ionic liquids. The following synthesis steps are involved in the preparation of 5-HMF: (i) hydrolysis of polysaccharide (cellulose) into glucose in the presence of a Brønsted acid catalyst, (ii) isomerization (conversion of glucose into fructose) in the presence of a Lewis acid or Brønsted base catalyst, and (iii) dehydration of fructose, catalyzed by a Brønsted acid catalyst to form 5-HMF. To increase the yield and stability of 5-HMF, acidic salts including ferrous chloride and aluminum chloride along with organic solvents (dimethyl sulfoxide) are added [30].

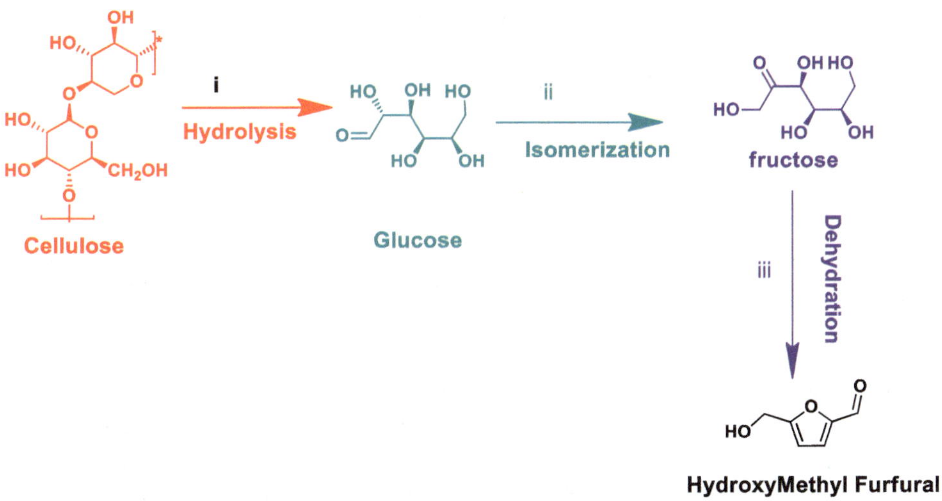

Figure 12.7: Synthesis pathway of 5-HMF from cellulose feedstock.

The presence of both the aldehyde and alcohol functional groups exhibits virtues of aromaticity in 5-HMF; thus, it can be explored for the production of myriad aromatic compounds including biofuels, 2,5-dimethylfuran, gasoline, and diesel through the hydrogenation process. Oxidation of alcohol and aldehyde groups leads to the formation of diol or dicarboxylic acid and can be subjected to the synthesis of polymers. Another compound, 2,5-furan dicarboxylic acid, is an important substitute for terephthalic acid and is utilized for the preparation of polyethylene furanoate. These derived components are widely employed as precursors for the manufacturing of polyurethane, polyesters, and polyamides. Amongst these, polyurethane shows good thermal resistance, photoreactive polyesters are utilized in ink products, and polyamide exhibits liquid crystal properties. LA, a key platform chemical, is also derived from 5-HMF, which is involved in the preparation of a range of biofuels, including 1,4-pentanediol, Y-valerolactone, 2-methyl tetrahydrofuran, and levulinate esters. In the food industry, 5-HMF is utilized as a biomarker and flavoring agent [32]. Several drug intermediates, such as 5-chloromethylfurfural, furan-2,5-dicarboxylic acid diamide, 2-(hydroxymethyl)-

5-(aminomethyl)-furan, and 5-[2-hydroxy-2-(5-hydroxymethyl-furan-2-yl)-acetyl]-furan-2-carbaldehyde, are synthesized from 5-HMF via several treatments (halogenation, condensation, amidation, and esterification). 5-Arylaminomethyl-2-furanmethanol, a photosensitizer, is derived from 5-HMF. It is employed in chemotherapy and photodynamic therapy as a natural precursor. Similarly, 1,2,6-hexanetriol, a viscosity and humidity controller, has been utilized in the pharmaceutical industry for the preparation of drugs and cosmetics [33]. Owing to the versatile inherent properties of 5-HMF, it is recognized as a sleeping giant in the domain of platform chemicals. Figure 12.8 depicts various key ingredients that have been derived from 5-HMF. Abdulmalik et al. investigated the efficiency of 5-HMF for binding intracellular sickle hemoglobin without disturbing associated healthy tissues. It rapidly absorbs in blood via the gastrointestinal tract, forms a Schiff base, allosterically shifts oxygen, and inhibits red cell sickling [34].

12.2.2 Furfural

Furfural, or furan-2-carbaldehyde, a heteroatomic aldehyde, is a platform chemical and a key ingredient for the preparation of carbohydrate biomass refineries. It is colorless and oily in appearance, with an almond-like odor. It is considered safe and naturally degrades into ascorbic acid. It is a key ingredient in bread, cocoa, coffee, fruit juice, and wine. The C5 sugars (arabinose and xylose) present in hemicellulose are utilized as raw material for the synthesis of furfural [35]. Figure 12.9 illustrates synthesis of furfural from xylose. Its favorable features, namely, being inexpensive, easily synthesizable, and readily bioavailable, allow it to be widely employed for the synthesis of organic molecules. Furan, an important intermediate heterocyclic molecule for the synthesis of myriad pharmaceuticals is developed from furfural. Interestingly, furfural acts as an agrarian nematicide and is often used to control plant parasites without affecting the yield and growth of plants [36]. In industry, furfural is used as a lubricating oil and in the preparation of chemical intermediates such as furfuryl alcohol and tetrahydrofuran. In the year 1921, the Quaker Oats Company first commercially synthesized furfural from oat hulls in the presence of sulfuric acid. However, low yield, limited demand, and extreme maintenance costs limited the production of furfural. Currently, furfural is synthesized from C5 polysaccharides (pentosan sugars) following several processes such as hydrolysis, dehydration, and double distillation. Acidic catalysts, halide ions, and organic solvents are employed during the treatment [37]. Literature envisages that the application of cocatalysts (transition metals – Cr-II and -III) and hydrochloric acid at 200 °C can transform C5 sugars (corn stalk, pine wood, and rice straw) into *N,N*-dimethylacetamide and improve the yield of furfural up to 63% [38].

It has gained popularity in organic synthesis and serves as a precursor for the preparation of furfurymaleic acid, nitrofurazone, and 5-(3-nitrophenyl)-4-(2-furyl)-3-buten-2-one. This platform chemical is highly demanded by the chemical industries to form petrochemical compounds, biorenewable green plastic (polyethylene furanoate), sealants,

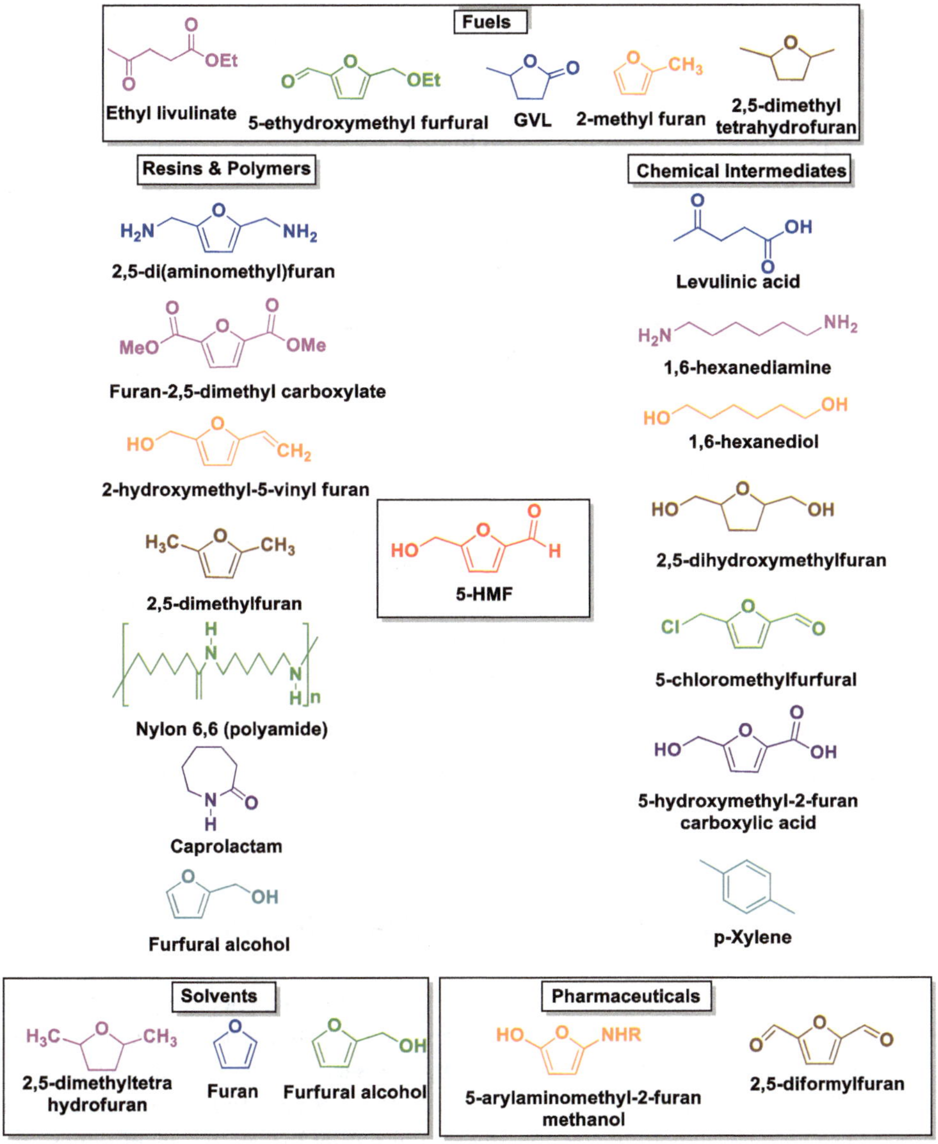

Figure 12.8: Various key ingredients derived from 5-HMF.

adhesives, polymers (polyurethane-polyurea copolymers, polyurethanes), automotive accessories, toys, and catering products [39]. Furfural, after oxidation, converts into furoic acid, a versatile component for the synthesis of myriad agricultural, pharmaceutical, and industrial fine chemicals. Furoic acid transforms into furoyl chloride, a precursor for insecticides. Similarly, furan, furfuryl alcohol, tetrahydrofurfuryl alcohol, 5-chloromethyl furfural, nitrofuran, and methylfuran are involved in the synthesis of several therapeu-

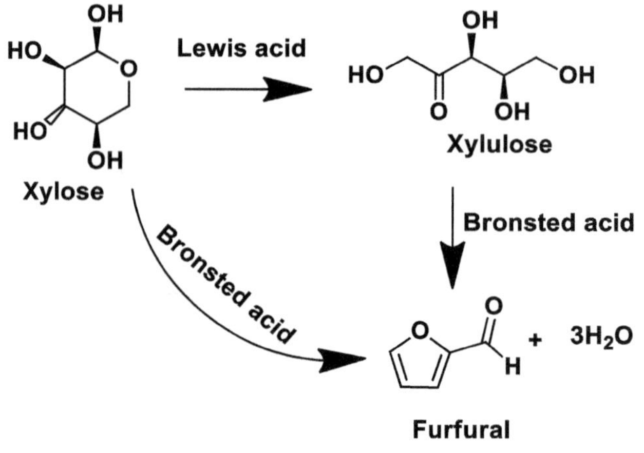

Figure 12.9: Synthesis of furfural from C-5 sugar.

tics. The heterocyclic furan ring is a key molecule for the synthesis of antibacterial (ceftiofur, furazolidone), antiprotozoal (diloxanide, nifuratel), anti-inflammatory (rofecoxib), analgesic (mirfentanil), diuretic (furosemide), antidepressant (siramesine), anti-Parkinson (preladenant), muscle relaxant (dantrolene), antiglaucoma (pilocarpine), antihypertensive (prazosin), and anticancer drugs (lapatinib). After hydrogenation, furfural is transformed into tetrahydrofuran, a commercial solvent and starting material of spandex fiber (polytetramethylene ether glycol), polyurethane, and nylon [40]. Furan-2,5-dicarboxylic acid is employed for the synthesis of polyester. Figure 12.10 compiles various bioactive agents derived from furan heterocyclic compounds.

12.3 Prebiotics from carbohydrate feedstock

Prebiotics are food supplements that are nondigestible and selectively promote the growth/flora of beneficial microorganisms in the colon region. Carbohydrates such as lactulose, inulin, fructo-oligosaccharides, soybean oligosaccharides, cyclodextrins, germinated barley, and galacto-oligosaccharides fall into this category. To fulfill the function of prebiotics, these carbohydrate components should not be hydrolyzed by enzymes and gastric acid. They should selectively stimulate the growth of intestinal microflora [41].

Sulfated fucans and polysaccharides are obtained from the cell wall of brown seaweeds including *Ascophyllum nodosum* and species of *Fucus*. These contain alginate and sulfated L-fucose, with a backbone of mannuronic acid. Besides their applications in the food industry, these brown algae can significantly augment the content of Bifidobacteria upon daily administration (10 g) for 2 weeks [42].

Ulvans, the green algae, have also gained therapeutic and nutraceutical values. They do not degrade in fecal microflora and have a great affinity for receptors present

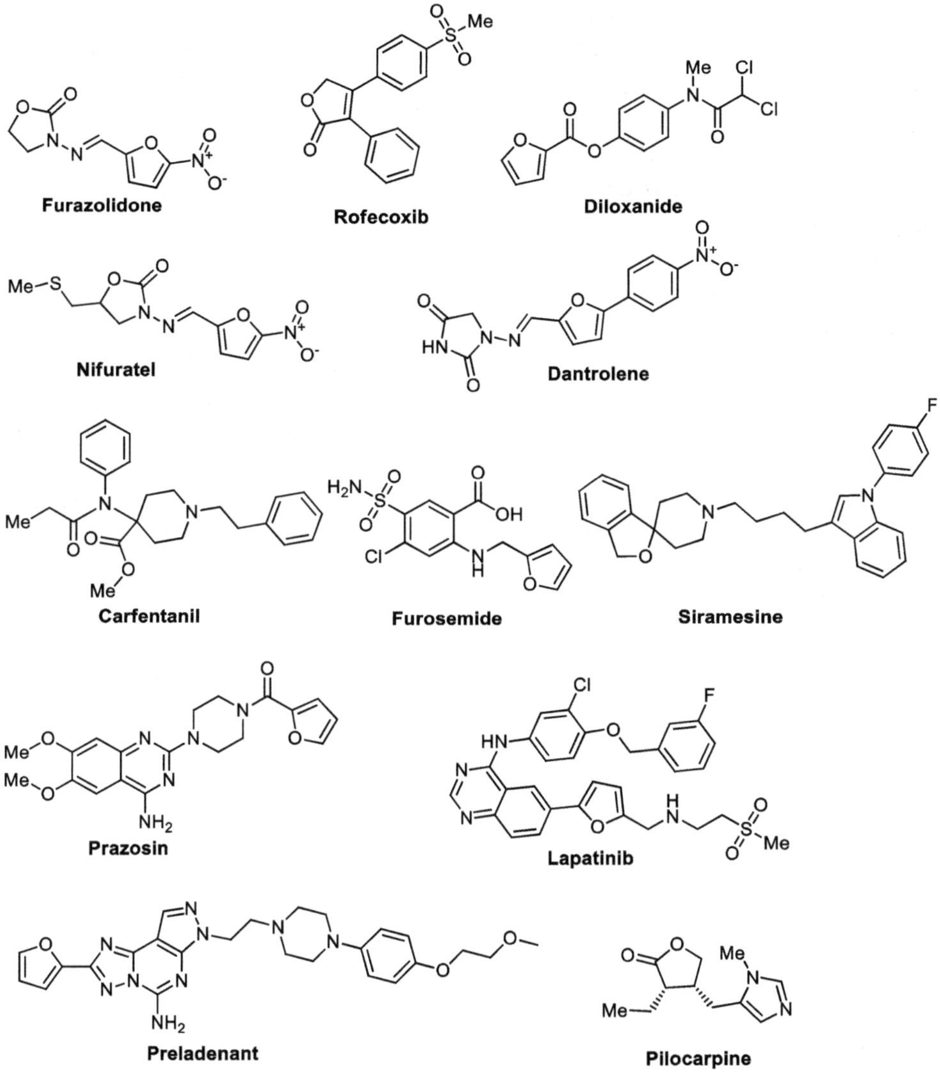

Figure 12.10: A myriad of bioactive agents derived from furan heterocyclic compounds.

in the gut region. Ulvans are growth promoters and stabilizers owing to their nondegradation property. The United States Food and Drug Administration (USFDA) has approved chitosan biopolymer obtained from naturally occurring chito-oligosaccharides for the preparation of food products. Primex®, a Norwegian company, has acquired GRAS (Generally Recognized as Safe) approval for chitosan as a functional food in the year 2011. The literature documents the benefits of chitosan in regulating lipid and lipoprotein components. Being biodegradable and biocompatible, chitosan also exhibits antibacterial and antifungal virtues that are required qualities for addition in food products [43]. Wang et al. administered chitosan in hyperlipidemic Wistar rats for 3 weeks and investigated its effect on lipid metabolism and related proteins. Outcomes gathered from plasma lipid level assays suggested approximately 31% decreased triglyceride and approximately 11% increased high-density lipoprotein after administration of chitosan oligosaccharide [44].

Raffinose oligosaccharide present in lupine seeds, that is, *Lupinus albus*, is employed for the manufacturing of bifidogenic prebiotics. It is added into the mixed culture of *Lactobacillus acidophilus* and *Bifidobacterium lactis* in a 1:1 ratio to modify the acidification rate, enhance culture growth, reduce fermentation time, and increase the effect of the prebiotic [45]. Another heteropolysaccharide prebiotic, that is, konjac mannan, pectin, and modified starches, is used to delay gastric emptying, resulting in increased satiety and reduced blood cholesterol and obesity as physiological responses. These fibrous heteropolysaccharides also reduce the incidence of hypertension. Furthermore, these therapeutic effects produced by the consumption of prebiotics lead to the generation of short-chain fatty acids in the colon that are involved in several metabolic processes in the body [46]. Figure 12.11 illustrates various prebiotics obtained through carbohydrate sources.

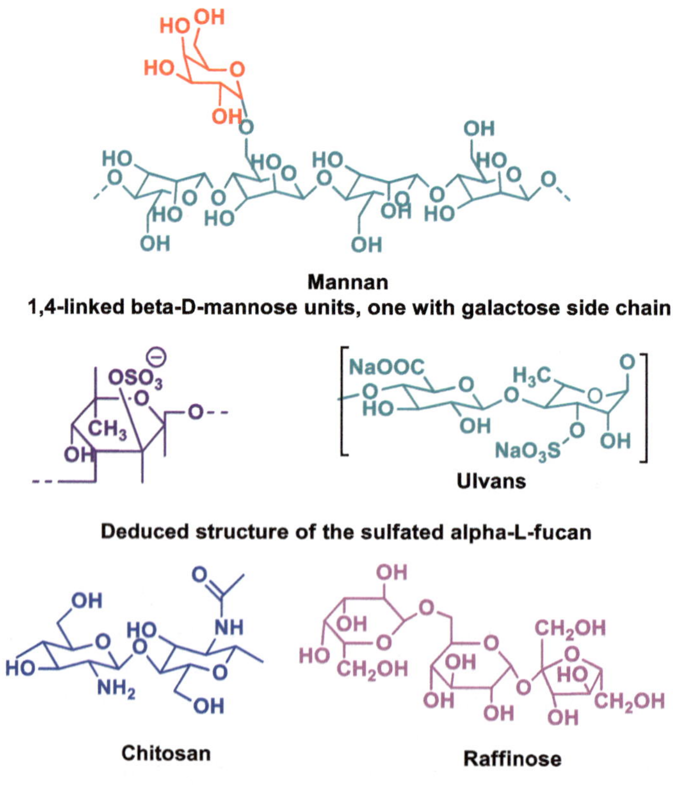

Figure 12.11: Various polysaccharide-derived prebiotics.

12.4 Biosurfactants (BSs)

An innovative approach to sustainable development involves the production of biosurfactants through the biodegradation of waste carbohydrate feedstocks. Science and technology are currently focused on ensuring the safety and sustainability of the environment, population, and ecosystems. This shift from "brown chemistry" to "green chemistry" requires simple changes to promote an eco-friendlier approach. Natural resources are being increasingly recognized as effective and reliable alternatives to accelerate this revolutionary change in chemistry. Promising sources of carbohydrates for biosurfactants include wastes of vegetables, vegetable oils, fruits, and starchy substances. These advantages include lower toxicity, greater environmental compatibility, high biodegradability, and the ability to be produced from biowaste. These virtues of biosurfactants make them suitable for the preparation of food products and skincare cosmetics. Furthermore, biosurfactants are gaining popularity in drug design and delivery and biotechnology products meant for the industrial and healthcare sectors. They are extensively explored as emulsifiers in the preparation of emulsions, deemulsifiers, wetting and foaming agents in fermenters, edible food additives in packaged products, cosmeceuticals, and as detergents. Biosurfactants also find applications in petrochemical industries, wastewater treatment, soil remediation, and agro-industries [47]. Despite numerous attractive features, the utilization of biosurfactants is hampered by high production costs. The production of biosurfactants is largely limited by expensive substrates and the need for effective bioprocessing methods. Additional challenges arise in terms of product recovery and purification strategies to produce biosurfactants that are environmentally and economically sustainable. One promising approach involves utilizing agro-industrial wastes as a carbohydrate source in microbial mediums. This environmentally conscious approach not only decreases manufacturing costs but also promotes the utilization of waste streams, contributing to a circular bioeconomy [48]. Moreover, it aligns with the pursuit of sustainable development goals. In 1999, Rosenberg and Ron categorized biosurfactants into two main groups: low-molecular-weight and high-molecular-weight (HMW) biosurfactants. The former group includes rhamnolipids, cellobiose, sophorolipids, lipopeptides, surfactin, iturin, phospholipids, glycolipids, fatty acids, and neutral lipids. Alternatively, the HMW group consists of polymeric surfactants (vesicles and fimbriae, emulsan polysaccharides, lipopolysaccharides, alasan, liposan, and lipoproteins) and particulate surfactants [49]. Popular biosurfactants prepared from carbohydrate sources are presented in Figure 12.12.

Various microorganisms, including yeasts, fungi, and bacteria, are utilized to produce diverse biosurfactants through the process of fermentation [50]. Carbohydrate sources, including sugarcane, date, beetroot, and their coproduct molasses, are used for the preparation of biosurfactants via fermentation using microorganisms. Corn, rice, wheat, potato, and cassava are extracted for starch, and a huge amount of husk and starchy water (high carbon source) is released that is used to produce surfactant through *Bacillus subtilis* [51]. Likewise, carbohydrate-rich cane and soy molasses have

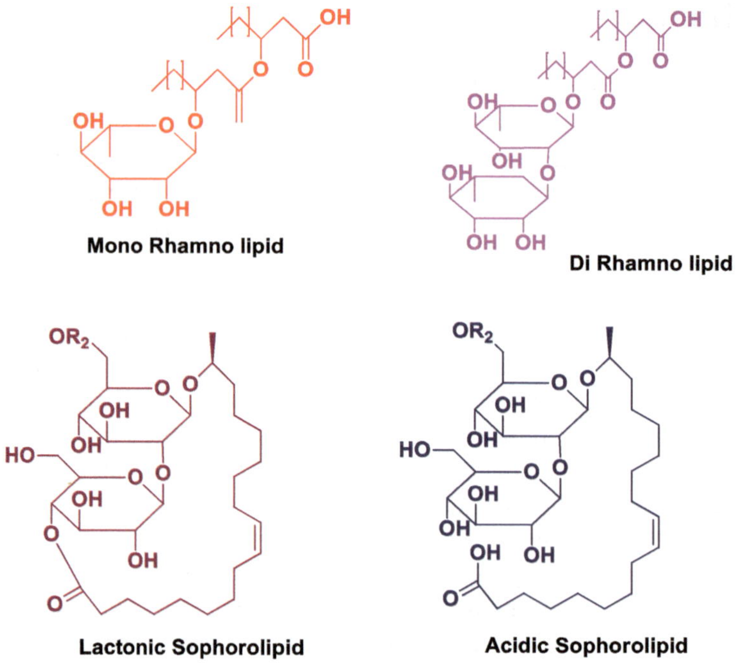

Figure 12.12: Popular biosurfactants derived from carbohydrate sources.

been explored to produce biosurfactants. Biosurfactants – rhamnolipids and lipopeptide – are prepared from carbon-rich molasses using *P. aeruginosa* and *Bacillus subtilis* through fermentation [52]. Cashew apple, a part of the cashew nut, is generally used to produce syrup, jam, and juice. It is an enriched carbohydrate source and employed for the preparation of biosurfactant [53]. Microorganisms including *Acinetobacter calcoaceticus, Bacillus subtilis, Yarrowia lipolytica,* and *Pseudomonas aeruginosa* are used for the fermentation of cashew apple. Fruits/vegetable peels (citrus, banana, orange, carrot, pineapple) and sugarcane bagasse enriched with carbon sources are subjected to submerged fermentation to prepare biosurfactants utilizing *Pseudomonas aeruginosa, Bacillus subtilis, Corynebacterium aquaticum,* and *Bacillus licheniformis* [54].

12.5 Carbohydrate feedstock-derived bioactive agents

Carbohydrates and their derived molecules play a crucial role in various biological functions within living organisms. As a result, research on carbohydrate-based compounds has gained momentum. Nitrogen-containing chitin from degraded carbohydrates is utilized for the synthesis of glycoconjugates and other heterocyclic molecules of medicinal value. Disubstituted imidazole containing a polyhydroxy group inside the chain is derived from D-glucose, formamidine acetate, and ammonium carbonate. The polysaccharide "hyaluronic acid" was investigated owing to its potential to alleviate arthritis [55]. Organophosphorus compounds are employed as insecticides, herbicides, and in some medical applications, even though these intermediates and products do not feature prominently among the most utilized substances in the chemical industry. Carbohydrate biomass-based 5-chloromethyl furfural is employed for the synthesis of prothrin, an insecticide. Furan-based prothrin has shown insecticidal activity against *Aedes* mosquitoes that cause yellow fever [56]. Over the past decade, numerous published scholarly contributions have provided updates on the advancements in carbohydrate-based drugs, diagnostic agents, and vaccines. These bioactive therapeutics are approved by esteemed regulatory authorities including the USFDA, EMA (European Medicines Agency), South Korea, India, and other countries. Table 12.1 enlists several therapeutics developed in two decades.

Table 12.1: List of newly carbohydrate-derived bioactive agents.

Therapeutic agent	Structure	Use	Origin country and year
Remdesivir	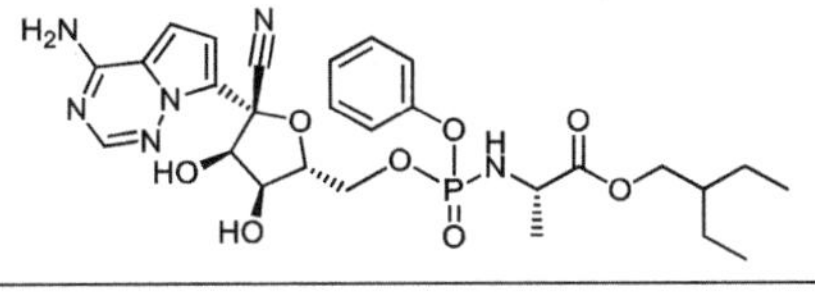	Antiviral	USA, 2020
Azvudine	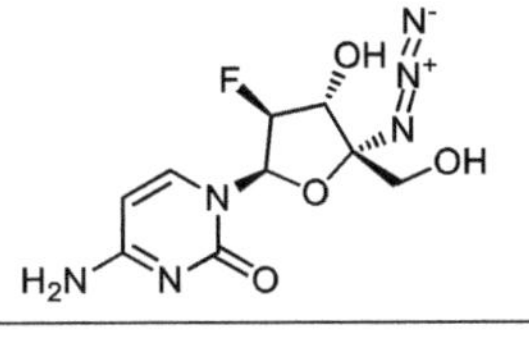	Antiviral	China, 2021
Maribavir	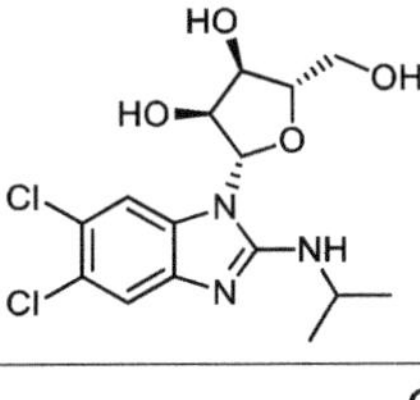	Antiviral	USA, 2021
Molnupiravir	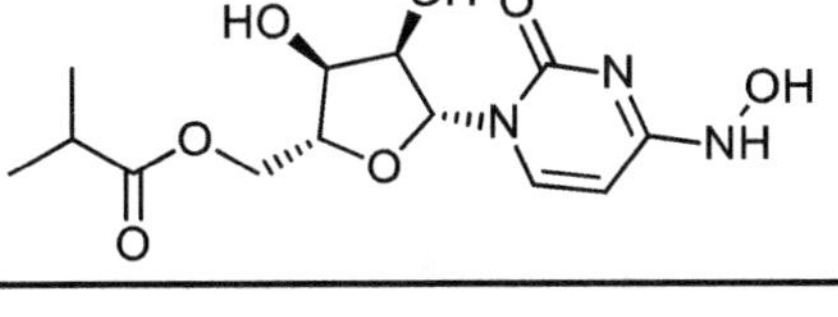	Antiviral	UK, 2021

Table 12.1 (continued)

Therapeutic agent	Structure	Use	Origin country and year
Remogliflozin etabonate	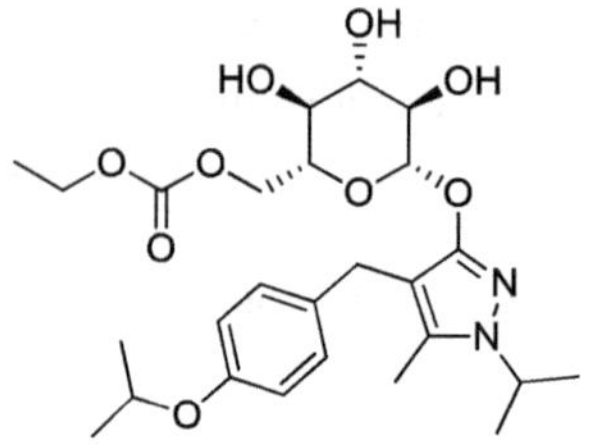	Antidiabetic	India, 2019
Shengjimycin	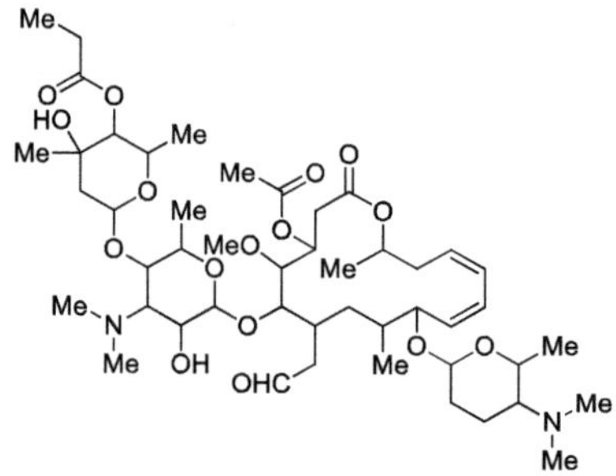	Antibacterial	China, 2019
Sotagliflozin	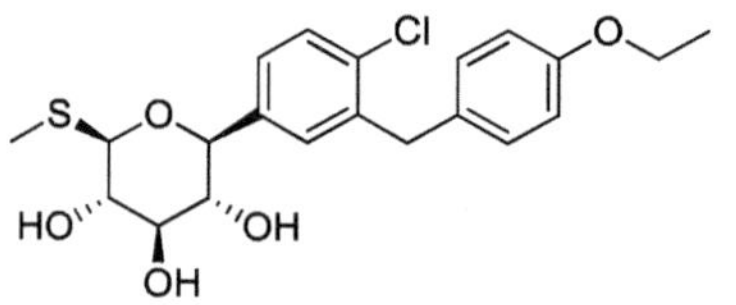	Antidiabetic	Europe, 2019
Plazomicin	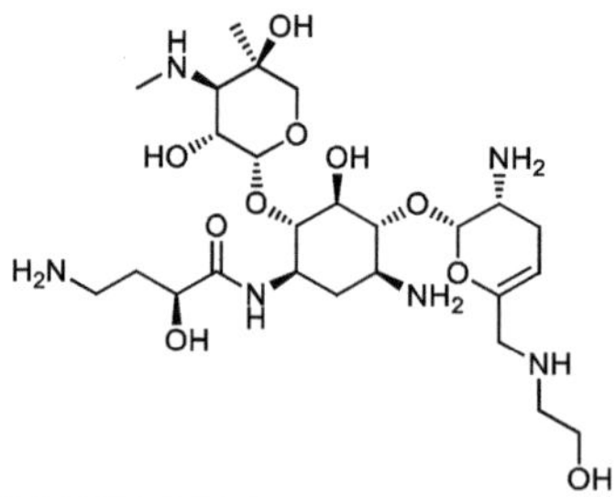	Antibacterial	USA, 2018
Ertugliflozin	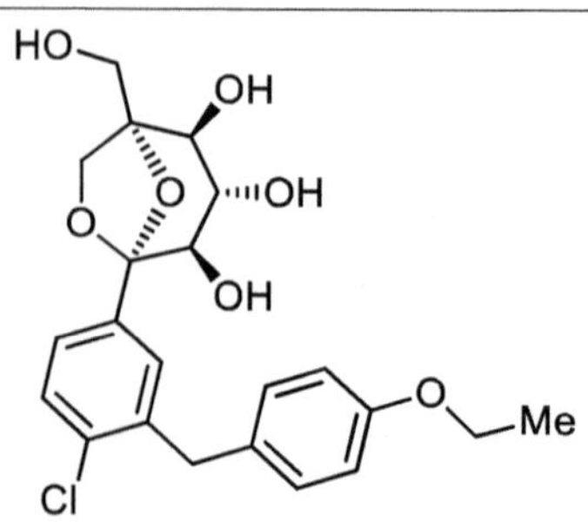	Antidiabetic	USA, 2017

Table 12.1 (continued)

Therapeutic agent	Structure	Use	Origin country and year
Immucillin H	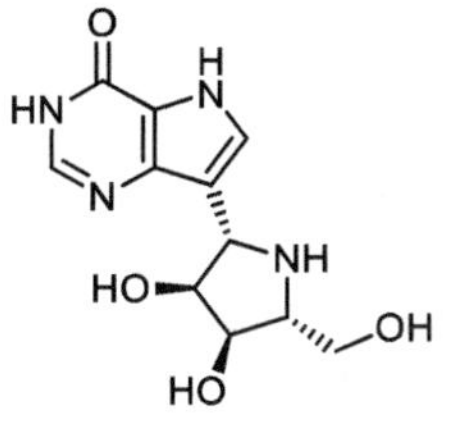	Anticancer	Japan, 2017
Midostaurin	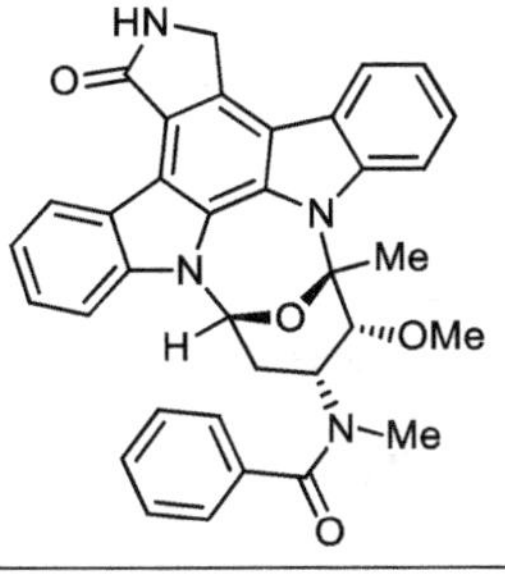	Anticancer	USA, 2017
Dalbavancin	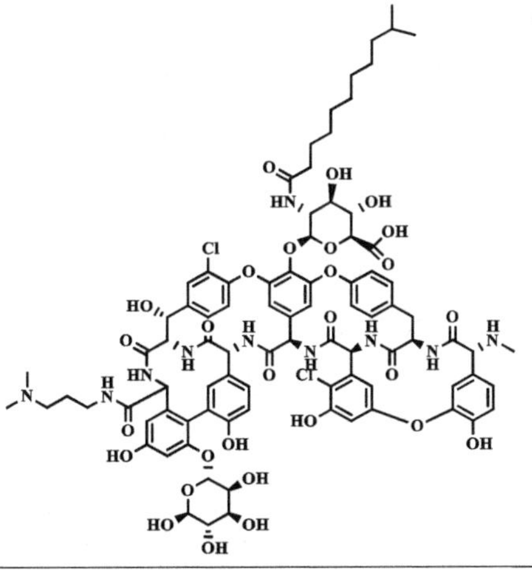	Antibacterial	USA, 2014
Empagliflozin	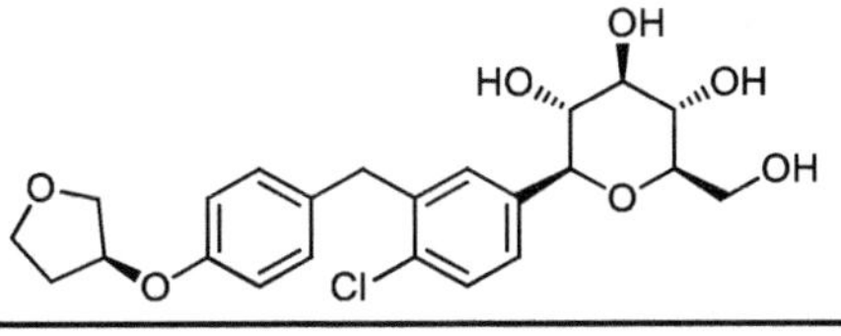	Antidiabetic	Europe, 2014

Table 12.1 (continued)

Therapeutic agent	Structure	Use	Origin country and year
Ipragliflozin	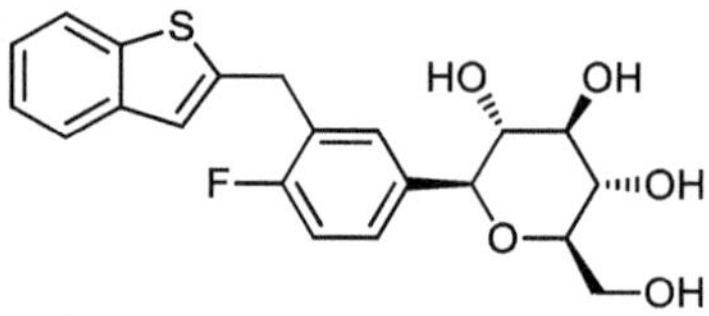	Antidiabetic	Japan, 2014
Luseogliflozin (diarylmethane)	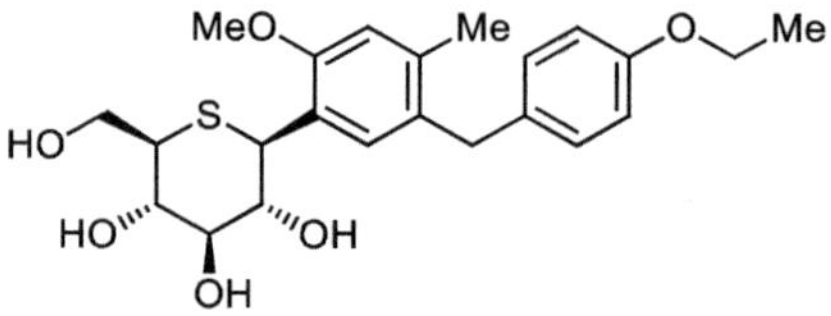	Antidiabetic	Japan, 2014
Tofogliflozin	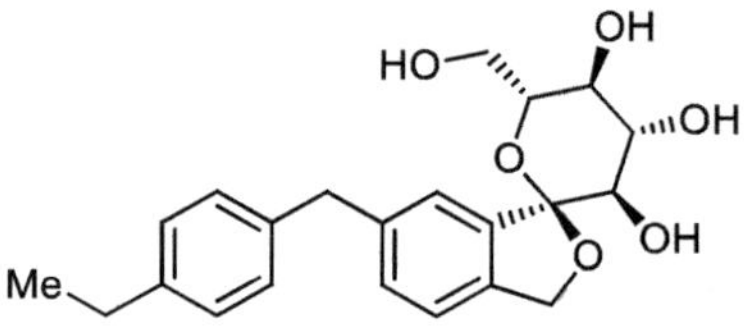	Antidiabetic	Japan, 2014
Canagliflozin	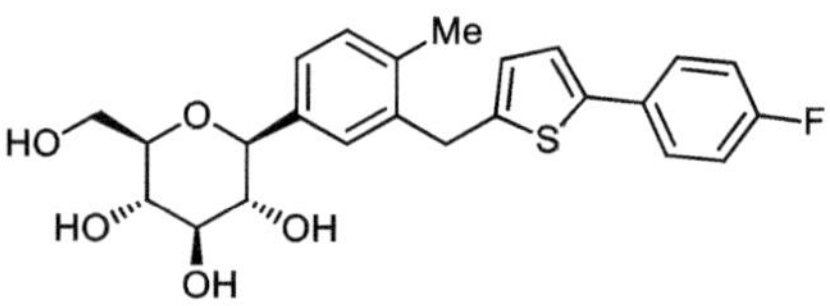	Antidiabetic	USA, 2013
Sofosbuvir	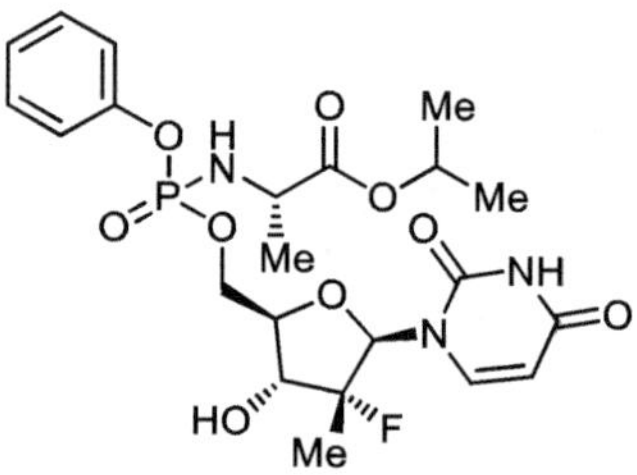	Antiviral	USA, 2013
Dapagliflozin	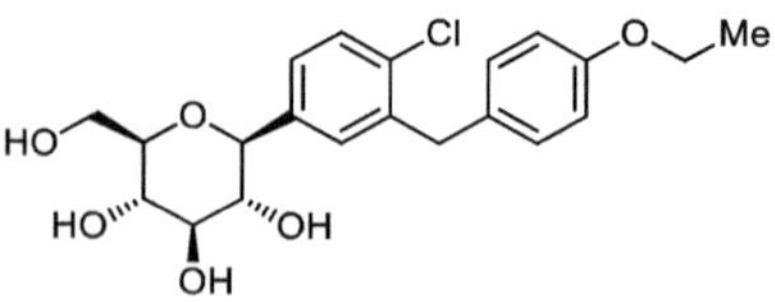	Antidiabetic	Europe, 2012
Fidaxomicin	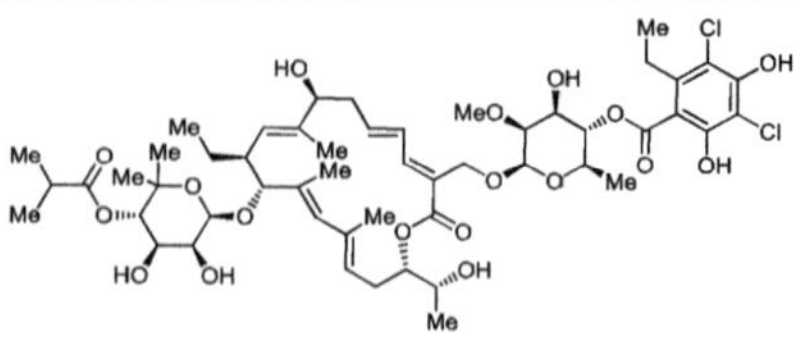	Antibacterial	USA, 2011

Table 12.1 (continued)

Therapeutic agent	Structure	Use	Origin country and year
Laninamivir octanoate	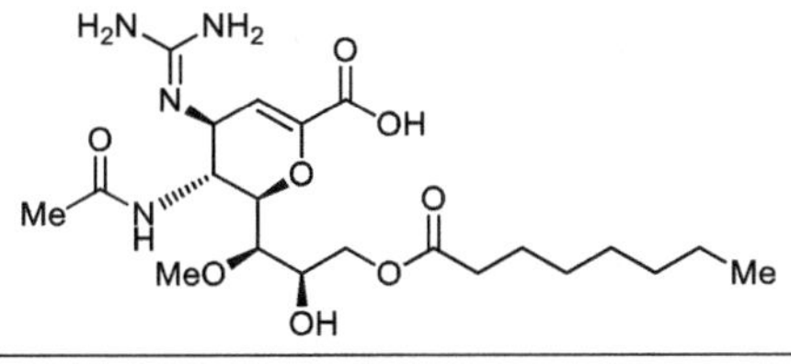	Antiviral	Japan, 2010
Peramivir	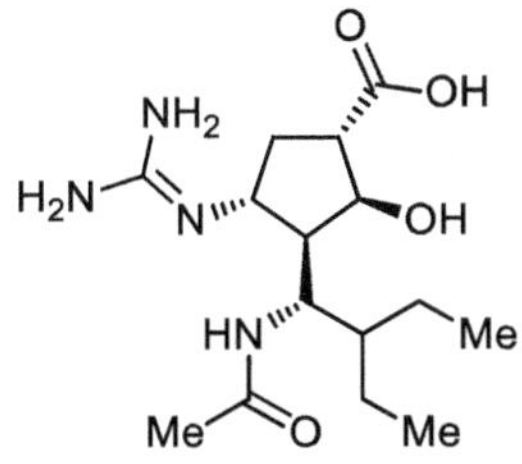	Antiviral	Japan, 2010

12.5.1 Antiviral agents and nucleoside analogues

Carbohydrate-derived nucleoside analogs are potential antiviral agents and are extensively utilized in chemotherapy. These chemically modified compounds contain a nucleobase connected to a furanose moiety, mimic endogenous nucleosides, and disrupt viral replication and cellular division by interfering with cellular DNA and RNA. There are several approved nucleotide analogue-derived antivirals (i.e., vidarabine and cytarabine) from D-arabinose (a 5-carbon monosaccharide) that are clinically safe and effective for the treatment of cancer. Similarly, sofosbuvir is a nucleoside analog that contains ribose and is phosphorylated [57]. It has the potential to treat specific types of hepatitis C virus (HCV) without the obligation of interferon. The USFDA granted its initial approval in 2013. Subsequently, it received approval for combination therapy in the treatment of HCV [58]. Figure 12.13 depicts antivirals developed from 5-carbon monosaccharides. Streptozotocin, a derivative of 2-deoxyglucose, is a powerful DNA-methylating antibiotic containing glucosamine. It has served as the primary treatment for most patients with pancreatic neuroendocrine tumors for many years [59].

12.5.2 Therapeutics for cardiovascular and nervous systems

A cardiotonic digoxin is composed of a carbohydrate-based pyranose sugar that influences the extent of digoxin binding, affinity, and selectivity for Na/K ATPase, which in turn enhances the intracellular concentration of sodium and calcium ions. It consists

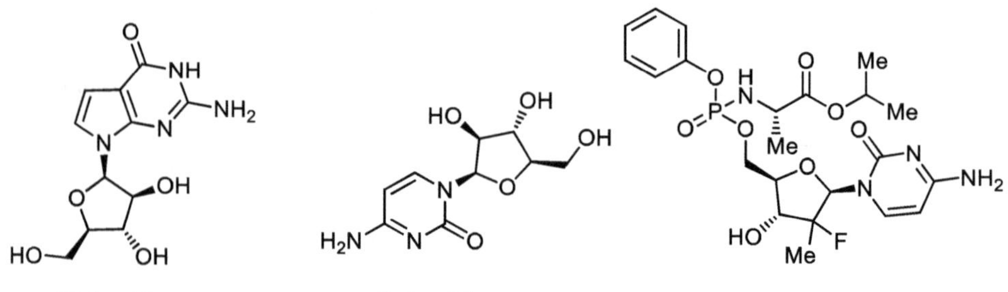

Figure 12.13: Antivirals developed from 5-carbon monosaccharides.

of cardenolide and *O*-2,6-dideoxy-β-D-ribo-hexapyranosyl sugar. Cangrelor is an ATP analog containing ribose sugar and is utilized in the realm of cardiovascular ailments [60]. It received FDA approval in 2015 for the purpose of decreasing the likelihood of periprocedural myocardial infarction during percutaneous coronary intervention [61]. Figure 12.14 depicts carbohydrate-derived cardiovascular drugs (digoxin, ouabain, and cangrelor). Heparin sulfate is a linear oligosaccharide glycosaminoglycan characterized by a high degree of sulfation and possesses a negative charge. It consists of 1,4-glycosidic bonds between D-glucosamine and D-glucuronic acid or L-iduronic acid units. Heparin shows anticoagulant action via the prevention of fibrin clotting in blood. Similar derivatives of heparin, that is, Tinzaparin and Enoxaparin, are developed by following processes including depolymerization, degradation in alkaline media, enzymatic hydrolysis, and fragmentation. Semisynthetic pentosan polysulfate, an oligosaccharide, contains β-D-xylopyranose residues and exhibits fibrinolytic and anticoagulant effects.

Sodium oligomannate and sugammadex are derived from carbohydrate feedstock and have been authorized for treating nervous system disorders. Sodium oligomannate (GV971) is employed for treating Alzheimer's disease, while sugammadex has been approved for the reversal of neuromuscular blockage. Sodium oligomannate is sourced from marine brown algae containing 2–10 sugar units. Research from ethological experiments has shown that sodium oligomannate exhibits a significant improvement in models of animals with cognitive disorders and multiple targeting mechanisms [62]. Further, it has the potential to cross the BBB in its original form through GLUT-1 (glucose transporter isoform 1/Slc2a1). It directly coordinates with Ab, hampers the synthesis of Ab-fibril, and enhances microglia-mediated Ab phagocytosis. Additionally, sodium oligomannate has a specific influence on gut microbiota, reducing the accumulation of amino acids (isoleucine and phenylalanine) to decrease the infiltration of the cells into the brain. Interaction with microglia may play a role in diminishing AD-related neuroinflammation and reversing cognitive impairment [63].

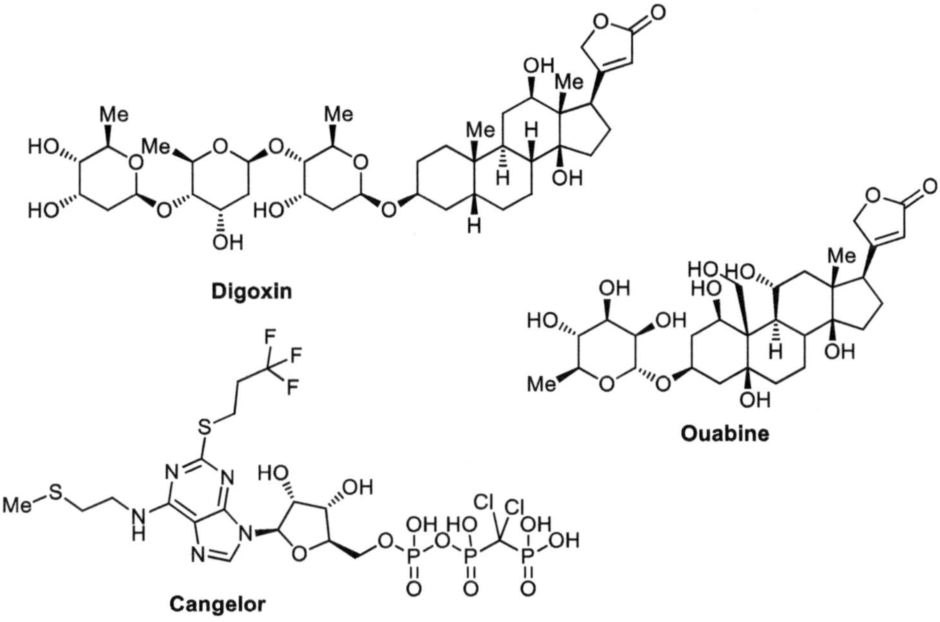

Figure 12.14: Carbohydrate-derived cardiovascular drugs.

12.6 Conclusion

Several carbohydrate feedstocks, including straw and stalk (rice, wheat, corn), wastes (onion and banana), and sugarcane bagasse, are utilized for the preparation of sustainable industrial and bioactive products such as biopolymers, bioplastics, biofuels, and oxygen and nitrogen-containing heterocyclic (furfural, LA, and 5-hydroxymethyl furfural) chemicals. Myriad therapeutics and bioactive molecules have been designed from carbohydrate sources owing to their active participation in energy metabolism, signal transduction, and immune defense mechanisms against pathogens. The abundance and easy accessibility of this biomass mediate the synthesis of several fine chemicals on a massive scale that are further utilized in the preparation of derived products. Although countless sustainable and cost-effective molecules can be synthesized from carbohydrate feedstocks, several challenges, including requirements of energy, microwave reactors, sophisticated chemical processes, and feedstock conversion, still need to be regulated to obtain high-yield products.

References

[1] Wang J, Azam W. Natural resource scarcity, fossil fuel energy consumption, and total greenhouse gas emissions in top emitting countries. Geosci Front, 2024, 15, 101757.

[2] Wassie SB. Natural resource degradation tendencies in Ethiopia: A review. Environ Syst Res, 2020, 9, 1–29.

[3] Manisalidis I, Stavropoulou E, Stavropoulos A, Bezirtzoglou E. Environmental and health impacts of air pollution: A review. Front Public Health, 2020, 8, 505570.

[4] Ayangbenro AS, Babalola OO. Reclamation of arid and semi-arid soils: The role of plant growth-promoting archaea and bacteria. Curr Plant Biol, 2021, 25, 100173.

[5] Lichtenthaler FW, Peters S. Carbohydrates as green raw materials for the chemical industry. C R Chimie, 2004, 7, 65–90.

[6] Schwach P, Pan X, Bao X. Direct conversion of methane to value-added chemicals over heterogeneous catalysts: Challenges and prospects. Chem Rev, 2017, 117, 8497–8520.

[7] Huber GW, Iborra S, Corma A. Synthesis of transportation fuels from biomass: Chemistry, catalysts, and engineering. Chem Rev, 2006, 106, 4044–4098.

[8] Li C, Zhao X, Wang A, Huber GW, Zhang T. Catalytic transformation of lignin for the production of chemicals and fuels. Chem Rev, 2015, 115, 11559–11624.

[9] Upton BM, Kasko AM. Strategies for the conversion of lignin to high-value polymeric materials: Review and perspective. Chem Rev, 2016, 116, 2275–2306.

[10] Ragauskas AJ, Beckham GT, Biddy, Davison BH, Dixon RA, Gilna P, Keller M, Langan P. Lignin valorization: Improving lignin processing in the biorefinery. Science, 2014, 344, 1246843.

[11] Kumar MNR. A review of chitin and chitosan applications. React Funct Polym, 2000, 46, 1–27.

[12] Chen X, Yang H, Yan N. Shell biorefinery: Dream or reality? Chem Eur J, 2016, 22, 13402–13421.

[13] Yan N, Chen X. Sustainability: Don't waste seafood waste. Nature, 2015, 524, 155–157.

[14] Jiménez-González C, Poechlauer P, Broxterman QB, Yang BS, Am Ende D, Baird J, Bertsch C, Hannah RE, Dell'Orco P, Noorman H, Yee S. Key green engineering research areas for sustainable manufacturing: A perspective from pharmaceutical and fine chemicals manufacturers. Org Process Res Dev, 2011, 15, 900–911.

[15] Priddy RD. Sustainability: The train has left the station. MRS Energy Sustain, 2017, 4, E3.

[16] Krane J. Climate change and fossil fuel: An examination of risks for the energy industry and producer states. MRS Energy Sustain, 2017, 4, E2.

[17] Sheldon RA. Green chemistry and resource efficiency: Towards a green economy. Green Chem, 2016, 18, 3180–3183.

[18] Llatance-Guevara L, Flores NE, Barrionuevo GO, Mullo Casillas JL. Waste biomass selective and sustainable photooxidation to high-added-value products: A review. Catalysts, 2022, 12, 1091.

[19] Gandini A. Furans as offspring of sugars and polysaccharides and progenitors of a family of remarkable polymers: A review of recent progress. Polym Chem, 2010, 1, 245–251.

[20] Morone A, Apte M, Pandey RA. Levulinic acid production from renewable waste resources: Bottlenecks, potential remedies, advancements and applications. Renew Energy, 2015, 51, 548–565.

[21] Wettstein SG, Alonso DM, Gürbüz EI, Dumesic JA. A roadmap for conversion of lignocellulosic biomass to chemicals and fuels. Curr Opin Chem Eng, 2012, 1, 218–224.

[22] Dashtban M, Gilbert A, Fatehi P. Production of furfural: Overview and challenges. J Sci Technol For Prod Process, 2012, 2, 44–53.

[23] Liu X, Zhu Z. Synthesis and catalytic applications of advanced Sn-and Zr-zeolites materials. Adv Sci, 2024, 11, 2306533.

[24] Deshan ADK, Moghaddam L, Atanda L, Wang H, Bartley JP, Doherty WO, Rackemann DW. High conversion of concentrated sugars to 5-hydroxymethylfurfural over a metal-free carbon catalyst: Role of glucose–fructose dimers. ACS Omega, 2023, 8, 40442–40455.

[25] Yuan Y, Macquarrie DJ. Microwave assisted acid hydrolysis of brown seaweed *Ascophyllum nodosum* for bioethanol production and characterization of alga residue. ACS Sustain Chem Eng, 2015, 3, 1359–1365.
[26] Guo X, Gu D, Wang M, Huang Y, Li H, Dong Y, Tian J, Wang Y, Yang Y. Characterization of active compounds from *Gracilaria lemaneiformis* inhibiting the protein tyrosine phosphatase 1B activity. Food Funct, 2017, 8, 3271–3275.
[27] Kwon OM, Kim DH, Kim SK, Jeong GT. Production of sugars from macro-algae *Gracilaria verrucosa* using combined process of citric acid-catalyzed pretreatment and enzymatic hydrolysis. Algal Res, 2016, 13, 293–297.
[28] Sweygers N, Somers MH, Appels L. Optimization of hydrothermal conversion of bamboo (*Phyllostachys aureosulcata*) to levulinic acid via response surface methodology. J Environ Manag, 219, 95–102.
[29] Kim B, Yang K, Park K, Cline J, Thorpe R, Kiely CJ, McIntosh S. Insights into the electrochemical oxidation selectivity of hydroxymethylfurfural over humin-layered Au nanoparticles. J Catal, 2024, 435, 115542.
[30] De Corato U, De Bari I, Viola E, Pugliese M. Assessing the main opportunities of integrated biorefining from agro-bioenergy co/by-products and agroindustrial residues into high-value added products associated to some emerging markets: A review. Renew Sustain Energy Rev, 2018, 88, 326–346.
[31] Climent MJ, Corma A, Iborra S. Conversion of biomass platform molecules into fuel additives and liquid hydrocarbon fuels. Green Chem, 2014, 16, 516–547.
[32] Li X, Xu R, Yang J, Nie S, Liu D, Liu Y, Si C. Production of 5-hydroxymethylfurfural and levulinic acid from lignocellulosic biomass and catalytic upgradation. Ind Crop Prod, 2019, 130, 184–197.
[33] Saikia K, Rathankumar AK, Kumar PS, Varjani S, Nizar M, Lenin R, George J, Vaidyanathan VK. Recent advances in biotransformation of 5-hydroxymethylfurfural: Challenges and future aspects. J Chem Technol Biotechnol, 2022, 97, 409–419.
[34] Oder E, Safo MK, Abdulmalik O, Kato GJ. New developments in anti-sickling agents: Can drugs directly prevent the polymerization of sickle haemoglobin in vivo? Br J Haematol, 2016, 175, 24–30.
[35] Jaswal A, Singh PP, Mondal T. Furfural – A versatile, biomass-derived platform chemical for the production of renewable chemicals. Green Chem, 2022, 24, 510–551.
[36] Rachamontree P, Douzou T, Cheenkachorn K, Sriariyanun M, Rattanaporn K. Furfural: A sustainable platform chemical and fuel. Appl Sci Eng Prog, 2020, 13, 3–10.
[37] Matsagar BM, Hossain SA, Islam T, Alamri HR, Alothman ZA, Yamauchi Y, Dhepe PL, Wu KCW. Direct production of furfural in one-pot fashion from raw biomass using Brønsted acidic ionic liquids. Sci Rep, 2017, 7, 13508.
[38] Zhang Z, Zhao ZK. Microwave-assisted conversion of lignocellulosic biomass into furans in ionic liquid. Bioresour Technol, 2010, 101, 1111–1114.
[39] Jiang Z, Hu D, Zhao Z, Yi Z, Chen Z, Yan K. Mini-review on the synthesis of furfural and levulinic acid from lignocellulosic biomass. Processes, 2021, 9, 1234.
[40] Saeid H, Al-sayed H, Bader M. A review on biological and medicinal significance of furan. AJMAS, 2023, 44–58.
[41] O'Sullivan L, Murphy B, McLoughlin P. Prebiotics from marine macroalgae for human and animal health applications. Mar Drugs, 2010, 8, 2038–2064.
[42] Chen Z, Liang W, Liang J, Dou J, Guo F, Zhang D, Xu Z, Wang T. Probiotics: Functional food ingredients with the potential to reduce hypertension. Front Cell Infect Microbiol, 2023, 13.
[43] Elsabee MZ, Abdou ES. Chitosan based edible films and coatings: A review. Mater Sci Eng C, 2013, 33, 1819–1841.
[44] Wang D, Han J, Yu Y, Li X, Wang Y, Tian H, Guo S, Jin S, Luo T, Qin S. Chitosan oligosaccharide decreases very-low-density lipoprotein triglyceride and increases high-density lipoprotein cholesterol in high-fat-diet-fed rats. Exp Biol Med, 2011, 236, 1064–1069.

[45] Romero-Espinoza AM, Vintimilla-Alvarez MC, Briones-García M, Lazo-Vélez MA. Effects of fermentation with probiotics on anti-nutritional factors and proximate composition of lupin (*Lupinus mutabilis* sweet). Lwt, 2020, 130, 109658.

[46] Rezende ESV, Lima GC, Naves MMV. Dietary fibers as beneficial microbiota modulators: A proposed classification by prebiotic categories. Nutrition, 2021, 89, 111217.

[47] Naughton PJ, Marchant R, Naughton V, Banat IM. Microbial biosurfactants: Current trends and applications in agricultural and biomedical industries. J Appl Microbiol, 2019, 127, 12–28.

[48] Kashif A, Rehman R, Fuwad A, Shahid MK, Dayarathne HNP, Jamal A, Aftab MN, Mainali B, Choi Y. Current advances in the classification, production, properties and applications of microbial biosurfactants – A critical review. Adv Colloid Interface Sci, 2022, 306, 102718.

[49] Mohanty SS, Koul Y, Varjani S, Pandey A, Ngo HH, Chang JS, Wong JW, Bui XT. A critical review on various feedstocks as sustainable substrates for biosurfactants production: A way towards cleaner production. Microb Cell Fact, 2021, 20, 120.

[50] Khanna A, Handa S, Rana S, Suttee A, Puri S, Chatterjee M. Biosurfactant from Candida: Sources, classification, and emerging applications. Arch Microbiol, 2023, 205, 149.

[51] Zhi Y, Wu Q, Xu Y. Production of surfactin from waste distillers' grains by co-culture fermentation of two *Bacillus amyloliquefaciens* strains. Bioresour Technol, 2017, 235, 96–103.

[52] Rodrigues MS, Moreira FS, Cardoso VL, de Resende MM. Soy molasses as a fermentation substrate for the production of biosurfactant using *Pseudomonas aeruginosa* ATCC 10145. Environ Sci Pollut Res Int, 2017, 24, 18699–18709.

[53] Abbasi H, Sharafi H, Alidost L, Bodagh A, Zahiri HS, Noghabi KA. Response surface optimization of biosurfactant produced by *Pseudomonas aeruginosa* MA01 isolated from spoiled apples. Prep Biochem Biotech, 2013, 43, 398–414.

[54] Das AJ, Kumar R. Utilization of agro-industrial waste for biosurfactant production under submerged fermentation and its application in oil recovery from sand matrix. Bioresour Technol, 2018, 260, 233–240.

[55] Huang H, Ding X, Xing D, Lin J, Li Z, Lin J. Hyaluronic acid oligosaccharide derivatives alleviate lipopolysaccharide-induced inflammation in ATDC5 cells by multiple mechanisms. Molecules, 2022, 27, 5619.

[56] Chang F, Dutta S, Becnel JJ, Estep AS, Mascal M. Synthesis of the insecticide prothrin and its analogues from biomass-derived 5-(chloromethyl) furfural. J Agric Food Chem, 2014, 62, 476–480.

[57] Jordheim LP, Durantel D, Zoulim F, Dumontet C. Advances in the development of nucleoside and nucleotide analogues for cancer and viral diseases. Nat Rev Drug Discov, 2013, 12, 447–464.

[58] Lawitz E, Jacobson IM, Nelson DR, Zeuzem S, Sulkowski MS, Esteban R, Brainard D, McNally J, Symonds WT, McHutchison JG, Dieterich D. Development of sofosbuvir for the treatment of hepatitis C virus infection. Ann NY Acad Sci, 2015, 1358, 56–67.

[59] Lenzen S. The mechanisms of alloxan-and streptozotocin-induced diabetes. Diabetologia, 2008, 51, 216–226.

[60] MacLeod-Glover N, Mink M, Yarema M, Chuang R. Digoxin toxicity: Case for retiring its use in elderly patients? Can Fam Physician, 2016, 62, 223–228.

[61] Chen DP, Xiong YJ, Tang ZY, Yao QY, Ye DM, Liu SS, Lin Y. Characteristics of deslanoside-induced modulation on jejunal contractility. World J Gastroenterol, 2012, 18, 5889.

[62] Wang T, Kuang W, Chen W, Xu W, Zhang L, Li Y, Li H, Peng Y, Chen Y, Wang B, Xiao J. A phase II randomized trial of sodium oligomannate in Alzheimer's dementia. Alzheimers Res Ther, 2020, 12, 1–10.

[63] Syed YY. Sodium oligomannate: First approval. Drugs, 2020, 80, 441–444.

Ali Hamza, Akbar Ali*, Muhammad Ibrahim, Muhammad Imran Irfan, and Azhar Abbas

Chapter 13 Applications of carbohydrate-based gelators

13.1 Introduction

Gelators are a fascinating class of multifunctional chemicals that create gels when dissolved in various solvents. Gels are soft materials with mechanical and viscoelastic qualities intermediate between solids and liquids, ranging from tough and hard to weak and spongy. A minute amount of the gelator is heated in the suitable solvent or aqueous solution and then allowed to attain room temperature, producing a gel of suitable characteristics. The gelator molecules combine to construct a 3D complex of fibers that can be micro- or nanoscale in length or thickness.

Different scientists have described various ways in which gelators self-assemble. It begins with one-dimensional interactions, which allow the molecules to combine at the molecular level to produce one-dimensional frameworks [1]. These 1D structures subsequently intertwine to form a spiral structure made up of vesicles, micelles, fibers, or tapes, forming the tertiary structure, which represents the gel's overall network [1]. In the same way, further intertwining causes a 3D structure. One of the most crucial transitions determining whether gelation occurs is that from the secondary to tertiary assembly, the fibers must be able to cross-link to create a continuous network with the solvent/mixture immobilized by capillary forces or surface tension [1–3]. If a continuous chain of fibers does not form, precipitation may occur instead of gelation.

Gels are generated in various organic solvents or aqueous solutions by immobilizing the solvent through molecular self-assembly, with the solvent serving as the primary component (Figure 13.1) [4]. They can be categorized in a variety of ways based on the solvent system in which they form a gel and the interactions that control their self-assembly.

The molecules that produce a gel in organic solvents are termed organogelators, however, the molecules that form a gel in water are called hydrogelators [5–8]. Other

***Corresponding author: Akbar Ali,** Department of Chemistry, Government College University, Faisalabad 38000, Faisalabad, Pakistan, e-mail: akbarchm@gmail.com
Ali Hamza, Department of Chemistry, Government College University, Faisalabad 38000, Faisalabad, Pakistan
Muhammad Ibrahim, Department of Applied Chemistry, Government College University, Faisalabad 38000, Faisalabad, Pakistan
Muhammad Imran Irfan, Instiitute of Chemistry, University of Sargodha, Sargodha 40100, Punjab, Pakistan
Azhar Abbas, Government Ambala Muslim Graduate College, Sargodha 40100, Punjab, Pakistan

https://doi.org/10.1515/9783111447094-013

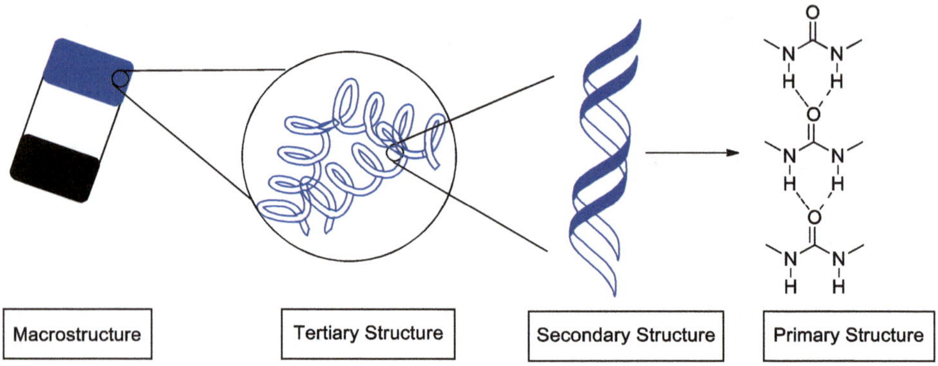

Figure 13.1: A microscopic view of the macrostructure of a gel formed from a simple molecule due to various kinds of linkages and certain conditions like pH, alterations in temperature, and enzymatic influence.

kinds include physical or chemical gels, depending on the gelation mechanism or interactions that form the cross-linked network [9–11]. Non-covalent interactions, such as hydrogen bonding, π–π interaction, CH–π interactions, dipole-dipole forces, Van der Waals interactions, and London dispersion forces, govern the self-assembling process for the formation of physical gels [1, 12, 13]. The latter interactions are insufficient for the gelation process; hence, certain functional groups are usually added to the gelator molecules, such as urea, amides, and long-chain alkanes for hydrophobic contacts, and nucleobases and carbohydrates with an abundance of O–H groups for hydrogen bonding [1].

13.2 Low-molecular-weight gelators (LMWGs)

Low-molecular-weight gelators (LMWGs), also known as supramolecular gelators, were formerly characterized as having a molecular weight of 300–1,000; however, this figure has recently been expanded to less than 3,000 [14, 15]. Smart materials with built-in functions in molecules are now being developed in all sectors of chemistry. Low-molecular-weight gelators (LMWGs) are molecular assemblies made from adaptable tiny molecules and have become the topic of interest owing to their extensive applications, which include cosmetics and biomedicine. A variety of ample LMWGs have been presented to provide gels appropriate for various purposes. Gelators form self-assembled fibrous structures in a solvent, making them unique materials with a wide range of uses. The hydrogel obtained using LMWGs has the potential to replace standard polymer hydrogels in applications requiring biocompatibility. The production of such hydrogels is reversible, increasing their usefulness in a variety of sectors [8].

13.3 Carbohydrate-based low-molecular-weight gelators

Carbohydrates are macromolecules that serve as a vital source of energy in living bodies [16]. They are biological units capable of mimicking the actions and structures of other physiologically relevant macromolecules, including cell signaling capabilities [17, 18]. They have been employed in supramolecular gelator systems because of their high hydrogen bonding potential, which is vital for supramolecular interactions [19]. Carbohydrates are naturally occurring renewable resources that are chiral with diverse functional groups. The hydroxyl groups enable sugars to generate broad hydrogen bonding connections, allowing them to self-assemble into gel networks [16, 17, 20]. Sugars are nontoxic, biodegradable, biocompatible, and environmentally benign, having affordable precursors that can be easily manipulated synthetically to generate new compounds [16, 17].

The massive number of citations regarding successful synthetic procedures for various monomer units of polysaccharides, disaccharides, and monosaccharides for gelation demonstrates how much the topic of molecular gelators has grown and gained interest over the years in research [4, 17]. As a result, carbohydrates are easily capable of forming a wide range of LMWGs, allowing their substantial investigation to be used as supramolecular gelators [17]. A study reported the synthesis of several derivatives (**1a, 1b, 2a, 2b**) of substituted monosaccharides (Figure 13.2) [21].

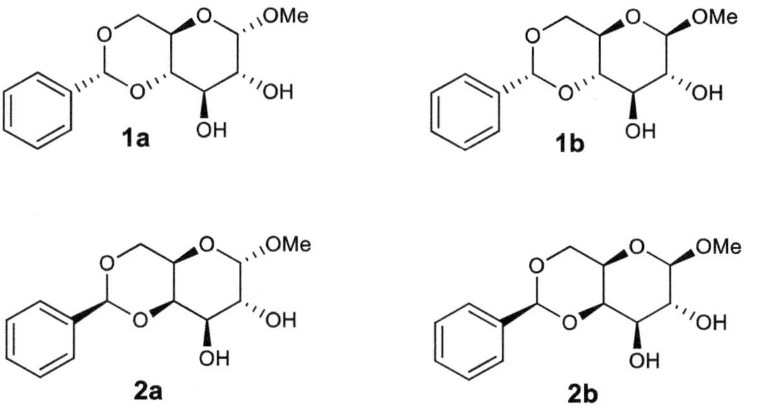

Figure 13.2: Monosaccharide-based gelators containing Methyl 4,6-Benzylidene.

Many carbohydrate compounds (**3–7**) (Figure 13.3) and (**8–10**) (Figure 13.4) have been modified with functional groups in order to rationally create new gelators and assess the features that influence gelation. Some of the functional groups introduced include, but are not limited to, amide, ester, urea, triazole, amine, carbamate, and others [22]. Some sugars have even been combined with other molecules, including lipids, diacetylenes, and other sugars, to create disaccharide-based gelators.

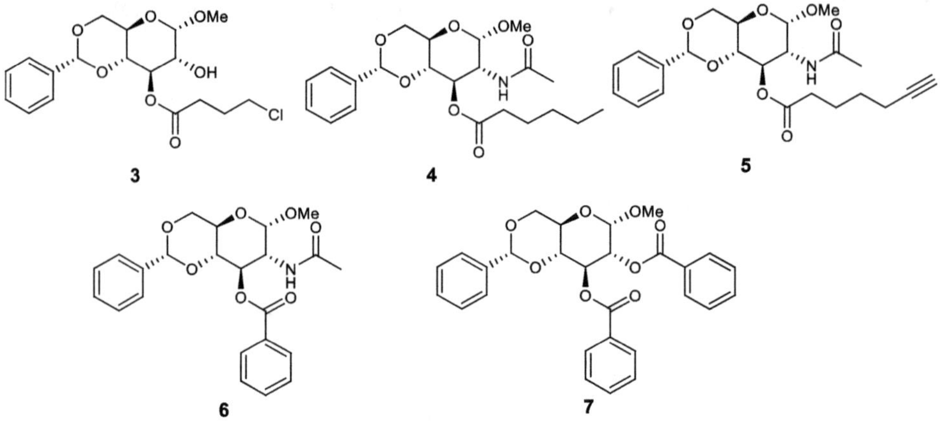

Figure 13.3: Some monoesters and diesters reported as gelators.

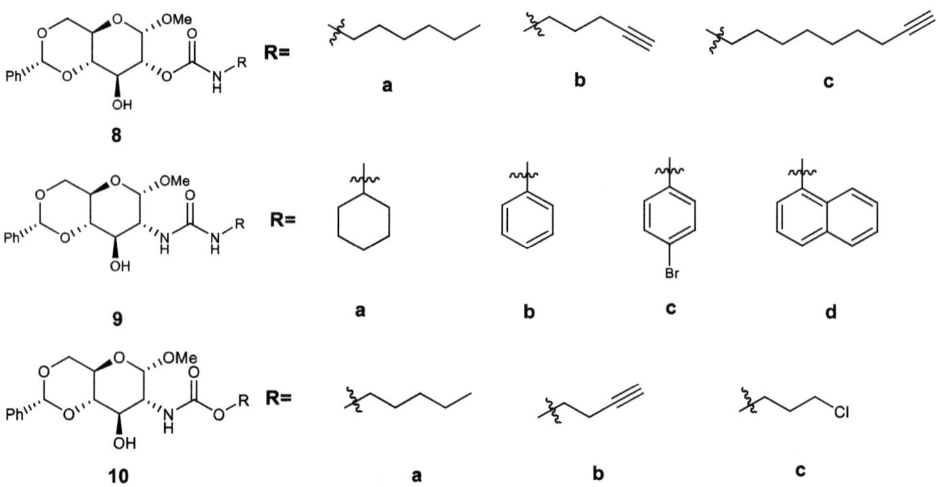

Figure 13.4: Carbamate (**8**), urea (**9**), and amide (**10**) containing derivatives of glucopyranosides with gelation properties.

Sugars are biocompatible and possess intrinsic chirality, making them superior for biomedical applications such as controlled release systems and biomolecule segregation. Recently, low-molecular-weight carbohydrates have received a lot of interest as gelators due to their supramolecular structures. Carbohydrate-based gelators are novel materials with several applications, including medication delivery, environmental cleanup, antibacterial agents, tissue engineering, thixotropy (the ability to reduce viscosity in response to applied stress), enzyme immobilization, nanomaterial fabrication, and wound healing. CBGs have various distinguishing characteristics, such as ease of availability, environmental friendliness, and cost-effectiveness, which make them the ideal choices for gelators over other traditional organic gelators.

Gelators derived from isosorbide, xylitol, and mannitol (**11–13**) (Figure 13.5) have also been described in the literature. Xylitol-based gelators have been designed particularly for oil phase gelation [23]. The compounds synthesized by Kesava [23] formed efficient gels in aromatic solvents but performed poorly in paraffinic solvents. Compounds (**11a–c**) revealed the natural aptitude of the gels for a variety of crude oils with varying densities, which supports their potential for oil spill cleanup.

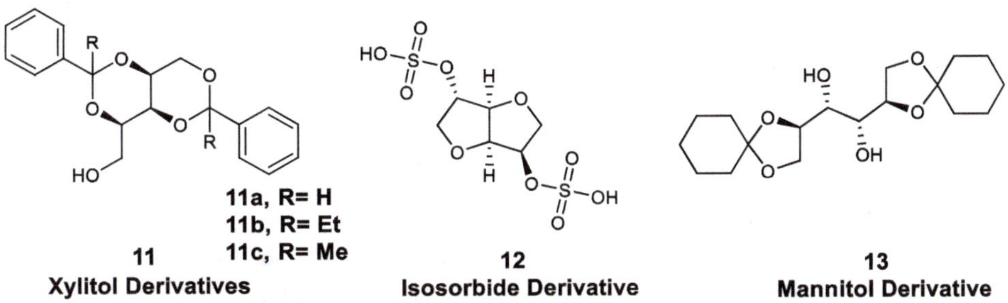

Figure 13.5: Gelators containing derivatives of xylitol, isosorbide, and mannitol.

13.3.1 Historical perspective

The earliest description of carbohydrate-containing organogelators is known to have existed in 1891, when Meunier [24] described the gel formation phenomenon of two benzaldehyde-D-sorbitols formed by condensing the combination of D-sorbitol and benzaldehyde using strong inorganic acids.

Yamasaki identified a 1,3:2,4-*Di-O*-benzylidene-D-sorbitol (DBS) molecule (**14**), in the 1990s (Figure 13.6), that formed a gel in ethylene glycol with optical anisotropy in

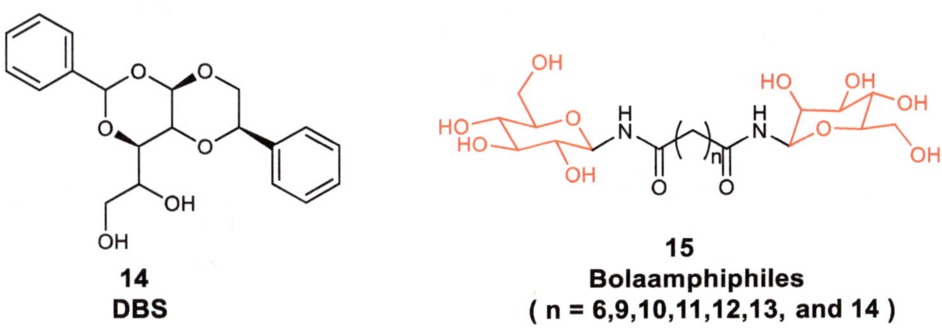

Figure 13.6: Representative carbohydrate gelators discovered in the 1990s.

crossed-polarized light [25]. A series of bola-amphiphiles (**15**) with 1-glucosamide as the head group on both ends of the molecule has been reported (Figure 13.6). Some of the molecules self-assembled into fibrous formations and various morphologies in water, depending on the length of their alkyl chain [26].

13.4 Applications

There is diversity in the applications of gels containing carbohydrate species as a necessary component in their molecular structure. Their applications lie in various fields, ranging from biomedical to cosmetics.

13.4.1 Antibacterial agents and wounds healing

As conventional antibacterial medications face significant microbial resistance, pharmaceutical innovation is required for the development of drugs with little resistance and fewer negative effects for safe and useful antibiotic usage. Among the several types of antibacterial medicines, carbohydrate-based gels are particularly helpful and trustworthy due to their low side effects, with the advantage of inhibiting the formation of biofilms.

Monosaccharides among the sugar-based compounds have shown promising results in breaking up biofilms. A report by Prasad and coworkers regarding monosaccharide derivatives having gelation properties has been compiled here. The synthesized compounds (**16**, **17**), which are effective gelators for a variety of oils and hydrophobic solvents, were known to be antibacterial agents. The solutions from both gelators were allowed to interact with biofilms from various bacterial strains. The chemical compound without the lengthy aliphatic chain (**16**) was capable of disrupting biofilms generated by gram-positive strains of several bacteria. The second gelator (**17**) was found more potent against bio-films of uropathogenic *Escherichia coli* and *Salmonella enterica typhimurium* (Figure 13.7) [27]. The gels were effective competently for oily entities, but they unexpectedly generated foam in water, which remained stable at ambient temperature for around 6 h [27].

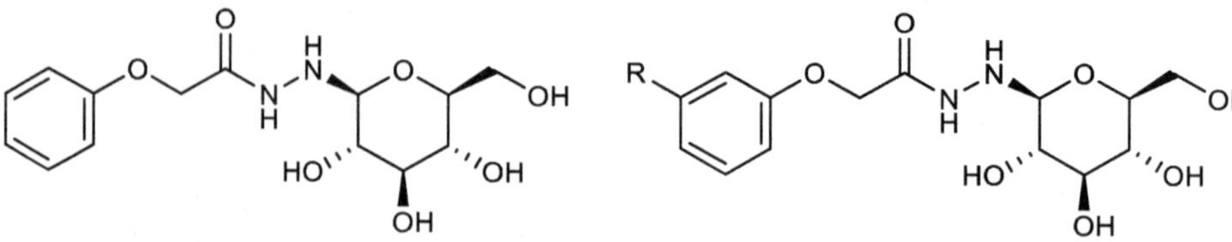

16

17

Figure 13.7: Monosaccharide-based gelators having antimicrobial potential.

Glycolipids acting as gelators have also been found to have wound healing action. The glycolipid (**18**) can form a gel both in DMSO:H_2O and organic solvent. The organogels of (**18**) have remarkable results in encapsulating curcumin. The morphological analysis has revealed fibrous networks throughout the gel matrix [28]. Furthermore, the gels have displayed thixotropic qualities, as well as injectability and self-healing capabilities that have allowed them to be used in vivo [28, 29]. The gels of the compound (**18**) **have** gone through a test to monitor their capacity to suppress the development of Methicillin-resistant *S. aureus* (MRSA), *Pseudomonas aeruginosa* (PA), and Methicillin-susceptible *S. aureus* (MSSA). The investigations found that it could prevent the development of MRSA and MSSA but poses a weaker influence on PA [28]. The gel of compound (**18**) efficiently prevented the formation of biofilm in all bacterial strains (Figure 13.8) [29]. The gel in composite form of the same compound, prepared containing curcumin and oleogel, was capable of assisting in closing wounds and clot formation in diabetics, as the results expressed by tests on Wistar rats [28, 29]. The worth of the research work was enhanced by the specialty of gels in boosting collagen production, normalizing cell proliferation and overlapping stages of inflammation, and managing the formation of free radicals from the site of the wound [28, 29].

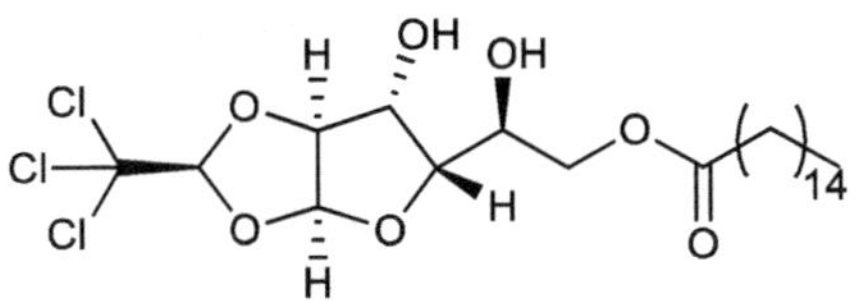

18

Figure 13.8: A glycolipid having gelation properties effective against various microbes.

There are also investigations into cytidine (**19**) analogues to be potentially applied as antimicrobial and wound healing agents. The synthesized gelator has created aqua gels by linking silver nitrate or boric acid [30, 31]. Actually, cytidine (**19**) forms a dimer with boric acid in the form of borate diester (**20**), which polymerizes to form a gel in the presence of silver ions from either source (Figure 13.9). The subsequent hydrogels (**21**) contain intrinsic antibacterial action due to cationic silver, which is a recognized antibiotic (Figure 13.10) [30, 31].

The hydrogel containing silver nitrate suppressed the growth of bacteria, healing the burned wounds by promoting the development of the hypodermis [30]. The thixotropic attributes make it an attractive option for use in various biological implementations, including culturing tissues. The efficiency of the hydrogel containing silver acetate is less against *S. pneumoniae* and *S. aureus* compared with a strain of *M. morganii, P. aeruginosa, K. pneumoniae*, and *E. coli*, but with no discoloration effect on the tissues [28, 29].

A few gluconamide-based amphiphiles (**22, 23**) have recently been tested for antibacterial activity against a variety of bacterial species (Figure 13.11) [28]. In vitro tests have shown the ability of the aforementioned chemicals to limit bacterial growth with significant potential. Further research demonstrated that these gelators might

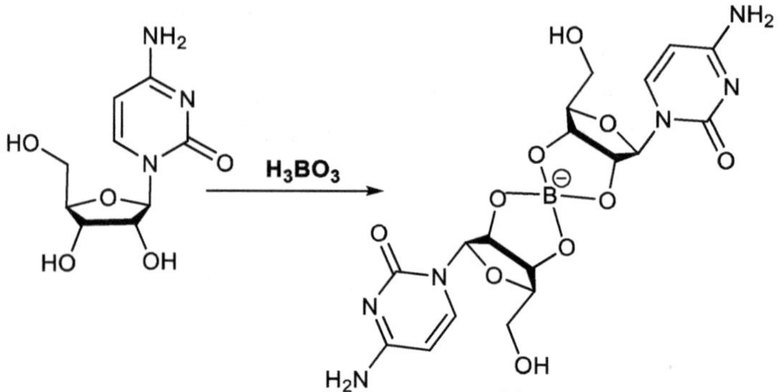

19
Cytidine

20
Borate diester

Figure 13.9: Mechanism of dimer formation in cytidine in the presence of boronic acid.

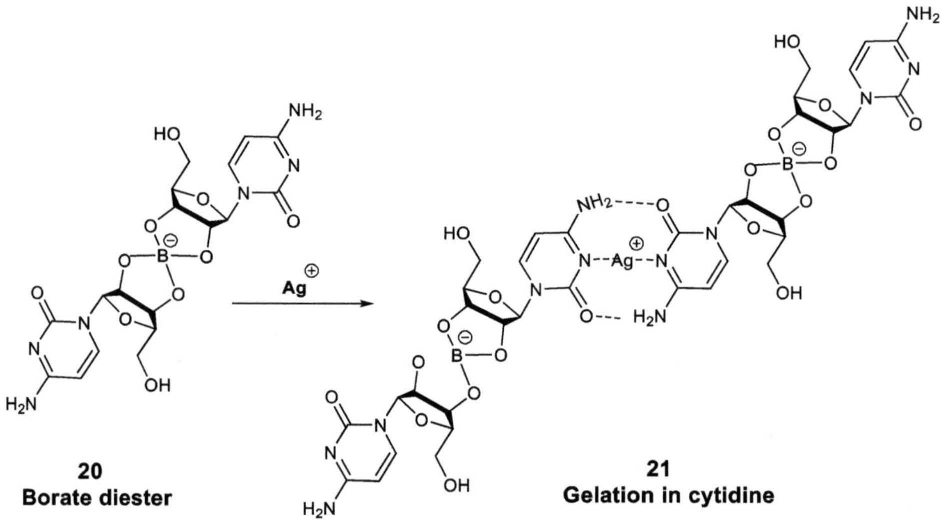

20
Borate diester

21
Gelation in cytidine

Figure 13.10: Gelation in cytidine from borate diester bridging with silver ions.

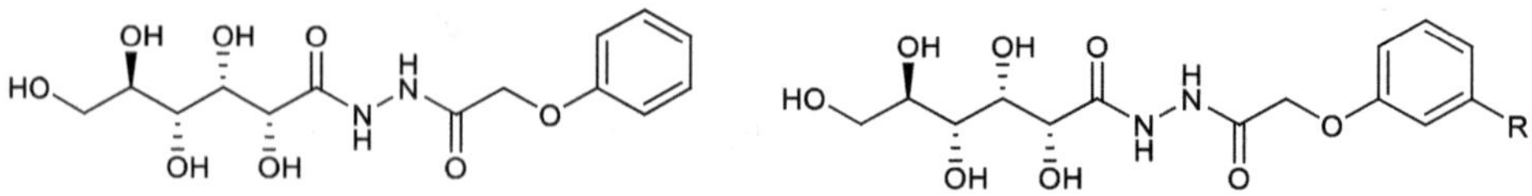

22

23

Figure 13.11: Some gluconamide-containing gelators.

limit the formation of biofilm, making them capable in various applications where the reduction of biofilm is needed [27, 29].

13.4.2 Drug delivery

Medication delivery is the technique for delivering a medication to a specific location in the body, being released and absorbed, along with the successful transfer of active ingredients across biological membranes and the bloodstream to the site of action of the drug. Carbohydrates have biocompatibility, affordability, and environmental friendliness. Extensive research on carbohydrate-based gels for utilization in medicinal delivery purposes has gained much interest [32]. Many of the sugar-based hydrogels are spongy, causing them to expand and shrink, making them ideal for drug delivery applications.

An example of drug delivery utilizing a polysaccharide-based hydrogel has been reported [33]. They used a framework of glycol chitosan (**24**) having a covalent linkage with glycidyl methacrylate. After this, doxorubicin hydrochloride (**27**) surrounded this assembly (Figure 13.12) [33]. After being subjected to the blue portion of visible light (430–485 nm), the combination produced a gel in an aqueous medium that contained doxorubicin hydrochloride (**27**). The researchers have concluded that the drug's release increased over the first 18 h, but then slowed and was maintained throughout 7 days, making it suitable for the release of drugs that need sustained release over a specified time period [33]. This has shown that the hydrogel holds pronounced prospects for medicinal delivery purposes.

The benzamide gelator (**28**) creates a hydrogel with potential biological uses. Drug entrapment and release studies using naproxen sodium (**29**) and chloramphenicol (**30**) as references to investigate possible uses in sustained drug release were effectively demonstrated by the prepared gel (Figure 13.13). For the first 40 h, almost 97% of the naproxen (**29**) diffuses into the aqueous phase [34]. Meanwhile, it has been studied that gel concentrations have no effect on the rate of diffusion [34]. The findings of these investigations suggest that the gel has potential uses in sustained-release medication delivery.

For the purpose of drug encapsulation and sustained release of chloramphenicol (**30**), Chen and colleagues tested a gel in an aqueous medium. The concentration of the gelator (**31**) (Figure 13.14) and its strength have a significant impact on the delayed disintegration of a medication tested in contrast with the benzamide gelator (**28**) [35].

Naproxen sodium (**29**) can likewise be encapsulated using the phenyl carbamate gelator (**32**) (Figure 13.15), and it gradually diffuses into the aquatic medium. For the preparation of this gel, one needs to dissolve naproxen sodium (**29**) in the preferred solvent. After 14 h of room temperature treatment, the gel is covered with phosphate-buffered saline (pH 7.47). The encapsulating process produces very good results in making the drug able to have sustainability in its release [36].

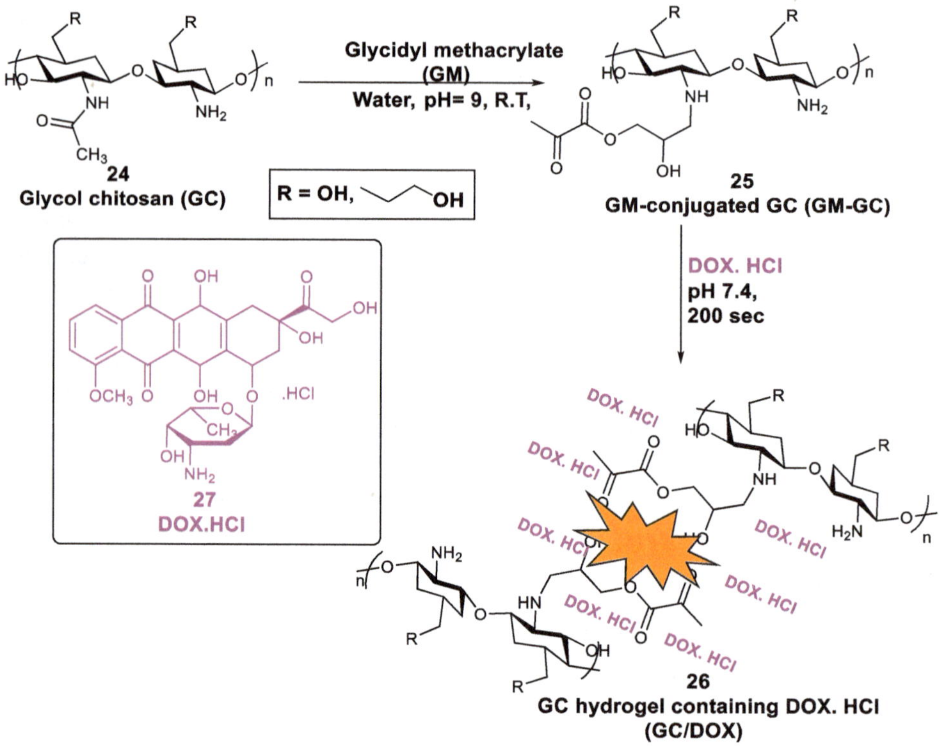

Figure 13.12: Preparatory sketch for GC/DOX.

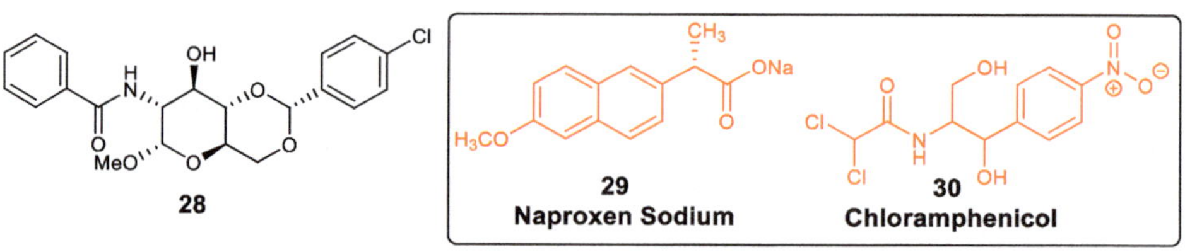

Figure 13.13: Benzamide gelator compound **(28)** for hydrogel formation and standard drugs, Naproxen Sodium **(29)** and Chloramphenicol **(30)**, which were employed for drug delivery experiments.

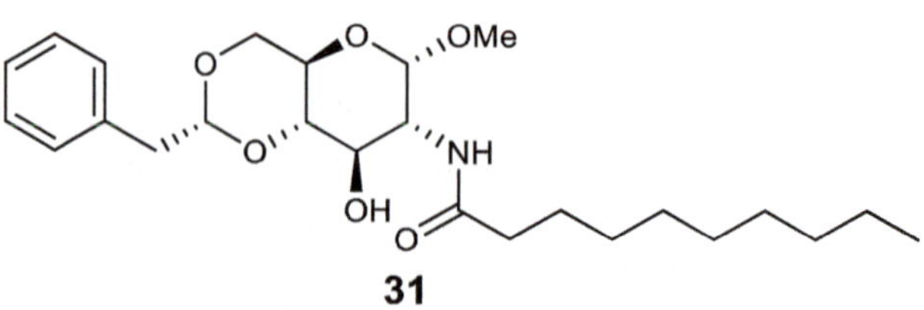

Figure 13.14: A hydrogel having a significant impact on drug release of chloramphenicol compared to a benzamide gelator (**31**).

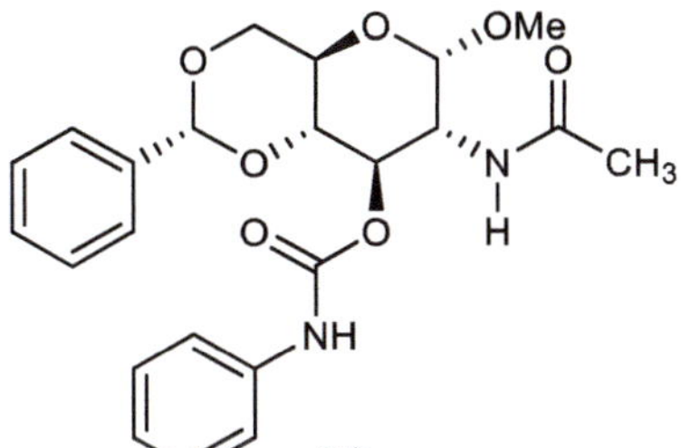

32

Figure 13.15: Derivative of phenyl carbamate used for gel formation.

Three gelator compounds (**33–35**) were prepared and investigated for their potential in delivering medications. In comparison to the others, compound (**33**) was discovered to be the most efficient in entrapping an anesthetic agent, prilocaine hydrochloride (**36**), and some hydrophilic dyes [37]. The diffusion rate is inversely associated with the concentration of gelator molecules (**33–35**), as evidenced by in vitro drug release experiments (Figure 13.16).

The release profile of prilocaine HCl (**36**) is consistent with the Peppas-Korsmeyer model (a model used to investigate the drug delivery capability of any molecule) [37]. In vitro cytotoxicity experiments have revealed that these gels are highly biocompatible, implying that dendron derivatives of poly(aryl ether) glucose core (**33–35**) are excellent contenders for targeted and sustained medication release in topical applications.

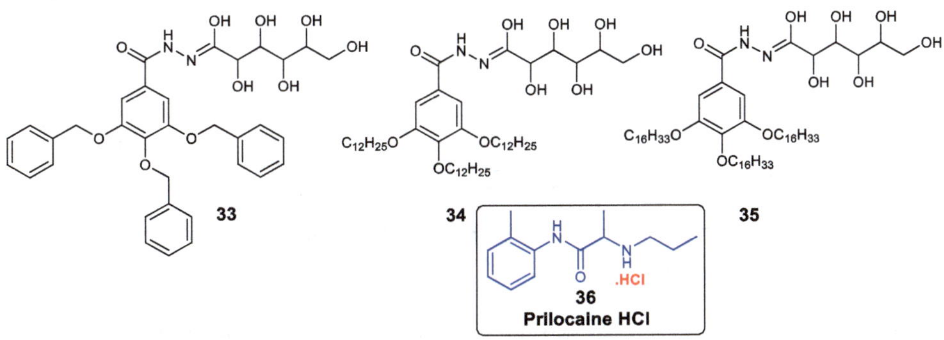

Figure 13.16: Glucose-imitative molecules having poly(aryl ether) Dendron (**33–35**) and a local anesthetic, prilocaine hydrochloride (**36**).

A new injectable hydrogel made of thiolated hyaluronic acid (HA) (**37**) has been synthesized [38]. Oxidation of the free thiol groups in HA-SH (**39**) is responsible for hydrogel formation, resulting in intermingled disulfide linkages (Figure 13.17a) [38]. Investigating the release of sorafenib (**42**), metformin (**43**), and doxorubicin (DOX) (**27**) from the compound (**39**) hydrogel, it has been concluded that drug-loaded hydrogels have injectability and stability under physiological environments, allowing for rapid and accurate drug discharge (Figure 13.17b).

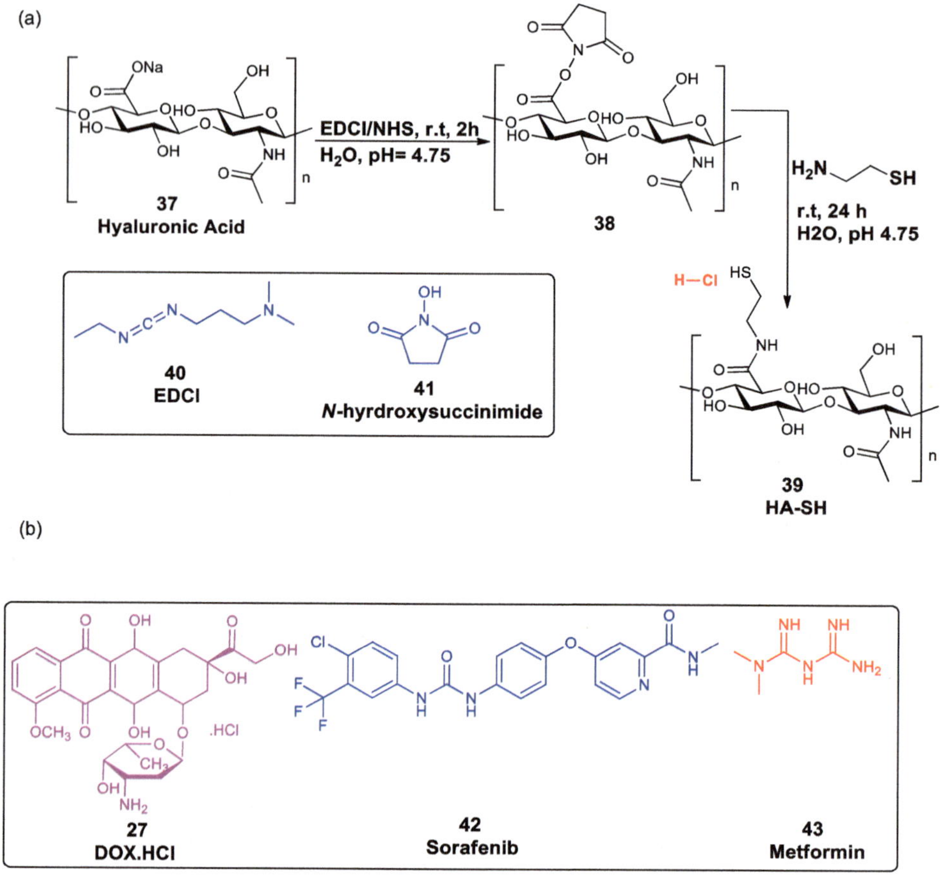

Figure 13.17: (a) Synthesizing a copolymer (HA-SH) (**39**) (b) Structures of the standard drugs for drug delivery experiments.

A recent study stated the biocompatible generation of potassium-coupled guanosine-quadruplex hydrogel (**47**) (Figure 13.18) having regeneration abilities, to be utilized in prolonged vitamin B2 and B12 delivery [39]. The vitamin release was much quicker in an acidic environment than at biological pH [39]. Furthermore, the release profile of doxorubicin (**27**) studied in an acidic pH revealed that the maximum amount of the entrapped antitumor drug was disintegrated in the first 15 h, indicating the potential implementation of the drug in pH-sensitive environments [39].

Therapeutic proteins and peptides are made possible in significant ways, allowing them to retain their biological activity and stability. A hydrogel based on polysaccharide has been successfully manufactured by Xu and colleagues, having biocompatibil-

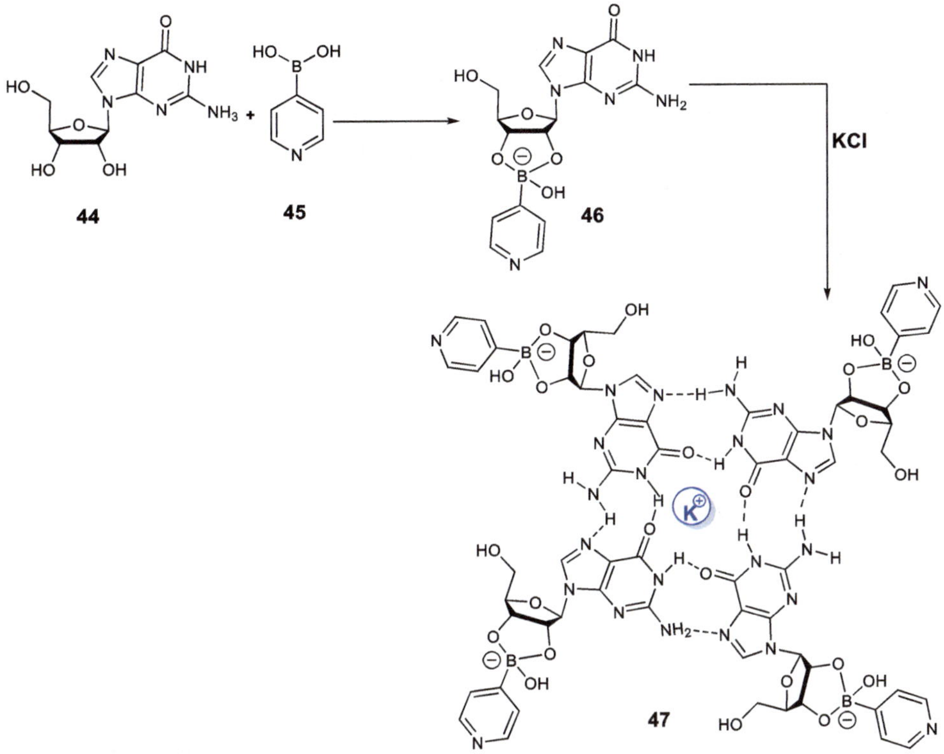

Figure 13.18: Preparation and self-assembling of guanosine-containing gelator.

ity and biodegradability [40]. The hydrogel containing epichlorohydrin cross-linked hydroxypropyl pachyman (ECH-HPP) (50) has high pH sensitivity, making it ideal for protein discharge in intestinal media. The created self-assembly of ECH-HPP (50) has exhibited encapsulation of different protein medicines up to 97.6% with uniform distribution (Figure 13.19) [40]. During encapsulation and release, the protein stability was not compromised. Furthermore, the insulin-loaded hydrogel reduced blood glucose levels in diabetics while demonstrating minimal damage in either in vitro or in vivo [40].

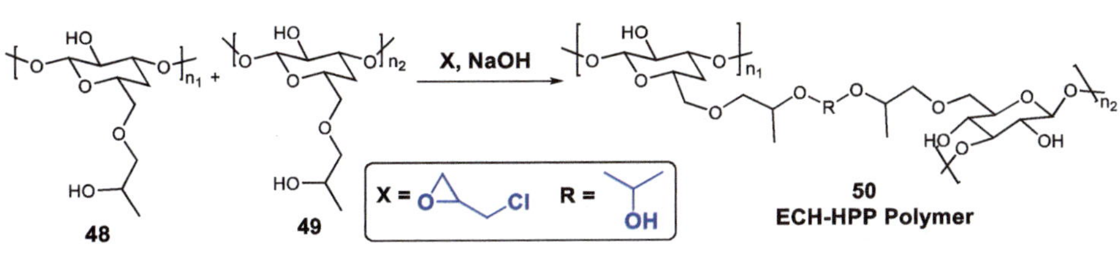

Figure 13.19: Synthesis of ECH-HPP (50) polymer having efficacy for the delivery of protein-based drugs.

13.4.3 Anticancer agents

A study has been explained in the previous section about drug delivery by Yoo and others. In the same research work, they also examined the anti-tumor effect of the GC/DOX gel **(26)** (Figure 13.20). The medication-bearing hydrogel **(26)** had a greater anticancer impact on thyroid tumor cells than the individual drug [33].

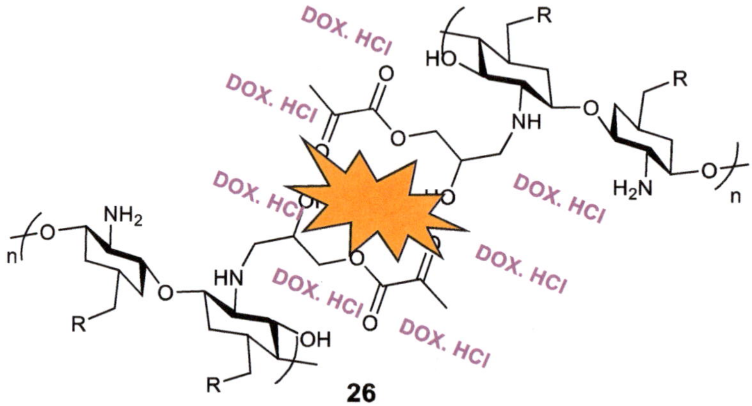

26
GC hydrogel containing DOX. HCl (GC/DOX)

Figure 13.20: GC/DOX hydrogel demonstrating efficacy in drug delivery and anti-tumor activity.

Polymeric gelators (HA-SH) **(39)** (Figure 13.21) have been synthesized and show good anticancer activity in addition to drug delivery properties for various renowned drugs. The hydrogel of HA-SH **(39)** has shown the ability to encapsulate many medicines at once, exhibiting a synergistic impact on tumor cells. The results indicate that HA-SH **(39)** is more efficient against cancer cells compared to GC/DOX **(27)** [38].

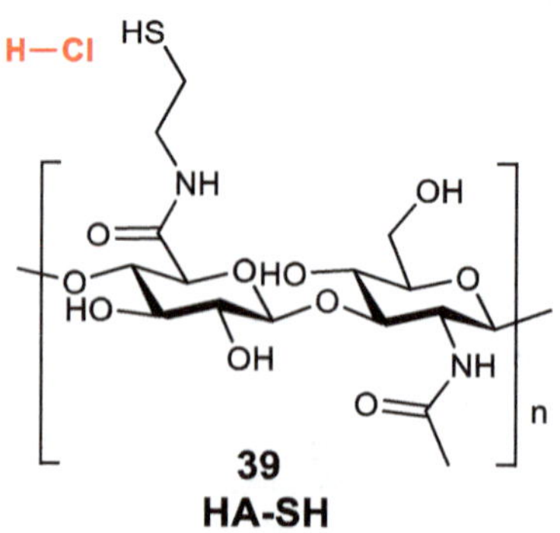

39
HA-SH

Figure 13.21: A polymeric gel (HS-SH) with anti-tumor activity.

13.4.4 Tissue engineering

The usefulness of supramolecular gels in biomedicine lies in the simulation of biological conditions. In stem cell culture, the stiffness of LMWGs has a major influence on cell differentiation [41, 42]. The field of tissue engineering, also called regenerative medicine, has been growing rapidly in the last 30 years. It is a very intriguing field of study, associated with numerous other fields, with the aim of helping in healing organ failure [43–45]. There are few studies in the literature regarding the use of carbohydrate-based LMWGs in tissue culturing.

The researchers have developed a glycopeptide (**51**) that is catabolized by alkaline phosphatase (ALP) for the purpose of gelation [46]. It does not gelate in water, but it does after being catalyzed by ALP (Figure 13.22) [46]. The gel's (**52**) capacity to serve as a matrix for cell culture to develop human umbilical vein endothelial cells (HUVECs) was investigated, and it was found that the adhesion of the gel to the cells also caused their proliferation [46]. The gel (**52**) was found to be highly biocompatible and injectable. The proliferation of cells was attributed to the glucose entities, making it useful in in vivo applications as the gel matrix was still capable of promoting the progression of blood capillaries, despite the decrease in the gel's volume progressively [46]. Overall, the glucose-containing hydrogel was capable of working in tissue culturing as an extracellular matrix.

Figure 13.23 shows a few structures made from *N*-alkyl-galactonamide (**53a-c**) (Figure 13.23), which is an outstanding example of a carbohydrate-based gelator for tissue engineering. It has been found that the neural stem cells differentiated into neuronal and glial cells when treated with biocompatible hydrogels, resulting in a completely interwoven neural network [47]. The gelator molecule (**53b**) clearly demonstrated the excellent properties required for tissue engineering.

13.4.5 Environmental remediation

In modern civilization, treating wastewater containing dangerous chemicals and oil spills is quite difficult. It has been discovered that organic synthetic dyes are to blame for a number of human health problems. Several synthetic materials were used in the past to remove pollutants from the environment, but some of those compounds were even more harmful than the contaminants they were meant to remove. Many researchers have resorted to soft materials like nontoxic, biodegradable gels in an effort to solve this global issue in a more ecologically acceptable manner [48]. Because of their significant surface area for contaminant absorption and ability to function over a wide pH range, gels have garnered a lot of attention for this application. Furthermore, it is simple to modify the gelator molecules to create new ones with durable and stable mechanical attributes. Designing effective candidates for sanitation re-

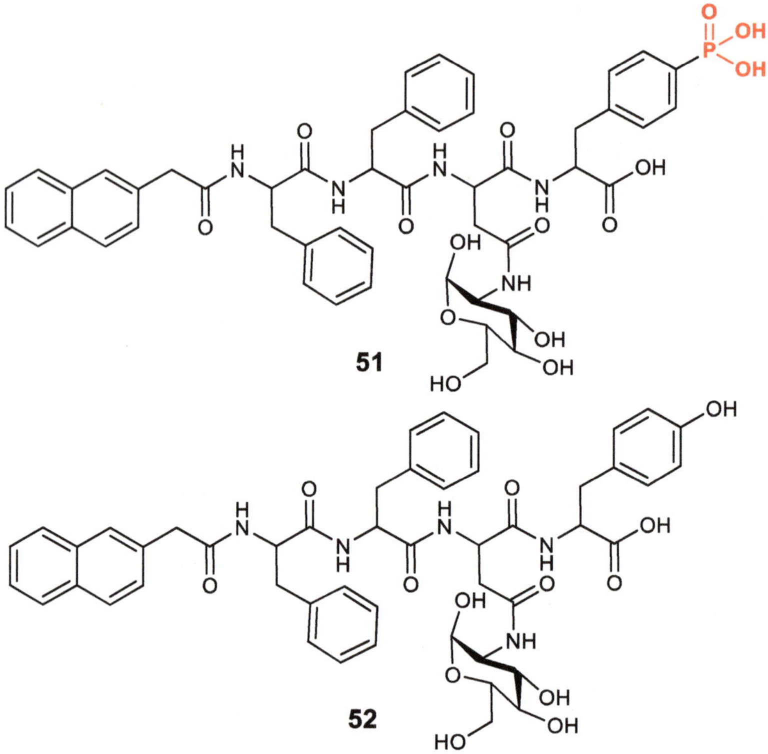

Figure 13.22: Glycopeptide-based gelator (51) forming a gel (52) upon treatment with an enzyme.

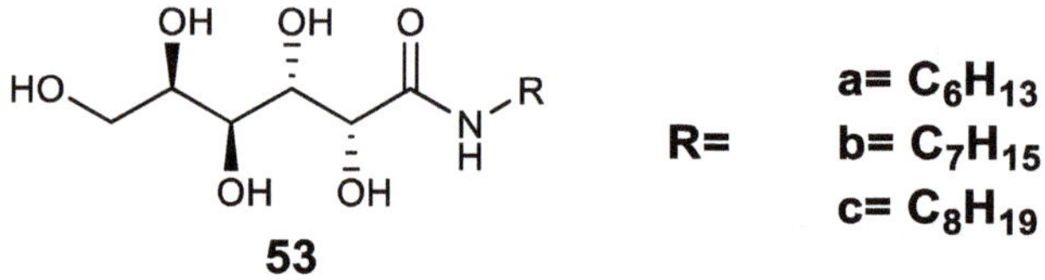

Figure 13.23: Gelators containing alkylglactonamide moiety employed in tissue culturing.

quires making sure the gelator has the chemical capabilities to engage with the specific pollutant more frequently.

N-Acetyl-D-glucosamine derivative (54) (Figure 13.24) capable of forming gels in a diversity of organic solvents, can coagulate fuels; therefore, its phase-selective gelation capabilities have been investigated [48]. The toluene gel of compound (54) can absorb nearly 96% of rhodamine B (55) (Figure 13.25) base from a solution poured on top of it [48]. Furthermore, it has been shown to generate both organic solvent gels and hydrogels. The findings revealed that this gelator was multifunctional and had significant potential for use in environmental remediation [48].

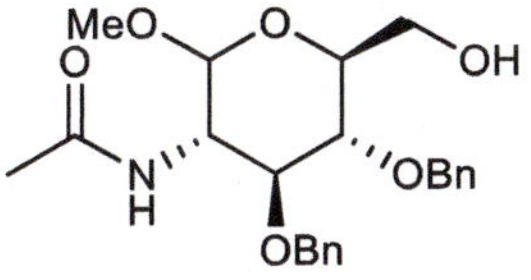

54

Figure 13.24: A gelator derived from *N*-acetyl-glucosamine.

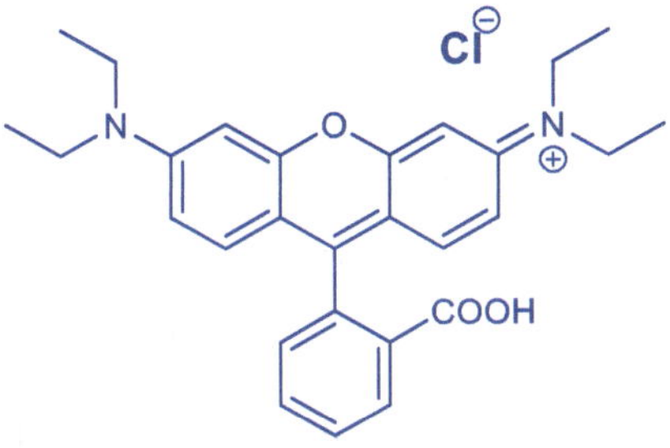

55
Rhodamine B

Figure 13.25: Structure of the pollutant rhodamine B (**55**).

The gelator compounds **(16, 17)** (Figure 13.26), as earlier described for antimicrobial applications, also have the potential to be used as cleaning agents in both solutions and gels. Overall, both the supramolecular materials are moderately efficient for cleaning purposes. They have the ability to gelate in both water and organic solvents [27].

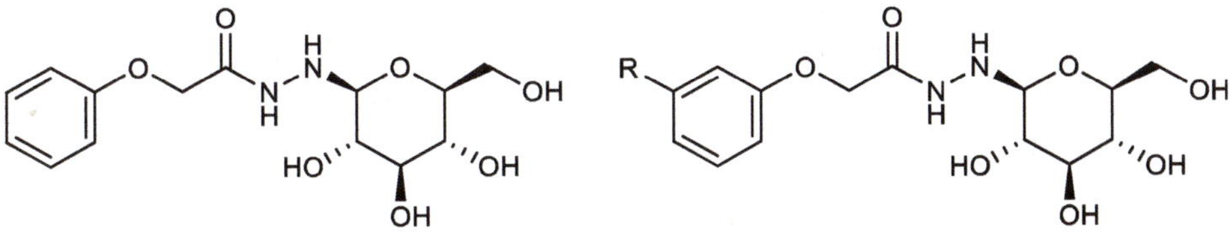

16

17

Figure 13.26: Monosaccharide-based gelators with cleaning action and antimicrobial potential.

The compound **(28)** (Figure 13.27) produced a gel in water, and its dye absorption capability was tested using toluidine blue (TBO) **(56)**. The gel was used to extract TBO **(56)** from a solution containing it. The collected aqueous phase demonstrated that enough stability was present in the blue gel for regeneration [34]. In this research, it has been estimated that about 89% of the TBO dye **(56)** was removed from the gel column, leaving less than 11% in the collected aqueous solution. It was interesting to note that the gel, even at extremely low concentrations, is efficient in eliminating hazardous pigments from water. A similar approach was used to make a gel of **(28)** using

EtOH/H_2O in the ratio (1:2), which removed the dye in bulk amount from the TBO (**56**) containing medium as well [34].

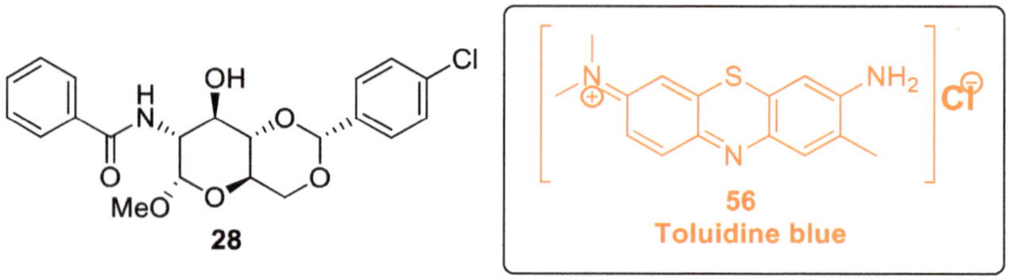

Figure 13.27: Benzamide gelator compound (**28**) for hydrogel formation and toluidine blue dye being utilized in cleaning experiments.

Compound (**57**), having a double bond at the *cis*-position, formed a phase-selective gelating porous amorphous powder for refined petroleum products, crude oils, and aromatic solvents. The derivatives of D-gluconamide (**58** and **59**), having different chain lengths, also have phase-specific gelation properties for oils. Gelation happened at room temperature in milliseconds (Figure 13.28). This phase-specific organogelator demonstrated good reclamation rates and is suitable for a wide variety of fuels and aromatic liquids [41]. This compound's (**57**) susceptibility to gelling in solvents at ambient temperature upon solidification is most likely due to the presence of the *cis* double bond, which produced an uneven structure in the alkyl chain.

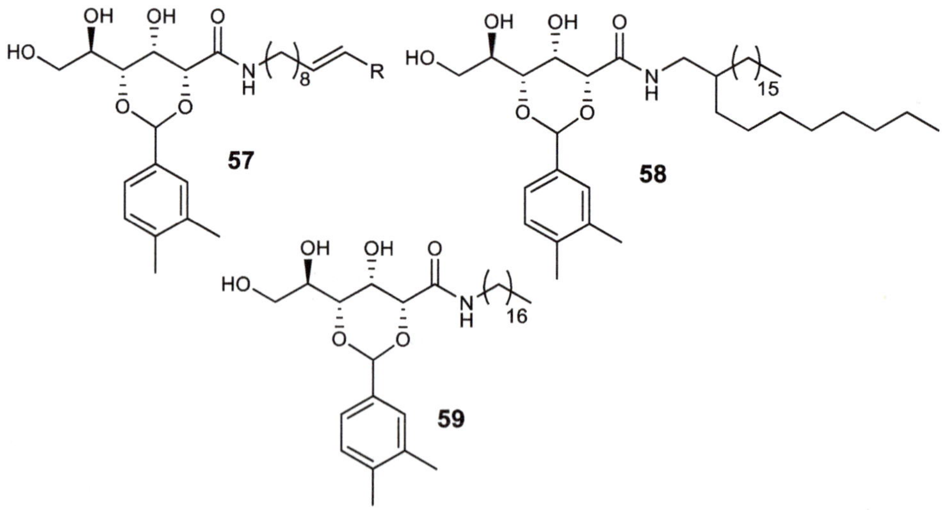

Figure 13.28: Some D-gluconic acetal-based gelators for refining purposes.

Among the arabinopyranoside-based gelators, the compound (**62**) has been found efficient for phase-specific crude oil gelation and oil extraction from ocean water (Figure 13.29). The starch composite gelator (**62**) solidifies the crude oil layer, resulting in a consistent dispersion of the gelator across the oil layer at ambient temperature to scoop away crude oil powder from the water, allowing the gelator to quickly and uniformly form a gel. Gelator dispersion is firmly supported by the starch. This has resulted in significant coagulation of crude oil from mixtures having oil and water combined [49]. The D-mannose derivative gelator (**60**) and two other members of arabinose form a gel in some organic solvents and separate crude products from oil and water mixtures.

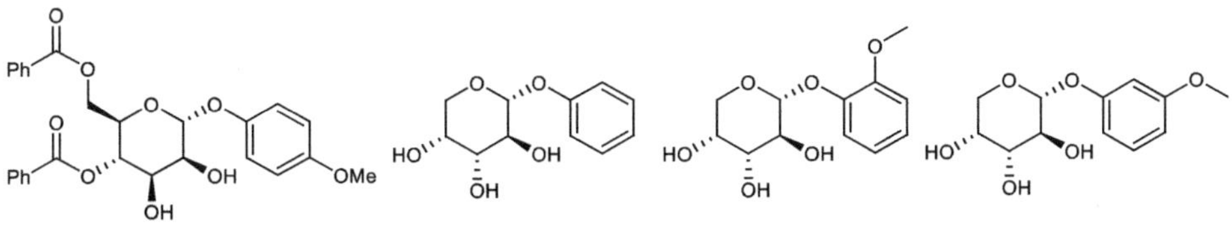

60 **61** **62** **63**

Figure 13.29: Gelation molecules of D-mannose derivative (**60**) and arabinose derivatives (**61–63**).

Recovering spilled oil, treating sewage, and the fabrication of stretchable optical devices have all been investigated as potential applications for (**64**) gel (Figure 13.30). The gel's fascinating feature of adsorbing crystal violet and iodine from water suggests that it might have potential uses in waste water treatment. After 24 h of submersion in an aqueous solution of crystal violet, the (**64**) gel successfully adsorbed the crystal violet molecules from water [50]. The compound **64** absorbed more crystal violet than previously reported in the field of LMWGs. The pump oil layer was watched to see if it gelled, which it did almost immediately, showing the possible use of it in actual spilled oil remediation.

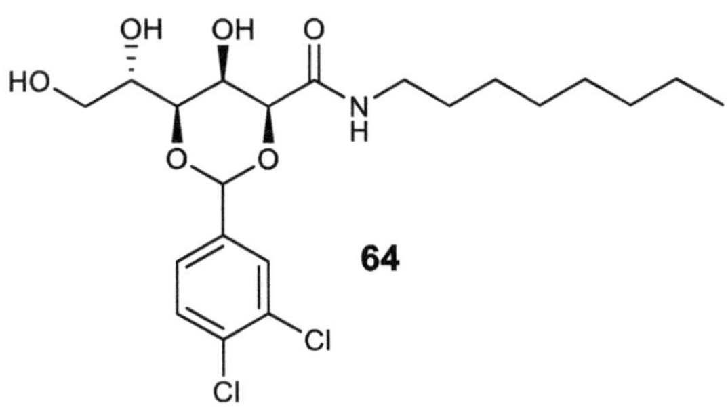

64

Figure 13.30: D-Gluconic acetal-based gelator in the field of water treatment.

Dibenzylidene acetal–protected sorbitol compounds (**65**) have been examined as a gelator entity in applications of solidification and elimination of an extensive range of oils by adding the gelator in powdered form to the oils. Gel formation is independent of the amount of water used. The synthesized LMWG (**65**) has greater potential than (**66**) of the same class in cleaning up marine oil spills (Figure 13.31) [51].

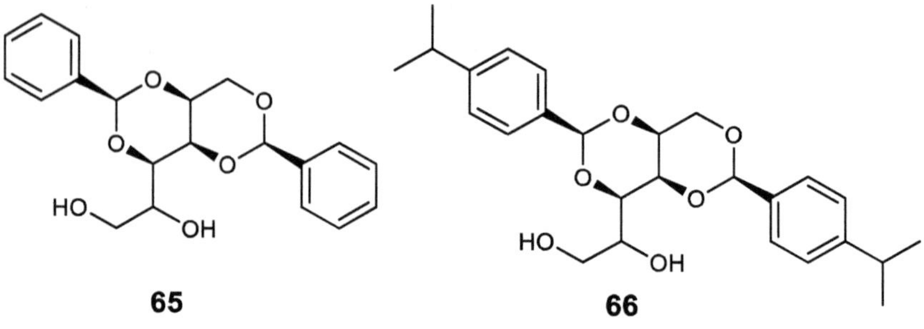

Figure 13.31: Dibenzylidene acetal–protected sorbitol compounds utilized for marine water cleanup.

Glucose-derived phase-specific organogelators (**67**) were synthesized, and when applied in powdered form, they may form gels at ambient temperature. Thioalkyl or thioaryl motifs were inserted at the anomeric position of the sugar head group in order to decrease the water solubility [52]. It was discovered that the resultant compounds selectively congealed oils in aqueous-oil biphasic solutions and were found to be supergelators with minimum gelation requirements. In subsequent research, they produced a number of *N*-glycosylamine variants (**68** and **69**) based on 3,4,5-tri-*O*-benzohydrazide and investigated their applications as LMWGs in environmental cleaning activities (Figure 13.32). Toluene, dichlorobenzene, benzene, mineral oil, coconut oil, gasoline, kerosene, and crude oil could be extracted from each sugar-based gelator as gel. The remarkable capability of the gelator was its recyclability after distillation of benzene from the gel. This gel has been potentially applied as supramolecular scrubbers for the successful removal of dyes, including rhodamine B (55), from aqueous environments [53].

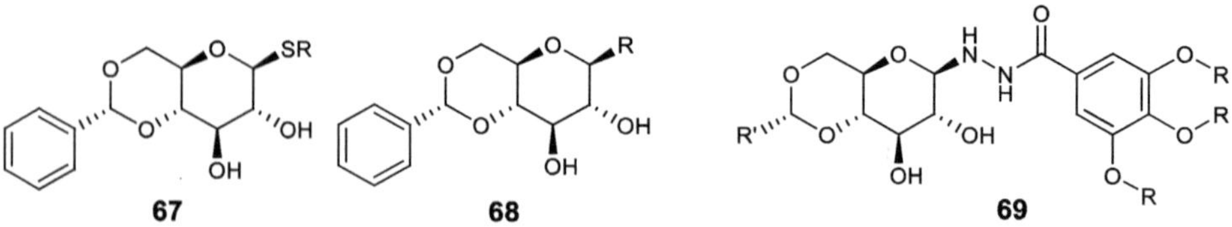

Figure 13.32: Representatives of *N*-glycosylamine-derived gelators for oil–water separation.

Variants of di-2,3-methylbenzylidene acetal-protected sorbitol (DMDBS) (**70**) with aliphatic associations of different chain lengths were formulated. It was discovered that compounds (**70a–c**) are phase-specific gelators, only gelling the organic phase in aqueous biphasic combinations (Figure 13.33). Compound (**70c**) was capable of gelling the gasoline layer within 3 min of addition while utilizing THF as a carrier solvent [54].

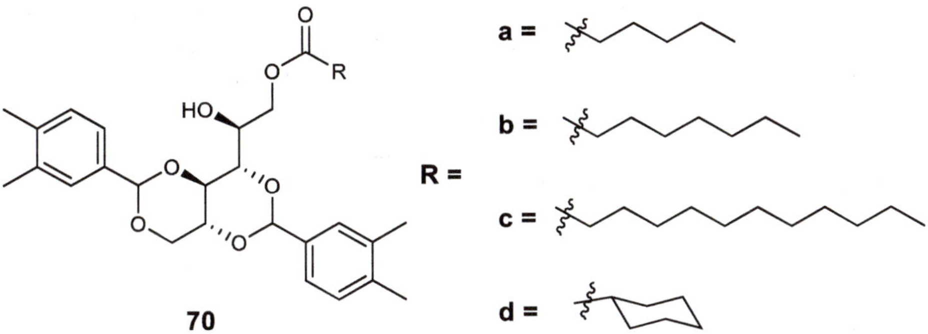

70

Figure 13.33: Sorbitol-protected di-2,3-methylbenzylidene acetal (DMDBS) **(76)** derivatives having gelation properties.

Imidazolium salt carbohydrate analogs **(71** and **72)** were developed and produced [55] as supramolecular ionic liquid gelators (Figure 13.34). Because of their lengthy alkyl chains, the imidazolium cations are the elements that contribute to both hydrophobic and aromatic–aromatic associations. The anionic components derived from carbohydrates play a role in hydrogen bonding interactions. Aqueous solutions, ionic liquids, and organic solvents were used to evaluate gelation. Sixty percent of the chromium was removed from all tested gels after 3 h of exposure. Exceptionally high adsorption capabilities were discovered when tiny samples of the gels were submerged in large chromium solutions. In comparison to the bromide gels, the gels made from carbohydrates demonstrated improved adsorption of chromium. The imidazolium salt carbohydrate hydrogel's adsorption/reduction capability can be very helpful for treating water [55].

71
Gluconate

72
Saccharate

Figure 13.34: Analogues of imidazolium salts of gluconate and saccharate for gel preparation for the removal of chromium ions.

In a recent study investigating the potential of employing the DPSC (75) hydrogel as an adsorption substance for the elimination of organic dyes that are water-soluble, researchers were encouraged by the molecular structure and design (Figure 13.35) [56]. To test the dye adsorption ability of DPSC (75) gel, the commonly used industrial effluents rhodamine B (55) and others were used. Although the direct application of DPSC (75) gel in the form of beads packed in a column produced good results in terms of dye removal, an attritional loss of beads was seen due to the absence of extraordinary strength throughout the dye recovery and regeneration processes. A cellulose-based cartridge has been produced via template-assisted assembly of the DPSC phenomena for potential commercial applications. The cartridge or container demonstrated the ability to effectively absorb dyes by adhering to the intercalation method. The established methodology is a user-friendly and economically feasible approach due to the cartridge's easy adsorption, desorption, regeneration, and material stability [56].

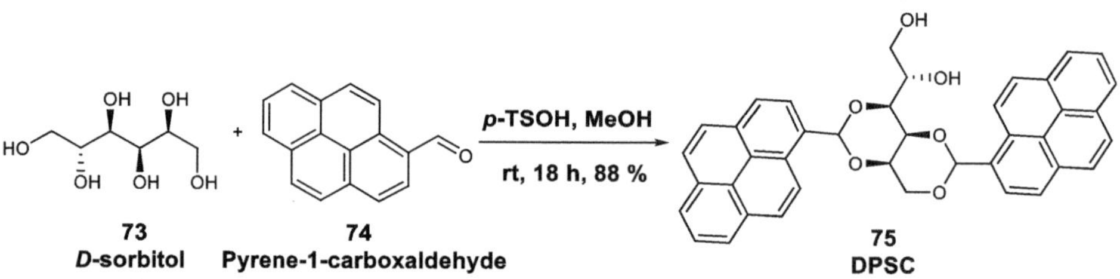

Figure 13.35: Synthetic scheme of DPSC **(75)**.

13.4.6 Shaping, thixotropy, self-healing, and printing

The task of producing gels for shaping purposes from LMWGs has proven to be quite challenging. Small molecules combine to generate supramolecular gels, which are usually delicate and challenging to work with without breaking the gels. Forming a gel in a mold is the simplest method for shaping it; this is how most LMWGs are discovered. The rheological characteristics of gel materials can be changed by varying the amounts of gelators and by strengthening procedures [41, 57]. Certain peptide-based LMWGs, as reported by Adams et al., have demonstrated the ability to endure the extrusion process used in 3D printing [58]. Though fewer instances of 3D printing using carbohydrate-containing LMWGs have been demonstrated, polysaccharides and their derivatives have been used with much greater success in tissue engineering and 3D printing [59, 60].

Thixotropic gels are proven to be self-healing and capable of carrying loads. The mechanical strength of the gel **(64)** is strong (Figure 13.30). Even when compressed by a heavy object, the gel held its shape, suggesting that it is capable of supporting considerably heavier loads on its own. An incredibly rare characteristic of the gel **(64)** was its exceptional load-bearing strength [61]. The gel's capacity for self-healing was investigated. After the gel was divided into two halves, it took only 10 min for the parts to fuse together to create a single, continuous block. The block became violet during the course of 12 h, signifying the active exchange of dissolved molecules across the fusion interface of the self-assembled fibrous network throughout the healing process [50].

The thixotropic features of supramolecular gels derived from LMWGs are necessary for the gels to be 3D printable. Thixotropy is a characteristic property in which a material becomes less viscous when a shear force is applied, and then it returns to a more viscous state over time. The gel needs to undergo shear thinning before it can be extruded via the printer's nozzle in an extrusion-based 3D printing process. After extrusion, the material must then become more viscous in order to transform back into a stable gel [58]. Rapid solvent exchange is a substitute for extrusion-based printing in the in situ gel formation process. This method involves dissolving a gelator in a solvent and injecting the gelator solution into a bath containing a solvent in which the gelator is insoluble. A gel filament or print is formed quickly as a result of a quick solvent exchange that takes place [62].

Fitremann and coworkers demonstrated that solvent exchange between DMSO/H_2O may create well-organized hydrogel filaments [62]. Like various supramolecular gels containing LMWGs, the aqueous gels of *N*-heptyl-D-galactonamide (**53b**) are brittle and suffer shortening when subjected to shear strain, making them unsuitable for effective extrusion (Figure 13.23). In this work, a continuous, homogeneous hydrogel filament was created by injecting a solution of *N*-heptyl-D-galactonamide (**53b**) containing DMSO in water in a specific ratio that was adjusted and managed by a syringe pump. This made it possible to 3D print non-thixotropic gels since the solvent exchange could allow the gel to form on a surface.

Fitremann and associates eventually demonstrated that this non-thixotropic supramolecular gel based on carbohydrates could be 3D printed by exchange of solvent [63]. In order to print a gel onto a glass slide covered with a hydrophobic polycarbonate membrane, a syringe pump was used to feed a DMSO solution of *N*-heptyl-D-galactonamide (**53b**) into a water bath. Until an acceptable adhesion to the polycarbonate membrane was achieved, the flow rate was modified. This technology allows for the customized printing of 3D biocompatible supramolecular gel scaffolds utilizing carbohydrate-containing LMWGs. The said research provides a technique for 3D printing based on non-thixotropic supramolecular gelators.

The derivative of methyl glucoside (**76**) (Figure 13.36) produced a transparent oil–silicone gel expertise in extrusion. The extruded gel's strength was determined utilizing rheological parameters, demonstrating that the molded soft substance preserved gel-like viscoelasticity. This investigation also included a study of the thixotropic properties of the gel formed from (**76**). By applying a vortex mixer, the gel of (**76**) in ethanol/water 1:1 was cleaved. It was found that the gel will reconstitute after an hour of lying at ambient temperature without being disturbed [64].

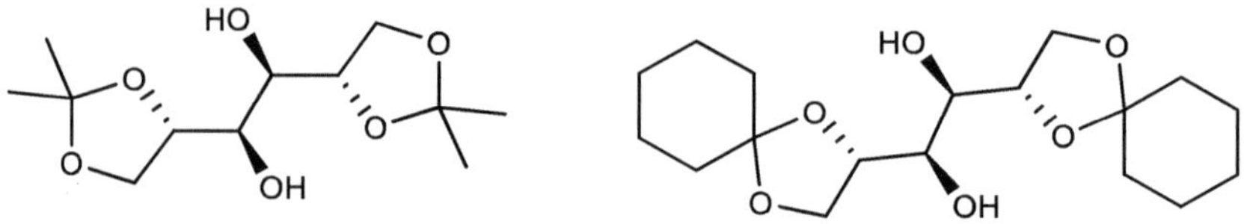

77 78

Figure 13.36: A gelator containing a methyl glucoside derivative (**76**).

In forming supramolecular gels, self-healing characteristics play a crucial role alongside thixotropy. Phase-specific aqueous-organic gelation and considerable self-healing were demonstrated by the mannose-containing organogelator (**60**) (Figure 13.29) [65]. This organogelator (**60**) has displayed remarkable self-healing capabilities besides phase-sensitive aqueous-organic gelation [65].

13.4.7 Development of soft-optical devices

The amphiphilic diols (77 and **78**) proved to be effective gelators for a variety of non-polar solvents and oils. A proposed method includes gelators self-assembling into fibrils by H-bonding, followed by lateral assembly of fibrils into fibers and 3D tangling of these fibers, resulting in a fibrous network. The gels generated by these gelators were found to be transparent with great transmittance in the visible range. Because the oils have refractive indices similar to those of glass and the gels generated by gelators (77 and **78**) in oils have good transmittance, numerous soft-optical devices have been constructed using these oil gels (Figure 13.37) [66]. It's interesting to note that these gels have self-healing qualities, which makes them the perfect soft material for optical devices that won't break or scratch. It was discovered that these constructed optical devices, including prisms and lenses, were very similar to standard glass-based optical devices.

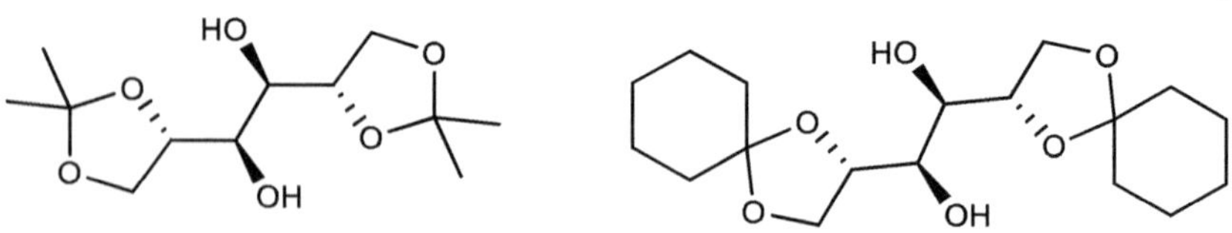

77 78

Figure 13.37: Diols having gelation properties applicable in optical devices.

13.4.8 Preparation of nanomaterials

For the creation of nanoparticles, carbohydrates are well recognized as efficient surface passivating agents. Numerous carbohydrates, including starch, fructose, glucose, sucrose, and chitosan, have been used as capping agents to keep the nanoparticles' development controlled. Carbohydrates have occasionally functioned as reducing agents in addition to capping molecules, particularly during the creation of metal nanoparticles.

Green production of gold nanoparticles using a buffered glucose and starch solution was described by Engelbrekt and colleagues. Gels have been shown to be an incredibly effective reaction medium for the controlled development of nanoparticles. When it comes to gels, the likelihood of manufactured nanoparticles clumping together is quite low since the particles are enmeshed in the gel matrix and cannot move freely. For a variety of biological and medicinal applications, nanoparticles trapped in a semisolid gel matrix have attracted a lot of interest. These hybrid systems of gel and nanoparticles are very helpful for a variety of applications in material science, chiral catalysis, biosensors, and nonlinear optics [67].

Narayana's research proved the effective usage of a novel *N*-acetyl glucosamine-based gelator (**54**) for a variety of applications, including oil/water separation, metals (Au and Ag) nanoparticle production, and dye removal from waterborne solutions (Figure 13.24). The organogel-forming capacity of the *N*-acetyl glucosamine-based gelator (**54**) was tested in various organic solvents. Aside from organogels, hydrogels have also been created with the same gelator. In addition to oil/water separation, the current paper describes the implementation of a carbohydrate-containing gelator for the synthesis of metal nanoparticles. Additionally, the *N*-acetyl glucosamine (GlcNAc)-based organogel (**54**) is also explored for the removal of rhodamine B dye (**55**) from water, as illustrated in Section 13.4.5 [48].

Even at extremely low concentrations, gelators have shown exceptional gelation capacity and can form gels with a variety of organic solvents. The compound **54** (Figure 13.24) was discovered to have dual functions as a gelator and a reducing agent, obviating the necessity for an external reducing agent in the reduction of gold salt into Au nanoparticles. The compound **54** is an attractive material for the creation of nanoparticles because of its dual functionality. It is possible to create Au nanoparticles with a size range of 5–10 nm and embed them in hydrogel fibers.

13.4.9 Chirality of resin

N-Palmitoylglucosamine (**79**) is a molecule whose chirality is inherited and is induced during gel formation in aqueous methanol (Figure 13.38). During the manufacture of polymeric resins, this property has been used as a tool to control the chirality [68]. As a result, left-handed twisted nanoribbons are formed. Eventually, the template is removed to reveal resin nanotubes, indicating the chirality of the organopolymeric resin nanotubes. These resin-based nanotubes underwent carbonization to produce equivalent M-helical carbonaceous nanotubes, which may find use in energy storage and electromagnetic wave absorption [69].

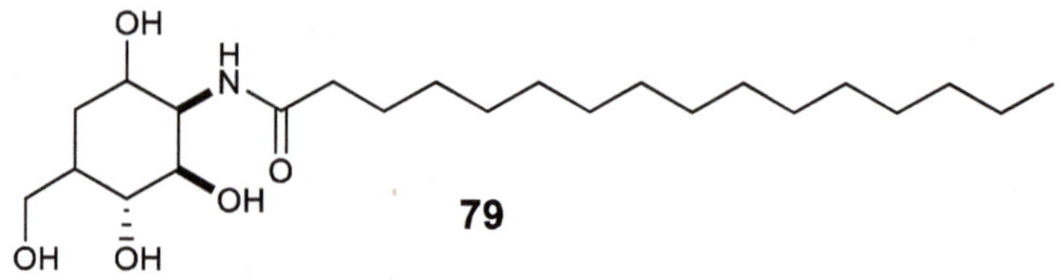

79

Figure 13.38: *N*-Palmitoylglucosamine organogelator.

13.4.10 Semiconductor fabrics

Previously, numerous technologies were used to make conducting fabrics [70–72], but they were associated with certain major problems [73, 74]. Diacetylene 4,6-*O*-benzylidene -β-D-galactopyranoside **(80)**, which congeals various organic liquids, was created by Sureshan and colleagues to manufacture a comparable proportion of the substance across the tightly bound fibers [75]. Photo-irradiated topological polymerization of the self-assembled gelator molecule **(80)** in the gel state produced polydiacetylenes (PDA) **(81)** (Figure 13.39). PDA **(81)** textiles were created by gelling the cloth and then photopolymerizing it. Surface resistance was promising in a PDA-cotton textile devoid of benzylidene and with semiconducting characteristics. This could prove to have an encouraging future utilization in developing semiconducting garments.

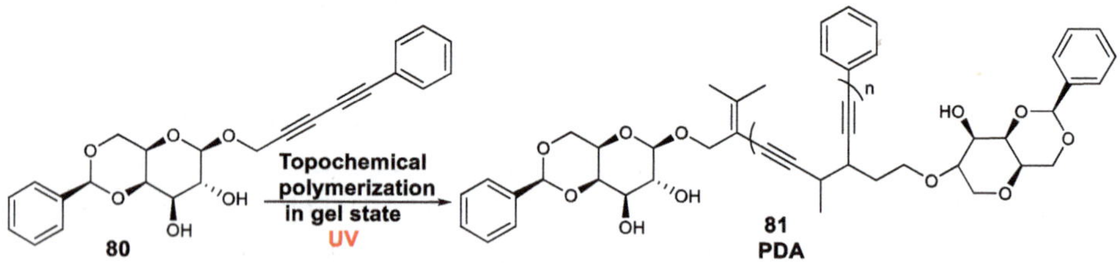

Figure 13.39: Polydiacetylene (**81**) derived from galactose as a semiconducting fabric.

13.4.11 Structuring agent

Interest in the development of oil-structuring agents to replace trans and saturated fats has increased due to the advantages for human health. John and colleagues used a biocatalyst to synthesize sugar-based ester organogelators in order to create value-added products. This was done in accordance with green principles [76]. Using Novozyme 435 and fatty acids, raspberry ketone glycosides (**82**) were esterified to produce the monosaturated raspberry ketone glycoside oleate (**83**) as well as the related ester amphiphiles caprylate and stearate (Figure 13.40). In contrast to the other two varieties, the gel based on the latter gelator exhibited significantly higher structural properties. A portion of these sugar esters self-assembled in edible oils and organic solvents. It has been suggested that the primary noncovalent interaction inside the supramolecular gel assembly is H-bonding. The fatty acid chain's gel packing was discovered to be similar to that of natural fats.

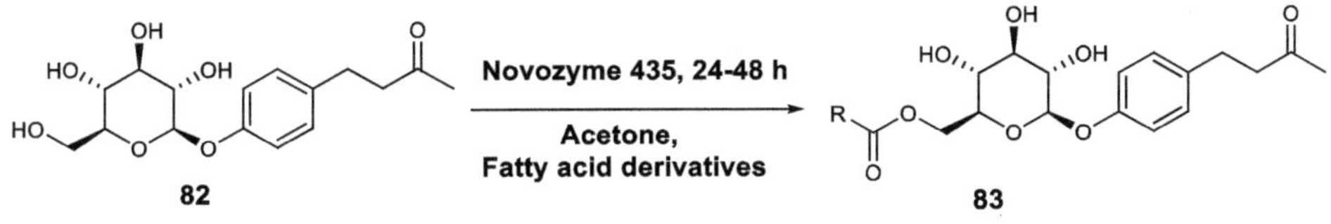

Figure 13.40: Preparation of raspberry ketone glucoside gelators.

13.4.12 Development of a multifunctional gelator library

Even while gelators have been shown to have a wide range of uses, a certain gelator is often only beneficial for a specific application. Synthesizing gelators for different purposes is practically challenging because of this one gelator for one application pattern. In order to overcome this limitation, combining the work of various researchers and the use of click chemistry to synthesize a library of gelators from a common gelator core, 4,6-*O*-benzylidene β-D-galactopyranosylazide (**84**), has been summarized here [77].

The researchers have shown how two similar sugar-based gelators (**84** and **85**), in their xerogel form, exhibit a topochemical reaction with complementary reactive motifs (i.e., azide and alkyne). After being mildly heated, a thermal topochemicalazide-alkyne cycloaddition (TAAC) reaction took place in a xerogel formed from this gel.

Numerous applications, like gas storage, drug encapsulation, catalysis [78], filtration/separation membranes, and energy-storage materials, use porous polymers [79–81]. A styrene gel (**86**) was made by Prathap, and the gel was polymerized to create polystyrene (Figure 13.41). Washing the gelator resulted in porous polystyrene [66]. The concentration may be changed to control the pore size. These porous polystyrenes are highly pursued in the biomedical area, hydrophobic coatings, composite materials, and catalysis.

By adding the anthracene moiety to the gelator core, a luminous gel was created, making it useful in organic electronics, photovoltaics, and sensors [82–84]. The gel generated by the compound (**87**) displayed red-shifted absorption in a non-gelling solvent, just like other known fluorescent gelators (Figure 13.41) [61, 85]. Fluorescence tests that are temperature and concentration-dependent provide evidence for gelation-induced fluorescence.

The gel of (**88**) was polymerized to produce silica corrugated rods with approximate widths of 200–300 nm. It has never before been reported that a silica rod with such a distinctive shape was synthesized using an organogel template [61].

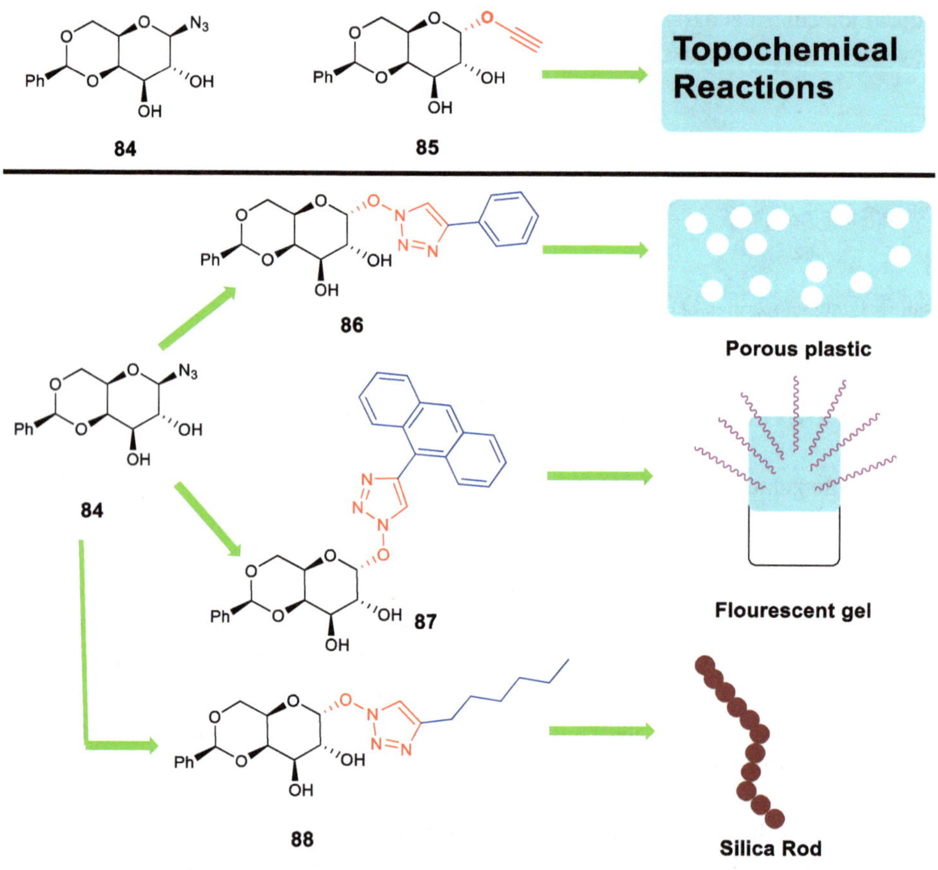

Figure 13.41: Structure of the gelator 4,6-*O*-benzylidene β-D-galactopyranosylazide (**84**) and its library of gelators from a common gelator molecule.

13.5 Conclusion

In light of the explanation of the above-mentioned applications, it can be concluded that carbohydrate-based gelators have gained much interest compared to ordinary gelators. The excellence of these gelators lies in their thixotropic characteristics, as they can bear stress without deformation and provide good usage facilities for their implementation in different fields. Another property that makes them superior to ordinary gels is their self-healing ability. They can return to their original shape after being used for any purpose. There is a large diversity in carbohydrates, and hence, many categories have been employed to synthesize gelators, providing us with great diversity in the field of gelators based on carbohydrate entities. They are also superior compared to other gels due to their preferable distinct applications with a common ancestor molecule from which they have been derived. There has been found the difference of only a group in two molecules, and they have been found applicable for different purposes.

References

[1] Skilling KJ, Citossi F, Bradshaw TD, Ashford M, Kellam B, Marlow M. Insights into low molecular mass organic gelators: A focus on drug delivery and tissue engineering applications. Soft Mater, 2019, 10(2), 237–256.

[2] Lloyd GO, Steed JW. Anion-tuning of supramolecular gel properties. Nat Chem, 2009, 1(6), 437–442.

[3] Draper ER, Adams DJ. Low-molecular-weight gels: The state of the art. Chem, 2017, 3(3), 390–410.

[4] Weiss RG. The past, present, and future of molecular gels. What is the status of the field, and where is it going? J Am Chem Soc, 2014, 136(21), 7519–7530.

[5] Vintiloiu A, Leroux JC. Organogels and their use in drug delivery-A review. J Control Release, 2008, 125(3), 179–192.

[6] Steed JW. Anion-tuned supramolecular gels: A natural evolution from urea supramolecular chemistry. Chem Soc Rev, 2010, 39(10), 3686–3699.

[7] Murdan S. Organogels in drug delivery. Expert Opin Drug Deliv, 2005, 2(3), 489–505.

[8] Du X, Zhou J, Shi J, Xu B. Supramolecular hydrogelators and hydrogels: From soft matter to molecular biomaterials. Chem Rev, 2015, 115(24), 13165–13307.

[9] Sangeetha NM, Maitra U. Supramolecular gels: Functions and uses. Chem Soc Rev, 2005, 34(10), 821–836.

[10] George M, Weiss RG. Molecular organogels. Soft matter comprised of low-molecular-mass organic gelators and organic liquids. Acc Chem Res, 2006, 39(8), 489–497.

[11] De Loos M, Feringa BL, van Esch JH. Design and application of self-assembled low molecular weight hydrogels. Eur J Org Chem, 2005, 2005(17), 3615–3631.

[12] Adarsh NN, Kumar DK, Dastidar P. Composites of N, N′-bis-(pyridyl) urea-dicarboxylic acid as new hydrogelators-a crystal engineering approach. Tetrahedron, 2007, 63(31), 7386–7396.

[13] Dong R, Pang Y, Su Y, Zhu X. Supramolecular hydrogels: Synthesis, properties and their biomedical applications. Biomater Sci, 2015, 3(7), 937–954.

[14] Van Esch J, Schoonbeek F, De Loos M, Marc Veen E, Kellogg RM, Feringa BL. Low molecular weight gelators for organic solvents: From serendipity towards design. J Supramol Struct, 1999, 233–259.

[15] Dastidar P. Designing supramolecular gelators: Challenges, frustrations, and hopes. Gels, 2019, 5(1), 15.

[16] Basu N, Chakraborty A, Ghosh R. Carbohydrate derived organogelators and the corresponding functional gels developed in recent time. Gels, 2018, 4(2), 52.

[17] Datta S, Bhattacharya S. Multifarious facets of sugar-derived molecular gels: Molecular features, mechanisms of self-assembly and emerging applications. Chem Soc Rev, 2015, 44(15), 5596–5637.

[18] Matson JB, Stupp SI. Self-assembling peptide scaffolds for regenerative medicine. Chem Com, 2012, 48(1), 26–33.

[19] Delbianco M, Bharate P, Varela-Aramburu S, Seeberger PH. Carbohydrates in supramolecular chemistry. Chem Rev, 2016, 116(4), 1693–1752.

[20] Hollingsworth RI, Wang G. Toward a carbohydrate-based chemistry: Progress in the development of general-purpose chiral synthons from carbohydrates. Chem Rev, 2000, 100(12), 4267–4282.

[21] Shinkai S, Ikeda M, Sugasaki A, Takeuchi M. Positive allosteric systems designed on dynamic supramolecular scaffolds: Toward switching and amplification of guest affinity and selectivity. Acc Chem Res, 2001, 34(6), 494–503.

[22] Smith DK. Dendritic supermolecules–towards controllable nanomaterials. Chem Comm, 2006, 1, 34–44.

[23] Raju CSK, Pramanik B, Ravishankar R, Rao PVC, Sriganesh G. Xylitol based phase selective organogelators for potential oil spillage recovery. RSC Adv, 2017, 7(59), 37175–37180.

[24] Meunier JA. Sur les composes que la mannite et la sorbiteforment avec les aldehydes. Gauthier-Villars, 1891.

[25] Yamasaki S, Tsutsumi H. Microscopic studies of 1, 3: 2, 4-di-O-benzylidene-D-sorbitol in ethylene glycol. Bull Chem Soc Jpn, 1994, 67(4), 906–911.

[26] Shimizu T, Masuda M. Stereochemical effect of even-odd connecting links on supramolecular assemblies made of 1-glucosamide bolaamphiphiles. J Am Chem Soc, 1997, 119(12), 2812–2818.

[27] Prasad YS, Miryala S, Lalitha K, Ranjitha K, Barbhaiwala S, Sridharan V, Nagarajan S. Disassembly of bacterial biofilms by the self-assembled glycolipids derived from renewable resources. ACS Appl Mater Interfaces, 2017, 9(46), 40047–40058.

[28] Prasad YS, Miryala S, Lalitha K, Saritha B, Maheswari CU, Sridharan V, Nagarajan S. An injectable self-healing anesthetic glycolipid-based oleogel with antibiofilm and diabetic wound skin repair properties. Sci Rep, 2020, 10(1), 18017.

[29] Prasad YS, Saritha B, Tamizhanban A, Lalitha K, Kabilan S, Maheswari CU, Nagarajan S. Enzymatic synthesis and self-assembly of glycolipids: Robust self-healing and wound closure performance of assembled soft materials. RSC Adv, 2018, 8(65), 37136–37145.

[30] Tang Q, Plank TN, Zhu T, Yu H, Ge Z, Li Q, Pei H. Self-assembly of metallo-nucleoside hydrogels for injectable materials that promote wound closure. ACS Appl Mater Interfaces, 2019, 11(22), 19743–19750.

[31] Bhattacharyya T, Chaudhuri R, Das KS, Mondal R, Mandal S, Dash J. Cytidine-derived hydrogels with tunable antibacterial activities. ACS Appl Bio Mater, 2019, 2(8), 3171–3177.

[32] Pan A, Roy SG, Haldar U, Mahapatra RD, Harper GR, Low WL, Hardy JG. Uptake and release of species from carbohydrate containing organogels and hydrogels. Gels, 2019, 5(4), 43.

[33] Yoo Y, Yoon SJ, Kim SY, Lee DW, Um S, Hyun H, Yang DH. A local drug delivery system based on visible light-cured glycol chitosan and doxorubicin hydrochloride for thyroid cancer treatment in vitro and *in vivo*. Drug Deliv, 2018, 25(1), 1664–1671.

[34] Bietsch J, Olson M, Wang G. Fine-tuning of molecular structures to generate carbohydrate based super gelators and their applications for drug delivery and dye absorption. Gels, 2021, 7(3), 134.

[35] Chen A, Adhikari SB, Mays K, Wang G. Synthesis and study of molecular assemblies formed by 4, 6-O-(2-phenylethylidene)-functionalized D-glucosamine derivatives. Langmuir, 2017, 33(32), 8076–8089.

[36] Wang D, Chen A, Morris J, Wang G. Stimuli-responsive gelators from carbamoyl sugar derivatives and their responses to metal ions and tetrabutylammonium salts. RSC Adv, 2020, 10(66), 40068–40083.

[37] Kannan R, Muthuvijayan V, Prasad E. *In vitro* study of a glucose attached poly (aryl ether) dendron based gel as a drug carrier for a local anaesthetic. New J Chem, 2017, 41(15), 7453–7462.

[38] He M, Sui J, Chen Y, Bian S, Cui Y, Zhou C, Zhang X. Localized multidrug co-delivery by injectable self-crosslinking hydrogel for synergistic combinational chemotherapy. J Mater Chem B, 2017, 5(25), 4852–4862.

[39] Ghosh T, Biswas A, Gavel PK, Das AK. Engineered dynamic boronate ester-mediated self-healable biocompatible G-quadruplex hydrogels for sustained release of vitamins. Langmuir, 2020, 36(6), 1574–1584.

[40] Xu W, He X, Zhong M, Hu X, Xiao Y. A novel pH-responsive hydrogel based on natural polysaccharides for controlled release of protein drugs. RSC Adv, 2015, 5(5), 3157–3167.

[41] Hu Y, Gao W, Wu F, Wu H, He B, He J. Low molecular weight gels induced differentiation of mesenchymal stem cells. J Mater Chem B, 2016, 4(20), 3504–3508.

[42] Latxague L, Ramin MA, Appavoo A, Berto P, Maisani M, Ehret C, Barthelemy P. Control of stem-cell behavior by fine tuning the supramolecular assemblies of low-molecular-weight gelators. Angew Chem, 2015, 127(15), 4600–4604.

[43] Khan F, Tanaka M. Designing smart biomaterials for tissue engineering. Int J Mol Sci, 2017, 19(1), 17.

[44] Guo JL, Kim YS, Mikos AG. Biomacromolecules for tissue engineering: Emerging biomimetic strategies. J Biol Macromol, 2019, 20(8), 2904–2912.

[45] Williams DF. Challenges with the development of biomaterials for sustainable tissue engineering. Front Bioeng Biotechnol, 2019, 7, 127.

[46] Qi J, Yan Y, Cheng B, Deng L, Shao Z, Sun Z, Li X. Enzymatic formation of an injectable hydrogel from a glycopeptide as a biomimetic scaffold for vascularization. ACS Appl Mater Interfaces, 2018, 10(7), 6180–6189.

[47] Chalard A, Vaysse L, Joseph P, Malaquin L, Souleille S, Lonetti B, Fitremann J. Simple synthetic molecular hydrogels from self-assembling alkylgalactonamides as Scaffold for 3D neuronal cell growth. ACS Appl Mater Interfaces, 2018, 10(20), 17004–17017.

[48] Narayana C, Upadhyay RK, Chaturvedi R, Sagar R. A versatile carbohydrate based gelator for oil water separation, nanoparticle synthesis and dye removal. New J Chem, 2017, 41(6), 2261–2267.

[49] Pathak NP, Yadav S. A gelator–starch blend for dry powder based instant solidification of crude oil at room temperature. Chem Comm, 2020, 56(20), 2999–3002.

[50] Guan X, Fan K, Gao T, Ma A, Zhang B, Song J. A novel multi-stimulus responsive gelator based on D-gluconic acetal and its potential applications. Chem Comm, 2016, 52(5), 962–965.

[51] Zhuan C, Li Y, Yuan X, Zhao J, Hou X. A sorbitol-based phase-selective organogelator for crude oil spills treatment. J Appl Polym Sci, 2019, 136(6), 47052.

[52] Vibhute AM, Muvvala V, Sureshan KM. A sugar-based gelator for marine oil-spill recovery. Angew Chem, 2016, 128(27), 7913–7916.

[53] Soundarajan K, Das TM. Sugar-benzohydrazide based phase selective gelators for marine oil spill recovery and removal of dye from polluted water. Carbohydr Res, 2019, 481, 60–66.

[54] Fan K, Ma Y, Wang X, Yu S, Han G, Yin Z, Song J. Sorbitol-based supramolecular organogelators with effective phase-selective gelation and significant self-healing property. Soft Mater, 2018, 16(1), 1–6.

[55] Rizzo C, Andrews JL, Steed JW, D'Anna F. Carbohydrate-supramolecular gels: Adsorbents for chromium (VI) removal from wastewater. J Colloid Interface Sci, 2019, 548, 184–196.

[56] Singh V, Lalitha K, Maheswari CU, Sridharan V, Pradhan D, Batra S, Nagarajan S. Remediation of dyes using supramolecular material derived from carbohydrate based π-gelator using the bottom-up assembly approach. ACS Omega, 2024.

[57] Panja S, Fuentes-Caparros AM, Cross ER, Cavalcanti L, Adams DJ. Annealing supramolecular gels by a reaction relay. Chem Mater, 2020, 32(12), 5264–5271.

[58] Nolan MC, Caparros AMF, Dietrich B, Barrow M, Cross ER, Bleuel M, Adams DJ. Optimizing low molecular weight hydrogels for automated 3D printing. Soft Mater, 2017, 13(45), 8426–8432.

[59] Ouyang L, Highley CB, Rodell CB, Sun W, Burdick JA. 3D printing of shear-thinning hyaluronic acid hydrogels with secondary cross-linking. ACS Bio Mater Sci Engr, 2016, 2(10), 1743–1751.

[60] Pekkanen AM, Mondschein RJ, Williams CB, Long TE. 3D printing polymers with supramolecular functionality for biological applications. Biomacromolecules, 2017, 18(9), 2669–2687.

[61] Sahoo P, Sankolli R, Lee HY, Raghavan SR, Dastidar P. Gel sculpture: Moldable, load-bearing and self-healing non-polymeric supramolecular gel derived from a simple organic salt. Chem Eur J, 2012, 18(26), 8057–8063.

[62] Chalard A, Joseph P, Souleille S, Lonetti B, Saffon-Merceron N, Loubinoux I, Fitremann J. Wet spinning and radial self-assembly of a carbohydrate low molecular weight gelator into well-organized hydrogel filaments. Nanoscale, 2019, 11(32), 15043–15056.

[63] Chalard A, Mauduit M, Souleille S, Joseph P, Malaquin L, Fitremann J. 3D printing of a biocompatible low molecular weight supramolecular hydrogel by dimethylsulfoxide water solvent exchange. Addit Manuf, 2020, 33, 101162.

[64] Ono F, Ichimaru K, Hirata O, Shinkai S, Watanabe H. Universal glucose-based low-molecular-weight gelators for both organic and aqueous solvents. Chem Lett, 2020, 49(2), 156–159.

[65] Pal KB, Mukhopadhyay B. Carbohydrate-basedsafe fuel gel with significant self–healing property. Chem Select, 2017, 2(3), 967–974.

[66] Prathap A, Sureshan KM. Sugar-based organogelators for various applications. Langmuir, 2019, 35(18), 6005–6014.

[67] Engelbrekt C, SØrensen KH, Zhang J, Welinder AC, Jensen PS, Ulstrup J. Green synthesis of gold nanoparticles with starch–glucose and application in bioelectrochemistry. J Mater Chem, 2009, 19(42), 7839–7847.

[68] Li J, Li Y, Li B, Yang Y. Left-handed helical polymer resin nanotubes prepared by using N-palmitoyl glucosamine. Chirality, 2018, 30(4), 439–444.

[69] Liu S, Duan Y, Feng X, Yang J, Che S. Synthesis of enantiopure carbonaceous nanotubes with optical activity. Angew Chem, 2013, 125(27).

[70] Cucchi I, Boschi A, Arosio C, Bertini F, Freddi G, Catellani M. Bio-based conductive composites: Preparation and properties of polypyrrole (PPy)-coated silk fabrics. Synth Met, 2009, 159(3–4), 246–253.

[71] Xia Y, Lu Y. Fabrication and properties of conductive conjugated polymers/silk fibroin composite fibers. Compos Sci Technol, 2008, 68(6), 1471–1479.

[72] Molina J, Del Río AI, Bonastre J, Cases F. Electrochemical polymerisation of aniline on conducting textiles of polyester covered with polypyrrole/AQSA. Eur Polym J, 2009, 45(4), 1302–1315.

[73] Gupta ND, Das S, Das NS, Banerjee D, Sarkar D, Chattopadhyay KK. Improvement of adhesion and continuity of polypyrrolethin films through surface modification of hydrophobic substrates. J Appl Polym Sci, 2014, 131(2).

[74] Kaynak A, Hakansson E. Short-term heating tests on doped polypyrrole-coated polyester fabrics. Synth Met, 2008, 158(8–9), 350–354.

[75] Krishnan BP, Mukherjee S, Aneesh PM, Namboothiry MA, Sureshan KM. Semiconducting fabrics by in situ topochemical synthesis of polydiacetylene: A new dimension to the use of organogels. Angew Chem, 2016, 128(7), 2391–2395.

[76] Silverman JR, John G. Biobased fat mimicking molecular structuring agents for medium-chain triglycerides (MCTs) and other edible oils. J Agric Food Chem, 2015, 63(48), 10536–10542.

[77] Krishnan BP, Sureshan KM. A library of multipurpose supramolecular supergelators: Fabrication of structured silica, porous plastics, and fluorescent gels. Chem Asian J, 2018, 13(2), 187–193.

[78] Cui J, Yan Y, Wang Y, Caruso F. Templated assembly of pH-labile polymer-drug particles for intracellular drug delivery. Adv Func Mater, 2012, 22(22), 4718–4723.
[79] Germain J, Svec F, Frechet JM. Preparation of size-selective nanoporous polymer networks of aromatic rings: Potential adsorbents for hydrogen storage. Chem Mater, 2008, 20(22), 7069–7076.
[80] Chen L, Honsho Y, Seki S, Jiang D. Light-harvesting conjugated microporous polymers: Rapid and highly efficient flow of light energy with a porous polyphenylene framework as antenna. J Am Chem Soc, 2010, 132(19), 6742–6748.
[81] Patra A, Scherf U. Fluorescent microporous organic polymers: Potential testbed for optical applications. Chem Eur J, 2012, 18(33), 10074–10080.
[82] Babu SS, Prasanthkumar S, Ajayaghosh A. Self-assembled gelators for organic electronics. Angew Chem Int Ed, 2012, 51(8), 1766–1776.
[83] Kartha KK, Babu SS, Srinivasan S, Ajayaghosh A. Attogram sensing of trinitrotoluene with a self-assembled molecular gelator. J Am Chem Soc, 2012, 134(10), 4834–4841.
[84] Stolley RM, Helm ML. Soft support for energy conversion. Nat Chem, 2014, 6(11), 949–950.
[85] Lekha PK, Prasad E. Aggregation-controlled excimer emission from anthracene-containing polyamidoamine dendrimers. Chem Eur J, 2010, 16(12), 3699–3706.

Ayushi Aggarwal, Aastha Palta, Anupama Parmar, and
Harish Kumar Chopra*

Chapter 14
Carbohydrate-based ionic liquids for biomass utilization

14.1 Introduction

Ionic liquids (ILs) have gained significant attention in various research areas in chemical sciences, food, and pharmaceutical sciences over the past decade [1, 2]. ILs, which are salts with melting temperatures lower than 100 °C, are well known for their exceptional solvation capabilities as well as their exceptional nonvolatile, low viscosity, high thermal stability, and recyclable characteristics [3]. ILs are often referred to as "solutions to the future" due to their nonflammability and negligible vapor pressure. They can easily be designed by altering the combination of different anions and cations, hence defined as "designer compounds" [4]. Generally, in ILs, the cationic part is made up of organic moieties such as ammonium, pyridinium, phosphonium, and sulfonium ions, whereas the anionic part is made up of inorganic or organic salts such as lactates, sugars, amino acids, and tartrates [5, 6]. ILs have been known for a long time, with the first reported ILs in 1888 [7] and true RTILs in 1914 [8]. However, the discovery of imidazolium-based tetrafluoroborate and hexafluorophosphate ILs in 1992 broadened the research of ILs and their applications [9]. The synthesis of novel ILs has been extensively studied for their potential applications in organic synthesis [10], cellulose processing [11], batteries [12], electrochemical synthesis [13], and as extracting agents [14]. The synthesis and applications of ILs are widely explored, now found in various subclasses of ILs such as RTILs, PILs, CILs, and MILs, each with specific applications in their respective fields, such as asymmetric chemistry and hydrogels [15]. Out of all these sub classes, CILs are emerging as promising compounds for separating enantiomers and enhancing analytical techniques (GC, HPLC, CE, etc.). CILs can be produced from both naturally occurring and synthetic compounds such as amino acids, alcohols, terpenes, and carbohydrates [16, 17].

This chapter focuses on carbohydrate-based ILs (CBILs), including those with an intact carbohydrate moiety as part of an anion or cation. Like ILs, CBILs are salts with melting points less than 100 °C, often acting as RTILs. First, D-fructose-derived ILs

***Corresponding author: Harish Kumar Chopra,** Department of Chemistry, Sant Longowal Institute of Engineering and Technology, Longowal, Sangrur 148106, Punjab, India, e-mail: hk67@rediffmail.com
Ayushi Aggarwal, Aastha Palta, Department of Chemistry, Sant Longowal Institute of Engineering and Technology, Longowal, Sangrur 148106, Punjab, India
Anupama Parmar, PG Department of Chemistry, M. M. Modi College, Patiala 147001, Punjab, India

https://doi.org/10.1515/9783111447094-014

were synthesized by S.T. Handy et al. in 2003 and further used as recyclable solvents for the Heck reaction of aryl iodides [18, 19]. CBILs naturally contain stereochemical information, which also defines them as carbohydrate-based chiral ILs (CCILs), and similar to CBILs, they can be utilized as additives or catalysts for asymmetric synthesis [20]. The versatility of CBILs enhances their ability to serve as both organocatalysts and as catalysts in biomass conversion.

14.1.1 Carbohydrate-based ionic liquids as organocatalysts

CBILs have shown great potential as organocatalysts due to their structural diversity, biocompatibility, and ability to participate in various catalytic processes [21–23]. Organocatalysis, which relies on small organic molecules to accelerate chemical reactions without the need for metals, is a field that aligns well with the principles of green chemistry. CBILs offer unique advantages in this space, thanks to the intrinsic chiral nature of carbohydrates, which can induce stereoselectivity in reactions [24–27].

The functionality of CBILs as organocatalysts is largely driven by the modification of their carbohydrate backbone, which can be tailored to enhance their catalytic activity [28–30]. For instance, the hydroxyl groups of carbohydrates can be strategically functionalized to provide hydrogen-bonding sites, proton-donating centers, or nucleophilic groups, enabling CBILs to mediate key reaction mechanisms such as aldol condensations, Michael additions, and Diels-Alder reactions. These properties make them especially valuable in asymmetric synthesis, where selectivity is critical [31–33].

Moreover, CBILs, being derived from natural sources, offer a low-toxicity, biodegradable alternative to conventional synthetic catalysts. Their ability to act under mild conditions further reduces the environmental footprint of chemical processes. The potential for these ILs to be recycled and reused also enhances their economic viability, making them attractive candidates for industrial applications [34–36].

14.1.2 Carbohydrate-based ionic liquids in biomass conversion

The shift toward renewable feedstocks in chemical manufacturing has spurred considerable interest in the conversion of biomass into value-added products [37–40]. CBILs, with their tunable chemical properties and environmental benefits, have emerged as highly effective catalysts in the realm of biomass conversion.

Biomass, comprising lignocellulosic materials, sugars, and polysaccharides, is a complex and heterogeneous resource. The breakdown and conversion of biomass into fuels, chemicals, and materials require catalysts that can operate efficiently in this challenging environment. CBILs have demonstrated their ability to act as both solvents and catalysts in biomass conversion processes, including the depolymerization of cellulose and hemicellulose, the hydrolysis of lignin, and the conversion of sugars into platform chemicals such as 5-hydroxymethylfurfural (HMF) and levulinic acid [33, 41–43].

One of the key advantages of CBILs in biomass conversion is their capacity to dissolve biomass components, which traditional solvents often struggle to achieve. This facilitates the catalytic breakdown of polysaccharides and lignin into smaller, more manageable molecules. Additionally, the tunable acidity and basicity of CBILs can be tailored to optimize reaction conditions, improving yields and selectivity in the production of biofuels and chemicals [44–46].

CBILs also play a pivotal role in facilitating cascade reactions, where multiple reaction steps occur in sequence without the need for isolation of intermediates. This is particularly useful in biomass conversion, where complex starting materials require several transformations to yield the desired products. By acting as both catalyst and solvent, CBILs streamline these processes, reducing the need for multiple solvents and simplifying reaction setups [47–50].

In summary, CBILs offer a versatile and sustainable approach to catalysis, functioning effectively as organocatalysts for stereoselective reactions and as catalysts in the conversion of biomass into valuable chemicals and fuels. Their unique combination of green chemistry principles, catalytic efficiency, and renewable origins positions them at the forefront of efforts to create more sustainable chemical processes. This chapter will further explore the design, synthesis, and application of CBILs in these two critical areas, highlighting recent advances and future opportunities for their broader industrial implementation.

14.2 Case study

14.2.1 Organocatalysts

O. N. Van Buu et al. synthesized isosorbide-based chiral ILs (CCILs, **4**–7) by beginning with commercially available carbohydrates known as isosorbide [51]. These CILs were obtained by selective monobenzylation of the hydroxyl group at the endo position (C3), and then the hydroxyl group present at the exo-position (C6) was treated with benzenesulfonyl chloride. The sulfonate group was further treated with an excess of primary aromatic and aliphatic amines at 160 °C to give the corresponding amino ethers via an S_N2 substitution reaction with complete inversion of configuration, followed by quarterization of the amine group. Lastly, these obtained CCILs were treated with TfOH and $LiNTf_2$ for anion metathesis, as shown in Figure 14.1. Furthermore, the produced isosorbide-based CILs were used as a chirality inducer in the asymmetric aza Diels-Alder reaction. As shown in Figure 14.2, they carried out the reaction between Danishefsky's diene and imine at room temperature; a good yield and high diastereoselectivity of the product were produced.

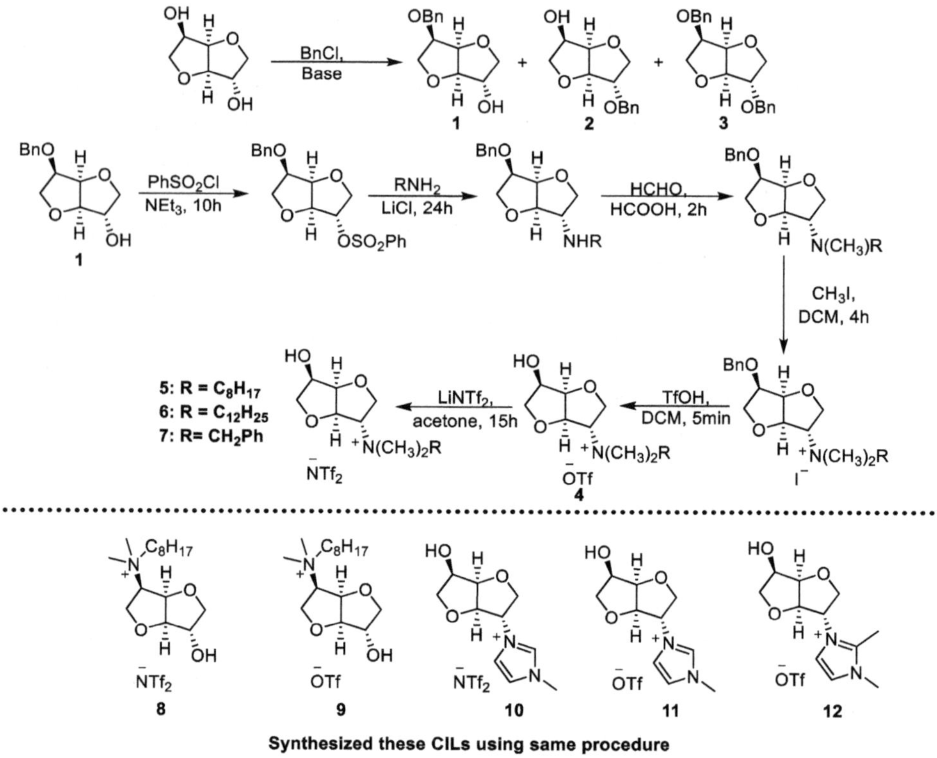

Figure 14.1: Synthesis of isosorbide-based chiral ionic liquids (CILs) [51].

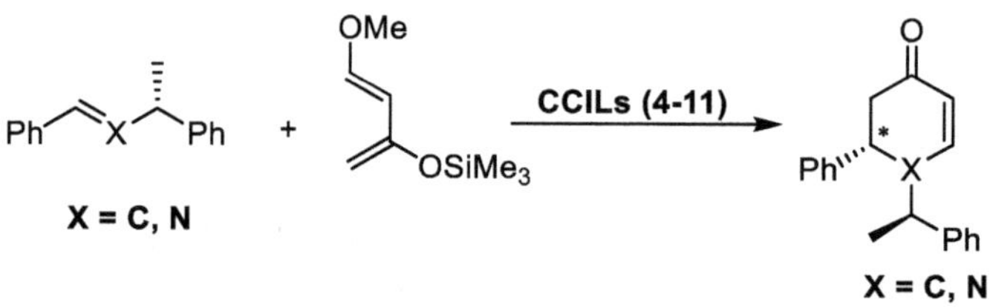

Figure 14.2: Asymmetric aza Diels-Alder reaction using isosorbide-based chiral ionic liquids (CILs) [51].

R. Jayachandra et al. synthesized D-xylose and D-galactose-based CILs (**13–17**) and examined their catalytic applications [52]. They started with the protection of hydroxyl groups of D-xylose by the tosylation process and then inserted the *N*-methylimidazolium moiety as the cationic part, followed by anion exchange reactions with various inorganic salts, as shown in Figure 14.3. After the synthesis of D-xylose-based CILs, they played a vital role as chiral catalysts in the Michael addition reaction between chalcone and diethyl malonate in chloroform (solvent) using K_2CO_3 (base) at room temperature, as seen in Figure 14.4. The authors repeated the same procedure three times at different temperatures (room temperature, 0 to −5 °C, and −40 °C) and found that among the five CCILs, the iodide counter ion (**13**) containing CCIL showed maximum enantiomeric excess (14%), CCIL **14** and **15** gave 2–3% *ee*, and CCILs **16** and **17** were racemic at room temperature. After repeating the same procedure at 0 to −5 °C, the percentage of enantiomeric excess (*ee*) increased in all CCILs (1–5) due to the reduced restricted rotation in molecules (as shown in Table 14.1) but CCIL **13** showed maximum ee (33%). Further, the reaction was performed at −40 °C with CCIL 13, but unfortunately, the reaction did not proceed at this temperature.

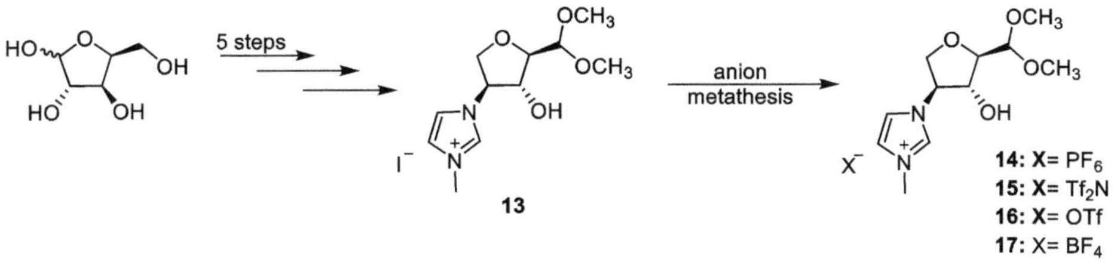

Figure 14.3: Synthesis of D-xylose and D-galactose-based CILs [52].

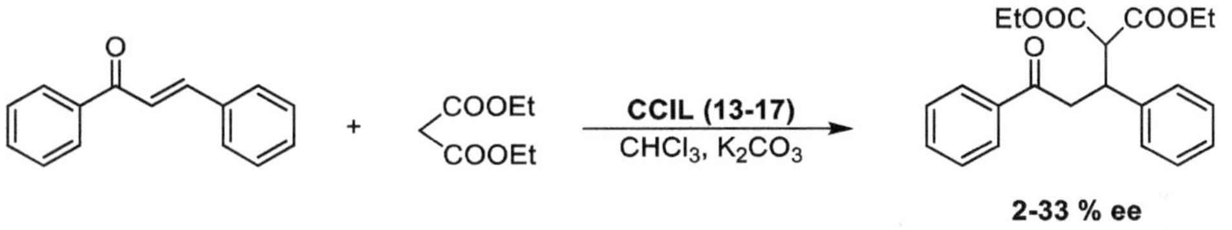

Figure 14.4: Asymmetric Michael addition reaction using D-xylose and D-galactose-based CILs as organocatalysts [52].

Table 14.1: Yields and enantiomeric excess (ee) values of Michael adducts at room temperature (rt) and 0 °C to −5 °C [52].

S. no.	CCIL	*T* (°C)	*t* (h)	Yield (%)	ee (%)
a	**13**	rt	7	81	14
b	**13**	0 to −5	12	73	33
c	**14**	rt	7	84	2
d	**14**	0 to −5	12	71	8
e	**15**	rt	8	76	3
f	**15**	0 to −5	13	69	9
g	**16**	rt	8	79	Racemic
h	**16**	0 to −5	12	70	4
I	**17**	rt	8	78	Racemic
j	**17**	0 to −5	13	72	4

Yun *et al.* synthesized a glucose-based CIL containing a pyridine moiety in the synthesis of chiral Tröger's base [53]. The reaction started with the acetylation of D-glucose using acetic anhydride (Ac_2O), substituting bromine, and reacting with pyridine, followed by an anion exchange reaction with sodium tetrafluoroborate to give glucose-containing pyridinium CIL, as shown in Figure 14.5. Furthermore, it was used as both a catalyst and a solvent in the synthesis of optically active Tröger's base, as shown in Figure 14.6. This helped the Tröger's base to maintain its optical activity, as other methods of synthesizing Tröger's base resulted in racemization of its enantiomers as shown in Table 14.2.

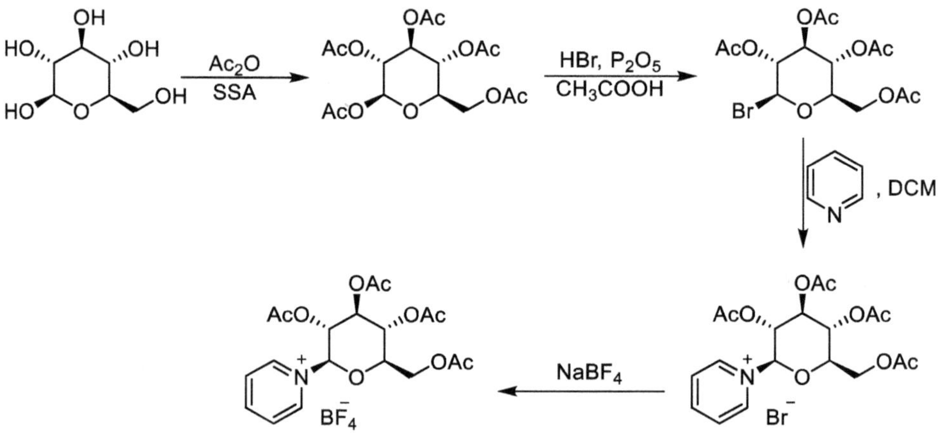

18

Figure 14.5: Synthesis of glucose-based CIL-containing pyridine [53].

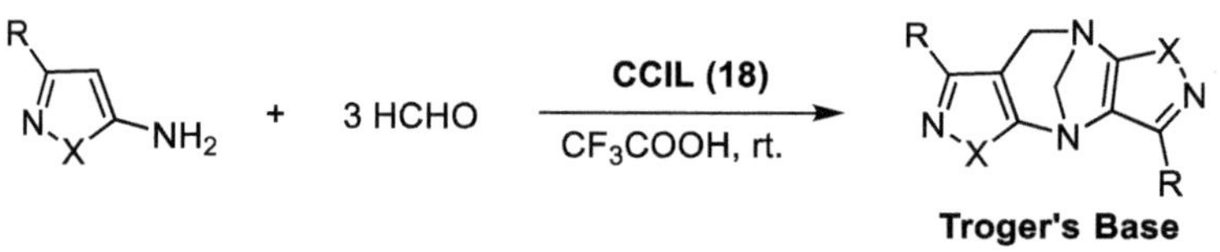

Figure 14.6: Synthesis of Tröger's base using glucose-based CIL-containing pyridine [53].

Kaur and Chopra synthesized six D-galactose containing DABCO (1,4-diazabicyclo[2.2.2]octane) CILs [54]. These were synthesized by protecting four OH groups of D-galactose with acetone in the presence of zinc chloride and H_2SO_4 followed by treating the primary OH group with iodine, which further quaternized with nucleophilic amine DABCO to give CCIL **19**. After that, CCIL **19** was reacted with various inorganic salts to obtain other CCILs (**20**–**24**) as shown in Figure 14.7. Furthermore, these obtained CILs (**19**–**24**) were used as chiral recognition agents for the enantioseparation of sodium salts of Mosher's acid and as organocatalysts in the reduction of aromatic prochiral ketones into second-

Table 14.2: Yield (%) and enantiomeric excess (%) of the various synthesized chiral Tröger's base [53].

R	X	t (h)	Yield (%)	ee (%)
CH_3	NPh	5	83	98
p-MePh	NPh	7	76	>99
p-OMePh	NPh	7	73	>99
CH_3	*p*-MePhN	6	71	90
Ph	NPh	7.5	70	84
H	NPh	9	79	44
p-ClPh	NPh	9	65	Racemic
CH_3	O	10	51	Racemic

ary alcohols. The secondary alcohols were obtained in low-to-moderate enantiomeric excess (5–17%) and high yields, as shown in Figure 14.8.

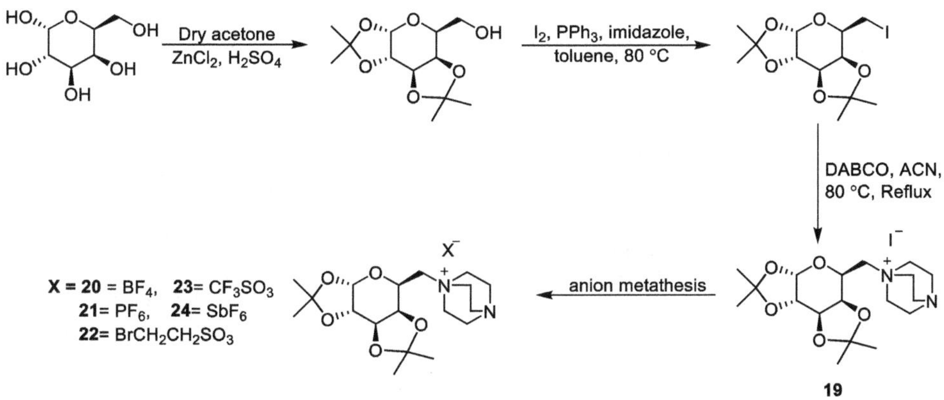

Figure 14.7: Synthesis of D-galactose containing DABCO (1,4-diazabicyclo[2.2.2]octane) CILs [54].

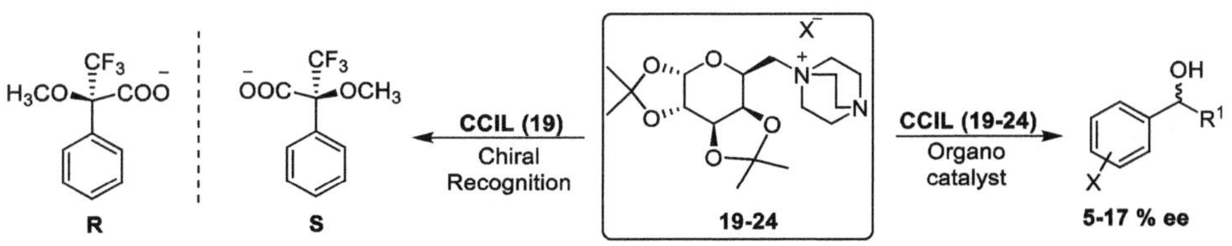

Figure 14.8: D-Galactose containing DABCO (1,4-diazabicyclo[2.2.2]octane) CILs were used as chiral recognition agents and organocatalysts [54].

Chrobok and colleagues published a significant study exploring the influence of CBILs on Diels-Alder reactions [55]. They synthesized new hydrogen-bond-rich ILs, including three CBILs derived from glucose. In these CBILs, glucose remains unmodified within the side chain of tetraalkylammonium ILs, synthesized through a three-step procedure (Figure 14.9). Each of these CBILs functions as a room-temperature IL (RTIL), with glass transition temperatures ranging from −30 to −14 °C and high thermal stability, decomposing at approximately 460 °C. Using a model reaction between cyclopentadiene and diethyl maleate (as shown in Figure 14.10), they systematically investigated a series of tetraalkylammonium ILs containing increasing numbers of hydroxyl groups in their side chains, with their glucose-based CILs featuring the highest hydroxyl content. While initial experiments employed equimolar quantities of ILs, the observed high reaction rates enabled the reduction of the catalyst loading to just 4%. Their investigation revealed a clear structure-activity relationship between the number of free hydroxyl groups and both reaction kinetics and selectivity. CBCILs containing two or more free hydroxyl groups exhibited superior endo/exo selectivity with ratios exceeding 12.2, while those with one or no hydroxyl groups showed significantly lower selectivity (ratios of 4.7 or below). Reaction rates also increased dramatically with higher hydroxyl content, with the glucose-based CIL demonstrating optimal performance.

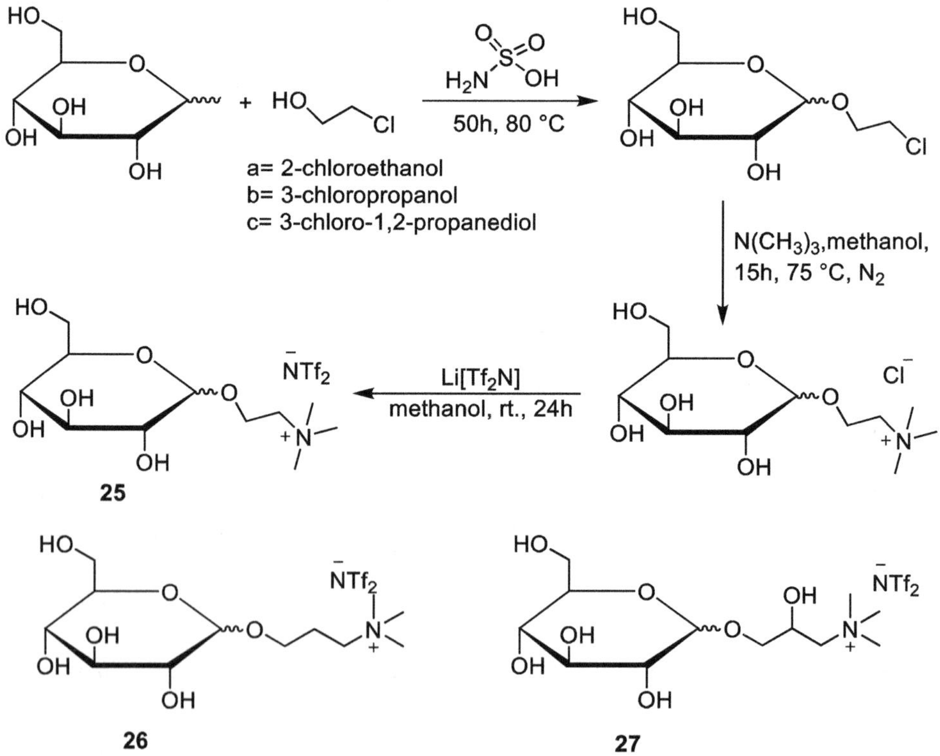

25

26

27

Figure 14.9: Synthesis of hydrogen-bond-rich ionic liquid CILs using D-glucose [55].

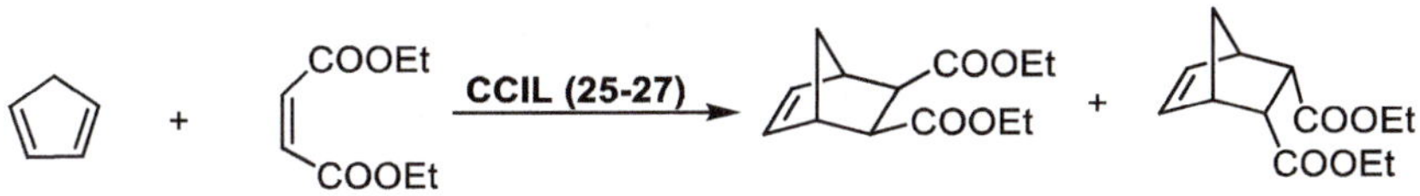

Figure 14.10: The model Diels-Alder reaction catalyzed by using hydrogen-bond-rich ionic liquid CILs using D-glucose (**25–27**) [55].

Gathergood and Bouquillon studied the selective catalytic hydrogenation of 1,5-cyclooctadiene using ILs derived from natural products [56]. Their work describes the one-step synthesis of various ILs derived from natural carboxylic acids through a reaction with tetrabutylammonium hydroxide, including derivatives of glucuronic and galacturonic acid (Figure 14.11). They conducted the reaction in water using $PdCl_2$ as a catalyst, as shown in Figure 14.11, under 1 atmosphere of H_2. In their initial run without ILs, they achieved complete conversion to cyclooctene. However, when they tried to reuse the catalyst, the selectivity decreased significantly, resulting in more cyclooctene formation. When they introduced ILs to the reaction system, they observed improved selective reduction of 1,5-cyclooctadiene to cyclooctene (Figure 14.12). They tested several tetrabutylammonium-based ILs with different natural anions: succinate, lactate, pyruvate, glucuronate, and galacturonate. Among these, the galacturonate-based CBIL performed best, maintaining both consistent conversion rates and high selectivity for cyclooctene production through 10 reaction cycles.

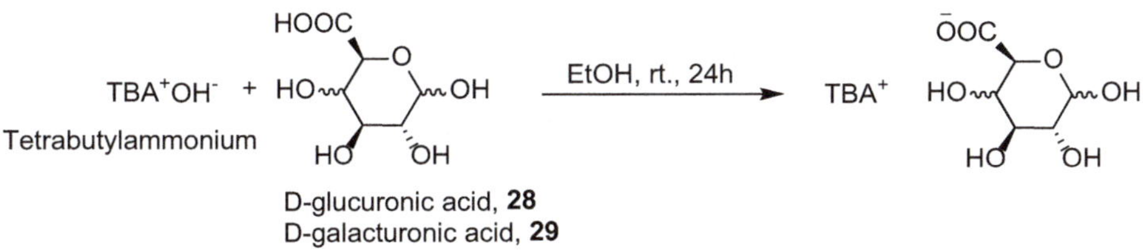

Figure 14.11: Synthesis of TBA-based CCILs [56].

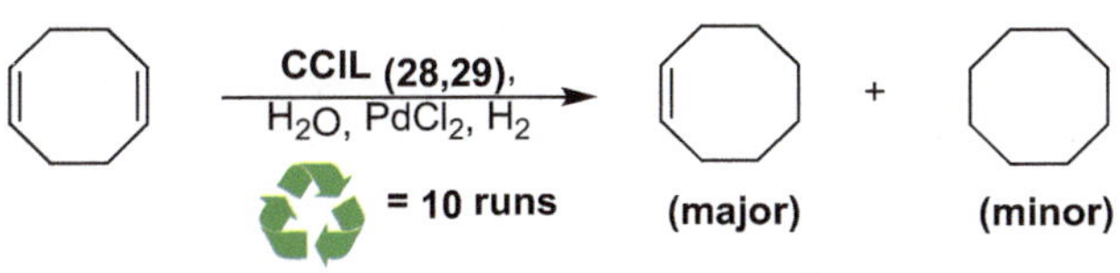

Figure 14.12: Selective hydrogenation of 1,5-cyclooctadiene in the presence of TBA-based CCILs [56].

Erfurt and the group synthesized quaternary ammonium salts with a modified D-glucose cation and bromide anion for this study. Various alkyl chain lengths were introduced at the carbon in the cation to assess the salts' lipophilicity effects, their toxicity, and biodegradability. For comparison, in one of the salts with a 16-carbon alkyl side chain, the Br^-

anion was replaced by NTf_2^-, resulting in an IL (Figure 14.13). Then they investigated the use of D-glucose-derived ILs as environmentally friendly catalysts for the industrial production of 2-chloro-1,3-butadiene (chloroprene), as shown in Figure 14.14, a key compound in the synthesis of cross-linked rubbers, which are widely used in applications ranging from adhesives to automotive components [57]. The study focused on promoting the use of carbohydrate-derived ILs, with an emphasis on developing catalysts that are less toxic and more biodegradable.

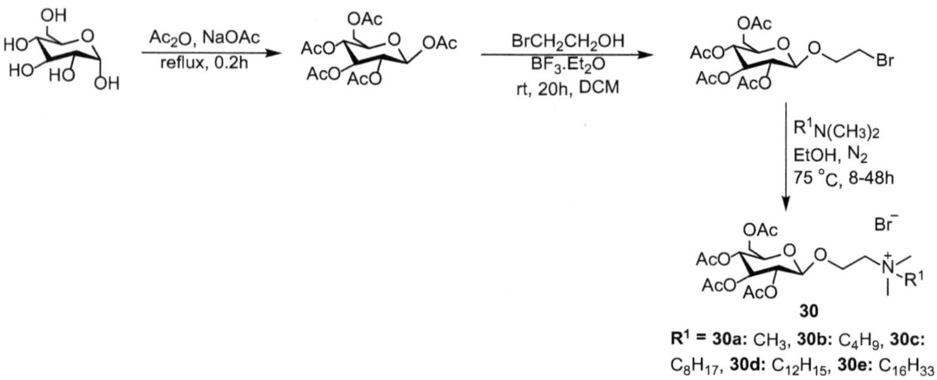

Figure 14.13: Synthesis of quaternary ammonium bromides using D-glucose as a substrate [57].

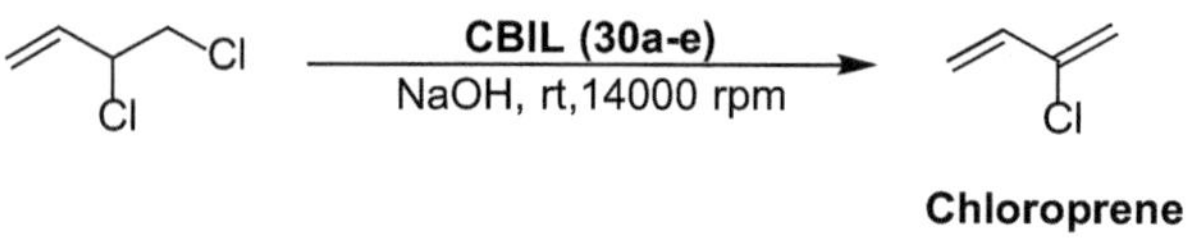

Figure 14.14: Dehydrohalogenation reaction of 3,4-dichlorobut-1-ene to yield chloroprene [57].

In another study exploring the versatility of CBILs, Singh and colleagues investigated the use of their arabinose-derived CBIL as an additive in the reaction between methylmagnesium chloride and benzaldehyde in THF [46]. Starting with methyl α-D-glucopyranoside, an Appel reaction was performed, and subsequently, *N*-substituted imidazoles were introduced at the 6-position, converting the sugars into glucosyl imidazolium iodide salts with yields ranging from 71% to 99%. These were then converted into a phosphate, which served as a leaving group to be displaced by *N*-methylimidazole (Figure 14.15). The resulting product is reported as a RTIL with a glass transition temperature of 16 °C. Although the reaction proceeded efficiently with excellent yields exceeding 95%, no enantioselectivity was observed. Nevertheless, the study demonstrated the notable stability of the CBILs under organometallic reaction conditions, as evidenced by its successful recovery post-reaction (Figure 14.16).

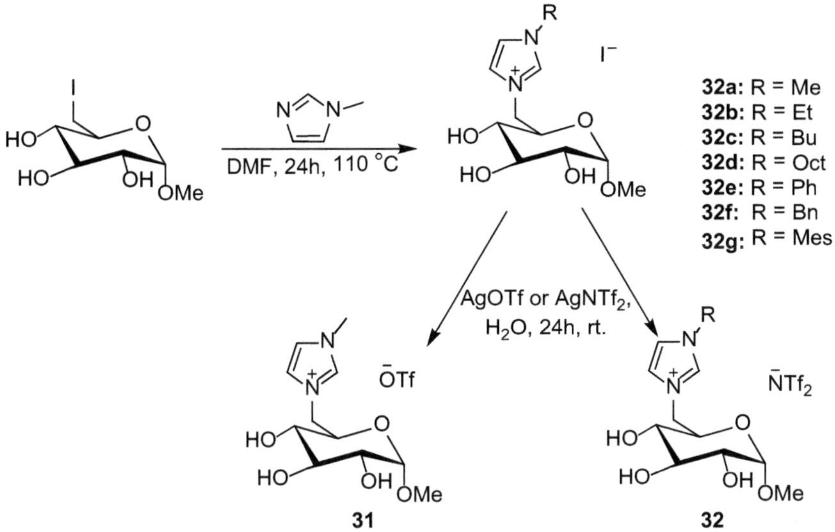

Figure 14.15: Synthesis of glucosyl imidazolium ionic liquids with varying *N*-substituted imidazoles and varying anions [46].

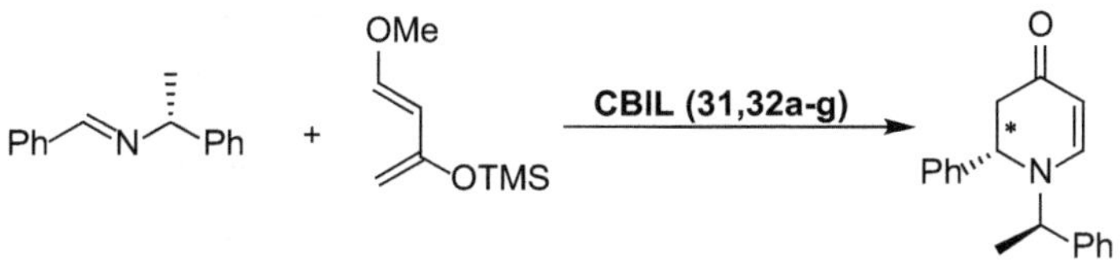

Figure 14.16: Model aza-Diels-Alder reaction catalyzed by glucosyl imidazolium ionic liquids [46].

Zhou et al. developed efficient glucopyranoside-substituted asymmetric bis(imidazolium) ILs to facilitate an environmentally friendly, efficient, and reusable Pd-catalyzed Suzuki reaction between arylboronic acids and 2-bromofluorene, as shown in Figure 14.18 [58]. Carbohydrate-containing mono(imidazolium) ILs were synthesized from *N*-carbohydrate-substituted imidazole and a haloarene or haloalkane. Symmetric bis(imidazolium) ILs, on the other hand, can be conveniently produced through a one-step method, as shown in Figure 14.17. The results demonstrated that the reaction could be conducted in air with a significantly reduced amount of Pd catalyst (0.01 mol%). The choice of ligand played a crucial role in the yield, with *bis*-imidazolium, featuring a more rigid linker between its cationic moieties, emerging as the most effective. The optimized conditions, $Pd(OAc)_2$/CIL, were successfully applied to synthesize a wide range of biaryl compounds. Finally, recycling tests showed favorable results, maintaining efficiency even after five cycles.

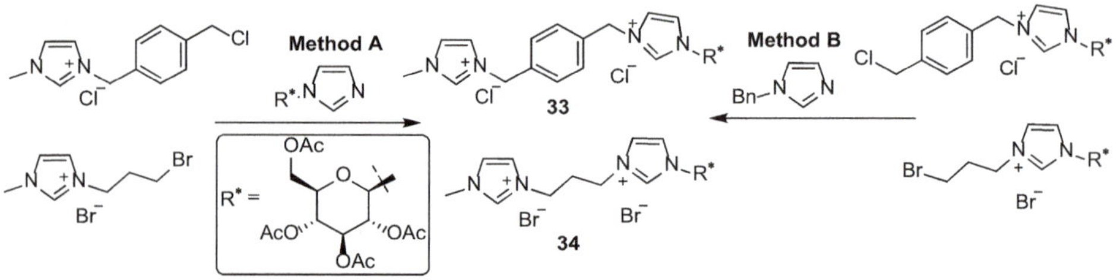

Figure 14.17: Synthesis of glucopyranoside-substituted asymmetric bis(imidazolium) ILs (33–34) [58].

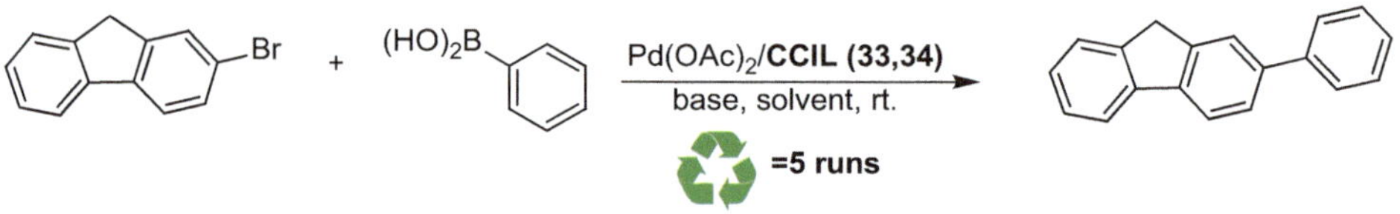

Figure 14.18: $Pd(OAc)_2$/CCIL (33–34)-catalyzed Suzuki reaction [58].

In 2020, Zhou and the group synthesized a series of Glu-IMSs, combining glucoside and imidazolium head groups with various substituents, with an isolated yield of 92–95% [59]. The synthetic method for Glu-IMSs is straightforward, using inexpensive, commercially available starting materials, as shown in Figure 14.19. When combined directly with $PdCl_2$ (molar ratio 1:2), Glu-IMSs form an efficient catalyst for Heck-Mizoroki reactions involving aryl iodides, bromides, and activated aryl chlorides, as shown in Figure 14.20. Additionally, the optimal $PdCl_2$/Glu-IMS catalytic system was applied to Suzuki reactions. A series of fluorine-centered organic functional materials (OFMs) were synthesized via Suzuki coupling between 2-bromofluorene, 2,7-dibromofluorene, 2,7-dibromo -9,9-dibutyl-9H-fluorene, 2,7-dibromo-9,9-dihexyl-9H-fluorene, and various aryl boronic acids (Figure 14.21). Among these, Glu-IMS featuring a coordinating oxygen functional group was found to be the most effective. High-resolution TEM analysis revealed the formation of Pd nanoparticles functionalized with IMS, with an average size of approximately 4.0 nm, which served as an efficient catalyst for promoting cross-coupling reactions in water. The Glu-IMS containing both –OH and NHC coordination sites exhibited the highest catalytic activity. Also, another study was conducted on a series of glucopyranoside-substituted imidazolium-based CILs (Glu-imi-CILs) as ligands, as shown in Figure 14.19, for Pd-catalyzed homo-coupling reactions of arylboronic acids in water (Figure 14.21) [60]. They synthesized symmetrical biphenyl compounds via $Pd_2(dba)_2$/Glu-imi-CIL-catalyzed homo-coupling of arylboronic acids. The ancillary ligand Glu-imi-CIL 1d, featuring a free OH group, was identified as the most effective ligand, and the reaction reached optimal productivity with a catalytic amount of NaClO. This new approach enabled the synthesis of 13 symmetric biaryl compounds in high isolated yields (85–99%) under green reaction conditions.

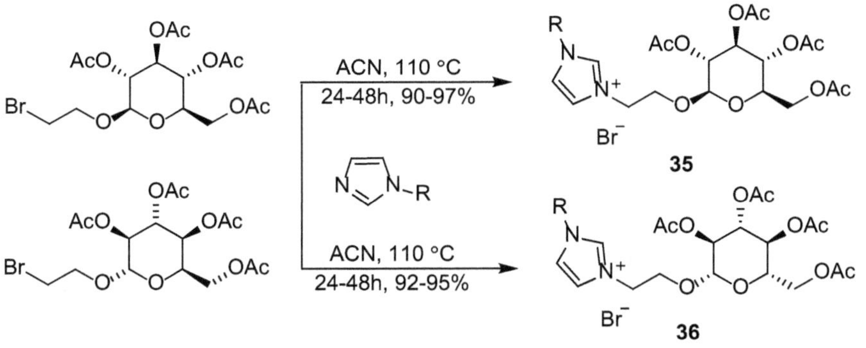

Figure 14.19: Synthesis of Glu-IMSs-based CILs [59, 60].

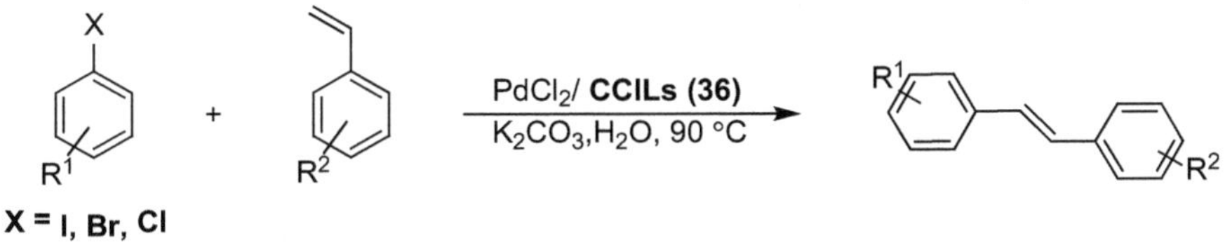

Figure 14.20: $PdCl_2$-catalyzed Heck-Mizoroki reactions in the presence of Glu-IMS-based CILs [60].

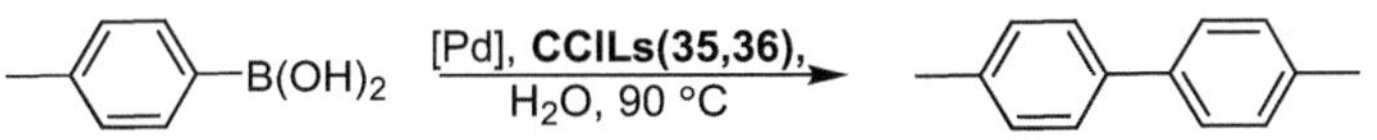

Figure 14.21: Pd-catalyzed Suzuki reaction in the presence of Glu-IMS-based CILs [59, 60].

Nemcsok *et al.* synthesized new carbohydrate-based ammonium and triazolium salts through a multi-step process [61]. The ammonium or triazolium functionalities were introduced at various positions on the altrose and glucose frameworks to investigate the structure–efficiency relationship of these new catalysts, as shown in Figure 14.22. The synthesized quaternary salts effectively promoted the alkylation of a glycine-derived Schiff base, although enantioselectivities were modest. However, the catalyst showed 5% asymmetric induction, indicating that such compounds hold potential as enantioselective catalysts, though further research and optimization are needed to enhance their efficiency, as shown in Figure 14.23. Additionally, ^{19}F NMR study of the sugar-based ammonium and triazolium Mosher's carboxylate salts confirmed that some of these compounds possess significant chiral recognition ability.

Jha and the group synthesized 1,2,3-triazolium ILs with sugar moiety at the 1,4-position using click chemistry, representing the first example of low-cost, carbohydrate-based ILs with a triazolium cation (Figure 14.24) [62]. Screening these ILs in Cu(I)-catalyzed amination showed that they serve as reusable coordinating solvents and ligands, effectively promoting the amination of iodo-, bromo-, and chlorobenzenes under

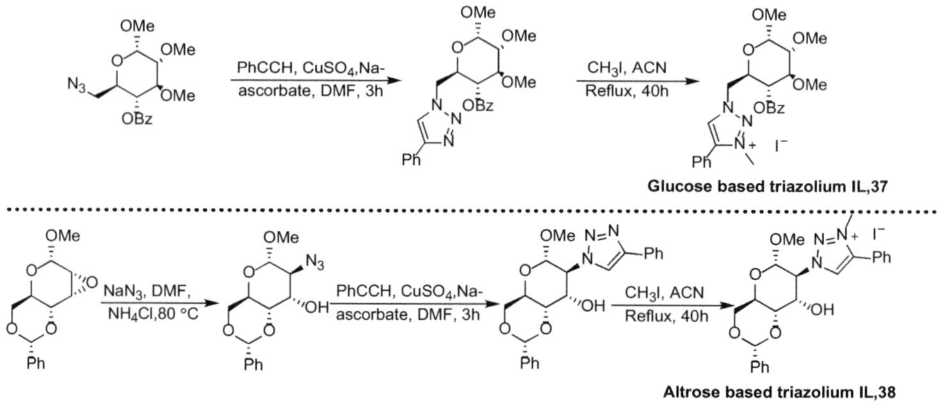

Figure 14.22: Synthesis of glucose- and altrose-based IL [61].

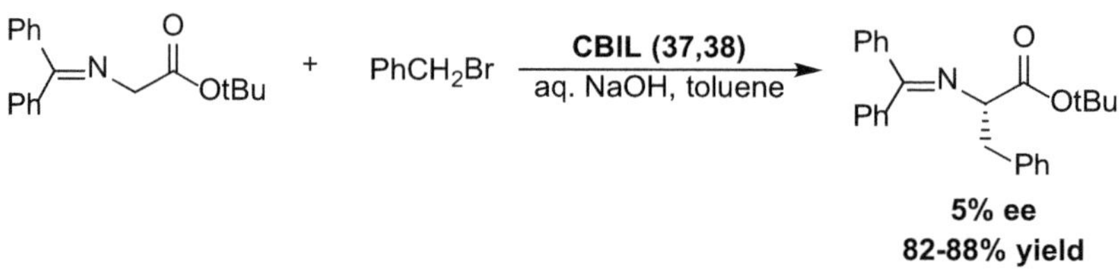

Figure 14.23: Asymmetric reaction catalyzed by glucose- and altrose-based IL [61].

mild conditions (Figure 14.25). The superior performance of IL-2 in amination may be attributed to free hydroxyl groups on the pyranose sugar, which enhance the stability of the Cu(I) species during the reaction.

14.2.2 Biomass utilization

CBILs have been widely utilized as organocatalysts for organic transformations, as mentioned above, because of their high solubility and low volatility. However, they may also be used as extractants and dissolving media for the processing of biomass.

In order to decrease the dependency on hydrocarbons from fossil fuels, it is highly desirable to value renewable biomass. In these procedures, waste materials from forestry or food production can be used to synthesize ILs, which can then be used as regenerable solvents. Therefore, the idea of a closed-loop biorefinery, which uses ILs generated from biomass feedstock to process biomass, was recently proposed [63]. Given this potential, it would be advantageous to create sugar-based ILs, which are far less toxic than the majority of other solvents used for the dissolution of biopolymers and can be utilized to break down difficult-to-dissolve polysaccharides [63].

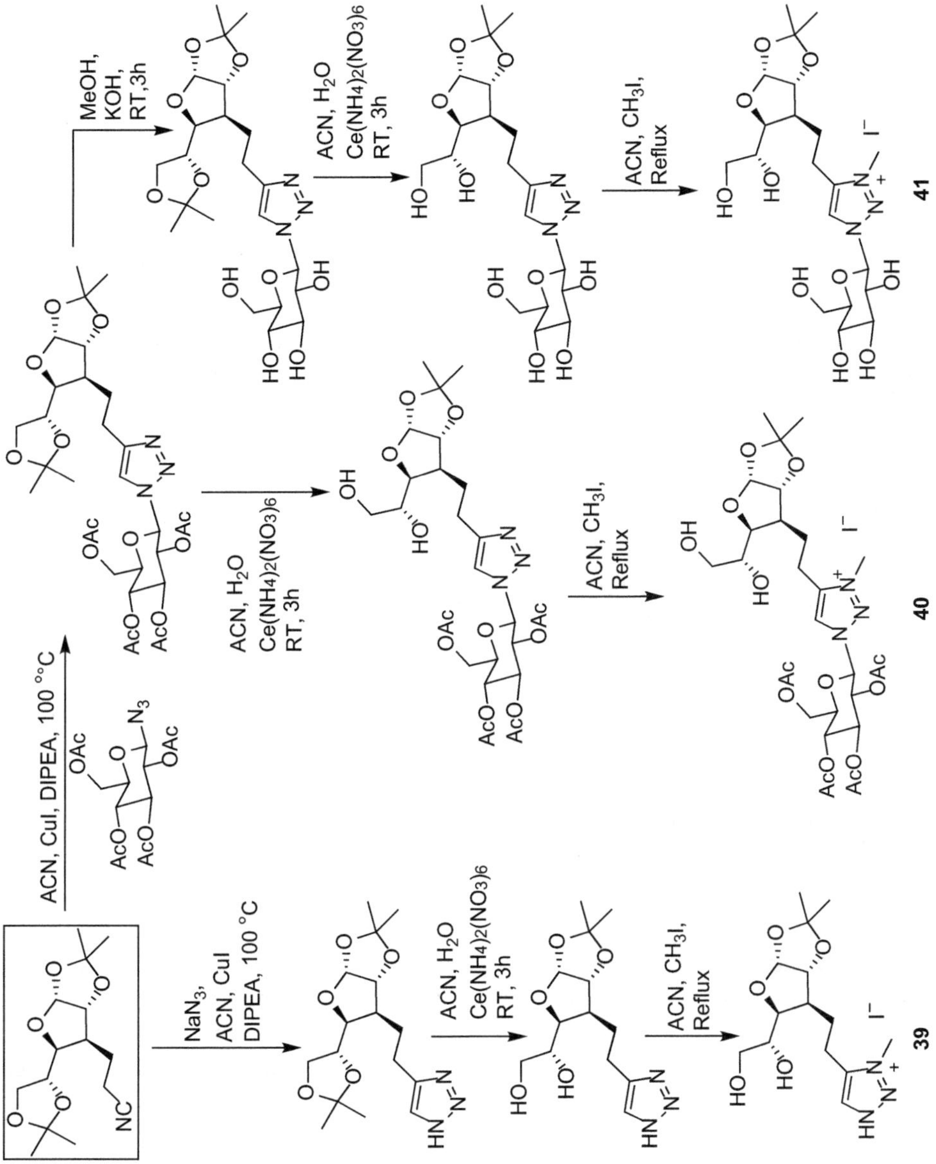

Figure 14.24: Synthesis of carbohydrate-based 1,2,3-triazolium IL [62].

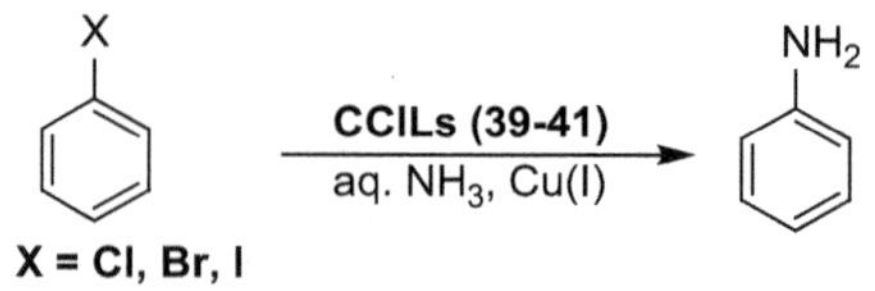

Figure 14.25: Amination reaction catalyzed by carbohydrate-based 1,2,3-triazolium IL [62].

Inspired by the ideas, Javed and coworkers synthesized *N,N*-diethyl-*N,N*-dimethylammonium gluconate by neutralizing diethyl dimethyl ammonium hydroxide with gluconic acid, as shown in Figure 14.26, and further utilized it to extract cellulose from oil palm lignocellulosic biomass [49]. Without any pre-treatment, more than half by weight (52%) of cellulose was extracted from the crude biomass utilizing *N,N*-diethyl-*N,N*-dimethylammonium gluconate **(42)** in 30 min at 25 °C. While conventional ILs have been extensively explored for their ability to dissolve polysaccharides [64], CBILs have received very less attention for biomass utilization.

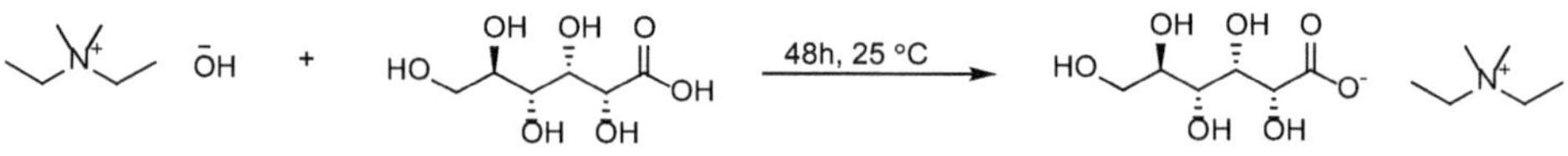

42

Figure 14.26: Synthesis of *N,N*-diethyl-*N,N*-dimethylammonium gluconate [49].

Additionally, CBILs can be used as solvents with strong hydrogen-bonding capabilities and few environmental effects. Furthermore, the prospective application of CBILs as extractants should be investigated in order to capitalize on the powerful coordinating properties of hydroxyl groups. Billeci and coworkers recently synthesized gluconate-based ammonium and phosphonium ILs, as shown in Figure 14.27, including gelators, which develop carbohydrate-based gels with a strong hydrogen-bond network that are then utilized for fuel desulfurization [65]. About 70–80% of benzothiophene (BT) and dibenzothiophene (DBT) were extracted from the fuel, which are often regarded as refractory chemicals in desulfurization.

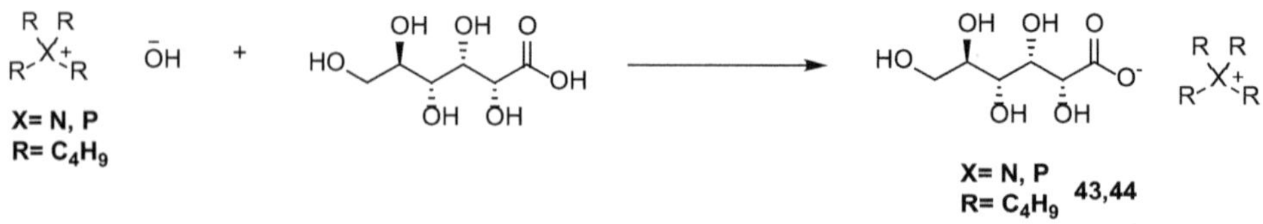

43,44

Figure 14.27: Synthesis of gluconate-based ammonium and phosphonium ILs [65].

14.3 Conclusion

Manufacturing ILs from abundant, affordable, and renewable carbohydrates offers a promising path toward sustainable chemical production amidst declining fossil resources. CBILs can be synthesized through a variety of methods that are increasingly appealing from both economic and sustainability perspectives. Although the purification of intermediates is a challenge in carbohydrate IL synthesis, recent methods that replace column chromatography with simple extraction using environmentally friendly organic solvents mark a significant step forward in green chemistry.

Carbohydrate-derived ILs, with their unique characteristics, provide an alternative to amino acid derivatives, which are more commonly used in bio-IL synthesis. Notable features of carbohydrate ILs include their hydrogen-bond-rich structure, which enhances their coordinating ability and solubility in polar solvents. Additionally, their chirality, low toxicity, and high biodegradability make them excellent candidates for applications in catalysis, biomedicine, and environmental technologies. The range of potential uses for carbohydrate-based ILs has broadened considerably, now encompassing roles as solvents, extractants, and materials for energy conversion and storage.

A key advantage of ILs is their tunability, allowing specific properties to be tailored by choosing particular cations, anions, side chains, or functional groups. The class of carbohydrate ILs and salts is still underexplored, emphasizing the need for more data on how structural variations impact their toxicological profiles. This knowledge will ultimately support more precise, targeted design within this class of bio-ILs.

Abbreviations

ILs	Ionic liquids
RTILs	Room-temperature ILs
PILs	Polymerized ILs
MILs	Magnetic ILs
CILs	Chiral ILs
CBILs	Carbohydrate-based ionic liquids
CCILs	Carbohydrate-based CILs

References

[1] Greaves TL, Drummond CJ. Protic ionic liquids: Properties and applications. Chem Rev, 2008, 108, 206–237.

[2] Zhao H, Malhotra SV. Applications of ionic liquids in organic synthesis, 2002.

[3] Aggarwal A, Singh A, Chopra HK. Merrifield resin supported ionic liquids: Catalytic applications in organic synthesis. Curr Org Chem, 2023, 27, 130–152.

[4] Aggarwal A, Chopra HK. Cyclodextrin functionalized ionic liquids: Synthesis and application in analytical chemistry. Crit Rev Anal Chem, 2024, 1–26.
[5] Welton T. Room-temperature ionic liquids. Solvents for synthesis and catalysis. Chem Rev, 1999, 99, 2071–2084.
[6] Chaturvedi D. Recent developments on task specific ionic liquids. Curr Org Chem, 2011, 15, 1236–1248.
[7] Gabriel S, Weiner J. Ueber einige abkömmlinge des propylamins. Berichte der deutschen chemischen Gesellschaft, 1888, 21, 2669–2679.
[8] Gibson HW. Perspective for a special issue of polymer reviews on: Ionic liquids and their derivatives in polymer science and engineering. Polym Rev, 2009, 49, 289–290.
[9] Wilkes JS, Levisky JA, Wilson RA, Hussey CL. Dialkylimidazolium chloroaluminate melts: A new class of room-temperature ionic liquids for electrochemistry, spectroscopy and synthesis. Inorg Chem, 1982, 21, 1263–1264.
[10] Qureshi ZS, Deshmukh KM, Bhanage BM. Applications of ionic liquids in organic synthesis and catalysis. Clean Technol Environ Policy, 2014, 16, 1487–1513.
[11] Wang H, Gurau G, Rogers RD. Ionic liquid processing of cellulose. Chem Soc Rev, 2012, 41, 1519–1537.
[12] Ray A, Saruhan B. Application of ionic liquids for batteries and supercapacitors. Materials, 2021, 14, 2942.
[13] Liu N, Luo F, Wu H, Liu Y, Zhang C, Chen J. One-step ionic-liquid-assisted electrochemical synthesis of ionic-liquid-functionalized graphene sheets directly from graphite. Adv Funct Mater, 2008, 18, 1518–1525.
[14] Stojanovic A, Keppler BK. Ionic liquids as extracting agents for heavy metals. Sep Sci Technol, 2012, 47, 189–203.
[15] Lei Z, Chen B, Koo Y-M, MacFarlane DR. Introduction: Ionic Liquids. ACS Publications, 2017, pp. 6633–6635.
[16] Singh A, Chopra HK. New benzimidazolium-based chiral ionic liquids: Synthesis and application in enantioselective sodium borohydride reductions in water. Tetrahedron Asymm, 2016, 27, 448–453.
[17] Singh A, Chopra HK. Synthesis, characterization and applications of some novel DMAP-based chiral ionic liquids. J Mol Liq, 2018, 266, 106–111.
[18] Gaida B, Brzęczek-Szafran A. Insights into the properties and potential applications of renewable carbohydrate-based ionic liquids: A review. Molecules, 2020, 25, 3285.
[19] Handy ST, Okello M, Dickenson G. Solvents from biorenewable sources: Ionic liquids based on fructose. Org Lett, 2003, 5, 2513–2515.
[20] Kaur N, Singh A, Chopra HK. Exploring low-cost natural precursors as chiral building blocks in synthesis: Chiral carbohydrate-ionic liquids. Mini Rev Org Chem, 2018, 15, 208–219.
[21] Jopp S. Carbohydrate based ionic liquids (CHILs): Synthesis and applications. Eur J Org Chem, 2020, 2020, 6418–6428.
[22] Lambrecht S, Schröter H, Pohle H, Jopp S. Swelling behavior of novel hydrogels produced from glucose-based ionic monomers with varying cross-linkers. ACS Omega, 2024, 9, 5418–5428.
[23] Correia DM, Fernandes LC, Fernandes MM, Hermenegildo B, Meira RM, Ribeiro C, Ribeiro S, Reguera J, Lanceros-Méndez S. Ionic liquid-based materials for biomedical applications. Nanomaterials, 2021, 11, 2401.
[24] Choudhary G, Dhariwal J, Saha M, Trivedi S, Banjare MK, Kanaoujiya R, Behera K. Ionic liquids: Environmentally sustainable materials for energy conversion and storage applications. Environ Sci Pollut Res, 2024, 31, 10296–10316.
[25] Hassanpour M, Shahavi MH, Heidari G, Kumar A, Nodehi M, Moghaddam FD, Mohammadi M, Nikfarjam N, Sharifi E, Makvandi P. Ionic liquid-mediated synthesis of metal nanostructures: Potential application in cancer diagnosis and therapy. J Ionic Liq, 2022, 2, 100033.

[26] Liu C, Raza F, Qian H, Tian X. Recent advances in poly (ionic liquid) s for biomedical application. Biomater Sci, 2022, 10, 2524–2539.
[27] Yang Q, Zhang Z, Sun X-G, Hu Y-S, Xing H, Dai S. Ionic liquids and derived materials for lithium and sodium batteries. Chem Soc Rev, 2018, 47, 2020–2064.
[28] Brzeczek-Szafran A, Erfurt K, Blacha-Grzechnik A, Krzywiecki M, Boncel S, Chrobok A. Carbohydrate ionic liquids and salts as all-in-one precursors for N-doped carbon. ACS Sustain Chem Eng, 2019, 7, 19880–19888.
[29] De Jesus SS, Maciel Filho R. Are ionic liquids eco-friendly? Renew Sust Energ Rev, 2022, 157, 112039.
[30] Zullo V, Iuliano A, Guazzelli L. Sugar-based ionic liquids: Multifaceted challenges and intriguing potential. Molecules, 2021, 26, 2052.
[31] Jopp S, Fleischhammer T, Lavrentieva A, Kara S, Meyer J. Synthesis, biocompatibility, and antimicrobial properties of glucose-based ionic liquids. RSC Sustain, 2023, 1, 1751–1764.
[32] Blaga AC, Tucaliuc A, Kloetzer L. Applications of ionic liquids in carboxylic acids separation. Membranes, 2022, 12, 771.
[33] Szelwicka A, Erfurt K, Jurczyk S, Boncel S, Chrobok A. Outperformance in acrylation: Supported D-glucose-based ionic liquid phase on MWCNTs for immobilized lipase B from candida antarctica as catalytic system. Materials, 2021, 14, 3090.
[34] Khorasani F, Ranjbar-Karimi R, Marra A. Recent advances in the synthesis and applications of ionic liquids derived from natural products. Synthesis, 2024.
[35] Wen X, Wang J, He L, Wei B, Xie Y. Synthesis of glucopyranoside benzimidazolium-based ionic liquids for Pd-catalyzed aqueous Suzuki reaction. J Carbohydr Chem, 2022, 41, 369–384.
[36] Sabbasani RR, Krishnaraj P, Bhulakshmi S, Cheng LC, Chacko TM. Sustainable synthesis of D-Glucose based IL: A renewable organo catalyst for the efficient synthesis of dihydropyrano coumarins. Asian J Org Chem, e202400202.
[37] Sheldon RA. Biocatalysis in ionic liquids: State-of-the-union. Green Chem, 2021, 23, 8406–8427.
[38] Grassiri B, Mezzetta A, Maisetta G, Migone C, Fabiano A, Esin S, Guazzelli L, Zambito Y, Batoni G, Piras AM. Betaine-and l-carnitine-based ionic liquids as solubilising and stabilising agents for the formulation of antimicrobial eye drops containing diacerein. Int J Mol Sci, 2023, 24, 2714.
[39] Baghersaei S, Mokhtari B, Soulgani BS, Pourreza N, Veiskarami S. Investigation on asphaltene dispersion activity of tetrabutylammonium carboxylate ionic liquids. Energy Fuels, 2023, 37, 7085–7093.
[40] Dong L, Xia J, Song H, Yang J, Sun Y, Li X, Zhu X. Chiral ionic liquids synthesis and their applications in racemic drug separation and analysis. J Mol Liq, 2023, 371, 120982.
[41] Lehmann P, Jopp S. Novel glucosylimidazolium ionic-liquid-supported novozym 435 catalysts–a proof of concept for an acrylation reaction. ChemistryOpen, 2022, 11, e202200135.
[42] Furuishi T, Taguchi S, Wang S, Fukuzawa K, Yonemochi E. The development and characterization of novel ionic liquids based on mono-and dicarboxylates with meglumine for drug solubilizers and skin permeation enhancers. Pharmaceutics, 2024, 16, 322.
[43] Garg P, Reddy SR. Biomass-derived sugar ionic liquid as a sustainable organocatalyst: An efficient synthesis of functionalized dihydropyrano coumarins. Asian J Org Chem, 2022, 11, e202200322.
[44] Ghirardello M, Costantini M, Vecchi A, Pacifico S, Pazzi D, Castiglione F, Mele A, Marra A. Synthesis of chiral ionic liquids from natural monosaccharides. Eur J Org Chem, 2022, 2022, e202200100.
[45] Komabayashi M, Okushiba S, Nokami T, Jopp S. Scope and limitations in the synthesis of glucosamine-based ionic liquids. Asian J Org Chem, 2023, 12, e202300093.
[46] Komabayashi M, Jopp S. Glucose-based ionic liquid organocatalysts for asymmetric aza-Diels-Alder reactions. ChemistryEurope, 2024, e202400052.
[47] Barteczko N, Grymel M, Erfurt K, Jakóbik-Kolon A, Bręczek-Szafran A, Chrobok A. Aqueous olefin metathesis with D-glucose-based bromides giving more efficient catalysis. J Mol Liq, 2023, 386, 122484.

[48] Hong M, Miao Z, Xu X, Zhang Q. Magnetic iron oxide nanoparticles immobilized with sugar-containing poly (ionic liquid) brushes for efficient trapping and killing of bacteria. ACS Appl Bio Mater, 2020, 3, 3664–3672.
[49] Javed F, Ullah F, Akil HM. Synthesis, characterization and cellulose dissolution capabilities of ammonium-based room temperature ionic liquids (RTILs). Pure Appl Chem, 2018, 90, 1019–1034.
[50] Liu Y, Nie Y, Lu X, Zhang X, He H, Pan F, Zhou L, Liu X, Ji X, Zhang S. Cascade utilization of lignocellulosic biomass to high-value products. Green Chem, 2019, 21, 3499–3535.
[51] Nguyen Van Buu O, Vo-Thanh G. Synthesis of novel chiral ammonium-based ionic liquids derived from isosorbide and their applications in an asymmetric aza diels-alder reaction. Lett Org Chem, 2007, 4, 158–167.
[52] Jayachandra R, Reddy SR, Balakrishna. Natural sugars derived chiral ionic liquids for asymmetric michael addition reaction. ChemistrySelect, 2016, 1, 2341–2343.
[53] Yuan R, Wang Y-J, Fang Y, Ge W-H, Lin W, Li M-Q, Xu J-B, Wan Y, Liu Y, Wu H. The first direct synthesis of chiral Tröger's bases catalyzed by chiral glucose-containing pyridinium ionic liquids. Chem Eng J, 2017, 316, 1026–1034.
[54] Kaur N, Chopra HK. Synthesis and applications of carbohydrate based chiral ionic liquids as chiral recognition agents and organocatalysts. J Mol Liq, 2020, 298, 111994.
[55] Erfurt K, Wandzik I, Walczak K, Matuszek K, Chrobok A. Hydrogen-bond-rich ionic liquids as effective organocatalysts for Diels–Alder reactions. Green Chem, 2014, 16, 3508–3514.
[56] Ferlin N, Courty M, Gatard S, Spulak M, Quilty B, Beadham I, Ghavre M, Haiß A, Kümmerer K, Gathergood N. Biomass derived ionic liquids: Synthesis from natural organic acids, characterization, toxicity, biodegradation and use as solvents for catalytic hydrogenation processes. Tetrahedron, 2013, 69, 6150–6161.
[57] Erfurt K, Markiewicz M, Siewniak A, Lisicki D, Zalewski M, Stolte S, Chrobok A. Biodegradable surface active D-glucose based quaternary ammonium ionic liquids in the solventless synthesis of chloroprene. ACS Sustain Chem Eng, 2020, 8, 10911–10919.
[58] Zhou Z, Yuan Y, Xie Y, Li M. Green, efficient and reusable bis (imidazolium) ionic liquids promoted Pd-catalyzed aqueous Suzuki reaction for organic functional materials. Catal Lett, 2018, 148, 2696–2702.
[59] Zhou Z, Xie Q, Zhou X, Yuan Y, Pan Y, Lu D, Du Z, Xue J. Synthesis of glucoside-based imidazolium salts for Pd-catalyzed cross-coupling reaction in water. Carbohydr Res, 2020, 496, 108079.
[60] Zhou Z, Li J, Wu Y, Yuan Y, Kong L, Xue J, Huang Z. Glucopyranoside-substituted imidazolium-based chiral ionic liquids for Pd-catalyzed homo-coupling of arylboronic acids in water. J Carbohydr Chem, 2020, 39, 288–299.
[61] Nemcsok T, Rapi Z, Bagi P, Bakó P. Synthesis and application of novel carbohydrate-based ammonium and triazolium salts. Synth Commun, 2019, 49, 2388–2400.
[62] Jha AK, Jain N. Synthesis of glucose-tagged triazolium ionic liquids and their application as solvent and ligand for copper (I) catalyzed amination. Tetrahedron Lett, 2013, 54, 4738–4741.
[63] Hulsbosch J, De Vos DE, Binnemans K, Ameloot R. Biobased ionic liquids: Solvents for a green processing industry? ACS Sustain Chem Eng, 2016, 4, 2917–2931.
[64] Zhang J, Wu J, Yu J, Zhang X, He J, Zhang J. Application of ionic liquids for dissolving cellulose and fabricating cellulose-based materials: State of the art and future trends. Mater Chem Front, 2017, 1, 1273–1290.
[65] Billeci F, D'Anna F, Gunaratne HN, Plechkova NV, Seddon KR. "Sweet" ionic liquid gels: Materials for sweetening of fuels. Green Chem, 2018, 20, 4260–4276.

Ali Ramazani*, Seyedmohammad Hosseininezhad, and Yavar Ahmadi

Chapter 15 Advances in carbohydrate-supported nanostructured material in organic transformations

15.1 Introduction

Sugars, starches, and natural fibers, which are primarily found in plants, consist largely of carbohydrates, making them one of the most prevalent organic materials found in the natural world. Carbohydrates, recognized as macromolecules, play a crucial role in providing structural integrity to the cells of plants and to arthropods, including crabs, lobsters, and shrimp [1]. Derived from chitin through a deacetylation process in an alkaline environment, chitosan is an environmentally friendly linear polysaccharide made up of glucosamine and is biodegradable [2, 3]. Alginate, a plentiful carbohydrate found in the cell walls of brown macroalgae, is composed of guluronic and mannuronic acid units. These units are connected through β-1,4-glycosidic bonds. As a versatile biopolymer, alginate is employed across various sectors such as drug delivery systems, the food industry as an additive, and in the textile industry for printing. Of late, it's gaining attention as a potential biofuel source aimed at generating beneficial compounds [4]. Pectin, a naturally occurring polysaccharide found in the primary cell walls of plants, is water-soluble and is made up of linear chains of α-D-galacturonic acid, which are connected by 1,4-linkages [5]. Cellulose stands out as a biogenic substance notable for its intriguing structure and properties. It ranks among the most significant and abundant sources of renewable organic polymers, with an annual production of approximately 1.5×10^{12} tons [6–8]. Gums are organic, water-soluble substances and non-starch-based polysaccharides. They typically exhibit a complex, heterogeneous structure and have substantial molecular weights. The primary gums sourced from trees include gum tragacanth, gum kondagogu, gum arabic, ghatti, and gum karaya [9]. Chitin is derived from the hard

***Corresponding author: Ali Ramazani,** The Organic Chemistry Research Laboratory (OCRL), Department of Chemistry, Faculty of Science, University of Zanjan, Zanjan 45371-38791, Iran; The Convergent Sciences and Technologies Laboratory (CSTL), Research Institute of Modern Biological Techniques (RIMBT), University of Zanjan, Zanjan 45371-38791, Iran, e-mail: aliramazani@znu.ac.ir
Seyedmohammad Hosseininezhad, The Organic Chemistry Research Laboratory (OCRL), Department of Chemistry, Faculty of Science, University of Zanjan, Zanjan 45371-38791, Iran
Yavar Ahmadi, Department of Chemistry Education, Farhangian University, P.O. Box 14665-889, Tehran, Iran

https://doi.org/10.1515/9783111447094-015

outer shells of marine creatures such as shrimps, crabs, and squids [2]. Starch, a cost-effective and sustainable polysaccharide, is made up of numerous glucose molecules connected through glycosidic linkages. It is produced by plants through the process of photosynthesis, utilizing water and CO_2. Common sources like wheat, rice, and potatoes are rich in starch content [10].

Catalytic organic transformations are widely employed in chemical manufacturing due to their essential role in numerous procedures. An expansive review of carbohydrate-based nanostructures, which facilitate a diverse array of organic transformations that contribute to eco-friendly methods, is both appropriate and necessary [11]. In this chapter, we aim to explore significant carbohydrates, including starch, cellulose, pectin, gum, alginate, chitin, chitosan, among others, that are integrated with nanostructures. We will also delve into the latest advancements in carbohydrate-supported nanostructures, such as nanoparticles, quantum dots, metal-organic frameworks, and so on, and their role in facilitating organic reactions, particularly in the synthesis of diverse heterocyclic compounds.

15.2 Chitosan-based nanostructures

In 2023, research outlined the extraction of carbon quantum dots (CQDs) from the fruits of *Momordica charantia* using microwave irradiation, adhering to the principles of green chemistry. It also detailed the development of three distinct catalysts: silver nanoparticles embedded in chitosan, a composite of silver nanoparticles with chitosan and *M. charantia* fruit powder, and a composite of silver nanoparticles with chitosan and CQDs. The support structures for the catalysts were created by embedding CQDs or *M. charantia* powder within a network of chitosan beads. Ag metallic particles were then attached to the chitosan beads, which had been cross-linked with glutaraldehyde, using an $AgNO_3$ solution. These composite Ag catalysts were subsequently employed in the reduction of harmful nitroaromatic compounds, including 1,2-diamino-4-nitrobenzene, 2-nitroaniline, 2,4-dinitrophenol, and 4-nitrophenol. Future research may explore the application of Ag(0)-chitosan-CQDs systems in organic synthesis processes. This includes testing their effectiveness in C–C coupling reactions, such as Suzuki-Miyaura and Heck reactions, as well as in oxidation and hydrogenation reactions, or in the degradation of organic pollutants [12].

A sustainable method was developed for the in situ immobilization of Pd NPs using ultrasound. This method involved the use of biodegradable chitosan/agarose-modified ferrite nanoparticles (Fe_3O_4@CS-Agarose/Pd). The catalytic efficiency of the corresponding nanocomposite was examined through the ultrasound-assisted oxidation process of various alcohol derivatives (Figure 15.1). This process successfully transformed a broad spectrum of primary alcohols (**1**) into aldehyde derivatives (**2**), achieving yields ranging from moderate to high. The durability of the corresponding nanocomposite was demonstrated through its ability to be reused successfully at least ten times, as well as through leaching and heterogeneity tests. The use of ultrasonic treatment in both the catalyst's synthesis and the oxidation process markedly increased the speed of reactions when compared to traditional heating methods [13].

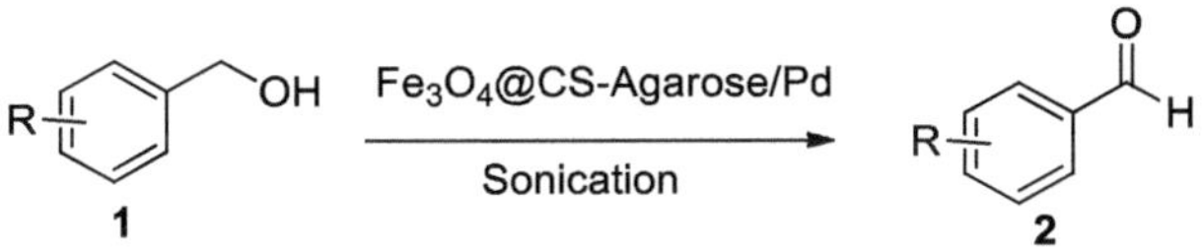

Figure 15.1: Oxidation of benzyl alcohols under sonication.

In a study detailing the development of a palladium-based catalyst anchored on magnetic chitosan (Pd@IO-Chitosan), designed for use in the Suzuki-Miyaura carbon-carbon coupling reaction (Figure 15.2), the refined reaction parameters, employing Pd@IOChitosan as the catalyst for the Suzuki coupling of iodobenzene (**3**) and phenylboronic acid (**4**), resulted in a high yield (over 99%) of the desired products (**5**) and significant catalytic efficiency, with a Turn Over Number (TON) of 1.85×10410^4 under gentle conditions. The catalyst Pd@IO-Chitosan illustrated a strong super paramagnetic characteristic, measuring 39.037 emu/g, which enabled its efficient retrieval and subsequent reuse from the reaction mixture. Following the completion of the 10th cycle, its magnetic strength diminished to 14.0 emu/g; however, this level remained adequate to allow for its magnetic separation [14].

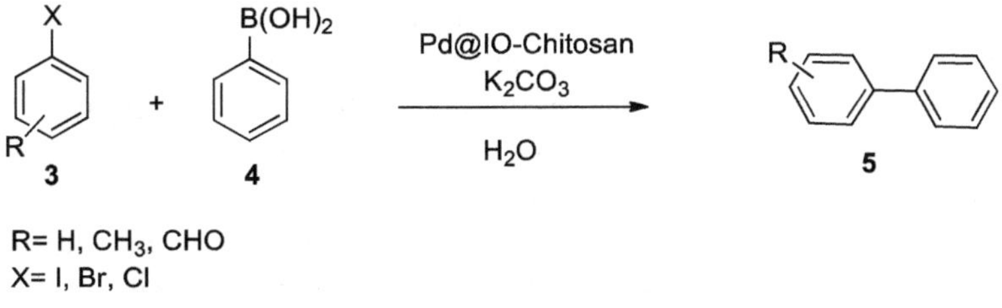

Figure 15.2: Suzuki cross-couplings of aryl halide derivatives (**3**) catalyzed by Pd@IO-chitosan.

A study introduced an eco-friendly and innovative approach for the synthesis of Fe_3O_4 MNPs (metallic nanoparticles). These were combined with chitosan (CS) and 5-amino-1H-tetrazole, which possesses an extended chain of ((3-chloropropyl)trimethoxysilane), and was employed to immobilize the palladium catalyst, referred to as Fe_3O_4–CS@tet-Pd(II). The Suzuki-Miyaura cross-coupling reaction was conducted to synthesize biaryls (**8**) by reacting aryl halides (**6**) with phenylboronic acid (**7**) (Figure 15.3). This process utilized the Fe_3O_4–CS@tet-Pd(II) catalyst, K_2CO_3, and a mixture of EtOH:H_2O as an environmentally friendly solvent at 90 °C. A variety of substituted aryl halides were tested to produce the desired biaryl compounds (**8**). The corresponding nanostructured catalyst offered multiple benefits: it yielded high product outputs and was created from natural, cost-effective, and harmless materials. Moreover, its remarkable performance in eco-friendly solvents and the straightforward recovery method using a magnetic field highlight its potential as an ideal option for sustainable chemistry practices [15].

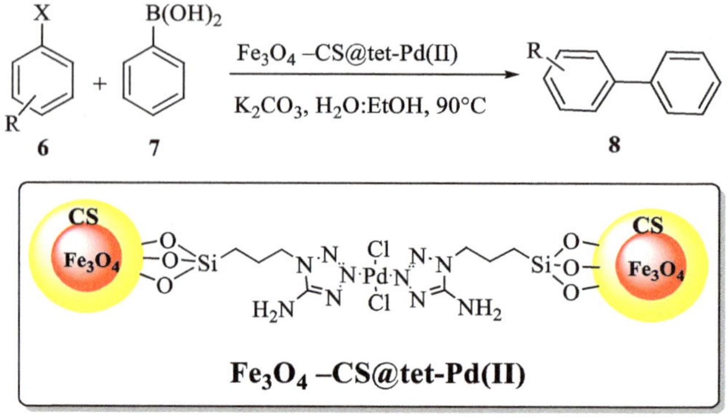

Figure 15.3: Suzuki-Miyaura cross-coupling reaction of aryl halide derivatives with phenylboronic acid catalyzed by Fe_3O_4–CS@tet-Pd(II).

A team innovated an in situ synthesized Pd NP, which is intricately adorned on a chitosan-coated SiO_2-Fe_3O_4 nanocomposite catalyst. This novel catalyst was specifically designed for the Suzuki-Miyaura coupling reaction and the reduction of 4-nitrophenol (9) (Figure 15.4). The authors evaluated the catalyst's effectiveness in producing a range of biaryls (**11**) through a straightforward process. In a similar vein, they successfully expedited the reduction of the organic pollutant 4-nitrophenol (**9**) to the less harmful 4-aminophenol (**10**). The corresponding catalyst boasted easy recovery with a magnet and demonstrated the ability to be reused up to eight times in both the coupling and reduction reactions, maintaining consistent reactivity throughout [16].

In a conducted research, Fe_3O_4/chitosan/pumice hybrid beads were innovatively crafted to serve as a novel stabilizing agent. On this newly formed stabilizer, Pd NPs were effectively affixed without the need for any harmful reducing agents. The study further explored the efficacy of PdNPs@Fe_3O_4/chitosan/pumice hybrid beads as a heterogeneous catalyst in the cyanation of various aryl halides (**12**) using $K_4[Fe(CN)_6]$ (Figure 15.5). These reactions displayed the high catalytic efficiency of the beads, successfully transforming aryl halides (**12**) into the targeted benzonitriles (**13**) with impressive yields ranging from 80% to 98%. Owing to their magnetically separable properties, the Pd NPs@Fe_3O_4/chitosan/pumice hybrid beads were able to be recycled multiple times, maintaining an 86% yield even after six consecutive uses [17].

Reyes-Mercado et al. developed $CuSO_4$-supported chitosan (CTS-$CuSO_4$) and chitosan-supported copper nanoparticles/copper(I) oxide microcrystals (CTS-CuNPs/Cu_2OMCs) as heterogeneous catalysts using a straightforward, green, and effective immobilization

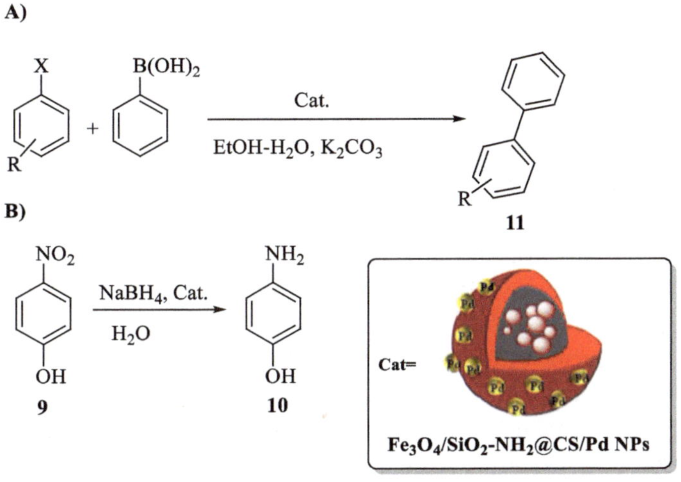

Figure 15.4: Suzuki-Miyaura coupling (A) and 4-nitrophenol reduction (B) reactions catalyzed by Fe_3O_4/SiO_2-NH_2@CS/Pd NPs [16].

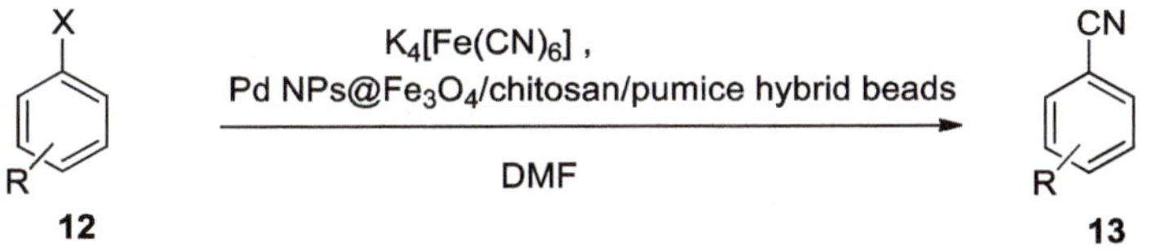

Figure 15.5: Cyanation reaction of aryl halides catalyzed by Pd NPs@Fe_3O_4/chitosan/pumice hybrid beads.

process that leverages the chelating ability of CTS. These catalysts, particularly CTS-$CuSO_4$ and CTS-CuNPs/Cu_2OMCs, enhanced the production of bis(indolyl)methanes (BIMs) **(16)** by enabling the catalysts to be recovered and reused for up to five cycles (Figure 15.6). They facilitated solvent-free reactions under gentle conditions, delivered high yields, required minimal catalyst quantities, and generated no metal waste due to the recyclability of the catalysts. In addition, the authors' findings indicated that the CTS-CuNPs/Cu_2OMCs catalyst is a viable substitute for CTS-$CuSO_4$ in the production of BIMs when using aliphatic aldehydes **(14)**. This study was the first report of employing CTS-$CuSO_4$ and CTS-CuNPs/Cu_2OMCs catalysts for BIM synthesis via a cost-effective, simple, environmentally friendly, and sustainable method [18].

A team successfully applied a chitosan (CS) coating to Fe_3O_4 particles. The resulting magnetic chitosan (Fe_3O_4@CS) showed significant catalytic activity in converting 4-nitrophenol (4-NP) **(17)** to 4-aminophenol (4-AP) **(18)** with the aid of $NaBH_4$ solution (Figure 15.7). To the authors' knowledge, this marked the first instance of employing magnetic CS for 4-NP reduction **(17)**. Chitosan's surface can be advantageously modified through the reaction of its amine groups with the safe and straightforward or-

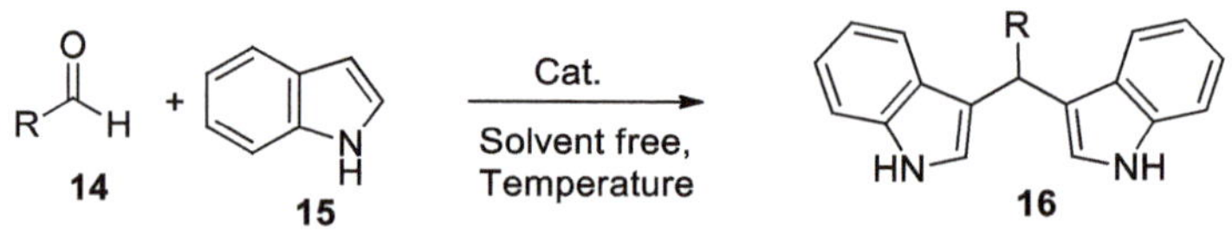

Figure 15.6: Synthesis of bis(indolyl)methanes catalyzed by CTS-$CuSO_4$ and CTS-CuNPs/Cu_2OMCs.

ganic compound, methyl salicylate, transforming them into amide groups that facilitate metal interaction. They effectively altered the surface of magnetic CS and anchored silver nanoparticles (Fe_3O_4@CS@MS@Ag), enhancing the reduction efficiency of 4-NP (**17**) to 4-AP (**18**) by a factor of four compared to the unmodified Fe_3O_4@CS. Importantly, this study was the inaugural demonstration of using methyl salicylate to functionalize the surface of magnetic CS, creating a magnetically separable, silver-nanoparticle-based heterogeneous catalyst. Also, the catalyst Fe_3O_4@CS@MS@Ag can be magnetically retrieved from the reaction mixture and reused up to five times, maintaining nearly 90% efficiency in reducing 4-NP (17) to 4-AP (**18**) [19].

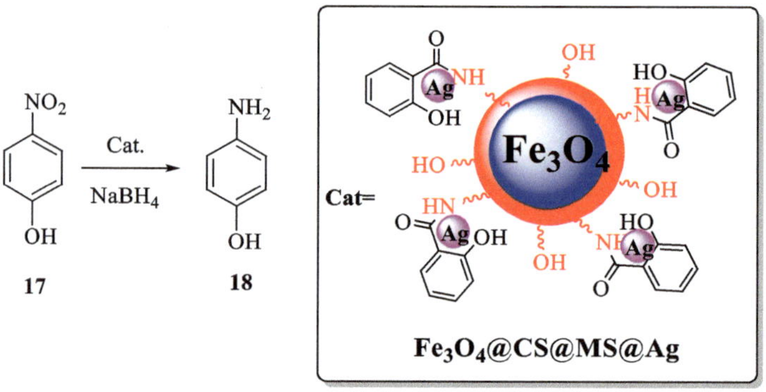

Figure 15.7: Reduction of 4-nitrophenol (4-NP) to 4-aminophenol (4-AP) in the presence of $NaBH_4$.

Researchers developed a straightforward and effective method for creating palladium catalysts anchored on chitosan-biguanidine coated magnetic Fe_3O_4 NPs, designated as [Pd(0/II)/CS-bigua@Fe_3O_4]. This innovative and recyclable magnetic nanocatalyst illustrated commendable catalytic performance in the Suzuki coupling reaction (Figure 15.8). Notably, the catalyst maintained its catalytic prowess across seven successive reaction cycles without significant degradation. The study showcased the transformation of various aryl iodides (**19**) and bromides (**19**) into their corresponding products (**21**) with high efficiency when reacted with phenyl boronic acid (**20**). However, the coupling of aryl chlorides (**19**) with phenyl boronic acid (**20**) required a longer reaction time compared to aryl bromides and iodides (**19**), yielding moderate amounts of the targeted products [20].

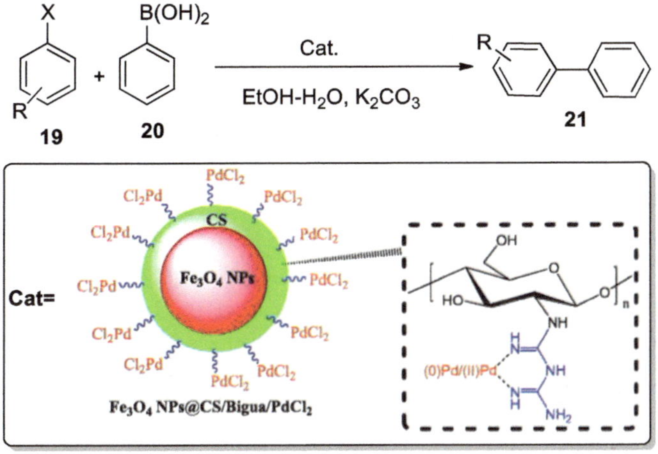

Figure 15.8: Suzuki-Miyaura reaction in the presence of Pd(0/II)/CS-bigua@Fe_3O_4 [20].

The developed magnetic interphase nanocatalyst, consisting of an oxo-vanadium Schiff base complex anchored on Fe_3O_4 NPs, referred to as [VO(CS)@MNPs], was synthesized by Veisi et al. Furthermore, it was demonstrated that the corresponding catalyst serves as an eco-friendly, efficient, and recyclable heterogeneous magnetic nanocatalyst for the selective oxidation of sulfides (**22**) into sulfoxides (**23**), employing hydrogen peroxide in the absence of solvents at room temperature (Figure 15.9). The benefits of this approach include the use of a readily available, environmentally benign, cost-effective, and chemically stable oxidant, along with mild reaction conditions, straightforward processing, practicality, rapid reaction times, high to outstanding yields, easy and efficient catalyst recovery using an external magnetic field, multiple reuses, and exceptional chemoselectivity [21].

In a study, researchers developed a self-doped nitrogen-enriched chitosan carbon (Ch) via activated carbonization. The authors then created a variety of CdS QDs/xCh composite photocatalysts using a hydrothermal process, which incorporated varying amounts of CdS QDs. The catalytic oxidation process of 5-hydroxymethyl furfural (HMF) was carried out to yield 2,5-diformyl furan (DFF). The findings indicated that with a 20% Ch loading (CC-20), a reaction duration of 10 h, and a raw material to catalyst ratio of 1/4, the authors attained a 78.73% conversion rate for HMF and a 97.31% selectivity for DFF [22].

The magnetic DAR-chitosan compound was created through a process where magnetic chitosan, denoted as [Fe_3O_4@CS], was combined with a newly made diacetylresorcinol, abbreviated as DAR, which served as a binding agent. This combination formed a Schiff base precursor on the resulting magnetic DAR-chitosan polymer, which was then able to bond with gold(III) ions, resulting in a magnetic DAR chitosan-based gold(III) Schiff base complex, represented as Fe_3O_4@CS/DAR-$AuCl_3$. This newly synthesized complex, Fe_3O_4@CS/DAR-$AuCl_3$, was employed as a renewable catalyst in A3 coupling reactions. These reactions are instrumental in producing propargylamine, a fundamental

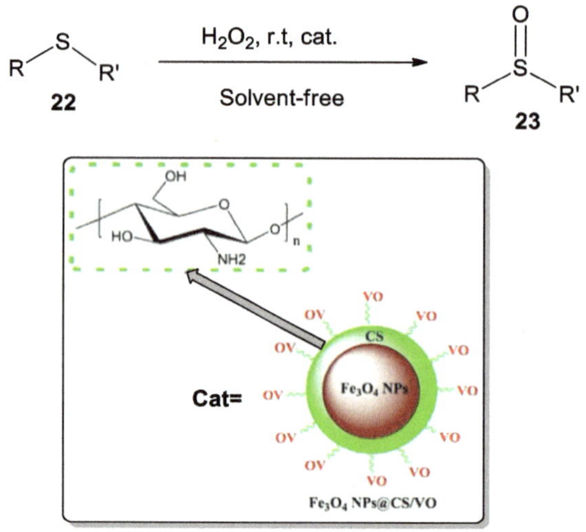

Figure 15.9: Selective oxidation of sulfides into sulfoxides in the presence of Fe_3O_4 NPs@CS/VO [21].

component for pharmaceuticals, yielding high results ranging from 80 to 92% under gentle conditions. In addition, it was illustrated that the magnetic Fe_3O_4@CS/DAR-$AuCl_3$ could be efficiently separated and retained up to 76% of its original effectiveness after being reused eight times, indicating its substantial capacity for reuse [23].

A emphasized the creation of a Fe_3O_4-modified chitosan composite and its assessment as a catalyst in the multicomponent Knoevenagel-Michael cascade reaction. The study's findings revealed that the corresponding composite is an effective catalyst, achieving a high yield of 92.3% under ideal conditions. Due to its reusability, consistent high performance, and straightforward separation process, the corresponding composite qualified as an eco-friendly catalyst [24].

15.3 Alginate-based nanostructures

A research study introduced a novel method for creating annulated pyrazole compounds **(27)** through a one-pot cyclocondensation process involving aromatic aldehyde derivatives **(25)**, malononitrile **(24)**, and phenylhydrazine **(26)**. This method aligned with the principles of green chemistry, utilizing a hybrid magnetic bio-based catalyst, Fe_3O_4@Alg@CPTMS@Arg, which boasted exceptional efficiency and sustainability due to the amino acid coordination sphere's influence (Figure 15.10). Moreover, this technique offered the benefits of using a non-toxic catalyst that can be easily recovered and reused up to seven times without significant loss of performance, thereby promoting the eco-friendly production of pyrazole variants (up to 97% yield) [25].

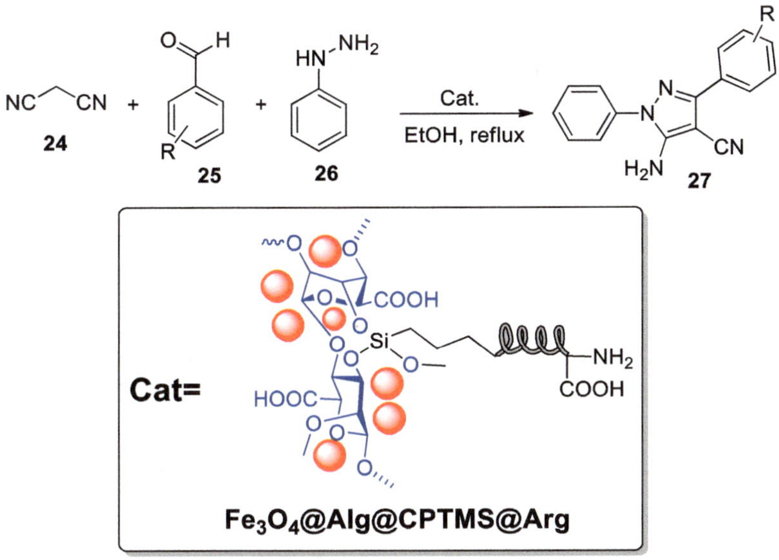

Figure 15.10: The synthesis of pyrazole derivatives in the presence of Fe_3O_4@Alg@CPTMS@Arg.

A study detailed a single-step method for creating AgNP within calcium alginate (CA) beads that form in situ. The CA beads, produced by the interaction of sodium alginate with calcium hydroxide, served multiple roles: they reduced and stabilized the nanoparticles and provided a matrix for their support. Calcium hydroxide supplied Ca^{2+} ions, leading to the formation of a Ca-alginate gel polymer that encapsulated the AgNP. Notably, the nanoparticle size remained relatively unchanged regardless of the alginate concentration. Alginate also effectively prevented the nanoparticles from clumping together. The catalytic performance was assessed through the reduction of para-NP to para-AP using an excess of sodium borohydride. TOF and the catalyst's reusability were the metrics used to gauge its effectiveness. The findings confirmed that the catalyst was environmentally friendly, could be reused, remained stable, and operated efficiently [26].

A study presented the fabrication of a nanocomposite where Au NPs were affixed to a sodium alginate (Alg) layer that coated magnetite (Fe_3O_4@Alg-AuNPs). The Alg, a natural anionic polysaccharide, formed a shell around the Fe_3O_4 microparticles, serving as a stabilizer for the Au NPs. The composite's catalytic efficacy was evaluated through the conversion of 4-nitrophenol **(28)** to 4-aminophenol **(29)** (Figure 15.11). This process was closely observed using UV-Vis spectroscopy, revealing outstanding yields (99%) within a brief reaction period (1.5–4 min) in an aqueous environment at ambient temperature, with a minimal catalyst quantity (1–3 mg). A kinetics analysis was also conducted, yielding a substantial rate constant. Moreover, the catalyst demonstrated excellent durability, being easily separable and reusable with no significant decline in its catalytic performance [27].

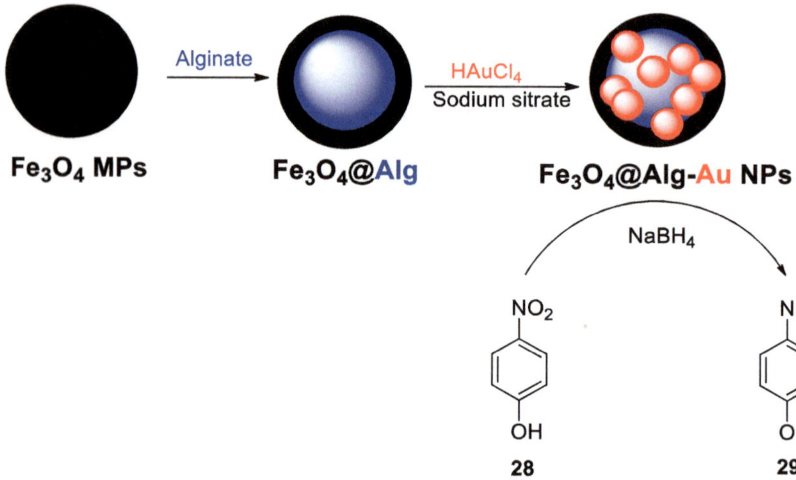

Figure 15.11: The reduction of 4-nitrophenol to 4-aminophenol in the presence of Fe_3O_4@Alg-AuNPs.

In a research study, a new and potent nanocatalyst composed of Fe_3O_4/g-C_3N_4/alginate-Ag was synthesized. The catalytic prowess of this newly developed nanocatalyst was assessed in the regioselective synthesis of triazole derivatives (**33** and **33′**), as well as A3 and KA2 coupling reactions in an aqueous setting (Figure 15.12). The corresponding catalyst could be effortlessly retrieved post-reaction with an external magnet and redeployed multiple times, exhibiting only a minor reduction in its catalytic efficiency [28].

In a research study, alginate (Alg) hydrogel beads were utilized as a base for iron oxide, resulting in the creation of iron oxide-embedded alginate beads (Io-alg). Alg, a cost-effective and harmless biopolymer, and iron, a widely available and safe element, were chosen for their beneficial properties. To enhance the beads' structural integrity, bentonite was incorporated, forming iron oxide-infused alginate-bentonite composite beads (Io-alg-ben). Both Io-alg and Io-alg-ben beads, in their hydrogel form, were employed as catalyst supports for transforming 4-NP into 4-AP within a borohydride setting. Notably, the hydrogel beads exhibited remarkable catalytic action in reducing 4-NP. The in situ-generated zero-valent iron (ZVI) during the 4-NP reduction in a NaBH4 medium served as the active catalyst. The reduction process was scrutinized at varying catalyst amounts and initial 4-NP concentrations, consistently following zero-order kinetics. The catalyst maintained its efficacy for up to five cycles of reuse. Its performance stood on par with other documented catalysts. Additionally, the influence of reaction temperature was explored to ascertain the activation energy [29].

The composite Alg@Cu_2O-Sb_2O_3, consisting of copper oxide-antimony oxide (Cu_2O-Sb_2O_3) encapsulated within Na alginate hydrogel, was synthesized and employed as a catalytic reactor. This reactor facilitated the reduction of various pollu-

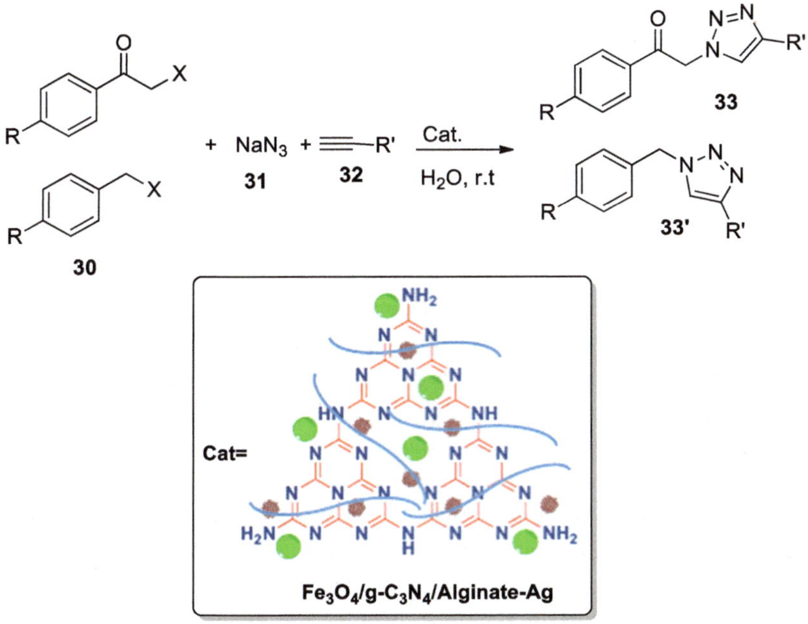

Figure 15.12: Synthesis of 1,4-disubstituted 1,2,3-triazoles in the presence of Fe_3O_4/g-C_3N_4/alginate-Ag [28].

tants, including 4-nitrophenol, 2-nitrophenol, 2,6-dinitrophenol, methyl orange, Congo red, acridine orange, methylene blue, and $K_3[Fe(CN)_6]$. Among these, the corresponding catalyst proved to be particularly selective and effective in reducing 2-NP. Additionally, the catalyst demonstrated remarkable stability and could be recycled with ease for the reduction of 2-NP. It achieved the reduction of 2-NP in just 1.0 min, with a rate constant of 3.8187 min^{-1} [30].

Sodium alginate (SA) and carboxymethyl cellulose (CMC) biopolymers were combined and modified with Fe_3O_4 nanoparticles, which served to anchor copper nanoparticles, creating Fe_3O_4@SA-CMC-CuNP nanocomposites during batch experiments. These nanocomposites were then employed as a heterogeneous catalyst for the conversion of 4-nitrophenol (4-NP) **(34)** to 4-aminophenol (4-AP) **(35)** in the presence of $NaBH_4$ (Figure 15.13). The reaction's progress was tracked using UV-visible spectrophotometry. Compared to the individual Fe_3O_4 and the Fe_3O_4@SA-CMC components, the Fe_3O_4@SA-CMC-CuNP nanocomposite demonstrated significantly enhanced catalytic efficiency in the reduction of 4-NP **(34)**. Upon examining variables like the catalyst quantity, it was found that using 30 mg of the catalyst led to a 95.0% reduction of 4-NP **(34)** within just 1.5 min [31].

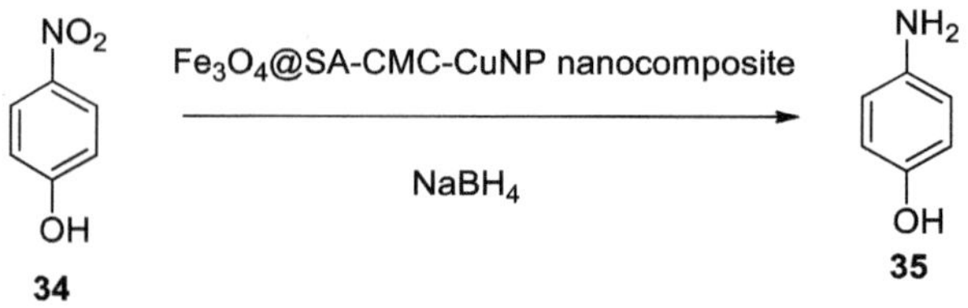

Figure 15.13: The reduction of 4-nitrophenol to 4-aminophenol in the presence of Fe_3O_4@SA-CMC-CuNP.

Innovative composite alginate beads were developed by integrating cyclodextrin nanosponges (CDNS), which acted as a molecular carrier, during the bead formation process. These composite beads proved to be an efficient platform for the stabilization of Pd NPs, thanks to the synergistic properties of alginate and CDNS that promote better Pd attachment. The synthesized composite, named Pd/Alg-CDNS, showcased superior catalytic performance in the hydrogenation of nitroarenes **(36)** within water-based solutions (Figure 15.14). By fine-tuning the reaction conditions, the most favorable outcomes were obtained using 0.03 g of Pd/Alg-CDNS for each mmol of nitrobenzene, conducted at 45 °C. Comparative analyses with baseline catalysts, Pd/Alg and Pd/CDNS, revealed that the inclusion of CDNS into the alginate beads significantly enhanced the catalyst's efficacy, especially for bulky substrates such as 1-nitronaphthalene. Impressively, Pd/Alg-CDNS proved to be highly reusable, maintaining its catalytic function across seven successive cycles without notable loss in performance. This method was adaptable to various nitroarenes, possessing different electronic characteristics. Remarkably, Pd/Alg-CDNS was selective in reducing the nitro groups [32].

CA-CMC/Fe_2O_3-NiO beads were crafted through a straightforward crosslinking technique, utilizing $AlCl_3$ as the crosslinking agent. These beads, in conjunction with $NaBH_4$, effectively reduced various nitrophenol isomers (4-NP, 2-NP, and 2,6-DNP) and organic dyes (methyl orange (MO) and eosin yellow (EY)) within 2 min. Additionally, $K_3[Fe(CN)_6]$ was reduced in a mere 30 s. The CA-CMC/Fe_2O_3-NiO beads demonstrated high efficiency against all tested pollutants, exhibiting rate constants of 2.50, 2.23, 1.74, and 4.82 (×10 − 2) s^{-1} for 4-NP, MO, EY, and $K_3[Fe(CN)_6]$, respectively. 4-NP was selected to fine-tune the beads' catalytic performance. Impressively, the beads maintained their catalytic prowess for up to four uses. The method's efficacy was also validated by reducing 4-NP in various real-world samples, achieving reductions of 88.5% in seawater over 8 min, 69% in wastewater over 7 min, 90.2% in orange juice over 4 min, and 73% in full-fat milk over 8 min [33].

In 2023, research focused on the development of an active hybrid bio-nanocomposite material, which is a combination of the biopolymer alginate with embedded TiO_2 and silver nanoparticles. Initially, a set amount of TiO_2 nanodispersion was blended into the alginate solution, followed by the formation of Alginate/TiO_2 bio-nanocomposite beads through ionic crosslinking with a $CaCl_2$ solution. These beads, while still moist, were in-

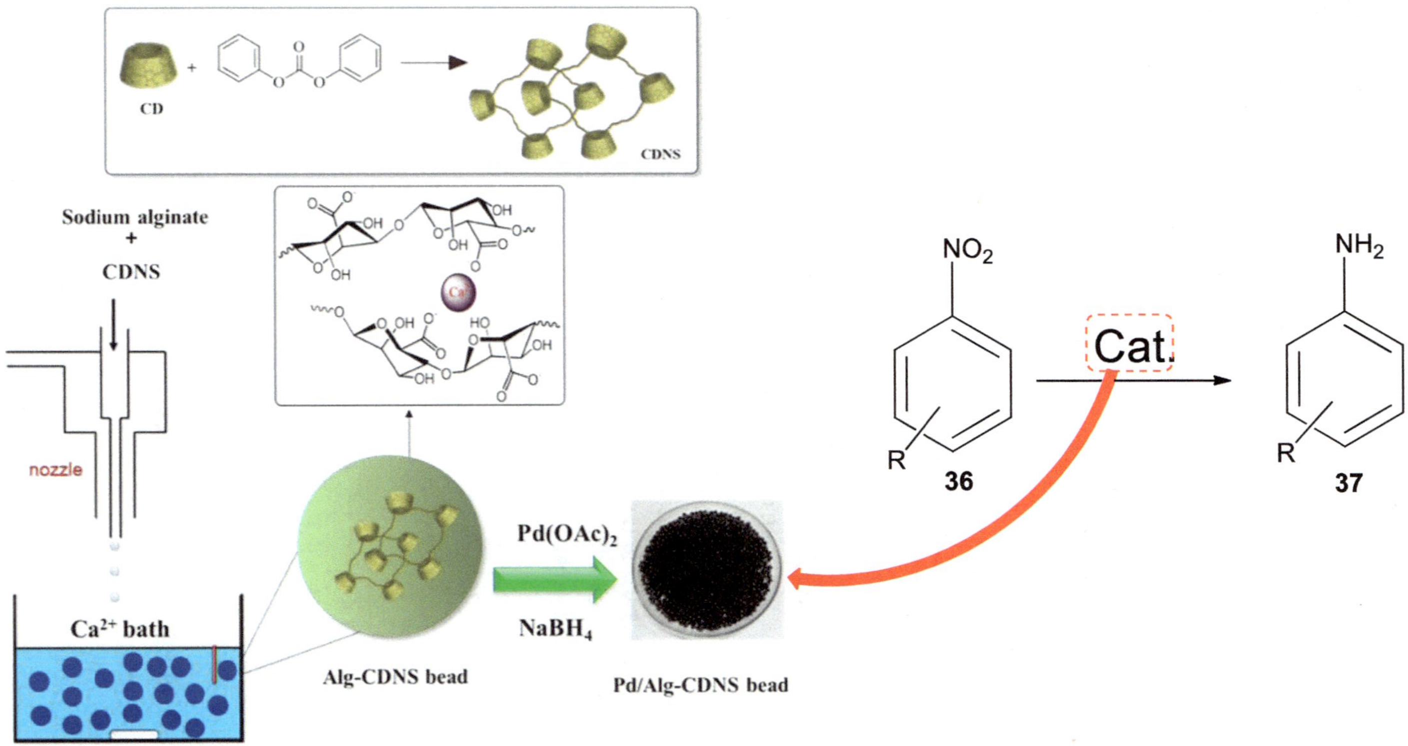

Figure 15.14: Synthetic route for Pd/Alg-CDNS and the reduction of nitrophenols to aminophenols in the presence of Pd/Alg-CDNS [32].

fused with Ag^+ ions using an adsorption method, and subsequently, the silver ions were converted into Ag NPs via an irradiation process. The alginate/TiO_2 and alginate/TiO_2-Ag bionanocomposite beads were thoroughly examined to evaluate the nanostructural morphology, specifically the size and shape, of the embedded TiO_2 and Ag nanoparticles. The catalytic capabilities of these beads were investigated by observing their effect on the conversion of 2-NP to 2-AP, considering various parameters such as the initial concentration of 2-NP, the duration of the reduction process, the quantity of the reducing agent, and the catalyst's mass. The inclusion of Ag nanoparticles was found to significantly enhance the catalytic process, with the hydrogenation of 2-NP to 2-AP achieving a 98% completion rate using alginate/TiO_2-Ag, in contrast to a 76% rate when using the alginate/TiO_2 bionanocomposite alone. Moreover, the hybrid bionanocomposite beads displayed a robust ability to reduce 2-NP concentrations up to 0.05 M. An optimal catalyst weight of 0.4 g was determined, which yielded a nearly 98% reduction efficiency [34].

Tummachote et al. synthesized an alginate-coated MNP and utilized it as a nanosorbent for Pd. By adding amino-rich components such as polyvinylamine and dimethylaminoethyl methacrylate to the alginate layer, we fostered advantageous interactions, notably the creation of carboxylate-ammonium salts. This modification enhanced the dispersal of Pd and markedly diminished Pd clustering on the nanoparticle's surface. Optimal catalytic conditions for 4-nitrophenol (4-NP) (**38**) reduction were achieved using 5 mol% Pd on G2-Pd nanoparticles, resulting in a turnover frequency (TOF) of 390 h^{-1} and a rate constant (k) of 2.95×10^{-2} s^{-1}. The surface alteration with amino groups led to catalysts that demonstrated superior catalytic efficacy in reducing 4-NP (**38**) in aqueous solutions (Figure 15.15). These modified particles also showed exceptional durability and maintained over 98% conversion efficiency after 20 reuse cycles, indicating minimal loss in performance [35].

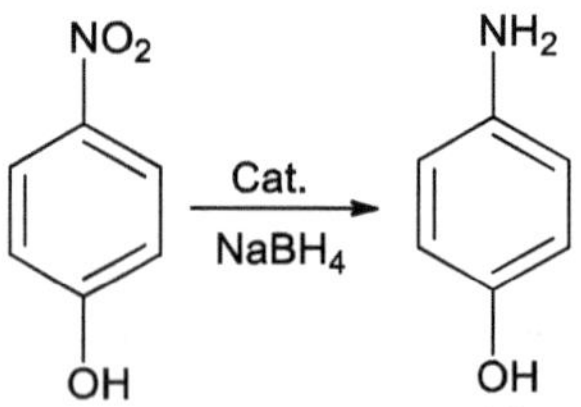

38 **39**

Figure 15.15: The reduction of 4-nitrophenol to 4-aminophenol in the presence of Pd-immobilized nanoadsorbents as catalysts.

15.4 Pectin-based nanostructures

A team created a new nanomagnetic interphasic material, designated as Fe_3O_4@Pectin/Pd, by integrating Pd NP with pectin-coated magnetic Fe_3O_4 nanoparticles. This innovative material showcased its potential as an eco-friendly, efficient, and reusable heterogeneous nanocatalyst suitable for carbon-carbon (Suzuki) and carbon-nitrogen (Buchwald-

Hartwig) cross-coupling reactions (Figure 15.16). The authors synthesized a diverse collection of biphenyl (**45**) and arylamine derivatives (**42**) under optimal conditions, achieving outstanding yields. In addition, the corresponding catalyst was successfully separated and reused multiple times without any notable loss in its catalytic efficacy [36].

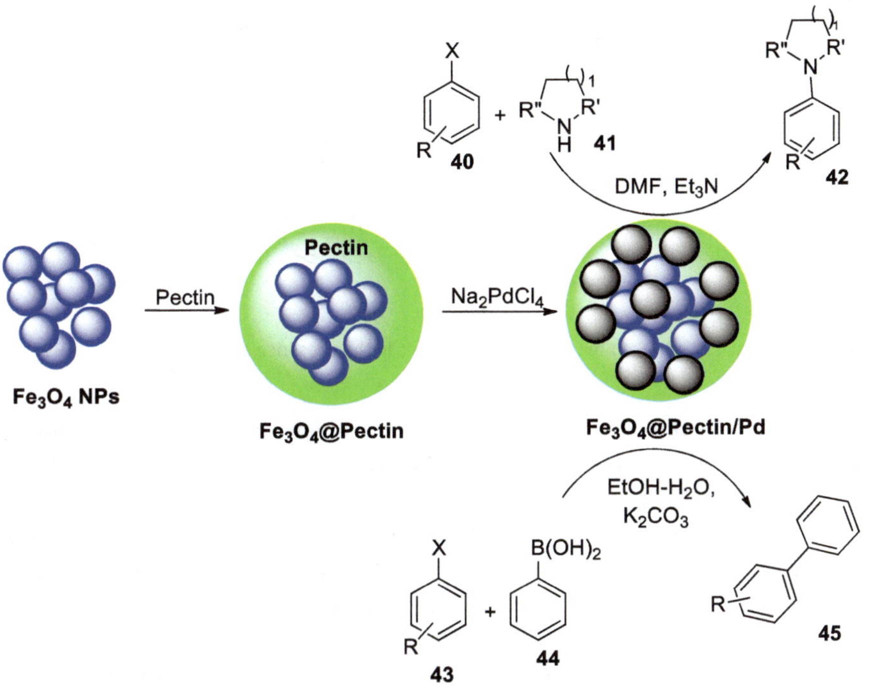

Figure 15.16: The Suzuki and Buchwald-Hartwig cross-coupling reactions catalyzed by Fe_3O_4@Pectin/Pd.

A research study culminated in the development of an innovative core-shell nanomaterial, termed Fe_3O_4/Pectin/Au nanocomposite, which features Au NPs embedded within and a pectin coating around magnetic nanoparticles. The pectin not only served as a protective layer for the Fe_3O_4 nanoparticles but also acted as an eco-friendly reducing agent for the on-site generation of Au NPs. Moreover, the biomolecule-created polar environment helped to evenly distribute the Au NPs on the surface, preventing them from clumping together. This nanocomposite was catalytically tested in the reduction of nitroaromatic compounds (**46**), delivering outstanding yields swiftly under gentle conditions (Figure 15.17). The magnetic nature of the catalyst facilitated its easy separation and enabled its reuse for 11 consecutive cycles with negligible decline in its catalytic function [37].

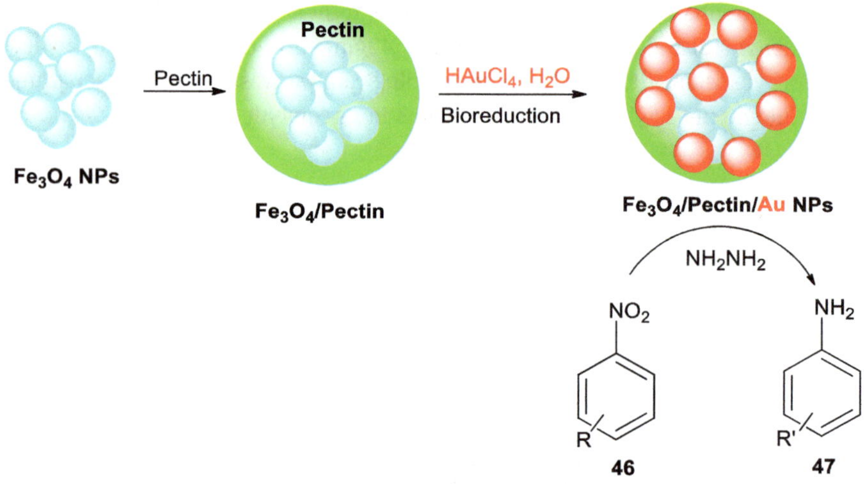

Figure 15.17: The reduction of nitrophenols to aminophenols in the presence of Fe_3O_4/pectin/Au NPs.

A study examined a catalyst composed of magnetic nanoparticles coupled with the naturally abundant pectin biopolymer (Fe_3O_4@Pectin), which supports nickel (II) acetate. This catalyst was employed to synthesize the precursor necessary for creating xanthene (**50**). Subsequently, the catalyst's efficacy was tested in alcohol (**48**) oxidation reactions (Figure 15.18), noted for their high efficiency and selectivity. The resulting aldehydes (**49**) were then utilized as precursors in the production of xanthans (**50**) (Figure 15.18). The key benefits of this approach include a straightforward work-up process, the ability to conduct reactions without solvents, and the capability to recover and reuse the catalyst multiple times using an external magnet without any loss in activity [38].

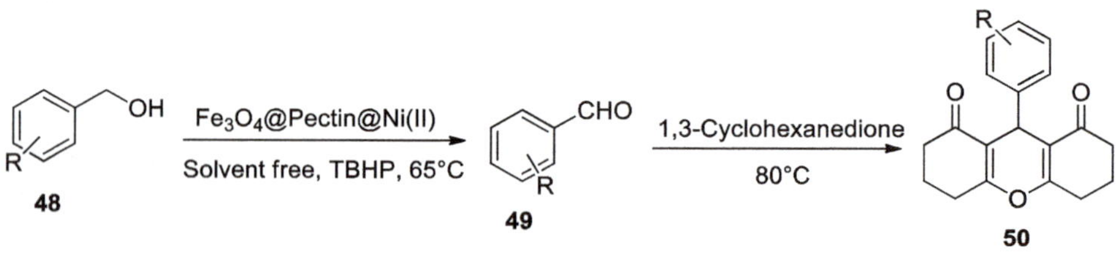

Figure 15.18: The alcohol oxidation reactions in the presence of Fe_3O_4@Pectin@Ni(II) and the synthesis of xanthans.

A research pioneered a sustainable, biogenic approach to crafting minute Cu NPs embedded within pectin-functionalized ferrite nanoparticles, forming a cutting-edge nanocomposite using the extract from Mentha pulegium flowers. The pectin@Fe_3O_4 composite's core-shell structure imparted substantial stability. In testing its catalytic capabilities, the authors discovered its aptness for the three-component assembly of various amines (**51**), triethyl orthoformate (**52**), and sodium azide (**53**), yielding a spectrum of 1-substituted-1H-tetrazole derivatives (**54**) (Figure 15.19). These reactions proceeded with high efficiency in the absence of solvents. The presence of Cu NPs was crucial, as the unadorned pectin@Fe_3O_4 catalyst showed limited effectiveness. Post-reaction, the catalyst was effortlessly recovered using a magnet and redeployed for nine cycles, maintaining consistent catalytic activity [39].

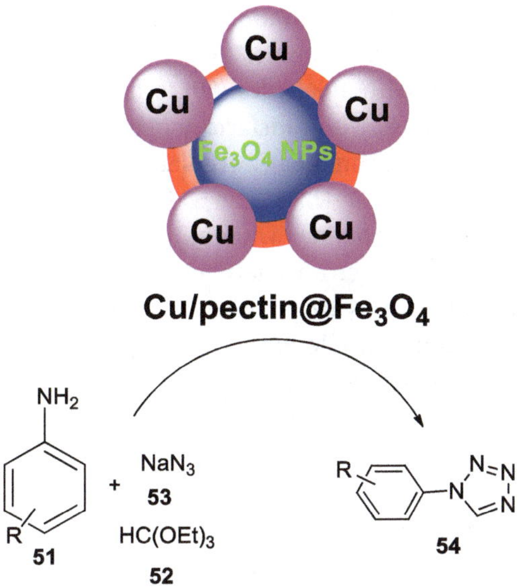

Figure 15.19: Synthesis of 1-substituted-1H-tetrazoles catalyzed by Cu/Pectin@Fe_3O_4.

A study presented a novel method for synthesizing PdNPs on-site using *Cyclea barbata* pectin, which acts as an environmentally friendly reducing and stabilizing agent. The PdNP@pectin complex demonstrated outstanding recyclable catalytic activity for the Heck coupling reaction in an aqueous medium (Figure 15.20). The study explored the kinetics and reusability of the nanoparticles in the catalytic reduction of *ortho*-, *meta*-, and *para*-nitrophenol, revealing efficient catalysis with up to five successful recycles without significant loss of performance. Notably, the nanocomposite served as an effective catalyst for transforming alkynes into *cis*-alkenes using KOH/DMF as a hydrogenation agent. Additionally, this reaction proved to be efficient for producing sex pheromones, achieving total yields of 70% for *Plutella xylostella* ((*Z*)-11-hexadecen-1-yl acetate) and 68% for *Cylas formicarius* ((*Z*)-3-dodecen-1-yl(*E*)-2-butenoate) [40].

In 2021, research showcased the effective creation of Ni-Fe_3O_4@Pectin ~ PPA ~ Piconal as a dual-function, heterogeneous nanomagnetic catalyst. This novel catalyst was adeptly employed for the reduction of nitroarenes, utilizing $NaBH_4$ as a safe hydrogen source in water-based solutions at room temperature. The catalytic efficiency

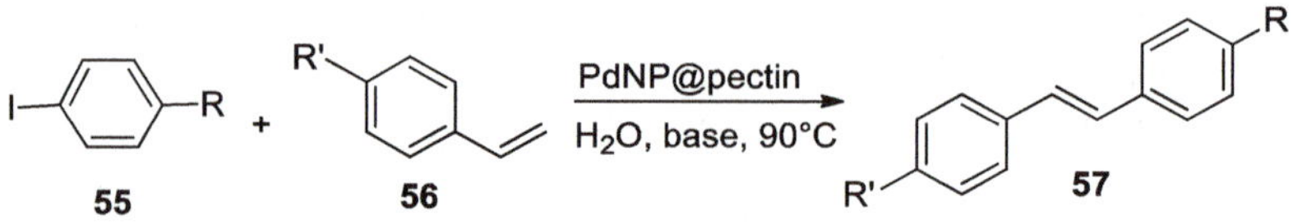

Figure 15.20: Heck coupling reaction in the presence of PdNP@pectin.

of Ni-Fe_3O_4@Pectin ~ PPA ~ Piconal was further demonstrated through a one-step method that included both reduction and Schiff base condensation reactions with various nitroarenes (**58**) (Figure 15.21). During this process, amines (**59**) were generated on-site with $NaBH_4$ and subsequently combined with aldehydes (**60**) to form imines (**61**). This approach yielded a diverse array of imines with satisfactory to excellent results. Thanks to its magnetic properties, the catalyst could be conveniently recovered with an external magnet. The dual-functionality of Ni-Fe_3O_4@Pectin ~ PPA ~ Piconal facilitated complex, multistage reactions in a single container without the need to isolate products after each phase, employing water as an eco-friendly solvent and $NaBH_4$ as a hydrogen source, offering a safer alternative to hydrogen gas. Furthermore, this cutting-edge catalyst proved to be separable and reusable for multiple cycles, maintaining its catalytic performance without notable degradation [41].

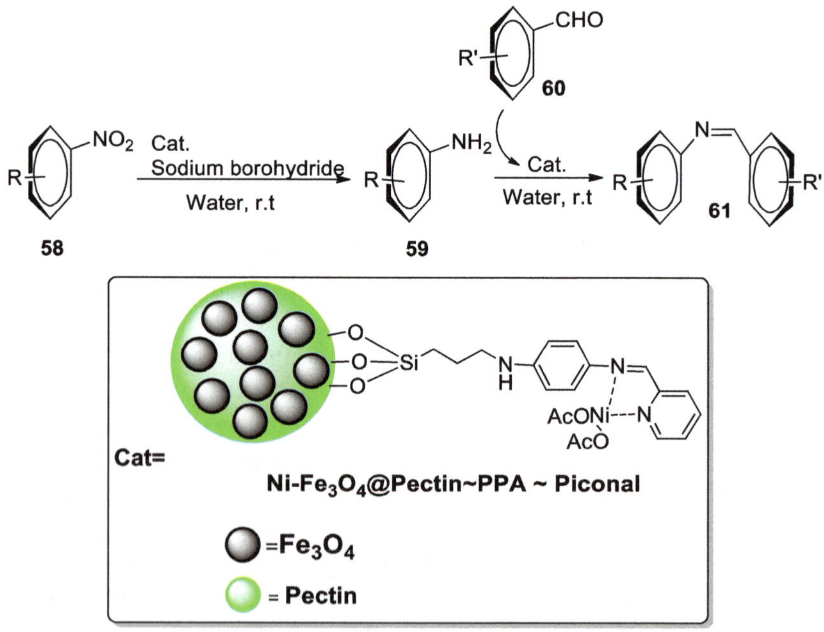

Figure 15.21: The tandem reduction-Schiff base condensation catalyzed by Ni-Fe_3O_4@Pectin ~ PPA ~ Piconal.

In an initiative, the innovative catalyst known as Fe_3O_4@Pectin@$(CH_2)_3$-Acetamide-Cu(II) was successfully synthesized. A key characteristic of the corresponding catalyst was its formulation from readily available materials, such as the naturally abundant biopolymer pectin. The catalytic prowess of the desired catalyst was put to the test in a traditional one-pot reaction involving terminal alkynes (**64**), alkyl halides (**63**), and sodium azide (**62**), all conducted in water (Figure 15.22). The reaction progressed seamlessly, yielding products (**65**) with both high yields and regioselectivity. The significant advantages of this method were manifold: it produced high yields, the catalyst can be recycled, the procedure was straightforward, it utilizes water – an environmentally benign solvent – and it accommodated a wide range of functional groups. These features contributed to both economic efficiency and environmental sustainability [42].

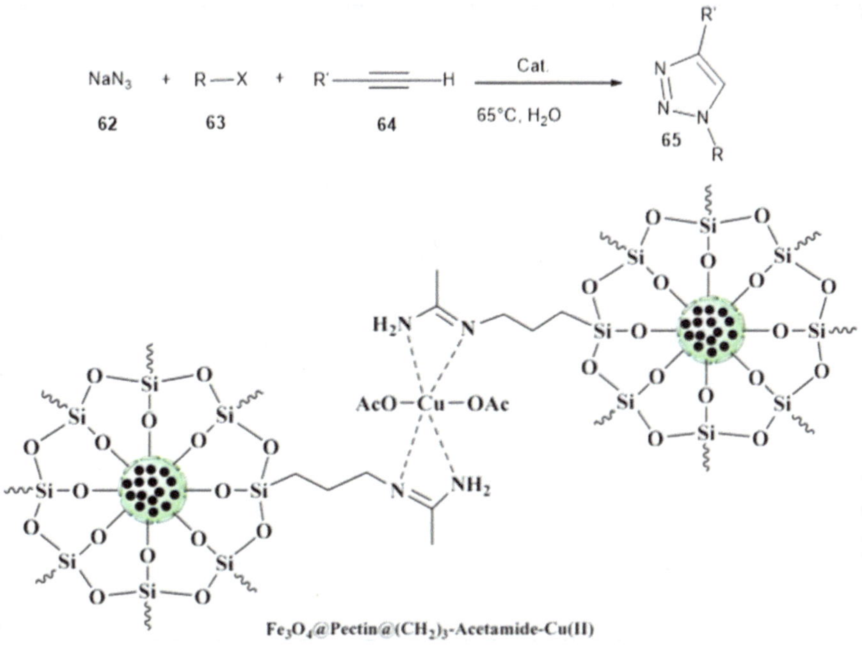

Figure 15.22: Multicomponent synthesis of 1,2,3-triazoles catalyzed by Fe_3O_4@Pectin@$(CH_2)_3$-Acetamide-Cu(II) [42].

A study successfully developed Fe_3O_4@Pectin ~ Imidazole ~ SO_3H-Cu(II), a potent magnetic catalyst derived from biopolymers, for the conversion of benzyl alcohols into aldehydes, employing tert-butyl hydroperoxide as the oxidizing agent in a solvent-free environment. The findings revealed that this catalyst was highly effective in transforming benzyl alcohols to their respective aldehydes using tert-butyl hydroperoxide. Noteworthy benefits of this catalyst include its high efficiency with minimal loading, straightforward processing, the attainment of uncontaminated products, the substitution of a cost-effective metal for more expensive alternatives, the absence of solvents, the employment of an environmentally friendly base, and the ability to be reused for a minimum of five consecutive cycles without experiencing a significant decrease in activity [43].

In a pioneering study, a group of researchers introduced a novel, recyclable deep eutectic solvent anchored onto pectin utilizing metformin as a hydrogen bond donor and choline chloride as the acceptor. Subsequently, these particles (Pec-Met-ChCl) were combined with magnetic Fe_3O_4 nanoparticles, resulting in the creation of the MNPs-Pec-Met-ChCl catalyst. The catalyst's stable and diverse nature, along with the presence of hydrogen bond donors, leads to efficient substrate interaction, defining its catalytic action. The inclusion of Fe_3O_4 magnetic nanoparticles not only simplified recovery but also enhanced catalytic performance. This innovative catalytic system boasted features such as environmental friendliness, versatility, sustainability, effortless recovery, and reusability, delivering outstanding results in the production of various aromatic and heteroaromatic 2,3-dihydro-4(1H)-quinazolinone derivatives (**69**) in a solvent-free process (Figure 15.23) [44].

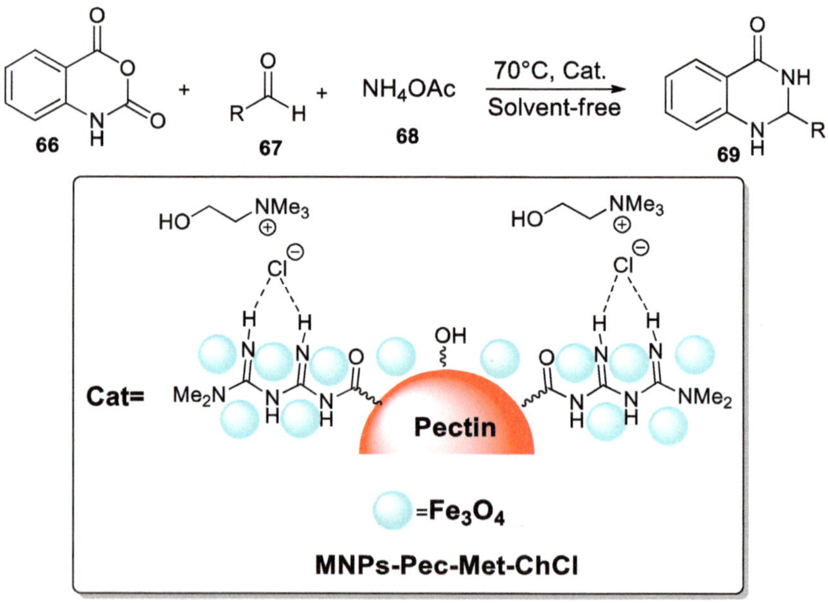

Figure 15.23: Synthesis of 2,3-dihydro-4(1H)-quinazolinone derivatives in the presence of MNPs-Pec-Met-ChCl.

An investigation introduced a bio-sourced magnetic nanocomposite, crafted by modifying pectin with 1-amino-8-naphthol-3,6-disulfonic acid through a two-stage process. Initially, DSA was integrated onto low-methoxy citrus pectin (Pec). Subsequently,

Fe_3O_4 nanoparticles were affixed, culminating in the magnetized nanocomposite (Fe_3O_4-Pec-DSA). This composite was utilized to facilitate the ultrasonic-assisted fabrication of biologically potent 1,4-dihydropyridines (**73**) using ethanol (Figure 15.24). An exhaustive exploration of variables, including the type of catalyst, choice of solvent, ultrasound power and frequency, temperature, duration, and catalyst quantity, was conducted to pinpoint the ideal conditions. The exceptional efficacy of the corresponding catalyst, characterized by abbreviated reaction durations (15–40 min) and substantial yields (84–95%), can be attributed to the combined effect emanating from the catalyst's structure, particularly the $-SO_3H$ group, Fe_3O_4 nanoparticles, and the application of ultrasound waves [45].

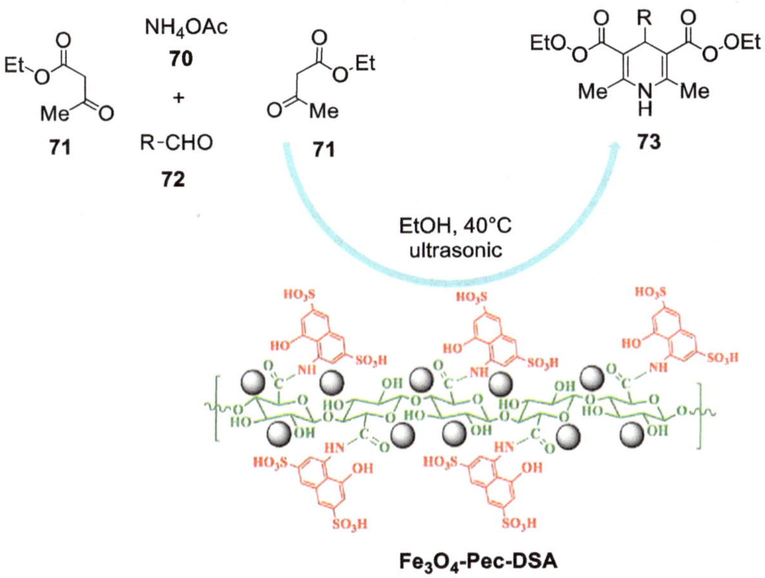

Figure 15.24: Synthesis of 1,4-dihydropyridines in the presence of Fe_3O_4-Pec-DSA [45].

Integrating nanoparticles within the hydrogel network fostered the creation of cutting-edge catalytic substances boasting enhanced characteristics. Chosen for its compatibility, pectin served as an ideal base. The pectin hydrogel was then synthesized through physical cross-linking in the presence of calcium chloride. Subsequently, iron and copper nanoparticles were embedded onto the hydrogel's surface, rendering it an effective and eco-friendly catalyst. The catalytic prowess of the γ-Fe_2O_3/Cu@HPECGs nanocomposite was assessed during the oxidative transformation of thiols into disulfides, utilizing H_2O_2 as an eco-conscious oxidizing agent. Moreover, the γ-Fe_2O_3/Cu@HPECGs nanocatalyst was capable of producing symmetrical sulfides via a one-step process, employing thiourea as the sulfur donor under gentle conditions. The inclusion of Cu nanoparticles, as opposed to solely iron nanoparticles, markedly amplifies the catalytic activity. These methods offer practical approaches, achieving product yields up to 90% with pronounced selectivity [46].

15.5 Cellulose-based nanostructures

A novel sulfonic acid-functionalized magnetic nanocomposite based on biopolymers was developed and utilized in two key acid-catalyzed organic heterocyclic reactions: the synthesis of tetrahydropyridines and Biginelli condensation (Figure 15.25). This new, environmentally friendly, and magnetically separable heterogeneous nanocatalyst boasted several benefits, including high product yields, rapid reaction completion, gentle conditions, straightforward processing, and solvent-free operations. Moreover, this nanocatalyst was cost-effective to produce from readily available materials, has been thoroughly characterized by standard methods, and can be recycled multiple times with negligible loss of efficacy. Its reusability was particularly valuable for bulk industrial applications. To highlight the superiority of $MgFe_2O_4$ over other ferrites, such as Fe_3O_4, due to its excellent high-temperature stability, $MgFe_2O_4$ was chosen for the creation of tetrahydropyridine and dihydropyrimidine compounds. The use of $MgFe_2O_4$, functioning as a Lewis acid, in conjunction with SO_3H, acting as a Brønsted acid, led to high product yields. Specifically, the synthesis of tetrahydropyridines using $MgFe_2O_4$/cellulose resulted in substantial yields, whereas the use of cellulose/SO_3H alone yielded lower quantities. This indicated a synergistic effect between the two, enhancing the reaction's efficiency [47].

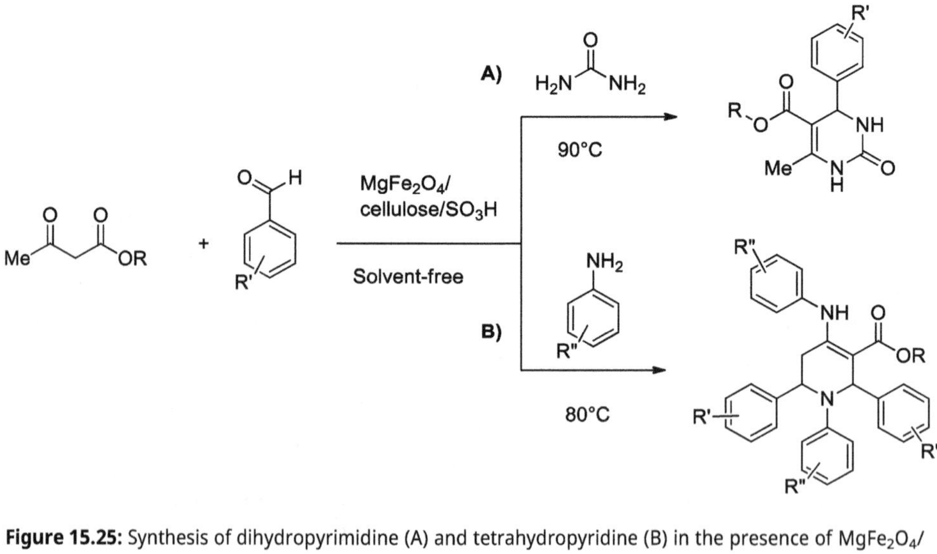

Figure 15.25: Synthesis of dihydropyrimidine (A) and tetrahydropyridine (B) in the presence of $MgFe_2O_4$/cellulose/SO_3H.

A study outlined a straightforward method for the immobilization of Fe–Cu bimetallic nanoparticles on microcrystalline cellulose, a substance that is safe for the environment. The resulting bimetallic cellulose nanocatalyst demonstrated excellent efficiency in the rapid reduction of a wide variety of nitroarenes (**74**) using $NaBH_4$ (Figure 15.26). The reduction processes were completed in water within 5–14 min, producing arylamines (**75**) with yields ranging from 89% to 95%. The Fe-Cu@MCC catalyst could be conveniently separated from the mixture using a magnet and was capable of being reused up to eight times without a noticeable decrease in its catalytic performance. The mild conditions required for the reaction, the simple post-reaction cleanup, the high yields, and the use of water as a safe, green, and cost-effective solvent are notable advantages that make this approach a valuable contribution to current methods in this field [48].

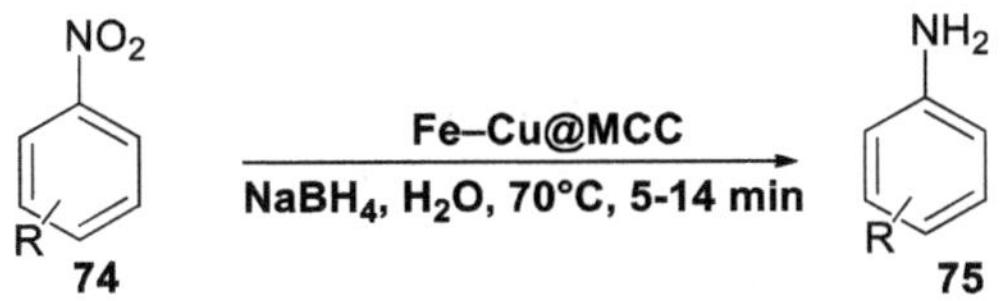

Figure 15.26: The reduction of nitroarenes in the presence of Fe-Cu@MCC.

The nanocellulose/BF_3/Fe_3O_4 catalyst, derived from biological sources and environmentally benign, exhibited remarkable catalytic efficiency. Its design allowed for simple magnetic separation and repeated use with minimal loss in performance. This catalyst was employed in the one-step synthesis of 4H-pyrimido[2,1-b]benzothiazole compounds (**79**) by condensing aldehydes (**76**), 2-aminobenzothiazole (**77**), and ethyl acetoacetate (**78**) (Figure 15.27). The process operated under solvent-free conditions at 100 °C, offering numerous benefits such as straightforward purification, outstanding yields, and reduced reaction times [49].

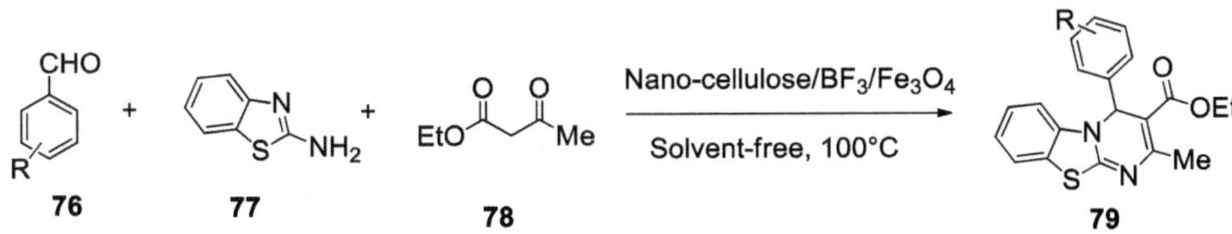

Figure 15.27: Synthesis of 4H-pyrimido[2,1-b]benzothiazoles in the presence of nanocellulose/BF_3/Fe_3O_4.

In a conducted study, carbon-doped CuO/Fe_3O_4 nanocomposites were effectively created through the direct calcination of Cu-BTC MOF (HKUST-1)/Fe_3O_4/cellulose microfibril composites in an atmospheric setting. This particular carbon-doped nanocomposite catalyst exhibited superior catalytic performance in the reduction of 4-nitrophenol, methylene blue, and methyl orange, in comparison to both the CuO/Fe_3O_4 catalyst derived from HKUST-1/Fe_3O_4 and the CuO obtained from HKUST-1, when $NaBH_4$ was present. The improved efficiency was credited to the carbon content, which hinders the CuO/ Fe_3O_4 particles from clumping together while also enhancing the surface area relative to volume and ensuring chemical robustness. Additionally, the nanocomposite catalyst can be easily separated for reuse with an external magnet, thanks to its superparamagnetic properties, and its potential for recycling was further explored [50].

A study introduced a Cu(II)-immobilized magnetic nano-fibrillated cellulose, synthesized by integrating $Cu(OAc)_2$ with cyclohexyldiamine-modified epichlorohydrin (Fe_3O_4@NFC/E-CHDA-Cu^{II}). The catalytic prowess of this novel magnetic nanocatalyst (Fe_3O_4@NFC/ECH-CHDA-Cu^{II}) was evaluated in a one-step, multicomponent reaction involving aromatic aldehydes (**80**), 1,3-cyclohexanediones (**81**), and malononitrile (**82**) to synthesize tetrahydrobenzo[b]pyrans (**83**) (Figure 15.28). This process was conducted under solvent-free conditions at room temperature. Notably, the nanocatalyst demonstrated effective reusability, being magnetically recoverable for a minimum of five cycles without a decline in performance, and the resulting product was easily purified through a straightforward procedure [51].

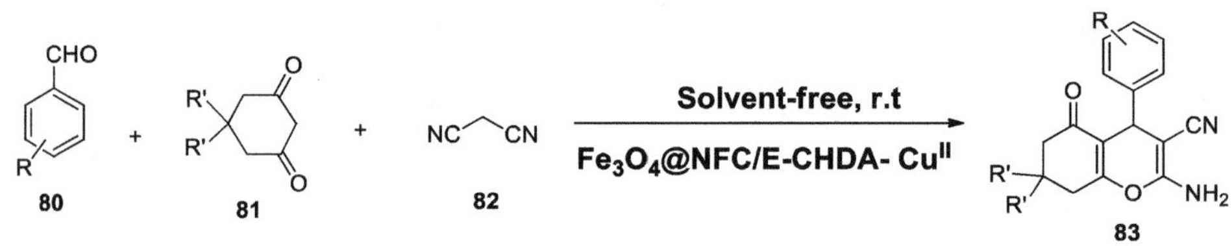

Figure 15.28: Synthesis of tetrahydrobenzo[b]pyrans in the presence of Fe_3O_4@NFC/E-CHDA-Cu^{II}.

The bio-based magnetic nano-catalyst Fe_3O_4@nanocellulose/Sb(V) was created through a straightforward in situ co-precipitation process. This involved combining Fe^{3+} and Fe^{2+} ions with NH_4OH within a nanocellulose aqueous suspension, followed by treating the resulting Fe_3O_4@nanocellulose with antimony pentachloride in a chloroform solution. The catalyst's effectiveness was evaluated during the synthesis of 4H-pyrimido[2,1-b]benzothiazoles (**87**) (Figure 15.29). This synthesis was conducted through a solvent-free, three-component reaction involving aromatic aldehydes (**84**), 2-aminobenzothiazoles (**85**), and ethyl acetoacetate (**86**) at 90 °C. Our findings demonstrate that these compounds were produced with high yields and purity in a brief period. Additionally, the results showed that the corresponding catalyst can be easily recovered and reused multiple times with minimal loss of efficiency [52].

A sustainable method was developed for creating Ni nanoparticles on a magnetized cellulose biopolymer. This biopolymer, which took on a core-shell nanocomposite structure (Fe_3O_4@CNF@Ni), was synthesized through sonication. The catalytic efficiency of the corresponding catalyst was tested in ultrasound-assisted Hiyama and

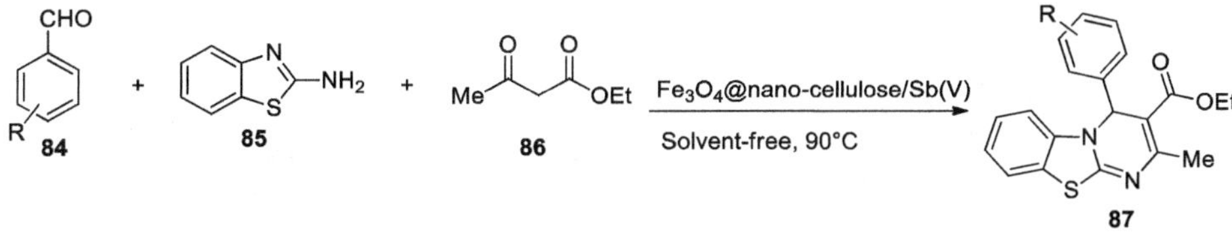

Figure 15.29: Synthesis of 4H-pyrimido[2,1-b]benzothiazoles in the presence of Fe_3O_4@nanocellulose/Sb(V).

Suzuki-Miyaura cross-coupling reactions, which are used to produce various biaryl derivatives. The results showed that sonication significantly enhanced the reaction speed, simplified the process, and increased the yields compared to traditional heating methods [53].

15.6 Gum-based nanostructures

A study described the creation of palladium nanoparticles achieved using gum arabic as a support, in an environmentally friendly manner without requiring any external reducing substances. These produced nanoparticles were then utilized as a catalyst for the Mizoroki-Heck reaction, which was conducted under solvent-free conditions (Figure 15.30). Also, the catalyst demonstrated its durability by being effectively recycled for five consecutive uses [54].

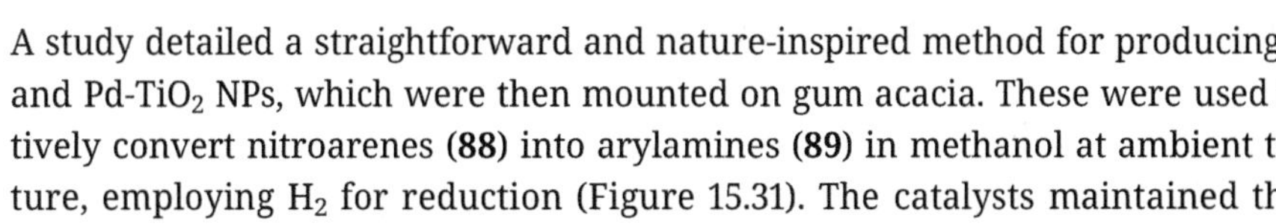

Figure 15.30: Mizoroki-Heck reaction in the presence of Pd nanoparticles supported on gum arabic.

A study detailed a straightforward and nature-inspired method for producing Pd-ZnO and Pd-TiO_2 NPs, which were then mounted on gum acacia. These were used to selectively convert nitroarenes (**88**) into arylamines (**89**) in methanol at ambient temperature, employing H_2 for reduction (Figure 15.31). The catalysts maintained their efficiency after being reused five times. This research highlighted a process that was kind to the environment, with rapid reaction times and substantial yields [55].

Xanthan gum is gaining increased interest in the field of catalysis science and technology. A study showed that a nanocatalyst made from Fe_3O_4 combined with xanthan gum biopolymer, which can be magnetically separated, was synthesized using an in situ approach. This nanocatalyst facilitated the creation of 2-amino-3-

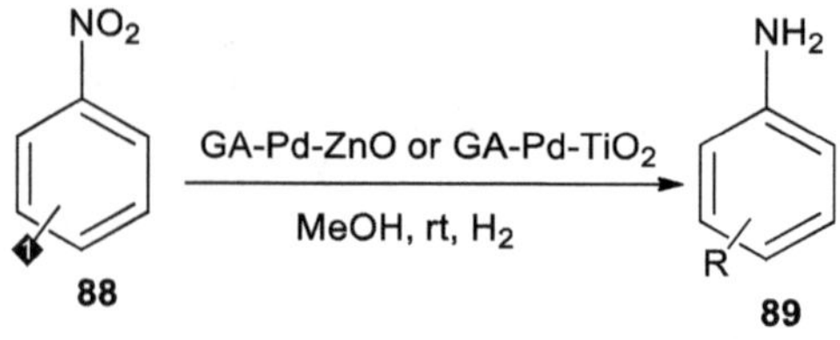

Figure 15.31: Selective hydrogenation of nitroarenes catalyzed by GA-Pd-ZnO or GA-Pd-TiO_2.

cyano-4H-pyran derivatives through a multicomponent reaction that occurred at room temperature with ethanol as the solvent [56]. A study outlined the development of a new eco-friendly polymer catalyst system for producing pyrazolopyridine compounds with biological activity. The nanocatalyst was created through a three-step process: (I) generating copper ferrite MNPs using the co-precipitation technique, (II) attaching acrylonitrile to the acacia gum (AG) framework through graft copolymerization, initiated by ammonium persulfate and crosslinked with N, N'-methylene bisacrylamide along with the copper ferrite, and (III) modifying the resulting hydrogel with hydroxylamine hydrochloride to form an acacia gum-grafted-polyamidoxime/$CuFe_2O_4$ (AG-g-PAO/$CuFe_2O_4$) hydrogel nanocatalyst. This process yielded the target products in impressive quantities (90–97%) swiftly (15–40 min) and without complex purification steps [57].

A study presented a catalytic process for efficiently producing pyrazolopyrazole derivatives. It involved a mix of hydrolyzed Arabic gum-g-polyacrylonitrile/$ZnFe_2O_4$ catalyst, malononitrile, aromatic aldehyde, hydrazine hydrate, and ethyl acetoacetate. The method utilized an eco-friendly, heterogeneous catalyst composed of zinc ferrite magnetic particles, Arabic gum, *N,N*-methylene bisacrylamide, ammonium persulfate, and an alkaline solution for hydrolysis. This organic catalyst, with its pliable, three-dimensional cross-linked structure featuring distinct acid and base sites, proved to be an effective catalyst for the one-step creation of pyranopyrazole derivatives. Using just 0.01 g of the catalyst at 80 °C in a water/ethanol solvent, the highest yield of dihydropyrano[2,3-c]pyrazole reached 96% in a mere 15 min [58].

15.7 Chitin-based nanostructures

An eco-friendly and economical method was created for the direct production of a Rh/N-C catalyst through a single-step pyrolysis process. This method utilized sustainable chitin as the source of both carbon and nitrogen to generate finely distributed Rh particles. By adjusting the pyrolysis temperature, it is possible to modify the oxidation state and size of the Rh particles within the nitrogen-doped carbon framework. These changes significantly affect the Rh/N-C catalyst's efficiency in selectively converting benzoic acid into cyclohexane carboxylic acid [59].

A study detailed an improved extraction technique for chitin (Ch) from the cuttlebone of cuttlefish found in the Persian Gulf. This α-chitin served as the biological basis for developing nanostructured Ch/ZnO through a hydrothermal process. The research aimed to enhance these materials with catalytic properties. Consequently, the nanostructured Ch/ZnO was employed as a recyclable nanocatalyst for the eco-friendly and effective production of benzo[a]pyrano[2,3-c]phenazine derivatives (**90**) (Figure 15.32), utilizing a four-component, microwave-assisted domino reaction [60].

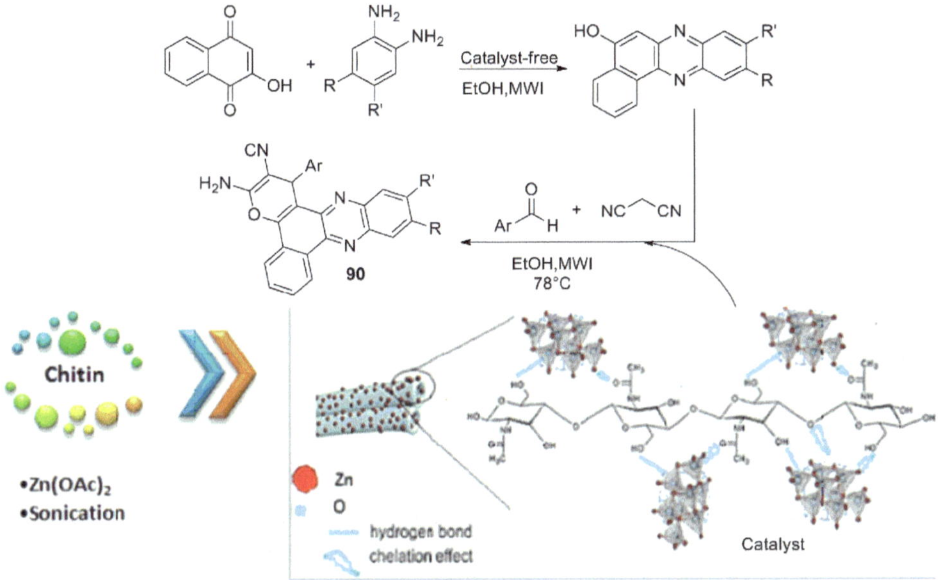

Figure 15.32: Synthesis of benzo[a]pyrano[2,3-c]phenazines in the presence of Ch/ZnO [60].

A study described the utilization of chitin and chitosan nanocrystals as natural supports for Pd NPs, which act as a heterogeneous catalyst in the Heck coupling reaction. A straightforward one-pot method was employed to reduce a Pd salt precursor and anchor it onto the nanocrystal supports. The Heck coupling reactions conducted with these catalysts achieved complete product formation under mild conditions, surpassing other biomass-supported catalysts like those based on cellulose nanocrystals. These findings highlighted the promising role of chitinous nanomaterials as efficient support structures in cross-coupling chemical reactions [61].

A study investigated the use of nanofibrous microspheres made from chitin solution as a base for creating three distinct Pd catalysts supported on chitin, each activated in a different way. This led to the identification of variations in their structures and catalytic behaviors. The Pd/chitin-Ar emerged as the superior catalyst, achieving complete CO conversion in the CO oxidation reaction with a reduced amount of Pd and maintaining remarkable stability over a 24-h reaction cycle. Notably, this catalyst was also tested in the Heck coupling reaction (Figure 15.33), where it demonstrated both high catalytic performance and stability, maintaining about 94% effectiveness over approximately six runs [62].

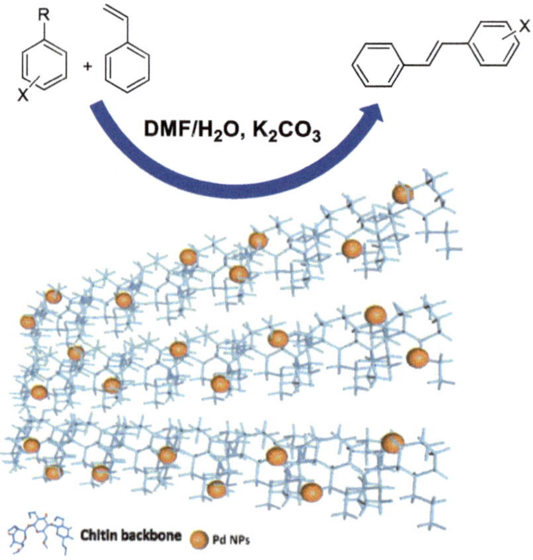

Figure 15.33: Heck coupling reaction in the presence of Pd/chitin-Ar [62].

Utilizing ultrasonic techniques, researchers developed a novel and straightforward method for azido pre-click modification of chitin, leading to the production of new water-soluble chitin derivatives through click chemistry. Additionally, the authors anchored a non-recyclable yet highly efficient homogeneous catalyst, the imino-isocyanide Pd(II) complex ([$PdCl_2$(N(p-tol) = CPh_2)(CNCy)]), onto the newly created triazole chitin derivatives for the Sonogashira reaction (Figure 15.34). The authors also established protocols for forming the Pd(II)-based chitin complex as either powder or nanoparticles. The nanoparticles, in particular, proved to be highly effective and reusable catalysts for the Sonogashira cross-coupling, operating without the need for anaerobic conditions or a copper cocatalyst, and uniquely, in water as the solvent. This research marked the inaugural demonstration of a chitin-supported Pd catalyst for cross-coupling reactions. Moving forward, the team plans to broaden the scope of cross-coupling reactions facilitated by these nanoparticles and to incorporate the azido chitin derivatives in click chemistry with various substrates to synthesize new water-soluble chitin derivatives for subsequent biological activity evaluations [63].

An article discussed the creation of chitin-derived carbon microspheres doped with nitrogen and oxygen (referred to as ChC) using straightforward methods like dissolution, sol-gel processing, and carbonization. Following this, a new type of magnetic catalyst composed of Ag-Fe_3O_4 attached to the chitin-based carbon microspheres

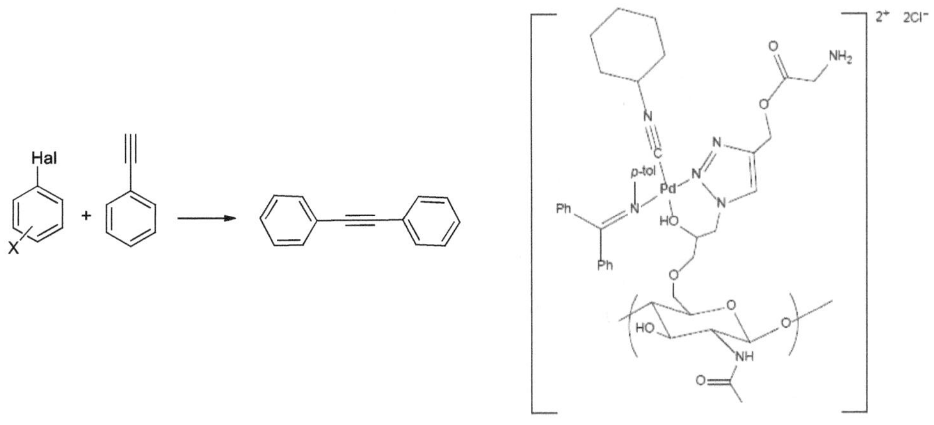

The structure of the Pd(II)-chitin based complex

Figure 15.34: The Sonogashira reaction and the structure of the Pd(II)-chitin-based complex [63].

(named MChC) was synthesized via an in situ redox reaction. The Ag-Fe_3O_4 NPs were anchored within the MChC structure by bonding with C, N, and O atoms. The catalytic efficiency of MChC was tested by its ability to reduce 4-nitrophenol. The substance 4-nitrophenol was completely converted to 4-aminophenol in just 5 min using the MChC-45 catalyst. Furthermore, MChC proved to be easily recoverable from the solution using a magnet in just 8 s and maintained a high level of catalytic performance even after being used for 10 cycles [64].

15.8 Starch-based nanostructures

A team created a catalyst that is both effective and eco-friendly, known as Pd@Fe_3O_4-NH_2/starch, which can be easily separated using magnets. This catalyst was tested for a range of organic chemical reactions, including the selective oxidation of benzyl alcohol derivatives **(91)**, the reduction of nitrocompounds **(92)**, and the reductive amination of aldehydes **(93)** with nitro compounds **(94)**, all under gentle conditions (Figure 15.35). The outcomes were outstanding [65].

A study was pioneering in its use of HO_3S-SNPs as an innovative and highly promising solid acid catalyst for crafting 14-aryl-14-H-dibenzo[a,j]xanthenes **(95)** (Figure 15.36). The use of starch-SO_3H stands out due to its high efficiency, affordability, simplicity in processing, and its status as a natural solid acid catalyst, making this method both economically viable and environmentally friendly. Additionally, the approach is aligned with the principles of green chemistry, thanks to its solvent-free nature, cost-effectiveness, nontoxicity, high yields of the targeted benzoxanthenes, and expedited reaction times [66].

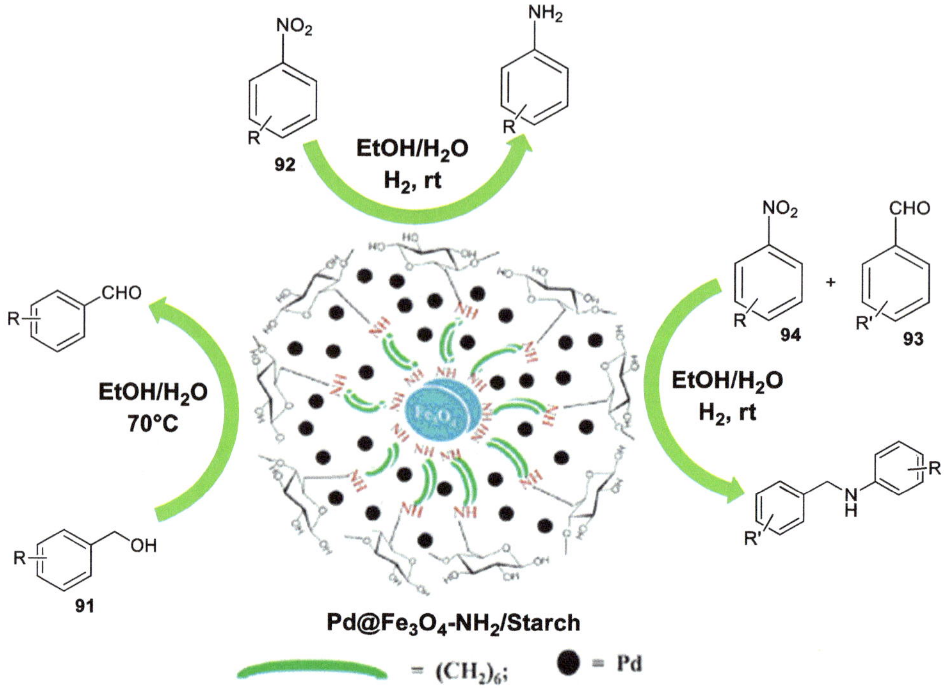

Figure 15.35: The selective oxidation of benzyl alcohols, the reduction of nitro compounds, and the reductive amination of aldehydes in the presence of Pd@Fe_3O_4-NH_2/starch [65].

A researcher synthesized a biomagnetic nanocomposite by functionalizing chitosan with cysteine and cross-linking it with starch oxide, then immobilizing silver ions onto it. They conducted a one-pot A3-coupling reaction involving both aliphatic and aromatic aldehydes (**96**), cyclic amines (**97**), and phenyl acetylene (**98**), facilitated by a small amount of the Fe_3O_4@CS-St_{OX}@Cys@Ag^+ nanocomposite catalyst (Figure 15.37). This process efficiently yielded a wide array of propargylamine derivatives (**99**) with good to excellent results in a short span of time. The methodology's benefits include not requiring additional bases or cocatalysts, avoiding side reactions like aldehyde/alkyne cyclization, and allowing for the catalyst's easy recovery and repeated use without diminishing its effectiveness [67].

An article introduced a cutting-edge catalyst created by integrating an ionic liquid with β-cyclodextrin, resulting in a compound known as βCD-IL@M-Starch. This catalyst was instrumental in producing imidazo[2,1-b][1,3,4]thiadiazol-5-amine and imidazo[1,2-a]pyridin-3-amine derivatives. It demonstrated excellent performance, particularly in generating diphenylimidazo[2,1-b][1,3,4]thiadiazol-5-amine derivatives from semicarbazide, isocyanides, and benzaldehyde. The reactions were carried out under gentle conditions, yielding high-quality products with the aid of βCD-IL@M-Starch. No-

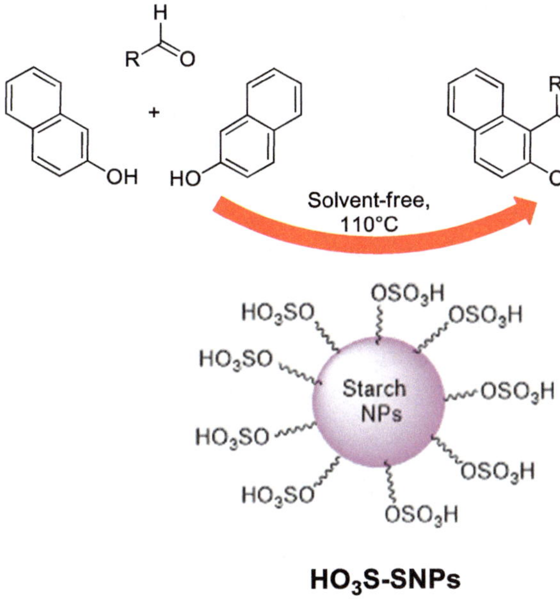

Figure 15.36: Synthesis of 14-aryl-14-H-dibenzo[a,j]xanthenes in the presence of HO_3S-SNPs [66].

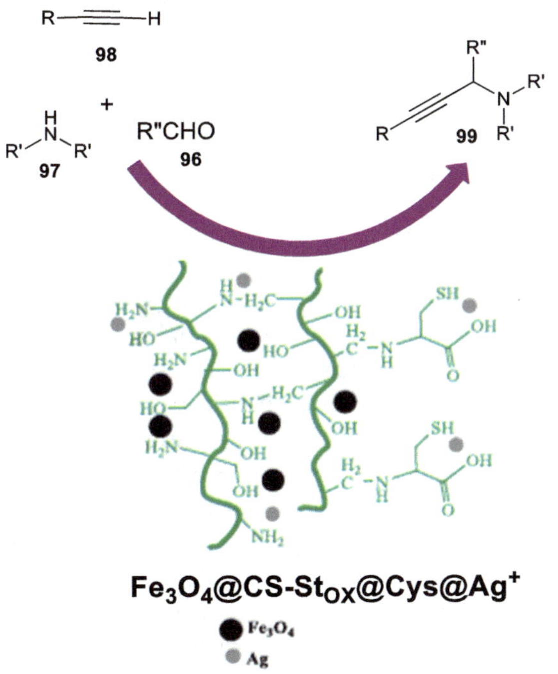

Figure 15.37: Synthesis of propargylamine derivatives in the presence of Fe_3O_4@CS-St$_{OX}$@Cys@Ag$^+$ [67].

tably, the catalyst proved to be magnetically recoverable and maintained its high efficiency across 10 consecutive uses [68].

A team devised a straightforward and effective method for synthesizing biologically significant imidazopyrimidine derivatives. This process was enhanced by ultrasound and uses starch-functionalized magnetite nanoparticles as a catalyst in water at ambient temperature. The method was characterized by its wide range of applicable substrates, high atom efficiency, and the ease with which both products and catalyst can be separated from the reaction mix. It also boasted excellent conversion rates, short reaction times, chemoselectivity, the use of environmentally friendly solvents, and the employment of a biocatalyst. These features render this protocol a superior alternative to those previously established. A notable advantage of this new methodology is the catalyst's ability to be recycled and used effectively for up to six cycles [69].

Magnetic nanoparticles with a core-shell structure, coated with a gold complex (Fe_3O_4@Starch-Au), were synthesized using a hydrothermal technique. These newly created catalysts demonstrated their effectiveness in facilitating cross-coupling reactions. They were used to combine nontoxic aryl tosylates (**100**) with phenols (**101**), resulting in the production of the targeted diaryl ethers (**102**) via an Ullmann-type diaryl ether formation process (Figure 15.38). This method enabled the successful coupling of various aryl tosylates and phenols, including those with aromatic and heterocyclic structures, in water, yielding significant amounts of the desired product [70].

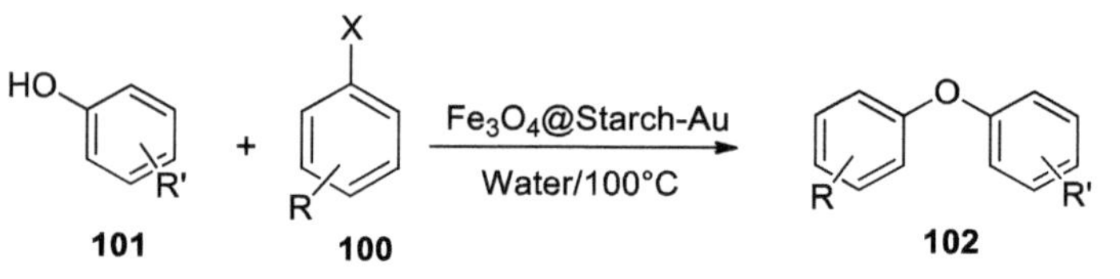

Figure 15.38: O-Arylation of substituted phenols with aryl tosylates catalyzed by Fe_3O_4@Starch-Au.

A research paper described the production of sulfonic acid-functionalized magnetic starch, labeled as starch/SPION@SO_3H, through the polymerization of 4-styrenesulfonic acid on magnetic starch's surface. The magnetic starch itself, referred to as starch/SPION, is created by attaching silica-encapsulated superparamagnetic iron oxide nanoparticles (SPION) to starch through covalent bonds. This was achieved by first modifying starch/SPION with allyltrimethoxysilane, followed by adding 4-styrenesulfonic acid to the material. The catalytic efficiency of starch/SPION@SO_3H was tested in a multicomponent reaction involving 4-hydroxycoumarin, benzaldehydes, dimedone, and ammonium acetate to produce chromeno[4,3-b]quinoline-6,8(9*H*)-dione derivatives. The catalyst was found to be highly effective in generating these derivatives. A variety of benzaldehydes, featuring either electron-donating or electron-withdrawing groups, were utilized as starting materials, and all reacted to form the end products with high isolated yields. Moreover, the catalyst proved to be highly reusable, maintaining its activity over 10 consecutive uses [71].

In a conducted research, sulfur-doped graphitic carbon nitride, referred to as S-g-C_3N_4, was synthesized in situ using thiourea through a straightforward thermal process and was utilized as an optimal support for catalysts. Ag nanoparticles were anchored onto the magnetic S-g-C_3N_4 that had been modified with starch, denoted as S-g-C_3N_4/starch-Fe_3O_4-Ag. This composite, S-g-C_3N_4/starch-Fe_3O_4-Ag, exhibited commendable catalytic performance in the reduction of nitro-aromatic compounds (**103**), with sodium borohydride ($NaBH_4$) serving as the reducing agent (Figure 15.39). Specifically, the reduction of 4-nitroaniline in water reached an efficiency of 99% when 5 mg of the catalyst was used at a temperature of 50 °C over a period of 90 min [72].

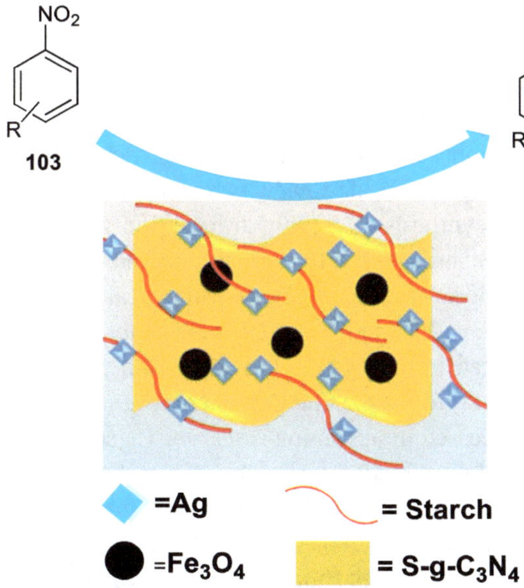

Figure 15.39: The reduction of nitro-aromatic compounds in the presence of S-g-C_3N_4/starch-Fe_3O_4-Ag [72].

An eco-friendly and gentle method was introduced for the selective oxidative esterification of a range of aldehydes using a new palladium-based chitosan-starch complex embedded within kaolin (Kaolin@CS-starch-Pd). This novel material served as a heterogeneous, recyclable, and biocompatible nanocatalyst. Molecular oxygen was employed as the oxidant, producing only water as a by-product. This technique successfully transformed numerous aldehydes into their corresponding methyl esters with high efficiency, all under mild conditions that eliminated the need for any preliminary activation steps [73].

15.9 Conclusion

Creating nanoparticles and quantum dots with the support of carbohydrates is a proficient approach in the field of nanotechnology. To synthesize potent catalysts, one can use carbohydrate-supported MNPs either on their own or in conjunction with other market-available substances. A wide array of carbohydrates is utilized in generating different nanocatalysts. However, there is still a considerable need for advancement in crafting eco-friendly catalysts. This is crucial to lessen the reliance on traditional toxic substances and to simplify as well as economize the production process required to produce the intended catalyst.

Carbohydrates serve as a base for stabilizing metal nanoparticles or compounds, including Pd, Ag, Cu, ZnO, TiO_2, and Au, among others. Recent advancements have highlighted the versatility of carbohydrate-supported various nanoparticles and quantum dots in diverse scientific inquiries. The use of these carbohydrate-based nanocatalysts has been particularly prominent in various chemical processes. These include Suzuki-Miyaura cross-coupling, Heck, Suzuki, Knoevenagel-Michael cascade reactions, nitro compound reductions, alcohol oxidations, and multicomponent reactions, which have gained considerable interest lately. Therefore, this review primarily aims to collate and discuss the latest findings on the synthesis of carbohydrate-supported nanoparticles and quantum dots, as well as their efficacy in the aforementioned chemical reactions. The synthesis of various compounds using carbohydrate catalysts supported with nanostructures has been economical, simple, environmentally friendly, and sustainable. The products are produced in high yield, and some catalysts were able to be reused for multiple loads with minimal impact on the reaction process.

Numerous obstacles remain to be addressed in upcoming research. These include the development of cost-effective, large-scale production techniques for carbohydrate-supported nanostructured materials, as well as enhancing the processes for creating carbohydrate-supported magnetic nanoparticles, which are favored for their straightforward removal from reaction mixtures. Also, studies should move toward commercialization, as conducting studies for commercialization in this field is limited and requires more detailed investigations.

References

[1] Khalilzadeh MA, Kim SY, Jang HW, Luque R, Varma RS, Venditti RA, Shokouhimehr M. Carbohydrate-based nanostructured catalysts: Applications in organic transformations. Mater Today Chem, 2022, 24, 100869. https://doi.org/10.1016/j.mtchem.2022.100869.

[2] Badawy MEI, Rabea EI. A biopolymer chitosan and its derivatives as promising antimicrobial agents against plant pathogens and their applications in crop protection. Int J Carbohydr Chem, 2011, 2011, 460381. DOI: 10.1155/2011/460381.

[3] Rinaudo M. Chitin and chitosan: Properties and applications. Prog Polym Sci, 2006, 31, 603–632. https://doi.org/10.1016/j.progpolymsci.2006.06.001.

[4] Jeon W, Ban C, Park G, Yu TK, Suh JY, Woo HC, Kim DH. Catalytic hydrothermal conversion of macroalgae-derived alginate: Effect of pH on production of furfural and valuable organic acids under subcritical water conditions. J Mol Catal A Chem, 2015, 399, 106–113. https://doi.org/10.1016/j.molcata.2015.01.011.

[5] Bonnin E, Garnier C, Ralet MC. Pectin-modifying enzymes and pectin-derived materials: Applications and impacts. Appl Microbiol Biotechnol, 2014, 98, 519–532. DOI: 10.1007/s00253-013-5388-6.

[6] Klemm D, Heublein B, Fink HP, Bohn A. Cellulose: Fascinating biopolymer and sustainable raw material. Angew Chem Int Ed, 2005, 44, 3358–3393. https://doi.org/10.1002/anie.200460587.

[7] Uyama H. Functional polymers from renewable plant oils. Polym J, 2018, 50, 1003–1011. DOI: 10.1038/s41428-018-0097-8.

[8] Chen P, Cho SY, Jin HJ. Modification and applications of bacterial celluloses in polymer science. Macromol Res, 2010, 18, 309–320. DOI: 10.1007/s13233-010-0404-5.

[9] Sashidhar RB, Raju D, Karuna R. Tree gum: Gum kondagogu. In Polysaccharides: Bioactivity and Biotechnology. Ramawat KG, Mérillon J-M. (Eds.). Springer International Publishing: Cham, 2015, pp. 185–217. DOI: 10.1007/978-3-319-16298-0_32.

[10] Ghaderi A, Gholinejad M, Firouzabadi H. Palladium deposited on naturally occurring supports as a powerful catalyst for carbon-carbon bond formation reactions. Curr Org Chem, 2015, 20, 327–348. DOI: 10.2174/1385272819666150423202719.

[11] Molnár Á, Papp A. ChemInform abstract: The use of polysaccharides and derivatives in palladium-catalyzed coupling reactions. Catal Sci Technol, 2014, 4. DOI: 10.1039/C3CY00699A.

[12] Salim FS, Sargin I, Arslan G. Carbon quantum dots and chitosan-based heterogeneous silver catalyst for reduction of nitroaromatic compounds. Turkish J Chem, 2023, 47, 148–163. DOI: 10.55730/1300-0527.3525.

[13] Shahriari M, Sedigh MA, Mahdavian Y, Mahdigholizad S, Pirhayati M, Karmakar B, Veisi H. In situ supported Pd NPs on biodegradable chitosan/agarose modified magnetic nanoparticles as an effective catalyst for the ultrasound assisted oxidation of alcohols and activities against human breast cancer. Int J Biol Macromol, 2021, 172, 55–65. DOI: 10.1016/j.ijbiomac.2021.01.037.

[14] Shaikh N, Pamidimukkala P. Magnetic chitosan stabilized palladium nanostructures: Potential catalysts for aqueous Suzuki coupling reactions. Int J Biol Macromol, 2021, 183, 1560–1573. DOI: 10.1016/j.ijbiomac.2021.05.103.

[15] Nasrollahzadeh M, Bakhshali-Dehkordi R, Kamali TA, Orooji Y, Shokouhimehr M. Preparation of magnetic chitosan-supported palladium-5-amino-1H-tetrazole complex as a magnetically recyclable catalyst for Suzuki-Miyaura coupling reaction in green media. J Mol Struct, 2021, 1244, 130873. DOI: 10.1016/j.molstruc.2021.130873.

[16] Veisi H, Ozturk T, Karmakar B, Tamoradi T, Hemmati S. In situ decorated Pd NPs on chitosan-encapsulated Fe3O4/SiO2-NH2 as magnetic catalyst in Suzuki-Miyaura coupling and 4-nitrophenol reduction. Carbohydr Polym, 2020, 235, 115966. DOI: 10.1016/j.carbpol.2020.115966.

[17] Baran T. Pd NPs@Fe3O4/chitosan/pumice hybrid beads: A highly active, magnetically retrievable, and reusable nanocatalyst for cyanation of aryl halides. Carbohydr Polym, 2020, 237, 116105. DOI: 10.1016/j.carbpol.2020.116105.

[18] Reyes-Mercado E, Rivas-Loaiza JA, García-Merinos JP, López Y, González-Campos JB. Chitosan-supported copper salt and copper metal nanoparticles/copper (I) oxide microcrystals: Efficient and recyclable heterogeneous catalysts for the synthesis of bis(indolyl)methanes. Chem Eng Process – Process Intensif, 2021, 159, 108201. DOI: 10.1016/j.cep.2020.108201.

[19] Hasan K, Shehadi IA, Al-Bab ND, Elgamouz A. Magnetic chitosan-supported silver nanoparticles :A heterogeneous catalyst for the reduction of 4-nitrophenol. Catalysts, 2019, 9, 839.

[20] Veisi H, Najafi S, Hemmati S. Pd(II)/Pd(0) anchored to magnetic nanoparticles (Fe3O4) modified with biguanidine-chitosan polymer as a novel nanocatalyst for Suzuki-Miyaura coupling reactions. Int J Biol Macromol, 2018, 113, 186–194. DOI: 10.1016/j.ijbiomac.2018.02.120.

[21] Veisi H, Sajjadifar S, Biabri PM, Hemmati S. Oxo-vanadium complex immobilized on chitosan coated-magnetic nanoparticles (Fe3O4): A heterogeneous and recyclable nanocatalyst for the chemoselective oxidation of sulfides to sulfoxides with H2O2. Polyhedron, 2018, 153, 240–247. DOI: 10.1016/j.poly.2018.07.013.

[22] Yuan C, Liu Q, Wei M, Zhao S, Yang X, Cao B, Wang S, Abomohra AE, Liu X, Hu Y. Selective oxidation of 5-hydroxymethylfurfural to furan-2,5-dicarbaldehyde using chitosan-based biochar composite cadmium sulfide quantum dots. Fuel, 2022, 320, 123994. https://doi.org/10.1016/j.fuel.2022.123994.

[23] Nourmohammadi M, Rouhani S, Azizi S, Maaza M, Msagati TA, Rostamnia S, Hatami M, Khaksar S, Zarenezhad E, Jang HW, Shokouhimehr M. Magnetic nanocomposite of crosslinked chitosan with 4,6-diacetylresorcinol for gold immobilization (Fe3O4@CS/DAR-Au) as a catalyst for an efficient one-pot synthesis of propargylamine. Mater Today Commun, 2021, 29, 102798. https://doi.org/10.1016/j.mtcomm.2021.102798.

[24] Liandi AR, Cahyana AH, Yunarti RT, Wendari TP. Facile synthesis of magnetic Fe3O4@Chitosan nanocomposite as environmentally green catalyst in multicomponent Knoevenagel-Michael domino reaction. Ceram Int, 2022, 48, 20266–20274. https://doi.org/10.1016/j.ceramint.2022.03.307.

[25] Amirnejat S, Nosrati A, Javanshir S. Superparamagnetic Fe3O4@Alginate supported L-arginine as a powerful hybrid inorganic–organic nanocatalyst for the one-pot synthesis of pyrazole derivatives. Appl Org Chem, 2020, 34, 1–13. DOI: 10.1002/aoc.5888.

[26] Supriya, Pal CK, Sengupta S, Basu JK. One pot synthesis of nano Ag in calcium alginate beads and its catalytic application in p-Nitrophenol reduction with kinetic parameter estimation and model fitting. Can J Chem Eng, 2021, 99, 359–373. DOI: 10.1002/cjce.23831.

[27] Ghorbani-Vaghei R, Veisi H, Aliani MH, Mohammadi P, Karmakar B. Alginate modified magnetic nanoparticles to immobilization of gold nanoparticles as an efficient magnetic nanocatalyst for reduction of 4-nitrophenol in water. J Mol Liq, 2021, 327, 114868. DOI: 10.1016/j.molliq.2020.114868.

[28] Daraie M, Heravi MM, Mohammadi P, Daraie A. Silver incorporated into g-C3N4/Alginate as an efficient and heterogeneous catalyst for promoting click and A3 and KA2 coupling reaction. Sci Rep, 2021, 11, 1–13. DOI: 10.1038/s41598-021-93239-z.

[29] Biswas S, Pal A. Iron oxide-loaded alginate-bentonite hydrogel beads as a green and sustainable catalyst for 4-nitrophenol reduction. Mater Today Commun, 2021, 28, 102588. DOI: 10.1016/j.mtcomm.2021.102588.

[30] Khan SB, Bakhsh EM, Akhtar K, Kamal T, Shen Y, Asiri AM. Copper oxide-antimony oxide entrapped alginate hydrogel as efficient catalyst for selective reduction of 2-nitrophenol. Polymers (Basel), 2022, 14, 458. DOI: 10.3390/polym14030458.

[31] Helmiyati H, Yunarti RT, Dini FW. Magnetic alginate–carboxymethyl cellulose to immobilize copper nanoparticles as a green and sustainable catalyst for 4-nitrophenol reduction. Heliyon, 2023, 9, e14111. DOI: 10.1016/j.heliyon.2023.e14111.

[32] Sadjadi S, Heydari A. Palladated cyclodextrin nanosponge-alginate dual bead as an efficient catalyst for hydrogenation of nitroarenes in aqueous solution. Polymers (Basel), 2023, 15, 3240. DOI: 10.3390/polym15153240.

[33] Bakhsh EM, Maslamani N, Akhtar K, Danish EY, Asiri AM, Khan SB. Fe2O3-NiO embedded calcium alginate-carboxymethyl cellulose composite as an efficient nanocatalyst for 4-nitrophenol reduction. Inorg Chem Commun, 2023, 156, 111157. https://doi.org/10.1016/j.inoche.2023.111157.

[34] Fadl FIAE, Taleb MFA. Hybrid Alginate/TiO2/Ag bio-nanocomposite beads for catalytic hydrogenation of 2-nitrophenol. J Inorg Organomet Polym Mater, 2023, 33, 2142–2153. DOI: 10.1007/s10904-023-02651-7.

[35] Tummachote J, Mahanitipong U, Inthanusorn W, Rutnakornpituk B, Rutnakornpituk M. Highly robust alginate-based magnetic nanosorbents of Pd catalysts for the 4-nitrophenol reduction. Trends Sci, 2024, 21, 1–17. DOI: 10.48048/tis.2024.7465.

[36] Zhang W, Veisi H, Sharifi R, Salamat D, Karmakar B, Hekmati M, Hemmati S, Zangeneh MM, Zhang Z, Su Q. Fabrication of Pd NPs on pectin-modified Fe3O4 NPs: A magnetically retrievable nanocatalyst for efficient C–C and C–N cross coupling reactions and an investigation of its cardiovascular protective effects. Int J Biol Macromol, 2020, 160, 1252–1262. DOI: 10.1016/j.ijbiomac.2020.05.247.

[37] Li Y, Li N, Jiang W, Ma G, Zangeneh MM. In situ decorated Au NPs on pectin-modified Fe3O4 NPs as a novel magnetic nanocomposite (Fe3O4/Pectin/Au) for catalytic reduction of nitroarenes and investigation of its anti-human lung cancer activities. Int J Biol Macromol, 2020, 163, 2162–2171. DOI: 10.1016/j.ijbiomac.2020.09.102.

[38] Khashei Siuki H, Bagherzade G, Ghamari Kargar P. A green method for synthesizing nickel nanoparticles supported by magnetized pectin: Applied as a catalyst for aldehyde synthesis as a precursor in xanthan synthesis. ChemistrySelect, 2020, 5, 13537–13544. DOI: 10.1002/slct.202002946.

[39] Xue W, Yang G, Karmakar B, Gao Y. Sustainable synthesis of Cu NPs decorated on pectin modified Fe3O4 nanocomposite: Catalytic synthesis of 1-substituted-1H-tetrazoles and in-vitro studies on its cytotoxicity and anti-colorectal adenocarcinoma effects on HT-29 cell lines. Arab J Chem, 2021, 14, 103306. DOI: 10.1016/j.arabjc.2021.103306.

[40] Le VD, Le TCH, Chau VT, Le TND, Dang CH, Vo TTN, Nguyen TD, Nguyen TD. Palladium nanoparticlesin situsynthesized onCyclea barbatapectin as a heterogeneous catalyst for Heck coupling in water, the reduction of nitrophenols and alkynes. New J Chem, 2021, 45, 4746–4755. DOI: 10.1039/d0nj05032f.

[41] Ghamari Kargar P, Ravanjamjah A, Bagherzade G. A novel water-dispersible and magnetically recyclable nickel nanoparticles for the one-pot reduction-Schiff base condensation of nitroarenes in pure water. J Chinese Chem Soc, 2021, 68, 1916–1933. DOI: 10.1002/jccs.202100172.

[42] Khashei Siuki H, Ghamari Kargar P, Bagherzade G. New acetamidine Cu(II) Schiff base complex supported on magnetic nanoparticles pectin for the synthesis of triazoles using click chemistry. Sci Rep, 2022, 12, 3771. DOI: 10.1038/s41598-022-07674-7.

[43] Ghamari Kargar P, Ghasemi M, Bagherzade G. Copper (II) supported on a post-modified magnetic pectin Fe3O4@Pectin~Imidazole~SO3H-Cu(II): An efficient biopolymer-based catalyst for selective oxidation of alcohols with aqueous TBHP. Sci Iran, 2022, 29, 1338–1350. DOI: 10.24200/sci.2021.58355.5689.

[44] Bakhtiarian M, Khodaei MM. Synthesis of 2,3-dihydro-4(1 H) quinazolinones using a magnetic pectin-supported deep eutectic solvent. Colloids Surf A Physicochem Eng Asp, 2022, 641, 128569. https://doi.org/10.1016/j.colsurfa.2022.128569.

[45] Bakhtiarian M, Khodaei MM. Sonochemical synthesis of 1,4-dihydropyridines using a bio-derived magnetic nanocomposite based on the pectin modified with the di-sulfonic acids under mild conditions. Mater Today Commun, 2021, 29, 102791. https://doi.org/10.1016/j.mtcomm.2021.102791.

[46] Yarmohammadi N, Ghadermazi M, Derikvand Z, Mozafari R. In situ synthesis of bimetallic γ-Fe2O3/Cu nanoparticles over pectin hydrogel obtained from biomass resource (orange peel) as a reusable green catalyst for oxidation and C-S cross-coupling reactions. Chem Pap, 2022, 76, 4289–4307. DOI: 10.1007/s11696-022-02174-4.

[47] Maleki A, Jafari AA, Yousefi S. MgFe2O4/cellulose/SO3H nanocomposite: A new biopolymer-based nanocatalyst for one-pot multicomponent syntheses of polysubstituted tetrahydropyridines and dihydropyrimidinones. J Iran Chem Soc, 2017, 14, 1801–1813. DOI: 10.1007/s13738-017-1120-2.

[48] Karami S, Zeynizadeh B, Shokri Z. Cellulose supported bimetallic Fe–Cu nanoparticles: A magnetically recoverable nanocatalyst for quick reduction of nitroarenes to amines in water. Cellulose, 2018, 25, 3295–3305. DOI: 10.1007/s10570-018-1809-0.

[49] Mirjalili BBF, Aref F. Nano-cellulose/BF3/Fe3O4: A magnetic bio-based nano-catalyst for the synthesis of pyrimido[2,1-b]benzothiazoles under solvent-free conditions. Res Chem Intermed, 2018, 44, 4519–4531. DOI: 10.1007/s11164-018-3401-5.

[50] Lu H, Zhang L, Ma J, Alam N, Zhou X, Ni Y. Nano-cellulose/mof derived carbon doped cuo/fe3o4 nanocomposite as high efficient catalyst for organic pollutant remedy. Nanomaterials, 2019, 9, 277. DOI: 10.3390/nano9020277.

[51] Salimi M, Nasseri MA, Jazi BN. Cu(II)-immobilized on functionalized magnetic nano-fibrillated cellulose (Fe3O4@NFC/E-CHDA-CuII): A novel, efficient and magnetically nanocatalyst for the one-pot synthesis of tetrahydrobenzo[b]pyran derivatives. J Iran Chem Soc, 2019, 16, 2221–2230. DOI: 10.1007/s13738-019-01689-0.

[52] Hosseinikhah SS, Mirjalili BBF. Fe3O4@NCs/Sb(V): As a cellulose based nano-catalyst for the synthesis of 4H-pyrimido[2,1-b]benzothiazoles. Polycycl Aromat Compd, 2022, 42, 1013–1022. DOI: 10.1080/10406638.2020.1764985.

[53] Ghamari Kargar P, Nayebi M, Parhizi Z, Varma RS. Nickel nanoparticles adorned on magnetized cellulose nanofibers: Ultrasound-facilitated cross coupling reactions. Cellulose, 2022, 29, 9183–9198. DOI: 10.1007/s10570-022-04823-z.

[54] Khazaei A, Rahmati S, Ghaderi A, Roshani L. Palladium nanoparticles supported on gum arabic as a reusable catalyst for solvent-free Mizoroki-Heck reaction. J Iran Chem Soc, 2014, 11, 263–269. DOI: 10.1007/s13738-013-0296-3.

[55] Supriya P, Srinivas BTV, Chowdeswari K, Naidu NVS, Sreedhar B. Biomimetic synthesis of gum acacia mediated Pd-ZnO and Pd-TiO2 – Promising nanocatalysts for selective hydrogenation of nitroarenes. Mater Chem Phys, 2018, 204, 27–36. DOI: 10.1016/j.matchemphys.2017.10.026.

[56] Esmaeili MS, Khodabakhshi MR, Maleki A, Varzi Z. Green, natural and low cost xanthum gum supported Fe3O4 as a robust biopolymer nanocatalyst for the One-Pot synthesis of 2-Amino-3-Cyano-4H-Pyran derivatives. Polycycl Aromat Compd, 2021, 41, 1953–1971. DOI: 10.1080/10406638.2019.1708418.

[57] Hassanzadeh-Afruzi F, Maleki A, Zare EN. Novel eco-friendly acacia gum-grafted-polyamidoxime@copper ferrite nanocatalyst for synthesis of pyrazolopyridine derivatives. J Nanostructure Chem, 2023, 13, 451–462. DOI: 10.1007/s40097-022-00471-8.

[58] Hassanzadeh-Afruzi F, Salehi MM, Heidari G, Maleki A, Zare EN. Hydrolyzed Arabic gum-grafted-polyacrylonitrile@ zinc ferrite nanocomposite as an efficient biocatalyst for the synthesis of pyranopyrazoles derivatives. J Mol Struct, 2023, 1274, 134490. https://doi.org/10.1016/j.molstruc.2022.134490.

[59] Cao Y, Tang M, Li M, Deng J, Xu F, Xie L, Wang Y. In situ synthesis of chitin-derived Rh/N-C Cataylsts: Efficient hydrogenation of benzoic acid and derivatives. ACS Sustain Chem Eng, 2017, 5, 9894–9902. DOI: 10.1021/acssuschemeng.7b01853.

[60] Abolghassem S, Molaei S, Javanshir S. Preparation of α-chitin-based nanocomposite as an effective biocatalyst for microwave aided domino reaction. Heliyon, 2019, 5, e02036. DOI: 10.1016/j.heliyon.2019.e02036.

[61] Jin T, Hicks M, Kurdyla D, Hrapovic S, Lam E, Moores A. Palladium nanoparticles supported on chitin-based nanomaterials as heterogeneous catalysts for the Heck coupling reaction. Beilstein J Org Chem, 2020, 16, 2477–2483. DOI: 10.3762/BJOC.16.201.

[62] Pei X, Li Y, Deng Y, Lu L, Li W, Shi R, Lei A, Zhang L. Chitin microsphere supported Pd nanoparticles as an efficient and recoverable catalyst for CO oxidation and Heck coupling reaction. Carbohydr Polym, 2021, 251, 117020. DOI: 10.1016/j.carbpol.2020.117020.

[63] Kritchenkov AS, Kletskov AV, Egorov AR, Kurasova MN, Tskhovrebov AG, Khrustalev VN. Ultrasound and click chemistry lead to a new chitin chelator. Its Pd(II) complex is a recyclable catalyst for the Sonogashira reaction in water. Carbohydr Polym, 2021, 252, 117167. DOI: 10.1016/j.carbpol.2020.117167.

[64] Wang G, Wang B, Chen X, Pei Y, Zhou S, Ding W. Magnetic porous nano-carbon catalysts supported silver nanoparticles derived from chitin and their application in catalytic reduction reactions. J Appl Polym Sci, 2021, 138, 1–9. DOI: 10.1002/app.51439.

[65] Sharma H, Bhardwaj M, Kour M, Paul S. Highly efficient magnetic Pd(0) nanoparticles stabilized by amine functionalized starch for organic transformations under mild conditions. Mol Catal, 2017, 435, 58–68. DOI: 10.1016/j.mcat.2017.03.019.

[66] Safari J, Aftabi P, Ahmadzadeh M, Sadeghi M, Zarnegar Z. Sulfonated starch nanoparticles: An effective, heterogeneous and bio-based catalyst for synthesis of 14-aryl-14-H-dibenzo[a,j]xanthenes. J Mol Struct, 2017, 1142, 33–39. DOI: 10.1016/j.molstruc.2017.02.095.

[67] Rafiee F, Karder FR. Bio-crosslinking of chitosan with oxidized starch, its functionalization with amino acid and magnetization: As a green magnetic support for silver immobilization and its catalytic activity investigation. Int J Biol Macromol, 2020, 146, 1124–1132. DOI: 10.1016/j.ijbiomac.2019.09.238.

[68] Bahadorikhalili S, Ansari S, Hamedifar H, Mahdavi M. The use of magnetic starch as a support for an ionic liquid-β-cyclodextrin based catalyst for the synthesis of imidazothiadiazolamine derivatives. Int J Biol Macromol, 2019, 135, 453–461. DOI: 10.1016/j.ijbiomac.2019.05.197.

[69] Verma P, Pal S, Chauhan S, Mishra A, Sinha I, Singh S, Srivastava V. Starch functionalized magnetite nanoparticles: A green, biocatalyst for one-pot multicomponent synthesis of imidazopyrimidine derivatives in aqueous medium under ultrasound irradiation. J Mol Struct, 2020, 1203, 127410. DOI: 10.1016/j.molstruc.2019.127410.

[70] Mousavi-Mashhadi SA, Shiri A. On-water and efficient Ullmann-Type O-Arylation cross coupling reaction of phenols and aryl tosylates in the presence of Fe3O4@Starch-Au as nanocatalyst. ChemistrySelect, 2021, 6, 3941–3951. DOI: 10.1002/slct.202004327.

[71] Sayahi MH, Shamkhani F, Mahdavi M, Bahadorikhalili S. Sulfonic acid functionalized magnetic starch as an efficient catalyst for the synthesis of chromeno[4,3-b]quinoline-6,8(9H)-dione derivatives. Starch/Staerke, 2021, 73, 1–8. DOI: 10.1002/star.202000257.

[72] Taheri S, Heravi MM, Mohammadi P. Sulfur-doped graphitic carbon nitride decorated with starch, Fe3O4, and Ag nanoparticles: As efficient and magnetic recoverable nanocatalyst for hydrogenation of nitroaromatics in aqueous media. Diam Relat Mater, 2022, 126, 109078. https://doi.org/10.1016/j.diamond.2022.109078.

[73] Veisi H, Ebrahimi Z, Karmakar B, Joshani Z, Ozturk T. Chitosan-starch biopolymer modified kaolin supported Pd nanoparticles for the oxidative esterification of aryl aldehydes. Int J Biol Macromol, 2021, 191, 465–473. https://doi.org/10.1016/j.ijbiomac.2021.09.106.

Arruje Hameed, Razia Noreen, Kulsoom Ghulam Ali, Bushra Parveen, Muhammad Afzal, Muhammad Abdul Qayyum, Hassam Mehmood, Tahir Farooq*, and Saher Gohar

Chapter 16 Bioconjugation of carbohydrates

16.1 Introduction

Among the various types of biomolecules, carbohydrates possess dynamic and multifaceted structures with hydroxyl groups showing different reactivity patterns, whereas proteins and nucleic acids have confined stereo-chemistries showing a narrow pattern of new combinations. The biomolecules connected with carbohydrate moieties play a variety of vital roles, including cell-to-cell interactions, ligand-receptor bindings, and pathogenic infections. The specific sequence of oligosaccharides conjugated to proteins and lipids acts as sugar codes for controlling a variety of biological functions. Therefore, the understanding of the structure-activity relationship has been considered crucial for carbohydrate-based therapy for controlled bio-functions [1]. Bioconjugation has emerged as a main research field in chemical biology with a vast scope in medical clinics, industrial synthesis, and research laboratories. The application of prepared bioconjugates heavily relies on the selected linkage strategy during bioconjugation. The optimal linkage controls the stability and desired properties of the bioconjugates. Naturally occurring vital glycoconjugates exist in small amounts and are difficult to isolate for chemotherapeutic applications. Therefore, over the years, there has been great interest in the development of synthetic bioconjugates of carbohydrates with tailored physio-biochemical properties [2]. In this connection, several ligation techniques have been developed for conjugating carbohydrates to proteins, lipids, and nucleic acids. Various efforts have been made to generate complex glycans with desired properties using functionalizable substituents and linkers [3]. Stereocontrolled glycosylation is achieved by using selectively protected glycosyl acceptors, promoters, and protecting groups. Generally, Lewis acids in stoichiometric amounts act as promoters activating the donor leaving groups. The oxocarbenium ions

***Corresponding author: Tahir Farooq**, Department of Applied Chemistry, Government College University, Faisalabad, Pakistan, e-mail: tahirfarooqfsd@gmail.com
Arruje Hameed, Razia Noreen, Muhammad Afzal, Department of Biochemistry, Government College University, Faisalabad, Pakistan
Kulsoom Ghulam Ali, Bushra Parveen, Department of Chemistry, Government College University, Faisalabad, Pakistan
Muhammad Abdul Qayyum, Department of Chemistry, Division of Science and Technology, University of Education, Lahore, Pakistan
Hassam Mehmood, Saher Gohar, Department of Applied Chemistry, Government College University, Faisalabad, Pakistan

https://doi.org/10.1515/9783111447094-016

couple with the glycosyl acceptor after the departure of LGs. The introduction of a non-carbohydrate acceptor leads to the formation of glycoconjugates. Group participation, anomeric effects, and steric hindrance largely influence the stereo-selectivity of the glycoconjugates (Figure 16.1) [4].

The list of glycosyl donors includes thioglycosides, imidates, and halides, each having an array of advantages and disadvantages. The oligosaccharides are generally modified with the introduction of functionalizable linkers. The mono and di-saccharides are generally equipped with linkers at the anomeric position. The oligosaccharides are synthesized chemically or result from the digestion of glycolipids and glycoproteins. The chemical treatments of oligosaccharides are usually challenging because of the vulnerable nature of glycosidic linkages. Since 1990, azide and alkene have become attractive functionalizable linkers for the facile preparation of glycoconjugates under mild conditions (Figure 16.2) [5, 6]. During further explorations, the Lewis acid was evaluated as a coupling promoter under metal-free conditions. It helps to activate the anomeric position directly during the glycoconjugation process.

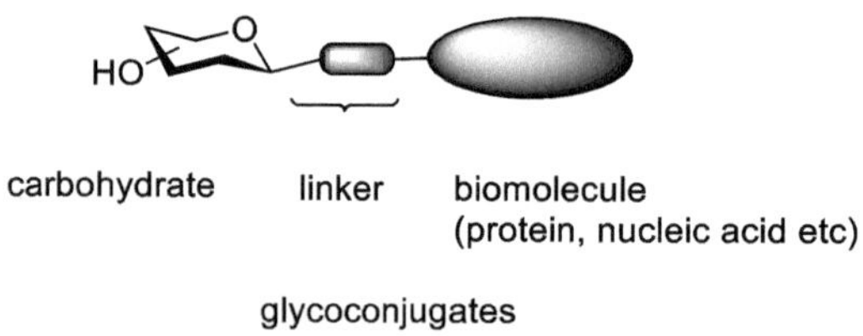

Figure 16.1: Bioconjugation of carbohydrates.

16.2 Classical bioconjugation strategies

The biopolymers and proteins regulate a countless number of physiobiochemical processes in living organisms. Over the years, there has been great interest in elucidating the functions of various biomolecules, including carbohydrates, proteins, and nucleic acids. The covalent derivatization of biomolecules, which is referred to as bioconjugation, has become a leading tool and strategy for understanding the diverse and complex functions of biomolecules [7]. The traditional protocols often resulted in heterogeneous reaction products due to a lack of control over the regiochemistry of reactions (Figure 16.3). Similarly, the target biomolecules often lose their functions due to poor control over undesirable off-site reactions. Over the years, covalent bioconjugation has undergone unprecedented developments, providing highly controlled and site-specific protocols [8]. The cysteine residues have become attractive points for site-specific conjugations due to their frequent occurrence in natural proteins. The thiolate group of cysteine has been variously exploited for most bioconjugation studies. The disulfides, maleimides, and iodoacetamides are the major reactive forms of thiol groups.

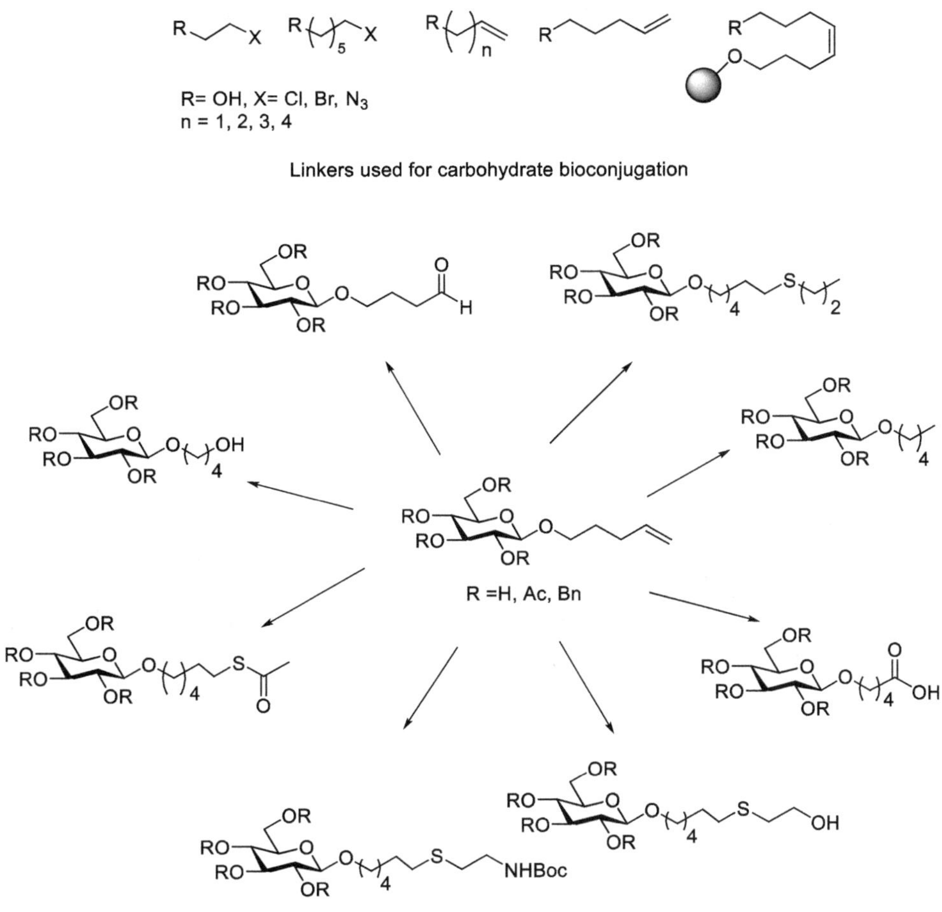

Figure 16.2: Saccharides functionalized with linkers for bioconjugation.

16.2.1 Bioconjugation based on amide bonds

Under neutral conditions, the amide bonds are highly stable and thus have received special attention for bioconjugation. The amino group of lysine side chains can easily be converted into amide bonds when proteins react with small molecules. Recently, Staudinger ligation and chemical ligation have emerged as the main techniques for site-specific bioconjugation of proteins. In chemical ligation, peptide ligation has become a mainstream method for protein bioconjugation. The amide bonds are formed after the reaction of thioester with *N*-terminal cysteine residues. However, the introduction of residual –SH is considered an undesirable aspect during bioconjugation. [7].

The Staudinger reaction has become a powerful ligation technique that proceeds with the generation of aza-ylide or iminophosphorane from the reaction of an azide

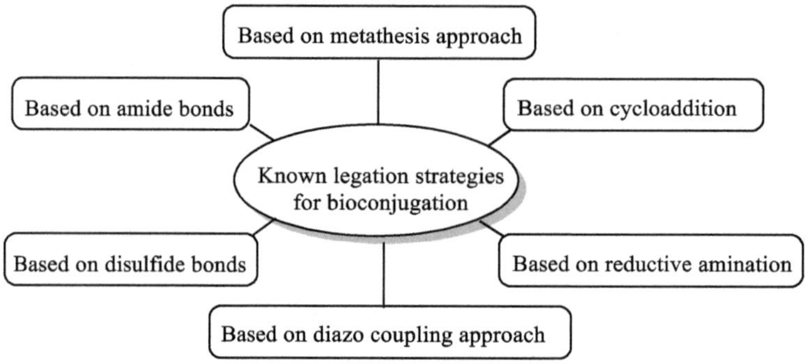

Figure 16.3: Classical bioconjugation strategies.

with trivalent phosphorus. The corresponding amines are produced after the hydrolysis of intermediates. The amides could also be produced from the reaction of acylating agents with intermediates. The Staudinger reactions of carbohydrates became possible with the introduction of glycosyl azides, which could lead to the formation of *N*-glycosylamide. The reaction proceeded from dialkyl-phosphane to phosphinothioester through different steps. Further, an asparagine derivative containing *N*-glycosylamide was received as a Staudinger-end product. The preparation of glycosylamides from azides in one step through the Staudinger reaction accelerated the opening of new horizons for the preparation of carbohydrate-based bioconjugates (Figure 16.4) [9].

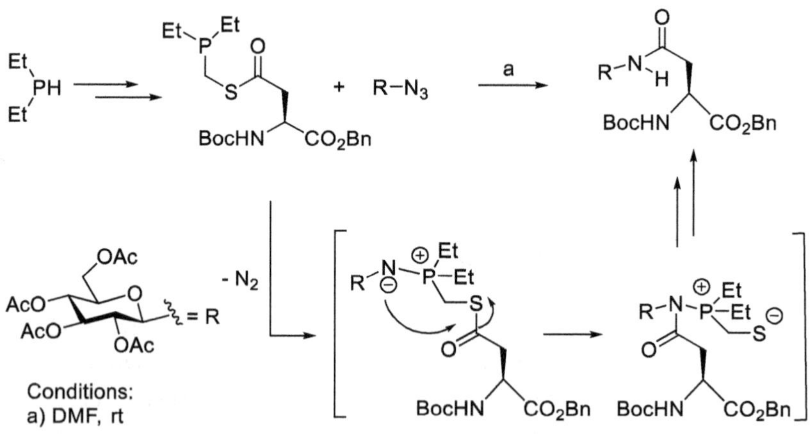

Figure 16.4: Phosphine reduces the azido group to the corresponding amide.

According to the protocol of Staudinger ligation, the azido-containing proteins, peptides, and carbohydrates show site-specific bioconjugation under gentle conditions. In this biochemical transformation, phosphine reduces the azido group to the corresponding amide with the liberation of nitrogen gas. The traceless version of Staudinger ligation produces an amidophosphonium salt, producing amide from phosphinothioester. This versatile ligation technique enabled the synthesis of *N*-glycopeptides and the introduction of azido-containing carbohydrates on cell surfaces. The azido transfer technique has been used for the installation of azido groups in proteins and subsequent site-specific bioconjugations [10].

The aldonic acid was prepared by the oxidation of sugar lactol and further subjected to amidation reaction using activating agents. An earlier report describes the conjugation of bovine serum albumin (BSA) protein to aldonic acid through amidation reaction. The mono-, di-, and tri-saccharide aldonic acids were studied using *N*-(3-dimethylaminopropyl)-*N*-ethylcarbodiimide as an activating agent under acidic conditions. In early years, great attention was focused on the development of linker-assisted amide-mediated glycoconjugations. The sugar moieties were subjected to glycosylation using 9-hydroxynonanoate ester. In subsequent steps, hydrazide and acylazide were formed, which finally reacted with BSA. The described methodology was favorable due to the formation of hydrazides from ester under milder conditions. Secondly, the conjugation reaction became attractive because of low temperature requirements for the reaction of acylazide with BSA. In 2009, the one-pot strategy was introduced for a direct amidation of lactols under green conditions. The iodine-promoted decarboxylative amidation of mono-, di-, and trisaccharides opened up a new and reliable approach for direct bioconjugations [11]. In 2011, 2-amino-2-deoxysugars were subjected to amidation using amine-protected amino acids with carbodiimides as activating agents (Figure 16.5) [12]. In lactol, the amine at C-2 was utilized for the amidation reaction. The large oligosaccharides show relatively slower amidation reactions due to steric hindrance issues. In 2014, the synthetic oligosaccharides were subjected to amide-mediated conjugation with BSA [13].

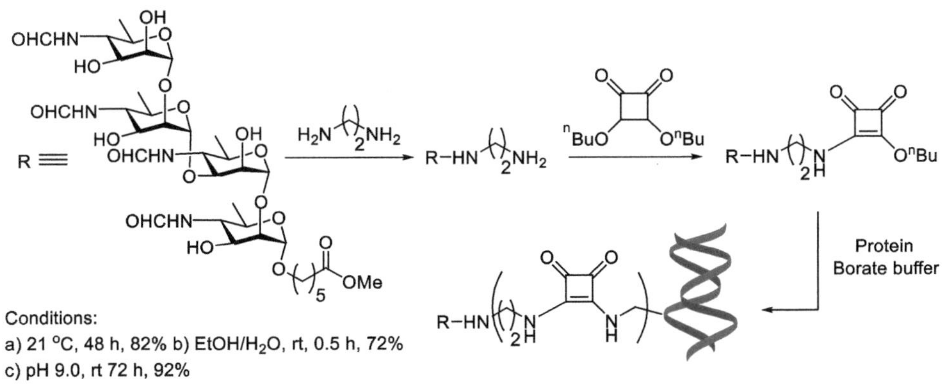

Figure 16.5: Amidation using carbodiimides as activating agents.

16.2.2 Bioconjugation based on reductive amination

The ease of formation of C=N from the condensation of ketones and aldehydes with nitrogen bases under normal conditions highlights its suitability for bioconjugations. The hydroxyl groups of carbohydrates are readily oxidized to aldehydes, thus could be used for the preparation of compounds with C=N. The cell surface carbohydrates could also be easily converted to ketones, enabling them for condensation reactions with nitrogen bases [7].

Reductive amination is one of the classical methods developed for the preparation of glycoconjugates. The known procedure proceeds with the reaction of an amine with a carbonyl functionality, generating an imine, which is subsequently reduced to amines. The reduction proceeds at room temperature under mild reaction conditions. The amino group of lysine generally serves as a target site for the generation of protein-based amines. The bioconjugation of proteins with carbohydrates is generally facilitated by introducing bi-functional linkers. The reductive amination method of bioconjugation facilitates coupling and covalent linkage, showing higher hydrolytic stability [14]. Commonly, three procedures are followed for the coupling of carbohydrates with amines:

1. The primary hydroxyl groups in carbohydrates are oxidized to aldehydic functionalities, which could generate corresponding imines.
2. The formation of dialdehydes from the oxidation of vicinal diols.
3. By reacting selective amines with anomeric carbonyl functionalities.

As described earlier, the amines and carbonyl functionalities are also connected by using bi-functional linkers, producing bioconjugates through reductive amination. According to one of the early known methods, the allyl glycosides were subjected to ozonolysis for the preparation of aldehydes, which produced bioconjugates with proteins through reductive amination (Figure 16.6) [15].

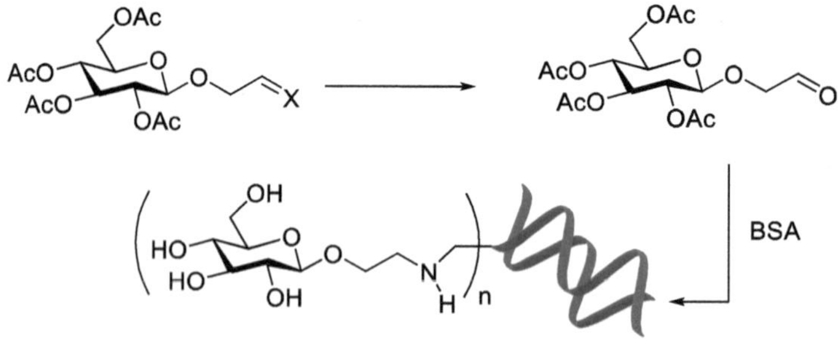

Conditions:
a) i) O_3, MeOH ii) Me_2S b) $NaCNBH_3$, pH 9.0, n = 38 - 45

Figure 16.6: Ozonolysis of allyl glycosides and bioconjugation through reductive amination.

16.2.3 Bioconjugation based on cycloadditions

Over the last two decades, the Huisgen 1,3-dipolar azide-alkyne cycloadditions catalyzed by Cu(I) have become a powerful strategy for various types of bioconjugations. The resulting 1,4-disubstituted-1,2,3-triazole has become a useful linkage for the bioconjugation of proteins, nucleic acids, and carbohydrates. This ligation technique has been widely exploited for the functionalization of biopolymers, DNA, and proteins with fluorescent molecules through bioconjugation. However, this ligation technique offers some issues related to Cu toxicity. Therefore, the Diels-Alder (DA) reaction, another version of cycloaddition, has gained attention for wide-scope applications in bioconjugations [16]. In 2002, a pioneering Cu-catalyzed cycloaddition reaction proved a great advancement for the preparation of various types of bioconjugates. The report described the conjugation of a monosaccharide with an alkyne-containing tetrapeptide. The 1,2,3-triazole glycoconjugates were obtained when 2-azido thiogalactoside reacted with alkyne in the presence of a Cu(I) catalyst (Figure 16.7) [17].

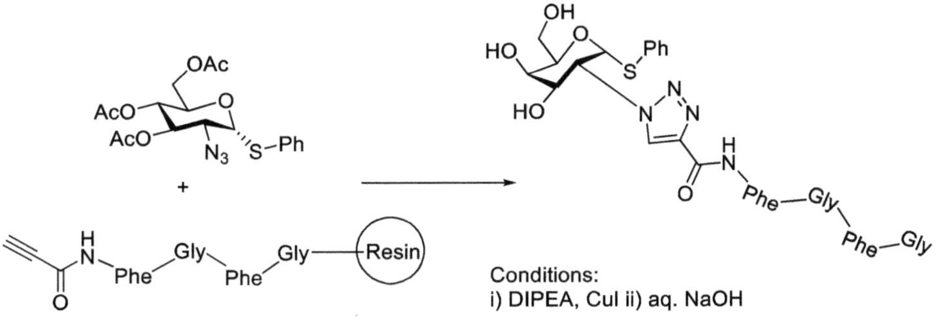

Figure 16.7: Cycloaddition of a monosaccharide with an alkyne-containing tetra-peptide.

Over the years, there has been great interest in the preparation of immunogenic materials from non-immunogenic molecules by covalently attaching proteins to small molecules. There is increasing interest in preparing anti-oligosaccharide antibodies from carbohydrates under this concept. The polysaccharide-protein conjugates have become attractive candidates for vaccine and pharmaceutical development against various human pathogens. The limited solubility, temperature, and pH sensitivity usually challenge the carbohydrate-protein conjugate formation under known low-yielding methods. Therefore, there has always been great interest in the development of high-yielding and less-demanding bioconjugation techniques [18].

In 2002, a report described the preparation of bioconjugates employing the Diels-Alder cycloaddition approach. The reported method described the synthesis of new glycoproteins using dienophile-equipped protein and diene-linked saccharide under mild conditions in aqueous media. The HSA protein was equipped with an electron-deficient double bond in phosphate buffer at pH 7.5. At the same time, the diene-derivatized saccharides were also prepared in mild to good yield. In the subsequent step, the diene and dienophile-containing molecules were made to undergo DA cycloaddition, producing carbohydrate-protein conjugates at ambient temperature. Advantageously, the recovery of uncoupled sugar derivatives could easily be managed by gel filtration or diafiltration. This new procedure provided an opportunity for the preparation of carbohydrate-protein conjugates under less demanding conditions and facile recovery of uncoupled saccharides (Figures 16.8–16.9) [19].

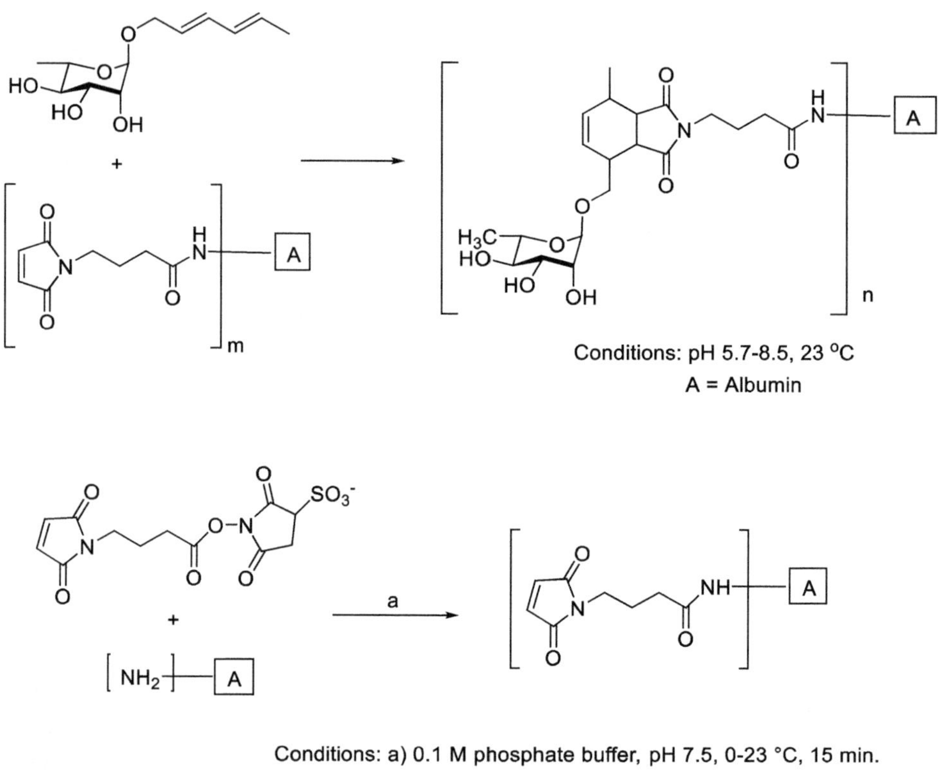

Figure 16.8: Diels-Alder cycloaddition approach for carbohydrate-protein conjugates.

16.2.4 Bioconjugation based on disulfide bonds

Over the years, the preparation of bioconjugates has evolved through various synthetic methodologies. The disulfide-based linkage has become a prominent approach for the preparation of a variety of functional glycoconjugates. In 2004, a procedure described the activation of the thiol moiety on protein to disulfide using phenylselenenyl sulfide [20]. In parallel, Lawesson's reagent was employed for the preparation of glycosyl-thiols. Subsequently, the selenenyl sulfide activation protocol was employed for the conjugation of mono-, tri-, and hepta-saccharide thiols with protein. In a recent report, the pyridyldisulfanyl substitution has also been explored for the activation of thiols and subsequent glycosyl-conjugation. The glucosyl-thiol was made to react with 21-mer RNA under pyridyldisulfanyl substitution activation, leading to good to higher yields of glycol-conjugates (Figure 16.10) [21].

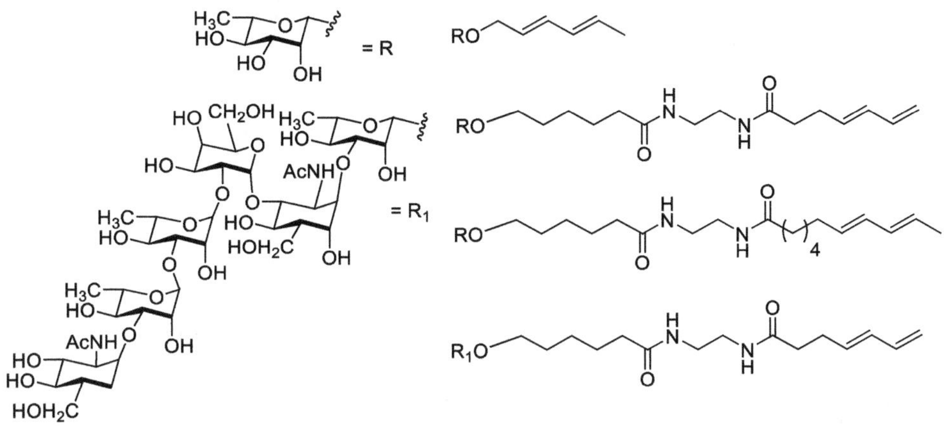

Figure 16.9: Diene-functionalized saccharides.

16.2.5 Bioconjugation based on diazo coupling approach

As described earlier, the detection of glycosylated DNA in T-even phages provided evidence of naturally occurring useful oligonucleotide-carbohydrate conjugates. Therefore, attention was diverted to the preparation of DNA-carbohydrate conjugates with a broad scope of applications. However, the lack of reliable methods for the synthesis of glycoconjugates challenges this emerging field. There was a dire need to develop novel methods for the introduction of multiple carbohydrates on one DNA chain. In the early 2000s, a report described the synthesis of DNA-carbohydrate conjugates using a diazo coupling mechanism. The simple synthetic procedure described the incorporation of oligosaccharide chains into guanine. The cellulose and lactose were converted into their diazonium derivatives and subsequently coupled with natural DNA via a diazo coupling mechanism. The prepared conjugates exhibited stronger resistance to endo- and exonucleases and a higher melting temperature compared to native DNA. The prepared conjugates show higher structural stability and nuclease resistance; thus, it could be a promising strategy for gene therapy applications (Figure 16.11) [22].

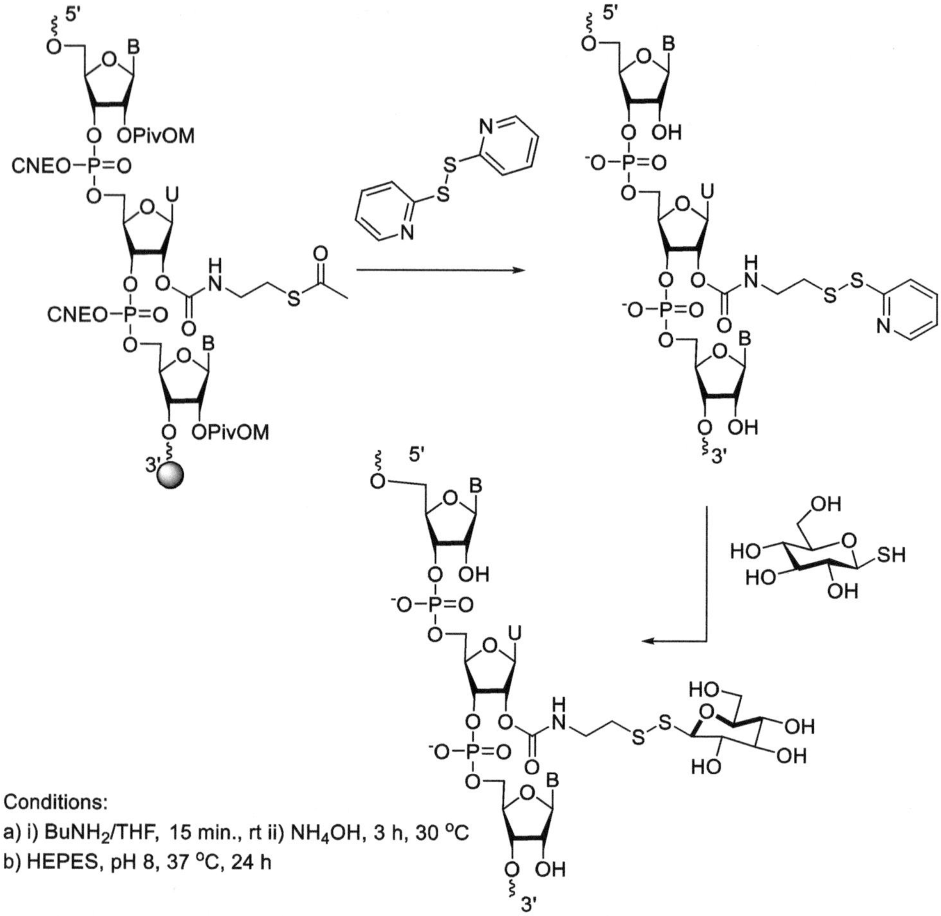

CNE = 2-cyanoethyl ; ● : controlled-pore glass support; PivOM: Pivaloyoxymethyl

Figure 16.10: Pyridyldisulfanyl-mediated conjugation of glucosyl-thiol.

16.2.6 Carbohydrate phosphoramidite building blocks

Over the years, there has been great interest in improving the design of nucleic acids for enhanced stability, better cellular uptake, and resistance against nucleases. Reports have already described the improved delivery of siRNA after conjugation with peptides, polymers, and lipids as delivery carriers [23]. The reports have also described the improved characteristic features and inhibited properties of conjugates of siRNA and DNA with cholesterol. The carbohydrate-oligonucleotide conjugates could have improved uptake owing to the sugar-binding membrane receptor. Glucose is vital for cellular metabolism; therefore, GLUT-protein facilitates its transportation across cellular membranes [24].

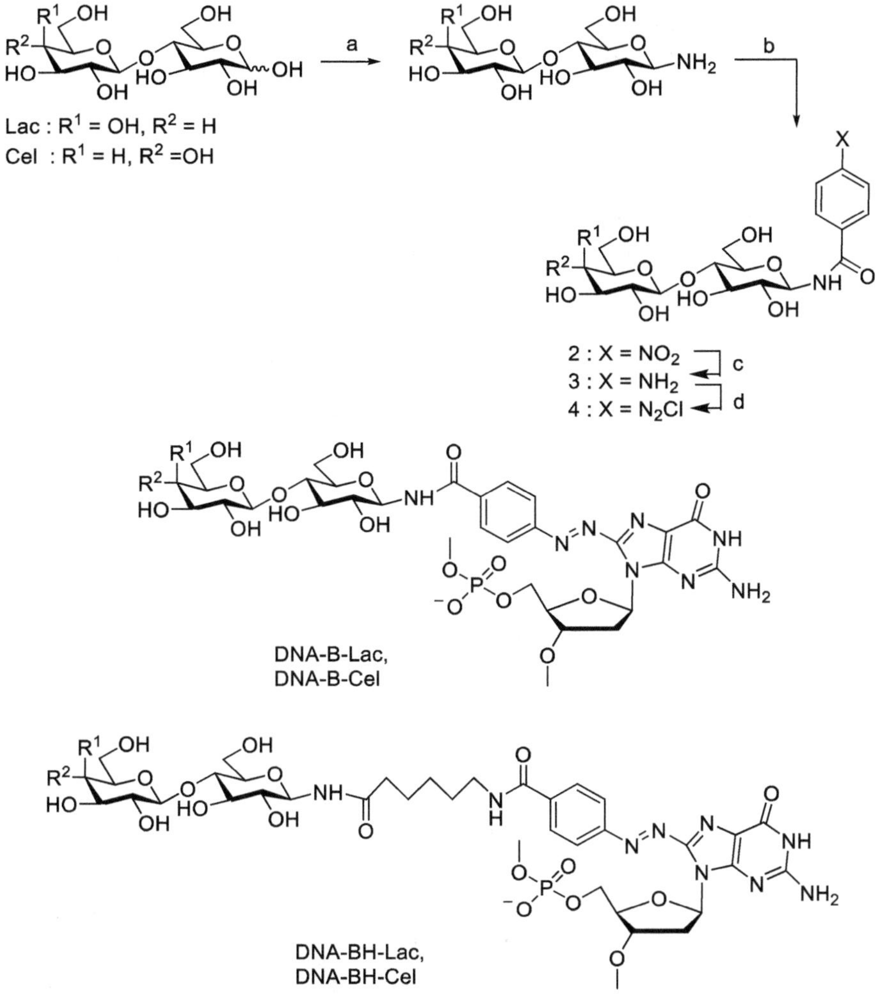

Figure 16.11: Diazonium derivatives of cellulose and lactose for bioconjugation.

Therefore, it was envisioned that GLUT receptors could facilitate the transportation of nucleotide conjugates with a glucose moiety. In this connection, in 2010, the phosphoramidite approach was employed for the conjugation of the non-reducing end of oligosaccharides to DNA strands. A library of conjugates was obtained using different dendrons and spacers. The conjugates with longer spacers showed better cellular uptake in target cells compared to the unmodified control (Figure 16.12) [25].

Phosphoramidite derivatives are synthesized for the incorporation of carbohydrates into oligonucleotides. This strategy has been well adopted for the preparation of oligonucleotide–carbohydrate conjugates with useful applications as primers for PCR [26]. Classically, mono- and di-saccharides with alcohol moieties undergo subsequent

Figure 16.12: Sugars with spacers.

phosphitylation. Efforts have also been made for the preparation of tri-saccharide-oligonucleotide conjugates using a chemo-enzymatic solid-phase synthesis approach (Figures 16.13–16.14) [26 27].

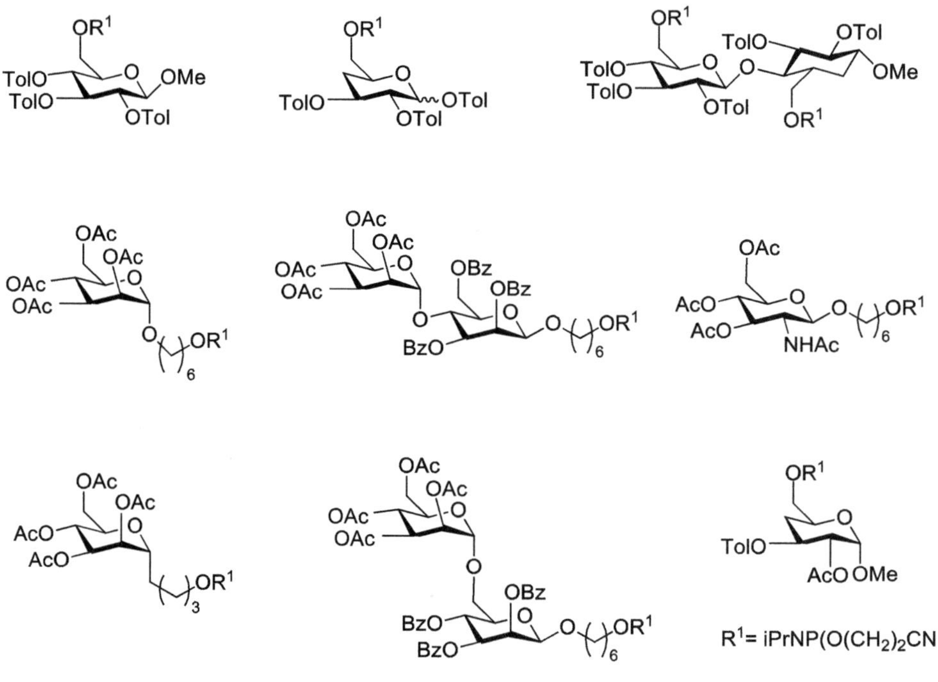

Figure 16.13: Phosphoramidite derivatives of carbohydrates.

16.2.7 Metathesis approach for bioconjugation

The thio-alkenylation was also introduced for the preparation of glycoconjugates. In this approach, cysteamine was reacted with alkenyl-functionalized carbohydrates to generate glycans with amines as terminal groups (Figure 16.15) [28].

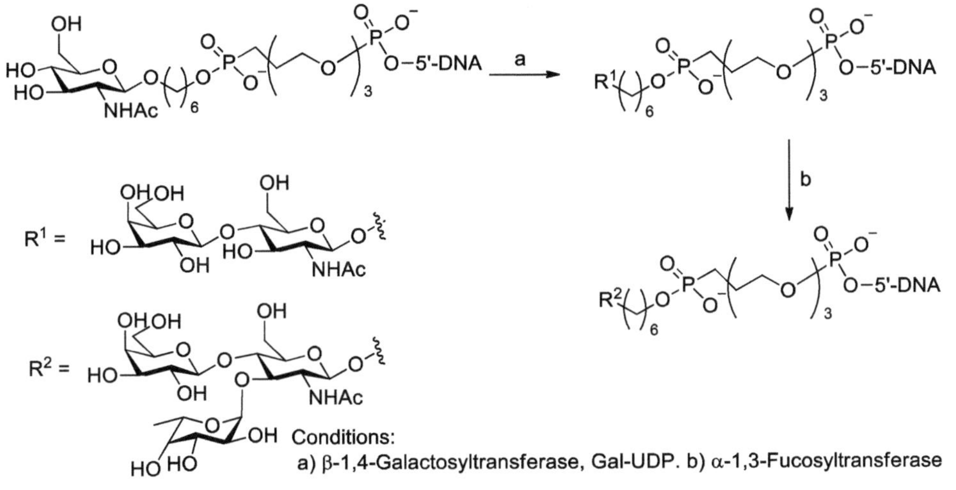

Figure 16.14: Carbohydrate-oligonucleotide conjugates via the phosphoramidite approach.

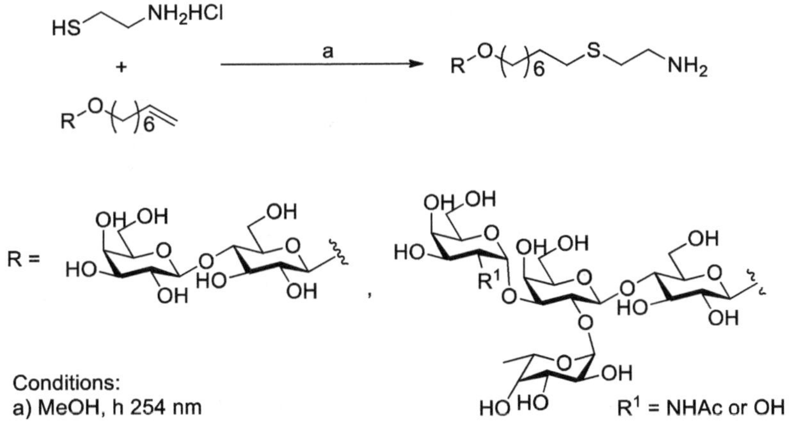

Figure 16.15: Reaction of cysteamine with alkenyl-functionalized carbohydrates.

16.3 Bioconjugates of carbohydrates with biomolecules

The site-specific covalent derivatization of biomolecules is referred to as bioconjugation. Over the years, bioconjugates have shown a wide scope of applications (Figure 16.16). The conjugates of carbohydrates with proteins and nucleic acids have also shown promising applications in disease diagnostics and imaging, biochemical assays, and in drug development. Therefore, various methods have been developed for the preparation of bioconjugates of carbohydrates with proteins, DNA, and RNA [9, 29].

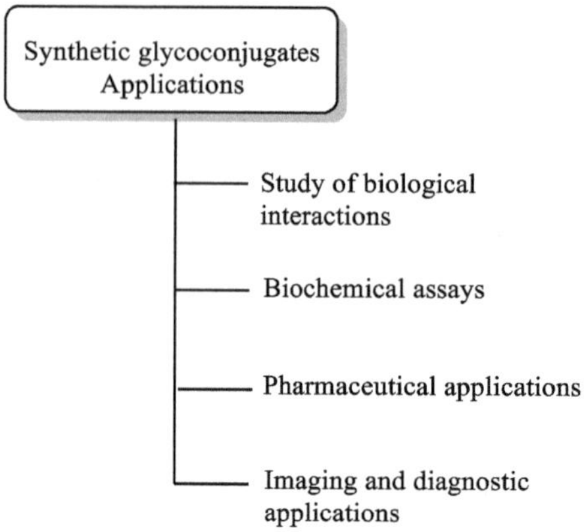

Figure 16.16: Applications of synthetic glycoconjugates.

16.3.1 Bioconjugation of carbohydrate with protein

Over the last few years, glycoproteins have emerged as potent candidates for the treatment of various diseases owing to their superior biological functions, low toxicity, and high specificity [30]. The covalently linked mono- and oligosaccharides play key roles in various processes like differentiation, cell growth, and fertilization. However, the complex structure and composition of saccharides cause some disadvantages for such biological macromolecules. Therefore, huge efforts have been made for the preparation of functional glycoconjugates using controlled chemistry of carbohydrate units. The proteins are linked to glycans using two common methods. The *N*- of asparagine participates in binding, resulting in the formation of *N*-linked glycans. The binding of the *O*- of threonine and serine provides *O*-linked glycans [31].

Over the years, there has been great interest in understanding the interaction and functions of natural glycans using new synthetic heterogeneous and homogeneous glycoconjugates. Accordingly, some synthetic strategies based on chemical and enzymatic approaches have been designed for the practical synthesis of carbohydrate-protein conjugates. The prepared bioconjugates based on the Maillard reaction could selectively recognize various cells, including cancer cells, depending on the nature and composition of oligosaccharide units. The structure of the sugar controlled the highly specific intracellular recognitions and lectin-carbohydrate interactions [32]. The synthetic oligosaccharides have been produced with allyl groups so they can undergo ozonolysis and subsequent bioconjugation with proteins through reductive amination. In one such report, the glycol-conjugate vaccine candidates were prepared from oligosaccharides with an *n*-pentenyl moiety, which underwent ozonolysis and reductive amination with a carrier protein (Figure 16.17). Carbohydrates tethered with aromatic aldehydes were made to undergo reductive amination with BSA. The amino acid analysis revealed the varying degree of modification of lysine residues in proteins after covalent linkage with the *N*-acetylglucosamine moiety at different time intervals (Figure 16.18).

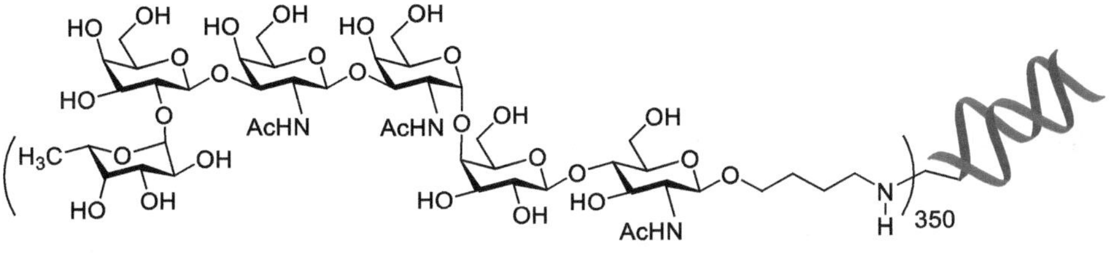

Figure 16.17: Glycol-conjugate vaccine candidate prepared via reductive amination.

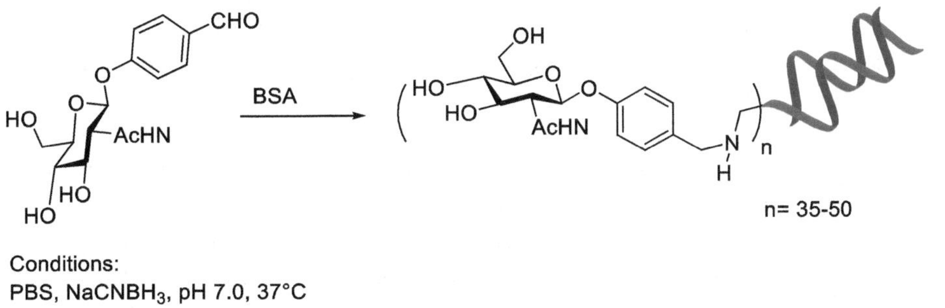

Figure 16.18: Saccharides tethered with aldehydes and reductive amination with BSA.

In 2014, the bioconjugates of BSA were prepared through reductive amination using synthetic oligosaccharides containing an alkylene chain terminated with an aldehydic group (Figure 16.19). The aldehydic functionality could selectively be prepared from the oxidation of the primary –OH group at C-6. The lysyl-lysine dipeptide was used to enhance the conjugation sites in a 14 amino acid residue peptide before coupling under reductive amination conditions (Figure 16.20) [9].

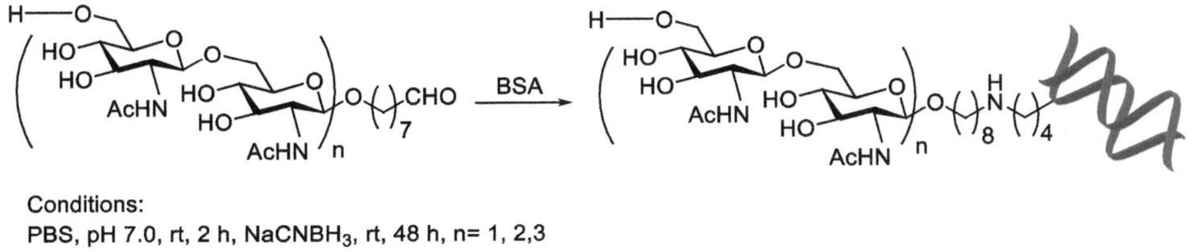

Figure 16.19: Synthetic oligosaccharides underwent reductive amination with BSA.

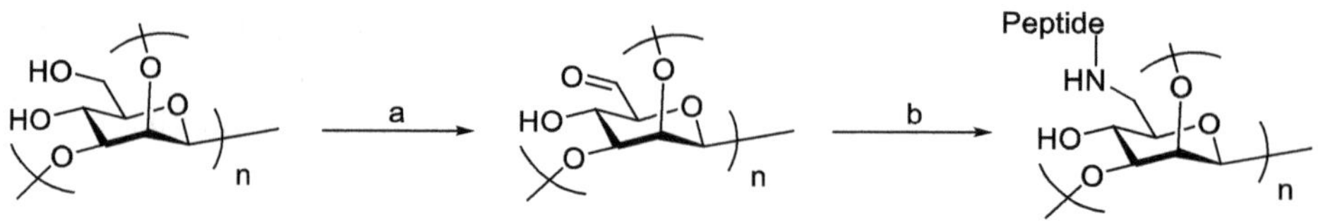

Conditions:
a) Dess-Martin periodinane, DMSO, 1 h, 30 °C
b) i) Peptide, PBS, pH 7.0, rt, 30 min ii) $NaCNBH_3$ PBS, pH 7.0, rt, 4 h

Figure 16.20: Sugars with aldehydes for reductive amination.

Carbohydrate-protein conjugates are vital for an array of biological functions. Over the last decade, they have also been a focus for the preparation of vaccines for various infectious diseases. Therefore, a number of methods have been developed for the covalent coupling of proteins to carbohydrates with varying success. Carbohydrates are generally modified with the introduction of reactive functionalities before coupling with proteins. The carbohydrates are subjected to oxidation of hydroxyls and/or addition of linkers before coupling reactions. The amino group on the protein could be attached to the reducing end of oligosaccharides employing the reductive amination method.

Over the years, reductive amination turned out to be a slow process, especially for *N*-linked sugars. In 2008, a study described an improved reductive amination method for the preparation of carbohydrate-protein conjugates. The reported procedure improved the efficiency of reductive amination using 500 mM sodium sulfate. The method involved the preparation of a carbohydrate microarray and was subsequently incorporated into mannose–BSA conjugates. This methodology proved to be a useful tool for the generation of glycoconjugates *via* reductive amination with a wide scope of applications (Figure 16.21) [33].

The amino functionalities of proteins can be conjugated to organic drugs, metal complexes, and other small biomolecules. There has been great interest in the development of biomaterials by immobilizing functional molecules on protein assemblies. In recent years, protein crystals have been exploited for the preparation of biohybrid materials using the bioconjugation concept. In 2015, a method disclosed the surface modification of polyhedra crystals (PhC) produced by cypovirus-infected insect cells. The cysteine residues on PhC were functionalized with alkynes, which subsequently coupled with an azido group on Lewis X carbohydrate under click conditions. The reported methodology enabled the preparation of hybrid biomaterials for application in drug delivery and bio-solid catalysis (Figure 16.22) [34].

The efficient bioconjugation methods are highly desirable for the economical synthesis of oligosaccharides. At the same time, the denaturation of protein should be avoided during conjugation with shorter reaction time, ambient temperature, physiological pH, and aqueous media. The saccharide chains with terminal linkers facilitate the conjugation process. The reactive groups on proteins and also on sugars should

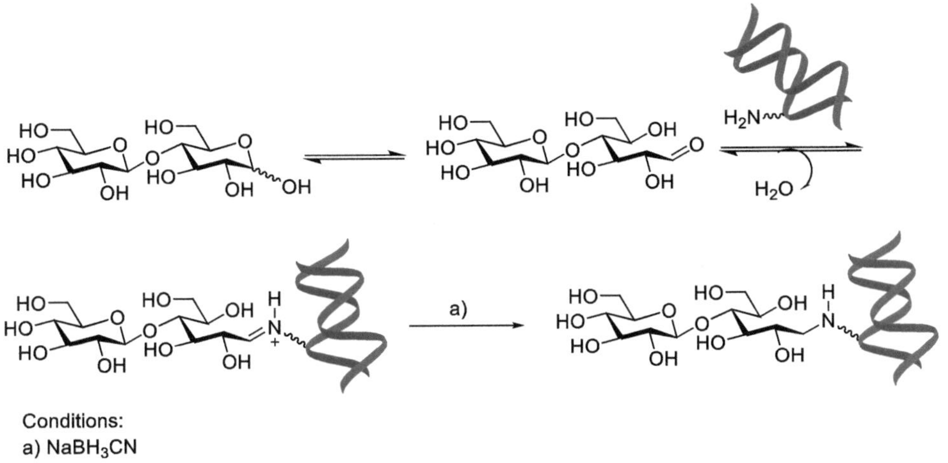

Figure 16.21: Improved reductive amination for carbohydrate-protein conjugates.

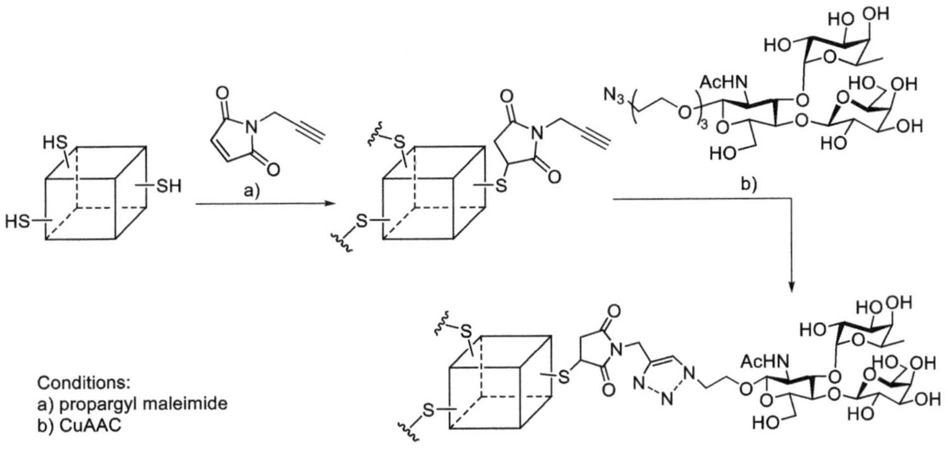

Figure 16.22: Site-selective bioconjugation of protein crystals with carbohydrates.

show resistance toward hydrolysis by keeping the structural morphology intact. Ideally, an equimolar ratio of reactive functionalities on protein should react with carbohydrates. However, the generally known bioconjugation methods show a number of limitations and fail to produce quantitative products.

In 2005, an economical coupling of proteins with carbohydrates was reported using a mild and efficient methodology. The reaction procedure reported the surface functionalization of protein with an aminooxy group-containing spacer through a 2-step method. Subsequently, it underwent a condensation reaction with ketonic or aldehydic functionalities of carbohydrates. Advantageously, the resulting bioconjugates could easily be separated from unreacted sugars. The report provided a detailed account of the coupling of aminooxylated BSA with mono and tetra saccharides under mild conditions. The reported procedure provided an economical and high-yielding strategy for the preparation of functional glycoconjugates (Figures 16.23–16.25) [35].

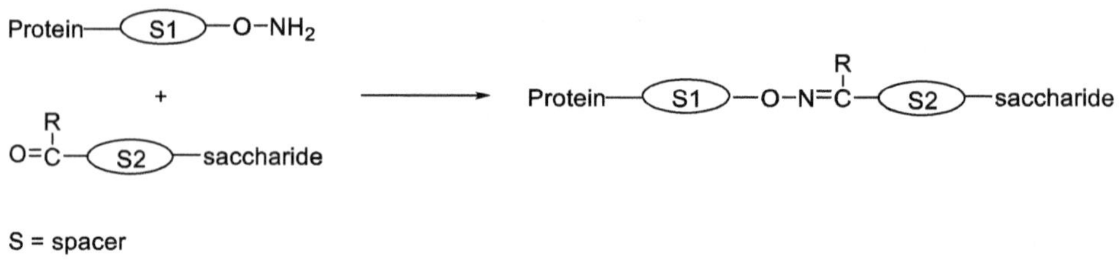

Figure 16.23: Surface functionalization of protein with an aminooxy group-containing spacer.

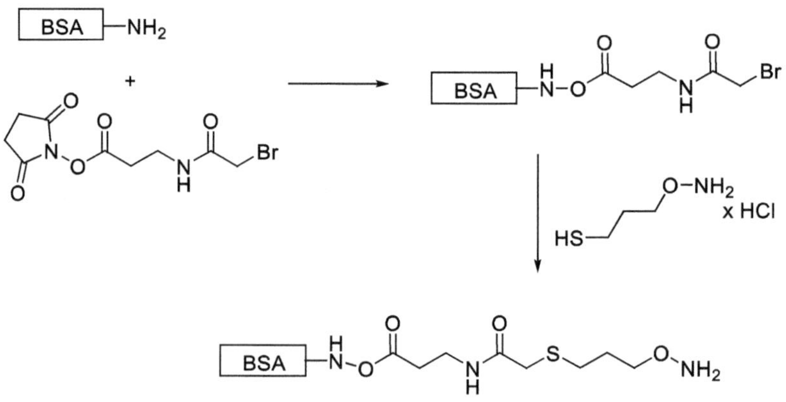

Figure 16.24: Coupling of aminooxylated BSA.

The anomeric positions in sugars were also functionalized with azido alcohol linkers using a Lewis acid as a promoter. The glycosylation was influenced by the length of the azido linker. The longer linkers exhibited low reactivity and resulted in poor yield. The 6-azidohexyl linkers turned out to be the best choice because of the higher C to N ratio. A recent study described the stereo-selective insertion of azide using the alpha-directing effect of butylsilylene along with a thiotolyl donor. The reported methodology utilized the de-protection of the protecting groups in the end product (Figure 16.26).

The site-specific ligation of tyrosine was performed after attaching a triazolinedione linker to the polysaccharide using a click chemistry approach. This strategy was later adopted to assess the number of conjugation sites and carbohydrates in potential vaccine candidates. Another approach described the attachment of a carboxybenzyl-protected amine as a linker on sugar moieties, which were subjected to glycosylation employing a one-pot ligation strategy. The azide was generated under azido transfer

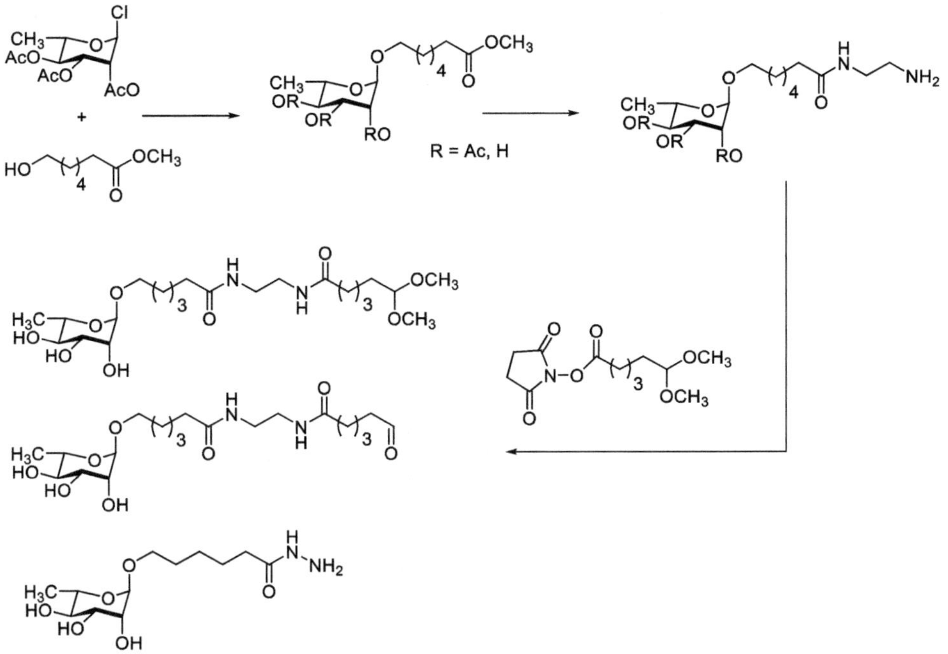

Figure 16.25: Saccharide chains with terminal linker.

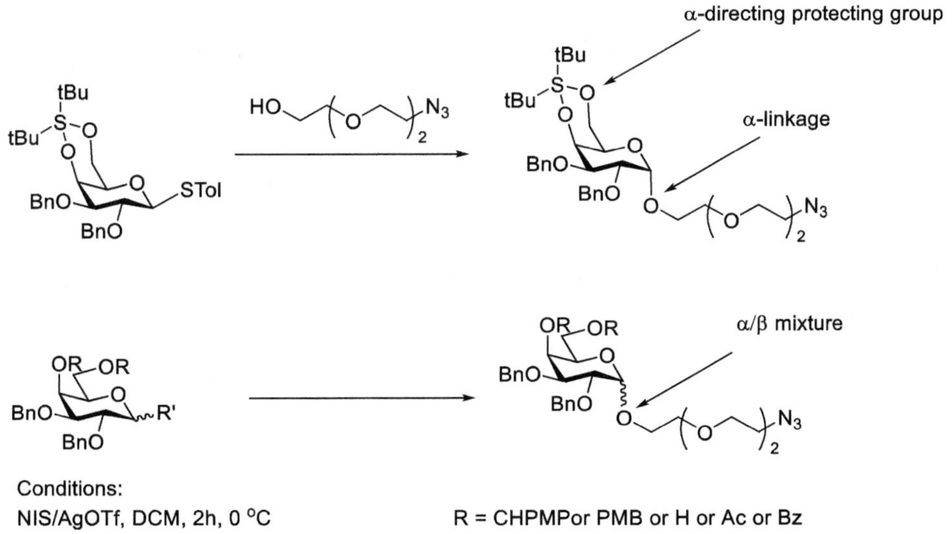

Figure 16.26: Sugars functionalized with azido alcohol linkers.

conditions and subsequently underwent a click reaction with disulfide-alkyne linker (Figures 16.27 and 16.28) [4].

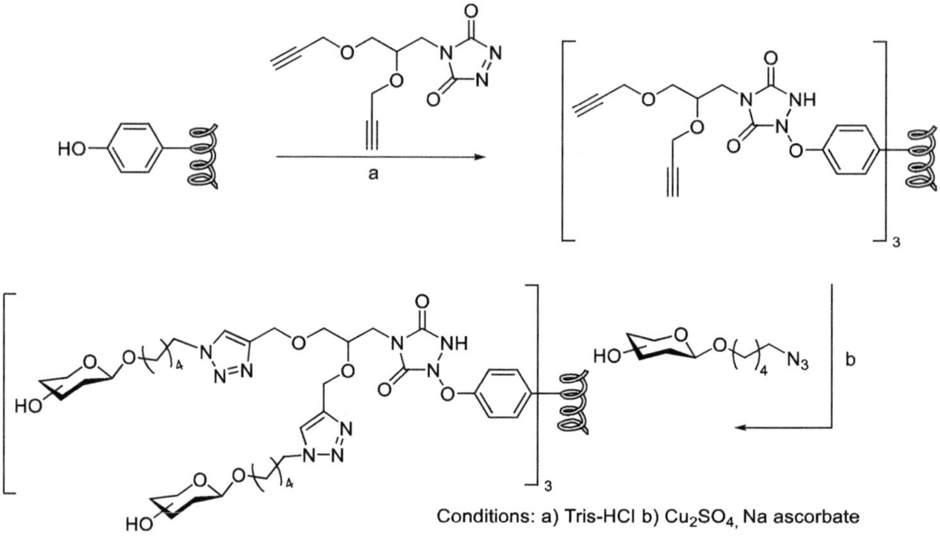

Figure 16.27: Site-selective ligation of tyrosine using a triazolinedione linker.

Carbohydrate functionalization has emerged as a promising synthetic tool in chemical biology. Many therapeutically potent compounds have saccharides as a basic bioactive motif [36]. It is an active structural unit in several antibiotics, vaccines, and candidates for anti-tumor agents. The oxygen-rich nature and structural diversity pose serious challenges to the selective labeling of carbohydrates. The multiple hydrophilic groups reduce their solubility and hinder the selective labeling of saccharides. Over the last few years, the preparation of bioconjugates of carbohydrates in ionic liquids has gained attention due to some advantages. The ionic liquids, owing to their greater dissolution power, have been widely exploited for the preparation of glycoconjugates. However, most of the carbohydrate conjugate preparation involved classical reaction approaches like simple substitution reactions and amide condensation. Very recently, a report described the bioconjugation of azide and amine-containing carbohydrates employing a phosphine-mediated chemical labeling strategy. The described azide–amine coupling reaction proceeded efficiently in ionic liquid. It described the application of ionic liquid and phosphine-mediated three-component coupling for labeling saccharides for bioconjugation formation. The loss of nitrogen from the azido group and the carbonyl group of CO_2 under phosphine-mediated coupling provided urea as the product. Consequently, several saccharides equipped with azido and amino groups were subjected to this urea-forming reaction. The mammalian cell lysate labeled with azide functionality exhibited bioconjugation in ionic liquid. The site-specific labeling technology has gained consider-

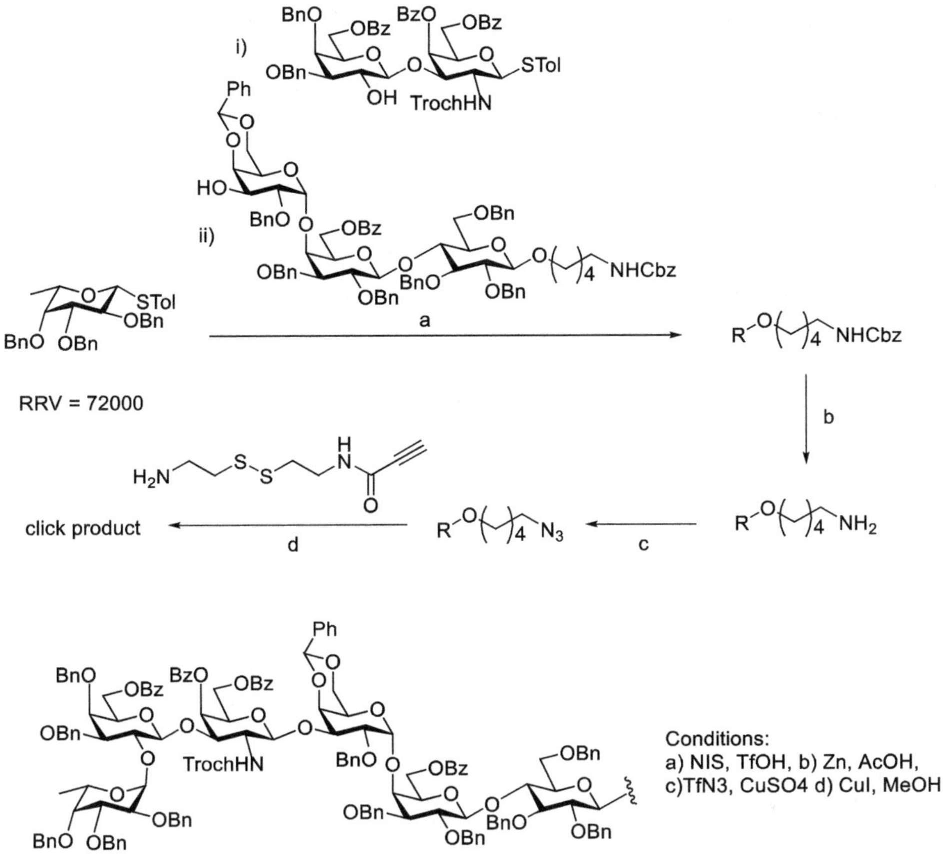

Figure 16.28: Carboxybenzyl-protected amine-linked saccharides and one-pot glycosylation.

able attention because of newly emerging amine-azide coupling methods. The versatility of the presented protocol was established from the chemical transformation of various mono-saccharides and polysaccharides in known traditional media (Figure 16.29) [37].

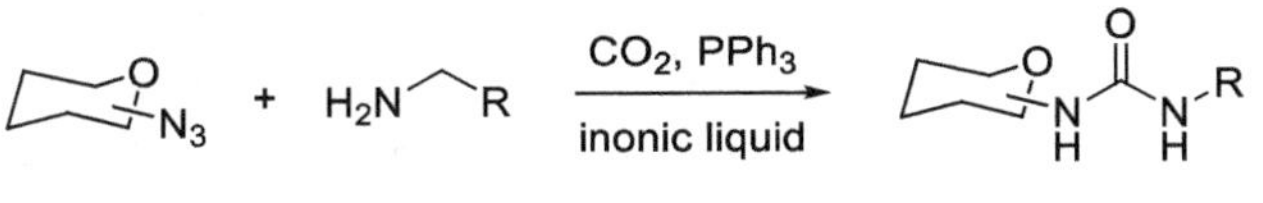

Figure 16.29: Bioconjugation in ionic liquids.

In recent years, small molecules like carbohydrates have been chemically grafted to gelatin to improve its bioactivity profile. The saccharides have gained attention because of their potent role in cell adhesion and cell surface interactions. In a few studies, the adhesion and proliferation of hepatocytes were improved using glycosylated derivatives [38]. The development of bioactive scaffolds from glycosylated collagen has shown promising applications in tissue engineering [39]. Similarly, chitosan modified with lactose has shown various applications in nerve, cartilage, and bone tissue engineering [40]. Very recently, a study disclosed the anchoring of saccharides on gelatin for 3D scaffold formation. The protocol described the formation of saccharide-gelatin conjugates as sponges for tissue engineering. The study described the grafting of gelatin with maltose and lactose for the improvement of adhesion and cellular interactions under mild conditions. The aldehydic moiety of disaccharides participated in a reductive amination reaction for the preparation of bioconjugates of gelatin. The model study provided an opportunity to understand the differentiation, proliferation, adhesion, and cellular interaction using different glycoconjugates. The disclosed simple and efficient protocol provided easy access to carbohydrate-gelatin as functional sponges for wide application in tissue engineering (Figure 16.30) [41].

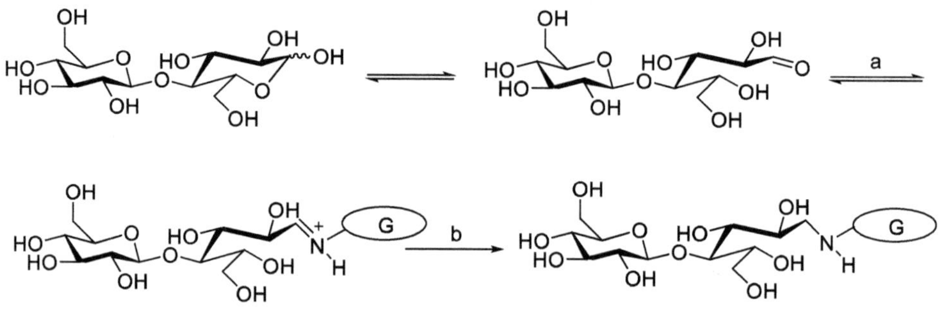

Conditions:
a) 1% (w/v) gelatin (G) in 100 mM BBS, Na_2SO_4 500 mM, pH 8.5, rt
b) $NaCNBH_3$ in 100 mM BBS, Na_2SO_4 500 mM, pH 8.5, rt

Figure 16.30: Carbohydrate-gelatin as functional glycoconjugates.

The cellular recognition is controlled by surface glycoproteins, which implicate lymphocyte trafficking, embryonic development, and cellular signaling. Therefore, a number of procedures have been developed for the functionalization of saccharides and proteins and their subsequent conjugation. However, most of the known methods generally require purified proteins. Over the last few years, strained alkenes and alkynes have been introduced using the aminoacyl-tRNA synthetase/tRNA pair, which is produced by a genetic encoding technique. The tetrazine-strained alkene reactions are promising for experiments involving living cells due to their bioorthogonality and fast kinetics [42]. In 2015, the inverse demand DA reaction of *trans*-cyclooctene and tetrazines provided saccharide-protein conjugates. The alkyne-tetrazine underwent a cycloaddition reaction with glycosyl azide, providing tetrazine-functionalized glycans. The genetic encoding technique was employed to generate proteins with *trans*-cyclooctene-substituted lysine. Further, the functionalized glycan and protein underwent chemo-selective and site-specific bioconjugation. The disclosed simple and efficient protocol established the biocompatibility of in-

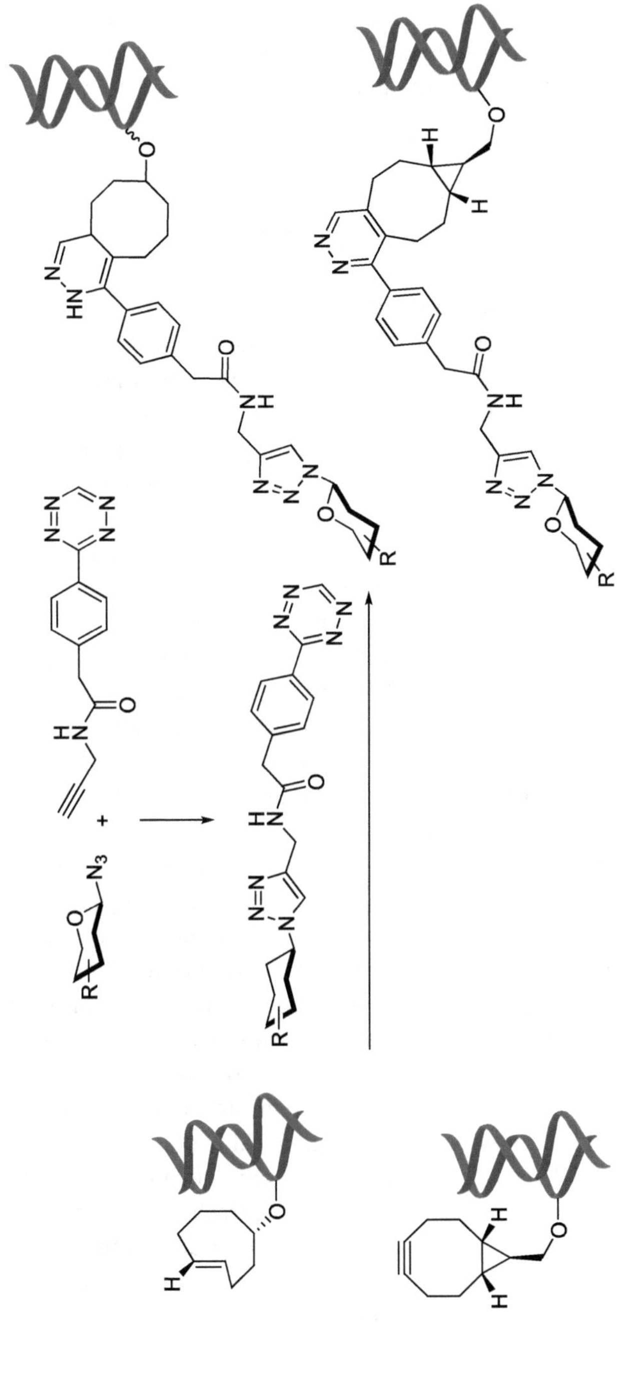

Figure 16.31: Inverse demand DA and CuAAC reaction for the preparation of bioconjugates.

verse demand DA and CuAAC reactions for the preparation of bioconjugates. Advantageously, simple as well as complex glycans can be functionalized conveniently with tetrazine. The facile accessibility of modified unnatural amino acids facilitates the formation of tailored bioconjugates (Figure 16.31) [42].

16.4 Bioconjugates of plant-based proteins

Proteins are key components of structural building blocks that stabilize several structures. They have also been considered potent nutrients maintaining the qualitative attributes of food. The functional properties of proteins are important for the development of food because they influence the bioavailability of bioactive compounds, minerals, and other nutrients. The potent functional properties of proteins include emulsification, foam, and stable gel formation, which maximize their nutritional value. However, their sensitive nature toward pH, high temperature, and proteolytic environments challenges their application in the food industry [43]. Over the last few decades, there has been great interest in the improvement of the functional properties of proteins by controlling their temperature and pH sensitivity. In this connection, the Maillard reaction has received huge attention because of the conjugation of carbohydrates with proteins with improved functional and biological properties. The carbohydrate-protein conjugates appear as the end product of the early-stage Maillard reaction, representing the condensation of $-NH_2$ of proteins with reducing sugars [44].

During the last few years, there has been an increasing interest in plant-based protein products. They have emerged as alternative sources of protein due to rising economic issues and health concerns. Several strategies have been developed to enhance the thermal stability and solubility of plant proteins. Accordingly, they are used in a range of products due to their improved physicochemical properties. Different chemical and enzymatic methods have improved the functional properties of proteins for the food industry (Figure 16.32). Recently, the physicochemical properties of plant proteins have also been improved by developing their conjugates with carbohydrates. The glycosylation of protein has become an attractive strategy for the preparation of protein-based value-added food products [45]. The carbohydrates and proteins are glycosylated using the Maillard reaction with minimal chemical additives. The reaction proceeds between the carbonyl group of reducing sugar and the amino group on proteins, leading to Schiff base formation. Subsequently, the rearrangement step provides the thermodynamically favored product. The reaction proceeds with the excess of reducing sugars at alkaline pH and high temperature. Both the wet and dry heating methods are employed for the conjugation reaction [46].

16.4.1 Dry heating method

The pH of the solution is adjusted by dissolving sugars and proteins in different ratios using the dry method. Later, the solution is freeze-dried at different temperatures for various time intervals, resulting in products with different degrees of conjugation. High molecular weight peanut protein was conjugated with dextran, and the glycosylated product formation increased with incubation time. Another study described the conjugate formation of dextran-whey protein at 50–55 °C after 144 h. A recent study described the effect of incubation time and weight ratio of the components on dextran-β-lactoglobulin conjugate. The weight ratio of components and temperature showed an inverse relationship for the preparation of the conjugate. The increasing amount of maltodextrin favored the degree of glycation of BSA. The conjugation was favored due to the available number of grafting sites on the protein under mild conditions.

16.4.2 Wet heating method

The conjugates of carbohydrates and proteins are also prepared using the wet heating method. In this method, dry conditions and higher temperatures are applied to freshly dissolved carbohydrates and proteins. The reaction is stopped by rapidly cooling the solution after reacting for a short time at an adjusted pH. The short reaction time is an advantage, but high temperatures could cause denaturing of proteins. The glycation of rice proteins with different saccharides, including dextran, malto-dextrin, maltose, lactose, and glucose, was carried out, showing different degrees of grafting. The conjugation process was influenced by the exposure of amino acids; thus, the unfolding and flexibility of proteins control the conjugate formation. During the Maillard reaction, proteins could also undergo partial denaturation; however, the protein surface characteristics are generally influenced by the extraction method [47].

Another study reported the influence of extraction method and pulse variety on the surface hydrophobicity of proteins. Proteins with higher hydrophobicity were achieved using isoelectric precipitation in alkaline extraction. The salt extraction was found to keep globulins and albumins in a more native state because of milder conditions [48].

16.4.3 Combined heating method

In recent years, classical heating has been replaced with pulsed electric fields, microwave, and ultrasonic-assisted heating techniques (Figure 16.33). They provide a higher degree of glycation in a shorter time. One such study described the microwave-assisted conjugation of albumin-maltodextrin in a short reaction time [49]. The ultrasonication technique promoted the Maillard reaction because of convenient protein unfolding and exposure of reactive amino groups. This resulted in a high degree of conjugation because more polysaccharides come in contact with the unfolded protein structure [50]. Another study explains the preparation of conjugates using gum acacia and soy protein, comparing classical and ultrasonic heating methods [51]. The biomacromolecules have also been modified using high-intensity short pulses for a short time at low temperature [52]. An increase in pulse intensity increased browning and the formation of intermediate products with a marked reduction of free amines.

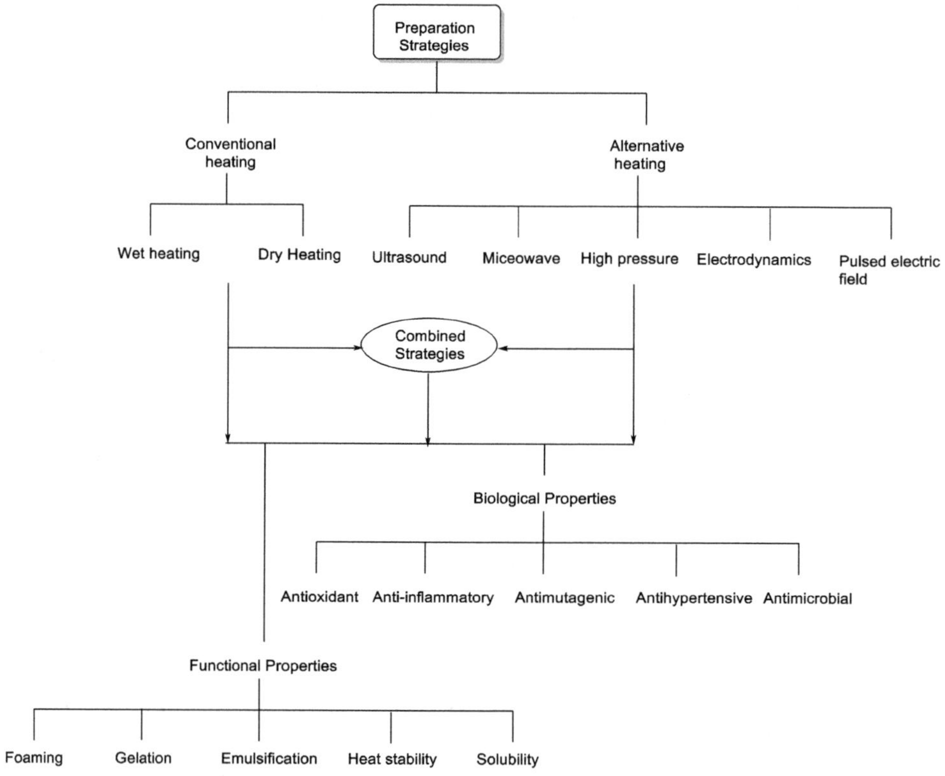

Figure 16.32: Classical and modern heating techniques improve functional properties.

16.4.4 Enzymatic methods

Over the last few decades, plant-based protein conjugates have gained considerable attention from the food processing industry. Generally, they follow the Maillard reaction for the processing of carbohydrate-protein conjugates as edible products. Over time, the focus has shifted to avoid the generally followed reaction because of undesirable advanced glycation end products and browning issues [53]. For this purpose,

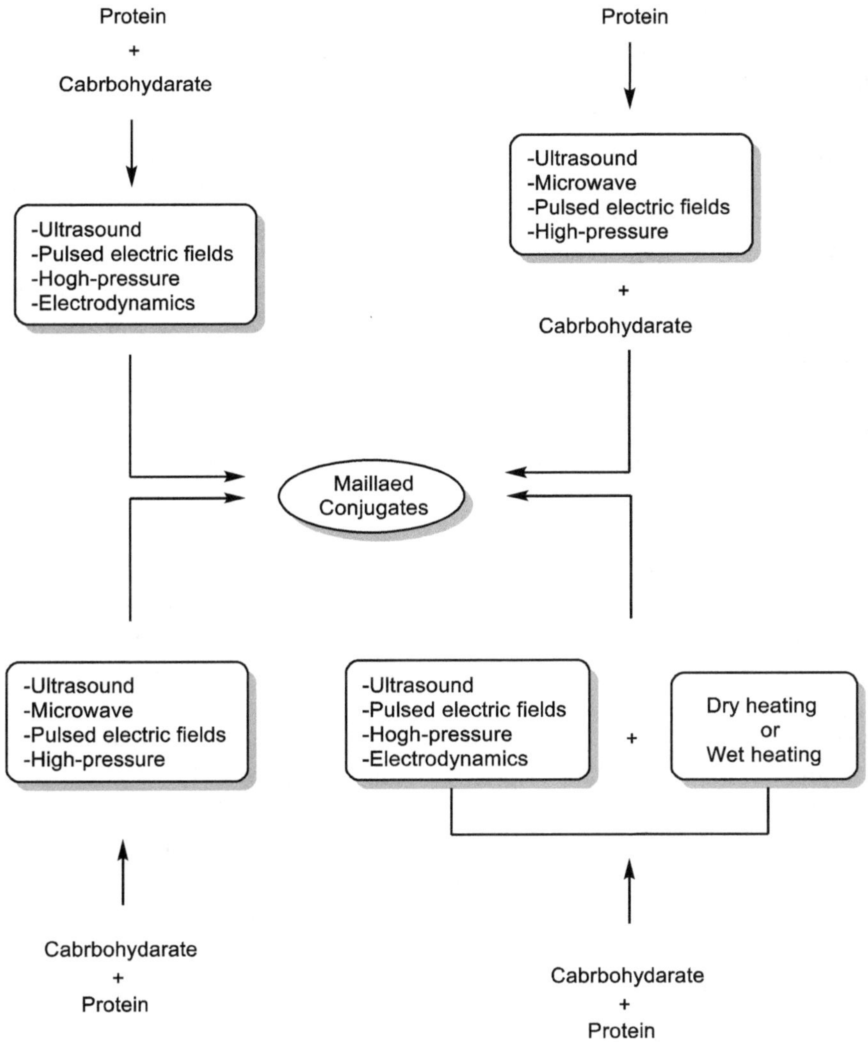

Figure 16.33: Maillard reaction for carbohydrate-protein conjugates.

enzyme-catalyzed reactions are considered an alternative strategy for the formation of carbohydrate-protein conjugates. The phenolic groups in proteins have been oxidized using polyphenol oxidases into reactive quinones, which react with the nucleophilic moiety of carbohydrates. The tyrosyl residues in gelatin were oxidized to *o*-quinones using tyrosinase, which subsequently reacted with the $-NH_2$ functionality of chitosan [54].

In another study, the conjugation between hydrolyzed oat spelt xylan and α-casein was effectively catalyzed by laccase. However, the tyrosinase was less effective for the same conjugation reaction [55]. According to another study, the binding of polysaccharide with protein could be explained using a protein-polyphenol-carbohydrate conjugate. In fact, the quinone produced from the oxidation of polyphenol does act as a coupler facilitating the binding of polysaccharides and proteins for conjugate formation [56]. The transglutaminase catalyzed the conjugation of glucosamine with soy protein, resulting in increased physiochemical properties of the prepared conjugate [57]. The oligochitosan–soy protein conjugates were received from a transglutaminase-catalyzed glycosylation reaction. The prepared conjugates exhibited better stability and emulsion activity [58]. The interaction of the conjugate with water controls its gelling, foaming, and emulsification properties. The covalent linking of saccharides with proteins increased the hydrophilic nature of the conjugates due to the extra exposure of –OH groups of saccharides. This results in increased conjugate–water interactions leading to higher solubility. Sometimes glycation unfolds the protein, which causes aggregation and reduces solubility. The pectin-soy protein conjugates exhibited improved solubility because of more –OH groups as a result of glycosylation. The polysaccharide chain caused steric hindrance and decreased surface hydrophobicity by reducing the aggregation of protein [59].

During food processing, the proteins can lose nutritional values, showing turbidity, low solubility, and aggregation due to thermal denaturation. The thermal stability of proteins increases with the formation of protein-sugar conjugates because sugar-mediated steric hindrance prevents protein-protein interaction. The glycoconjugate of dextran and rapeseed protein exhibited a significantly high denaturation temperature because of conjugation. The increasing size of carbohydrates positively influences the denaturation temperature of dextran-β-conglycinin conjugates [60]. The same effect was observed for glucose-phaseolin conjugates with 50:1 ratio [61]. The glycation of proteins not only improve their functional properties but the conjugates show an array of applications as nutraceuticals, encapsulation of lipids and drug coatings. Dry heating method was used for maltodextrin-casein conjugate formation for the encapsulation of hydrophilic EGCG and hydrophobic vitamin D. The conjugate encapsulation decreases the turbidity and improves the solubility of encapsulated nutraceuticals. It also enhances their shelf life and bioavailability when added to beverages [62].

The BSA-dextran conjugates were used for encapsulation and controlled delivery of ibuprofen [63]. The enzyme-mediated conjugation of sugar-protein provided conjugates with enhanced antioxidant activities for wide application in food products [64]. The allergenicity of proteins decreases significantly as a result of the glycosylation reaction. It lowers the immune responses due to potential changes in protein structure. The fructo-oligosaccharides–soy protein conjugates exhibited a marked decrease in immuno reactivity to native soy protein [65]. Similarly, the lactose-soybean conjugates showed a 43% decrease in allergenicity [66]. Glycation of plant-based proteins improved the functional properties of the conjugates with wide applications in drug delivery, food, and the pharmaceutical industry. The alteration of physiochemical attributes of the glycoconjugates depends on the reaction conditions, heating methods, and the structural chemistry of carbohydrates and proteins [67].

16.5 Bioconjugation of carbohydrates with nucleic acids

In 2011, a growing nucleotide was made to undergo conjugation with two different glycosyl components. The azido-containing nucleo-base underwent a click reaction with propargyl glycosyl under mild conditions. The resulting oligonucleotide conjugate exhibited higher stability in ionic solutions (Figure 16.34).

Sonogashira coupling approach was adopted to incorporate alkynyl galactose at C-5 uridine, resulting in the stabilization of the DNA duplex. The oligonucleotide-carbohydrate conjugates have also been synthesized using glycosylated solid supports (Figures 16.35 and 16.36) [68]. The *tris*-galactose derivatives of choline were attached to solid support using bi-functional linkers. The nucleotide synthesis followed the standard procedure under mild reaction conditions. In another remarkable study, guanine in plasmids and oligonucleotides was conjugated with diazo derivatives of disaccharides. The in situ preparation of diazo derivatives reduces the yield of conjugate products (Figure 16.37) [26].

Over the last two decades, siRNAs have been considered potent therapeutic agents; however, factors like poor delivery and low stability challenge their scope of applications [69]. Tremendous efforts have been made to increase their cellular uptake and resistance against nuclease degradation. To this end, various attempts have been made to modify the oligonucleotide and to conjugate siRNAs with peptides, polymers, lipids, and antibodies. Their bioconjugates with lipophilic molecules induced hydrophobic character, resulting in enhanced membrane permeability and cellular uptake. As described earlier, the attachment of sugar moieties at the 5′-end of the DNA duplex improves the cellular uptake and nuclease resistance of oligonucleotide-carbohydrate conjugates. Considering these facts, Empar *et al.*, in 2013, employed phosphoramidite chemistry to prepare carbohydrate-siRNA conjugates to target tumor necrosis factor. The 5′-end of the passenger strand carried one to four galactose and glucose moieties in the prepared conjugates. The attachment of siRNAs to carbohydrate moieties was supported with long (dodecanediol) and short (ethylene glycol) spacers. Such chemical modifications improved resistance against nucleases without altering the machinery of RNAi. The reported solid-phase automatic synthesis of carbohydrate–RNA conjugates used phosphoramidites and apolar carbohydrates. The siRNA duplexes connected with apolar carbohydrates improved the cellular uptake of oligoribonucleotides by providing hydrophobic platforms. The reported methodology provided an idea about the development of nucleic acid-based drugs using apolar carbohydrates as new modifying tools for bioconjugation (Figures 16.38–16.39) [70].

In 2011, a report described the carbohydrate-lectin interactions using peptide nucleic acids-sugar conjugates. The biological recognition events highly rely on carbohydrate-lectin interactions. Generally, proteins have weaker affinities for carbohydrate ligands. Therefore, there has always been great interest in developing affinity en-

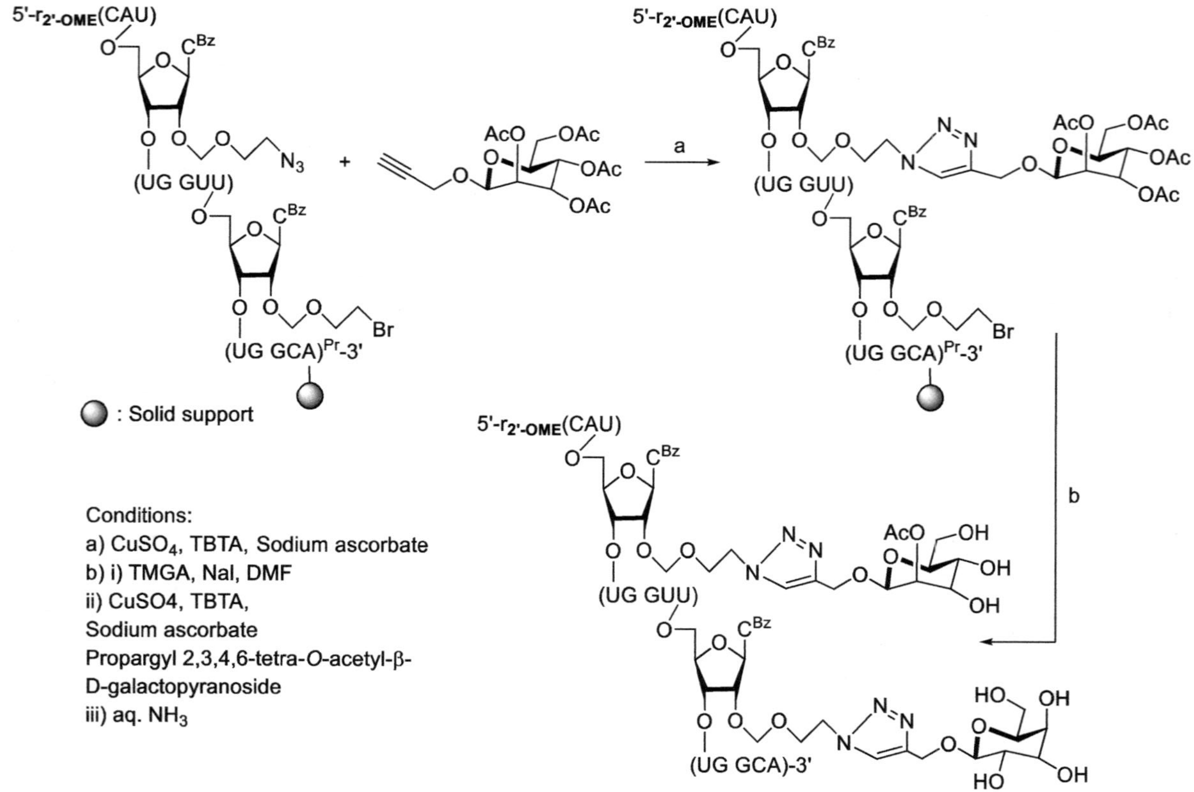

Figure 16.34: Nucleo-base with azide underwent a click reaction with propargyl glycosyl.

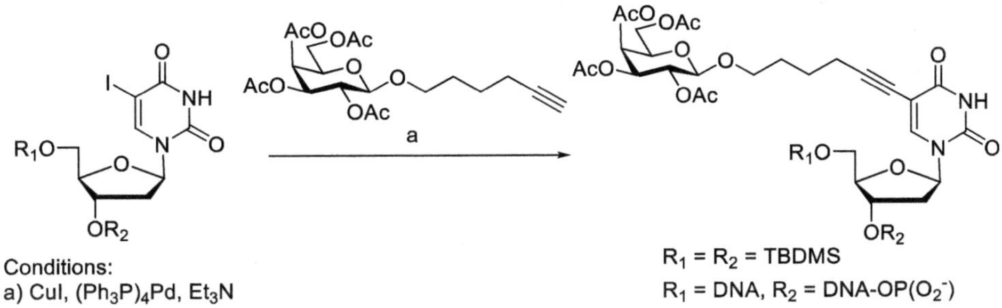

Figure 16.35: Sonogashira coupling approach to incorporate alkynyl galactose at uridine.

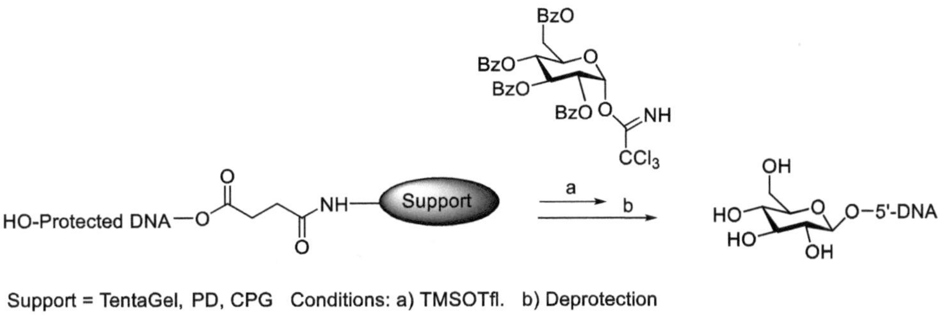

Figure 16.36: Solid support-assisted synthesis of oligonucleotide-carbohydrate conjugates.

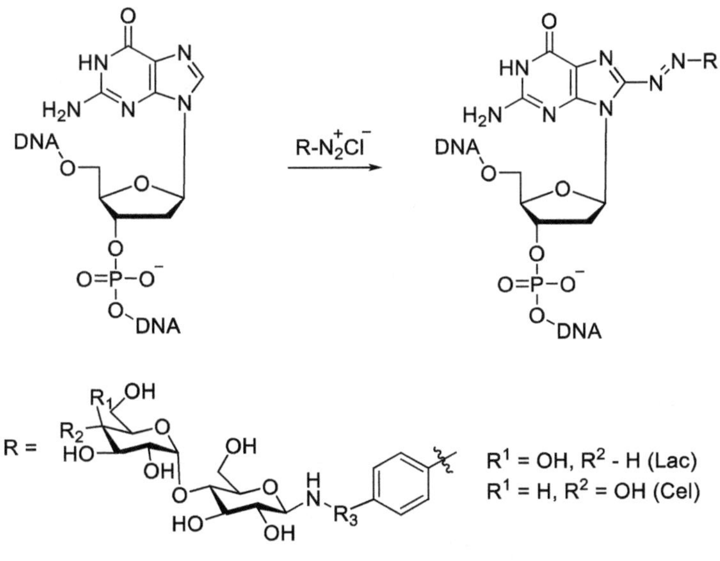

Figure 16.37: Bioconjugation of diazo derivatives of disaccharides with guanine.

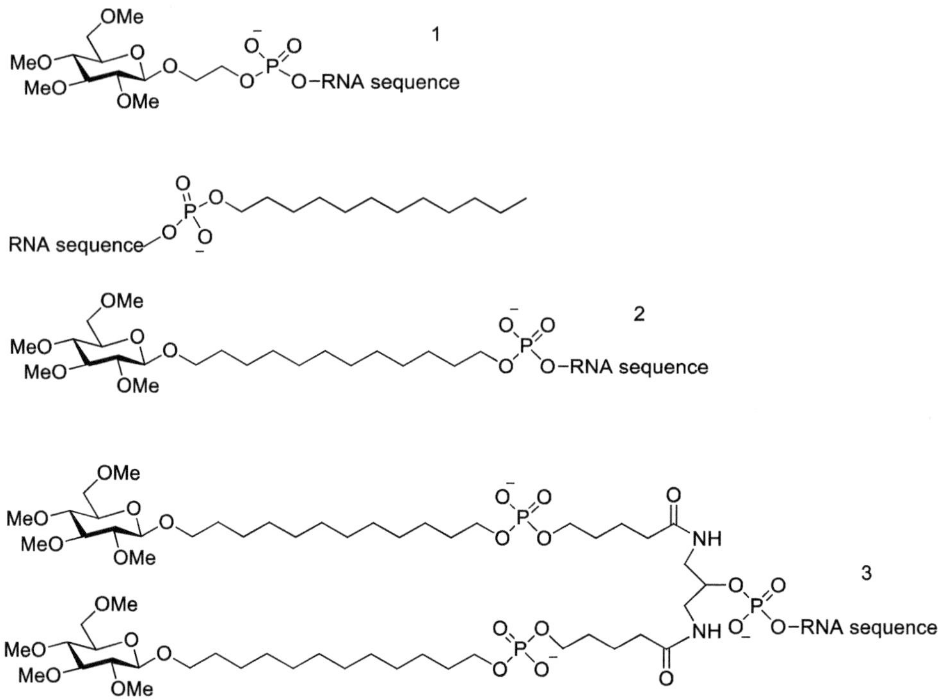

Figure 16.38: Carbohydrate-siRNA conjugates.

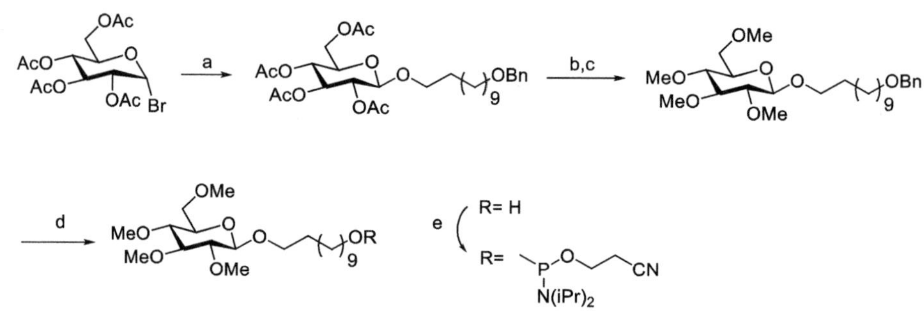

Conditions:
a) 1,12-dodecanediol, monobenzyl ether, Ag_2CO_3, CH_2Cl_2
b) Na_2CO_3, MeOH; c) MeI, NaH, DMF; d) H_2, $Pd(OH)_2$, THF/MeOH
e) 2-cyanoethyl-N,N'-diisopropylamino-chlorophosphoramidite, DIPEA, CH_2I_2

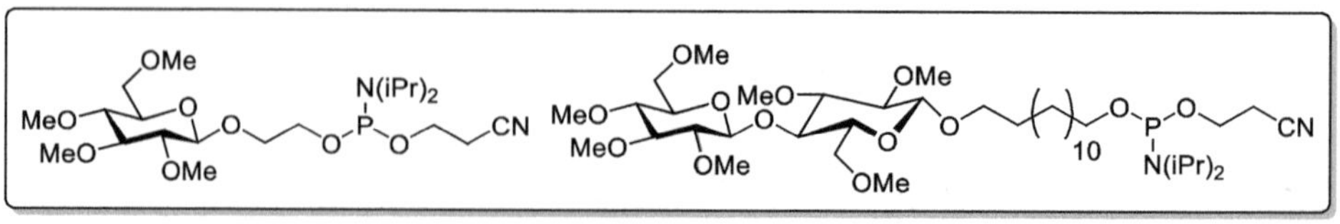

Figure 16.39: Carbohydrates attached with spacers.

hancement measures. The multivalent presentation of carbohydrates was made possible by preparing carbohydrate–PNA–DNA duplexes. These molecular rulers estimated the number and spatial alignment of carbohydrate ligands. It could help to determine lectin structure for drug development. It also provides insight into the binding affinity of monosaccharides with lectin (Figure 16.40) [71].

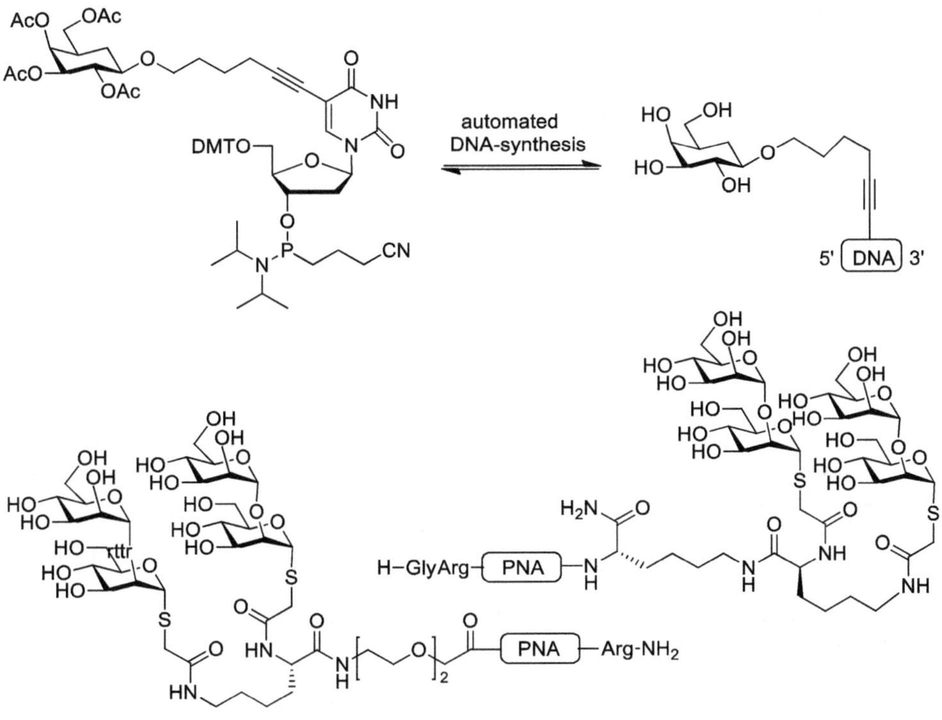

Figure 16.40: Preparation of carbohydrate-PNA-DNA conjugates.

Thus, the conjugates of carbohydrates with proteins and nucleic acids have also shown wide applications in drug development, disease diagnostics, biochemical assays, and imaging. Herein, we have presented some of the methods developed for the preparation of bioconjugates of carbohydrates with proteins and nucleic acids, which have promising applications.

16.6 Conclusion

Bioconjugation, the covalent derivatization of biomolecules, has become a leading strategy for understanding the diverse and complex functions of biomolecules. The bioconjugates have emerged as attractive candidates with a vast scope in medical clinics, industrial synthesis, and research laboratories. The application of prepared bioconjugates mainly relies on the selected linkage strategy during the bioconjugation process. The optimal linkage controls the stability and desired properties of the bioconjugates. Naturally occurring vital glycoconjugates exist in small amounts and are difficult to isolate for chemotherapeutic applications. Therefore, over the years, there has been great interest in the development of synthetic bioconjugates of carbohydrates with tailored physio-biochemical properties. Herein, we have explained the known classical methods employed for the bioconjugation of carbohydrates with proteins and nucleic acids. Further, the promising applications of such bioconjugates have been shown in drug development, disease diagnostics, biochemical assays, and imaging.

References

[1] Chizhov AO. Complex carbohydrates and glycoconjugates: Structure, functions and applications. Int J Mol Sci, 2021, 22, 12219.

[2] Crucho CIC, Correia-da-silva P, Petrova KT, Barros MT. Recent progress in the field of glycoconjugates. Carbohydr Res, 2015, 402, 124–132.

[3] Shivatare SS, Shivatare VS, Wong CH. Glycoconjugates: Synthesis, functional studies, and therapeutic developments. Chem Rev, 2022, 122, 15603–15671.

[4] Park SS, Wu HH, Jacquelyn GH. Anomeric O-functionalization of carbohydrates for chemical conjugation to vaccine constructs. Molecules, 2018, 23, 1742.

[5] Zhang X, Zhang Y. Applications of azide-based bioorthogonal click chemistry in glycobiology. Molecules, 2013, 18, 7145–7159.

[6] Villadsen K, Martos-Maldonado MC, Jensen KJ, Thygesen MB. Chemoselective reactions for the synthesis of glycoconjugates from unprotected carbohydrates. Chem Bio Chem, 2017, 18, 574–612.

[7] Kalia JR, Ronald T. Advances in bioconjugation. Curr Org Chem, 2010, 14, 138–147.

[8] Doelman W, Van Kasteren SI. Synthesis of glycopeptides and glycopeptide conjugates. Org Biomol Chem, 2022, 20, 6487–6507.

[9] Sarkar BJ. Narayanaswamy, Glycoconjugations of biomolecules by chemical methods. Front Chem, 2020, 8, 570185.

[10] Bianchi A, Bernardi A. Traceless staudinger ligation of glycosyl azides with triaryl phosphines: Stereoselective synthesis of glycosyl amides. J Org Chem, 2006, 71, 4565–4577.

[11] Cho CC, Liu JN, Chien CH, Shie JJ, Chen YC, Fang JM. Direct amidation of aldoses and decarboxylative amidation of α-keto acids: An efficient conjugation method for unprotected carbohydrate molecules. J Org Chem, 2009, 74, 549–1556.

[12] Zhang J, Zhao M, Peng S. Synthesis of mimetic peptides containing glucosamine. Carbohydr Res, 2011, 346, 1997–2003.

[13] Ganesh NV, Sadowska JM, Sarkar S, Howells L, McGiven J, David R. BundleMolecular recognition of Brucella A and M antigens dissected by synthetic oligosaccharide glycoconjugates leads to a disaccharide diagnostic for brucellosis. J Amer Chem Soc, 2014, 136, 16260–16269.

[14] Yang B, Jinnouchi A, Usui K, Katayama T, Fujii M, Suemune H, Aso M. Bioconjugation of oligodeoxynucleotides carrying 1, 4-dicarbonyl groups via reductive amination with Lysine Residues. Bioconj Chem, 2015, 26, 1830–1838.

[15] Bernstein MA, Hall LD. A general synthesis of model glycoproteins: Coupling of alkenyl glycosides to proteins, using reductive ozonolysis followed by reductive amination with sodium cyanoborohydride. Carbohydr Res, 1980, 78, C1–C3.
[16] Palomo JM. Diels–Alder cycloaddition in protein chemistry. Eur J Org Chem, 2010, 33, 6303–6314.
[17] Tornøe CW, Christensen C, Meldal M. Peptidotriazoles on solid phase:[1, 2, 3]-triazoles by regiospecific copper (I)-catalyzed 1, 3-dipolar cycloadditions of terminal alkynes to azides. J Org Chem, 2002, 67, 3057–3064.
[18] Ferreira Soares DC, Ramos Oda CM, Fonseca Monteiro LO, Branco de Barros AL, Tebaldi ML. Responsive polymer conjugates for drug delivery applications: Recent advances in bioconjugation methodologies. J Drug Target, 2019, 27, 355–366.
[19] Pozsgay VV, Alfred NEY. A method for bioconjugation of carbohydrates using Diels– Alder cycloaddition. Org Lett, 2002, 4, 3191–3194.
[20] Gamblin DP, Garnier P, Van Kasteren S, Oldham NJ, Fairbanks AJ, Davis BG. Glyco-SeS: Selenenylsulfide-mediated protein glycoconjugation – A new strategy in post-translational modification. Angew Chem, 2004, 116, 846–851.
[21] Gauthier F, Malher A, Vasseur JJ, Dupouy C, Debart F. Conjugation of small molecules to RNA using a reducible disulfide linker attached at the 2′-OH position through a carbamate function. Europe J Org Chem, 2019, 33, 5636–5645.
[22] Matsuura KA, Hibino T, Kazukiyo MK. Facile synthesis of stable and lectin-recognizable DNA-carbohydrate conjugates via diazo coupling. Bioconj Chem, 2000, 11, 202–211.
[23] Tai W. Current aspects of siRNA bioconjugate for in vitro and in vivo delivery. Molecules, 2019, 24, 2211.
[24] Pragallapati S, Manyam R. Glucose transporter 1 in health and disease. J Oral Maxillofac Pathol, 2019, 23, 443–449.
[25] Ugarte-Uribe BPR, Lucas S, Avino R, Reina A, Alkorta JJ, Eritja I, Morales R, Juan C. Synthesis, cell-surface binding, and cellular uptake of fluorescently labeled glucose– DNA conjugates with different carbohydrate presentation. Bioconj Chem, 2010, 21, 1280–1287.
[26] Zatsepin TS, Oretskaya TS. Synthesis and applications of oligonucleotide-carbohydrate conjugates. Chem & Biodivers, 2004, 1, 1401–1417.
[27] Akhtar S, Routledge A, Patel R, Gardiner JM. Synthesis of mono-and dimannoside phosphoramidite derivatives for solid-phase conjugation to oligonucleotides. Tetrahedron Lett, 1995, 36, 7333–7336.
[28] Daskhan GC, Ton Tran HT, Meloncelli PJ, Lowary TL, West LJ, Cairo CW. Construction of multivalent homo-and heterofunctional ABO blood group glycoconjugates using a trifunctional linker strategy. Bioconjug Chem, 2018, 29, 343–362.
[29] Burridge KM, Page RC, Konkolewicz D. Bioconjugates–From a specialized past to a diverse future. Polymer, 2020, 211, 123062.
[30] Ho WL, Hsu WM, Huang MC, Kadomatsu K, Nakagawara A. Protein glycosylation in cancers and its potential therapeutic applications in neuroblastoma. J Hematol Oncol, 2016, 9, 1–15.
[31] Wilkinson H, Saldova R. Current methods for the characterization of O-glycans. J Proteome Res, 2020, 19, 3890–3905.
[32] Kurbangalieva AZ, Nasibullin R, Yamada I, Katsunori KT. Homo-and heterogeneous glycoconjugates on the basis of N-glycans and human serum albumin: Synthesis and biological evaluation. Molecules, 2022, 27, 1285.
[33] Gildersleeve JCO, Simpson O, Benjamin JTA. Improved procedure for direct coupling of carbohydrates to proteins via reductive amination. Bioconj Chem, 2008, 19, 1485–1490.
[34] Abe ST, Pal Y, Komura R, Akihiro NI, Matsumoto KI, Sanghamitra H, Tabe NJM, Hiromune HA. Surface functionalization of protein crystals with carbohydrate using site-selective bioconjugation. Chem Lett, 2015, 44, 29–31.

[35] Kubler-Kielb JPV. A new method for conjugation of carbohydrates to proteins using an aminooxy-thiol heterobifunctional linker. J Org Chem, 2005, 70, 6987–6990.
[36] Dimakos V, Taylor MS. Site-selective functionalization of hydroxyl groups in carbohydrate derivatives. Chem Rev, 2018, 118, 11457–11517.
[37] Hall YDU, Nizam CP, Ishizawa ZM, El-Shaffey S, Jun HMO. Phosphine-mediated three-component bioconjugation of amino-and azidosaccharides in ionic liquids. Chem Commun, 2022, 58, 10568–10571.
[38] Wang BH, Wan Q, Yang T, Cui F, Hu L, Gong S, Li B, Zheng M, Chang Q. Porous lactose-modified chitosan scaffold for liver tissue engineering: Influence of galactose moieties on cell attachment and mechanical stability. Int J Polym Sci, 2016, 1, 2862738.
[39] Russo LB, Secchi C, Magnano V, Nappini E, Taraballi S, Gabrielli F, Comelli L, Papagni F, Barbara AC. Thiol–ene mediated neoglycosylation of collagen patches: A preliminary study. Langmuir, 2014, 30, 1336–1342.
[40] Tan HL, Wu L, Gong J, Changyou YG. Biomimetic modification of chitosan with covalently grafted lactose and blended heparin for improvement of in vitro cellular interaction. Polym Adv Technol, 2008, 19, 15–23.
[41] Pepe AL, Ostuni A, Alessandra SA, Laurita AB. Brigida bioconjugation of carbohydrates to gelatin sponges promoting 3D cell cultures. Biomimetics, 2023, 8, 193.
[42] Machida T, Lang K, Xue L, Chin JW, Winssinger N. Site-specific glycoconjugation of protein via bioorthogonal tetrazine cycloaddition with a genetically encoded trans-cyclooctene or bicyclononyne. Bioconj Chem, 2015, 26, 802–806.
[43] Kim W, Wang Y, Selomulya C. Dairy and plant proteins as natural food emulsifiers. Trends Food Sci Technol, 2020, 105, 261–272.
[44] Dumitrașcu LB, Iuliana DA. Alternative processing options for improving the proteins functionality by Maillard conjugation. Foods, 2023, 12, 3588.
[45] Karabulut G, Goksen G, Khaneghah AM. Plant-based protein modification strategies towards challenges. J Agri Food Res, 2024, 15, 101017.
[46] De Oliveira FC, Reis Coimbra JSD, De Oliveira EB, Zuñiga ADG, Rojas EEG. Food protein-polysaccharide conjugates obtained via the Maillard reaction: A review. Critical Rev Food Sci Nutr, 2016, 56, 1108–1125.
[47] Li YL, Luo F, Chen C, Mao Z, Shoemaker J, Fang ZC. Functional properties of the Maillard reaction products of rice protein with sugar. Food Chem, 2009, 117, 69–74.
[48] Karaca ACL, Michael NN. Emulsifying properties of chickpea, faba bean, lentil and pea proteins produced by isoelectric precipitation and salt extraction. Food Res Int, 2011, 44, 2742–2750.
[49] Nasrollahzadeh FV, Koocheki M, Farzin HA. Effect of microwave and conventional heating on structural, functional and antioxidant properties of bovine serum albumin-maltodextrin conjugates through Maillard reaction. Food Res Int, 2017, 100, 289–297.
[50] Shi WHS, Yu WW, Zhao SJ, Ming M. Study on the characteristic of bovine serum albumin-glucose model system, treated by ultrasonic. Food Res Int, 2010, 43, 2115–2118.
[51] Mu LZ, Yang M, Zhao B, Cui H, Qiangzhong CZ. Effect of ultrasonic treatment on the graft reaction between soy protein isolate and gum acacia and on the physicochemical properties of conjugates. J AgriFood Chem, 2010, 58, 4494–4499.
[52] Zeng XAHJ. Modification of plant biopolymers by pulsed electric fields. In Handbook of Electroporation. Springer International Publishing: Cham, 2016, pp. 1–16.
[53] Song CLZ, Huai X. Rheological, gelling and emulsifying properties of a glycosylated and cross-linked caseinate generated by transglutaminase. Int J Food Sci Technol, 2013, 48, 2595–2602.
[54] Wang YW, Guo Y, Xiong X, Guo Y, Xia MW. Microbial transglutaminase and tyrosinase modified gelatin-chitosan material. Soft Mat, 2015, 13, 32–38.

[55] Selinheimo EL, Mattinen P, Johanna MLB. Formation of protein– oligosaccharide conjugates by laccase and tyrosinase. J Agri Food Chem, 2008, 56, 3118–3128.
[56] Liu FM, Gao C, McClements Y, Julian D. Food-grade covalent complexes and their application as nutraceutical delivery systems: A review. Comp Rev Food Sci Food Saf, 2017, 16, 76–95.
[57] Jiang SJZ, Huai X. Transglutaminase-induced cross-linking and glucosamine conjugation in soybean protein isolates and its impacts on some functional properties of the products. Europ Food Res Technol, 2010, 231, 679–689.
[58] Song CLR, Chen J, Sun JP, Kopparapu XH, Xue NK, Guo Y. Effect of glycosylation and limited hydrolysis on structural and functional properties of soybean protein isolate. J Food Meas Charact, 2018, 12, 2946–2954.
[59] Ma XC, Yan W, Wang T, Hou D, Miao F, Donghong SL. Comparison of citrus pectin and apple pectin in conjugation with soy protein isolate (SPI) under controlled dry-heating conditions. Food Chem, 2020, 309, 125501.
[60] Chy X, Yu XQ, Qi SJ, Guo JR, Sun R, Yao WW, Zhao YJ, Ming M. The effect of glycosylation with dextran chains of differing lengths on the thermal aggregation of β-conglycinin and glycinin. Food Res Int, 2010, 43, 2270–2276.
[61] Tang CHS, Foegeding X, Allen E. Modulation of physicochemical and conformational properties of kidney bean vicilin (phaseolin) by glycation with glucose: Implications for structure–function relationships of legume vicilins. J Agri Food Chem, 2011, 59, 10114–10123.
[62] Markman GL, Function YD. Maillard-conjugate based core–shell co-assemblies for nanoencapsulation of hydrophobic nutraceuticals in clear beverages. Food, 2012, 3, 262–270.
[63] Ping LJY. Self-assembly of ibuprofen and bovine serum albumin– dextran conjugates leading to effective loading of the drug. Langmuir, 2009, 25, 6385–6391.
[64] Huang XT, Xiao Z, Wang H, Lan ZH, Hu YZ, Peipei QN. Characteristics and antioxidant activities of ovalbumin glycated with different saccharides under heat moisture treatment. Food Res Int, 2012, 48, 866–872.
[65] Van de Lagemaat JS, Moreno JM, Olano FJ, Castillo AD, Dolores M. In vitro glycation and antigenicity of soy proteins. Food Res Int, 2007, 40, 153–160.
[66] Palupi NSP, Faridah E, Asyhari DN, Hasriandy M. Reducing allergenicity of soy protein isolate from several varieties of soybean through glycation with lactose. Curr Res Nutr Food Sci J, 2020, 8, 268–280.
[67] Higa FAN, Michael T. Plant protein-carbohydrate conjugates: A review of their production, functionality and nutritional attributes. Food Rev Int, 2023, 39, 750–771.
[68] Maier MAY, Nazim CGM, Roland AF, Mohan H, Just V, Muthiah MG. Synthesis of antisense oligonucleotides conjugated to a multivalent carbohydrate cluster for cellular targeting. Bioconj Chem, 2003, 14, 18–29.
[69] Chernikov IV, Vlassov VV, Chernolovskaya EL. Current development of siRNA bioconjugates: From research to the clinic. Front Pharmacol, 2019, 10, 444.
[70] Vengut-Climent ET, Lucas M, Arévalo-Ruiz R, Eritja M, Morales R, Carlos J. Synthesis, RNAi activity and nuclease-resistant properties of apolar carbohydrates siRNA conjugates. Bioorg Med Chem Lett, 2013, 23, 4048–4051.
[71] Scheibe CSO. PNA–sugar conjugates as tools for the spatial screening of carbohydrate–lectin interactions. Pure App Chem, 2011, 84, 77–85.

Sultana Yeasmin and Pranab Ghosh*

Chapter 17
Carbohydrate polymers and their applications

17.1 Introduction

Carbohydrate polymers or polysaccharides are the most abundant naturally occurring macromolecules, which possess immense importance and utilization in versatile fields. Basically, these macromolecular substances consist of a number of monosaccharide units joined by glycosidic links. The biggest advantage of these bimolecular species is that they can be suitably designed with different monosaccharide units having different stereostructures and molecular weights, and thus can be modified to achieve desired properties required in various fields such as food, medicine, and other daily needs. These are of huge importance in biomedical applications like targeted drug delivery, controlled release, gene therapy, tissue engineering, preparation of functional biomimetic scaffolds, and so on [1, 2]. Besides, carbohydrate-based glycopolymers have been successfully employed in lectin recognition, bacterial adhesion, bioimaging, drug delivery, and gene delivery. Carbohydrateprotein interactions can be understood and effectively manipulated by mimicking the cell surface, which is possible due to the polyvalent nature of carbohydrate polymers [3]. Carbohydrate polymers such as cellulose, chitosan, dextrin, etc., can act as biocompatible heterogeneous catalysts for a number of organic syntheses under mild reaction conditions. These can simply be termed as renewable biodegradable polymeric feedstocks. Still, their complex structure is a matter of much intriguing area, and hence exploration of their physical properties, functions, and new applications is still expanding.

Carbohydrate polymers have conquered huge applications in various fields of our day-to-day life since they are

i. abundant in nature and hence easily available,
ii. nontoxic,
iii. biocompatible,
iv. biodegradable, and
v. cost-effective raw materials which can be modified via simple chemical reactions as required.

*Corresponding author: **Pranab Ghosh**, Natural Product and Polymer Chemistry Laboratory, Department of Chemistry, University of North Bengal, Darjeeling–734013, India, e-mail: pizy12@yahoo.com
Sultana Yeasmin, Department of Chemistry, Balurghat College (Affiliated to the University of Gour Banga), Balurghat, Dakshin Dinajpur 733101, West Bengal, India

https://doi.org/10.1515/9783111447094-017

Their chemical composition, i.e., the type of monosaccharide component, macromolecular weight, the anomeric configuration of glycosidic linkages, and the type of linkage, have been found to be the key factors responsible for versatile functionalities [4], for example, free amine and hydroxyl groups present in cellulose and chitosan are responsible for metal ion chelation via coordinate bonding utilizing the lone pairs at the metal-solution interface [5].

17.2 Different types of carbohydrate polymers

Depending on the functionalities performed by carbohydrate polymers in living cells, they can be classified into the following categories [6]:
1. Reserve polysaccharides such as starch
2. Structural polysaccharides, such as cellulose and
3. Protective polysaccharides such as pectin and hemicelluloses

Depending on the source, carbohydrate polymers can be divided into the following classes [7]:
a) Natural polysaccharides: These consist of homopolymers or heteropolymers, for example, cellulose, dextrin, and agarose.
b) Synthetic polysaccharides: These are derived from nature, and some chemical modification, such as ring-opening polymerization or other methods, is required on the carbohydrate backbone.
c) Glycopolymers: These are a kind of synthetic polysaccharides where monosaccharides or oligosaccharides are attached as pendant groups.

17.2.1 Marine carbohydrates

Polysaccharides obtained from living organisms residing under seawater, such as marine plants, animals, or bacteria, are termed marine carbohydrates [8]. Major examples are uronic acids, amino sugars, and neutral sugars. Marine carbohydrates such as agar, glucosamine glycon, chitin, chitosan, fucoidan, carrageenan, and alginic acid have found various applications, especially in the food [9], cosmetics [10], and pharmaceutical industries [11]. Most of them are bioactive enough as well and hence can be used as antioxidative, antibacterial, antiviral, antitumor, immunostimulatory, and anticoagulant agents.

17.2.2 Self-assembled carbohydrate polymers

Due to having polyfunctional groups, a variety of modifications can be done on carbohydrate polymers with utmost simplicity, whether it is chemical modification, physical, or even enzymatic modifications. One of the simplest methods is to manipulate the electrostatic and hydrogen bonding interactions, which produce self-assembled carbohydrate polymers in the form of capsules, coatings, and films mostly [12]. These have huge applications in the food industry. They can be used in the production of food materials with advanced properties that are responsive to multiple stimuli, whether temperature, pH, enzymes, etc. Another important application of these self-assembled moieties is the controlled release of bioactive substances. Since these are colloidal in nature, they can be treated as potential edible emulsions.

Some common carbohydrate polymers are illustrated below:

17.2.2.1 Cellulose

Cellulose, the most abundant carbohydrate polymer, is essentially a linear homopolymer consisting of 700–25,000 glucose units linked at the β-1,4 position. Besides being the major component of plant cell walls, some bacterial species also produce it via fermentation [13].

Starting from the paper, film, textile, wood product, and fiber industries to the recently developed sustainable and biodegradable fuel production, cellulose has been hugely utilized. According to requirements, it is often modified via various chemical reactions such as esterification [14, 15], etherification [16, 17] and acetylation [18].

17.2.2.2 Starch

Starch is a giant carbohydrate consisting of two polysaccharides, amylose and amylopectin, whose weight percentage varies depending on the source [5, 19]. Starch has found a variety of applications, starting from being a raw material to film production, food packaging, loose-fill products, disposables, and composting bags [20]. Various physical and chemical treatments, like heating, oxidation, and alkali or acid treatment, can lead to desired modifications to improve solubility issues, surface adhesion strength, and to break open the tightly bound starch granules [4].

17.2.2.3 Dextrin and cyclodextrins

During the catalytic hydrolysis reaction by the enzyme amylase, starch produces dextrin in the human mouth. Structurally, it is a low-weight polymer and consists of glucose units linked by glycosidic bonds [5]. In nature, dextrin exists as α,β-dextrin, amylodextrin, maltodextrin, cyclic, and highly branched cyclic dextrin compounds. Generally, three types of dextrins are found depending on their synthetic process, namely white dextrins, canary or yellow dextrins, and British gums.

17.2.2.4 Chitin and chitosan

Chitin, the second most abundant natural polysaccharide after cellulose, is composed of 2-acetamido-2-deoxy-β-D-glucose monomers, which are joined by β-(1→4) linkages [4]. Chitin, which is similar to cellulose in many aspects, is obtained mostly from the outermost shells of arthropods like shrimps, insects, lobsters, and crabs, as well as the cell walls of certain fungal species [21, 22]. After some chemical modification, chitin can be used as a thickening and stabilizing agent in the food processing industry [23]. It is also used in the production of edible films. Its unusual healing nature makes it a sought-after candidate toward the preparation of artificial skin, sutures, and drug delivery systems [24]. However, the biggest application of chitin lies in wastewater treatment worldwide [8].

Chitosan is another linear polysaccharide containing D-glucosamine monomers with *N*-acetylglucosamine substituents at random positions joined by β-(1→4) linkages. Chitosan can be produced from chitin simply by an *N*-deacetylation process [25]. The degree of deacetylation determines the characteristics of chitosan, such as reactivity, cell response, and solubility [26]. This crystalline polymer is basically hydrophilic in nature. Hence, it is quite good at gelation and film formation. In the textile industry, it provides an efficient and cost-effective method to remove alkaline effluent. Due to the presence of active functional groups like amine and hydroxyl groups, chitosan can undergo simple chemical modifications to produce a variety of materials with desired properties [22]. From clinical applications to the paper, food, and textile industries, chitosan occupies a renowned position mainly due to its biocompatibility [27].

Both chitin and chitosan exhibit excellent biological activity, with special mention of their antibacterial and antifungal properties. They are promising agents in drug delivery systems as well.

17.2.2.5 Gums

Gums are another group of carbohydrates that are capable of forming gels upon the introduction of counter-ions into their solution [8], such as alginic acid obtained from the cell wall of brown algae [28], guar gum from the seeds of *Cyamopsis tetragonolobus* [29], gellan gum from the bacterium *Sphingomonas elodea* [30], xanthan gum from the bacterium *Xanthomonas campestris* [31], and exudate gums excreted by plants upon external injury or fungal attack [31]. The parameters responsible for the degree of cross-linking during gel formation include the nature of the counter-ion, the charge density of the functional group, and the acidity or basicity of the solution [32].

Alginic acid, one of the most common gums, comprises a linear, anionic copolymer of D-mannuronic acid and L-guluronic acid. The characteristics of alginate polymers depend upon the percentage composition of D-mannuronic acid and L-guluronic acid. It has been reported that higher L-guluronic acid content renders better clinical applications of alginate polymers [33].

17.2.2.6 Agar

Agar or agar-agar is one of the biggest representatives of marine carbohydrates. It is obtained from an alga consisting of agarose and agaropectin. Being a gelatinous substance, it can be used as a laxative, vegetarian gelatin substitute, a thickener in soups, and so on. Mostly, it is used in the food industry, such as in fruit preserves, ice cream, and other desserts, as a clarifying agent in brewing, and also in the paper and textile industry for sizing paper and fabrics [34].

17.3 Applications of carbohydrate polymers

It is quite impossible to opt out of the application of carbohydrate polymers from a single area of this ever-developing world. Starting from the food industry to pharmaceuticals, the automotive industry to chemical laboratories, and even to our household utilities, they occupy almost everywhere to help civilization walk the greener runway toward sustainable development. Some applications are given below.

17.3.1 Biofuels

Sustainable development is a burning issue in today's world. Without an adequate energy source, the world will just cease to move. Hence, keeping an eye toward a greener yet developed society, the search for and establishment of bio-based energy sources is on the verge. In this regard, carbohydrate polymers are the ultimate answer, especially due to their easy availability, huge abundance, low cost, biodegradability, non-toxicity, and easy-going nature toward various kinds of chemical or physical modification to enrich or input desired properties:

- **Fuel cell**

Some common carbohydrate polymers, such as starch, cellulose, and chitosan, have been effectively utilized as raw materials to produce cost-effective, eco-friendly polymer electrolytes. Chitosan, having polyfunctional groups, can be modified very easily, whether physically or chemically, to act accordingly in the proton-exchange membrane in fuel cells. After modification, it can exhibit required characteristics like low methanol permeability and hydrophobicity very well.

17.3.2 Pharmaceutical uses

Almost all carbohydrate polymers exhibit some kind of biological activity. Many of them, such as cyclodextrin, dextran, alginate, chitosan, carrageenan, and heparin, play a significant role against diseases caused by viral and bacterial infections, cardiovascular disorders, oxidative stress, or metabolic disorders [35]. Their biocompatible nature also accounts for their vast application in the pharmaceutical industry, such as

- Coatings for tablets
- Binder for tablets, dispersant for medicines, controlled-release agent
- Blended sweeteners for medicines, for example, cold remedies
- Coating spheres for capsule medicines
- **Drug delivery:** The biggest pharmaceutical application of carbohydrate polymers lies in the realm of drug delivery systems, including targeting and even controlled release. Since they are found in living systems, they inherently possess the property of biocompatibility, which plays a crucial role behind such applications. Again, after delivery, these carriers are easily excreted from the system. Certain carbohydrate polymers are capable enough in their raw form without any modification, but most of them require some simple chemical or physical modification. Some examples are
  - Hyaluronic acid and chitosan are widely used as carriers for drugs via the dermal or oral route.
  - Xanthan gum and alginates are used in controlled release.
  - Dextran has found application in those unstable drugs that cannot pass through the cellular membrane easily.
  - Gellan is utilized as a mucoadhesive to fix the drug at the target when the drug is administered via the nasal route.
  - Pullulans are another class of carbohydrate polymers which were first utilized for drug delivery in microcapsules.

Moreover, carbohydrate polymers resolve the solubility issue faced by various nonpolar medicines, which do not dissolve well in our body fluids [36]. In such cases, the carbohydrate polymers are capable enough to solubilize the medicine and carry it to the target site, where they simply release the medicine.

- **Cell recognition and tissue engineering:** Synthetic polymers with carbohydrate moieties are capable of cell attachment at their surface. This property is applica-

ble to the cell recognition process by antiviral or antimicrobial agents and also in tissue engineering [3].
- **Role against pathogen infection:** Lectin-binding carbohydrate units play a significant role against pathogen infection [3]. Hence, synthetic polymers having such moieties are promising drug candidates, which can form biolayers via multivalent interaction against pathogen infection.
- **Determination of blood group:** Oligosaccharide chains present on RBCs actually determine the blood group [36]. Red blood cells are covered by the oligosaccharide chains where glucose is attached at the cell surface, followed by another four sugars, namely galactose, *N*-acetylglucosamine, then again galactose, and fructose. Such an arrangement of oligosaccharide chains labels the blood group as O-type. When one more monosaccharide, namely *N*-acetylgalactosamine, is attached to the chain, the blood group is A, and when the other group is galactose, the blood group is B. When both types of chains are present, the blood group is AB.
- **Vaccines:** For the synthesis of vaccines against viral and bacterial diseases like influenza, pneumonia, and meningitis, polysaccharides are treated as the key component [36]. Basically, a polysaccharide-based capsular cover called an antigen protects the inner cells of these bacterial species against the surroundings inside the host. Vaccines are prepared from these particular polysaccharides.

17.3.3 Nutraceutical role

Some carbohydrate polymers and their derivatives impart extra nutritional value to human health when taken as supplements, such as prebiotics and dietary fibers [36]. Certain oligosaccharides, like fructo-oligosaccharides, gluco-oligosaccharides, oatmeal starch, wheat bran, and almond shells, act as prebiotics to improve digestion and reduce the risk of intestinal infection.

17.3.4 Role in food industry

It is well known that carbohydrates are the best calorie reservoirs in living systems. Biocompatibility and nontoxicity are two major factors behind their vast and safe application in the food industry:
- **Food additives:** To maintain the quality and lifespan, especially of packaged food, a number of polysaccharides are used as food additives with very little or no side effects [4]. The additives act as thickeners, stabilizers, preservatives, and even maintainers of the perfect color of the food product during its life span. Some examples are given below.

Starch: Starch is a very good thickening additive and is used in sauces, soups, gravies, custard, and mayonnaise. Starch, after some modification, can exhibit better thermal stability. Hence, modified starch is utilized while processing frozen food. It renders better solubility and maintains the texture throughout. Furthermore, modified starch enhances the quality of baked food items and snacks.

Carboxymethyl cellulose (CMC): Although cellulose is the most abundant carbohydrate polymer obtained from the cell walls of plants, its application in its raw form is inhibited due to its insolubility in water. However, its derivative, CMC, has found enormous application as a food additive in desserts, cakes, cottage cheese, dressings, toppings, etc. It thickens the food item and binds it well, furnishing an enhanced life span. Furthermore, CMC is used as a laxative since it can be easily excreted from the body.

Apart from starch and cellulose derivatives, many more carbohydrate polymers exhibit similar additive properties, for example, carrageenan, pullulan, guar gum, and xanthan gum.

– **Artificial sweeteners**
  Since diabetes has become the most common threat to every human being, awareness has also spread at its peak. Hence, the demand for artificial sweeteners with lower calories as substitutes for extra added sugar is increasing day by day. Some commonly available examples are saccharin, aspartame, sucralose, and acesulfame potassium. But truly speaking, almost all the artificial sweetening agents available in the market are associated with multiple side effects on prolonged use, such as malignancy, gastrointestinal symptoms, hypertriglyceridemia, bloating, and weight gain. Hence, the search is on to obtain the safest artificial sweetening species with no or minimal side effects. In this respect, some monosaccharides, namely erythritol, mannitol, and xylitol, have emerged as promising candidates that are quite effective as alternative sweetening agents with no side effects [36]. For example, erythritol does not get absorbed in the body and neither increases calories nor blood sugar. Further, erythritol rather prevents tooth decay since it cannot be consumed by the responsible bacteria *Streptococcus mutans* due to structural differences from sucrose.

17.3.5 Cosmetics and household items

Most of the cosmetic products and household items we use in daily life are a kind of colloidal solutions or, better said, emulsions. For example, cream, shampoo, toothpaste, paints, ointments, hand wash, etc., are basically oil-water emulsions. Here, polysaccharides play a pivotal role. They stabilize the emulsion by keeping the oil and water layers intact and thus enhance the lifespan of the product. Some common examples are given below:

- Corn fiber gum and beet root pectin are utilized as emulsifiers in hydrating moisturizers applied on the skin.
- CMC, a magic derivative of cellulose, plays the role of a binding, emulsifying, and viscosity-controlling agent in toothpaste, laundry detergents, water-based paints, hand cream, moisturizers, shampoo, soaps, hand lotions, face wash, and so on.
- Cationic polymer is also used as an emulsifier in hair conditioners, hydrating and anti-aging face masks, hair tonic, deodorants, mouthwash, perfumes, and so on.
- Various skin care products are prepared utilizing the polysaccharides obtained from red, brown, and green algae due to their anti-wrinkle, skin-whitening, UV protective, anti-inflammatory, and anti-allergic properties.

17.3.6 Automotive manufacture

Carbohydrate polymers occupy a significant area in the automotive industry, especially as

- Corrosion inhibitor: In the automotive industry, corrosion inhibition is one of the major requirements. Corrosion is a kind of degradation of the surface caused by environmental factors and excessive friction between the surfaces or the presence of impurities. Many carbohydrate polymers, termed as chemical inhibitors, have the potential to protect the metal surface against environmental factors such as temperature and moisture, and even against internal factors of the device like the battery system where electrolyte change may have an effect. They have the ability to absorb the harmful molecules which cause the metal to corrode. For example, anticorrosion activities of exudate gums, cellulose, starch, pectin, pectate, chitosan, carrageenan, alginate, and dextrin have been reported with special emphasis on the effect of halide additives [5]. It is supposed that polymers and halide ions are bound by coulombic force, which leads to better protective films against corrosion.

17.3.7 Textile and paper industry

Textile and paper industries cannot be imagined without polysaccharides, especially cellulose and starch. The applications are outlined below:

- Starch is applied in textile finishing, warp sizing, and printing.
- In the paper industry, starch is also employed in sizing, feminine care products, babies' nappies, filler retention, and finishing surfaces.
- Starch polymers are used for loose fill, bags, wrap film, for example, Biolice, Biomax, Biopar, Solanyl, Cerestarch, and Mater-Bi.
- PLA is used for rigid barriers, barrier coatings, and films, for example, Ecovia.

- Sodium CMC forms a water-soluble film or coating and is used for sizing wool, yarn, laminating fabrics, and paper. This coating is also used in making grease-proof paper, which is used as butter paper for packaging oily foodstuffs. Moreover, sodium CMC increases the tensile strength, flexibility, tear-resistance, and ink penetration properties of paper. Besides, CMC produces a highly viscous product on reaction with epichlorohydrin. This viscous moiety is used in
  - Lubricants,
  - printing inks, and
  - fabric colors for dyeing denim, cotton, cellulose wool, linen cotton, artificial silk, and mixed fabrics.
- Chitosan and its derivatives (such as chitosan-guanidine and chitosan-cellulose) are widely used carbohydrate polymers in the paper and textile industries. Chitosan is used in printing inks as
  - binder and
  - thickener.

Its derivatives are used in paper coating since they have the ability to act as a barrier against water, grease, and even bacterial species. Cellulose can produce textile fiber. It can be used as a dye for cotton, rayon, polyester, silk, nylon, and wool, and also in sizing or desizing yarn.

17.3.8 Catalysis

In various organic reactions, carbohydrate polymers, especially chitosan, cellulose, and starch, have been employed successfully as heterogeneous catalysts, such as in solid-phase peptide synthesis. Here, they provide a suitable catalytic surface to the reactants due to the following reasons:

i. low or nontoxicity,
ii. high thermal stability,
iii. easy availability,
iv. the presence of polyfunctional groups helps to interact with the reactants as required,
v. easily tunable chemical and physical properties. and
vi. the separation of the product is easy after the completion of the reaction.

17.3.9 Essential items in chemical laboratories

Carbohydrate polymers are widely utilized in the preparation of multiple essential items required in chemical laboratories:

- Separation: Filter papers, like Whatman filter papers used in laboratories for separation, are made of 98% cellulose. Different kinds of filter papers are prepared depending upon the nature of the solute and solvent. Almost all of them are cellulose-based, like cellulose-EDTA, cellulose acetate, and cellulose nitrate. These biodegradable filter papers are nontoxic as well and hence are safer to use in food as tea bags, coffee filters, and also in various separating techniques applied in the pharmaceutical industry.
- Labwares: Variety of labwares required for different purposes in labs are prepared from carbohydrate polymers such as
  - absorbent papers,
  - blotting papers,
  - chromatography papers,
  - extraction cartridges,
  - cellulose crucibles, and
  - ion exchange resin paper.

17.3.10 Formation of nanocomposites

When carbohydrate polymers are inserted into metal and metal oxide nanoparticles, some unique kinds of nanocomposites are produced, which have been found to increase the conductivity and catalytic activities of electrochemical sensors [37]. The carbohydrate polymers are supposed to inhibit the self-aggregation process of the nanoparticles and thus furnish them with greater stability. Besides, they help to render an optimized current signal by the redox cycling process. These metal or metal oxide-carbohydrate polymer nanocomposites can be prepared from a number of carbohydrate polymers such as cellulose, starch, chitosan, alginate, pectin, dextran, gums, locust bean gum, latex, and mucilage. Due to the presence of bio-active moieties, these electrochemical sensors have potential applications with optimized selectivity in the detection of various biological analytes like hormones, enzymes, drugs, natural oxidants, aflatoxins, and even some water pollutants like metal ions, dyes, pesticides, and phenolics.

17.3.11 Production of new materials from waste resources

Comprising the biggest biomass resource, carbohydrates have been regarded as the best-suited sustainable substitute for conventional energy sources, as well as the raw materials for the production of various new products with unique characteristics in the chemical industry. Common carbohydrate polymers like starch, cellulose, hemicellulose, chitin, xyloglucan, and pectins possess all the equivalent qualities of fossil fuels and hence are potential candidates toward the independent production of biofuel, which is both economically and environmentally friendly. To achieve this goal, mass production of carbohydrate polymers is required, and their functionalities should be explored more and more. For example, polymannose, which is an important biomaterial required in antifungal drug delivery systems, has been produced on a mass level by an automated oligosaccharide synthesizer, THE GLYCONEER®, developed by the Max Planck Institute of Colloids and Interfaces [38, 39].

17.3.12 Adhesives or glues

Carbohydrate polymers are well utilized in the preparation of adhesives, such as
- hot melt glues,
- glues used in stamps, book binding, envelopes,
- wood adhesives and laminates,
- pressure-sensitive adhesives,
- corrugation (e.g., boxes), and
- paper sacks.

17.3.13 Construction industry

In the construction industry, carbohydrate polymers fulfill some important applications. For example, starch is used as a binder for concrete block binding, asbestos, clay or limestone binding, fire-resistant wall board, paint filler, and as a gypsum (plaster board) binder.

17.3.14 Explosive manufacture

In the preparation of explosives or matches required for household work, carbohydrate polymers, especially starch, are utilized as binders.

17.3.15 Printing inks

Some carbohydrate polymers and their derivatives are used in the manufacture of printing inks; for example, sucrose benzoate is used in water-based inks.

17.3.16 Application in agriculture

Some studies suggest that carbohydrate polymers may strengthen the immune systems of plants against pathogenic fungi [13]. It is reported that they can act as elicitors to inhibit spore germination and mycelial growth. Moreover, they provide a substitutive option for traditional fungicides by virtue of their ability to control the release of agrochemicals and different active essential ingredients. For example, chitosan, commonly used as a natural biocontrol agent, improves plant health and increases the yield of crops.

17.3.17 Water purification

The principle of water purification systems adopted by various water purifiers is based on adsorption theory, which is simple, cost-effective, highly efficient, and eco-friendly as well [40]. Activated carbon and silver nanoparticles have been excellent examples of effective adsorbents, but their utilization for edible purposes cannot be afforded due to their toxicity and high cost. In this regard, carbohydrate polymer-based adsorbents are gaining much popularity mainly due to their non-toxicity and low cost. Chitin, chitosan, and starch are some of the most abundant and efficient examples utilized for this purpose. Adsorbent systems based on such biopolymers can be prepared by crosslinking and coupling methods. In the first method, the hydroxyl or amino groups of the polysaccharides react with coupling reagents to produce a water-insoluble crosslinked network [41]. In the second method, polysaccharides go through a coupling reaction on insoluble supports to produce composite or hybrid materials [42].

17.3.18 Wastewater treatment

Wastewater produced from various domestic and industrial activities containing different kinds of organic, inorganic, and biological contaminants may cause serious health hazards and environmental threats. Hence, wastewater treatment has become one of the major steps toward a cleaner and greener world. Chitosan, a cationic biopolymer, has been successfully used in the treatment of wastewater produced in dairy factories [43]. For industrial effluents, chelation is the preferred process for the removal of dyes. The functional groups (amine group) of chitosan are responsible for chelate formation with metal cations from wastewater such as Cu^{2+}, Pb^{2+}, and Hg^{2+}. In an acidic medium, chitosan gets protonated and then it can attach to the anionic pollutant species via electrostatic attraction [44], and thus it can be effective for the removal of both kinds of effluents from wastewater. Another study also reported that chitin and chitosan both can be used for the removal of phenol and other effluents from wastewater via the adsorption method [45]. Some chemical modifications on both chitin and chitosan can make them more effective for wastewater treatment.

17.4 Future perspectives

As the world increasingly focuses on renewable and sustainable materials, it is likely that more applications of carbohydrate polymers will emerge, particularly in fuel cells and drug delivery systems, where they have already made a significant impact. Given the availability and biocompatibility of carbohydrate polymers, it can be expected that many more medical and healthcare applications of this type of material will be developed in the coming years. Pharmaceutical companies are thoroughly engaged in developing novel glycomimetic drugs for 24 × 7 urgent medical crises [13]. These natural biopolymers have the potential to produce nanofibrous scaffolds for both soft and hard tissue engineering [8]. Besides, until now, only plant, animal, and microbe-based carbohydrate polymers have been extensively explored to establish and commercialize their applications in various fields. However, there are many more such biopolymers from distinct environments, such as phytoplankton, hydrothermal vents, and benthic flora and fauna, etc. to be identified and explored for the benefit of mankind [36].

17.5 Conclusion

Carbohydrate polymers offer numerous applications in diverse fields, from medicine to cosmetics, from food to textiles, from glues to stitching threads, and so on, which is attributed mainly to their non-toxicity, easy availability, and biodegradability. Rather than just being energy reservoirs in our body system, they have expanded their utility to become almost a next-generation "smart" vehicle toward sustainable development. However, this area has not been fully explored yet, and more research is required on their complex structure to achieve more applications, especially in the field of nano-based medical science.

References

[1] Arima H, Kihara F, Hirayama F, Uekama K. Enhancement of gene expression by polyamidoamine dendrimer conjugates with α-, β-, and γ-Cyclodextrins. Bioconjugate Chem, 2001, 12, 476–484.

[2] Fulton DA, Stoddart JF. Neoglycoconjugates based on cyclodextrins and calixarenes. Bioconjugate Chem, 2001, 12, 655–672.

[3] Wang Q, Dordick JS, Linhardt RJ. Synthesis and application of carbohydrate-containing polymers. Chem Mater, 2002, 14, 3232–3244.
[4] Azeez MA, Olatunde OC. Recent developments on the application of carbohydrate polymers. IOSR J Appl Chem, 2018, 11, 68–80.
[5] Umoren SA, Eduok UM. Application of carbohydrate polymers as corrosion inhibitors for metal substrates in different media: A review. Carbohydr Polym, 2016, 140, 314–341.
[6] Singh AV. Biopolymers in drug delivery: A review. Pharmacologyonline, 2011, 1, 666–674.
[7] Chen G. The past ten years of carbohydrate polymers in ACS macro letters. ACS Macro Lett, 2021, 10, 1145–1150.
[8] Sudha PN, Aisverya S, Nithya R, Vijayalakshmi NK. Industrial applications of marine carbohydrates. Adv Food Nutr Res, 2014, 73, 145–181.
[9] Dhargalkar VK, Verlecar XN. Southern ocean seaweeds: A resource for exploration in food and drugs. Aquaculture, 2009, 287, 229–242.
[10] Stolz P, Obermayer B. Manufacturing microalgae for skin care. Cosmet Toi-letries, 2005, 120, 99–106.
[11] d'Ayala GG, Malinconico M, Laurienzo P. Marine derived polysaccharides for biomedical applications: Chemical modification approaches. Molecules, 2008, 13, 2069–2106.
[12] Valencia GA, Zare EN, Makvandi P, Gutiérrez TJ. Self-assembled carbohydrate polymers for food applications: A review. Compr Rev Food Sci Food Saf, 2019, 0, 1–16.
[13] Rao GM, Bharathi P, Akila RM. A comprehensive review on biopolymers. Scientific Rev Chem Commun, 2014, 4, 61–68.
[14] Spinella S, Maiorana A, Qian Q. Concurrent cellulose hydrolysis and esterification to prepare a surface-modified cellulose nanocrystal decorated with carboxylic acid moieties. ACS Sustain Chem Eng, 2016, 4, 1538–1550.
[15] Li D, Henschen J, Ek M. Esterification and hydrolysis of cellulose using oxalic acid dihydrate in a solvent-free reaction suitable for preparation of surface-functionalised cellulose nanocrystals with high yield. Green Chem, 2017, 19, 5564–5567.
[16] Fox SC, Li B, Xu D. Regioselective esterification and etherification of cellulose: A review. Biomacromolecules, 2011, 12, 1956–1972.
[17] Heinze T, Seoud O, Koschella A Etherification of Cellulose, Cellulose Derivatives. Springer, U.S., 2018, 429–477
[18] Cetin NS, Tingaut P, Ozmen N. Acetylation of cellulose nanowhiskers with vinyl acetate under moderate conditions. Macromol Biosci, 2009, 9, 997–1003.
[19] Brown WH, Poon T. Introduction to Organic Chemistry. 3rd ed., Wiley, Hoboken, New Jersey, U.S., 2005.
[20] Xiong H, Tang SH, Tang HAZ. The structure and properties of a starch-based iodegradable film. Carbohydr Polym, 2008, 71, 263–268.
[21] Tanase EE, Rapa M, Popa O. Biopolymers based on renewable resoures- A review. Scientific Bull Ser F Biotechnol, 2014, 18, 188–195.
[22] Aravamudhan A, Ramos DM, Nada AA. Natural polymers: Polysaccharides and Thier derivatives for biomedical applications. In Natural and Synthetic Biomedical Polymers. Elsevier, Amsterdam, Netherlands, 2014, pp. 67–83.
[23] Shahidi F, Arachchi JKV, Jeon YJ. Food applications of chitin and chitosans. Trends Food Sci Technol, 1999, 10, 37–51.
[24] Lee SH. Effect of chitosan on emulsifying capacity of egg yolk. Korean Society J Food Science Nutr, 1996, 25, 118–122.
[25] Parios I, Acosta N, Heras A. New drug delivery sysems based on chitosan. Curr Drug Discov Technol, 2008, 5, 333–341.

[26] Shukla RK, Tiwari A. Carbohydrate polymers: Aplications and recent advances in delivering drugs to the colon. Carbohydr Polym, 2012, 88, 399–416.

[27] Pachence JM, Bohrer MP, Kohn J. Biodegradable Polymers. Principles of Tissue Engineering, Elsevier, Amsterdam, Netherlands, 2007, 323–399.

[28] Mano JF, Silva GA, Azevedo HS. Natural origin biodegradable systems in tissue engineering and regenerative medicine: Present status and some moving trends. Journal of the Royal Society Interface, 2007, 4, 999–1030.

[29] Tiwari A, Prabaharan M. An amphiphilic nanocarrier based on guar gum-graft-poly (epsilo-caprolactone) for potential drug-delivery applications. J Biomater Sci Polym Ed, 2010, 21, 937–949.

[30] Coutinho DF, Saint SV, Shin H. Modified Gelan Gm hydrogels with tunable physical and mechanical properties. Biomaterials, 2010, 31, 7494–7502.

[31] Izdorczyk M, Cui SW, Wang Q. Polysaccharide Gums: Structures, Functional Properties, and Applications. Taylor & Francis, United Kingdom, 2005.

[32] Chandra R, Rustgi H. Biodegradable polymers. Prog Polym Sci, 1998, 23, 1273–1335.

[33] Christensena BE. Alginates as biomaterials in tissue engineering. Carbohydr Chem, 2011, 37, 227–258.

[34] Cregut M, Rondags E. New insights in agar biorefinery with arylsulphatase activities: Review. Process Biochem, 2013, 48, 1861–1871.

[35] Kar B, Pradhan D, Halder J, Rai VK, Ghosh G, Rath G. Antiviral application of carbohydrate polymers: A review. Current Pharmaceutical Design - Bentham Science Publisher, UAE, 2023, 29, 1441–1458.

[36] Nagar S. Hidden roles of carbohydrates. Science Reporter, 2020, 45–48.

[37] Zahran M. Carbohydrate polymer-supported metal and metal oxide nanoparticles for constructing electrochemical sensors. Mater Adv, 2024, 5, 68–82.

[38] Vargas PA, Delbianco M, Seeberger PH. Automated glycan assembly as an enabling technology. Curr Opin Chem Biol, 2018, 46, 48–55.

[39] Francis AP, Gurudevan S, Jayakrishnan AJ. Carbohydrates as sustainable materials for the future biomat. Science, 2018, 29, 529–1548.

[40] Nassar R, Browne E, Chen J. Removing human immune deficiency virus (HIV) from human blood using immobilized heparin. Biotechnol Lett, 2012, 34, 853–856.

[41] Varma AJ, Deshpande SV, Kennedy JF. Metal complexation by chitosan and its derivatives: A review. Carbohydr Poly, 2004, 55, 77–93.

[42] Berger J, Reist M, Mayer JM. Structure and interactions in chitosan hydrogels formed by complexation or aggregation for biomedical applications. Eur J Pharm Biopharm, 2004, 57, 35–52.

[43] Tokura S, Nishi N. Specification and Characterization of Chitin and Chitosan. Bangi: Universiti Kebangsaan Malaysia, 1995, pp. 68–86.

[44] Guibal E. Interactions of metal ions with chitosan-based sorbents: A review. Sep Purif Technol, 2004, 38, 43–74.

[45] Milhome MAL, Keukeleire D, Carvalho TV, Queiroz DC, Ribeiro JP, Nascimento RF. Removal of phenol and conventional pollutants from aqueous effluent by chitosan and chitin. Quimica Nova, 2009, 32, 2122–2127.

Surabhi Gupta*

Chapter 18
Preparation of biofuels from carbohydrates: an overview

18.1 Introduction

The global population keeps increasing and is anticipated to reach around 8 billion by 2025 [1]. Energy is the primary requirement in each aspect of human life, including food, water, transport, health, medicines, agriculture, and industry sectors [2]. The fast growth of the global population directly or indirectly demands energy to meet daily requirements in the form of electricity, heat, and fuels. It is well known that the economic growth of a country largely depends on energy resources [3]. Around 90% of the global energy needs are met by fossil fuels such as coal, natural gas, and petroleum oil [4]. Fossil fuels have some advantages, such as being cheap, reliable, and easily transportable from one place to another (Figure 18.1) [5].

Around 40% of the electricity requirement is derived from coal. On the other hand, the transportation sector accounts for 30% of the world's total produced energy in the form of petroleum oil and gas products [6]. However, non-renewability is one of the major limitations of fossil fuels, and world oil reserves are expected to run out by 2050 [7, 8]. Moreover, fossil fuels cause severe environmental contamination and climate change due to excessive carbon dioxide emissions during combustion [9]. Besides the greenhouse effect, acid rain, and soil and water pollution are also caused by fossil fuels. These drawbacks demand alternatives to fossil fuel energy. In this context, developing energy from green and renewable materials has become an important target for many countries [10]. In this aspect, biofuel, solar energy, wind energy, hydropower energy, and so on have been developed to meet daily requirements [11]. Among these energies, biofuels have received considerable attention in the past decades owing to their reliability.

Acknowledgments: The author thanks Dr. Jeyakumar Kandasamy, IIT (BHU), Varanasi, India, for his continuous support and discussion throughout the writing of the manuscript. The author also thanks Prof. Charles Gauthier, INRS, Laval, Canada, for providing a National Institute of Health (NIH) fellowship from the Research Grant Program R01 granted to him. To express her humble tribute, the author has dedicated this book chapter to Dr. Jeyakumar Kandasamy on his birthday.

***Corresponding author: Surabhi Gupta,** Institut National de la Recherche Scientifique (INRS), QC, H7V1B7 Canada, e-mail: guptasurabhipg2008@gmail.com

https://doi.org/10.1515/9783111447094-018

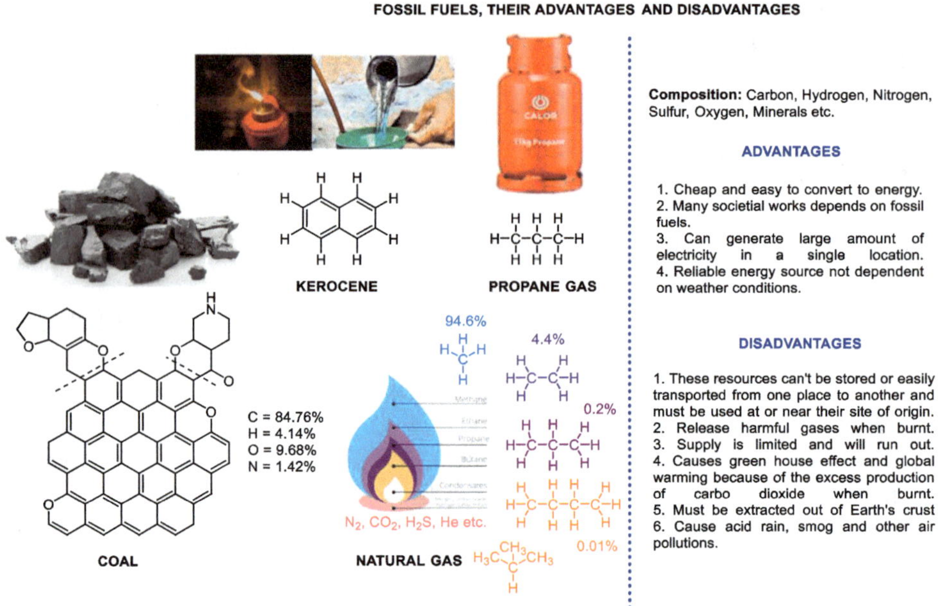

Figure 18.1: Fossil fuels: advantages and disadvantages.

18.2 Biofuels and their feedstocks

Biofuel is an important class of energy that is produced from biomass feedstocks [12]. Biofuels are efficient alternatives to fossil fuels that have received much attention due to their positive impact on the environment, ecosystem, economy, and biodiversity (Figure 18.2) [13, 14]. Table 18.1 displays various types of fuels, including petroleum-based fuels, bio-based fuels, and furan-based fuels with their structures. Biofuels exist in different physical states, but mainly as liquids. Types of biofuels include bioethanol, biobutanol, biodiesel, biogas, biogasoline, and bioethers [15].

Biofuel is renewable and produced from biological sources such as food crops, oil extracted from vegetables, plants, animals, industrial wastes, and microbial origin (Figure 18.3) [16, 17]. Biofuel has been considered a suitable solution to the increasing energy demands globally [18]. Besides being renewable, biofuel has many advantages, including utilizing wastes from industries, providing less expensive feedstocks, contributing to positive climatic changes, creating employment opportunities, and offering reliability (Figure 18.4) [19, 20]. Biofuel feedstocks are easy to store and transport on demand and would help developing countries increase their energy security as per requirements.

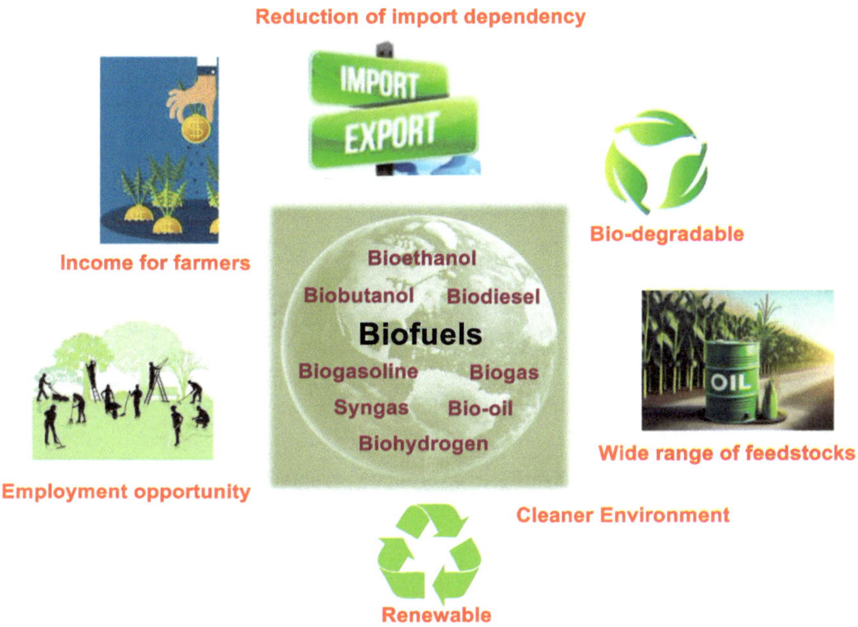

Figure 18.2: Biofuels and their advantages.

Table 18.1: Types of fuels and their chemical structures.

S. no.	Types of fuels	Name of fuels	Chemical structure
1)	**Petroleum-based fuel oxygenates**		
		Dimethyl ether (DME)	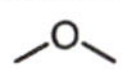
		Di-*n*-butyl ether (DBE)	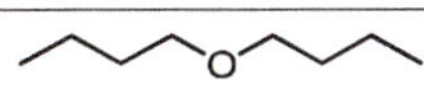
		Methyl tertiary-butyl ether (MTBE)	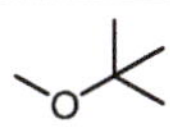
		Ethyl tertiary butyl ether (ETBE)	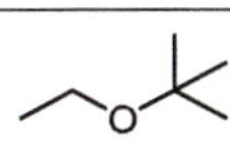
		Methanol	—OH
		Ethanol	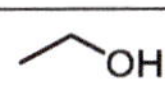
		n-Propanol	
		n-Butanol	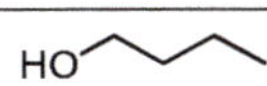
		iso-Butanol	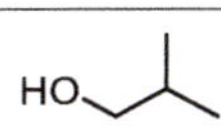

Table 18.1 (continued)

S. no.	Types of fuels	Name of fuels	Chemical structure
		sec-Butanol	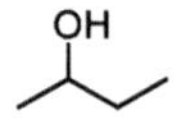
		tert-Butanol	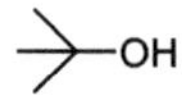
		tert-Amyl methyl ether (TAME)	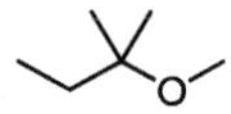
		Ethyl tertiary amyl ether (ETAE)	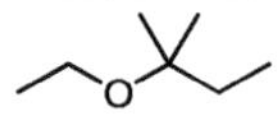
		Diisopropyl ether (DIPE)	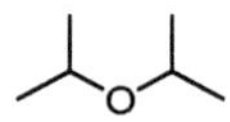
2)	**Bio-based fuel oxygenates (synthesized through microbial fermentation)**		
		Bioethanol	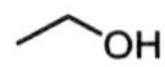 after distillation
		Biodiesel	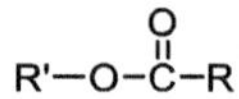 after distillation
		Biobutanol	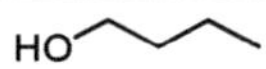 after distillation
3)	**Furanic fuel oxygenates (synthesized through catalytic conversion)**		
		2-Methylfuran (2-MF)	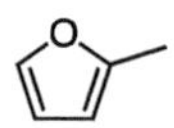
		2,5-Dimethylfuran (DMF)	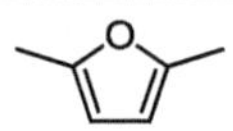
		γ-Valerolactone (GVL)	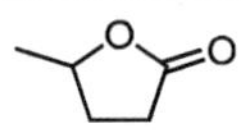
		5-Ethoxymethylfurfural (EMF)	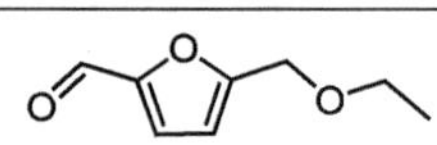
		2-Methyltetrahydrofuran (MTHF)	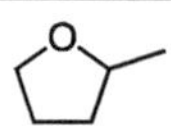
		Methyl levulinate (ML)	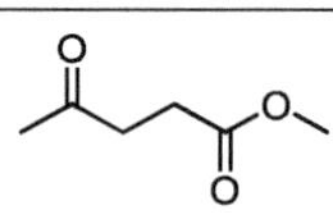
		Ethyl levulinate (EL)	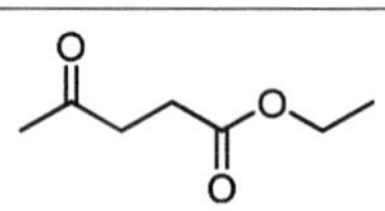
		Butyl levulinate (BL)	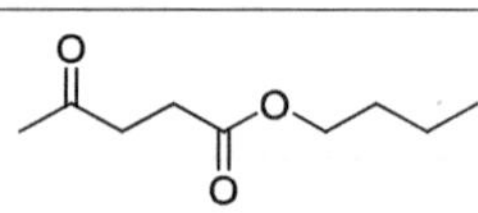

Table 18.1 (continued)

S. no.	Types of fuels	Name of fuels	Chemical structure
		Furan	
		Tetrahydrofuran (THF)	
		Tetrahydrofurfuryl alcohol (THFA)	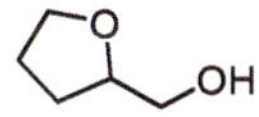
		Ethyl tetrahydrofurfuryl ether (ETE)	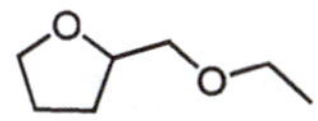
		Ethylfurfuryl ether (EFE)	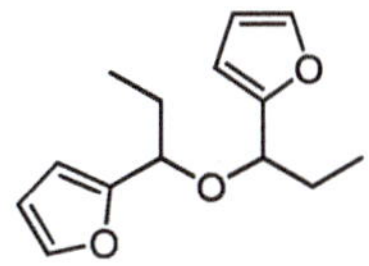
		Bisethoxy methyl furan (BEMF)	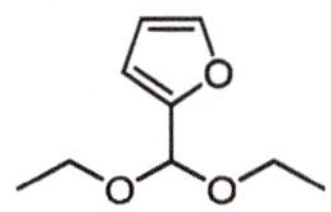

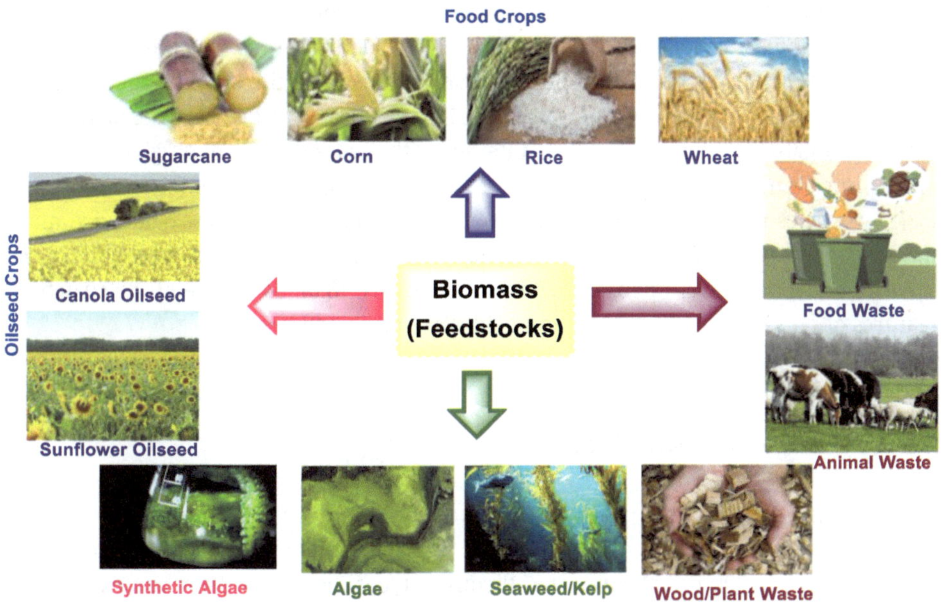

Figure 18.3: Feedstocks or biomass for biofuels.

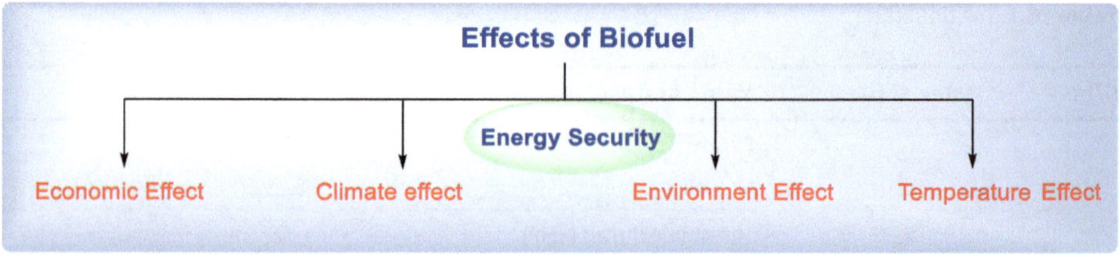

Figure 18.4: Effects of biofuels.

18.3 Biomass and carbohydrates

Biomass is a renewable organic material that comes from different biological sources [21]. Biomass is derived from various sources, but predominantly from plants, that is, 82.4%. The other sources of biomass are bacteria (12.8%), fungi (2.2%), and animals (0.47%) [22]. Carbon (~45%), oxygen (~45%), and hydrogen (~6%) are the main elements present in these feedstocks [23, 24]. Every year, around 200 billion metric tons of biomass are produced by nature, in which carbohydrates account for 75% [25]. The remaining 25% is polymer complexes and triglycerides [26]. In general, carbohydrates of biomass are present in different forms, which can be classified as simple and complex carbohydrates (Figure 18.5) [27]. Carbohydrates play a crucial role in building structural and non-structural units of biomass [28, 29]. The structural units of biomass are made up of complex polysaccharides (cellulose, hemicelluloses, chitin, chitosan, pectin, fucoidan, carrageenan, alginate, etc.), oligosaccharides, or disaccharides (lactose, maltose, sucrose, etc.) [30–32]. These sugar units interact with each other through glycosidic linkage or inter- and intra-molecular hydrogen bonding, Van der Waals forces, and so on [33]. Two types of natural sugars are present in biomass: (i) hexoses, which are made up of six-carbon (glucose, mannose, galactose, etc.), and (ii) pentoses, which are made up of five-carbon sugars (xylose, fructose, etc.) (Figure 18.6) [34]. Some of the structural features of polysaccharides and disaccharides are discussed below.

18.4 Polysaccharides

Polysaccharides are sugar polymers that are the major content of various biomass feedstocks. Polysaccharides are made up of multiple units of monosaccharides (ranging from 200 to 2,500 units) connected through glycosidic bonds. They can exist in linear or branched structural forms, which are known as simple and complex carbohydrates. The structures of polysaccharides vary depending on the biomass feedstocks, such as edible feedstocks, non-edible feedstocks, and algae. The structural features of some of the polysaccharides are described below.

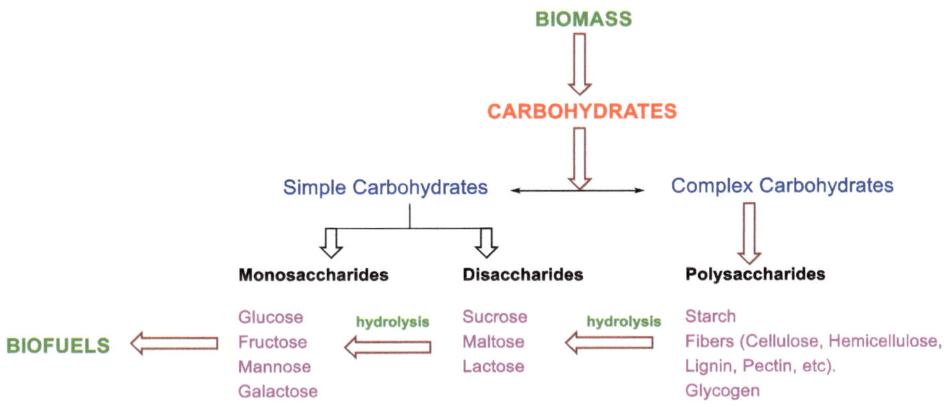

Figure 18.5: Biomass and carbohydrates.

18.4.1 Starch

Starch is a polysaccharide obtained from edible biomass feedstocks like rice, corn, wheat, potatoes, cassava, different fruits, and vegetables [35]. It is a soft, white, tasteless, and odorless polymer that is made up of glucose monomers. It is a chiral and biodegradable material that has an average melting point of 258 °C. Starch exists in two forms, namely, amylose and amylopectin (Figure 18.7). Amylose is a linear polysaccharide composed of D-glucose monosaccharides bonded through α-(1-4) glycosidic linkages. On the other hand, amylopectin is a branched polymer that is also made up of D-glucose units having both α-(1-4) and α-(1-6) glycosidic linkages. In starch, amylose is present in 20–30% while amylopectin is present in 70–80%.

18.4.2 Cellulose

Cellulose is a major carbohydrate polymer present in the cell walls of many plants and crops (Figure 18.8) [36]. It is a linear polysaccharide that is made up of D-glucose units linked through β-(1–4)-glycosidic bonds. The usual size of cellulose contains approximately 7,000–15,000 glucose monomers. Like starch, cellulose is also a tasteless and odorless compound that is insoluble in water and organic solvents. It is a chiral and biodegradable material that has an average melting point of 467 °C. Unlike starch, the structure of cellulose does not exist in a branched form.

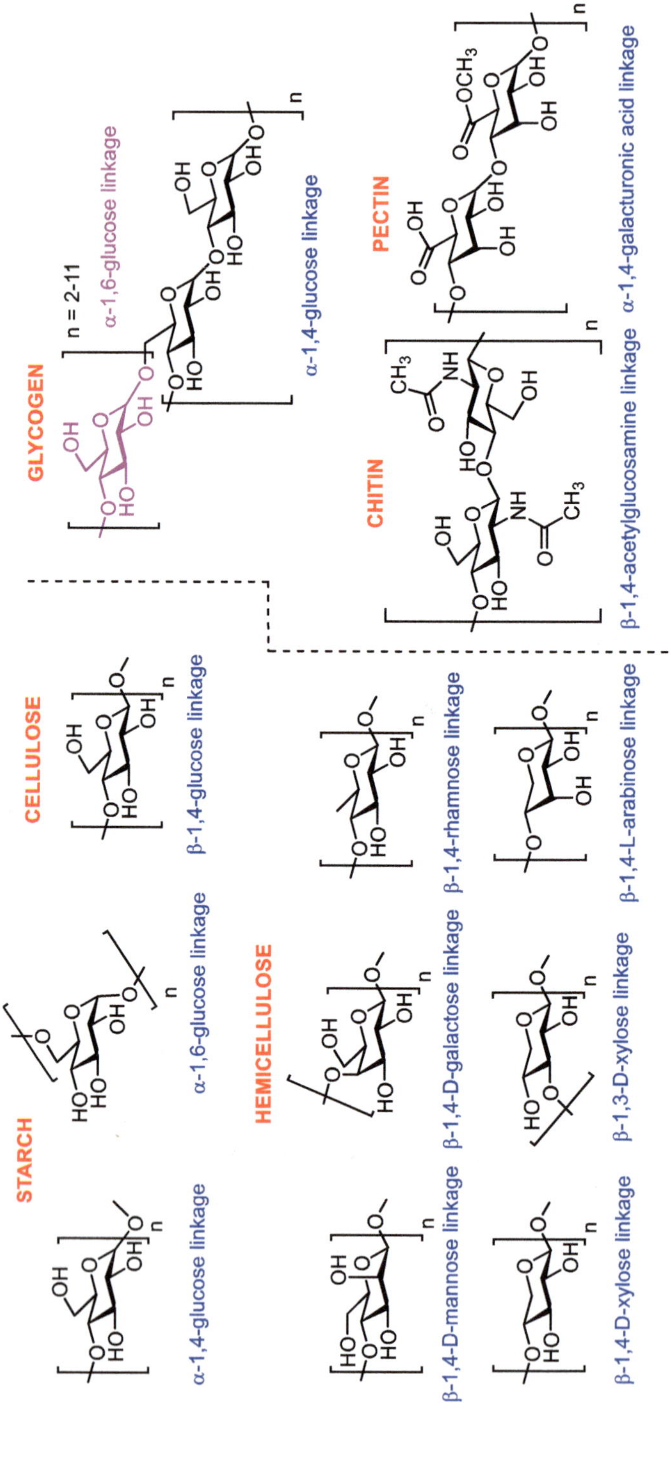

Figure 18.6: Polysaccharides (fermentable sugars) from different biomass.

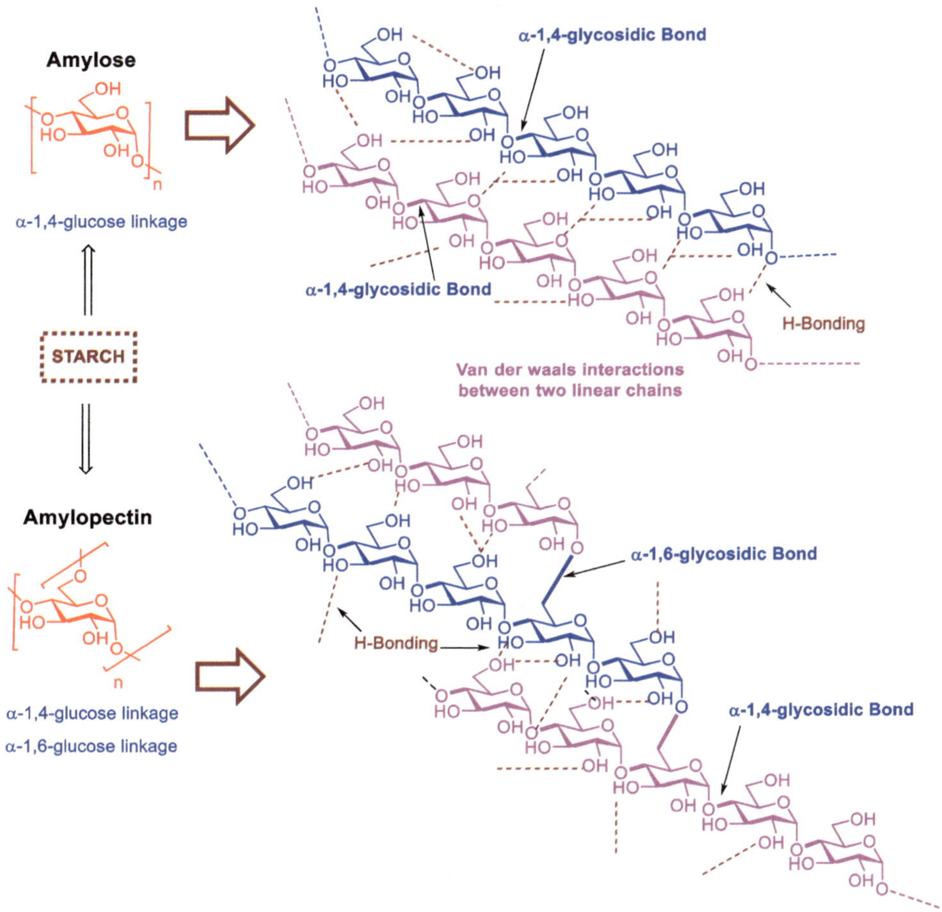

Figure 18.7: Different chemical bonding in a polysaccharide, for example, "starch".

18.4.3 Hemicellulose

Hemicellulose is also a polysaccharide predominantly present in the cell walls of plants and crops, along with cellulose and lignin (Figure 18.9) [37]. Unlike cellulose, hemicellulose is a heteropolysaccharide made up of five- and six-carbon sugars such as xylose, arabinose, mannose, galactose, and glucose. In the polymeric structure of hemicellulose, D-sugars are found predominantly, while L-sugars are found occasionally. The size of the hemicellulose polysaccharide is mainly based on the feedstock, usually made up of 500–3,000 sugar units. The glycosidic linkages in hemicelluloses are majorly found in the β-form.

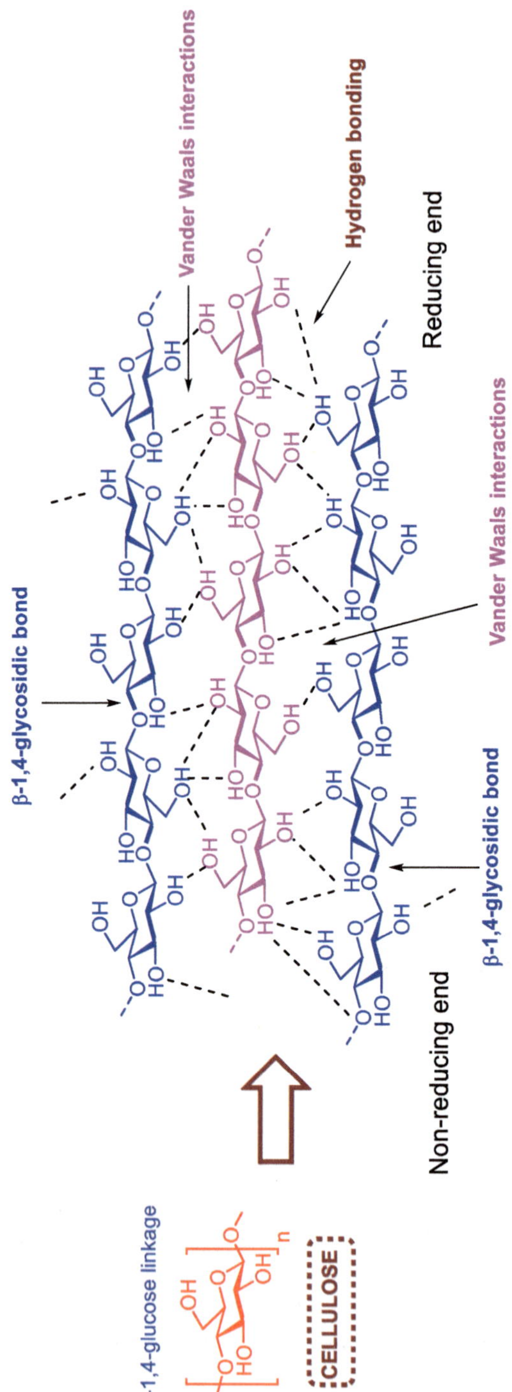

Figure 18.8: Different chemical bonds in cellulose.

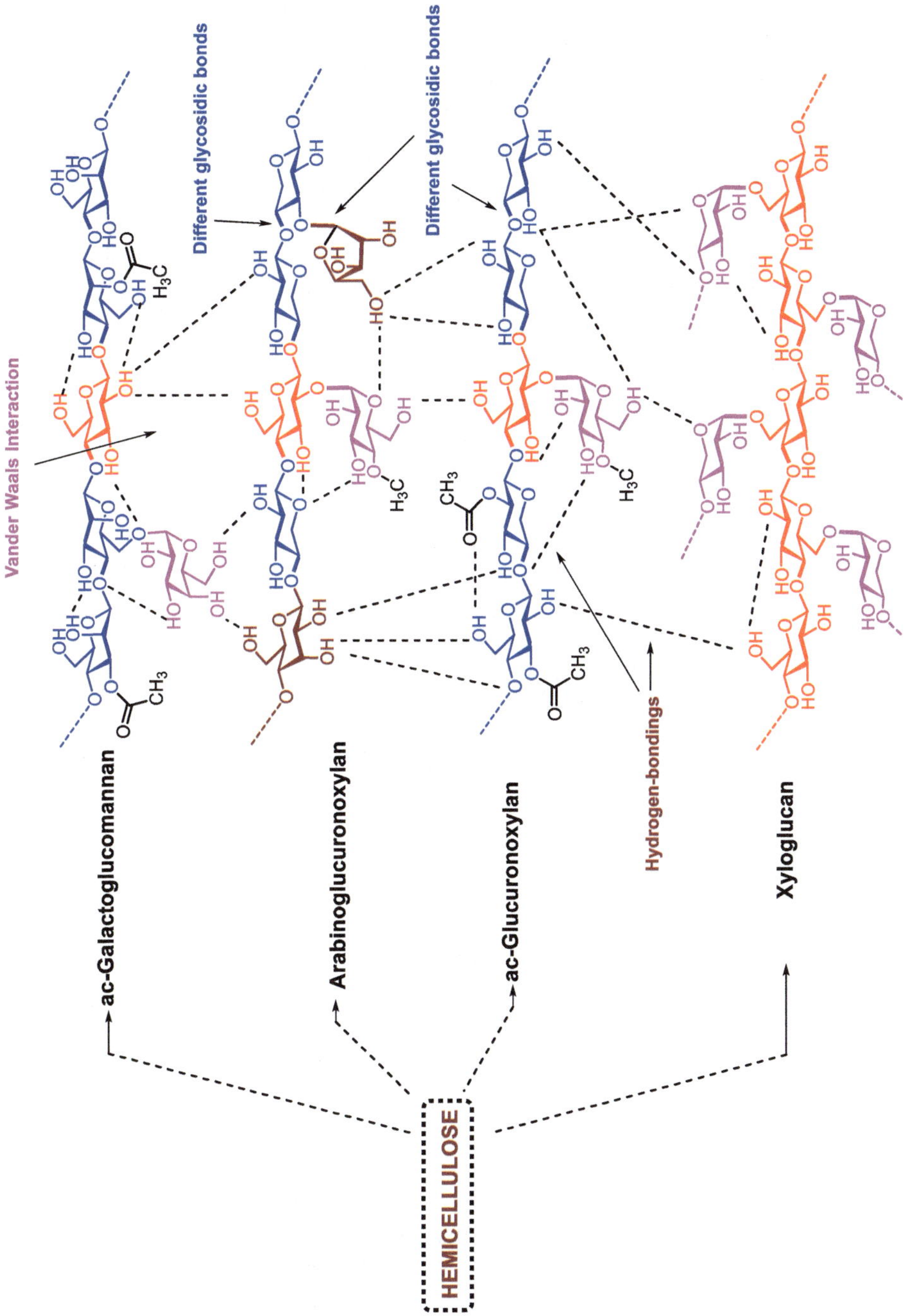

Figure 18.9: Structure of hemicelluloses and different chemical bonds.

18.4.4 Pectin

Like hemicellulose, pectin is a heteropolysaccharide typically found in the cell walls of terrestrial plants, vegetables, and fruits (Figure 18.10) [38]. The structural composition of pectin differs by source, although galacturonic acid or galacturonic acid methyl ester derivatives (~60%) are found predominantly. Besides galacturonic acid, rhamnogalacturonans, xylogalacturonans, arabinogalactans, and arabinans are also commonly found in pectin polysaccharides. In the homogalacturonan part of the polymer, galacturonic acid is linked through α-(1-4)-glycosidic linkages. Pectin is part of the human diet and acts as fiber. Various vegetables and fruits, including apples, guavas, and oranges, contain high amounts of pectin. Pectin is highly soluble in water and provides many health benefits.

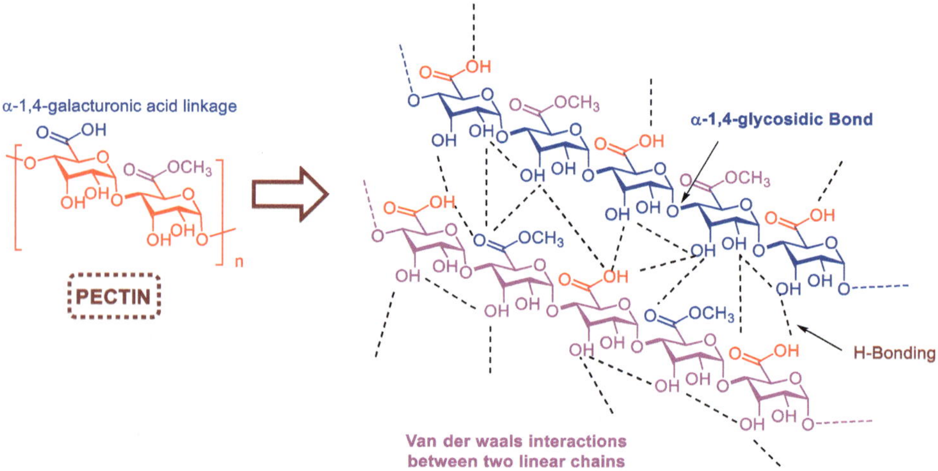

Figure 18.10: Structure of pectin and different chemical bonds.

18.4.5 Glycogen

Glycogen is an analog of starch and is formed in living organisms like humans, animals, bacteria, and fungi (Figure 18.11) [39]. It provides energy to living organisms. It is a branched homopolysaccharide made up of D-glucose residues. Like amylopectin, glucose units are linearly connected through α-(1-4) and α-(1-6) glycosidic linkages. The average length of glycogen is made up of 50,000 glucose units. Glycogen is richly found in cyanobacteria, an important feedstock that can be converted into biofuels.

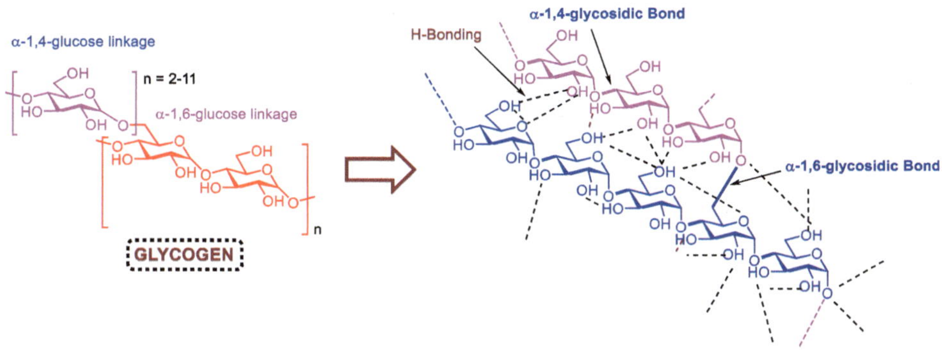

Figure 18.11: Structure of glycogen and different chemical bonds.

18.4.6 Chitin and chitosan

Chitin is a linear homopolysaccharide made up of *N*-acetylglucosamine (Figure 18.12) [40]. These monosaccharides are connected through β-(1,4) glycosidic bonds. Chitin is predominantly found in the cell walls and exoskeletons of seafood, fungi, invertebrates, crustaceans, and insects. Like cellulose, chitin is a highly abundant polysaccharide isolated as a by-product in the seafood industry. It has been extracted from the shells of crabs, shrimps, shellfish, and lobsters. Chitosan is a partially deacetylated form of chitin, a linear polysaccharide containing β-D-glucosamine and *N*-acetyl-D-glucosamine (Figure 18.13).

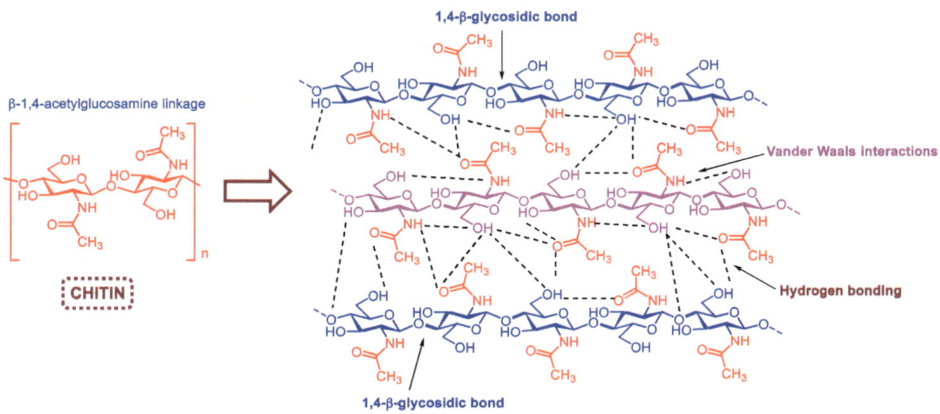

Figure 18.12: Structure of chitin and different chemical bonds.

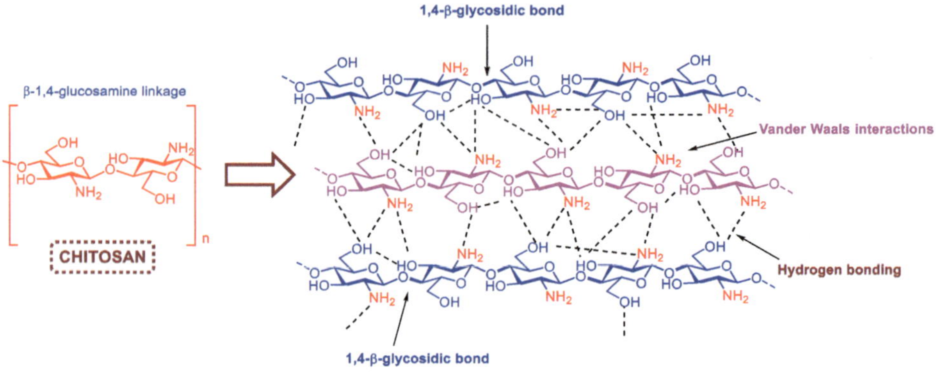

Figure 18.13: Structure of chitosan and different chemical bonds.

18.4.7 Fucoidan

Fucoidan is an L-fucose sugar-containing sulfated marine-based complex polysaccharide (Figure 18.14) [41]. The extracellular matrix of brown algae and marine invertebrates is a major source of fucoidan [42]. It is water-soluble and made up of L-fucose as the major part, along with different sugars, including galactose, glucose, and mannose. The fucoidan polymer is predominantly made up of fucose disaccharide repeating units with α-(1,3) and α-(1,4) linkages (Figure 18.15). Fucoidanases and sulfatases are enzymes that perform the sulfation or glycosylation to build the fucoidan polymer [43].

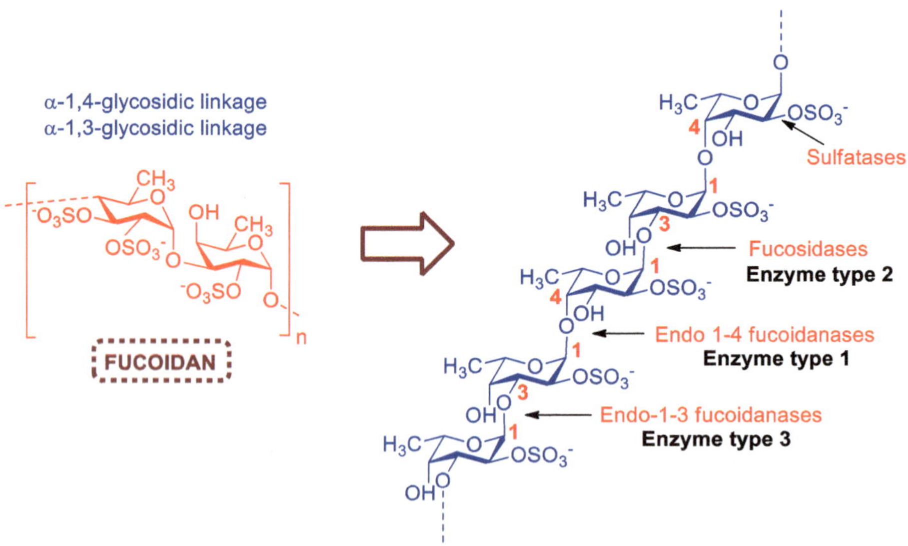

Figure 18.14: Structure of fucoidan.

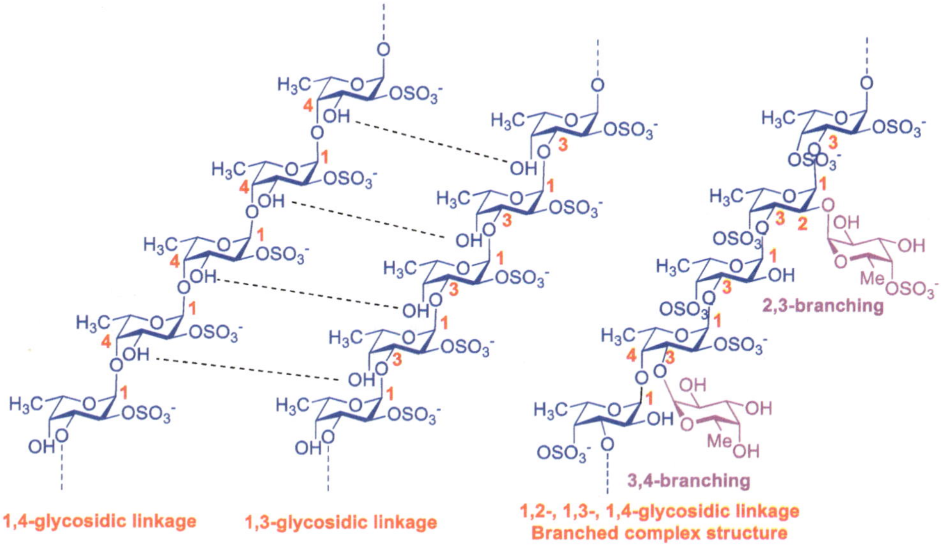

Figure 18.15: Types of fucoidan and different chemical bonds.

18.4.8 Carrageenans

Carrageenans are linear sulfated polysaccharides made up of disaccharide repeating units containing alternating β-D-galactopyranose and α-D-galactopyranose or 3,6-anhydro-α-D-galactopyranose (AnGal) (Figure 18.16) [44]. Red edible seaweeds are the major source of carrageenan [45]. Three important types of carrageenans are known, namely κ-(kappa), ι-(iota), and λ-(lambda) carrageenans. This classification is based on their structural features, including sulfation patterns and monosaccharide units. In κ- and ι-carrageenans, 3-linked β-D-galactose is coupled with 4-linked 3,6-anhydro-α-D-galactopyranose units. The λ-carrageenans contain alternating repeating units of 3- and 4-linked β-D-galactose units. In the case of κ-carrageenans, one sulfate group is found in the fourth –OH group of 3-linked β-D-galactose, while an additional sulfate group is found in ι-carrageenans at the second –OH of AnGal. The λ-carrageenans have three sulfate groups per disaccharide: (i) second –OH group of 3-linked β-D-galactose, and (ii) second –OH and sixth –OH group of 4-linked β-D-galactose. Carrageenans from industrial feedstocks have potential applications in biofuel production.

18.4.9 Alginate

Algin (i.e., alginic acid) is a natural linear anionic polysaccharide predominantly found in the brown algae cell walls as Ca, Mg, and Na salts of alginic acid (Figure 18.17)

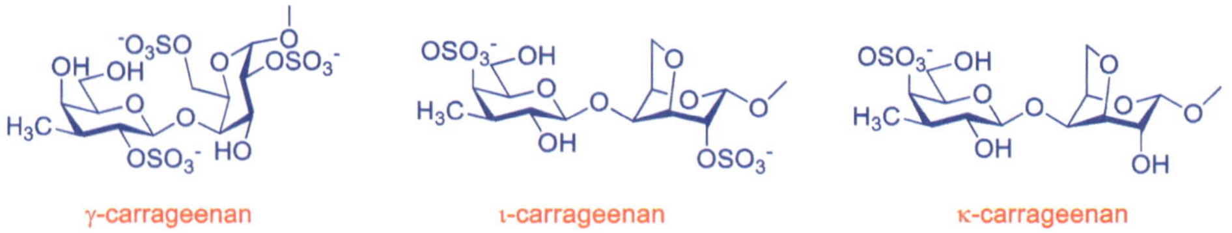

Figure 18.16: Types and structures of carrageenan.

[46]. It is made up of β-D-mannuronate (M) and α-L-guluronate (G) units via (1–4) glycosidic linkages in different blocks (e.g., GG, MM, and MG or GM blocks). The recent advancements in biotechnologies allow the production of biofuels from alginate using microbes.

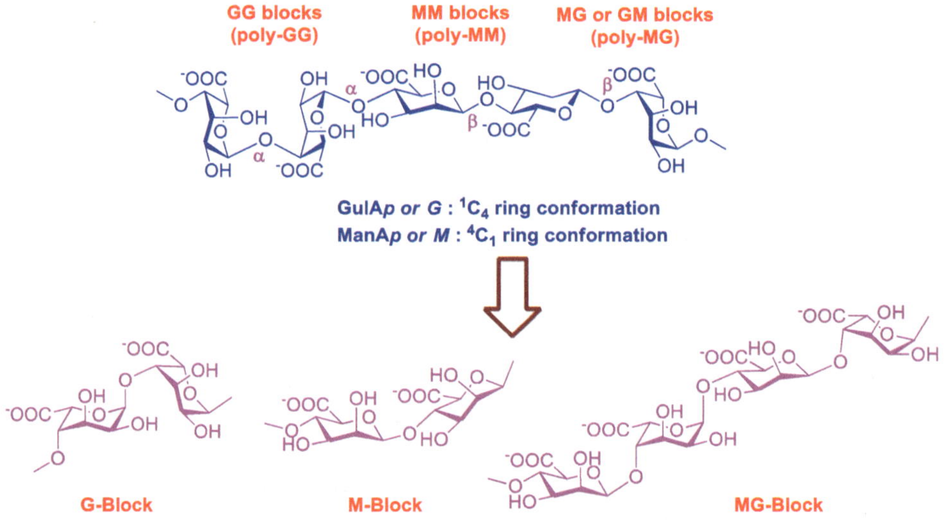

Figure 18.17: Structure of alginate.

18.5 Disaccharides

Disaccharides are an important class of compounds formed by connecting two monosaccharides through glycosidic bonds (Figure 18.18). Several disaccharides are known for their role in chemistry and biology. Besides the natural disaccharides, chemical or enzymatic hydrolysis of polysaccharides also provides disaccharides. Disaccharides are further hydrolyzed into monosaccharides with yeast and related microbes. Sucrose and lactose are the naturally occurring disaccharides, while maltose is obtained from starch via hydrolysis.

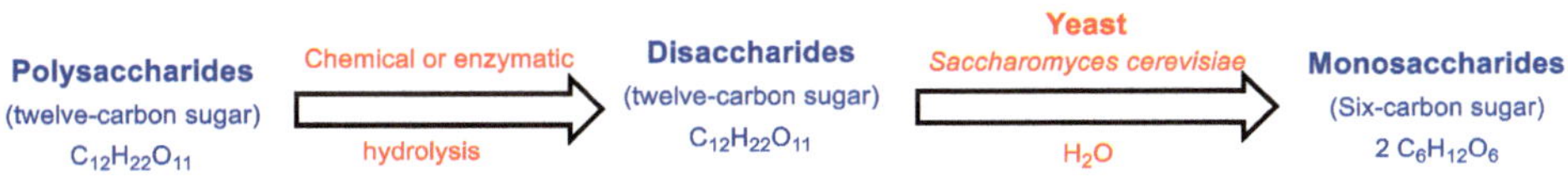

Figure 18.18: Formation of monosaccharides from disaccharides.

18.5.1 Sucrose

Sucrose is a disaccharide, which is known as table sugar. Sucrose is composed of glucose and fructose units (Figure 18.19) [47]. In sucrose, the monomers glucose and fructose are linked via a glycosidic bond between C1 on the glucosyl subunit and C2 on the fructose unit. Glucose exists as α anomers, and fructose is in β-form. Sucrose is a natural compound obtained predominantly from sugarcane and sugarbeets. Sucrose is one of the important biomass carbohydrates and is hydrolyzed to glucose and fructose using enzymes like sucrase or invertase. The monomers are used to produce various biofuels like bioethanol, biobutanol, and hydroxymethylfurfural (HMF) through fermentation methods or chemical methods using catalysts.

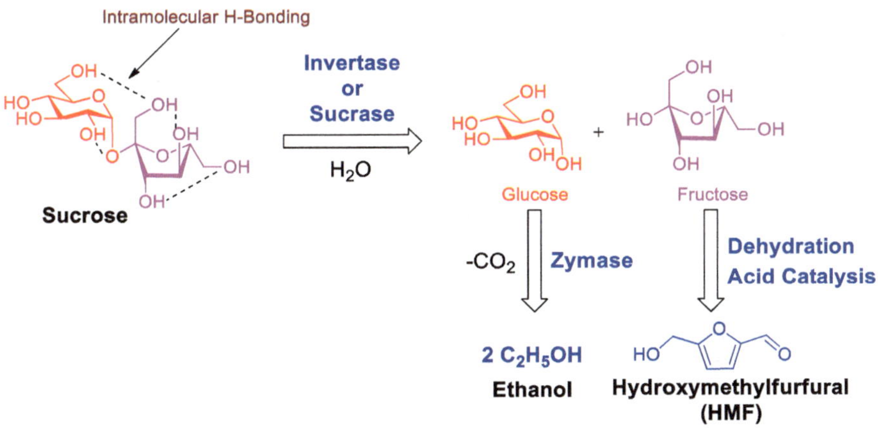

Figure 18.19: Synthesis of bioethanol and HMF from sucrose.

18.5.2 Lactose

Lactose is a sugar that is naturally occurring in milk and dairy products [48]. Like sucrose, it is a disaccharide composed of galactose and glucose units linked through a β-(1–4) glycosidic linkage. Food and dairy wastes contain various amounts of lactose compounds, which can be used for the production of biofuels. Lactose is hydrolyzed to galactose and glucose using the enzymes galactosidase or lactase. Glucose can be converted to bioethanol using zymase, while galactose can be converted to bioethanol

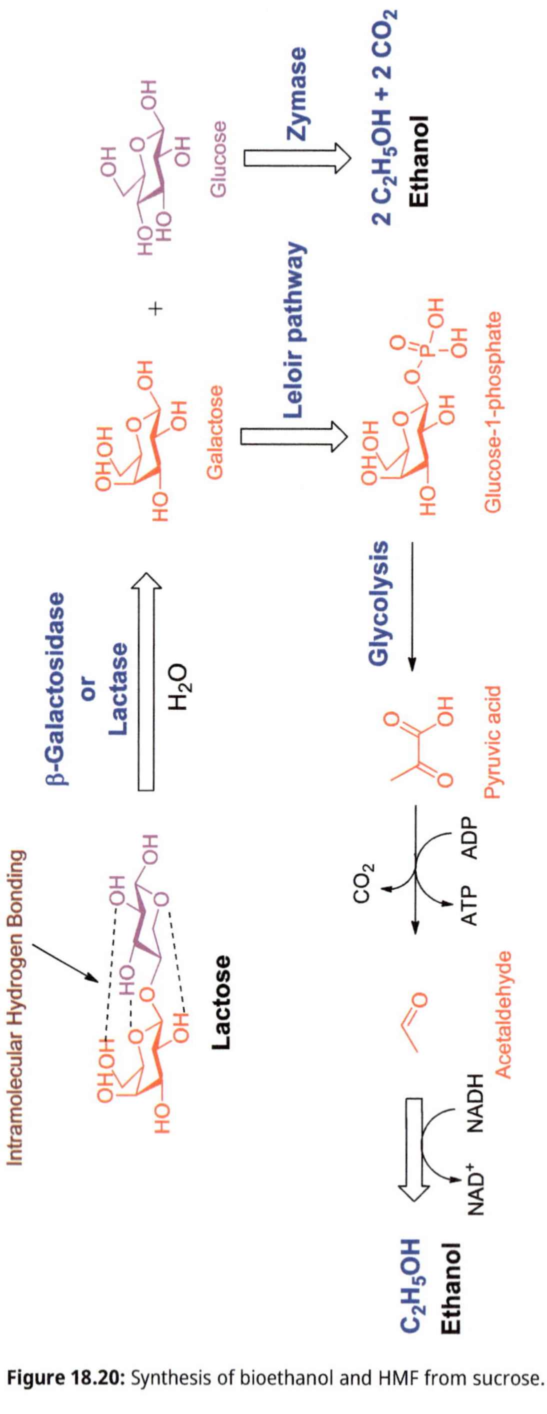

Figure 18.20: Synthesis of bioethanol and HMF from sucrose.

under the Leloir pathway through galactose-1-phosphate, using the glycolysis process to synthesize pyruvic acid and acetaldehyde (Figure 18.20).

18.5.3 Maltose

Maltose is a disaccharide, a lower unit of starch, obtained via hydrolysis of amylose using diastase or amylase [49]. In maltose, glucose is connected through an α-(1,4)-glycosidic bond. Maltose can be further hydrolyzed to glucose using the maltase enzyme and converted into various biofuels, including bioethanol, under the fermentation method (Figure 18.21).

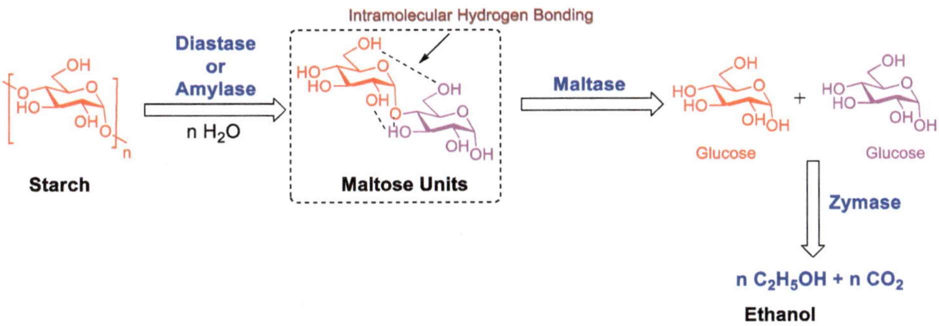

Figure 18.21: Synthesis of bioethanol and HMF from starch.

18.6 Monosaccharides

Monosaccharides are the simplest carbohydrates, which cannot be hydrolyzed further. Glucose, galactose, and fructose are examples of monosaccharides, usually obtained via hydrolysis (chemical and enzymatic) of polysaccharides and disaccharides. The monosaccharides are used as precursors for biofuel production (Figure 18.22).

18.7 Production of biofuels from carbohydrates-enriched biomass

Biomass is a renewable biomaterial used as a feedstock to produce biofuels. Carbohydrates are the major constituent of biomass and are in high demand due to their applications in biofuel production. The extraction of major polysaccharides from the biomass is an important step in biofuel production [50]. However, such extraction is

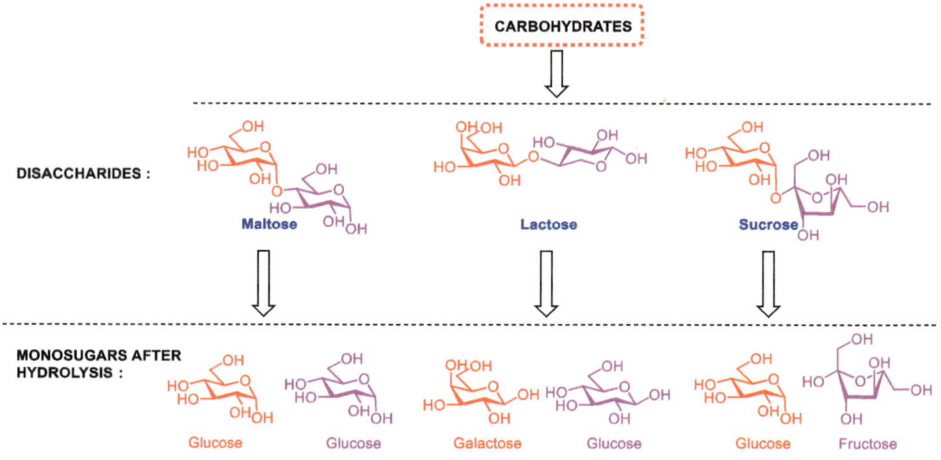

Figure 18.22: Different disaccharides and monomers for biofuel formation.

rather difficult owing to the complex structural features of polysaccharides. For instance, polysaccharides are made up of monosaccharides through glycosidic linkages. Moreover, polysaccharides have intra-molecular and inter-molecular H-bonding (due to the presence of many hydroxyl groups), van der Waals interactions, and so on. Therefore, the processing of raw materials is necessary to achieve a plethora of carbohydrates from biomass [51]. Such processing weakens the bonding that exists in the polysaccharides and allows easy extraction of carbohydrates. Hence, biomass is initially subjected to pretreatment methods under different conditions such as physical, chemical, physicochemical, and biological methods for the extraction of carbohydrates (Figure 18.23) [52].

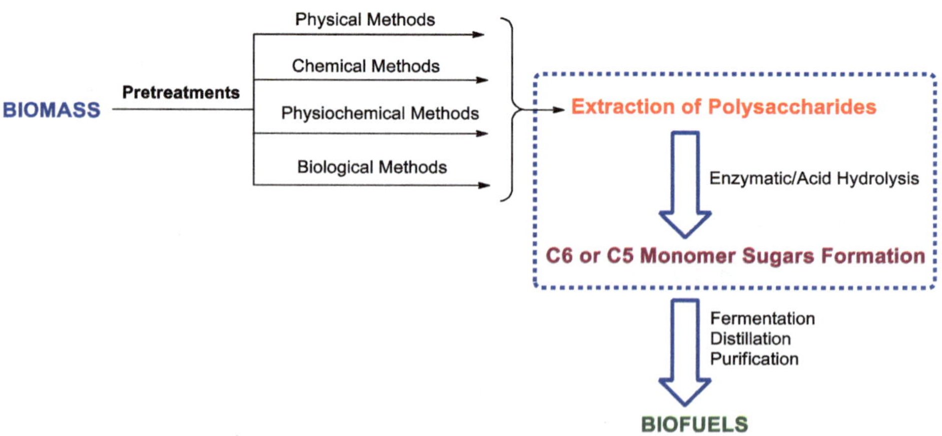

Figure 18.23: Synthetic route for the formation of biofuels from carbohydrates.

18.8 Biomass pretreatment

Pretreatment is an essential step for the extraction of chemical compounds present in the biomass [53]. Pretreatment provides structural alteration of core chemicals to overcome the rough nature of biomass. Some of the important pretreatment methods include the physical pretreatment method, chemical pretreatment method, physicochemical method, and biological pretreatment method, which are outlined below (Figure 18.24) [54].

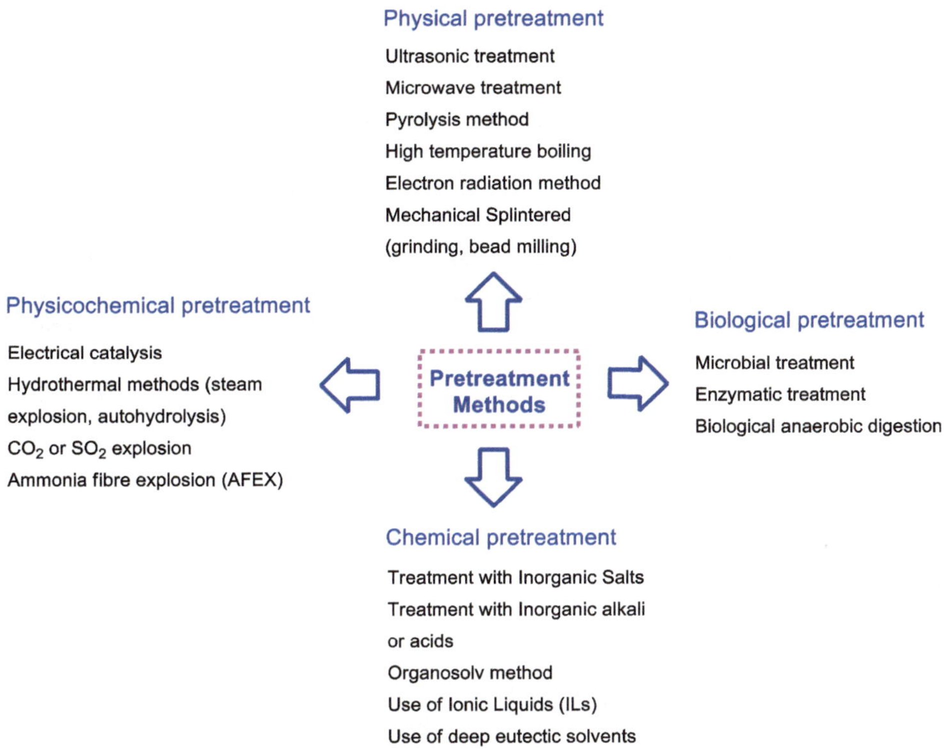

Figure 18.24: Different types of pretreatment methods.

18.8.1 Physical pretreatment method

The objective of the physical pretreatment of biomass is to decrease the particle size and increase the biomass's surface area and pore size [55]. In lignocellulosic biomass, physical pretreatment reduces cellulose's crystalline nature and degree of polymerization. In this physical treatment, biomass is subjected to ultrasonic treatment, microwave treatment, ball milling, pyrolysis method, high-temperature boiling, electron radiation method, mechanical splintering (grinding and bead milling), and so on, after which the extraction process takes place (Figure 18.25).

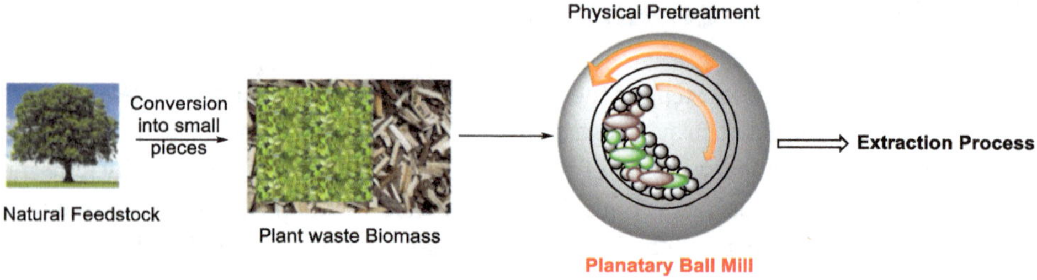

Figure 18.25: Physical pretreatment methods using a ball mill.

18.8.2 Chemical pretreatment method

In chemical pretreatment, organic or inorganic compounds are used for the hydrolysis of intra- and inter-polymer bonds present in biomass [56]. In this treatment, ionic liquids and deep eutectic compounds are also used as solvents. The biomass is subjected to heat under high temperatures through radiation or conventional methods. The soluble fractions are extracted using different solvents from the biomass residue (Figure 18.26). Lignocellulosic biomass is, in general, resistant to chemical breakdown due to its recalcitrant nature, such as structural complexity, heterogeneity, and crystalline nature.

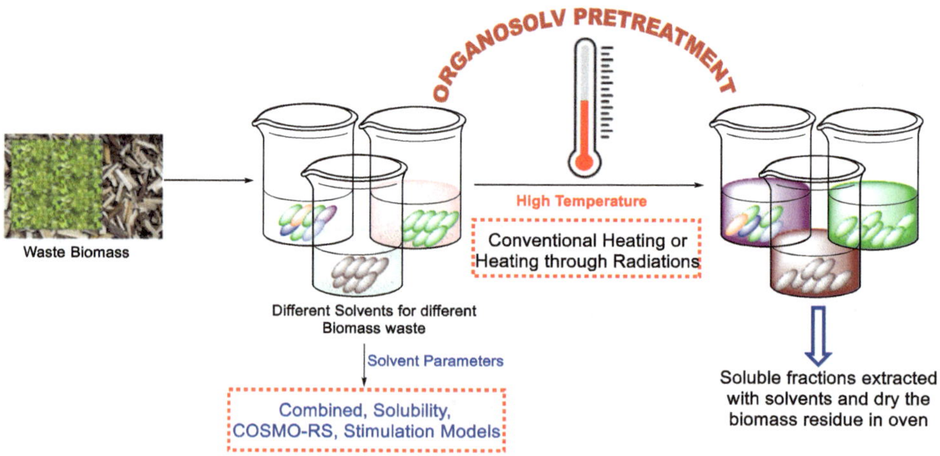

Figure 18.26: Chemical pretreatment methods using different solvents.

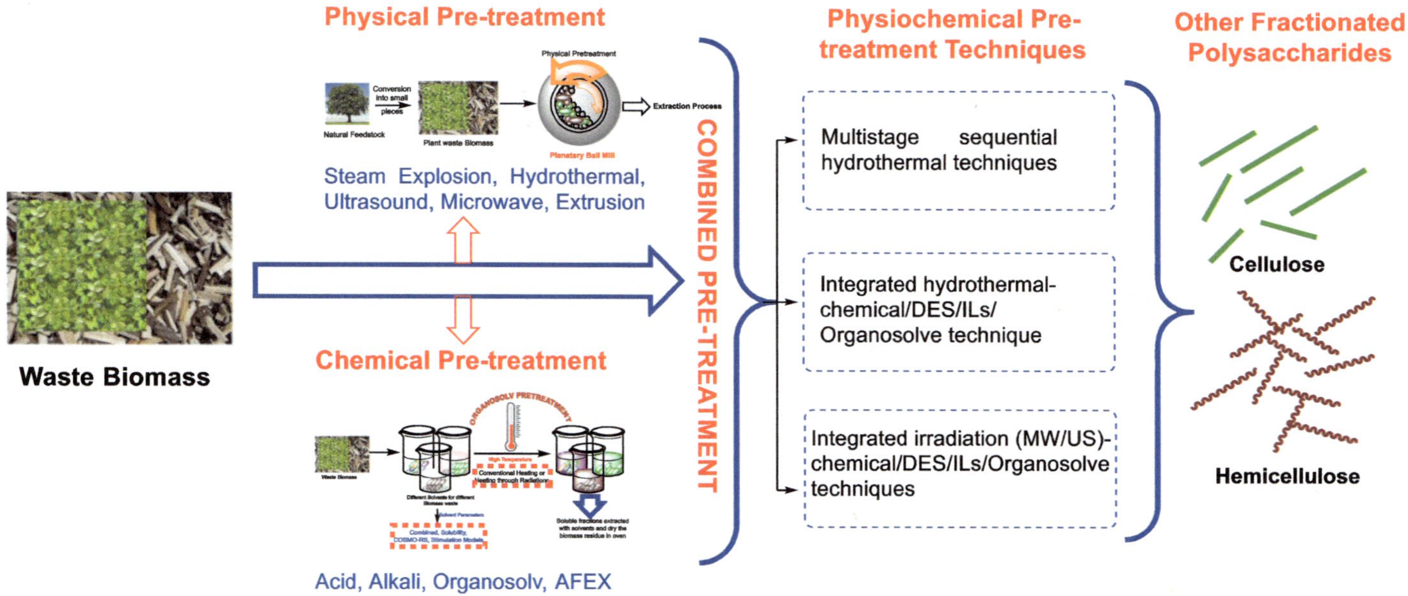

Figure 18.27: Physicochemical pretreatment methods under different physical and chemical conditions.

18.8.3 Physiochemical method

The physicochemical pretreatment method combines physical and chemical processes to ensure the efficient fragmentation of lignocellulosic biomass materials [57]. The well-known physicochemical pretreatments include "ammonia fiber explosion", "steam explosion", "carbon dioxide explosion", and "liquid hot water" (LHW) methods. The steam explosion employs a combination of mechanical forces and chemical reactions, which is a frequently used method for the fragmentation of lignocellulosic biomass materials. In this method, the biomass is treated under high pressure and temperature in the presence of steam. In the ammonia fiber explosion method, liquid ammonia is used as the solvent, and the biomass is treated at 60–100 °C under high pressure. In the CO_2 explosion method, carbon dioxide is dissolved in water to form carbonic acid. The resulting carbonic acid catalyzes the biomass hydrolysis step. The liquid hot water approach employs water as the solvent to fragment the biomass at high temperatures and pressure in place of steam. Cellulose and hemicellulose polysaccharides are obtained by the process of multi-stage sequential hydrothermal technique, integrated hydrothermal technique, and microwave/ultrasound techniques, ionic liquids, and deep eutectic and organosolv techniques (Figure 18.27).

18.8.4 Biological pretreatment method

In biological pretreatment, biological substances like fungi, bacteria, and enzymes are used to fragment, depolymerize, and break the bonds in biomass [58]. This pretreatment method is considered to have advantages when compared with other pretreatment methods. For instance, this method uses less energy, involves fewer toxic substances, and yields more chemicals (Figure 18.28). However, this process is relatively slow and requires proper experimental conditions.

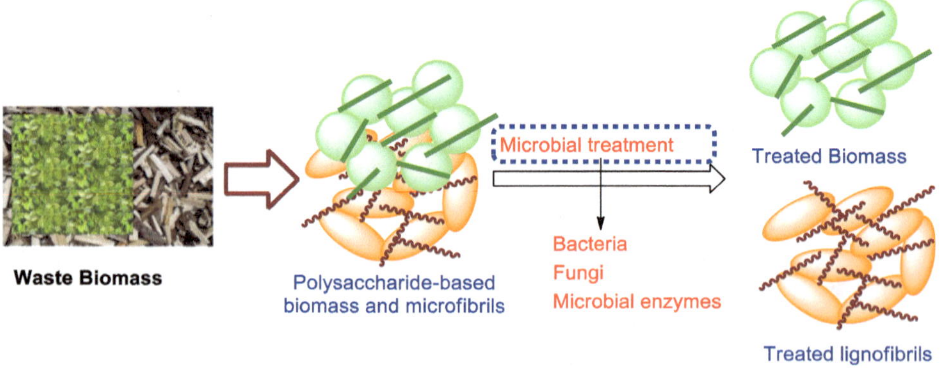

Figure 18.28: Biological pretreatment methods using different microbes.

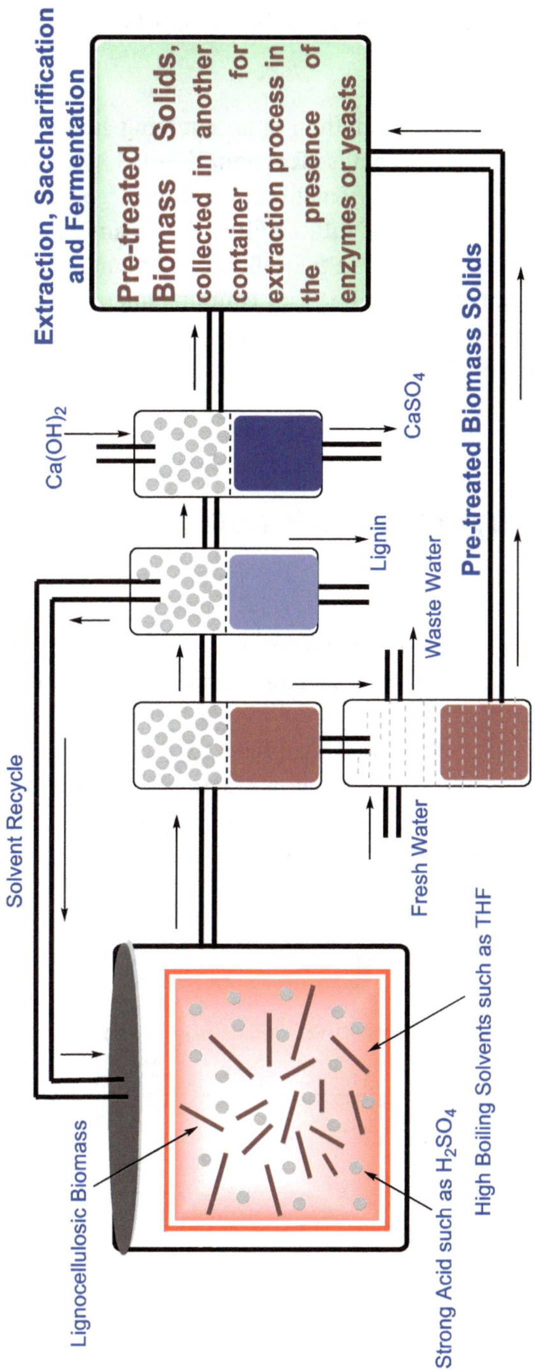

Figure 18.29: CELF pretreatment methods for lignocellulosic biomass.

18.8.5 Cosolvent-enhanced lignocellulosic fractionation (CELF) pretreatment

Cosolvent-enhanced lignocellulosic fractionation method is an evolving solvolysis pretreatment method used for lignocellulosic biomass fractionation [59, 60]. This method uses renewable tetrahydrofuran (THF) to fractionate cellulosic biomass under homogeneous conditions in the presence of dilute sulfuric acid and water. After the reaction, THF is removed by distillation, and the solid biomass (mainly xylose and arabinose) is transferred to another chamber that contains microbes like yeast or enzymes for the fermentation process (Figure 18.29). Using this method, one can achieve high yields of carbohydrates, furfural, 5-hydroxymethylfurfural, levulinic acid, and so on.

18.9 Extraction of carbohydrates (fermentable sugars)

After the pretreatment methods, the extraction of carbohydrates is required to produce biofuels [61]. There are two types of methods available, namely conventional methods and advanced methods (Figure 18.30). The conventional methods include liquid–liquid extraction, liquid–solid extraction, and acid-base extraction. On the other hand, advanced methods include "microwave-aided extraction", "ultrasound-aided extraction", "supercritical fluid extraction", "enzyme-utilized extraction", and "pressurized liquid extraction". A brief introduction to these methods is described below.

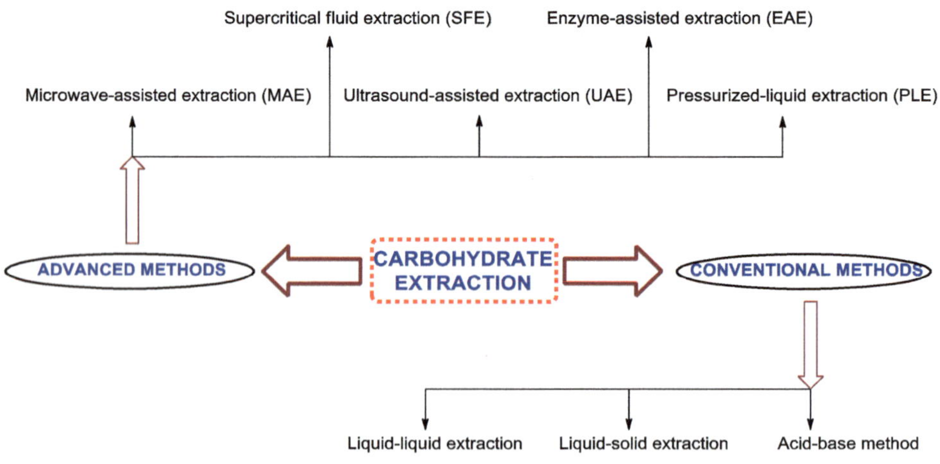

Figure 18.30: Different extraction methods for carbohydrates from pretreated biomass.

18.9.1 Conventional methods

18.9.1.1 Liquid-liquid extraction (LLE)

Liquid-liquid extraction (or solvent extraction) follows the partition/distribution coefficient principle (Figure 18.31) [62]. Two immiscible solvent systems are used for the extraction of carbohydrates based on their relative solubilities. In one of the solvents, carbohydrates are soluble, while insoluble in the other solvent. The carbohydrate-containing liquid will be evaporated to obtain the carbohydrate polymer, which will be subjected to fermentation to produce biofuels. During this procedure, various toxic and unwanted compounds are separated.

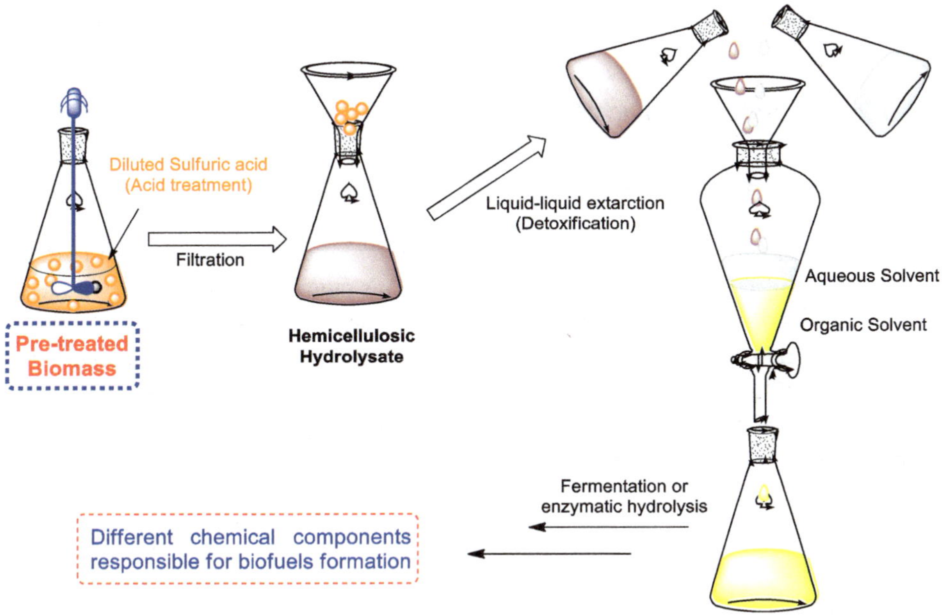

Figure 18.31: Liquid-liquid extraction method for carbohydrates.

18.9.1.2 Liquid-solid extraction (LSE)

Liquid-solid extraction of carbohydrates is very similar to liquid-liquid extraction, but instead of two liquids, the solute is spread on a solid material (Figure 18.32) [63]. The liquid phase is filtered from the solid, which contains toxic and unwanted chemicals. The sugar-containing solid phase will be dried and processed to obtain chemicals and biofuels.

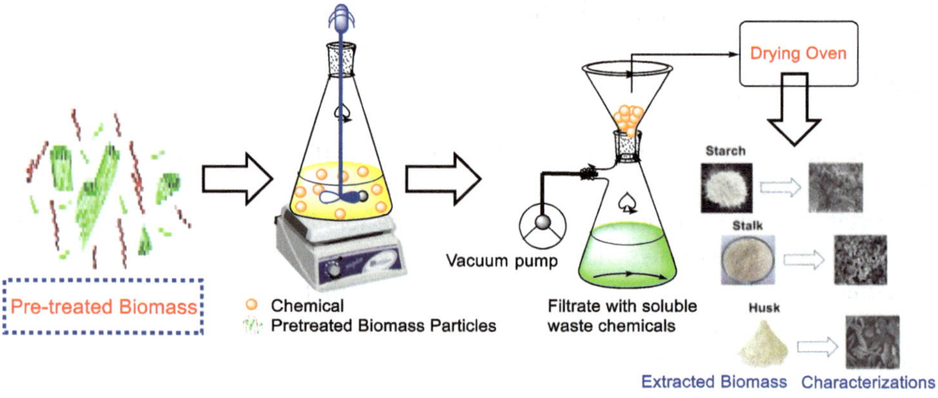

Figure 18.32: Liquid-solid extraction method for carbohydrates.

18.9.2 Advanced methods

18.9.2.1 Microwave-assisted extraction (MAE)

Microwave-assisted extraction helps in the separation of carbohydrates from biomass easily (Figure 18.33) [64]. In this technique, microwave energy is applied to the biomass-containing chamber. Water and ethanol are used as solvents. MAE offers quick and effective extraction due to the microwave properties. Microwaves heat the biomass uniformly without heating the container and allow carbohydrate extraction in a cheap and efficient manner in a short time. After the irradiation, carbohydrates are obtained using the centrifuge technique.

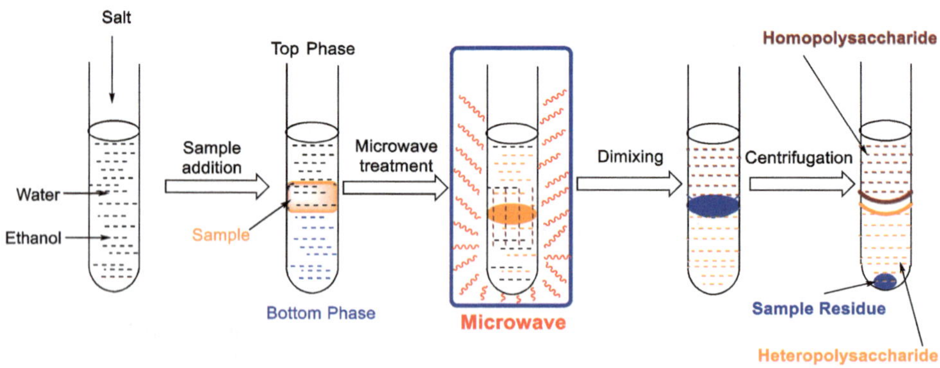

Figure 18.33: Microwave-assisted extraction (MAE) method of carbohydrates.

18.9.2.2 Ultrasound-assisted extraction (UAE)

Ultrasound is a mechanical wave that is widely used for the extraction of carbohydrates from biomass (Figure 18.34) [65]. In particular, high-intensity ultrasounds (i.e., frequency of 20–100 kHz) have been employed for carbohydrate extraction. Under this method, carbohydrate structural properties are preserved. These carbohydrate compounds are further subjected to enzymatic extraction to afford the carbohydrates.

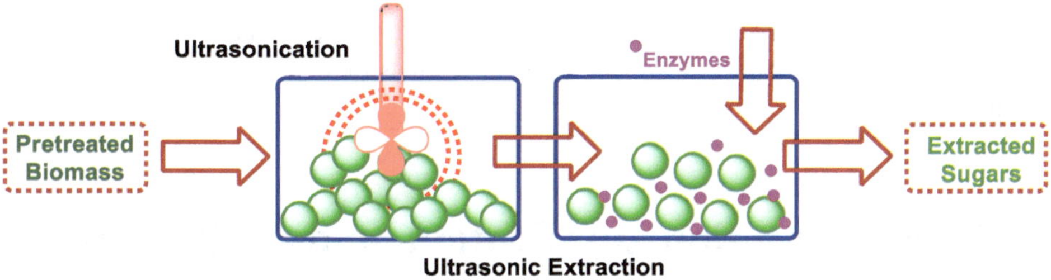

Figure 18.34: Ultrasound-assisted extraction (UAE) method of carbohydrates.

18.9.2.3 Pressurized-liquid extraction (PLE)

Pressurized liquid extraction (PLE) relies on the separation of carbohydrates under the solid-liquid extraction method at elevated temperatures and pressures (Figure 18.35) [66]. Under high pressure and temperature, the extraction of the solvent becomes easier, allowing deeper diffusion to extract the material from the solids. It offers quick extraction of carbohydrates in higher yields. In this setup, pre-treated biomass is transferred to a filtration chamber using a vacuum pump through an extractor under high temperature and pressure.

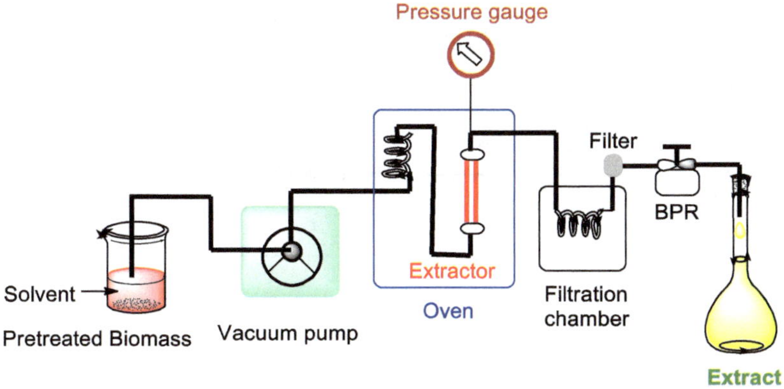

Figure 18.35: Pressurized-liquid extraction (PLE) method of carbohydrates.

18.10 Acid and enzymatic hydrolysis of polysaccharides

After the extraction of carbohydrates from biomass, the polysaccharides are supposed to be hydrolyzed to small units, particularly as monosaccharides [67, 68]. Two types of hydrolysis are known: (i) acid hydrolysis and (ii) enzymatic hydrolysis. In acid hydrolysis, polysaccharides like starch, cellulose, and pectin are hydrolyzed to monosaccharides using various mineral acids, such as sulfuric acid and nitric acid. On the other hand, enzymatic hydrolysis employs enzymes to hydrolyze the polysaccharides into monosaccharides that can subsequently be converted to biofuels by microorganisms via the fermentation process.

18.11 Fermentation, purification, and distillation

Fermentation is a process that aids the conversion of organic substances into valuable chemicals through the action of enzymes [69]. The fermentation process is widely used in biofuel production. For instance, the conversion of biomass carbohydrates into ethanol is achieved via fermentation using microorganisms like bacteria

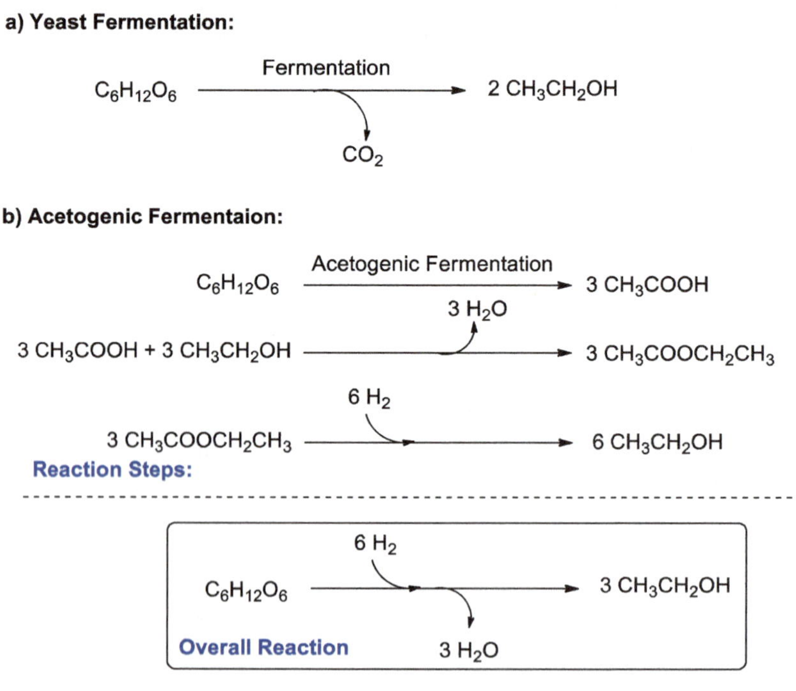

Figure 18.36: Ethanol formation through different fermentation techniques.

and yeast, which metabolize plant sugars into ethanol. Similarly, acetogenic fermentation aids the production of ethanol via acetic acid and ethyl acetate from glucose in three steps (Figure 18.36).

Distillation is an important purification step widely used in the industry to obtain pure biofuels from the fermentation step [70]. For instance, the separation of ethanol and water after the fermentation process is achieved via the distillation step using their boiling points [71].

18.12 Classification of biofuels

Based on the types of feedstocks, biofuels have been classified into four generations (Figure 18.37) [12, 72]. First-generation biofuels employ edible biomass feedstocks, whereas non-edible biomass has been utilized to produce second-generation biofuels. On the other hand, third-generation biofuel uses microalgae and other aquatic biomass as feedstock, while fourth-generation biofuel is produced from genetically modified plant microorganisms. The first two generations have been fairly explored, while the last two generations are in the initial stages.

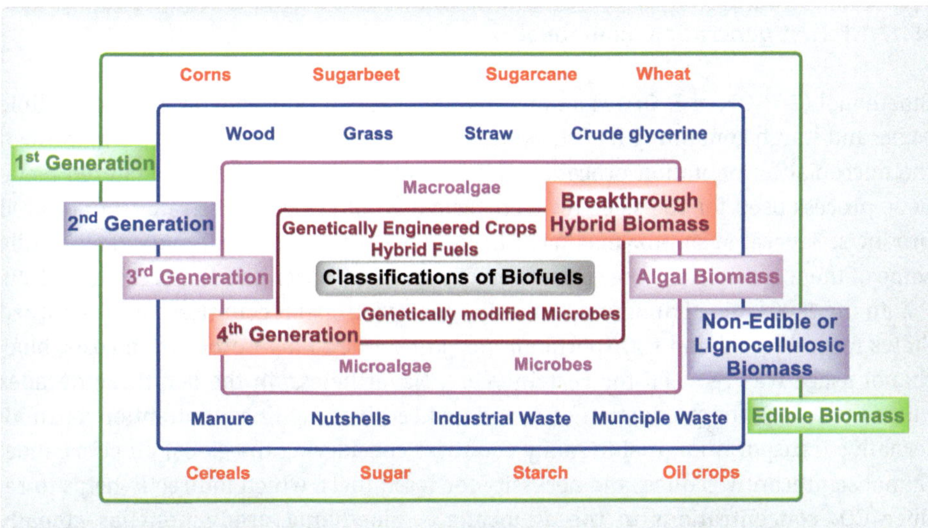

Figure 18.37: Classification of biofuels.

18.12.1 First-generation biofuels

First-generation biofuels are produced from starch and sugar-containing edible biomass feedstocks (Figure 18.38) [73]. Various vegetables and grains, including potatoes, wheat, corn, barley, sugarcane, and sugarbeet, are rich in starch and sucrose. This approach initially showed potential to minimize fossil fuel usage and offered positive environmental benefits. First-generation biofuels, including biodiesel, bioethanol, and biogas, are produced at around 50 billion liters per year. These fuels are used directly or blended with petroleum-based fossil fuels in the transportation sector.

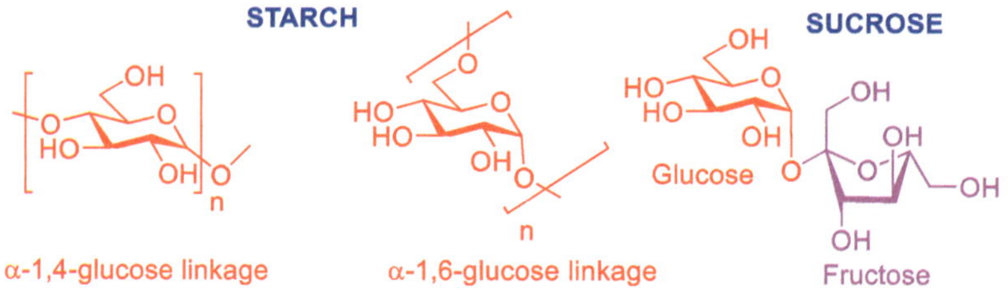

Figure 18.38: Carbohydrates for first-generation biofuels.

18.12.1.1 First-generation bioethanol

Bioethanol (C_2H_5OH or EtOH) is a liquid biofuel and is produced from different edible sugar and starch-containing feedstocks, including wheat, sugarbeet, and corn, through the microbial fermentation process [74]. Fermentation is the enzyme-catalyzed metabolic process used for the selective transformation of organic substrates into desired products. Several such enzymes have been identified as fermentative agents, while some of them are used for the conversion of sugar and starch to ethanol (Figure 18.39).

In the 1900s, bioethanol was utilized in many European countries and the United States for industrial and transportation purposes. Due to high production costs, bioethanol usage was reduced for certain years. Nevertheless, in the last three decades (after the 1970s), the production of bioethanol received significant attention as an alternative transportation fuel in many countries considering the global oil crisis. Bioethanol significantly reduces the necessity for fossil fuels, which indirectly helps to reduce CO_2 concentrations in the atmosphere. Bioethanol production has globally increased in recent decades not only from edible feedstocks but also from non-edible and microalgae feedstocks.

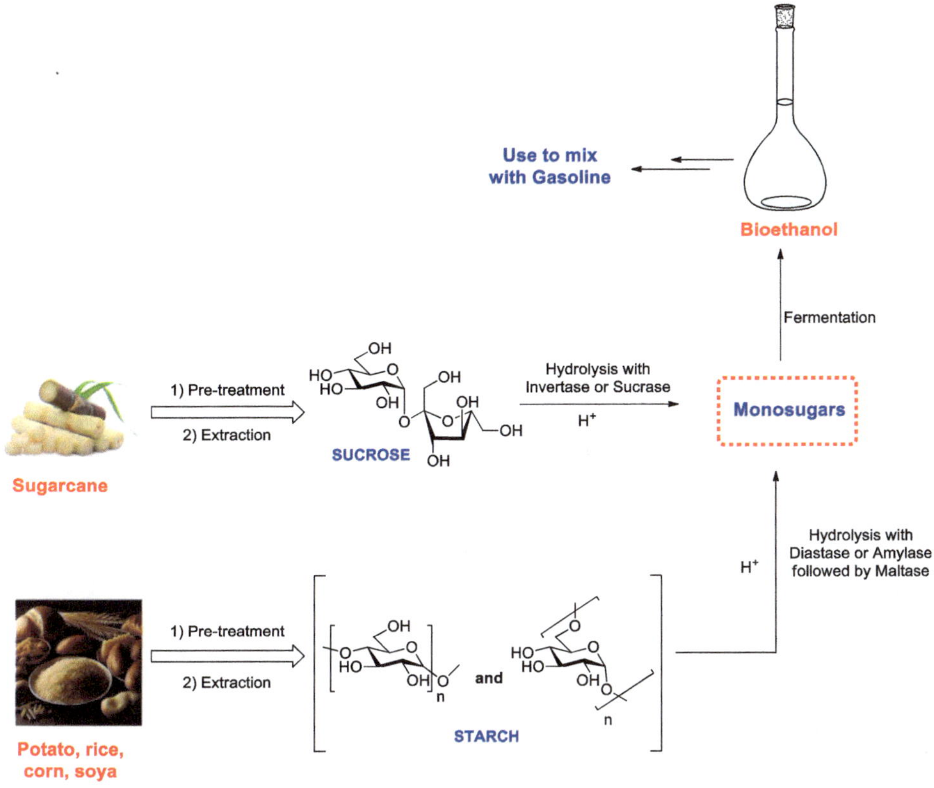

Figure 18.39: Synthesis of bioethanol from first-generation biomass.

18.12.1.2 First-generation biodiesel

Biodiesel is also a liquid fuel with a chemical formula ranging between $C_{10}H_{20}$ and $C_{15}H_{28}$ [75]. Biodiesel has been produced from different feedstocks using well-established technologies and is widely used for transportation (Figure 18.40). Biodiesel fuel production is mainly obtained from edible oil crops. Corn, rice bran, rapeseed, soybean, sunflower, peanut, safflower, coconut, olive, castor, and so on are the major sources of feedstocks used to produce biodiesel. Besides these edible sources, some common non-edible oil sources (cotton seed oil, tobacco, neem oil, etc.) are also used to produce biodiesel. Transesterification of oils such as triglycerides in the presence of alcohol is one of the reliable approaches for biodiesel production. In this method, glycerin (glycerol) is obtained as a by-product.

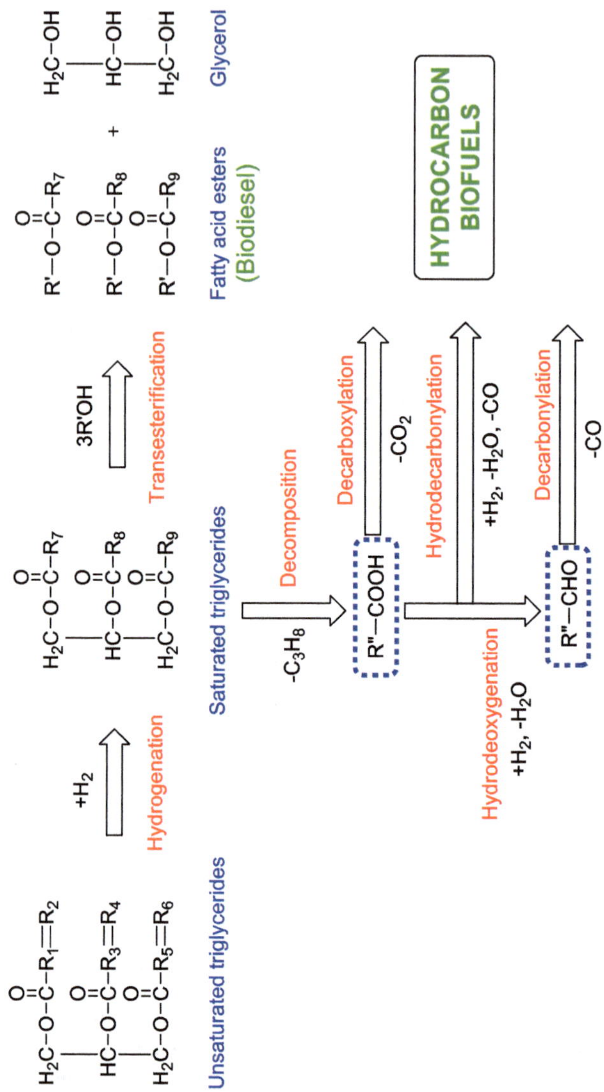

Figure 18.40: Synthesis of biodiesel and hydrocarbon biofuels from first-generation biomass.

18.12.1.3 First-generation bioethers

Bioether is a chemical compound in which an oxygen atom is bonded to two alkyl groups. Bioethers are usually produced from wheat or sugar beets through different methods. Bioethers can also be produced from glycerol, the by-product resulting from the production of biodiesel (Figure 18.41). Bioethers help in enhancing engine performance and reducing engine wear and toxic emissions. Bioalcohol can be converted into bioethers, which have been used as fuel additives. For instance, well-known fuel additives such as ethyl *tert*-butyl ether (ETBE), glycerol tertiary butyl ether (GTBE), or glycerol di-butyl ether (GDBE) have been prepared by the reaction of bioalcohols and isobutylene via an etherification reaction [76].

18.12.1.4 The drawbacks of first-generation biofuels

First-generation biofuels face some major drawbacks since they are produced from edible biomass feedstocks [77]. The extensive use of edible biomass feedstocks leads to competition with food resources. Besides, large areas of cropland, water, fertilizers, pesticides, and so on. are required for growing feedstocks. When compared with fossil fuels, first-generation biofuel production costs are high. The requirement of manpower for cultivation practices and soil erosion are additional problems with this generation.

18.12.2 Second-generation biofuels

Considering the drawbacks of the first generation, the development of second-generation biofuels was established [78, 79]. The non-edible lignocellulosic biomass, which is renewable, has been used as the feedstock for second-generation biofuels (Figure 18.42). Many feedstocks, including switch grass, sawmill dust, low-cost woods, and crop and municipal wastes, have been employed. The lignocellulosic biomass contains three chemical components: lignin, cellulose, and hemicellulose. Cellulose and hemicellulose bind with lignin, a non-carbohydrate biopolymer, and provide stiffness to plants. On the other hand, cellulose is the linear carbohydrate polymer made of β-D-glucose with β-1,4-glycosidic bonds. Hemicellulose is a branched heteropolymer consisting of a variety of carbohydrate monomers connected through glycosidic bonds. These lignocellulosic biomasses have high sugar content, which is an ideal choice to produce biofuels. The second-generation biofuels do not require agricultural land for the growth of feedstocks, and they do not compete with human food resources. The overall carbon emission from second-generation biofuels is neutral or negative. These are the major advantages of second-generation biofuels. Nevertheless, second-generation biomass feedstocks are relatively more complex than first-generation in terms of chemical content. Hence, additional steps are required for the fermentation process, which increases the production costs.

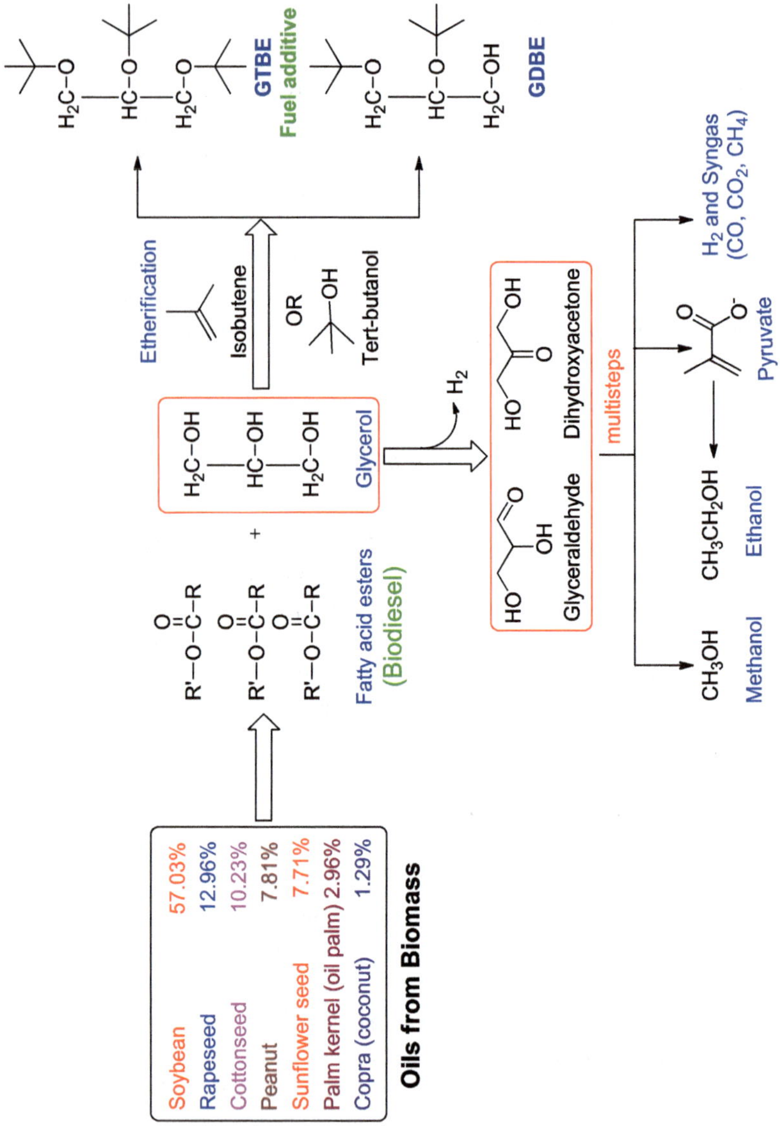

Figure 18.41: Synthesis of bioethers and other biofuel additives from glycerol.

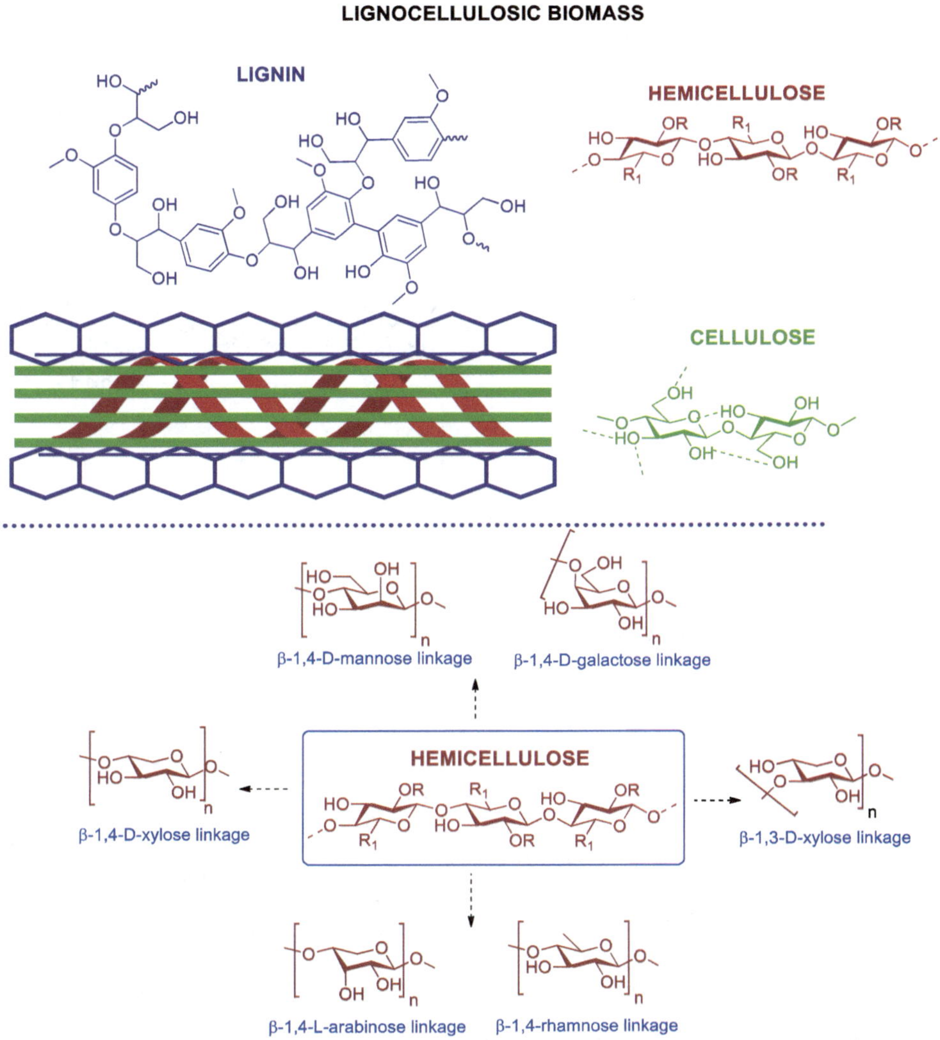

Figure 18.42: Structure of lignocellulosic biomass.

Cellulose has been obtained through pretreatment followed by extraction. Cellulose will be subjected to enzymatic hydrolysis using the cellulase enzyme to obtain glucose (Figure 18.43) [80]. Acid dehydration of glucose will lead to biofuel intermediates such as hydroxymethyl furfural (HMF) and levulinic acid. On the other hand, hydrogenolysis provides a sorbitol intermediate. Like cellulose, hemicellulose is hydrolyzed to arabinose and xylose by the action of the xylanase enzyme (Figure 18.44) [81]. The acid dehydration of xylose directly provides the biofuel intermediate furfural. Isomerization of xylose leads to the formation of xylulose, which can be converted to furfural through acid dehydration.

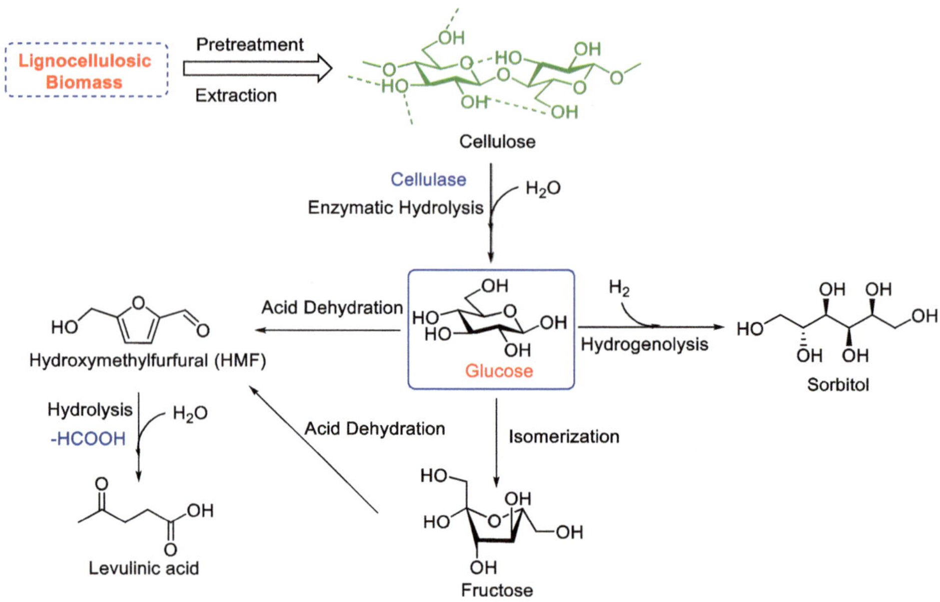

Figure 18.43: Biofuel formation from cellulose.

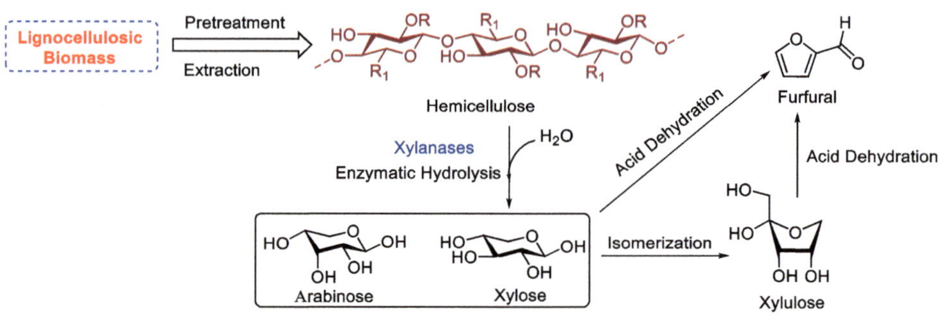

Figure 18.44: Biofuel formation from hemicellulose.

18.12.2.1 Second-generation bioethanol

Bioethanol has been obtained from lignocellulosic biomass carbohydrates via successive hydrolysis and fermentation processes [82]. The transformation of lignocellulosic biomass to bioethanol requires pretreatment, hydrolysis, fermentation, and product isolation steps. In one of the approaches, glucose is converted into pyruvate, followed by ethanol using the pyruvate kinase enzyme (Figure 18.45). This reaction proceeds through erythrose, glycoaldehyde, and ethylene glycol [83]. On the other hand, bioethanol can also be obtained from cellulose under catalytic conversion [84]. Thermochemical processes can also be utilized to produce bioethanol from lignocellulosic biomass via gasification followed by fermentation. However, this process has some drawbacks: (i) the difficulty of lignocellulosic biomass hydrolysis, (ii) identifying a suitable fermentation agent due to the heterogeneous mixture of sugars that results from lignocellulosic biomass hydrolysis, and (iii) the gathering and storage.

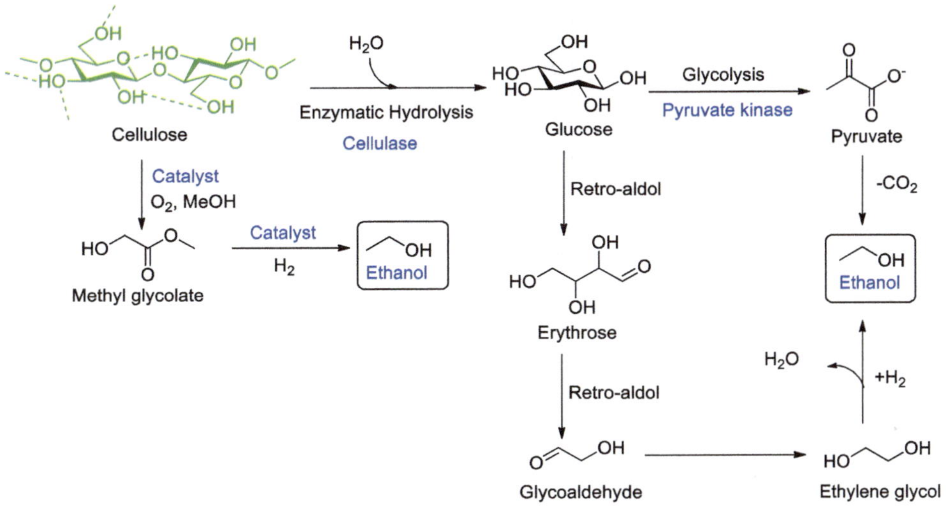

Figure 18.45: Synthesis of second-generation bioethanol from lignocellulosic biomass.

18.12.2.2 Second-generation biodiesel

Many different second-generation biomass feedstocks have been utilized to produce biodiesel [85]. In particular, energy crops, agricultural and wood wastes, and so on have been used. In addition, "salmon oil", "rubber tree", "Jatropha", and so on are the most common energy crops used to produce biodiesel. Furthermore, cooking oil waste, non-edible oil crops, and animal wastes can also be utilized as biodiesel feedstocks. Waste biomass is converted to cellulose and hemicellulose and further hydrolyzed to different monosaccharides. These monosaccharides were subjected to fermentation with *L. starkeyi* to obtain biodiesel (Figure 18.46) [86].

18.12.2.3 Second-generation biobutanol

Biobutanol is a renewable biofuel with the chemical formula $C_4H_{10}O$. It is produced from waste non-edible biomass feedstocks (Figure 18.47) [87, 88]. There are four types of butanol known, including *n*-butanol, *sec*-butanol, *iso*-butanol, and *tert*-butanol. Biobutanol has more heat energy, high volatility, low water miscibility, low flammability,

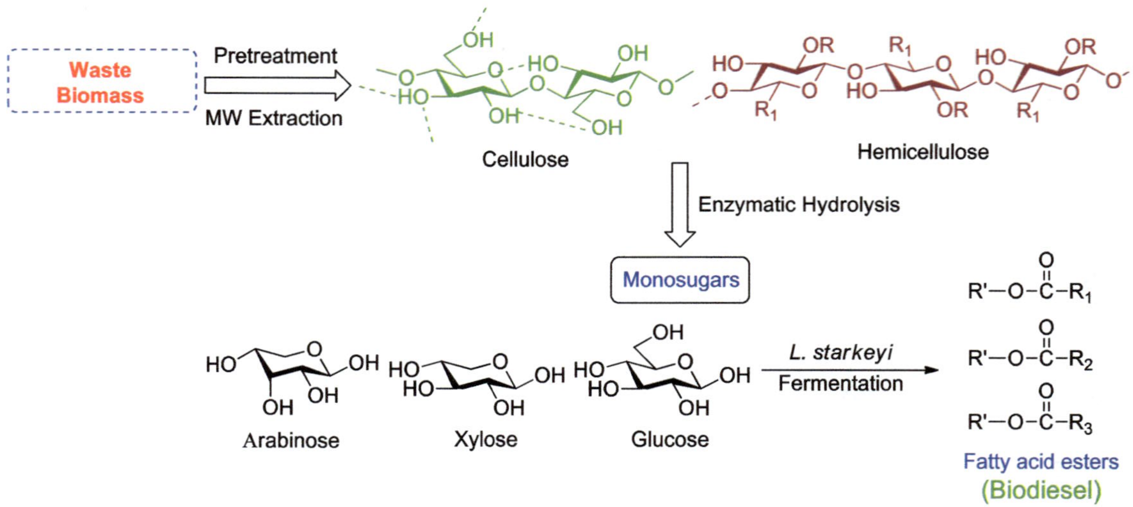

Figure 18.46: Synthesis of second-generation biodiesel from lignocellulosic biomass.

and low corrosiveness than ethanol, which are important advantages. Production of biobutanol from second-generation feedstocks can be achieved via the acetone-butanol-ethanol (ABE) fermentation process, which has received great interest globally.

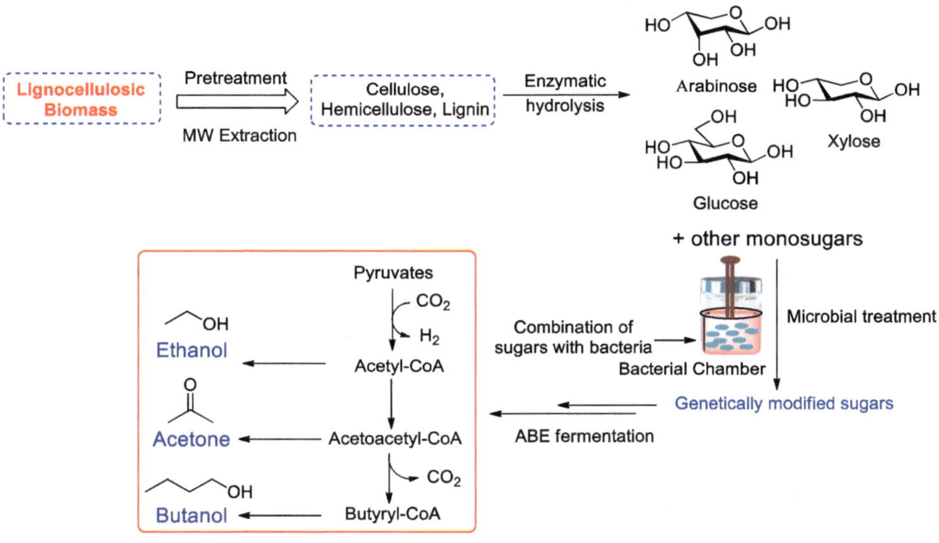

Figure 18.47: Synthesis of second-generation biobutanol from lignocellulosic biomass.

18.12.2.4 Limitations of second-generation biofuels

The agricultural farming of second-generation biofuels requires a large amount of water, free land, and manpower for harvesting, storage, and transport. Hence, its production cost is high when compared with fossil fuels. Second-generation biofuels often release nitrates that significantly contribute to environmental pollution and acid rain.

18.12.3 Third-generation biofuels

Third-generation biofuels have been produced from algae and other aquatic biomass feedstocks (Figure 18.48). This generation of biofuel is considered the best alternative renewable resource that overcomes the issues faced by first and second-generation biofuels [89]. Algae are categorized into autotrophic, heterotrophic, or mixotrophic based on their groups, like prokaryotic and eukaryotic organisms. Different types of biofuels, including biodiesel, gasoline, butanol, propanol, and ethanol, can be obtained from algae in high yields. Transformation of algal biomass to energy sources uses biochemical, chemical, and thermochemical methods. Cultivating algae is very fast when compared with first and second-generation feedstock growth. Additionally, algae do not require large free lands or pure water and can be grown on waste as well as salt water. Most importantly, algae cultivation utilizes excessive CO_2 which can reduce the green house effect. Hence, it is cost-efficient and environmentally friendly.

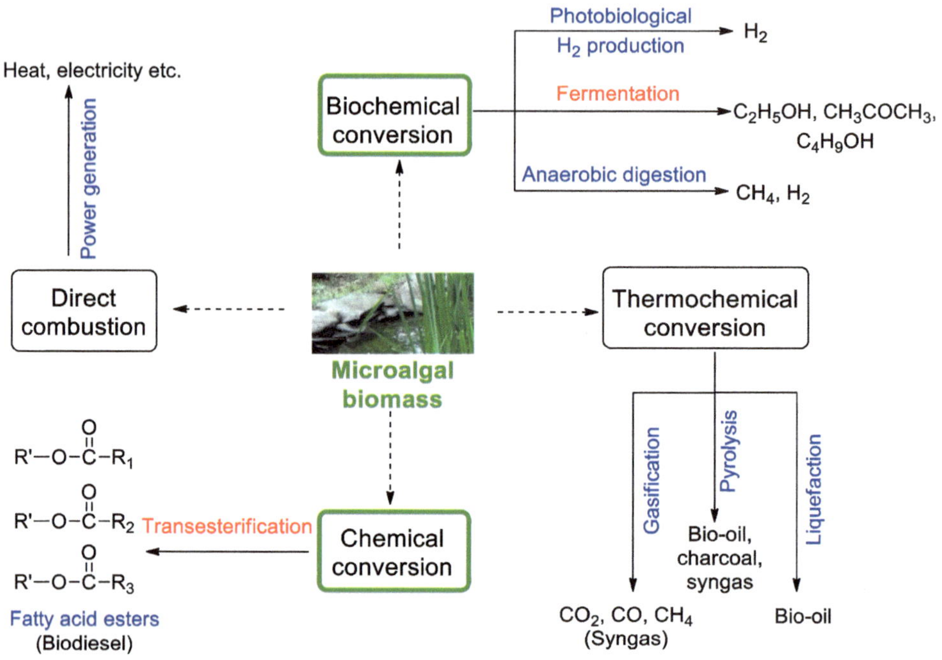

Figure 18.48: Third-generation biofuels formation from microalgal biomass.

18.12.3.1 Third-generation bioethanol

Algae have received significant attention over other biofuel feedstocks. Bioethanol can be produced from different algae (mixotrophic, heterotrophic, and autotrophic) carbohydrates by overcoming the issues associated with first and second-generation biofuels [90]. Initially, algae are subjected to pretreatment and hydrolysis to obtain starch, cellulose, and other carbohydrates (Figure 18.49). The resulting sugars are subjected to fermentation to produce ethanol. It is worth noting that algae-based bioethanol production is easier than that of the second generation of biomass, such as lignocellulosic biomass. Separation of algae-based carbohydrates does not require any chemical or physical pretreatment steps.

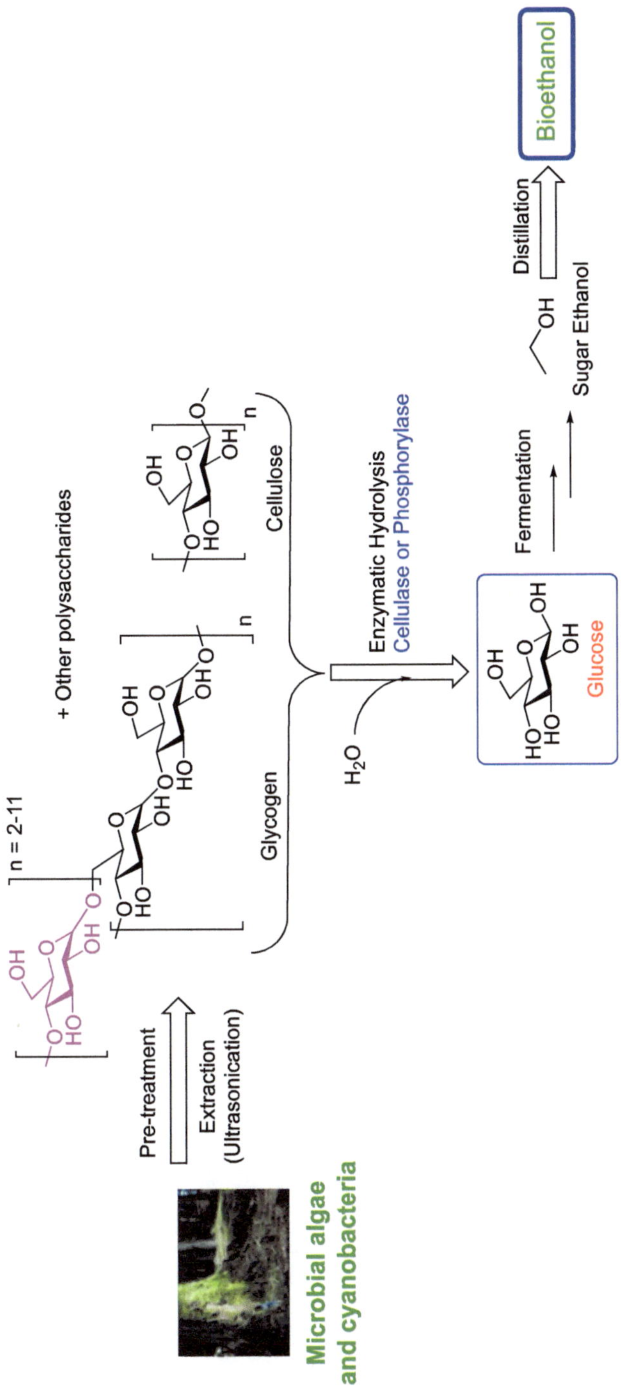

Figure 18.49: Synthesis of third-generation bioethanol from microalgal biomass.

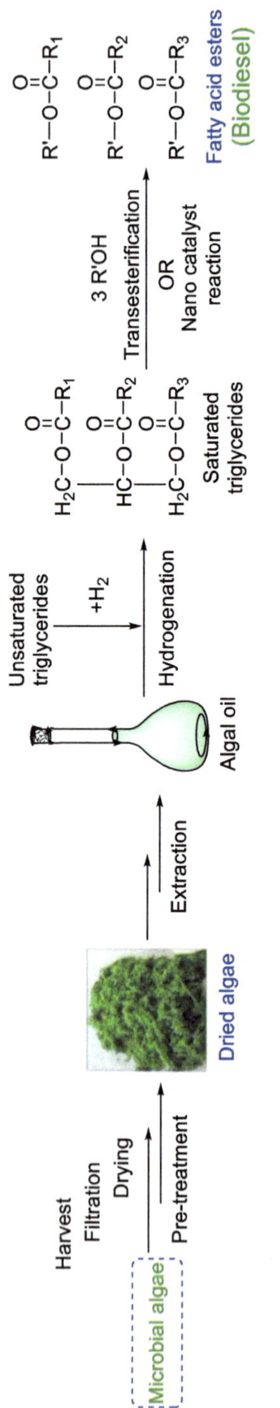

Figure 18.50: Synthesis of third-generation biodiesel from microalgal biomass.

18.12.3.2 Third-generation biodiesel

Biodiesel production from microalgae using a variety of microorganisms is a promising approach (Figure 18.50) [91]. Biodiesel can be produced from microalgae autotrophs and heterotrophs in different yields based on the microorganisms used. Microalgae provide better yields of biodiesel when compared with other generation feedstocks. Like other biodiesel production methods, the transesterification method is used to make biodiesel from microalgae oil.

18.12.4 Fourth-generation biofuel

Genetically modified feedstocks such as microalgae, yeast, fungi, and cyanobacteria are used for the preparation of fourth-generation biofuels (Figure 18.51) [92]. In this case, biomass is subjected to pyrolysis followed by carbon sequestration and fuel upgradation to obtain biohydrogen, bioethanol, and so on. Fourth-generation feedstocks are designed to utilize more atmospheric carbon dioxide (CO_2) from the environment, suppressing the emission of various poisonous gases. Different biological methods are

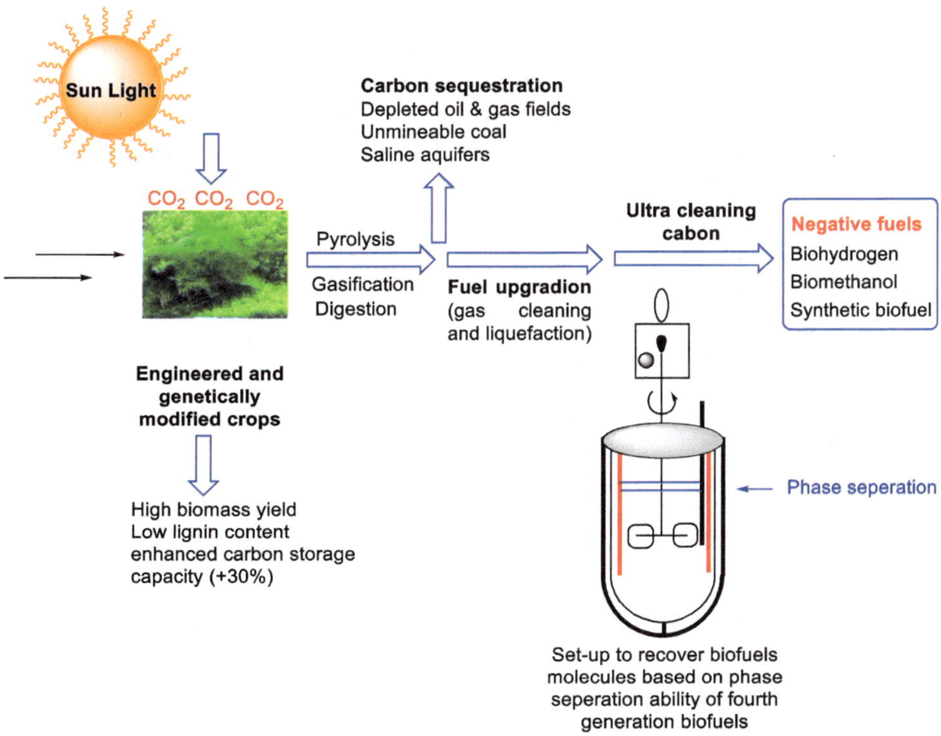

Figure 18.51: Fourth-generation biofuels synthetic route from engineered and genetically modified crops.

utilized for the extraction of sugars from fourth-generation feedstocks. It is in the early stages while merits and demerits are being investigated.

18.13 Furan and related biofuels from carbohydrate biomass

Biofuels possess various advantages over fossil fuels, including renewability, promising energy security, less pollution to the environment, cost-effectiveness, and easy transportation on demand. Carbohydrates are the major component of biomass and are being explored as important precursors to produce biofuels and value-added chemicals. As discussed in the previous section, biofuels are classified into four generations based on the origin of feedstock and the types of carbohydrates present in the biomass. Each generation has advantages and disadvantages in terms of cost, productivity, and efficiency. Among these four generations, the second-generation biofuels, that is, biofuels from non-edible lignocelluloses, are more promising in terms of economic and environmental benefits. These second-generation feedstocks are relatively cheap and do not cause food scarcity. Hence, second-generation biofuels are attractive for fulfilling the increasing global energy demands.

Carbohydrate-enriched biomasses have been used to produce various biofuels, including biomethanol, bioethanol, biobutanol, biodiesel, and biogas. Although these biofuels have many advantages over fossil fuels, a few demerits have also been observed. For instance, biodiesel has poor temperature performance and heating value, low volatility and oxidation stability, high nitroxide emissions, and viscosity. Similarly, bioethanol absorbs water and moisture easily. Moreover, it has high volatility and heat vaporization properties. In this context, furan-based biofuels that are derived from lignocellulosic biomass (and also from other carbohydrate biomass) have received considerable interest in the energy sector owing to their high content of oxygenation, better heating value and volatility, and low water miscibility (Figure 18.52) [93].

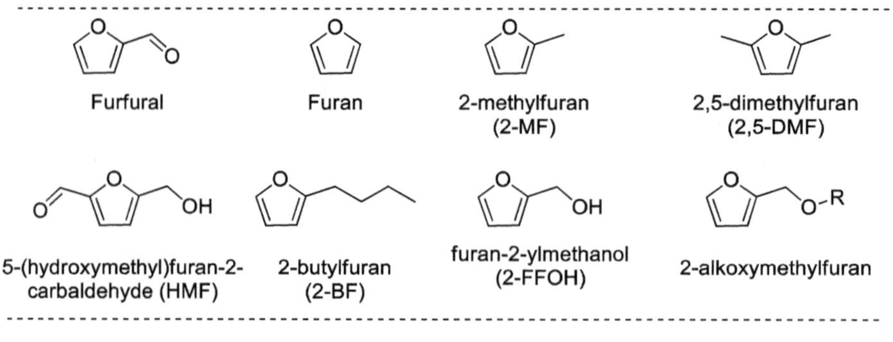

Figure 18.52: Furan and related biofuels from biomass carbohydrates.

These furanic compounds have useful properties that can be used as fuel or fuel additives with fossil fuels in an efficient manner. Furan-based biofuels can be produced from different hexose and pentose sugars through chemical transformation. The compounds 2,5-dimethylfuran (DMF) and 2-methylfuran (2-MF) are highly studied as biofuels, among others. Some of them are described below.

18.13.1 2,5-Dimethylfuran (DMF)

2,5-Dimethylfuran (DMF) is a heterocyclic compound that has ideal biofuel properties compared to bioethanol and biodiesel [94]. The energy density of DMF is comparable to petrol and is higher than that of bioethanol. DMF is highly stable and immiscible with water. Moreover, DMF does not absorb moisture. DMF can be produced from various carbohydrate-containing biomass, including lignocelluloses and starch. Among the various methods, the conversion of polysaccharides into monomers like glucose or fructose followed by DMF is one of the straightforward approaches for DMF production (Figure 18.53). DMF is produced from various lignocellulosic biomass feedstocks like crops and related wastes, industrial wastes from paper and cotton mills, and municipal solid wastes. In DMF production, initially, biomass feedstock is pretreated under different conditions. During this pretreatment process, the polysaccharides (e.g., cellulosic, starch, or algae) undergo hydrolysis to obtain glucose or fructose after isomerization under acidic conditions. In the next step, fructose undergoes dehydration under acidic conditions to produce the important intermediate 5-hydroxymethylfurfural (HMF). Various heterogeneous and homogeneous acid catalysts have been used for this transformation. Finally, hydrodeoxygenation (HDO) of HMF to DMF takes place in the presence of Brønsted acids, Lewis acids, or bifunctional catalysts. Various catalysts have been designed to enhance the efficiency of hydrolysis, isomerization, and dehydration reactions [95].

18.13.2 2-Methylfuran (MF)

2-MF has a chemical structure similar to DMF, with only one alkyl group. The fuel properties of 2-MF are comparable to DMF [96]. The synthesis of 2-MF is also achieved using steps similar to those of DMF. Cellulose is initially converted into HMF via glucose formation. The resulting product can be further converted into 2-MF via 5-methyl furfural or DMF. Hydrogenolysis and decarboxylation steps have been used for these transformations (Figure 18.54) [97]. On the other hand, hemicellulose is converted into xylose, which is transformed into 2-MF via furfural and furfuryl alcohol (FAL) (Figure 18.55).

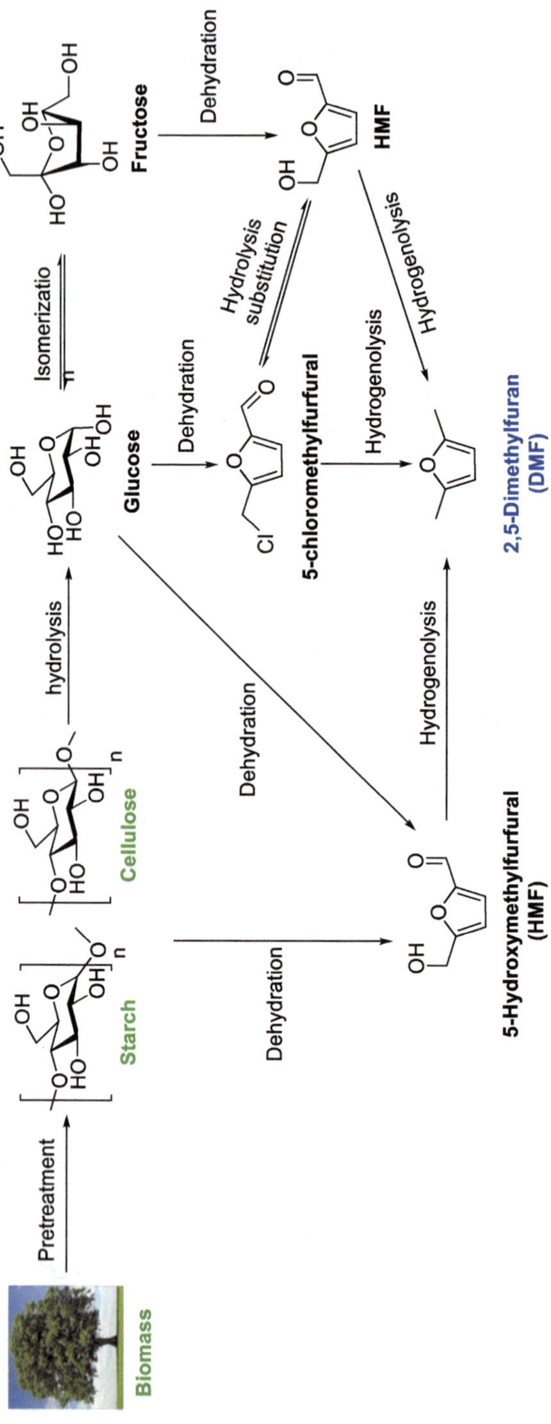

Figure 18.53: Formation of 2,5-dimethylfuran from biomass.

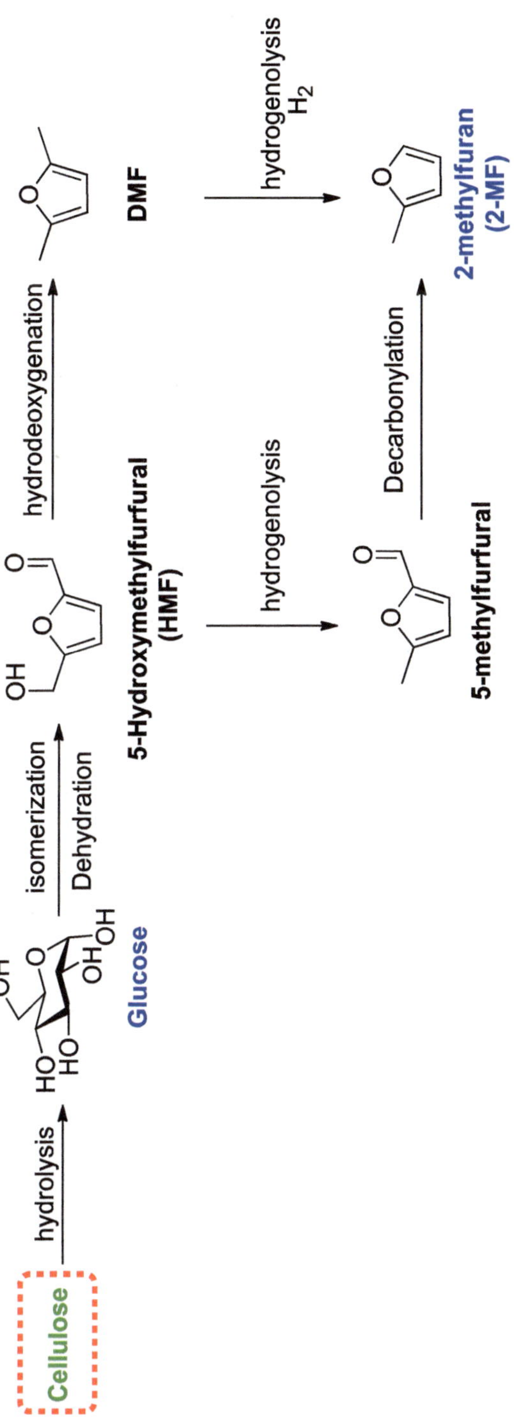

Figure 18.54: Formation of 2-methylfuran from cellulosic biomass.

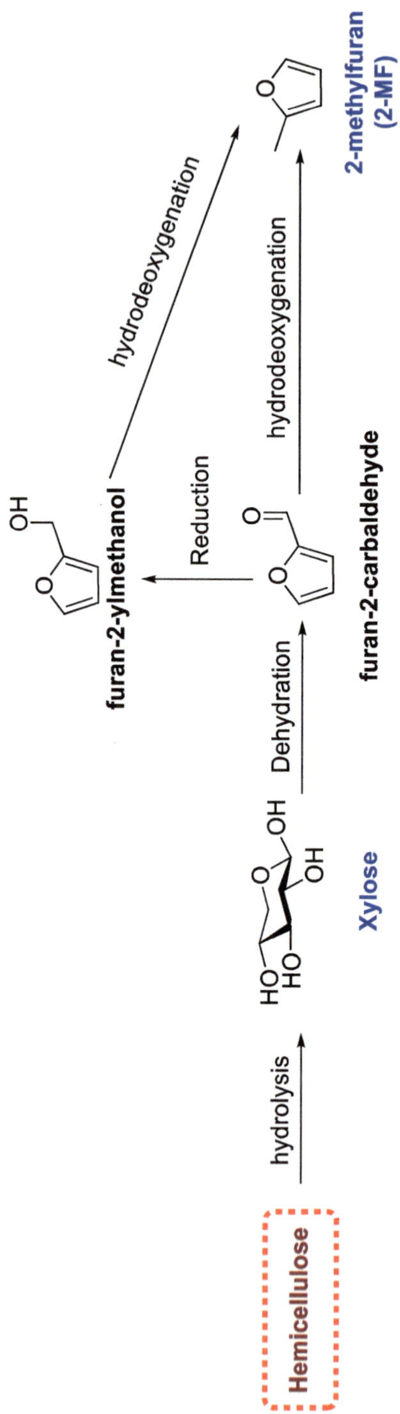

Figure 18.55: Formation of 2-methylfuran from hemicellulosic biomass.

18.13.3 γ-Valerolactone (GVL)

γ-Valerolactone is a cyclic ester compound formed from various biomass feedstocks via *a* levulinic acid intermediate [98]. γ-Valerolactone is typically derived from lignocellulosic biomass such as corn stover, sawgrass, or wood [99]. The production of GVL involves a few steps: (i) the hydrolysis of cellulosic biomasses into glucose, (ii) the dehydration of glucose to hydroxymethylfurfural, (iii) the ring opening of hydroxymethyl furfural to levulinic acid, and (iv) cyclization of levulinic acid to γ-valerolactone (Figure 18.56). Similarly, GVL can be obtained from hemicellulose through furfural, furfuryl alcohol, and alkyl levulinate (Figure 18.57). GVL has potential applications as a liquid fuel and fuel additive. In the past few decades, GVL has been used as a solvent that is renewable, safe to handle and store, and easily transportable [100]. It has high boiling and low melting points (−31 °C and 207 °C, respectively), which is ideal for fuel applications.

18.13.4 5-Ethoxymethylfurfural (EMF)

5-Ethoxymethylfurfural is a class of furan compounds considered to be a promising biofuel, gasoline alternative, and additive for diesel [101]. Synthesis of 5-ethoxymethylfurfural has been achieved through chemical synthesis from glucose (Figure 18.58). In this process, glucose is converted into 5-hydroxymethylfurfural initially and then converted into 5-EMF via an etherification reaction. On the other hand, 5-hydroxymethylfurfural is converted into 5-bromomethylfurfural and 5-chloromethylfurfural and subsequently converted into 5-EMF through the action of ethanol. 5-Ethoxymethylfurfural has high-energy density, high stability, and is environment-friendly, which makes it ideal for biofuel. 5-EMF has been typically prepared from fructose, glucose, or 5-hydroxymethylfurfural (5-HMF) in high yields [102].

18.13.5 2-Methyltetrahydrofuran

2-Methyltetrahydrofuran (2-MTHF) is a furan-based compound, which is a highly flammable liquid. 2-MTHF has been used as a green solvent and an alternative to fossil-based THF [103]. MTHF has various ideal fuel properties, which can be mixed up to 70% in gasoline [104]. The synthesis of 2-MTHF has been obtained in two ways from levulinic acid [104]. In one of the approaches, levulinic acid is converted into γ-valerolactone, followed by 2-MTHF via hydrogenation (Figure 18.59). In another approach, levulinic acid is converted into α and β-angelica lactone, followed by 2-MTHF via catalytic hydrogenation reactions (Figure 18.60).

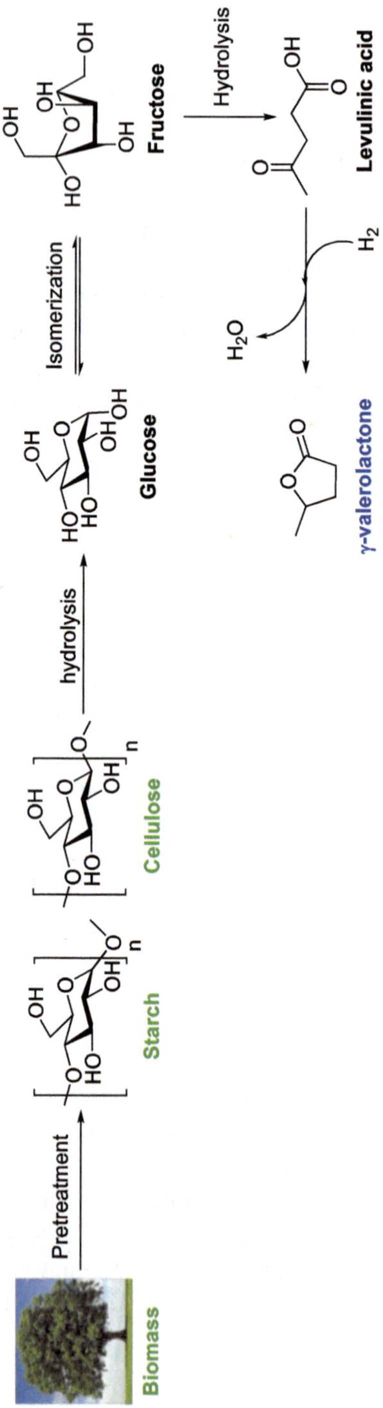

Figure 18.56: Formation of γ-valerolactone from cellulosic biomass.

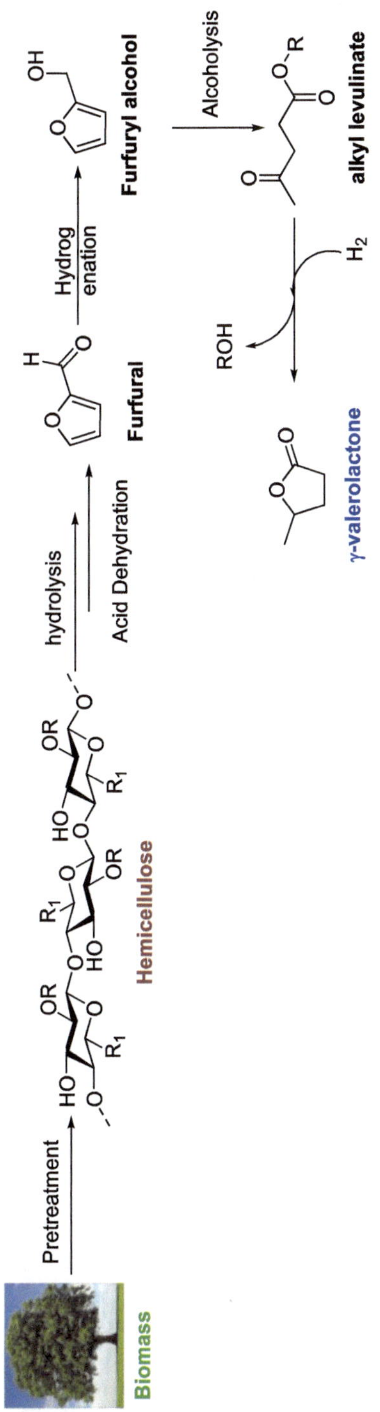

Figure 18.57: Formation of γ-valerolactone from hemicellulosic biomass.

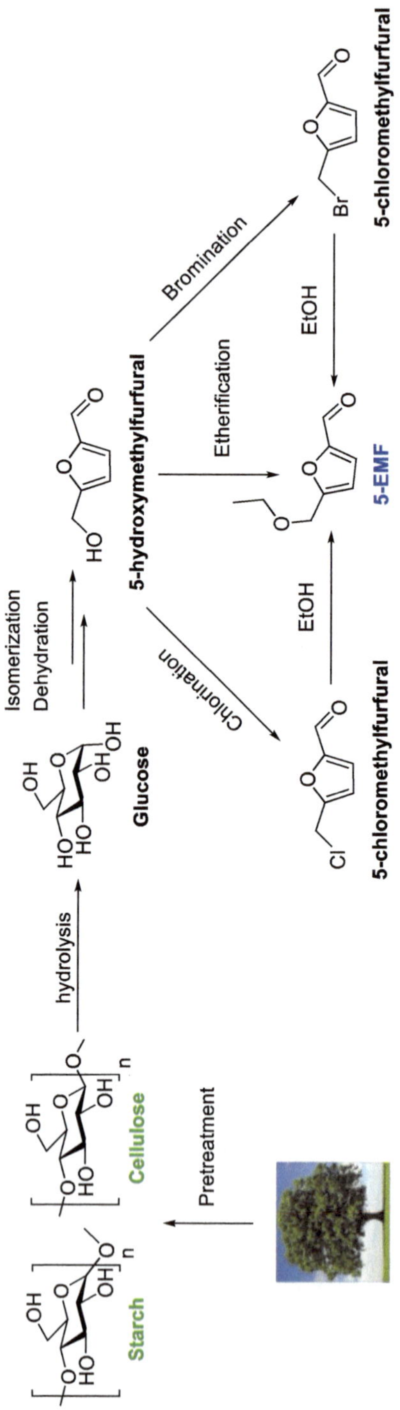

Figure 18.58: Formation of 5-ethoxymethylfurfural from biomass.

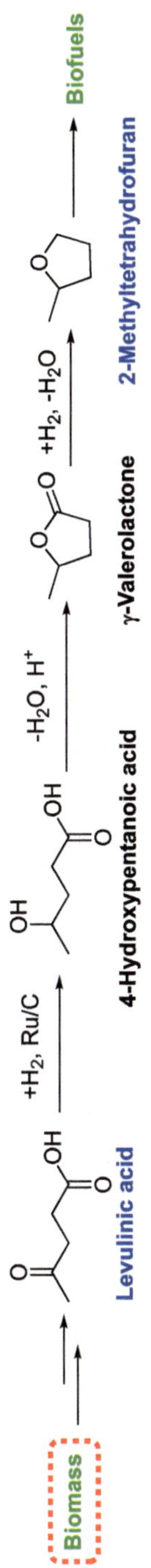

Figure 18.59: Formation of 2-methyltetrahydrofuran from levulinic acid derived from biomass.

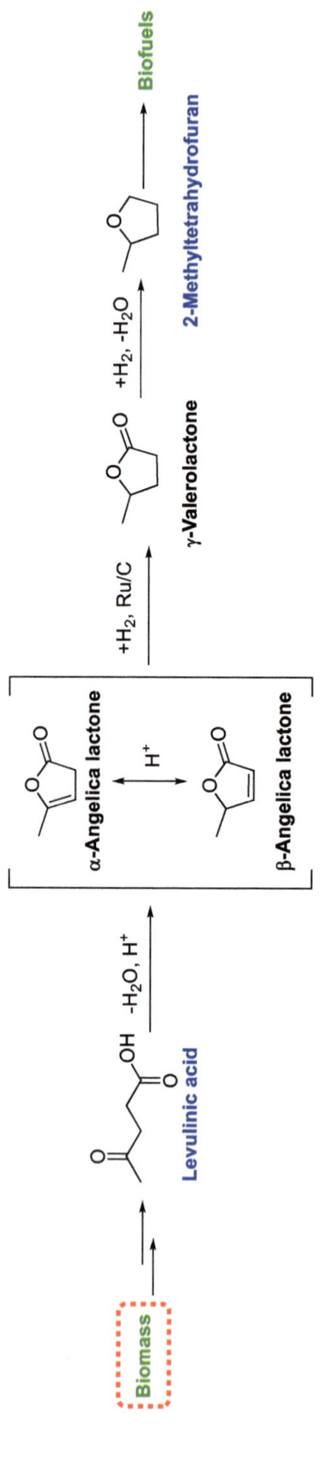

Figure 18.60: Another method for the formation of 2-methyltetrahydrofuran from levulinic acid derived from biomass.

18.13.6 Alkyl levulinate biofuels (EL, ML, and BL)

Alkyl levulinates, such as ethyl levulinate (EL), methyl levulinate (ML), and butyl levulinate (BL), are a few other types of biofuels that are obtained from lignocellulosic biomass (cellulose and hemicellulose) via furan derivatives (Figures 18.61 and 18.62) [105]. The applications of alkyl levulinates (AlkLevs) have been investigated in different fields, including biofuel additives. All of them have the potential to be used as fuel additives for petroleum products [106]. AlkLevs have been found in applications as solvents and in the food industry.

18.13.7 Tetrahydrofurfuryl alcohol (THFA)

Tetrahydrofurfuryl alcohol (THFA) is an important biofuel derivative from the furan family with the chemical formula $HOCH_2C_4H_7O$. A hydroxymethyl group is substituted in the 2-position of THF [107]. THFA is obtained from biomass through furan-2-carboxaldehyde (furfural) and furan-2-yl methanol using sequential hydrogenation reactions (Figure 18.63).

18.13.8 Ethylfurfuryl ether (EFE) and ethyl tetrahydrofurfuryl ether (ETE)

Ethylfurfuryl ether (EFE) and ethyl tetrahydrofurfuryl ether (ETE) are furanic-based fuels with ideal fuel properties. Both of them have lower heating values, which is an important fuel property. The synthesis of EFE has been achieved from biomass through xylose and furfural (Figure 18.64) [108]. Reductive etherification, that is, reduction of aldehyde or hydrogenation of aldehyde followed by etherification of the resulting furfuryl alcohol, provides direct access to EFE. On the other hand, acetal production of aldehyde with ethanol followed by reduction provides EFE. ETE is a suitable diesel fuel additive. ETE can be derived from EFE via a hydrogenation reaction.

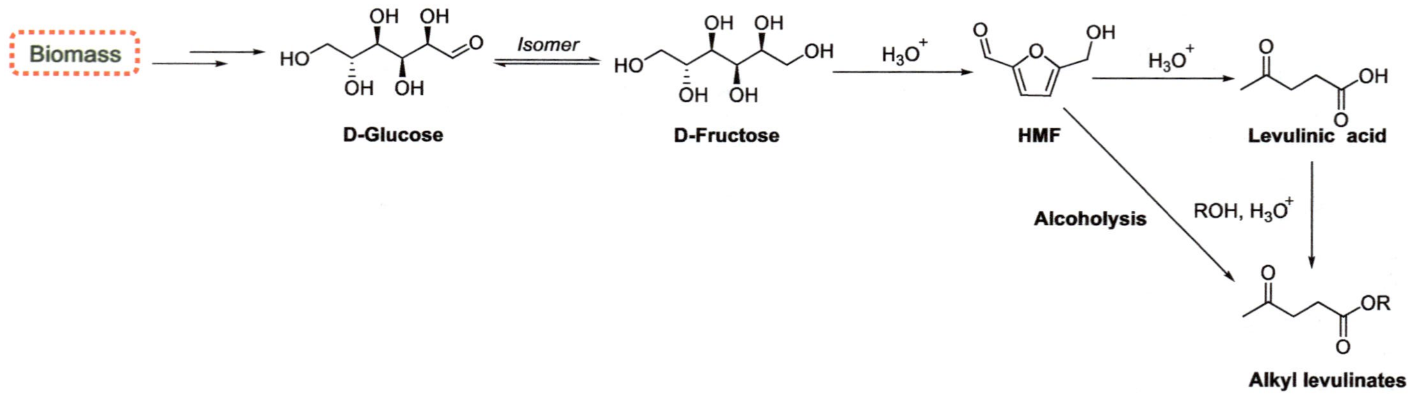

Figure 18.61: Formation of alkyl levulinates from levulinic acid derived from biomass.

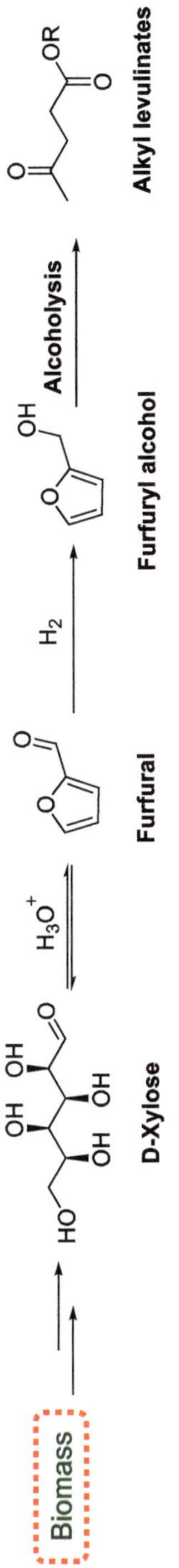

Figure 18.62: Formation of alkyl levulinates from furfuryl alcohol derived from biomass.

Figure 18.63: Formation of (tetrahydrofuran-2-yl)methanol from biomass.

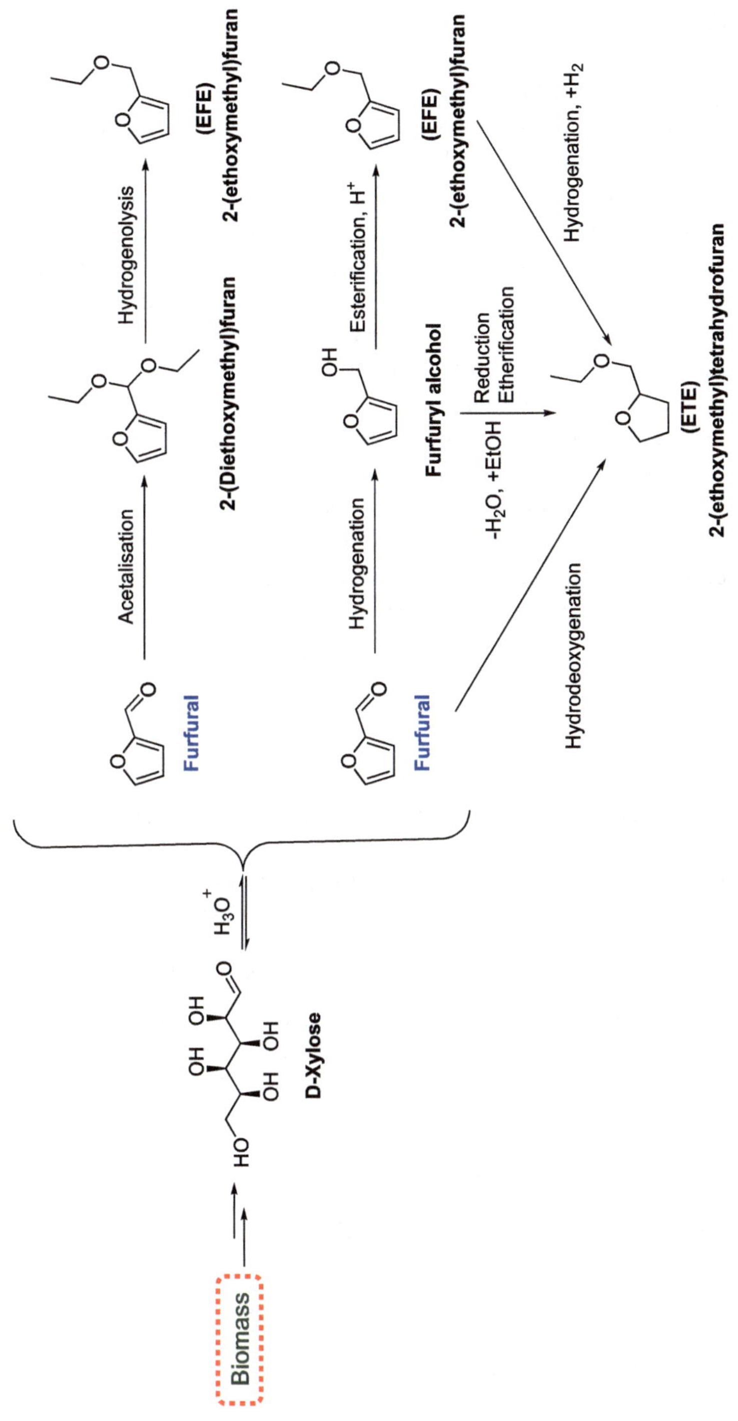

Figure 18.64: Formation of EFE and ETE from biomass.

18.14 Conclusion

Biofuels are very important to meet the long-term energy requirements globally. Biofuels can be obtained from various sources of carbohydrate-enriched biomass feedstocks. Based on the feedstocks and carbohydrates, biofuels have been classified into four types. First-generation biofuels help in decreasing carbon dioxide emissions (CO_2 emissions) along with cost and environmental issues. The use of edible biomass feedstock is one of the major disadvantages. The second-generation biofuels overcome this issue by using non-edible biomass feedstocks such as lignocellulosic biomass. Second-generation biofuels provide economic and environmental benefits. However, the production cost of second-generation biofuels is high due to the complex nature of feedstocks. The third generation of biofuels is derived from algae and microorganisms. However, like second-generation biofuels, the production cost of the third generation of biofuels is expensive. The fourth generation of biofuels relies on the use of genetically modified crops, which is in the early stages. Nevertheless, one common thing in all these generations is carbohydrates. The high content of carbohydrates in biomass is highly useful in terms of cost and production. More technologies are yet to come for the efficient conversion of carbohydrates into biofuels, which will solve the energy crisis in the present era.

References

[1] Gu D, Andreev K, Dupre ME. Major trends in population growth around the world. China CDC Wkly, 2021, 3, 604–613.

[2] Jaiswal KK, Chowdhury CR, Yadav D, Verma R, Dutta S, Jaiswal KS, et al. Renewable and sustainable clean energy development and impact on social, economic, and environmental health. Energy Nexus, 2022, 7, 100118–100132.

[3] Ockwell DG. Energy and economic growth: Grounding our understanding in physical reality. Energ Policy, 2008, 36, 4600–4604.

[4] She ZW, Kibsgaard J, Dickens CF, Chorkendorff I, Nørskov JK, Jaramillo TF. Combining theory and experiment in electrocatalysis: Insights into materials design. Science, 2017, 355, eaad4998–5011.

[5] Abas N, Kalair A, Khan N. Review of fossil fuels and future energy technologies. Futures, 2015, 69, 31–49.

[6] Helm D. The future of fossil fuels – is it the end?. Oxf Rev Econ Policy, 2016, 32, 191–205.

[7] Maibach E. Accelerating the end of the fossil fuel era: A human and planetary health imperative and opportunity. Front Sci, 2024, 2, 1383921–1383924.

[8] Utama NA, Fathoni AM, Kristianto MA, McLellan BC. The end of fossil fuel era: Supply-demand measures through energy efficiency. Procedia Environ Sci, 2014, 20, 40–45.

[9] Martins F, Felgueiras C, Smitkova M, Caetano N. Analysis of fossil fuel energy consumption and environmental impacts in European countries. Energies, 2019, 12, 964–975.

[10] Megia PJ, Vizcaino AJ, Calles JA, Carrero A. Hydrogen production technologies: From fossil fuels toward renewable sources. A Mini Review Energ Fuel, 2021, 35, 16403–16415.

[11] Ang T-Z, Salem M, Kamarol M, Das HS, Nazari MA, Prabaharan N. A comprehensive study of renewable energy sources: Classifications, challenges and suggestions. Energy Strat Rev, 2022, 43, 100939–100966.

[12] Mahapatra S, Kumar D, Singh B, Sachan PK. Biofuels and their sources of production: A review on cleaner sustainable alternative against conventional fuel, in the framework of the food and energy nexus. Energ Nexus, 2021, 4, 100036–100048.

[13] Jeswani HK, Chilvers A, Azapagic A. Environmental sustainability of biofuels: A review. Proc R Soc A, 2020, 476, 20200351–20200388.

[14] Liew WH, Hassim MH, Dks N. Review of evolution, technology and sustainability assessments of biofuel production. J Clean Prod, 2014, 71, 11–29.

[15] Malik K, Capareda SC, Kamboj BR, Malik S, Singh K, Arya S, Bishnoi DK. Biofuels production: A review on sustainable alternatives to traditional fuels and energy sources. Fuel, 2024, 5, 157–175.

[16] Williams CL, Westover TL, Emerson RM, Tumuluru JS, Chenlin L. Sources of biomass feedstock variability and the potential impact on biofuels production. Bioenerg Res, 2016, 9, 1–14.

[17] Panja A, Paul S, Jha P, Ghosh S, Prasad R. Waste and their polysaccharides: Are they worth bioprocessing?. Bioresour Technol Rep, 2023, 24, 101594–101616.

[18] Cherwoo L, Gupta I, Flora G, Verma R, Kapil M, Arya SK, Ravindran B, Khoo KS, Bhatia SK, Chang SW, Ngamcharussrivichai C, Ashokkumar V. Biofuels an alternative to traditional fossil fuels: A comprehensive review. Sustain Energy Technol Assessments, 2023, 60, 103503–103515.

[19] Mat Aron NS, Khoo KS, Chew KW, Show PL, Chen W-H, Nguyen THP. Sustainability of the four generations of biofuels – A review. Int J Energy Res, 2020, 44, 9266–9282.

[20] Kalak T. Potential use of industrial biomass waste as a sustainable energy source in the future. Energies, 2023, 16, 1783–1808.

[21] Amjith LR, Bavanish B. A review on biomass and wind as renewable energy for sustainable environment. Chemosphere, 2022, 293, 133579–133587.

[22] Tursi A. A review on biomass: Importance, chemistry, classification, and conversion. Biofuel Res J, 2019, 6, 962–979.

[23] Giudicianni P, Gargiulo V, Grottola CM, Alfe M, Ferreiro AI, Mendes MAA, Fagnano M, Ragucci R. Inherent metal elements in biomass pyrolysis: A review. Energ Fuels, 2021, 35, 5407–5478.

[24] Liu W-J, Li -W-W, Jiang H, Yu H-Q. Fates of chemical elements in biomass during its pyrolysis. Chem Rev, 2017, 117, 6367–6398.

[25] Alper K, Tekin K, Karagoz S, Ragauskas AJ. Sustainable energy and fuels from biomass: A review focusing on hydrothermal biomass processing. Sustain Energ Fuels, 2020, 4, 4390–4414.

[26] Corma A, Iborra S, Velty A. Chemical routes for the transformation of biomass into chemicals. Chem Rev, 2007, 107, 2411–2502.

[27] Luis HSG. Carbohydrates from biomass: Sources and transformation by microbial enzymes. In Carbohydrates: Comprehensive Studies on Glycobiology and Glycotechnology. Chuan-Fa C. (Ed.). IntechOpen: Carbohydrates, 2012, 441–458.

[28] Trouvelot S, Heloir M-C, Poinssot B, Gauthier A, Paris F, Guillier C, Combier M, Trda L, Daire X, Adrian M. Carbohydrates in plant immunity and plant protection: Roles and potential application as foliar sprays. Front Plant Sci, 2014, 5, 592–606.

[29] Hartmann H, Trumbore S. Understanding the roles of nonstructural carbohydrates in forest trees-from what we can measure to what we want to know. New Phytol, 2016, 211, 386–403.

[30] Benalaya I, Alves G, Lopes J, Silva LR. A review of natural polysaccharides: Sources, characteristics, properties, food, and pharmaceutical applications. Int J Mol Sci, 2024, 25, 1322–1354.

[31] Zhang J, Li P, Yu Y, Xu Y, Jia W, Zhao S. A review of natural polysaccharides-based flocculants derived from waste: Application efficiency, function mechanism, and development prospects. Ind Eng Chem Res, 2023, 62, 15774–15789.

[32] Pauly M, Keegstra K. Cell-wall carbohydrates and their modification as a resource for biofuels. Plant J, 2008, 54, 559–568.

[33] Yadav H, Karthikeyan C. 1-Natural polysaccharides: Structural features and properties. In Polysaccharide Carriers for Drug Delivery. Maiti S, Jana S. (Eds.). Woodhead Publishing Ltd, 2019, pp. 1–17, Sawston, United Kingdom.

[34] Mark QG, Xinzhong H, Changlu W, Lianzhong A. Polysaccharides: Structure and solubility. In Solubility of Polysaccharides. Zhenbo X. (Ed.). IntechOpen, 2017, pp. 7–21, London, United Kingdom.

[35] Apriyanto A, Compart J, Fettke J. A review of starch, a unique biopolymer-Structure, metabolism and in planta modifications. Plant Sci, 2022, 318, 111223–111231.

[36] Etale A, Onyianta AJ, Turner SR, Eichhorn SJ. Cellulose: A review of water interactions, applications in composites, and water treatment. Chem Rev, 2023, 123, 2016–2048.

[37] Rodionova MV, Bozieva AM, Zharmukhamedov SK, Leong YK, Chi-Wei Lan J, Veziroglu A, Veziroglu TN, Tomo T, Chang J-S, Allakhverdiev SI. A comprehensive review on lignocellulosic biomass biorefinery for sustainable biofuel production. Int J Hydrogen Energy, 2022, 47, 1481–1498.

[38] Roman-Benn A, Contador CA, Li M-W, Lam H-M, Ah-Hen K, Ulloa PE, Ravanal MC. Pectin: An overview of sources, extraction and applications in food products, biomedical, pharmaceutical and environmental issues. Food Chem Adv, 2023, 2, 100192–100205.

[39] Liu X, Gilbert RG. Normal and abnormal glycogen structure-A review. Carbohydr Polym, 2024, 338, 122195–122204.

[40] Zargar V, Asghari M, Dashti A. A review on Chitin and Chitosan polymers: Structure, chemistry, solubility, derivatives, and applications. Chem Bio Eng Reviews, 2015, 2, 204–226.

[41] Li B, Lu F, Wei X, Zhao RF. Structure and bioactivity. Molecules, 2008, 13, 1671–1695.

[42] Li J, He Z, Liang Y, Peng T, Hu Z. Insights into Algal polysaccharides: A review of their structure, depolymerases, and metabolic pathways. J Agri Food Chem, 2022, 70, 1749–1765.

[43] Zayed A, Cao HTT, Trang VTD, Ulber R. Structural tailoring of fucoidan backbones for maximizing their benefits: Enzymatic, chemical, and physical strategies. J Appl Phycol, 2023, 35, 2445–2462.

[44] Guo Z, Wei Y, Zhang Y, Xu Y, Zheng L, Zhu B, Yao Z. Carrageenan oligosaccharides: A comprehensive review of preparation, isolation, purification, structure, biological activities and applications. Algal Res, 2022, 61, 102593–102607.

[45] Krishnan L, Ravi N, Kumar Mondal A, Akter F, Kumar M, Ralph P, Kuzhiumparambil U. Seaweed-based polysaccharides-review of extraction, characterization, and bioplastic application. Green Chem, 2024, 26, 5790–5823.

[46] Lee KY, Mooney DJ. Alginate: Properties and biomedical applications. Prog Polym Sci, 2012, 37, 106–126.

[47] Ni D, Chen Z, Tian Y, Xu W, Zhang W, Kim B-G, Mu W. Comprehensive utilization of sucrose resources via chemical and biotechnological processes: A review. Biotechnol Adv, 2022, 60, 107990–108006.

[48] Lee B, Szilagyi A, Lactose: Uses in industry and biomedical importance lactose a functional disaccharide. In Dietary Sugars: Chemistry, Analysis, Function and Effects. Vr P. (Ed.). Royal Society of Chemistry, 2012, pp. 806–820, Milton Road, Cambridge.

[49] Khan R. The Chemistry of Maltose. In Advances in Carbohydrate Chemistry and Biochemistry. vol. 39, Tipson RS, Horton D. (Eds.). Elsevier, 1981, pp. 213–278, Academic press, Oxford, United Kingdom.

[50] Smit AT, Van Zomeren A, Dussan K, Riddell LA, Huijgen WJJ, Dijkstra JW, Bruijnincx PCA. Biomass pre-extraction as a versatile strategy to improve biorefinery feedstock flexibility, sugar yields, and lignin purity. ACS Sustain Chem Eng, 2022, 10, 6012–6022.

[51] Anthony A, Jonas B. Biomass pretreatment and characterization: A review. In Biotechnological Applications of Biomass. Thalita Peixoto B, Thiago Olitta B, Luiz Carlos B. (Eds.). IntechOpen, 2021, 1–17.

[52] Kumar AK, Sharma S. Recent updates on different methods of pretreatment of lignocellulosic feedstocks: A review. Bioresour Bioprocess, 2017, 4, 7–26.
[53] Singh JK, Vyas P, Dubey A, Upadhyaya CP, Kothari R, Tyagi VV, Kumar A. Assessment of different pretreatment technologies for efficient bioconversion of lignocellulose to ethanol. Front Biosci, 2018, 10, 350–371.
[54] Taylor MJ, Alabdrabalameer HA, Skoulou V. Choosing physical, physicochemical and chemical methods of pre-treating lignocellulosic wastes to repurpose into solid fuels. Sustain, 2019, 11, 3604–3631.
[55] Manna B, Das M, Patra P, Ghosh A. Insight into various conventional physical and chemical methods for the pretreatment of lignocellulosic biomass. In Thermochemical and Catalytic Conversion Technologies for Future Biorefineries. vol. 2, Verma P. (Ed.). Springer Nat Singapore: Singapore, 2022. pp. 31–57.
[56] Sai Bharadwaj AVSL, Dev S, Zhuang J, Wang Y, Yoo CG, Jeon B-H, Aggarwal S, Park SH, Kim TH. Review of chemical pretreatment of lignocellulosic biomass using low-liquid and low-chemical catalysts for effective bioconversion. Bioresour Technol, 2023, 368, 128339–128353.
[57] Basak B, Kumar R, Bharadwaj AVSLS, Kim TH, Kim JR, Jang M, Oh S-E, Roh H-S, Jeon B-H. Advances in physicochemical pretreatment strategies for lignocellulose biomass and their effectiveness in bioconversion for biofuel production. Bioresour Technol, 2023, 369, 128413–128425.
[58] Saritha M, Arora A, Lata. Biological pretreatment of lignocellulosic substrates for enhanced delignification and enzymatic digestibility. Indian J Microbiol, 2012, 52, 122–130.
[59] Zhao Z-M, Meng X, Scheidemantle B, Pu Y, Liu Z-H, Li B-Z, Wyman CE, Cai CM, Ragauskas AJ. Cosolvent enhanced lignocellulosic fractionation tailoring lignin chemistry and enhancing lignin bioconversion. Bioresour Technol, 2022, 347, 126367–126377.
[60] Nguyen TY, Cai CM, Kumar R, Wyman CE. Co-solvent pretreatment reduces costly enzyme requirements for high sugar and ethanol yields from lignocellulosic biomass. Chem Sus Chem, 2015, 8, 1716–1725.
[61] Mena-Garcia A, Ruiz-Matute AI, Soria AC, Sanz ML. Green techniques for extraction of bioactive carbohydrates. TRAC-Trends Anal Chem, 2019, 119, 115612–115622.
[62] Roque LR, Morgado GP, Nascimento VM, Ienczak JL, Rabelo SC. Liquid-liquid extraction: A promising alternative for inhibitors removing of pentoses fermentation. Fuel, 2019, 242, 775–787.
[63] Szparaga A, Kocira S, Kapusta I, Zaguła G. Solid-liquid extraction of bioactive compounds as a green alternative for developing novel biostimulant from Linum usitatissimum L. Chem Biol Technol Agric, 2023, 10, 108–123.
[64] Solarte DA, Ruiz-Matute AI, Chito-Trujillo DM, Rada-Mendoza M, Sanz ML. Microwave assisted extraction of bioactive carbohydrates from different morphological parts of Alfalfa (Medicago sativa L.). Foods, 2021, 10, 346–360.
[65] Florez-Fernandez N, Gonzalez, Munoz MJ. Ultrasound-assisted extraction of bioactive carbohydrates. In Water Extraction of Bioactive Compounds. Dominguez Gonzalez H, Gonzalez Munoz MJI. (Eds.). Elsevier, 2017, pp. 317–331, Academic press, Oxford, United Kingdom.
[66] Barp L, Visnjevec AM, Moret S. Pressurized liquid extraction: A powerful tool to implement extraction and purification of food contaminants. Foods, 2023, 12, 2017.
[67] Sakamoto Y, Imamura K, Onda A. Hydrolysis of oligosaccharides and polysaccharides on sulfonated solid acid catalysts: Relations between adsorption properties and catalytic activities. ACS Omega, 2020, 5, 24964–24972.
[68] Amandio MST, Rocha JMS, Xavier AMRB. Enzymatic hydrolysis strategies for cellulosic sugars production to obtain bioethanol from eucalyptus globulus bark. Ferment, 2023, 9, 241.
[69] Lin Y, Tanaka S. Ethanol fermentation from biomass resources: Current state and prospects. Appl Microbiol Biotechnol, 2006, 69, 627–642.
[70] Huang H-J, Ramaswamy S, Liu Y. Separation and purification of biobutanol during bioconversion of biomass. Sep Purif Technol, 2014, 132, 513–540.

[71] Huang H-J, Ramaswamy S. Overview of biomass conversion processes and separation and purification technologies in biorefineries. In Seperation and Purification Technology in Biorefineries. Ramaswamy S, Huang H-J, Ramarao BVI. (Eds.). John Wiley & Sons Ltd, 2013, pp. 1–36, New Jersey, United State.

[72] Cavelius P, Engelhart-Straub S, Mehlmer N, Lercher J, Awad D, Bruck T. The potential of biofuels from first to fourth generation. PLoS Biol, 2023, 21, e3002063–83.

[73] Naik SN, Goud VV, Rout PK, Dalai AK. Production of first and second generation biofuels: A comprehensive review. Renew Sustain Energy Rev, 2010, 14, 578–597.

[74] Bertrand E, Dussap C-G. First generation bioethanol: Fundamentals-definition, history, global production, evolution. In Liquid Biofuels: Bioethanol. vol. 12, CR S, et al. (Ed.). Springer, 2022, pp. 1–12, New York, United State.

[75] Mahmudul HM, Hagos FY, Mamat R, Adam AA, Ishak WFW, Alenezi R. Production, characterization and performance of biodiesel as an alternative fuel in diesel engines – A review. Renew Sustain Energy Rev, 2017, 72, 497–509.

[76] Liu H, Wang Z, Zhang J, Wang J, Shuai S. Study on combustion and emission characteristics of Polyoxymethylene Dimethyl Ethers/diesel blends in light-duty and heavy-duty diesel engines. Appl Energy, 2017, 185, 1393–1402.

[77] Mohr A, Raman S. Lessons from first generation biofuels and implications for the sustainability appraisal of second generation biofuels. Energ Policy, 2013, 63, 114–122.

[78] Ben Hassen Trabelsi A, Zaafouri K, Baghdadi W, Naoui S, Ouerghi A. Second generation biofuels production from waste cooking oil via pyrolysis process. Renew Energy, 2018, 126, 888–896.

[79] Nygaard I, Dembele F, Daou I, Mariko A, Kamissoko F, Coulibaly N, Borgstrom RL, Bruun TB. Lignocellulosic residues for production of electricity, biogas or second generation biofuel: A case study of technical and sustainable potential of rice straw in Mali. Renew Sustain Energy Rev, 2016, 61, 202–212.

[80] Shi N, Zhu Y, Qin B, Zhu T, Huang H, Liu Y. Conversion of Cellulose into 5-Hydroxymethylfurfural in a biphasic system catalyzed by aluminum sulfate and byproduct characterization. ACS Sustain Chem Eng, 2022, 10, 10444–10456.

[81] Bao Y, Du Z, Liu X, Liu H, Tang J, Qin C, Liang C, Huang C, Yao S. Furfural production from lignocellulosic biomass: One-step and two-step strategies and techno-economic evaluation. Green Chem, 2024, 26, 6318–6338.

[82] Aditiya HB, Mahlia TMI, Chong WT, Nur H, Sebayang AH. Second generation bioethanol production: A critical review. Renew Sustain Energy Rev, 2016, 66, 631–653.

[83] Gong Z, Lv X, Yang J, Luo X, Shuai L. Chemocatalytic conversion of lignocellulosic biomass to ethanol: A mini-review. Catal, 2022, 12, 922.

[84] Wu Y, Dong C, Wang H, Peng J, Li Y, Samart C, Ding M. One-pot ethanol production from cellulose transformation over multifunctional Pt/WOx and hollow Pt@HZSM-5 catalysts. ACS Sustain Chem Eng, 2022, 10, 2802–2810.

[85] Bhuiya MMK, Rasul MG, Khan MMK, Ashwath N, Azad AK. Prospects of 2nd generation biodiesel as a sustainable fuel-Part: 1 selection of feedstocks, oil extraction techniques and conversion technologies. Renew Sustain Energy Rev, 2016, 55, 1109–1128.

[86] Zhang L, Lee JTE, Ok YS, Dai Y, Tong YW. Enhancing microbial lipids yield for biodiesel production by oleaginous yeast Lipomyces starkeyi fermentation: A review. Bioresour Technol, 2022, 344, 126294–126306.

[87] Lapuerta M, Hernandez JJ, Fernandez-Rodriguez D, Cova-Bonillo A. Autoignition of blends of n-butanol and ethanol with diesel or biodiesel fuels in a constant-volume combustion chamber. Energy, 2017, 118, 613–621.

[88] Wilbanks B, Trinh CT. Comprehensive characterization of toxicity of fermentative metabolites on microbial growth. Biotechnol Biofuel, 2017, 10, 262–273.

[89] Leong W-H, Lim J-W, Lam M-K, Uemura Y, Ho Y-C. Third generation biofuels: A nutritional perspective in enhancing microbial lipid production. Renew Sustain Energy Rev, 2018, 91, 950–961.

[90] John RP, Anisha GS, Nampoothiri KM, Pandey A. Micro and macroalgal biomass: A renewable source for bioethanol. Bioresour Technol, 2011, 102, 186–193.

[91] Milano J, Ong HC, Masjuki HH, Chong WT, Lam MK, Loh PK, Vellayan V. Microalgae biofuels as an alternative to fossil fuel for power generation. Renew Sustain Energy Rev, 2016, 58, 180–197.

[92] Moravvej Z, Makarem MA, Rahimpour MR. The fourth generation of biofuel. In Second and Third Generation of Feedstocks. Basile A, Dalena F. (Eds.). Elsevier, 2019, pp. 557–597, Elsevier BV, Oxford, United Kingdom.

[93] Eldeeb MA, Akih-Kumgeh B. Recent trends in the production, combustion and modeling of furan-based fuels. Energies, 2018, 11, 1–47.

[94] Qian Y, Zhu L, Wang Y, Lu X. Recent progress in the development of biofuel 2,5-dimethylfuran. Renew Sustain Energy Rev, 2015, 41, 633–646.

[95] Saha B, Abu-Omar MM. Current technologies, economics, and perspectives for 2,5-dimethylfuran production from biomass-derived intermediates. Chem Sus Chem, 2015, 8, 1133–1142.

[96] Jezak S, Dzida M, Zorebski M. High pressure physicochemical properties of 2-methylfuran and 2,5-dimethylfuran – Second generation biofuels. Fuel, 2016, 184, 334–343.

[97] Zheng H-Y, Zhu Y-L, Teng B-T, Bai Z-Q, Zhang C-H, Xiang H-W, Li Y-W. Towards understanding the reaction pathway in vapour phase hydrogenation of furfural to 2-methylfuran. J Mol Catal A Chem, 2006, 246, 18–23.

[98] Horvath IT, Mehdi H, Fabos V, Boda L, Mika LT. γ-Valerolactone – A sustainable liquid for energy and carbon-based chemicals. Green Chem, 2008, 10, 238–242.

[99] Lopez-Aguado C, Del Monte DM, Paniagua M, Morales G, Melero JA. Techno-economic assessment of conceptual design for γ-valerolactone production over a bifunctional Zr-Al-beta catalyst. Ind Eng Chem Res, 2022, 61, 5547–5556.

[100] Cordova-Perez GE, Cortez-Elizalde J, Silahua-Pavon AA, Cervantes-Uribe A, Arevalo-Perez JC, Cordero-Garcia A, De Los Monteros AEE, Espinosa-Gonzalez CG, Godavarthi S, Ortiz-Chi F, Guerra-Que Z, Torres-Torres JG. γ-valerolactone production from levulinic acid hydrogenation using Ni supported nanoparticles: influence of Tungsten loading and pH of synthesis. Nanomater, 2022, 12, 2017.

[101] Liu X, Wang R. 5-Ethoxymethylfurfural – A remarkable biofuel candidate. In Biomass, Biofuels, Biochemicals. Saravanamurugan S, et al. (Ed.). Elsevier, 2020, pp. 355–375.

[102] Zuo M, Lin L, Zeng X. The synthesis of potential biofuel 5-ethoxymethylfurfural: A review. Fuel, 2023, 343, 127863–127884.

[103] Soszka E, Jedrzejczyk M, Keller N, Ruppert AM. High yield production of 2-methyltetrahydrofuran biofuel with reusable Ni-Co catalysts. Fuel, 2023, 332, 126118–126128.

[104] Alorku K, Shen C, Li Y, Xu Y, Wang C, Liu Q. Biomass-derived 2-methyltetrahydrofuran platform: A focus on precious and non-precious metal-based catalysts for the biorefinery. Green Chem, 2022, 24, 4201–4236.

[105] Yamanaka N, Shimazu S. Conversion of biomass-derived molecules into Alkyl levulinates using heterogeneous catalysts. React, 2023, 4, 667–678.

[106] Nelson Appaturi J, Andas J, Ma Y-K, Lee Phoon B, Muazu Batagarawa S, Khoerunnisa F, Hazwan Hussin M, Ng E-P. Recent advances in heterogeneous catalysts for the synthesis of alkyl levulinate biofuel additives from renewable levulinic acid: A comprehensive review. Fuel, 2022, 323, 124362–124387.

[107] Tran L-S, Carstensen -H-H, Lamoureux N, Foo KK, Gosselin S, El Bakali A, Gasnot L, Desgroux P. Exploring the flame chemistry of C5 tetrahydrofuranic biofuels: tetrahydrofurfuryl alcohol and 2-methyltetrahydrofuran. Energy Fuel, 2021, 35, 18699–18715.

[108] Cao Q, Guan J, Peng G, Hou T, Zhou J, Mu X. Solid acid-catalyzed conversion of furfuryl alcohol to alkyl tetrahydrofurfuryl ether. Catal Commun, 2015, 58, 76–79.

Hina Anwar, Akbar Ali*, Muhammad Ibrahim, Musrat Shaheen, and Tanveer Husssain Bukhari

Chapter 19 Industrial applications of carbohydrates

19.1 Introduction

Carbohydrates include polyhydroxy aldehydes, polyhydroxy ketones, and their derivatives. They are a wide and diverse class of organic molecules composed of carbon, hydrogen, and oxygen in the ratio *n*2*n*–*n*, respectively. They are the most abundant group of organic molecules in plants and provide a continually renewable source of raw materials [1]. They are inexpensive, and features good chemical and enantiomeric purity. They primarily function as energy transporters and structural elements of living organisms. They are generally known as saccharides meaning sugar forming. Carbohydrates are categorized into different categories based on their structure or function. They are classified as mono, di, oligo and polysaccharides depending upon their structure. Moreover, there are three other types of carbohydrates based on their role: 1) structural polysaccharides (cellulose and chitin), 2) storage polysaccharides (starch and glycogen), and 3) gel-forming polysaccharides (alginic acid and mucopolysaccharides). They are vital for humanity's survival, as food components or as medicinal medication constituents. Their physical, chemical, and biological properties are diverse and varied. In addition to being important elements of numerous biologically active chemicals for example, glycoproteins proteoglycans, glycolipids, heparin and nucleic acids, etc., they also act as a primary energy reserve for a variety of essential physiological processes. They may make up to 80–90% of caloric energy in third World countries diets which are totally grain-based. However, in a typical American diet they make up to 46% of calories [2]. A monosaccharide, often known as a simple sugar, is a carbohydrate that cannot be further hydrolyzed, for example, glucose (**1**) and fructose (**2**). A disaccharide contains two units of monosaccharides join together by a glycosidic bond, for example, sucrose (**3**) (Figure 19.1). Many of these elements are in equilibrium between an open-chain form and one or more ring forms, resulting in a hemiacetal or hemiketal connection between the aldehyde (aldose) or ketone (ketose) moiety of the open-chain form, with two potential stereochemical configurations [3].

***Corresponding author: Akbar Ali,** Department of Chemistry, Government College University Faisalabad, Faisalabad 38000, Pakistan, e-mails: akbarchm@gmail.com; akbarali@gcuf.edu.pk
Hina Anwar, Musrat Shaheen, Tanveer Husssain Bukhari, Department of Chemistry, Government College University, Faisalabad, Faisalabad 38000, Pakistan
Muhammad Ibrahim, Department of Applied Chemistry, Government College University, Faisalabad, Faisalabad 38000, Pakistan

https://doi.org/10.1515/9783111447094-019

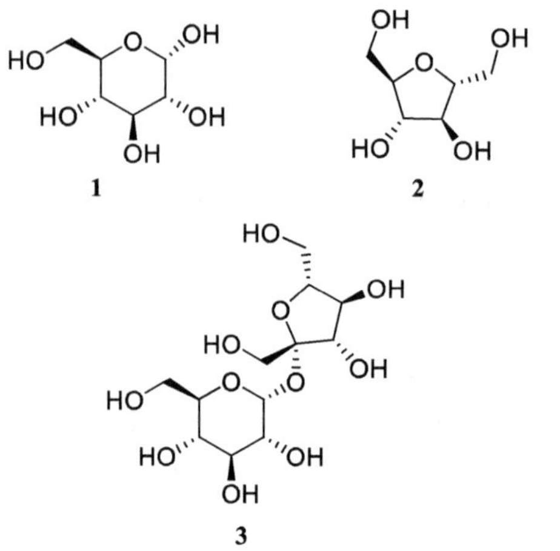

Figure 19.1: Structures of mono- and disaccharides.

Polysaccharides are complex carbohydrate derivatives made up of chains of monosaccharide subunits connected by intermediate connections [4]. They occur naturally in plants, microorganisms, and animal sources (Figure 19.2). Plant biomass is seen as a possible bio-factory for producing various types of polysaccharides due to its widespread availability and sustainability [5]. They can be classified based on structure, chemical content, solubility, viscosity, and source. Polysaccharides can be divided into two types based on their chemical composition [4]:
1) Homopolysaccharides consist of a single type of monosaccharide unit, such as cellulose, starch, dextrin, or chitin.
2) Hyaluronic acid, tragacanth gum, and heparin are examples of heteropolysaccharides, which are made up of many types of monosaccharide units.

19.1.1 Cellulose

Cellulose (**4**) (Figure 19.3) is a polysaccharide and constitutes a major portion of plant cell wall. It makes up to half of the biomass of plants existing on earth therefore it is considered to be the most abundant polymers on the planet. The major sources of cellulose include, cotton, jute, linen, kenaf, sugar cane, cereal straws, wood, and hemp. The non-plant sources include algae (*Valonia* and *Microdictyon*), bacteria (*Acetobacter*), and marine organisms (Ascite family) [6]. Cellulose polymeric chains contain straight chains of glucose subunits and contain three hydroxyl groups per anhydroglucose unit. These OH groups are responsible for intra- and inter-hydroxyl H-bonds, making cellulosic chains a supramolecular polymer with crystallographic sections [7].

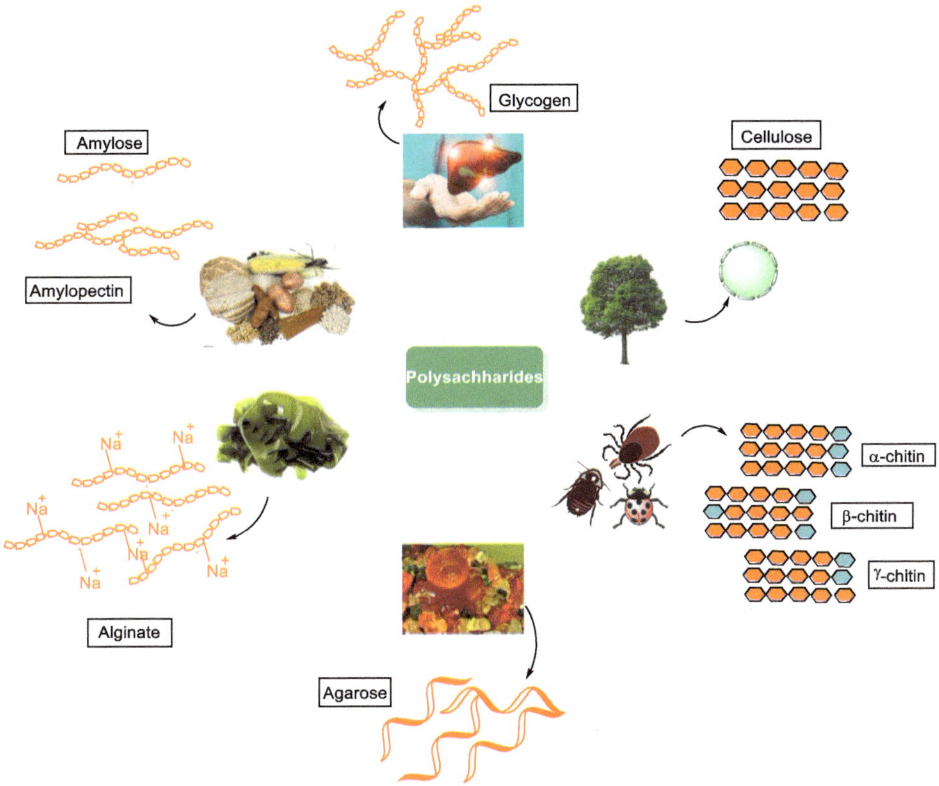

Figure 19.2: Different types of polysaccharides obtained from natural sources.

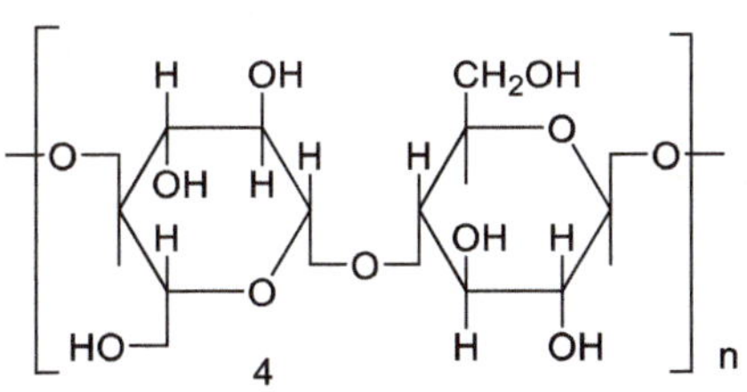

Figure 19.3: Cellulose polymer chain.

19.1.2 Hemicelluloses

Plant cell wall contains a polysaccharide molecule namely hemicelluloses (5) (Figure 19.4) that forms a matrix around cellulose microfibrils. Chemically, it consists of (1–3)-(1–4)-glucans, xylans, mannans, and xylogens [8–10]. H-bonding interaction and weak Van der Waals forces may help them to link with cellulosic microfibrils. In plant cells, the hemicelluloses act as signaling molecules and metabolic reserves [11]. They are considered as the second largest renewable source of lignocellulosic biomass with global annual production of approximately 60 billion tons [12, 13].

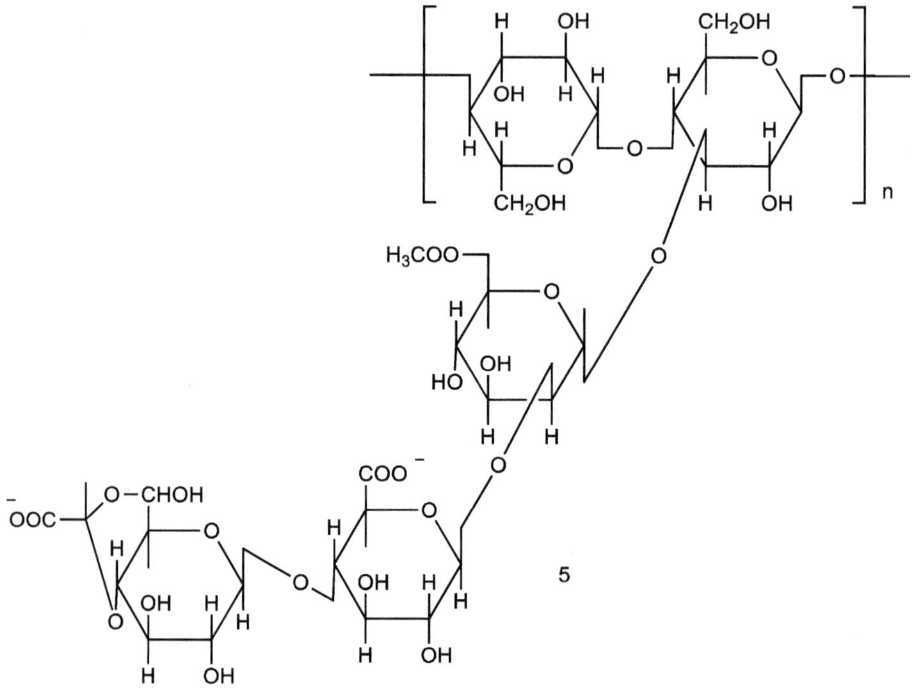

5

Figure 19.4: Hemicellulose polymer chain.

19.1.3 Pectin

In addition to cellulose, hemicellulose and lignin plant cell walls also contain pectin (**6**) (Figure 19.5). It is a gel-forming polysaccharide that forms gel with mucopolysaccharides and agar [14]. At industrial scale, pectin is known to be the most important hydrocolloids, and it finds its applications in various food products. Both high and low methoxy pectin on industrial scale can be obtained from plants such as sunflower is the major source of low-methoxyl (50%) pectin while apple and citrus are sources of high-methoxyl (>50%) pectin [15].

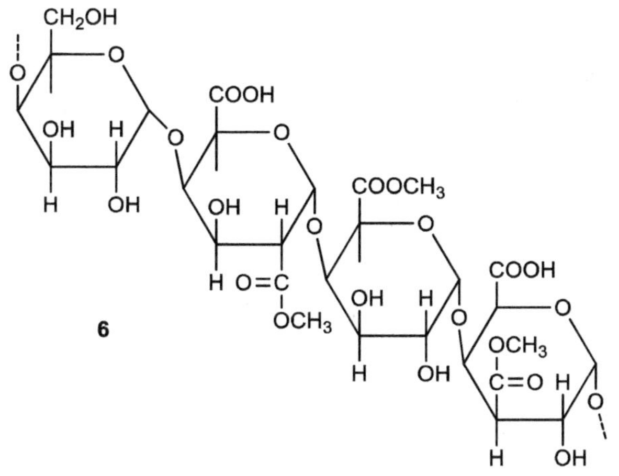

6

Figure 19.5: Subunit of pectin polymer.

19.1.4 Starch

Pure starch is a white odorless powder comprising of amylose (7) and amylopectin (**8**) chains (Figure 19.6). Amylose and amylopectin are two forms of starch. Amylose consists of linear chains of monosaccharide joined by α-1,4-glycosidic linkages, while amylopectin is found in the form of branching chains joined by α-1,4-glycosidic linkages and α-1,6-glycosidic bonds with 20–24 glucose subunits. Both amylose and amylopectin have different concentration in starch which depends on the source and nature of starch. Usually, amylose makes 20–25% and amylopectin make 75–80% of total starch [16]. Starch is considered to be the major energy reserves in plants and found in seeds, stem core, tubers, rhizomes, roots of plants, and fruits [17–19]. In a typical human diet, starch is the bulk source of energy; for example, it is found in potatoes, wheat, maize, and rice [20].

19.1.5 Glycogen

It is another form of energy storage carbohydrates. It is also known as "animal starch" because its functions are same as amylopectin. Glycogens have similar structural backbones as amylopectin, but its polymer chains are more compact and branched. Excess glucose in animal body stored in the form of glycogen in the cytoplasm of cells. Glycogen provides energy when required hence it functions in physiological metabolism [21].

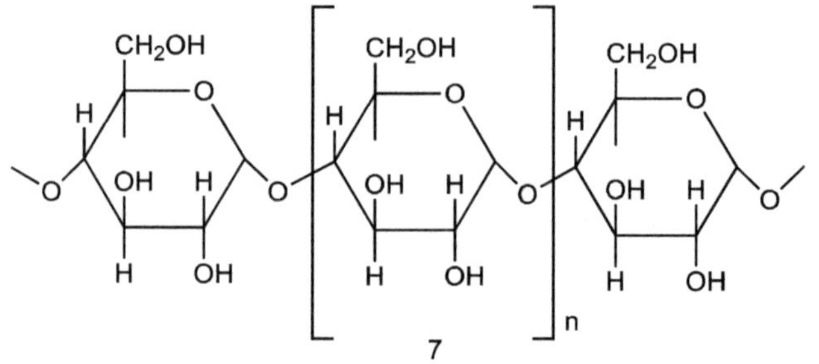

7

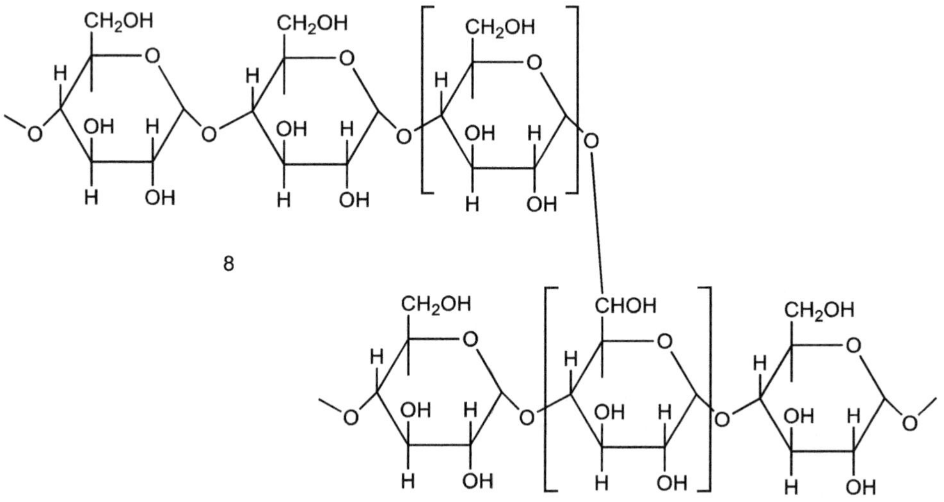

8

Figure 19.6: Structural representation of amylose and amylopectin chains.

19.1.6 Chitin

Chitin (**9**) (Figure 19.7) is another type of structural polysaccharide derived from exoskeletons of insects, crustaceans, and other invertebrates. It is highly hydrophobic polysaccharides with linear chains. Chemically it contains glucose derivatives of acetyl and amino groups. Due to its hydrophobicity, chitin shows insolubility in water and typical organic solvents, but it is soluble in dimethylacetamide with 5% lithium chloride and chloroalcohols in aqueous mineral acid solutions. It is also found in it can be found in the mycelia and spores of fungi [22, 23].

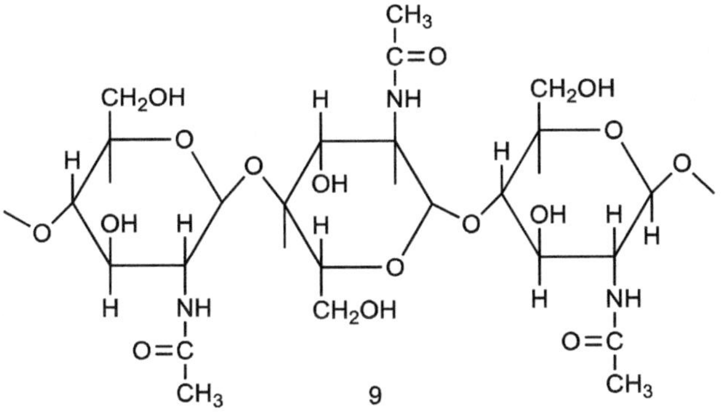

9

Figure 19.7: Chitin polymer subunit.

19.1.7 Hyaluronic acid

Hyaluronic acid **(10)** (Figure 19.8), also known as mucopolysaccharides or mucins, is a glycoprotein. It is a hybrid of protein and polysaccharides formed by the covalent bonds with proteins. Hyaluronic acid is one of the proteoglycans consisting of straight chains of polysaccharide [24]. It has a wide range of applications including medicinal, cosmetics, and food [25, 26].

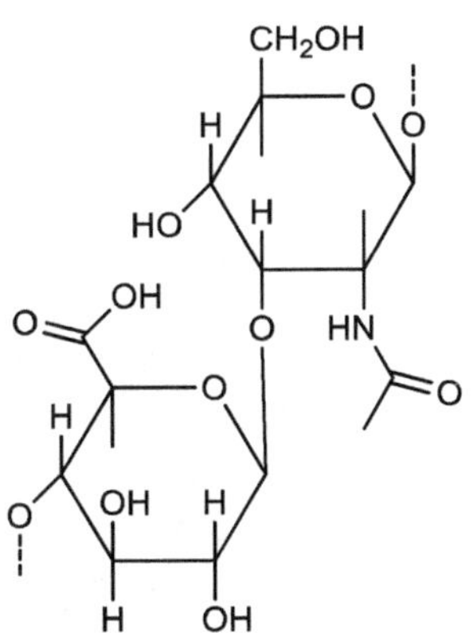

10

Figure 19.8: Subunit of hyaluronic acid.

19.1.8 Alginate

The main constituent of algae cell wall is alginate (**11**) (Figure 19.9) which provides strength and flexibility due to its hydrophilic properties [23]. Alginate is a naturally occurring and safe food component. Alginate has been widely employed in the food and nutraceutical industries because of its superior biological qualities, which include ion cross-linking, pH sensitivity, biocompatibility, and biodegradability, as compared to other seaweed polysaccharides. The only polymer that has carboxyl groups spontaneously occurring in every component residue is alginate [27].

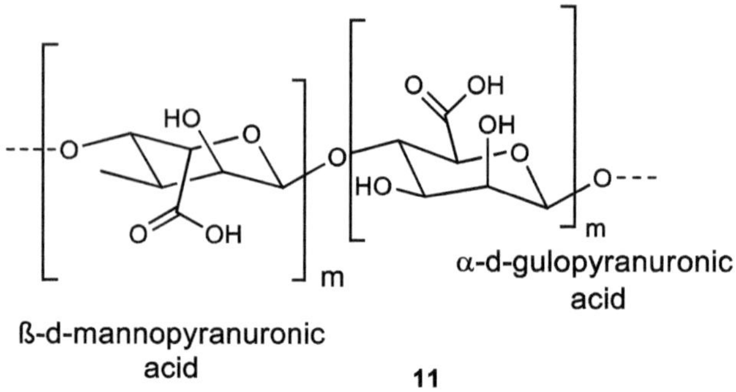

11

Figure 19.9: Structure of alginic acid.

A lot of focus has been placed on investigating the possible uses of carbohydrates as industrial raw materials in recent decades. So, here are the applications of carbohydrates in various industries (Figure 19.10).

19.2 Food industry

The food sector has largely employed carbohydrates due to their ability to enhance food quality by modifying texture and consistency, prevents crystallization, and is used as flavoring agents, coating agents and sweetening agents. Carbohydrate-based polymers are majorly used in food applications as they are safe because of their biodegradability, vast mechanical properties and natural sources. They functioned in human diet and are used as key energy providers in living beings hence the functionality of vital organs and physical activities in humans as well as all other living organisms depends upon the carbohydrates [28]. Because of their complex and unique structures, some carbohydrate molecules impart healthy benefits to their consumers [29].

19.2.1 Sweetening agent

Sweeteners are essential components of the everyday diet. One of the most widely used sweeteners among them is sucrose, a disaccharide that occurs naturally in large

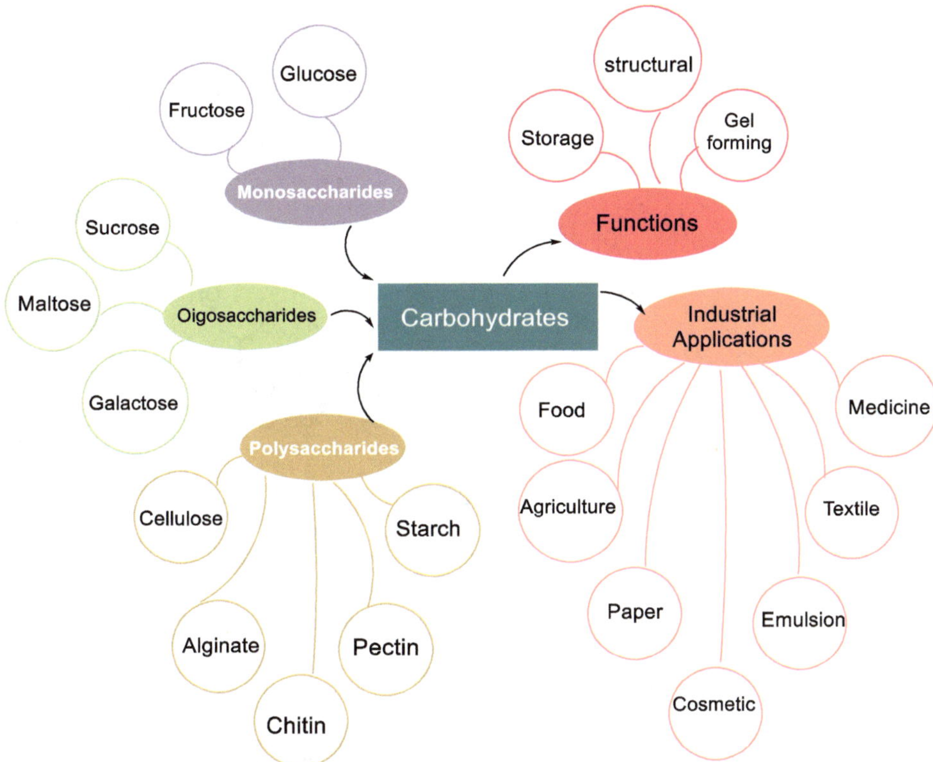

Figure 19.10: Overview of carbohydrate: types, function, and applications.

quantities. According to a study, sucrose (2-*O*-α-D-glucopyranosyl-D-fructose), commonly referred to as table sugar, is found in many plants, particularly sugar cane and sugar beet, and is therefore thought to be the most readily available low-molecular-weight carbohydrate in nature [30]. But consuming sugar can raise blood-glucose levels quickly, which can result in many health issues. The isomers of sucrose include trehalulose (1-*O*-α-D-glucopyranosyl-D-fructose) **(12)**, turanose (3-*O*-α-D-glucopyranosyl-D-fructose) **(13)**, maltulose (4-*O*-α-D-glucopyranosyl-D-fructose) **(14)**, leucrose (5-*O*-α-D-glucopyranosyl-D-fructose) **(15)**, and isomaltulose (6-*O*-α-D-glucopyranosyl-D-fructose) **(16)** (Figure 19.11). Trehalulose, turanose, leucrose, and isomaltulose are all promising substitute sweeteners that are gaining popularity because of their feasible production processes, appropriate sweetness, and possible physiological advantages.

These isomers have been used as sweeteners in many industries. For instance, in the Cargill company's Xtend™ sucromalt sweetener blend, leucrose was used as sweetening agent. The FDA (Food and Drug Administration) of the United States approved Xtend™ in 2009 as generally regarded as safe (GRAS) (no. 258). Xtend™ has been utilized as a sweetener in certain blood glucose-management products, like Glucerna® Shake. Leucrose also helps to manage obesity and related chronic diseases. Recent re-

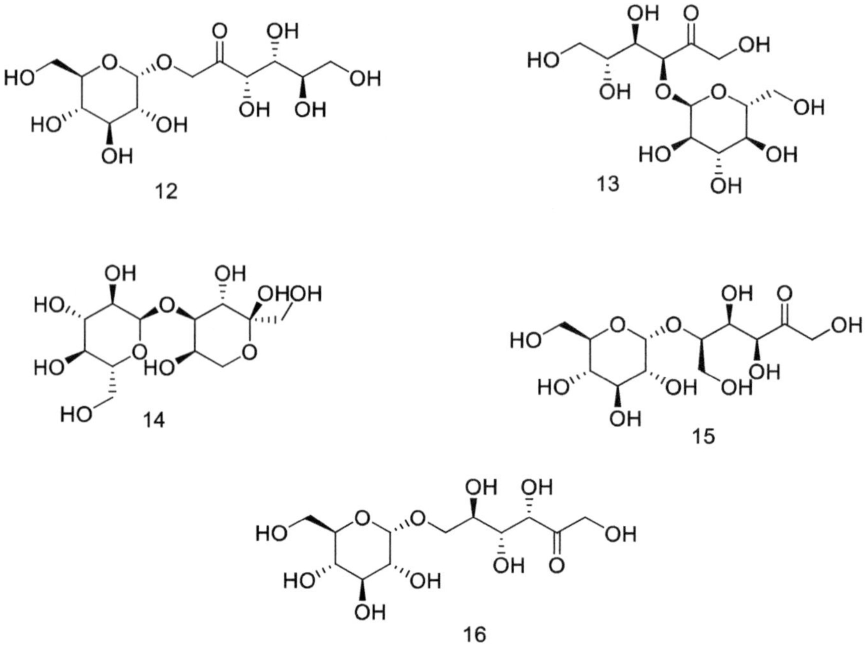

Figure 19.11: Isomers of sucrose used as sweetening agent.

search has shown that it may also be used as a new medicinal sweetener [31, 32]. Isomaltulose (content >98%) was declared GRAS (no. 184) by the FDA in 2006. FDA authorized dried isomaltulose syrup (content ≥80%) as GRAS (no. 681), in 2017 and stated that it could be used as a food and beverage sweetener at the same levels as sucrose. Food-grade isomaltulose is cost friendly, as compared to sucrose. Isomaltulose's application in yoghurt [33], jam [34], juice [35], and beer has been thoroughly investigated. It has also been utilized as a sweetener in certain foods made for dietary nutritional supplements and sports energy supplements, such as Granite™ carb supplement and Honey Badger® vegan hydration electrolyte powder [36].

19.2.2 Food quality enhancers

In food packaging, as well as in foods and bioproducts, carbohydrates have been used extensively as stabilizing agents, and fat substitutes. Turanose **(13)** can be added to rice flour to improve its viscoelasticity, stabilize rice dough [37]. Self-assembling of carbohydrate polymers (SCPs) have resulted in their useful characteristics when added to food [38]. They can interact with food products to create self-assembled structures by applying pressures. Thus, carbohydrate polymers (CPs) could bind and encapsulate a wide variety of biomaterials that are both hydrophilic and hydrophobic. The intrinsic functions, temperature stability, edibility, and other characteristics of carbohydrate-based polymers make them ideal candidates for use in food applications. The physical and chemical characteristics of CPs have a strong correlation with their ability to self-assemble [38].

In the food sector, starch is utilized as a gelling agent, emulsion stabilizer, thickener, and water binder [39]. Gum-bio-based compounds are primarily utilized as thickeners and emulsifiers in food and bioproduct formulations, as well as in pharmaceutical and cosmetic formulations, to regulate or enhance the products' texture, microstructure, viscosity, and stability. Gums are commonly categorized as food hydrocolloids when they are utilized in food [40, 41]. In order to extend the shelf life of foods, the layer-by-layer (L-b-L) technique to create food coatings from self-assembled chitosan (Cs) cellulose combinations was used [42]. Fruits and vegetables may ripen as a result of these materials' decreased oxygen permeability. A study emphasized that electrospun self-assembled cellulosic nanofibers can be used to mask off unpleasant flavors and odors in food as well as to thicken and texturize foods like sauces, dairy products, beverages, and meat [43]. For food applications, self-assembled cellulose gels were created that were sensitive to pH and ionic strength [44].

19.2.3 Dairy products

Many carbohydrates find their applications in dairy products to enhance their quality and shelf life. For instance, an investigation demonstrated that adding 3% oat β-glucan (**17**) (Figure 19.12) along with carrageen an improves the viscosity of milk, increase the sensory contents in chocolate milk and adds dietary fiber to the product, enhancing the drink quality [45]. In order to create a high-protein milk beverage, oat flour was used as a source of β-glucan and as a stabilizer. It was shown that adding oat flour in amounts of 2.5% and 1.9% (0.75 g of oat β-glucan per serving) resulted in strong suspension stability for the beverage (>80%) and the viscosity that is typical of liquid drinks (<50 mPa · s) when stabilized [46]. Viscosity of protein drinks increased significantly (from 51 to 100 mPa · s) when the amount of oat flour was increased, which has a detrimental effect on how they function. Scientists have frequently reminded producers that when developing recipes for novel beverages, especially milk drinks, that contain dietary fiber like β-glucan, a methodical approach to figuring out the appropriate dosage of the ingredient is necessary [47]. Manufacturers are quickly influenced by the popularity of dietary fiber due to its functional qualities and marketing promotion, which makes them incorporate these fibers into their products and present them as enriched or healthy [48]. On the other hand, if the amount of β-glucan in cereals is higher than normal, the product will have a sandy consistency, an oatmeal aftertaste, or dry mouth [49].

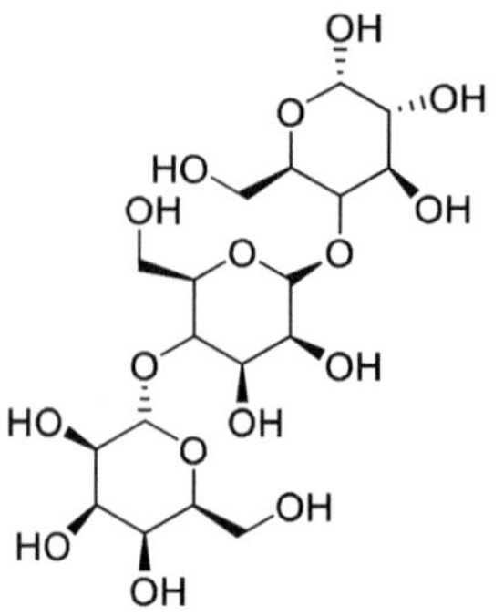

17

Figure 19.12: Structural representation of β-glucan.

19.2.4 Food packaging

The quality of food ingredients and products can deteriorate from farm to table due to external contamination or damage, primarily from oxidation and pathogenic microorganisms [50]. To preserve the food's original quality, appropriate precautions must be taken. Among these, food packaging can preserve quality, prevent spoiling, increase shelf life, and cut down on waste. One significant environmental problem over time has been the usage of synthetic polymers for food packaging has been seen. This issue is connected to both the depletion of fossil fuels and the pollution they produce. Additionally, these issues have been brought up by the current epidemic, as food delivery and the demand for packaged food consumption have been increased significantly [51, 52].

Several biopolymers, including polysaccharides, protein and lipid, can be utilized to create a packaging material [53]. To date, lipids (like beeswax and lauric acid) [54, 55], proteins (like whey, soy, and silk proteins) [56, 57], polysaccharides (like starch, chitosan, and cellulose) [58, 59], and so forth have been utilized to create fully biodegradable films. Due to its low cost, plentiful supply, edibility, and degradability, starch is one of the natural polymers that can be used for food packaging [60]. One new technology in the food packaging industry is multilayer films. The capacity of this kind of material to enhance performance using a single packaging system is the main reason of this type of packaging [55].

A study developed multilayer film for food packaging using chitosan and polylactic acid (PLA) [61]. The production of active bilayer films utilizing polyvinyl acetate and cassava starch as biopolymers is another example [62]. Furthermore, this kind of material is frequently developed using additional biopolymers, such as soy protein isolate, zein, gelatin, and potato starch [63–66]. Apart from the potential to enhance the mechanical and barrier qualities by employing various polymers, the multilayer system permits the regulated release of active substances that can be incorporated into the polymer matrix, hence enhancing the effectiveness of active packaging [67].

The antibacterial qualities of PLA-starch multilayer films were assessed leading to the demonstration that PLA monolayer films, and PLA-starch films that included ferulic and cinnamic acids have antibacterial property [68].

19.3 The pharmaceutical and medical industries

Carbohydrate molecules have extensively been utilized in the medicinal industry to treat various diseases because of their bioactive potential. Leucrose which is an isomer of sucrose has been shown in recent research to inhibit adipogenic genes in adipocytes and decrease lipid accumulation. It has also been shown to control hepatic lipogenesis and β-oxidation in obese mice [32]. D-Allose's ability to suppress cancer and tumors is thought to be its most significant characteristic [69]. Prostate [70, 71], hepatocellular [72, 73], cervical and skin [74], and ovarian [75] cancers are among the human tumors that D-allose has shown to be effective against.

D-Allose also inhibits the growth of leukemia [76], lung cancer [77], pancreatic [78], and head and neck cancer [79]. Additionally, when combined with D-allose, radiation dramatically increases the amount of reactive oxygen species (ROS) produced by cancer cells and has a five-fold influence on apoptosis [80]. D-Allose can also lessen the nephrotoxicity caused by *cis*-platin because of its anti-inflammatory properties [81]. It also has an inhibitory impact on ischemic lesions of the inner retina [82], strengthens the immune system [83], and prevents neutrophils from producing ROS. It can be utilized to prevent osteoporosis by preventing osteoclast development [84] and has a protective effect against liver ischemia reperfusion injury [85, 86]. Moreover, studies have shown that it has positive effects against obesity [69], hypertension, and stroke [87]. Rats used in preclinical studies have also shown D-allose to be safe as it showed no toxic effect [88]. With the exception of the trace amounts that stayed in the cecum and small intestine, D-allose was rapidly absorbed into the bloodstream and primarily eliminated through the urine [89]. The hydrophilic polymer chitosan (Cs) possesses antioxidant and antibacterial properties. Alkaline deacetylation of chitin, or more precisely, poly(*N*-acetyl glucosamine), is typically used to produce Cs, a cationic CP [90].

19.3.1 Drug targeting

Due to the higher stability, enhanced aqueous solubility and lower toxicity, mono and disaccharides are widely used as target molecules for prodrug synthesis. In addition to this, these sugar base prodrugs can release the drug molecule by the glycosidase activity in specific tissues. There is no other toxic compound released during hydrolysis of these molecules [91]. They can be used to make glycoconjugates which ultimately make a way to selectively localize a drug molecule to or near the specific target site. This approach makes extensive use of sugars in drug design, development, and delivery.

Mostly, hexoses such as D-glucose and D-galactose serve as the source of energy for brain and are transported through blood-brain barrier using GLUT1 and GLUT3. GLUT1 and GLUT3 are glucose carriers present on the endothelial surface of cell membrane [92–94]. Because of their ability to cross blood-brain-barrier these hexoses are considered as the vital conjugates for the delivery of important drugs related to brain such as dopamine [95–99], arginine [100, 101], and other medications [102, 103]. These drugs form conjugates with the hexose sugar and transport across central nervous system via specialized carrier-mediated transport. Moreover, these conjugates containing monosaccharides have shown to greater permeability, high bioavailability, and less side effects. Due to these properties these glycoconjugates have widely been used to improve the efficacy of the parent medicine [93].

19.3.2 Anti-inflammatory activity

Plant- and animal-derived dietary supplements have gained immense attention in the design and development of new pharmaceuticals due to their therapeutic effects [104, 105]. Nanoparticles of natural polysaccharide have frequently been investigated for their anti-inflammatory activity [106, 107]. They can potentially inhibit the activity of inflammatory enzymes and block the chemotactic and adhesion factor expression [108]. In addition, they have the ability to inhibit mediators such as cytokines (IL-1b, IL-6, and TNF-α), NO (nitric oxide), and inflammatory cell infiltration that are directly linked to cell inflammation. Sulfur containing polysaccharides obtained from algae show anti-inflammatory activity by blocking transport of leukocytes to site of inflammation [109]. Some plant-derived polysaccharides have the bioactive potential to act as nonsteroidal anti-inflammatory drug (NSAID) and they are used to treat bone and muscle disease such as osteoarthritis [110].

19.3.3 Hypoglycemic and hypocholesterolemic activities

Certain polysaccharide obtained from *Ganoderma atrum* species has the ability to treat type-2 diabetes as well as kidney diseases by treating insulin resistance of cell, hyperinsulinemia, hyperlipidemia, and hyperglycemia. The polysaccharides based nanocarrier can be used to enhance therapeutic potential and stability of loaded proteins. Polyelectrolyte of dextran-chitosan nanocarrier loaded with insulin given orally showed longer hypoglycemic potential and higher bioavailabilty [111]. *Bullacta exarate* derived chitosan, kefiran, and sulfated polysaccharides also exhibit hypocholestrolemic and hypoglycimaproperties [112–114]. Diabetes treatment by inhibiting key enzyme alpha amylase and alpha glucosidase is another effective method, because they play a key role in producing high blood glucose level. Hence inhibiting these enzymes can delay the breakdown of carbohydrates to glucose absorption in blood [115].

Inhibiting these enzymes has been shown to impair carbohydrate breakdown and delay glucose absorption [115]. Acarbose (**18**) (Figure 19.13) and plant extracts can lower blood glucose levels by inhibiting α-glucosidase [116]. Intake of certain food comprising carbohydrates can significantly regulate the synthesis, absorption, and secretion of plasma cholesterol level specifically LDL-c [117–119].

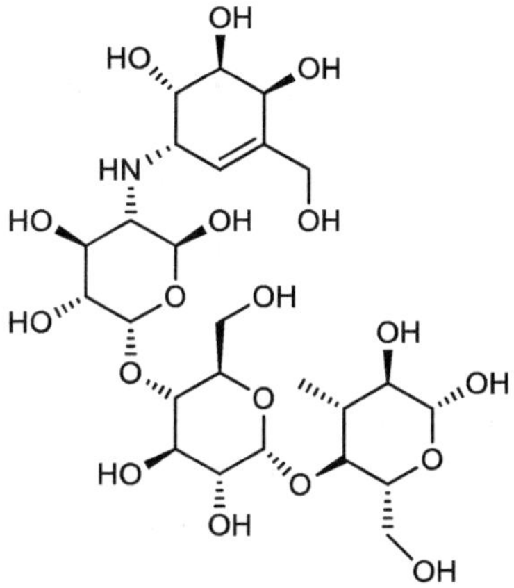

18

Figure 19.13: Structure of standard antidiabetic drug acarbose.

19.3.4 Anticoagulant activity

Sulfated polysaccharide such as low molecular weight and unfractionated heparins has been used as anticoagulants [120]. The anticoagulant activity of various polysaccharides has got significant attention. These biopolymers possess bioactive potential because of the high concentration of sulfate [121]. These polysaccharides also showed a variety of other bioactivities such as antitumor, antioxidant, and anticoagulant properties [112]. Other natural polysaccharides obtained from marine organism also have anticoagulant activity [122, 123].

19.3.5 Antiviral activity

Sulfated polysaccharides have extensively been used against several viruses such as herpes simplex virus (HSV), human immunodeficiency virus (HIV), dengue virus, respiratory syncytial virus, and human cytomegalovirus since the 1950s [108]. These antiviral polysaccharides inhibit the replication of enveloped viruses and obtained from many microalgae species [121]. Traditional Chinese medicines derived polysaccharides have largely been used as antiviral agents by stimulating and enhancing the immune system. They activate macrophagocytes to improve phagocytic activity and cause the release of antibodies [109]. The biochemical characteristics of polysaccharides obtained from microalgae include antibacterial, antiviral, and antioxidant activity [124].

19.4 Agriculture industry

The agriculture industry is currently facing challenges due to the pests and viruses attack on crops. Hence it has become difficult to meet the global demand of food for growing population [125]. Among plant pathogens, the most devastating species are fungi and oomycetes [126, 127]. Infections caused by fungal or oomycete constantly harm the crops and make food security hazardous. These infections grow through transit and trade or because of environmental change. In this way, the plant resistance for pathogenic fungi decreases which favor the resistance of infections to the synthetic fungicides [128]. The first organic fungicide was introduced in the 1940s resulting in plant protection that continuously expands over the next three years. Significant families of novel fungicides were developed during that time and applied to horticulture and cereal crops [129]. However, extensive usage of fungicide results in serious environmental impacts including both terrestrial as well as aquatic ecosystem targeting animals and harmful effects on human health [130, 131]. The hazardous impacts of these chemicals have resulted in the formulations of policies and laws regarding the usage of these agrochemicals and the registration of the active components of newly synthesized fungicides [132]. To mitigate the issue concerning the use of synthetic fungicides, carbohydrate-based biopolymers have drawn attention, because of their natural origin, non-toxicity, biodegradability, and easy availability. These properties make carbohydrates a suitable candidate to be used in agrochemicals. In addition, they can interact both as hydrophobic and hydrophilic chemicals in complex formulations [133]. These biopolymers can perform mainly three functions in protecting plants against pathogenic fungi: 1) they may change the fungal growth by inhibiting spore germination and mycelial growth; 2) they also act as a stimulator of plant immune system; and 3) they can also act as carriers for controlled-release agrochemicals or other active substances [133–136].

19.4.1 Carbohydrates as fungicide

Polysaccharides such as cellulose derivatives have extensively been used in protecting plants by acting as biocontrol agents, binders, and transporter of active substances. They also formed a coating around plant cells which is biodegradable and act as physical barrier to the microorganism or environment. Other carbohydrate biopolymers such as chitosan showed a wide range of bioactivities against viruses, bacteria, and fungi to protect plant from theses microorganisms. Specifically, it showed higher potential against fungi as compared to bacteria and viruses. However, it also showed antibacterial potential against gram-positive bacteria [137]. Chitosan derivatives have been widely used as agrochemical in crops before and after harvest to lower microbial growth [135]. An enhanced antifungal activity was seen in the compounds or composites especially metal containing compounds when chitosan was added [138]. For instance, chitosan was mixed into solution of silver nanoparticles and the nanocomposite formed was tested for its antifungal potential against *Alternaria alternata*, *Rhizoctonia solani*, and *Aspergillus flavus* which were isolated from chickpea seeds. An economic standpoint of using chitosan as fungicide is that it distorted the germination and hyphal shape of fungi (e.g., *Botrytis cinerea* and *Rhizopus stolonifer*) that harm the harvested crops [139]. This biopolymer also inhibits the growth of several other pathogens of plants and mycoparasitic fungi. Chitosan has the ability to penetrate the plasma membrane of certain sensitive fungi in an energy-dependent manner [140]. It has been shown to act as fungicide against broad spectrum of plant fungi. In addition, it can limit the growth of pathogenic fungal species at various stages of their life cycle [141].

19.4.2 Carbohydrates found in agrochemicals

Nature's most common carbohydrates include cellulose, chitosan, and alginate. Furthermore, chitosan and cellulose can be derived from agricultural waste, wood processing waste, or food-industry crustaceans. Furthermore, because these are natural polymers, many microorganisms in the environment can degrade them, making them extremely biodegradable and appropriate for organic agriculture.

Furthermore, these biopolymers are safer to use by consumers environment due to their low toxicity and biodegradability. Carbohydrate-derived polymers can be employed in the encapsulation procedure to shield microorganisms which act as biocontrol agents in biological labs to protect from adverse environmental stress and extend their vitality [142, 143]. In the same way, they can be employed to distribute chemicals and active volatile components used in agricultural processes. This approach leads the biopolymer to enhance the efficiency of encapsulated compounds and delay their release into crops ultimately enhancing the efficiency of plant protection and minimizing the agrochemical impact on environment [136, 144, 145]. In addition, gel-forming polysaccharide gives an advantage to soil by forming hydrogel which increases water retention in the soil that may help the plants to face the drought stress [133].

The enzymatic degradation or hydrolysis products (mono or oligosachharides) of carbohydrate base biopolymer may be used in stimulating plant defense mechanism accounting as another application of these molecule [136]. The film-forming properties of cellulose, chitosan, and alginate make them a suitable candidate for seed and harvested fruits coating which ultimately inhibit fungal growth even if they act as passive inhibitors [146].

19.4.3 Carbohydrates in plant immunity

Plants produce carbohydrates through photosynthesis, and they play a vital role as energy reserves, storage system of organic compounds in plants. In addition to this, there are other essential functions which become apparent is to act as signaling molecule in the same way as hormones [147, 148] and is nowadays largely investigated. In this way carbohydrates-based sugars can interact with hormone signaling, biotic and abiotic stresses and diurnal changes, because they act as complex communication system that is essential for the metabolism coordination, environmental responses, stress, and growth development [149]. Both disaccharides (sucrose and trehalose) as well as oligosaccharides (raffinose and fructans) act as antioxidants against ROS that is produced in plant in response to abiotic stresses. In addition, the antioxidant compounds in plants are both enzyme inhibitors including dismutase, ascorbate, superoxide, glutathione peroxidase, and superoxide, as well as nonenzymatic inhibitors that include glutathione, ascorbate, and α-tocopherol. Sugar acts as a key factor in integrated cellular redox network. The sugars also provide fuel energy that is required for the defense against microbial attack and to regulate the defense genes [150, 151]. Sugars are known as “sweet immunity” and “sugar enhanced defense” because of their vital role in plant immunity [152].

Rare sugars act as both growth regulators in plants and herbicide which may help to address the importance of synthetic plant growth regulators [153]. For instance, a rare monosaccharide sugar 7-deoxy-sedoheptulose **(19)** (Figure 19.14) was shown to be act as potential herbicide against Arabidopsis [153]. The sugar also shows activity at a very small concentration (260 µM) and has no harmful effects on mammalian cells. The sugar also shows bioactive potential by inhibiting shikimate pathway’s 3-dehydroquinate synthase. Unlike the synthetic inhibitors of 3-dehydroquinate synthase which include D-gluco-heptulosonate 7-phosphonate, 3-deoxy-D-arabino-heptulosonate 7-phosphonate, carbaphosphonate, and its cyclohexenyl derivatives) the sugar shows high-potential herbicidal properties in vivo [153].

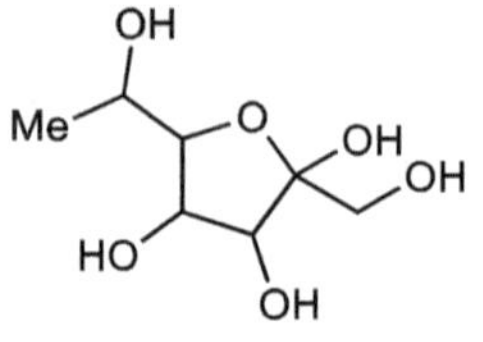

19

Figure 19.14: Structure of 7-deoxy-sedoheptulose.

19.5 Textile industry

Fibers serve as the foundation for all textile goods that meet society's basic necessities. Short-length fibers, often known as staple fibers, range in length from 3 to 20 cm, as do most natural fibers. A filament is an indefinitely long fiber, of which silk is the only naturally occurring example. Mostly synthetic and recycled fibers are produced as filaments [154]. Natural fibers, including animal (protein) and vegetable fibers, account for up to 40% of global textile fiber production each year [155]. Cotton crop being the only major source of cellulose-based textiles which have been used for thousands of years because of it excellent properties [156]. As a biodegradable, renewable, biocompatible, and inexpensive polymer, it has a variety of applications, including textiles, diapers, foams, coatings, and composites [157]. Although the cotton crop has a wide range of applications in the textile industry, but the production of this crop needs a large irrigation system and a considerable amount of expensive inorganic fertilizers [158]. In this regard, the researchers are focusing on two directions; 1) To identify the changes which can lead to beneficial production strategies of cotton [159], and 2) the alternative way of obtaining cellulose from other natural sources [160]. Cellulose derivatives account for over 90% of all natural fibers, and is the most commonly utilized in the production of clothing, home furnishings, and industrial products [161]. Cellulosic fiber utilization in textile industry is anticipated to increase from 3.7 kg per capita today to 5.4 kg by 2030 [162]. Cotton fiber contains about 90% cellulose, dried hemp has 40–50%, and wood has 40–55%, which is usually mixed with other components such as hemicelluloses and lignen [163, 164]. Along with sisal, flax, cotton, ramie, kenaf, and jute are commonly utilized. Cotton hybrid cultivation will expand, resulting in a harvest output increase from 800 kg/ha (2010) to 925 kg/ha (2030).

19.5.1 Nanocelluloses

In recent years, many nanocelluloses (NCs) derived from plant and microbial sources have been studied as a source of textile fibers. The interest in NCs is primarily centered on taking advantage of their increased crystallinity, because they promote considerable mechanical resistance. NC is a family of materials having significantly unique properties, owing to various sources and synthesis processes [165].

19.5.2 Plant nanocelluloses

The cellulosic chain hydroxyl group binds with the other, forming rigid and stable molecules that provide the plant stiffness and strength. The H-bonding among cellulose strands renders it impermeable in water. Fibrils are created by connecting cellulose molecules together that aggregate into bundles, which the plant uses to construct the cell wall alongside hemicelluloses and lignin. The cellulose fibril morphology shows a structure that looks like a mesh [166].

19.5.3 Bacterial nanocellulose

Bacterial nanocellulose (BNC) is a homopolysaccharide extracted from gram-negative species of the genera *Komagataeibacter, Acetobacter, Rhizobium, Agrobacterium, Pseudomonas, Salmonella, Alcaligenes*, and the only gram-positive bacterial genus, Sarcina [167]. Different bacteria generate cellulose with unique morphology, structure, characteristics, and yields [164]. To obtain high BNC yields, the highest cellulose producing species, such as Komagataeibacterxylinus, must be used [167]. In addition to using high BNC producers, additional ways to boost BNC production include the use of advanced reactors, complicated growth mediums, and even the development of cell-free enzyme systems [168].

The naturally occurring biopolymer obtained from microbes has the potential to promote the creation of novel textile and sustainable materials, as well as manufacturing processes in the future. Suzanne Lee, a British fashion designer, pioneered the usage of bacterial cellulose by launching BioCouture, a sustainable fashion research initiative. Experimenting with kombucha, she made a bacterial cellulose textile directly in a rectangular jar, as well as a microbial cellulose jacket and gloves using traditional garment manufacturing processes such as cutting and sewing clothing panels [169, 170]. An effort named "Malai" creates vegan alternatives to leather using bacterial cellulose derived from a culture medium including coconut water and industrial leftovers, which are then utilized to create fashion accessories in a variety of colors and tones. Manufacturers believe that the material is resilient, biodegradable, and can have a long life, as long as care is taken to preserve it.

19.6 Cosmetic industry

Consumers are increasingly interested in cosmetics incorporating natural components due to their desire for high efficacy and safety. Cosmetics from natural components are compounds extracted and separated from plants, animals, and microbes found in nature using chemical or physical methods, as opposed to synthetic active equivalents, which are efficient and sustainable [171]. In comparison to many synthetic products, natural cosmetics are becoming known for their low toxicity, ease to use, and high activity levels [172]. In vitro investigations have shown that seaweed oligosaccharides could be used as cosmetics for skin whitening [173–175]. They have been shown to prominently reduce melanin synthesis [174]. Furthermore, these oligosaccharides have been evaluated in vitro, indicating their potential as skin moisturizers [173, 174].

Many investigations have shown that agar oligosaccharides and carrageenan oligosaccharides (COS) as “generally recognized as safe (GRAS)” [176]. COS absorbs better and is less poisonous than carrageen a making it efficient and as safe (GRAS) [177]. COS had no effect on the growth of normal endothelia ECV304 cells and caused no direct harm to ECV304 cells [178]. As the sales price for cosmetics originated from natural sources grows year after year, the expansion of its activity is critical. It is vital to investigate the mechanism of action of polysaccharide-based cosmetics that ultimately increase product utilization rates and meet market demands [179].

With the expansion of polysaccharide applications, the cosmetic actions of polysaccharides include the use as moisturizing, brightening, and anti-aging [180]. Complex long chains of polysaccharides are frequently used to carry out a variety of functions. More polar groups, such as hydroxyl and carboxylic groups can capture and bind water molecules through H-bonding, forming network structures to exhibit moisturizing activity [181]. Polysaccharides block tyrosinase, scavenge free radicals, and have been shown to promote skin whitening and generate antioxidants [182, 183]. Skin creams produced with polysaccharides had considerable moisturizing and anti-aging properties in volunteer skin testing, confirming polysaccharides’ efficacy in cosmetic application [184]. Polysaccharides also have a mending impact on injured skin, comparable to synthetic cosmeceuticals, and could be used as an active component in restorative cosmetics [185]. It is clear that developing numerous polysaccharide activities can be a useful method of producing more effective and secure cosmetics.

19.7 Emulsion

Emulsions are often thermodynamically unstable colloidal systems [186]. They are made up of small oil droplets that are suspended in water. An emulsion is a colloidal material in which the dispersion media and its dispersed state are both liquids. Emulsifiers form emulsions by adhering onto droplet faces and reducing interfacial tension. Once created, it becomes a simple disintegration of the droplet. Simple emulsions can exist in either oil-in-water (o/w) or water-in-oil (w/o) systems [187]. Emulsification capacity of polysaccharides enhanced when they are covalently bound to glycoproteins or glycolipids. However, polysaccharides are amphiphilic in nature, so they can stabilize emulsions [188] by adsorbing onto the outermost layer of oil droplets [189, 190]. Sorbitan ester was the first emulsifying agent created throughout the process of producing polysaccharide-based biosurfactants. The intriguing fact about sorbitan esters is that around 20,000 tons of these surfactants are manufactured and put on the market each year.

Sorbitan ester is primarily used as an emulsifier in food manufacturing, beauty products, polymerization, and pharmaceutical industries. Unlike sorbitan esters, sucrose esters are relatively mild and only used in dermatological goods. Many countries have approved it as a dietary processing emulsifier. Sucrose-fatty acid esters are generally accepted as non-ionic surfactants that are commonly used as emulsifiers in the food and beverage industry. Sucrose is an extremely temperature-sensitive chemical; therefore, esterification requires an optimal process and precise control. For these reasons, sucrose emulsifier manufacturing is far more complicated than sorbitan esters [187, 191].

19.8 Paper industry

Paper is a thin sheet of cellulose made by mechanical or chemical processing of cellulosic fibers. The cellulose fibers for paper making can be obtained from wood, grasses, herbivore dung, or other vegetable sources. The cellulose fiber dissolved in water and drained through a fine mesh to obtain a finely distributed sheet which is followed by pressing and drying. To enhance the whiteness, durability, printing capacity, transparency, brightness, and writing capability of different paper goods, mineral-based fillers and pigments are used [192, 193]. These substances may be applied as additives in the paper making processes for example, super-calendered (SC) paper or pigments for the manufacturing of coated papers [194, 195]. There is a focus on designing biodegradable biopolymer-based organic fillers and pigments from sustainable natural resources. In this regard, the polysaccharide base organic fillers and pigments have gained significant attention for the paper industry globally [196, 197]. For instance, starch acetate, which is water insoluble starch ester have been synthesized as pigment or filler in paper making. The acetylation of starch enhances its temperature durability as compared to starch when the product formation required an elevated temperature. Sone and Matsunaga in 1992 patented the use of cellulosic fibers (or beads) as a pigment in coatings. The beads containing cellulose can be developed with viscose as the initial ingredient, and the procedure followed by three steps: (1) two-particle spout spray, (2) treatment with acid to reproduce cellulose, and (3) drying with hot air to create particles. The aforesaid procedure may create cellulose particles with grain sizes ranging from 0.1 to 1,000 m, with fine pellets with grain sizes ranging from 1 to 20 m being preferred as a covering pigment, according to the inventors. Furthermore, some natural disaccharide derivatives may be employed as promising agents for bio-based pigments for manufacturing paper. For instance, lactose octaacetate which is a lactose derivative has been utilized as fillers or pigment in paper industry [198].

19.9 Conclusion

Carbohydrates are abundantly found in nature and have extensive applications in various industries. In plants and animals, carbohydrates play different roles depending upon their chemical structure and nature of origin. Naturally, they can be found in the form of mono, di, and oligo polysaccharide form. They can be used in the food industry both for eating purposes and other applications such as packaging, stabilizing, sweetening, and flavoring agents. They show high bioactive potential against many diseases therefore they have extensively been used in medicinal industry for the design and development of newer drugs. They also have a role in plant immunity and signaling pathway and hence find applications in agriculture industry. Carbohydrate-based molecules are also used in many other industries such as cosmetic, textile, paper, and emulsion industry. These are promising agents in the synthesis of novel organic compounds which have a wide range of applications

References

[1] Honnavally PR, Rudrapatnam NT. Carbohydrates the renewable raw materials of high biotechnological value. Crit Rev Biotechnol, 2003, 23(2), 149–173.

[2] Hinde S. Understanding the role of carbohydrates in optimal nutrition. Nurs Stand, 2019, 34, 76–82.

[3] Monice MF, Wilma FB, Donald VB, Ronald AH, Curtis DK, Daniel CL, James G, Marks J, Ronald CS, Thomas JS, Paul WS, Lillian JG, Bart H. Safety assessment of monosaccharides, disaccharides, and related ingredients as used in cosmetics. Int J Toxicol, 2019, 38, 55–385.

[4] Li Q, Wang S, Jin X, Huang C, Xiang Z. The application of polysaccharides and their derivatives in pigment, barrier, and functional paper coatings. Polymers, 2020, 12(8), 1837.

[5] Gopinath V, Saravanan S, Al-Maleki AR, Ramesh M, Vadivelu J. A review of natural polysaccharides for drug delivery applications: Special focus on cellulose, starch and glycogen. Biomed Pharmacol, 2018, 107, 96–108.

[6] Liu R, Yu H, Huang Y. Structure and morphology of cellulose in wheat straw. Cellulose, 2005, 12, 25–34.

[7] Kulasinski K, Keten S, Churakov SV, Derome D, Carmeliet J. A comparative molecular dynamics study of crystalline, paracrystalline and amorphous states of cellulose. Cellulose, 2014, 21, 1103–1106.

[8] Chang C, Duan B, Cai J, Zhang L. Superabsorbent hydrogels based on cellulose for smart swelling and controllable delivery. Eur Polym J, 2010, 46, 92–100.

[9] Zielke C, Stradner A, Nilsson L. Characterization of cereal β-glucan extracts: Conformation and structural aspects. Food Hydrocoll, 2018, 79, 218–227.

[10] Hu X, Zhao J, Zhao Q, Zheng J. Structure and characteristic of β-glucan in cereal: A review. J Food Process Preserv, 2015, 39, 3145–3153.

[11] Kuki H, Yokoyama R, Kuroha T, Nishitani K. Xyloglucan is not essential for the formation and integrity of the cellulose network in the primary cell wall regenerated from *Arabidopsis* protoplasts. Plants, 2020, 9, 629.

[12] Li Z, Pan X. Strategies to modify physicochemical properties of hemicelluloses from biorefinery and paper industry for packaging material. Rev Environ Sci Bio, 2018, 17, 47–69.

[13] Rao J, Lv Z, Chen G, Peng F. Hemicellulose: Structure, chemical modification, and application. Prog Polym Sci, 2023, 140, 101675.

[14] Chandel V, Biswas D, Roy S, Vaidya D, Verma A, Gupta A. Current advancements in pectin: Extraction, properties and multifunctional applications. Foods, 2022, 11, 2683.
[15] Basak S, Annapure US. Trends in "Green" and novel methods of pectin modification. A review. Carbohydr Polym, 2022, 278, 118967.
[16] Kajiwara K, Miyamoto T. ChemInform abstract: Progress in structural characterization of functional polysaccharides. Chem In, 2010, 30, 1–55.
[17] Zhang J, Zhai A. Microstructure, thermodynamics and rheological properties of different types of red adzuki bean starch. Qual Assur Saf Crops Foods, 2020, 12, 89–99.
[18] Farooq U, Dimattia C, Faieta M, Flamminii F, Pittia P. Colloidal properties and stability of olive oil-in water emulsions stabilized by starch particles. Ital J Food Sci, 2021, 33, 4.
[19] Bojarczuk A, Skąpska S, Khaneghah AM, Marszałek K. Health benefits of resistant starch: A review of the literature. Joffunc Foods, 2022, 1(93), 105094.
[20] Tan L, Kong L. Starch-guest inclusion complexes: Formation, structure, and enzymatic digestion. Crit Rev Food Sci Nutr, 2020, 8(60), 5, 780–790.
[21] Zhu Y, Delbianco M, Seeberger PH. Automated assembly of starch and glycogen polysaccharides. J Am Chem Soc, 2021, 11(143), 26, 9758–9768.
[22] Jolly G, Kuscu MC, Kokate P, Younis M. A low-energy key management protocol for wireless sensor networks. In Proceedings of the Eighth IEEE Symposium on Computers and Communications vol. 3. ISCC, 2003, pp. 335–340.
[23] Kulkarni V, Butte K, Rathod S. Natural polymers a comprehensive review. Int J Res Pharm Biomed Sci, 2012, 3, 1597–1613.
[24] Guizzardi S, Uggeri J, Belletti S, Cattarini G. Hyaluronate increases polynucleotides effect on human cultured fibroblasts. J Cosmet Dermatol Sci Appl, 2013, 1(3), 124.
[25] Salwowska NM, Bebenek KA, Żądło DA, Wcisło-Dziadecka DL. Physiochemical properties and application of hyaluronic acid: A systematic review. J Cosmet Dermatol, 2016, 15(4), 520–526.
[26] Oerther S, Le Gall H, Payan E, Lapicque F, Presle N, Hubert P, Dexheimer J, Netter P, Lapicque F. Hyaluronate-alginate gel as a novel biomaterial: Mechanical properties and formation mechanism. Biotechnol Bioeng, 1999, 63, 206–215.
[27] Bi D, Yang X, Yao L, Hu Z, Li H, Xu X, Lu J. Potential food and nutraceutical applications of alginate: A review. Mar Drugs, 2022, 4(20), 9, 564.
[28] Fathi M, Martin A, Mcclements DJ. Nanoencapsulation of food ingredients using carbohydrate based delivery systems. Trends Food Sci Technol, 2014, 1(39), 1, 18–39.
[29] Gutiérrez TJ. Chitosan applications for the food industry. Chitosan: Deriv Compos Appl, 2017, 18, 183–232.
[30] Peters S, Rose T, Moser M. Sucrose: A prospering and sustainable organic raw material. Carbohydr Sustain Dev, 2010, 1–23.
[31] Lee D, Lee J, Hong MG, Lee BH, Kim YM, Chang PS, Kim Y, Yoo SH. Optimization of leucrose production by dextransucrase from *Streptococcus mutans* and its application as an adipogenesis regulator. J Funct Foods, 2017, 1(39), 238–244.
[32] Lee J, Kim E, Kim Y, Yoo SH. Leucrose, a sucrose isomer, suppresses hepatic fat accumulation by regulating hepatic lipogenesis and fat oxidation in high-fat diet-induced obese mice. J Cancer Prev, 2018, 23(2), 99.
[33] Guggisberg D, Piccinali P, Schreier K. Effects of sugar substitution with Stevia, Actilight™ and Stevia combinations or Palatinose™ on rheological and sensory characteristics of low-fat and whole milk set yoghurt. Int dairy J, 201(1), 21, 9, 636–644.
[34] Kim HW, Han SH, Lee SW, Suh HJ. Effect of isomaltulose used for osmotic extraction of Prunus mume fruit juice substituting sucrose. Food Sci Biotechnol, 2018, 27, 1599–1605.
[35] Peinado I, Rosa E, Heredia A, Escriche I, Andrés A. Influence of processing on the volatile profile of strawberry spreads made with isomaltulose. Food Chem, 2013, 138(1), 621–629.

[36] Tian Y, Deng Y, Zhang W, Mu W. Sucrose isomers as alternative sweeteners: Properties, production, and applications. Appl Microbiol Biotechnol, 2019, 103, 8677–8868.

[37] Park Y, Park SW, Ryu K, Lee S. Elucidation of rheological, microstructural, water mobility, and noodle-making properties of rice flour affected by turanose. Food Chem, 2019, 15(276), 9–14.

[38] Roy A, Shrivastava SL, Mandal SM. Self-assembled carbohydrate nanostructures: Synthesis strategies to functional application in food. In novel approaches of nanotechnology in food. Acad Press, 2016, 1, 133–164.

[39] Kaisangsri N, Kerdchoechuen O, Laohakunjit N. Biodegradable foam tray from cassava starch blended with natural fiber and chitosan. Ind Crop Prod, 2012, 37(1), 542–546.

[40] Gularte MA, Rosell CM. Physicochemical properties and enzymatic hydrolysis of different starches in the presence of hydrocolloids. Carbohydr Polym, 2011, 85(1), 237–244.

[41] Samutsri W, Suphantharika M. Effect of salts on pasting, thermal, and rheological properties of rice starch in the presence of non-ionic and ionic hydrocolloids. Carbohydr Polym, 2012, 87(2), 1559–1568.

[42] Li F, Biagioni P, Finazzi M, Tavazzi S, Piergiovanni L. Tunable green oxygen barrier through layer-by-layer self-assembly of chitosan and cellulose nanocrystals. Carbohydr Polym, 2013, 92(2), 2128–2134.

[43] Rezaei A, Nasirpour A, Fathi M. Application of cellulosic nanofibers in food science using electrospinning and its potential risk. Compr Rev Food Sci Food Saf, 2015, 14(3), 269–284.

[44] Roversi T, Piazza L. Supramolecular assemblies from plant cell polysaccharides: Self-healing and aging behavior. Food Hydrocol, 2016, 54, 189–195.

[45] Chatterjee BA, Patel T. Increased sensory quality and consumer acceptability by fortification of chocolate flavored milk with oat beta glucan. Int J Clin Biomed Res, 2016, 2, 25–28.

[46] Vasquez-Orejarena E, Simons CT, Litchfield JH, Alvarez VB. Functional properties of a high protein beverage stabilized with oat-β-glucan. J Food Sci, 2018, 83(5), 1360–1365.

[47] Lumaga RB, Azzali D, Fogliano V, Scalfi L, Vitaglione P. Sugar and dietary fibre composition influence, by different hormonal response, the satiating capacity of a fruit-based and a β-glucan-enriched beverage. Food Funct, 2012, 3(1), 67–75.

[48] Chakraborty P, Witt T, Harris D, Ashton J, Stokes JR, Smyth HE. Texture and mouth feel perceptions of a model beverage system containing soluble and insoluble oat bran fibres. Food Res I, 2019, 120, 62–72.

[49] Lyly M, Ohls N, Lähteenmäki L, Salmenkallio-Marttila M, Liukkonen KH, Karhunen L, Poutanen K. The effect of fibre amount, energy level and viscosity of beverages containing oat fibre supplement on perceived satiety. Food Nutr Res, 2010, 54(1), 2149.

[50] Ncube LK, Ude AU, Ogunmuyiwa EN, Zulkifli R, Beas IN. An overview of plastic waste generation and management in food packaging industries. Recy, 2021, 6, 12.

[51] Ahmad AA, Sarbon NM. A comparative study: Physical, mechanical and antibacterial properties of bio-composite gelatin films as influenced by chitosan and zinc oxide nanoparticles incorporation. Food Biosci, 2021, 43, 101250.

[52] Alias AR, Wan MK, Sarbon NM. Emerging materials and technologies of multi-layer film for food packaging application: A review. Food Control, 2022, 136, 108875.

[53] Andrade J, González-Martínez C, Chiralt A. Antimicrobial PLA-PVA multilayer films containing phenolic compounds. Food Chem, 2022, 375, 131861.

[54] Ochoa TA, Almendárez BE, Reyes AA, Pastrana DM, López GF, Belloso OM, González CR. Design and characterization of corn starch edible films including beeswax and natural antimicrobials. Food Bioproc Tech, 2017, 10, 103–114.

[55] Amin U, Khan MU, Majeed Y, Rebezov M, Khayrullin M, Bobkova E, Shariati MA, Chung IM, Thiruvengadam M. Potentials of polysaccharides, lipids and proteins in biodegradable food packaging applications. Int J Biol Macromol, 2021, 183, 2184–2198.

[56] Calva-Estrada SJ, Jiménez-Fernández M, Lugo-Cervantes E. Protein-based films: Advances in the development of biomaterials applicable to food packaging. Food Eng Rev, 2019, 11, 78–92.
[57] Xu YP, Wang Y, Zhang T, Mu GQ, Jiang SJ, Zhu XM, Tuo YF, Qian F. Evaluation of the properties of whey protein films with modifications. J Food Sci, 2021, 86(3), 923–931.
[58] Zhu F. Polysaccharide based films and coatings for food packaging: Effect of added polyphenols. Food Chem, 2021, 359, 129871.
[59] Kocira A, Kozłowicz K, Panasiewicz K, Staniak M, Szpunar-Krok E, Hortyńska P. Polysaccharides as edible films and coatings: Characteristics and influence on fruit and vegetable quality-A review. Agronomy, 2021, 11(5), 813.
[60] Agarwal S. Major factors affecting the characteristics of starch based biopolymer films. Eur Polym J, 2021, 160, 110788.
[61] Cheikh D, Majdoub H, Darder M. An overview of clay-polymer nanocomposites containing bioactive compounds for food packaging applications. Appl Clay Sci, 2022, 216, 106335.
[62] Azmir J, Zaidul IS, Rahman MM, Sharif KM, Mohamed A, Sahena F, Jahurul MH, Ghafoor K, Norulaini NA, Omar AK. Techniques for extraction of bioactive compounds from plant materials: A review. J Food Eng, 2013, 117(4), 426–436.
[63] Carocho M, Barreiro MF, Morales P, Ferreira IC. Adding molecules to food, pros and cons: A review on synthetic and natural food additives. Compr Rev Food Sci Food Saf, 2014, 13(4), 377–399.
[64] Chen C, Li C, Yang S, Zhang Q, Yang F, Tang Z, Xie J. Development of new multilayer active packaging films with controlled release property based on polypropylene/poly (vinyl alcohol)/ polypropylene incorporated with tea polyphenols. J Food Sci, 2019, 84(7), 1836–1843.
[65] Chi W, Cao L, Sun G, Meng F, Zhang C, Li J, Wang L. Developing a highly pH-sensitive κ-carrageenan-based intelligent film incorporating grape skin powder via a cleaner process. J Clean Prod, 2020, 20(244), 118862.
[66] da Rosa GS, Vanga SK, Gariepy Y, Raghavan V. Development of biodegradable films with improved antioxidant properties based on the addition of carrageenan containing olive leaf extract for food packaging applications. J Polym Envir, 2020, 28, 123–130.
[67] De Villiers MM, Otto DP, Strydom SJ, Lvov YM. Introduction to nanocoatings produced by layer-by-layer (LbL) self-assembly. Adv Drug Deliv Rev, 2011, 63(9), 701–715.
[68] Avila LB, Schnorr C, Silva LF, Morais MM, Moraes CC, Da Rosa GS, Dotto GL, Lima ÉC, Naushad M. Trends in bioactive multilayer films: Perspectives in the use of polysaccharides, proteins, and carbohydrates with natural additives for application in food packaging. Foods, 2023, 12, 1692.
[69] Chen Z, Chen J, Zhang W, Zhang T, Guang C, Mu W. Recent research on the physiological functions, applications, and biotechnological production of D-allose. Appl Microbiol Biotechnol, 2018, 102, 4269–4278.
[70] Naha N, Lee HY, Jo MJ, Chung BC, Kim SH, Kim MO. Rare sugar D-allose induces programmed cell death in hormone refractory prostate cancer cells. Apoptosis, 2008, 13, 1121–1134.
[71] Jeong RU, Lim S, Kim MO, Moon MH. Effect of D-allose on prostate cancer cell lines: Phospholipid profiling by nanoflow liquid chromatography–tandem mass spectrometry. Anal Bioanal Chem, 201(401), 689–698.
[72] Yamaguchi F, Kamitori K, Sanada K, Horii M, Dong Y, Sui L, Tokuda M. Rare sugar d-allose enhances anti-tumor effect of 5-fluorouracil on the human hepatocellular carcinoma cell line HuH-7. J Biosci Bioeng, 2008, 106, 248–252.
[73] Yokohira M, Hosokawa K, Yamakawa K, Saoo K, Matsuda Y, Zeng Y, Kuno T, Imaida K. Potential inhibitory effects of d-allose, a rare sugar, on liver preneoplastic lesion development in F344 rat medium-term bioassay. J Biosci Bioeng, 2008, 105, 545–553.
[74] Sui L, Dong Y, Watanabe Y, Yamaguchi F, Hatano N, Tsukamoto I, Izumori K, Tokuda M. The inhibitory effect and possible mechanisms of D-allose on cancer cell proliferation. Int J Oncol, 2005, 27, 907–912.

[75] Sui L, Dong Y, Watanabe Y, Yamaguchi F, Hatano N, Izumori K, Tokuda M. Growth inhibitory effect of D-allose on human ovarian carcinoma cells in vitro. Anticancer Res, 2005, 25, 2639–2644.
[76] Hirata Y, Saito M, Tsukamoto I, Yamaguchi F, Sui L, Kamitori K, Dong Y, Uehara E, Konishi R, Janjua N, Tokuda M. Analysis of the inhibitory mechanism of D-allose on MOLT-4F leukemia cell proliferation. J Biosci Bioeng, 2009, 107(5), 562–568.
[77] Kanaji N, Kamitori K, Hossain A, Noguchi C, Katagi A, Kadowaki N, Tokuda M. Additive antitumour effect of D-allose in combination with cisplatin in non-small cell lung cancer cells. Oncol Rep, 2018, 39, 1292–1298.
[78] Malm SW, Hanke NT, Gill A, Carbajal L, Baker AF. The anti-tumor efficacy of 2-deoxyglucose and D-allose are enhanced with p38 inhibition in pancreatic and ovarian cell lines. J Exp Clin Cancer Res, 2015, 34, 1–12.
[79] Indo K, Hoshikawa H, Kamitori K, Yamaguchi F, Mori T, Tokuda M, Mori N. Effects of d-allose in combination with docetaxel in human head and neck cancer cells. Int J Oncol, 2014, 45, 2044–2050.
[80] Hoshikawa H, Indo K, Mori T, Mori N. Enhancement of the radiation effects by d-allose in head and neck cancer cells. Cancer Lett, 2011, 306, 60–66.
[81] Miyawaki Y, Ueki M, Ueno M, Asaga T, Tokuda M, Shirakami G. D-allose ameliorates cisplatin-induced nephrotoxicity in mice. Tohoku J Exp Med, 2012, 228, 215–221.
[82] Hirooka K, Miyamoto O, Jinming P, Du Y, Itano T, Baba T, Tokuda M, Shiraga F. Neuroprotective effects of D-allose against retinal ischemia-reperfusion injury. Invest Ophthalmol Vis Sci, 2006, 47, 1653–1657.
[83] Hossain MA, Wakabayashi H, Goda F, Kobayashi S, Maeba T, Maeta H. Effect of the immunosuppressants FK506 and D-allose on allogenic orthotopic liver transplantation in rats. Transplant Proc, 2000, 32, 2021–2023.
[84] Noguchi Y, Kawate H, Nomura M, Takayanagi R. Eldecalcitol for the treatment of osteoporosis. Clin Interv Aging, 2013, 8, 1313–1321.
[85] Akram HM, Izuishi K, Maeta H. Protective effects of D-allose against ischemia reperfusion injury of the rat liver. J Hepatobiliary Pancreat Sci, 2003, 10, 218–225.
[86] Hossain M, Wakabayashi H, Izuishi K, Okano K, Yachida S, Maeta H. The role of prostaglandins in liver ischemiareperfusion injury. Curr Pharm Des, 2006, 12, 2935–295.
[87] Gao D, Kawai N, Nakamura T, Lu F, Fei Z, Tamiya T. Anti-inflammatory effect of D-allose in cerebral ischemia/reperfusion injury in rats. Neurol Med Chir, 2013, 53, 365–374.
[88] Iga Y, Nakamichi K, Shirai Y, Matsuo T. Acute and sub-chronic toxicity of D-allose in rats. Biosci Biotechnol Biochem, 2010, 74, 1476–1478.
[89] Iga Y, Matsuo T. D-allose metabolism in rats. J Jpn Soc Nutr Food Sci, 2010, 63, 17–19.
[90] Quinones JP, Peniche H, Peniche C. Chitosan based self-assembled nanoparticles in drug delivery. Polymers, 2018, 10(3), 235.
[91] Melisi D, Curcio A, Luongo E, Morelli E, Grazia RM. D-galactose as a vector for prodrug design. Curr Top Med Chem, 2011, 11(18), 2288–2298.
[92] Daci A, Bozalija A, Jashari F, Krasniqi S. Individualizing treatment approaches for epileptic patients with glucose transporter type1 (GLUT-1) deficiency. Int J Mol Sci, 2018, 19(1), 122.
[93] Yuan SS, Li ML, Chen JS, Zhou L, Zhou W. Application of mono- and disaccharides in drug targeting and efficacy. Chem Med Chem, 2018, 13(8), 764–778.
[94] Tabernero A, Medina JM, Giaume C. Glucose metabolism and proliferation in glia: Role of astrocytic gap junctions. J Neurochem, 2006, 99(4), 1049–1061.
[95] Bonina F, Puglia C, Rimoli MG, Melisi D, Boatto G, Nieddu M, Caprariis PD. Glycosyl derivatives of dopamine and L-dopa as anti-Parkinson prodrugs: Synthesis, pharmacological activity and in vitro stability studies. J Drug Deliv, 2003, 11(1), 25–36.

[96] Fernández C, Nieto O, Fontenla JA, Rivas E, De Ceballos ML, Fernández-Mayoralas A. Synthesis of glycosyl derivatives as dopamine prodrugs: Interaction with glucose carrier GLUT-1. Org Biomol Chem, 2003, 1(5), 767–771.

[97] Ruocco LA, Viggiano D, Viggiano A, Abignente E, Rimoli MG, Melisi D, Sadile AG. Galactosylated dopamine enters into the brain, blocks the mesocorticolimbic system and modulates activity and scanning time in Naples high excitability rats. Neuroscience, 2008, 152(1), 234–244.

[98] Ruocco LA, Viggiano D, Pignatelli M, Iannaccone T, Rimoli MG, Melisi D, Sadile AG. Galactosilated dopamine increases attention without reducing activity in C57BL/6 mice. Behav Brain Rese, 2008, 187(2), 449–454.

[99] Giordano C, Siniscalco D, Melisi D, Luongo L, Curcio A, Soukupova M, De Novellis V. The galactosylation of Nω-nitro-L-arginine enhances its anti-nocifensive or anti-allodynic effects by targeting glia in healthy and neuropathic mice. Eur J Pharma, 2011, 656(1–3), 52–62.

[100] Melisi D, Rosso F, Curcio A, Tortora C, Nieddu M, Marino G, Rimoli MG. Galactosyl derivative of Nω-nitro-l-arginine: Study of antiproliferative activity on human thyroid follicular carcinoma cells. J Cell Physiol, 2009, 221(2), 440–447.

[101] Chen Q, Gong T, Liu J, Wang X, Fu H, Zhang Z. Synthesis, in vitro and in vivo characterization of glycosyl derivatives of ibuprofen as novel prodrugs for brain drug delivery. J Drug Target, 2009, 17(4), 318–328.

[102] Gynther M, Ropponen J, Laine K, Leppanen J, Haapakoski P, Peura L, Rautio J. Glucose promoiety enables glucose transporter mediated brain uptake of ketoprofen and indomethacin prodrugs in rats. J Med Chem, 2009, 52(10), 3348–3353.

[103] Shelke SV, Gao GP, Mesch S, Gäthje H, Kelm S, Schwardt O, Ernst B. Synthesis of sialic acid derivatives as ligands for the myelin-associated glycoprotein (MAG). Bioorg Med Chem, 2007, 15(14), 4951–4965.

[104] Li G, Lou HX. Strategies to diversify natural products for drug discovery. Med Res Rev, 2018, 38, 1255–1294.

[105] Newman DJ, Cragg GM. Natural products as sources of new drugs from 1981 to 2014. J Nat Prod, 2016, 79, 629–661.

[106] Ding F, Zhang Y. Biomedical applications of glycoconjugates. Mini Rev Med Chem, 2018, 18, 1508–1523.

[107] Mzoughi Z, Abdelhamid A, Rihouey C, Le Cerf D, Bouraoui A, Majdoub H. Optimized extraction of pectin-like polysaccharide from suaedafruticosa leaves: Characterization, antioxidant, anti-Inflammatory and analgesic activities. Carbohydr Polym, 2018, 185, 127–137.

[108] Chen Y, Yao F, Ming K, Wang D, Hu Y, Liu J. Polysaccharides from traditional Chinese medicines: Extraction, purification, modification, and biological activity. Molecule, 2016, 21, 1705.

[109] Jiao G, Yu G, Zhang J, Ewart HS. Chemical structures and bioactivities of sulfated polysaccharides from marine algae. Mar Drugs, 2011, 9, 196–233.

[110] Bost J, Maroon A, Maroon J. Natural anti-inflammatory agents for pain relief. Surg Neurol Int, 2010, 1, 80.

[111] Liu Z, Jiao Y, Wang Y, Zhou C, Zhang Z. Polysaccharides-based nanoparticles as drug delivery systems. Adv Drug Deliv Rev, 2008, 60, 663–702.

[112] Yu Y, Shen M, Song Q, Xie J. Biological activities and pharmaceutical applications of polysaccharide from natural resources: A review. Carbohydr Polym, 2018, 183, 91–101.

[113] Wang PC, Zhao S, Yang BY, Wang QH, Kuang HX. Anti-diabetic polysaccharides from natural sources: A review. Carbohydr Polym, 2016, 148, 86–97.

[114] Moradi Z, Kalanpour NK. A branched polysaccharide: Preparation, properties and applications: A review. Carbohydr Polym, 2019, 223, 115100.

[115] Bhandari MR, Jong-Anurakkun N, Hong G, Kawabata J. α-Glucosidase and α-amylase inhibitory activities of Nepalese medicinal herb pakhanbhed (Bergenia ciliata, Haw.). Food Chem, 2008, 106, 247–252.
[116] Preetha PP, Devi VG, Rajamohan T. Hypoglycemic and antioxidant potential of coconut water in experimental diabetes. Food Func, 2012, 3, 753–757.
[117] Lovati MR, Manzoni C, Gianazza E, Arnoldi A, Kurowska E, Carroll KK, Sirtori CR. Soy protein peptides regulate cholesterol homeostasis in Hep G2 Cells. J Nutr, 2000, 130, 2543–2549.
[118] Benalaya I, Alves G, Lopes J, Silva LR. A review of natural polysaccharides: Sources, characteristics, properties, food, and pharmaceutical applications. Int J Mol Sci, 2024, 25(2), 1322.
[119] Kalita P, Ahmed AB, Sen S, Chakraborty R. A comprehensive review on polysaccharides with hypolipidemic activity: Occurrence, chemistry and molecular mechanism. Int J Biol Macromol, 2022, 206, 681–698.
[120] Zhao S, Rong C, Liu Y, Xu F, Wang S, Duan C, Wu X. Extraction of a soluble polysaccharide from *Auricularia polytricha* and evaluation of its anti-hypercholesterolemic effect in rats. Carbohydr Polym, 2015, 122, 39–45.
[121] De Jesus Raposo MF, de Morais AMMB, De Morais RMSC. Bioactivity and applications of polysaccharides from marine microalgae. In Polysaccharides. Springer International Publisher Cham: Switze, 2014, pp. 1–38.
[122] Mohan K, Ravichandran S, Muralisankar T, Uthayakumar V, Chandirasekar R, Seedevi P, Rajan DK. Potential uses of fungal polysaccharides as immunostimulants in fish and shrimp aquaculture: A review. Aquaculture, 2019, 500, 250–263.
[123] Olennikov DN, Kashchenko NI, Chirikova NK, Koryakina LP, Vladimirov LN. Bitter gentian teas: Nutritional and phytochemical profiles, polysaccharide characterisation and bioactivity. Molecules, 2015, 20, 20014–20030.
[124] Peng X, Usman B, Kaushik N, Hoffman J, Wang D, Saenko K, Vis D. The visual domain adaptation challenge. arXiv, 2017, 1710, 06924.
[125] Savary S, Willocquet L, Pethybridge SJ, Esker P, McRoberts N, Nelson A. The global burden of pathogens and pests on major food crops. Nat Ecol Evol, 2019, 3, 430–439.
[126] Bebber DP, Gurr SJ. Crop-destroying fungal and oomycete pathogens challenge food security. Fungal Genet Biol, 2015, 74, 62–64.
[127] Fisher MC, Henk DA, Briggs CJ, Brownstein JS, Madoff LC, McCraw SL, Gurr SJ. Emerging fungal threats to animal, plant and ecosystem health. Nature, 2012, 484, 186–194.
[128] Fones HN, Bebber DP, Chaloner TM, Kay WT, Steinberg G, Gurr SJ. Threats to global food security from emerging fungal and oomycete crop pathogens. Nat Food, 2020, 1, 332–342.
[129] Russell PE. A century of fungicide evolution. J Agric Sci, 2005, 143, 11–25.
[130] Carvalho FP. Pesticides, environment, and food safety. Food Energy Secur, 2017, 6, 48–60.
[131] Zubrod JP, Bundschuh M, Arts G, Bruhl CA, Imfeld G, Knabel A, Payraudeau S, Rasmussen JJ, Rohr J, Scharmuller A. Fungicides: An overlooked pesticide class? Environ Sci Technol, 2019, 53, 3347–3365.
[132] Russell PE. The development of commercial disease control. Plant Pathol, 2006, 55, 585–594.
[133] Campos EVR, de Oliveira JL, Fraceto LF, Singh B. Polysaccharides as safer release systems for agrochemicals. Agron Sustain Dev, 2015, 35, 47–66.
[134] Lopez-Moya F, Suarez-Fernandez M, Lopez-Llorca LV. Molecular mechanisms of chitosan interactions with fungi and plants. Int J Mol Sci, 2019, 20, 332.
[135] Xing K, Zhu X, Peng X, Qin S. Chitosan antimicrobial and eliciting properties for pest control in agriculture: A review. Agron Sustain Dev, 2015, 35, 569–588.
[136] Zheng F, Chen L, Zhang PF, Zhou JQ, Lu XF, Tian W. Carbohydrate polymers exhibit great potential as effective elicitors in organic agriculture: A review. Carbohydr Polym, 2020, 230, 115637.
[137] Malerba M, Cerana R. Chitosan effects on plant systems. Int J Mol Sci, 2016, 17, 996.

[138] Kaur P, Thakur R, Choudhary A. An in vitro study of the antifungal activity of silver/chitosan nanoformulations against important seed borne pathogens. Int J Sci Technol Res, 2012, 1, 83–86.
[139] Verlee A, Mincke S, Stevens CV. Recent developments in antibacterial and antifungal chitosan and its derivatives. Carbo hydr Polym, 2017, 164, 268–283.
[140] Palma-Guerrero J, Huang IC, Jansson HB, Salinas J, Lopez-Llorca LV, Read ND. Chitosan permeabilizes the plasma membrane and kills cells of *Neurospora crassa* in an energy dependent manner. Fungal Genet Biol, 2009, 46, 585–594.
[141] Rabea EI, Badawy MET, Stevens CV, Smagghe G, Steurbaut W. Chitosan as antimicrobial agent: Applications and mode of action. Biomacromolecules, 2003, 4, 1457–1465.
[142] Saberi-Rise R, Moradi-Pour M. The effect of *Bacillus subtilis* Vru1 encapsulated in alginate bentonite coating enriched with titanium nanoparticles against *Rhizoctonia solani* on bean. Int J Biol Macromol, 2020, 152, 1089–1097.
[143] John RP, Tyagi RD, Brar SK, Surampalli RY, Prevost D. Bio-encapsulation of microbial cells for targeted agricultural delivery. Crit Rev Biotechnol, 2011, 31, 211–226.
[144] Tleuova AB, Wielogorska E, Talluri V, Stepanek F, Elliott CT, Grigoriev DO. Recent advances and remaining barriers to producing novel formulations of fungicides for safe and sustainable agriculture. J Control Release, 2020, 326, 468–481.
[145] Ma HH, Zhao Y, Lu ZJ, Xing RL, Yao XM, Jin ZX, Wang YY, Yu F. Citral-loaded chitosan/carboxymethyl cellulose copolymer hydrogel microspheres with improved antimicrobial effects for plant protection. Int J Biol Macromol, 2020, 164, 986–993.
[146] Panahirad S, Dadpour M, Peighambardoust SH, Soltanzadeh M, Gullon B, Alirezalu K, Lorenzo JM. Applications of carboxymethyl cellulose and pectin-based active edible coatings in preservation of fruits and vegetables: A review. Trends Food Sci Technol, 2021, 110, 663–673.
[147] Rolland F, Baena-Gonzalez E, Sheen J. Sugar sensing and signaling in plants: Conserved and novel mechanisms. Annu Rev Plant Biol, 2006, 57, 675–709.
[148] Smeekens S, Ma J, Hanson J, Rolland F. Sugar signals and molecular networks controlling plant growth. Curr Opin Plant Biol, 2010, 13, 274–279.
[149] Rolland F, Moore B, Sheen J. Sugar sensing and signaling in plants. Plant Cell, 2002, S185–S205.
[150] Roitsch T, Balibrea ME, Hofmann M, Proels R, Sinha AK. Extracellular invertase: Key metabolic enzyme and PR protein. J Exp Bot, 2003, 54, 513–524.
[151] Bolton MD. Primary metabolism and plant defense–fuel for the fire. Mol Plant Microbe Interact, 2009, 22, 487–497.
[152] Bolouri-Moghaddam MR, Le Roy K, Xiang L, Rolland F, Van Den Ende W. Sugar signalling and antioxidant network connections in plant cells. FEBS J, 2010, 277, 2022–2037.
[153] Brilisauer K, Rapp J, Rath P, Schöllhorn A, Bleul L, Weiß E, Stahl M, Grond S, Forchhammer K. Cyanobacterial antimetabolite 7-deoxy-sedoheptulose blocks the shikimate pathway to inhibit the growth of prototrophic organisms. Nat Commun, 2019, 10.
[154] Felgueiras C, Azoia NG, Gonçalves C, Gama M, Dourado F. Trends on the cellulose-based textiles: Raw materials and technologies. Front Bioeng Biotechnol, 2021, 9, 1–20.
[155] Dunne R, Desai D, Sadiku R, Jayaramudu J. A review of natural fibres, their sustainability and automotive applications. J Reinf Plast Compos, 2016, 35, 1041–1050.
[156] Novakovic M, Popovic DM, Mladenovic N, Poparic GB, Stankovic SB. Development of comfortable and eco-friendly cellulose based textiles with improved sustainability. J Clean Prod, 2020, 267, 1–13.
[157] Kafy A, Kim HC, Zhai L, Kim JW, Hai LV, Kang TJ, Kim J. Cellulose long fibers fabricated from cellulose nanofibers and its strong and tough characteristics. Sci Rep, 2017, 7(1), 17683.
[158] Garcia S, Cordeiro A, de Alencar NI, de Oliveira Costa Neto PL. The sustainability awareness of Brazilian consumers of cotton clothing. J Clean Prod, 2019, 215, 1490–1502.

[159] Hadeyati M, Brock PM, Nachimuthu G, Schwenke G. Farm-level strategies to reduce the life cycle greenhouse gas emissions of cotton production: An Australian perspective. J Clean Prod, 2019, 212, 1–13.
[160] Basit A, Latif W, Baig SA, Afzal A. The mechanical and comfort properties of sustainable blended fabrics of bamboo with cotton and regenerated fibres. Cloth Text Res J, 2018, 36, 267–280.
[161] Martínez-Sanz M, Pettolino F, Flanagan B, Gidley MJ, Gilbert EP. Structure of cellulose microfibrils in mature cotton fibres. Carbohydr Polym, 2017, 175, 450–463.
[162] Häemmerle FM. The cellulose gap (the future of cellulose fibres). Lenzinger Ber, 2011, 89, 12–21.
[163] Sfiligoj SM, Hribernik S, Stana KK, Kreže T. Plant fibres for textile and technical applications. Adv Agrophy Res, 2013, 31, 369–398.
[164] Wang J, Tavakoli J, Tang Y. Bacterial cellulose production, properties and applications with different culture methods: A review. Carbohydr Polym, 2019, 1(219), 63–76.
[165] Clemons C. Nanocellulose in spun continuous fibers: A review and future outlook. J of Renew Mater, 2016, 4(5), 327–339.
[166] Thomas P, Duolikun T, Rumjit NP, Moosavi S, Lai CW, Johan MR, Fen LB. Comprehensive review on nanocellulose: Recent developments, challenges and future prospects. J Mech Behav Biomed Mater, 2020, 110, 103884.
[167] Dourado F, Ryngajllo M, Jedrzejczak-Krzepkowska M, Bielecki S, Gama M. Taxonomic review and microbial ecology in bacterial nanocellulose fermentation. Bacterial Nanocell, 2016, 1, 1–7.
[168] Ul-Islam M, Ullah MW, Khan S, Park JK. Production of bacterial cellulose from alternative cheap and waste resources: A step for cost reduction with positive environmental aspects. Korean J Chem Eng, 2020, 37, 925–937.
[169] Chan CK, Shin J, Jiang SX. Development of tailor-shaped bacterial cellulose textile cultivation techniques for zero-waste design. Clothing Text Rese J, 2018, 36(1), 33–44.
[170] Wood J. Bioinspiration in fashion – A review. Biomimetics, 2019, 4(1), 16.
[171] Dini I, Laneri S. The new challenge of green cosmetics: Natural food ingredients for cosmetic formulations. Molecules, 2021, 26(13), 3921.
[172] Chemat F, Abert Vian M, Ravi HK, Khadhraoui B, Hilali S, Perino S, Fabiano Tixier AS. Review of alternative solvents for green extraction of food and natural products: Panorama, principles, applications and prospects. Molecules, 2019, 24(16), 3007.
[173] Kobayashi R, Takisada M, Suzuki T, Kirimura K, Usami S. Neoagarobiose as a novel moisturizer with whitening effect. Biosci Biotechnol Biochem, 1997, 61(1), 162–163.
[174] Yun EJ, Lee S, Kim JH, Kim BB, Kim HT, Lee SH, Pelton JG, Kang NJ, Choi IG, Kim KH. Enzymatic production of 3, 6-anhydro-L-galactose from agarose and its purification and in vitro skin whitening and anti-inflammatory activities. Appl Microbiol Biotechnol, 2013, 97, 2961–2970.
[175] Jang MK, Lee DG, Kim NY, Yu KH, Jang HJ, Lee SW, Jang HJ, Lee YJ, Lee SH. Purification and characterization of neoagarotetraose from hydrolyzed agar. J Microbiol Biotechnol, 2009, 19(10), 1197–1200.
[176] Chi WJ, Chang YK, Hong SK. Agar degradation by microorganisms and agar-degrading enzymes. Appl Microbiol Biotechnol, 2012, 94, 917–930.
[177] Stephanie B, Eric D, Sophie FM, Christian B, Yu G. Carrageenan from Solieriachordalis (Gigartinales): Structural analysis and immunological activities of the low molecular weight fractions. Carbohydr Polym, 2010, 81, 448–460.
[178] Yao Z, Wu H, Zhang S, Du Y. Enzymatic preparation of κ-carrageenan oligosaccharides and their anti-angiogenic activity. Carbohydr Polym, 2014, 101, 359–367.
[179] Chen SK, Wang X, Guo YQ, Song XX, Yin JY, Nie SP. Exploring the partial degradation of polysaccharides: Structure, mechanism, bioactivities, and perspectives. Compr Rev Food Sci Food Saf, 2023, 22(6), 4831–4870.

[180] Erginer M, Gökalsin B, Tornaci S, Sesal C, Öner ET. Exploring the potential of halomonaslevan and its derivatives as active ingredients in cosmeceutical and skin regenerating formulations. Int J Biol Macromol, 2023, 240, 124418.

[181] Shao P, Chen X, Sun P. Improvement of antioxidant and moisture-preserving activities of *Sargassum horneri* polysaccharide enzymatic hydrolyzates. Int J Biol Macromol, 2015, 74, 420–427.

[182] Chen Q, Kou L, Wang F, Wang Y. Size-dependent whitening activity of enzymedegraded fucoidan from *Laminaria japonica*. Carbohydr Polym, 2019, 225, 115211.

[183] Deng R, Wang F, Wang L, Xiong L, Shen X, Song H. Advances in plant polysaccharides as antiaging agents: Effects and signaling mechanisms. J Agric Food Chem, 2023, 71(19), 7175–7191.

[184] Barreto SM, Maia MS, Benicá AM, de Assis HR, Leite-Silva VR, da Rocha-Filho PA, de Negreiros MM, de Oliveira Rocha HA, Ostrosky EA, Lopes PS, De Farias Sales VS. Evaluation of in vitro and in vivo safety of the by-product of *Agave sisalana* as a new cosmetic raw material: Development and clinical evaluation of a nanoemulsion to improve skin moisturizing. Ind Crops Prod, 2017, 108, 470–479.

[185] Morganti P, Morganti G, Colao C. Biofunctional textiles for aging skin. Biomedicines, 2019, 7(3), 51.

[186] Maphosa Y, Jideani VA. Factors affecting the stability of emulsions stabilised by biopolymers. Sci Technol behind Nanoemul, 2018, 22, 65.

[187] Pal A, Mondal MH, Adhikari A, Bhattarai A, Saha B. Scientific information about sugar-based emulsifiers: A comprehensive review. RSC Adv, 2021, 11(52), 33004–33016.

[188] Dickinson E. Hydrocolloids at interfaces and the influence on the properties of dispersed systems. Food Hydrocoll, 2003, 17(1), 25–39.

[189] Patel S, Goyal A. Applications of natural polymer gum arabic: A review. Int J Food Prop, 2015, 18(5), 986–998.

[190] Nie SP, Wang C, Cui SW, Wang Q, Xie MY, Phillips GO. A further amendment to the classical core structure of gum arabic (*Acacia senegal*). Food Hydrocoll, 2013, 31(1), 42–48.

[191] Yang Y, Gupta VK, Du Y, Aghbashlo M, Show PL, Pan J, Tabatabaei M, Rajaei A. Potential application of polysaccharide mucilages as a substitute for emulsifiers: A review. Int J Biol Macromol, 2023, 242, 124800.

[192] Hubbe MA, Pawlak JJ, Koukoulas AA. Paper's appearance: A review. Bio Resour, 2008, 3(2), 627–665.

[193] Song D, Dong C, Ragauskas AJ, Deng Y. Filler modification with polysaccharides or their derivatives for improved paper properties. J Biobas Mat Bioeng, 2009, 34, 321–334.

[194] Dai Z, Ni Y, Court G, Li Z. Mitigating pitch-related deposits at a TMP based speciality paper mill. Tappi J, 2011, 10.

[195] Hu K, Ni Y, Zou X. Substitution of hardwood bleached kraft pulp with aspen high-yield pulp in LWC wood-free papers, part 2: Impact on coated paper quality. Tappi J, 2007, 6(1), 26–32.

[196] Koivunen K, Alatalo H, Silenius P, Paulapuro H. Starch granules spotcoated with aluminum silicate particles and their use as fillers for papermaking. J Mater Sci, 2010, 45, 3184–3189.

[197] Patel SV, Venditti RA, Pawlak JJ. Dimensional changes of starch microcellular foam during the exchange of water with ethanol and subsequent drying. Bio Resour, 2010, 5(1), 121–134.

[198] Shen J, Song Z, Qian X, Ni Y. Carbohydrate-based fillers and pigments for papermaking: A review. Carbohydr Polym, 2011, 85(1), 17–22.

Ali Ramazani*, Seyedmohammad Hosseininezhad,
Maryam Manafi Moghadam, and Amit Kumar Nayak

Chapter 20
Advances in carbohydrates-based metal-organic frameworks

20.1 Introduction

Carbohydrates (CHs), including oligosaccharides and polysaccharides, are widely recognized as edible materials. Together with lipids, proteins, and nucleic acids, they form one of the four primary categories of biomolecules [1]. CHs constitute the primary components of sugars, starches, and natural fibers present in plants [2]. They are among the most abundant organic compounds found in nature. Broadly, CHs can be categorized into four primary types: monosaccharides, disaccharides, oligosaccharides, and polysaccharides [3, 4]. CHs function as macromolecules providing structural support to plant cells and various arthropods, such as crabs, shrimp, and lobsters. Typical monosaccharides are galactose, glucose, and fructose, whereas disaccharides include lactose, maltose, and sucrose [5, 6]. Glycogen and starch serve as examples of polysaccharides. In animals, glycogen acts as the storage form of glucose, while in plants, starch fulfills this role [7]. Additionally, cellulose, an abundant unbranched polysaccharide, plays a crucial structural role [8]. These polysaccharides not only provide energy for living organisms but also find applications in diverse fields, including the preparation of adhesives, agricultural chemicals, paper, and catalysts [9–11]. Moreover, their sustainable characteristics and scalability make them valuable as renewable raw materials for the applications mentioned earlier.

Metal-organic frameworks (MOFs) form through the coordination of organic ligands with metal ions (also known as metal ion clusters) [12–15]. MOFs exhibit remarkable features distinct from zeolites, active carbons, and mesoporous silicas. These include tunable structures, easy functionalization, and the ability to incorporate active sites [16–20]. Due to these distinctive advantages, MOFs find applications

***Corresponding author: Ali Ramazani**, The Organic Chemistry Research Laboratory (OCRL), Department of Chemistry, Faculty of Science, University of Zanjan, Zanjan 45371-38791, Iran, e-mail: aliramazani@znu.ac.ir; The Convergent Sciences and Technologies Laboratory (CSTL), Research Institute of Modern Biological Techniques (RIMBT), University of Zanjan, Zanjan 45371-38791, Iran
Seyedmohammad Hosseininezhad, Maryam Manafi Moghadam, The Organic Chemistry Research Laboratory (OCRL), Department of Chemistry, Faculty of Science, University of Zanjan, Zanjan 45371-38791, Iran
Amit Kumar Nayak, Department of Pharmaceutics, School of Pharmaceutical Sciences, Siksha 'O' Anusandhan (Deemed to be University), Bhubaneswar 751003, Odisha, India

https://doi.org/10.1515/9783111447094-020

in gas storage/separation [21–24], sensing [25–29], catalysis [30–33], proton conduction [34–37], environmental remediation [38–42], energy storage [43–45], and even biomedical contexts [46–49]. However, the widespread adoption of MOFs for commercial purposes faces significant challenges due to their inherent fragility, mechanical instability, and impracticality [50]. Researchers have tackled the challenges of pure MOFs by developing hybrid materials. They combined MOF particles with flexible and processable substrates to improve their practical applications. Over time, they have incorporated various materials, including graphene [51, 52], mesoporous silica [53, 54], and polymers [55, 56], as suitable substrates. Nevertheless, choosing substrate materials with long-term stability, good processability, high mechanical strength, and affordability remains a significant hurdle, as these qualities are fundamental for successful commercial applications [50].

In this review, we highlight recent advancements in different CHs, including cellulose, chitosan, sodium alginate, pectin, chitin, and cyclodextrin (CD) within the context of MOFs. Our focus is on the distinctive features of purposefully designed carbohydrate-MOFs, such as their chemico-thermal stability and customized porosity. Additionally, we delve into the synthesis methods and practical applications of these carbohydrate-MOF materials.

20.2 Cellulose-based MOF

Waste paper cellulose, characterized by its high length-to-diameter ratio and abundant –OH groups, served as a template for synthesizing a hybrid material composed of Cu-MOF and PSS. In this hybrid material, Cu-MOF was intricately threaded by PSS polymers. Following the initial synthesis, the WP@PSS@Cu-MOF was employed for extracting Li^+ in a brine solution. The findings revealed distinct selectivity and remarkable adsorption capacity for Li^+ ions. Furthermore, incorporating PSS into the Cu-MOF framework significantly enhanced the adsorption capacity and selectivity for Li^+. This enhancement is due to the combined effects of the size-filtering properties of Cu-MOF and the specific binding affinity of the PSS polymer. Even after five cycles of recycling, the Li^+ adsorption capacity of WP@PSS@Cu-MOF remained at approximately 88.3%. In the case of the synthesis of the desired hybrid absorbent, a mixture of WP@$Cu(OH)_2$ and PSS in deionized water was gently stirred for 60 min. The PSS^{n-} ions naturally merged with the WP@$Cu(OH)_2$ layer on the cellulose surface. Next, the WP@$Cu(OH)_2$@PSS solution was added to H_3BTC water-ethanol solution and stirred for 3 h. Lastly, by filtration and washing, the WP@PSS@Cu-MOF absorbent was generated (Figure 20.1) [57].

Scientists performed a study to compare the potential of cotton fabrics and bacterial cellulose nanofibers (NBC) as suitable substrates for MOF-199. The objective was to create micro and nanocomposites of cellulose@MOF-199, specifically tailored for

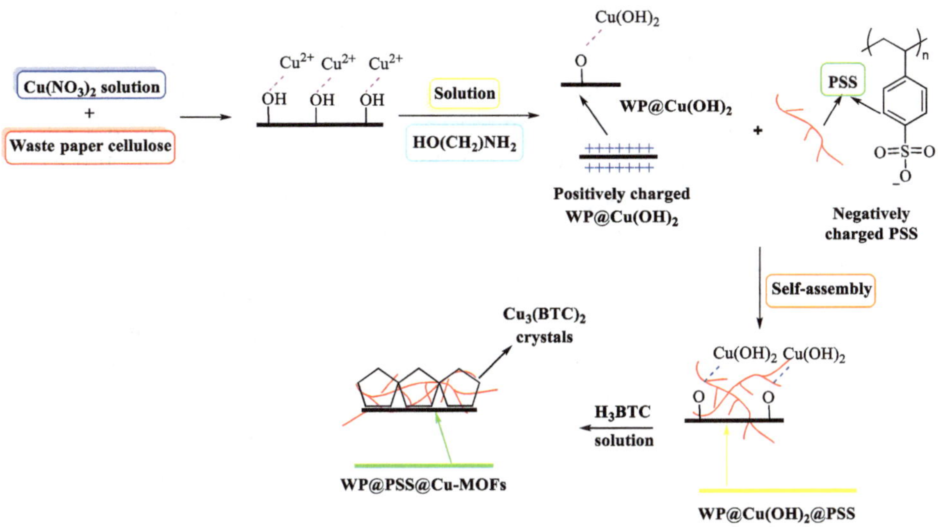

Figure 20.1: Manufacturing method for a PSS-integrated Cu-MOF composite material utilizing waste paper cellulose.

CO_2 capture. NBC acted as an effective substrate, showing complete coverage and even distribution of MOF crystals throughout the NBC@MOF-199 nanocomposite. The specific surface area for NBC@MOF-199 was 553.4 m^2 g^{-1}, while for Cotton@MOF-199, it measured 168.9 m^2 g^{-1}. Notably, NBC@MOF-199 demonstrated a high adsorption capacity of 2.9 mmol g^{-1} under ambient temperature and pressure, surpassing the performance of Cotton@MOF-199 (which achieved 1.2 mmol g^{-1}) [58].

20.3 Chitosan-based MOF

A group of researchers formulated a novel injectable hydrogel that promotes healing in chronic wounds commonly seen in diabetic patients. This advanced self-healing gel is composed of chitosan, K-γ-CD-MOFs, and hyaluronic acid. It is infused with alpha-lipoic acid (α-LA), increasing its antibacterial and antioxidant capabilities. The chitosan scaffold the authors engineered demonstrates exceptional qualities, including enhanced injectability, self-repairing capabilities, robust adhesive mechanical traits, effective antibacterial properties, and outstanding biocompatibility. To enhance the uptake and effectiveness of α-LA, a new hydrogel was created using a base of quaternary ammonium chitosan (QCS) and oxidized hyaluronic acid (OHA). This base served as a scaffold that, when combined with K-γ-CD-MOFs, facilitates the targeted and prolonged delivery of α-LA directly to the wound's periphery, thereby optimizing its healing potential. The total release of α-LA reached 83.4% from the QCS-OHA hydrogel, while the QCS-OHA scaffold integrated with K-γ-CD-MOFs exhibited a release of 72.4% [59]. A new ternary composite material, consisting of the nanoparticles of Ag housed within the pores of MOF-5 nanocrystals, which in turn are firmly bonded to the surface of a chitosan non-woven fabric, has been successfully synthesized by Liu et al. Research revealed that the Ag@MOF-5@chitosan composite is capable of purifying toxic molecules. This process involves the toxic molecules being first adsorbed into the MOF-5 channels and then undergoing hydrolytic breakdown, a reaction facilitated by the Ag nanoparticles embedded within the composite, all occurring under normal environmental conditions. In addition, studies indicated that the incorporation of Ag nanoparticles and MOF-5 onto the nonwoven chitosan fabric does not notably impede the intrinsic blood-clotting properties of the chitosan. The current hypothesis suggests that the Ag/Ag_2O components enhance the absorption of CEES/HD molecules into the MOF pores, likely due to the affinity or interaction between the S/Cl atoms and the Ag/Ag_2O. The interaction between Ag/Ag_2O and S/Cl leads to a transient state that significantly aids in the liberation of Cl − ions and the creation of highly reactive cyclic sulfonium ions. Then, they rapidly hydrolyze in the presence of water within the MOF host (Figure 20.2) [60].

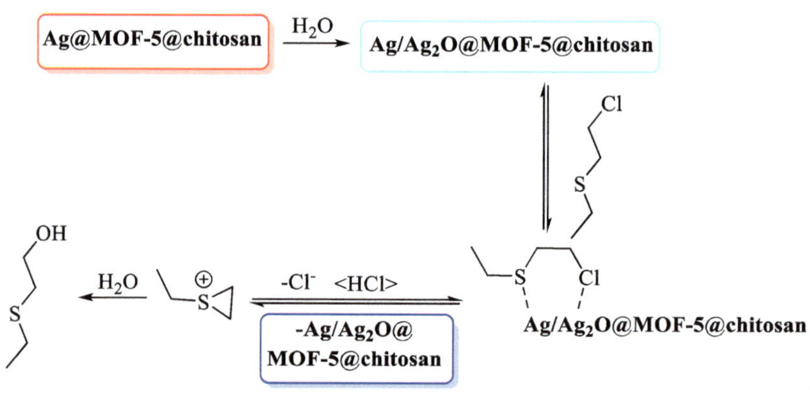

Figure 20.2: The process by which 2-CEES undergoes hydrolysis when catalyzed by Ag@MOF-5@chitosan.

An innovative catalyst, consisting of a Cu-based MOF combined with modified chitosan (CS-CA), referred to as CS-CA/Cu-MOF, was synthesized by Yousefian and Rafiee. This mesoporous catalyst demonstrates remarkable efficiency and cost-effectiveness for facilitating the Knoevenagel condensation reaction under mild conditions. Products resulting from the Knoevenagel reaction were obtained with high efficiency (86–100%). This catalyst has robust stability and has proven to be highly effective, with significant recuperation, for the production of benzylidenemalononitrile compounds. The current study also has other advantages, such as the simple and economical production of the catalyst, the ability to conduct reactions at ambient temperature, the use of a solvent-free environment, and the minimal use of the catalyst. The superior performance of the corresponding catalyst over either CS or Cu-MOF individually can be ascribed to the synergistic impact of the two constituents. The synthesis pathway for the CS-CA/Cu-MOF hybrid was reported (Figure 20.3). Initially, CA is transformed into its anhydride form by combining it with $(CH_3CO)_2O$ within a CH_3COOH solution and applying heat at 40 °C while stirring. Subsequently, CS undergoes modification through citric anhydride to yield the citrate ester derivative of CS. The final step involves the fabrication of Cu-MOF on the CS-CA matrix by connecting $Cu(NO_3)_2 \cdot 3H_2O$ with 1,4-BDC using a solvothermal method, culminating in the creation of the CS-CA/Cu-MOF composite [61].

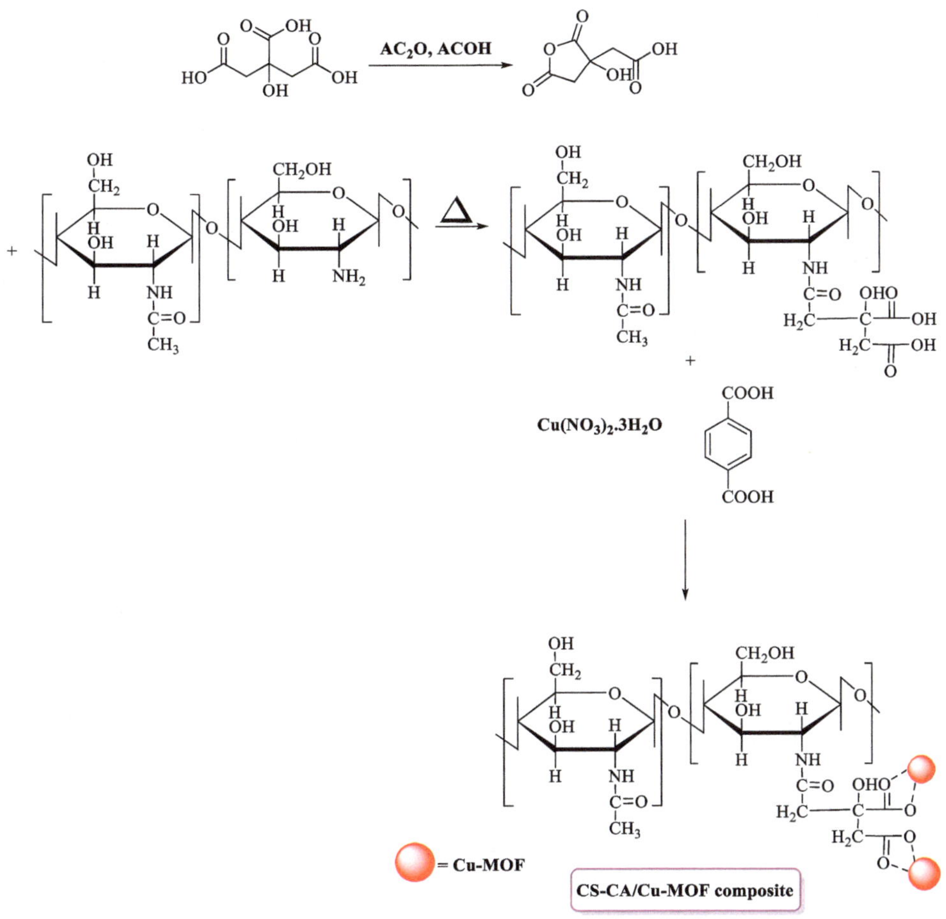

Figure 20.3: Synthesis of CS-CA/Cu-MOF.

In a research study conducted, magnetic nanoparticles of Fe_3O_4 were used to coat ZIF-67, which was subsequently integrated into an aminated chitosan (AmCs) framework, resulting in the creation of Fe_3O_4/ZIF-67@AmCs composite beads with a core-dual shell structure. The corresponding magnetic composite beads that were synthesized underwent an evaluation to determine their adsorptive removal aptitude in comparison with Cr(VI). It was found that the corresponding magnetic composite beads achieved a peak adsorption capability of 119.05 mg g^{-1} at a temperature of 25 °C. The adsorption process conformed to the principles of both the Langmuir and Freundlich isotherm. Moreover, the pseudo-second-order kinetic model was identified as the most appropriate representation for this process. Also, the desired composite illustrated outstanding reusability, effectively extracting Cr(VI) from their water-based solutions across seven successive cycles. According to the results of the temperature influence, elevating the temperature from 25 °C to 35 °C clearly enhances the efficiency of the process, as evidenced by the rise in both the adsorption capacity, from 71.35 mg g^{-1} to 75.56 mg g^{-1}, and the removal percentage, from 77.20% to 82.35%. In the case of the synthesis, under intense agitation, a specified quantity of AmCs was dissolved in an acetic acid solution. In a separate vessel, precise proportions of ZIF-67 and Fe_3O_4 were thoroughly mixed to create a uniform Fe_3O_4-ZIF-67 composite. Subsequently, the Fe_3O_4-ZIF-67 composite was incrementally introduced into the AmCs solution and agitated for a designated duration. The mixture, now containing the Fe_3O_4-ZIF-67, was methodically injected, dropwise, into 250 mL of a pre-prepared NaOH solution using a syringe. Following a 15 min setting period, the resultant magnetic Fe_3O_4/ZIF-67@AmCs wet beads were separated using an external magnet, rinsed with distilled water, and allowed to dry at ambient temperature throughout the night. This process resulted in the creation of the composite beads of magnetic Fe_3O_4/ZIF-67@AmCs [62]. A ternary nanocomposite consisting of POAP, chitosan, and DMOF-1 was synthesized using a straightforward electrochemical method. At a current density of 1 mA cm^{-2}, the ternary nanocomposite electrode illustrated a specific capacitance value of 3,150 mF cm^{-2}. The superior electrical storage capability of the corresponding nanocomposite electrode can be ascribed to the combined influence of chitosan/DMOF-1 and the Faradaic processes provided by DMOF-1, which together enhance the electrode's electrochemical behavior. The corresponding nanocomposite electrode demonstrated enhanced specific capacitance when contrasted with the previously documented specific capacitance values of POAP-based electrodes from other research. Also, the developed corresponding nanocomposite illustrated outstanding durability over numerous cycles. After 1,000 cycles, the apparatus maintained 93% of its original specific capacitance at a current density of 10 mA cm^{-2}, indicating its significant potential as an electrode material to use as an electrochemical supercapacitor [63]. Chitosan (CS) was used to modify MOFs (MIL-Fe composites) after their synthesis through solvothermal reactions. On the chitosan-MOF (CS-MIL-Fe), β-glucosidase was immobilized. The corresponding composites were characterized with several techniques. The desired composite showed 85% immobilization yield and 74% recovered activity. It also maintained 69% of its original activity, and this amount was obtained after ten successive cycles and 30 days of storage at 4 °C, while this amount for the free enzyme was obtained in 32%. Compared to the free enzyme, the immobilized enzyme illustrated greater stability across various pH levels and temperatures. In addition, the optimal pH and temperature of the immobilized enzyme were 6.0 and 60 °C, and these amounts for the free enzyme were 5.0 and 50 °C. The K_m value of the immobilized enzyme was obtained as 13.4 mM, while this amount for the free beta-glucosidase was obtained as 6.98 mM. In addition, the V_{max} values were determined to be 3.96 U mL^{-1} and 1.72 U mL^{-1}, respectively. In the case of the synthesis of the desired composite, $Fe(NO_3)_3{\cdot}9H_2O$ and $C_6H_3(CO_2H)_3$ were dissolved in DMF:EtOH. This solution was stirred, and trimethylamine was added. After autoclaving and cooling, the powder of MIL-Fe was isolated, washed, and dried. Separately, chitosan was immersed in acetic acid, combined with MIL-Fe, and added to NaOH. The final CS-MIL-Fe composite was washed until neutral pH was achieved [64]. Utilizing a microwave irradiation method, an active biological compound was effectively created. This compound consists of a CS biopolymer matrix embedding a zinc-based 1,3,5-benzenetricarboxylic acid framework (Zn-BTC). The BET analysis data indicated a significant decrease in the specific surface area of Zn-BTC, which dropped from 995.15 $m^2\ g^{-1}$ to 15.16 $m^2\ g^{-1}$ following the application of a chitosan coating. The distribution of pore sizes and the total pore volume for Zn-BTC and Zn-BTC@CS were predominantly at 37.26 nm and 22.5 nm, in that order. The compound Zn-BTC@CS illustrated an inhibition of growth in both DLD-1 and A549 cancer cell lines, showing a dose-responsive effect. The values of IC_{50} were determined to be 13.2 µg/mL for colon cancer cells and 19.8 µg mL^{-1} for lung cancer cells. The research demonstrated that Zn-BTC@CS affects vital proteins that play a role in metastasis, apoptosis, and proliferation. Consequently, Zn-BTC@CS exhibited promising capabilities as a chemotherapy agent in reducing the survival and proliferation of lung and colon cancer cells [65]. An array of composites, integrating zirconium-based MOFs with CS (denoted as MOF-808/chitosan), was developed to effectively adsorb and remove Cr(VI) ions from liquid solutions. Kinetic analysis confirmed that the adsorption of Cr(VI) ions using the corresponding composite adheres to the pseudo-2nd-order kinetic model. The adsorption isotherms conformed to the Langmuir isotherm model, indicating a uniform adsorption process of Cr(VI) ions on the corresponding composite's surface. The Langmuir model indicated that the highest adsorption capacity achieved was 320.0 mg g^{-1} at a pH level of 5. The thermodynamic analysis suggested that the adsorption process was spontaneous ($\Delta G° < 0$), disordered ($\Delta S° > 0$), and was endothermic ($\Delta H° > 0$).In addition, the corresponding composite was effectively reusable, maintaining its capacity to remove Cr(VI) from aqueous solutions across six consecutive cycles [66]. Composite foams made of chitosan and Zr-based MOF (UiO-66), referred to as CSF@UiO-66, were easily synthesized. This was achieved by integrating UiO-66 into the polymerized chitosan matrix, utilizing epichlorohydrin to serve as a connector. The composite foams, featuring hierarchical pores and numerous functional groups, were highly efficient in extracting ketoprofen from water. They reached a peak adsorption capacity of 209.7 mg g^{-1} at a pH level of 4, surpassing the majority of previously documented adsorbents in performance. Significantly, the composite foams demonstrated superior efficacy, maintaining their performance amidst various ions, and exhibited commendable recyclability for use in engineering contexts [67]. In an unprecedented approach, a novel Fe_3O_4@Cd-MOF was synthesized through eco-friendly methods. Subsequently, it was coated with CS (Fe_3O_4@Cd-MOF@CS). The performance of the Fe_3O_4@Cd-MOF@CS was investigated for its ability to eliminate amoxicillin (AMX) from various environments. The results of the adsorption research indicated that CS influences the efficacy of removal, achieving maximum effectiveness at a pH level of 8 and after a duration of 240 min. Following the refinement of conditions, the process succeeded in eliminating more than 75% of the AMX. The adsorption kinetics and isotherm were modeled using the pseudo-2nd-order model and the Langmuir isotherm, respectively. Ultimately, the adsorption process achieved a maximum capacity of approximately 103.09 mg g^{-1} [68].

20.4 Sodium alginate-based MOF

Newly developed hydrogels composed of alginate and nanocomposite materials, incorporating MOF-5, have been synthesized for the effective removal of Pb(II) and tetracycline (TC) contaminants from H_2O sources. The enhanced MOF-5/PNaSS/SA formulation exhibited a more effective adsorption capacity when contrasted with the SA and PNaSS/SA variants. According to laboratory experiments, the substance demonstrated its highest adsorption capabilities, achieving 189 mg g^{-1} for Pb(II) and 181 mg g^{-1} for TC, specifically at a pH level of 6. The adsorption behavior of MOF-5/PNaSS/SA was accurately described using the model of 2nd-order kinetics and the model of Langmuir isotherm. MOF-5/PNaSS/SA showcased remarkable selectivity in its adsorption process and maintained its efficiency despite the presence of salt ions. Furthermore, the material proved to be recyclable, retaining a high adsorption capacity even after ten repeated adsorption–desorption cycles. The existence of groups containing O/S and the structure of benzene rings were pivotal in the adsorption process, as evidenced by FTIR and XPS analyses [69]. Nanocomposite beads composed of NaAlg, carboxy-methyl cellulose, and a Cu-based MOF (SA/CMC/Cu-BDC) were fabricated through an ex-situ method that involved blending natural biopolymers with Cu-BDC powder, followed by a freeze-drying process. The beads composed of a composite material, which included 30 wt% Cu-BDC particles, exhibited a significantly large surface area characterized by both micro and mesoporous structures. The findings indicated that the optimal adsorption efficiency for TC was achieved at a pH level of 7 with the use of the corresponding composite beads containing 30 wt% Cu-BDC powder. Furthermore, the corresponding beads with 30 wt% concentration demonstrated a superior maximum adsorption capacity of 180.5 mg g^{-1}, in contrast to the 127 mg g^{-1} capacity of SA/Cu-BDC beads at the same weight percentage. The equilibrium and kinetic analysis of TC adsorption on the corresponding beads adhered to the model of Langmuir isotherm and the model of pseudo-2nd-order kinetic, respectively. The thermodynamic parameters indicated that the adsorption of TC occurs spontaneously and it was endothermic [70]. A green method for the direct creation of new antibacterial Cu-based

MOFs embedded in a sodium alginate hydrogel matrix was reported by Bradai et al. They synthesized two distinct types of composite beads made of Cu-based MOF and sodium alginate by treating copper-ion-linked alginate hydrogels with two separate ligand solutions – tartaric acid and oxalic acid – at 100 °C over a period of 24 h. According to thermal analysis, the newly created composite beads exhibit improved thermal stability when contrasted with the alginate beads cross-linked with Cu ions. The antibacterial assays demonstrated a significant efficacy of the Cu-based MOFs/sodium alginate composite beads in halting the proliferation of all tested bacterial strains, achieving a minimum inhibition zone of 23 mm. This underscores the synthesized materials' promise as exceptionally effective antibacterial substances [71]. In 2023, a zinc-based MOF (Zn-BTC), utilizing $C_6H_3(CO_2H)_3$, was combined with the derivatives of carbon, namely graphene quantum dots (GQDs) and reduced graphene oxide (rGO). The components were interconnected within an Alg hydrogel framework, resulting in the creation of a hydrogel of Zn-BTC@GQDs-rGO@Alg. The carefully designed beads of the hydrogel composite of Zn-BTC@GQDs-rGO@Alg were employed to improve the adsorption and extraction of U(VI) from aqueous solutions, taking advantage of the strong attraction of U(VI) to the O functional groups in the composite. The composite was distinguished by its multiple active functionalities, crystalline nature, and an average particle size of 19.43 nm. The adsorption isotherms for U(VI) demonstrated a strong correlation with the Langmuir adsorption model, both in linear and nonlinear representations. A novel nanohybrid composite, the hydrogel of Zn-BTC@GQDs-rGO@Alg, was successfully employed for U(VI) removal from wastewater, tap water, and seawater. The uptake percentages were impressive: 93.52% for seawater, 90.51% for tap water, and 92.68% for wastewater. Notably, this composite showed exceptional efficiency and impressive stability in absorbing U(VI), outperforming other adsorbents. Additionally, the composite was recycled five times without compromising its performance [72]. Alginate (Alg) is used as a base to integrate four distinct MOFs, namely Zn-BTC, Cu-BTC, Cr-BTC, and Fe-BTC, resulting in a variety of hydrogels designed for the adsorption of the vapor of *n*-hexane. The Alg composites with Zn-BTC and Cu-BTC exhibited a mesoporous structure, whereas those combined with Cr-BTC and Fe-BTC illustrated microporous structures. Among the various hydrogels, the Fe-BTC/Alg demonstrated the greatest capacity for adsorbing the vapor of *n*-hexane. In contrast, the Cr-BTC/Alg showed a lower adsorption capability, which can be attributed to the varying degrees of affinity that the metal ions have toward the Alg. The capacity of M-BTC/Alg to adsorb is influenced by a synergistic effect arising from both the surface area and the pore size. To assess the reproducibility of the M-BTC/Alg, five cycles of adsorption followed by thermal desorption were conducted, revealing that the Fe-BTC/alginate possesses excellent regenerative properties [73]. A group of researchers proposed a new one-step technique for preparing a 3D Alg-based MOF, which they then used for tetracycline removal. Instead of forming the hydrogel of ZIF-67-Alg (AM-M) using Co-Alg as the precursor, they allowed Co^{2+} to chelate with Alg and 2-methylimidazolate simultaneously, resulting in a distinct MOF hydrogel (MA-M). In terms of tetracycline adsorption, the MA-M hydrogel achieved equilibrium more quickly than the AM-M hydrogel. The highest adsorption capacity for tetracycline on MA-M reached 364.89 mg g^{-1}, exceeding the 302.32 mg g^{-1} capacity of AM-M. Additionally, MA-M exhibited outstanding regeneration capabilities even after being used for 10 cycles [74]. Eltaweil et al. developed a novel composite beads adsorbent based on MOF materials. Specifically, they integrated UiO-66 MOF and carboxylated graphene oxide (GOCOOH) into NaAlg. This composite was created to effectively remove methylene blue dye and Cu^{2+} ions from water solutions. Integrating both UiO-66 and GOCOOH into SA beads greatly improved their ability to adsorb MB and Cu^{2+} ions. Specifically, the highest adsorption capacities were 490.72 mg g^{-1} for MB and 343.49 mg g^{-1} for Cu^{2+}. The elimination process for both MB and Cu^{2+} follows pseudo-2nd-order kinetics and aligns with the Freundlich adsorption isotherm. Moreover, the regeneration experiments revealed that the removal efficiencies for MB and Cu^{2+} ions were maintained at over 87% even after five successive cycles [75]. A new composite of MOF(Fe)@NaAlg aerogels was reported using a straightforward ion cross-linking method. These aerogels, loaded with NH^{4+}, were designed as slow-release fertilizers. The adsorption capacity and swelling of the corresponding composite (with a ratio of 2:10) were 29.4 mg g^{-1} and 73 g g^{-1}, respectively. MOF(Fe) had an effective role in the slow-release properties of the prepared formulation [76].

20.5 Pectin-based MOF

In 2021, a modified Cu-based MOF infused with folic acid was crafted and integrated into appropriate pectin-based electrospun nanofibers. This enhancement not only refined the release dynamics of copper ions but also fortified the nanofiber mat, endowed it with antibacterial properties, and promoted angiogenesis as well as the migration and proliferation of fibroblasts, attributed to the presence of Cu^{2+} and folic acid. The study revealed that the incorporation of Cu-based MOFs increased the Young's modulus of the pectin nanofibers, elevating it to 19.13 MPa. In addition, the nanocomposite pectin nanofibers illustrated antibacterial properties and biocompatibility. These findings suggested that pectin nanofibers containing MOFs hold significant potential for use in biomedical applications [77]. The proton exchange membrane fuel cell (PEMFC), which utilized Ni-based pectin MOF as the anode catalyst and was denoted as PG-NaNi/Nf/Pt, was successfully created and underwent testing. The maximum current density of 5.2 mA cm^{-2}, maximum power density of 1.5 mW cm^{-2}, and an open circuit voltage of 0.96 V are considered sufficiently high for a PEMFC with an anode that does not contain a Pt group metal catalyst and a cathode that includes Pt [78]. A new hydrogel polymer, composed of a Cu/dipicolinic acid-MOF interlinked with oxidized pectin and chitosan (Cu/DPA-MOF/OP/CS), was crafted using an environmentally friendly and controllable method. This innovative material incorporates biodegradable polymers like pectin and chitosan within its architecture. The synthesized hydrogel was characterized by its nanometric particle size, extensive specific surface area, and remarkable thermal stability, which were its key attributes. The assessment of the antimicrobial efficacy of the newly created compounds involved a comparative analysis with cefazolin and terbinafine, two established pharmaceuticals. The study revealed that the antimicrobial capabilities of the synthetic compounds surpassed those of the commercial drugs. This enhanced antimicrobial activity is likely due to the compounds' intricate network structure, their nano-scale dimensions, and their considerable specific surface area coupled with consistent porosity [79]. Scientists developed an advanced dual-responsive pesticide delivery system by incorporating pectin with the nanoparticles of FeMOF. The nanoparticles of FeMOF were infused with pyraclostrobin, creating the nanoparticles of PYR@FeMOF-pectin. The system responded to both redox and pectinase stimuli. The study findings revealed that the corresponding nanoparticles had a loading capacity of almost 20.6% for pyraclostrobin. The pectin coating on the nanoparticles of PYR@FeMOF protected pyraclostrobin from photolysis and improved its spread on rice leaves. The bioactivity survey results showed that the nanoparticles of PYR@FeMOF-pectin exhibited better fungicidal effectiveness and lasted longer against Magnaporthe oryzae than the pyraclostrobin suspension concentrate. Furthermore, the nanocarriers of FeMOF-pectin showed no significant phytotoxicity and effectively increased both the shoot and root lengths in rice plants. Also, the nanoparticles of PYR@FeMOF-pectin illustrated an eightfold decrease in acute toxicity to zebrafish with respect to the pyraclostrobin suspension concentrate [80].

20.6 Chitin- and chitosan-based MOF

Since their inception, MOFs have experienced rapid development. Their remarkable properties, including expansive surface areas, customizable pore dimensions, robust thermal and chemical stability, and adaptable composition, have expanded their utility across diverse research domains such as gas storage, delivery systems, catalysis, sensing, adsorption, and separation. However, the relatively high production costs of MOFs have impeded their widespread practical implementation [81, 82].

Chitin and chitosan, distinguished by their inherent versatility and processability, have garnered significant attention. Their surface chemistry, amenable to tailored modification, positions them as promising candidates for diverse applications. Driven by the imperative for sustainable solutions, researchers have dedicated efforts to transform these biopolymers into eco-friendly alternatives to conventional materials across biomedical, environmental, and energy domains [83]

Chitin, a linear polysaccharide and precursor to chitosan, comprises repeating units of *N*-acetyl-D-glucosamine and D-glucosamine linked by 1–4-β-glycosidic bonds (Figure 20.4). This structural foundation underpins the material's unique characteristics and potential applications [84]. Primarily sourced from the exoskeletons of crustaceans, mollusks, insects, and certain fungi, chitin is a polysaccharide predominantly composed of *N*-acetyl-D-glucosamine. Characterized by its degree of deacetylation, chitin differs from chitosan, which is primarily D-glucosamine and defined by its degree of acetylation. Renowned for their antibacterial properties, nontoxicity, and biodegradability, these biopolymers offer a promising foundation for a wide array of composite applications. Their potential to serve as eco-friendly alternatives to synthetic counterparts in energy and environmental sectors has stimulated considerable research interest [85].

The structural foundation differentiating chitin from chitosan lies in the composition of their monomeric units. Chitin is primarily composed of *N*-acetyl-D-glucosamine residues, while chitosan predominantly consists of D-glucosamine units. This compositional variance significantly impacts their solubility behavior. Chitin, due to the presence of *N*-acetyl groups, exhibits limited solubility in most solvents, including aqueous solutions. Conversely, chitosan, with its abundant amino groups, readily dissolves in acidic aqueous environments owing to protonation. A critical factor distinguishing chitin from chitosan is the degree of deacetylation (DD). This parameter quantifies the extent to which acetyl groups have been removed from the chitin backbone to form chitosan. Generally, a DD value exceeding 50% is characteristic of chitosan, while lower values correspond to chitin. The DD significantly influences chitosan's properties, including solubility, viscosity, and biocompatibility [86]. Beyond compositional differences, the molecular architecture of these biopolymers contributes to their distinct characteristics. Both chitin and chitosan possess two hydroxyl groups per monomeric unit: a primary hydroxyl at the C-6 position and a secondary hydroxyl at the C-3 position. The presence of *N*-acetyl groups in chitin facilitates the formation of intramolecular and intermolecular hydrogen bonds, resulting in a more rigid and crystalline structure compared to chitosan. This structural arrangement impacts factors such as solubility, molecular weight, and mechanical properties [87].

Chitin, a natural polymer, exists in three primary structural forms: alpha, beta, and gamma, as illustrated in Figure 20.5. These forms exhibit distinct properties due to their unique microstructures. Alpha chitin, the most common form, has antiparallel chains and the highest crystallinity. It is primarily found in the outer skeletons of crustaceans, insect cuticles, fungi, yeast cell walls, and marine sponges. Beta chitin, with parallel chains, is primarily extracted from squids, diatom extracellular fibers, and the spines and bristles of certain annelids. Gamma chitin, a hybrid form, has a mixed arrangement of parallel and antiparallel chains. In contrast, chitosan is a heteropolysaccharide composed of glucosamine and *N*-acetylglucosamine units connected by a linear beta-1,4 linkage [88]. The distribution and arrangement of *N*-acetyl -D-glucosamine and D-glucosamine units within a polymer can significantly impact its material properties under various conditions. In heterogeneous environments, chitosan often exhibits an irregular distribution of these units, with a predominantly random arrangement and a minor degree of block-like distribution. The physicochemical

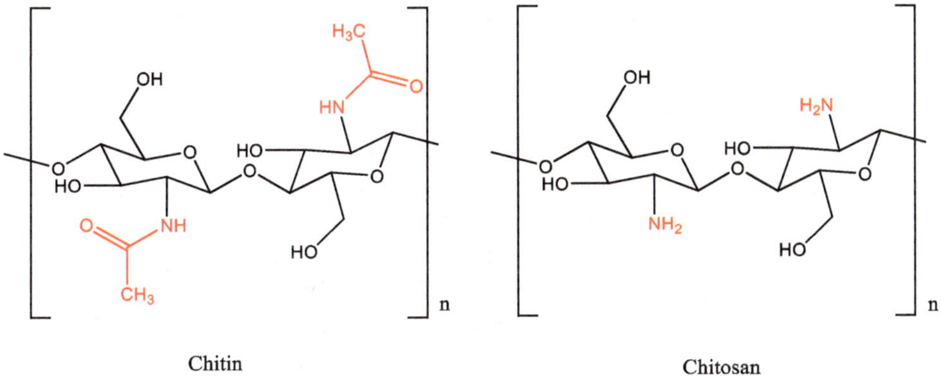

Figure 20.4: Structure of chitin and chitosan.

properties of this chitosan differ from those of chitosans that are randomly acetylated [89]. The presence of various surface functional groups, primarily hydroxyl, acetyl, and amino groups, contributes to a net positive charge on the material's surface [85].

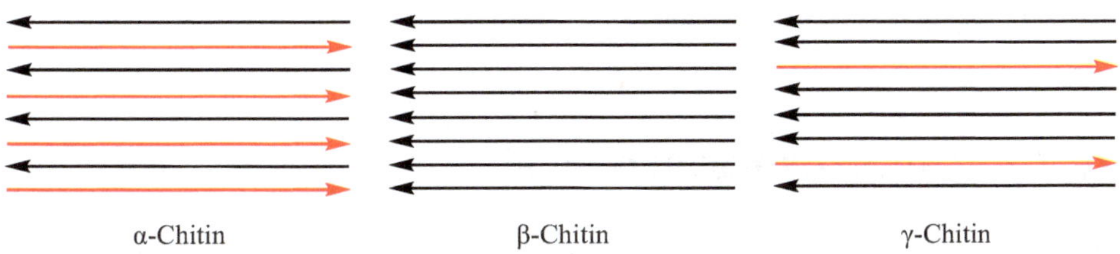

Figure 20.5: Schematic representation of the three polymorphic forms of chitin.

A novel composite, comprising chitin (CH) gel and the nanoparticles of MOF-808, was successfully synthesized by Sadjadi et al. The integration of the nanoparticles of MOF-808 into the CH matrix created a material with a unique hierarchical pore structure, encompassing micro-, meso-, and macropores. This multi-porous architecture significantly enhanced the surface area available for adsorption, making the composite an effective adsorbent for a wide range of pollutants. The CH/MOF-808 composite leveraged the individual properties of its components while demonstrating synergistic effects in certain cases. For instance, it outperformed both MOF-808 and CH in the adsorption of methylene blue, arsenic (AsV), Polymyxin B, and erythrosine. This enhanced performance was attributed to the compartmentalization of MOF-808 nanoparticles within the CH macropores and the synergistic interactions at the MOF-808/CH interface. The composite exhibited excellent adsorption kinetics and capacity for AsV removal from real-world water samples. Adsorption isotherm and kinetic studies, along with dielectric response analysis, confirmed that AsV was captured through chemisorption at the nanoparticles of MOF-808. The CH/MOF-808 composite represents a promising technology for addressing various water pollution challenges. Its versatility, derived from the structural diversity of MOFs and the biocompatibility of CH, makes it well-suited for applications where pore size, surface area, and pore chemistry are critical factors [89].

A novel covalent composite, consisting of MOFs (MOF)-Fe and Chitin (CH), was synthesized by Tovar et al. via a reaction between MOF-Fe and Cl-functionalized Ch. The composite was thoroughly characterized and demonstrated exceptional catalytic activity for the photodegradation of methylene blue (MB) under visible light irradiation. To optimize degradation efficiency, the influence of various reaction parameters, including initial MB concentration, pH, and catalyst dosage, was systematically investigated. Results indicated that alkaline conditions and a catalyst loading of 0.01 g L^{-1} yielded the most favorable degradation performance. Kinetic studies revealed that the composite exhibited significantly superior catalytic activity compared to its individual components, MOF-Fe and Ch. Scavenger experiments confirmed that photogenerated holes (h+) played a pivotal role in the photocatalytic degradation of MB. Furthermore, the MOF-Ch composite demonstrated excellent reusability, maintaining its catalytic activity after five consecutive reaction cycles. Notably, this versatile catalyst was not only effective for MB degradation but also exhibited promising performance in the degradation of other dyes [90].

20.7 Cyclodextrin-based MOF

CDs, a class of cyclic oligosaccharides, have gained significant attention as building blocks for biocompatible MOFs materials. These natural compounds were first isolated by a French scientist in 1891 using enzymes derived from Bacillus to catalyze amylose. Regulatory bodies like the European Medicines Agency (EMA) and the US Food and Drug Administration (FDA) have acknowledged their safety, approving both natural CDs and their derivatives for use as additives (Figure 20.6) [91].

CDs are cyclic oligosaccharides produced enzymatically from amylose. Classified as alpha-CD, beta-CD, or gamma-CD based on the number of glucose units (6, 7, or 8), CDs possess a truncated cone-like structure with a hydrophobic cavity and hydrophilic exterior. This amphiphilic nature enables CDs to form stable inclusion complexes with hydrophobic molecules, driven by various intermolecular forces and the release of water molecules from the CD cavity. The formation of host-guest complexes can significantly enhance the properties of the guest molecules, including stability, solubility, bioavailability, and anti-volatility. While alpha-CD, beta-CD, and gamma-CD share a similar molecular structure, their cavity sizes differ, influencing their inclusion capabilities. Beta-CD, with its intermediate cavity size and lower solubility, is commonly used for complexation. Researchers have synthesized CD-based MOFs by incorporating CDs as organic ligands and coordinating them with alkali metal salts.

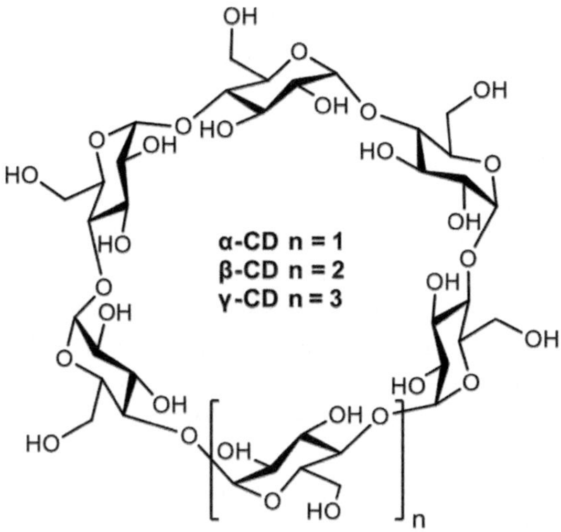

Figure 20.6: Chemical tridimensional structure and geometric dimensions of cyclodextrins.

The type of CD-MOF depends on the specific CD and metal ion combination, resulting in unique spatial structures [92, 93].

The formation of CD-MOFs involves a nucleation and growth process driven by the assembly of nucleation building units (NBUs). NBUs, smaller than secondary building units (SBUs), are soluble metal complexes that condense with metal ions, organic ligands, or other NBUs to form SBUs. These SBUs are then interconnected with organic ligands to create the final CD-MOF structure. The growth process and nucleation of CD-MOFs are influenced by various factors, including reaction reagents, solvents, regulators, heating methods, heating rate, and reaction time. These factors can change the thermodynamic and kinetic conditions during synthesis, thus controlling the formation of CD-MOF crystals. Researchers have used various techniques to regulate the size of MOF crystals, such as microwave-assisted, ultrasonic-assisted, and surfactant-assisted techniques. Surfactants, amphiphilic molecules with hydrophobic and hydrophilic groups, can selectively inhibit the growth of certain crystal surfaces, leading to changes in crystal morphology. Common surfactants used in MOF synthesis include cetyltrimethylammonium bromide, polyethylene glycol, and polyvinylpyrrolidone. Microwave and ultrasonic treatments offer simple, cost-effective, and efficient alternatives to traditional heating methods. These technologies create favorable conditions for CD-MOF crystal formation through electromagnetic waves and ultrasound, respectively, making them attractive approaches for researchers [94–96].

CD-MOFs offer promising applications as fragrance and flavor delivery systems. Their ability to protect volatile molecules from degradation and premature evaporation is particularly valuable in these industries. Menthol, a volatile terpene alcohol, was successfully encapsulated within beta-CD-MOFs, resulting in a 30.6% encapsulation efficiency and resistance to high temperatures. This encapsulation effectively prevented menthol volatilization and provided a controlled release profile. Zhang et al. loaded various ester and terpene fragrances into gamma-CD-MOFs, demonstrating that the encapsulation and release behavior were influenced by fragrance type. Ester fragrances exhibited higher loading and a linear release pattern, while terpene fragrances displayed an initial burst release. The results indicated that gamma-CD-MOFs could provide controlled release of fragrances over extended periods, making them suitable for applications in cosmetics, personal care, and food flavor industries. D-Limonene, a volatile citrus oil, was also encapsulated within gamma-CD-MOFs, effectively inhibiting its volatilization. This formulation showed potential as an inhalable dry powder inhaler, increasing the bioavailability of D-Limonene by 2.23-fold compared to oral administration. This highlights the potential of CD-MOFs as pulmonary drug delivery carriers [97–99].

CD-MOFs, with their porous structure and abundance of hydroxyl groups, offer promising applications in gas adsorption, storage, and separation. Their ability to physically or chemically adsorb various gases is particularly valuable for environmental and industrial purposes. Ethylene, a plant growth regulator used in postharvest ripening management, presents challenges due to its flammability and explosiveness. CD-MOFs, with their adjustable porosity, offer potential for long-term ethylene storage. Studies have demonstrated the superior ethylene encapsulation capabilities of CD-MOFs compared to other materials, such as V-type starch. The encapsulation and controlled release of ethylene within CD-MOFs can effectively delay the ripening of fruits and vegetables, offering novel solutions for the food and agriculture industries. Beyond ethylene, CD-MOFs can be used to encapsulate and deliver various functional factors. Their porous structure allows for the encapsulation of liquids, solids, and gases, while their high loading capacity makes them suitable for applications in biomedicine, agriculture, and food industries [100–102].

20.8 Future scope and summary

CHs-based MOFs represent a new and promising category of hybrid materials. Incorporating CHs into MOF structures enhances their structural robustness and flexibility, as well as their physicochemical properties. These CHs-MOFs are also biocompatible and provide additional mechanical strength. In the design of MOFs, the choice of ligand is crucial for controlling the size and shape of cavities and constructing functional MOFs. This selection depends on the specific CHs used, considering their charge, geometry, subunits, functional groups, and bioactive sites. These unique properties of CHs-MOFs enable their application in various fields, including (i) catalysis, (ii) air filtration, (iii) environmental remediation, and (iv) drug delivery.

Despite significant advancements in the preparation techniques of CHs-based MOFs, many limitations remain regarding their stability and practical applications. These limitations include:

(1) Challenges in preserving the open active site and porosity in CHs-based MOFs.
(2) Maintaining stability in an aqueous environment.
(3) Challenges in creating ordered crystalline materials and predicting their global structures.

We are confident that CHs-MOFs hold great potential for future applications in nanomedicine, bio-nanotechnology, material sciences, bio-electrochemistry, synthesis, and catalysts. It should be noted that CHs-MOFs will become valuable assets in the field of smart materials.

References

[1] Qiu J, Li X, Gref R, Vargas-Berenguel A. Carbohydrates in metal organic frameworks: Supramolecular assembly and surface modification for biomedical applications. In Metal-Organic Frameworks for Biomedical Applications. Woodhead Publishing, 2020, 445–465.
[2] Mika LT, Cséfalvay E, Németh Á. Catalytic conversion of carbohydrates to initial platform chemicals: Chemistry and sustainability. Chem Rev, 2018, 118(2), 505–613.
[3] Ayoub A, Treasure T, Hansen L, Nypelö T, Jameel H, Khan S, . . . Venditti RA. Effect of plasticizers and polymer blends for processing softwood kraft lignin as carbon fiber precursors. Cellulose, 2021, 28, 1039–1053.
[4] Varma RS. Journey on greener pathways: From the use of alternate energy inputs and benign reaction media to sustainable applications of nano-catalysts in synthesis and environmental remediation. Green Chem, 2014, 16(4), 2027–2041.
[5] Hardy JJ, Hubert S, Macquarrie DJ, Wilson AJ. Chitosan-based heterogeneous catalysts for Suzuki and Heck reactions. Green Chem, 2004, 6(1), 53–56.
[6] Huang G, Mo LQ, Wei YX, Zhou H, Guo YA, Wei SJ. Effect of mesoporous chitosan action and coordination on the catalytic activity of mesoporous chitosan-grafted cobalt tetrakis (p-sulfophenyl) porphyrin for ethylbenzene oxidation. Catalysts, 2018, 8(5), 199.
[7] Delbianco M, Bharate P, Varela-Aramburu S, Seeberger PH. Carbohydrates in supramolecular chemistry. Chem Rev, 2016, 116(4), 1693–1752.
[8] Zhang F, Wang D, Qin H, Feng L, Liang X, Qing G. Chemoselectivity of pristine cellulose nanocrystal films driven by carbohydrate–carbohydrate interactions. ACS Appl Mater Interfaces, 2019, 11(14), 13114–13122.
[9] Moon RJ, Martini A, Nairn J, Simonsen J, Youngblood J. Cellulose nanomaterials review: Structure, properties and nanocomposites. Chem Soc Rev, 2011, 40(7), 3941–3994.
[10] Moon RJ, Martini A, Nairn J, Simonsen J, Youngblood J. Cellulose nanomaterials review: Structure, properties and nanocomposites. Chem Soc Rev, 2011, 40(7), 3941–3994.
[11] Trache D, Tarchoun AF, Derradji M, Hamidon TS, Masruchin N, Brosse N, Hussin MH. Nanocellulose: From fundamentals to advanced applications. Front Chem, 2020, 8, 392.
[12] Yan Y, Li C, Wu Y, Gao J, Zhang Q. From isolated Ti-oxo clusters to infinite Ti-oxo chains and sheets: Recent advances in photoactive Ti-based MOFs. J Mater Chem A, 2020, 8(31), 15245–15270.

[13] Wang K, Li Q, Ren Z, Li C, Chu Y, Wang Z, . . . Zhang Q. 2D metal–organic frameworks (MOFs) for high-performance BatCap hybrid devices. Small, 2020, 16(30), 2001987.

[14] Yang B, Wang H, Zhang DW, Li ZT. Water-Soluble three-dimensional polymers: Non-covalent and covalent synthesis and functions. Chin J Chem, 2020, 38(9), 970–980.

[15] Zhang X, Wang X, Fan W, Sun D. Pore-Environment Engineering in Multifunctional Metal-Organic Frameworks. Chin J Chem, 2020, 38(5), 509–524.

[16] Cheng D, Li P, Zhu X, Liu M, Zhang Y, Liu Y. Enzyme-free Electrochemical Detection of Hydrogen Peroxide Based on the Three-Dimensional Flower-like Cu-based Metal Organic Frameworks and MXeneNanosheets. Chin J Chem, 2021, 39(8), 2181–2187.

[17] Gao X, Pei L, Xue W, Huang H, Gao Z, Zhao X. MonodentateAIEgen Anchored on Metal-Organic Framework for Fast Fluorescence Sensing of Phosphate. Chin J Chem, 2021, 39(1), 99–105.

[18] Huang H, Sun Y, Jia X, Xue W, Geng C, Zhao X, . . . Zhong C. Air-steam etched construction of hierarchically porous metal-organic frameworks. Chin J Chem, 2021, 39(6), 1538–1544.

[19] Lin X, Wang L, Han Z, Chen Z. Chiral spirocyclic phosphoric acids and their growing applications. Chin J Chem, 2021, 39(4), 802–824.

[20] Liu M, Ren X, Meng X, Li H. Metal-Organic frameworks-based fluorescent nanocomposites for bioimaging in living cells and in vivo. Chin J Chem, 2021, 39(2), 473–487.

[21] Li B, Zhang Z, Li Y, Yao K, Zhu Y, Deng Z, . . . Li J. Enhanced binding affinity, remarkable selectivity, and high capacity of CO2 by dual functionalization of a rht-type metal–organic framework. Angew Chem Int Ed, 2012, 51(6), 1412–1415.

[22] Liu K, Li B, Li Y, Li X, Yang F, Zeng G, . . . Song D. An N-rich metal–organic framework with an rht topology: High CO 2 and C 2 hydrocarbons uptake and selective capture from CH 4. Chem Commun, 2014, 50(39), 5031–5033.

[23] Wang H, Liu Y, Li J. Designer metal–organic frameworks for size-exclusion-based hydrocarbon separations: Progress and challenges. Adv Mater, 2020, 32(44), 2002603.

[24] Wang H, Warren M, Jagiello J, Jensen S, Ghose SK, Tan K, . . . Li J. Crystallizing atomic xenon in a flexible MOF to probe and understand its temperature-dependent breathing behavior and unusual gas adsorption phenomenon. J Am Chem Soc, 2020, 142(47), 20088–20097.

[25] Li C, Wang K, Li J, Zhang Q. Recent progress in stimulus-responsive two-dimensional metal–organic frameworks. ACS Mater Lett, 2020, 2(7), 779–797.

[26] Chang LM, An YY, Li QH, Gu ZG, Han YF, Zhang J. N-Heterocyclic carbene as a surface platform for assembly of homochiral metal–organic framework thin films in chiral sensing. ACS Appl Mater Interfaces, 2020, 12(34), 38357–38364.

[27] He J, Xu J, Yin J, Li N, Bu XH. Recent advances in luminescent metal-organic frameworks for chemical sensors. Sci China Mater, 2019, 62(11), 1655–1678.

[28] Gao LL, Gao EQ. Metal–organic frameworks for electrochemical sensors of neurotransmitters. Coord Chem Rev, 2021, 434, 213784.

[29] Lakshmi PR, Nanjan P, Kannan S, Shanmugaraju S. Recent advances in luminescent metal–organic frameworks (LMOFs) based fluorescent sensors for antibiotics. Coord Chem Rev, 2021, 435, 213793.

[30] Zhang Y, Jiao L, Yang W, Xie C, Jiang HL. Rational fabrication of low-coordinate single-atom Ni electrocatalysts by MOFs for highly selective CO2 reduction. Angew Chem Int Ed, 2021, 60(14), 7607–7611.

[31] Li Y, Lu M, Wu Y, Ji Q, Xu H, Gao J, . . . Zhang Q. Morphology regulation of metal–organic framework-derived nanostructures for efficient oxygen evolution electrocatalysis. J Mater Chem A, 2020, 8(35), 18215–18219.

[32] Li B, Leng K, Zhang Y, Dynes JJ, Wang J, Hu Y, . . . Ma S. Metal–organic framework based upon the synergy of a Brønsted acid framework and Lewis acid centers as a highly efficient heterogeneous catalyst for fixed-bed reactions. J Am Chem Soc, 2015, 137(12), 4243–4248.

[33] Liu X, Li J, Li N, Li B, Bu XH. Recent advances on metal-organic frameworks in the conversion of carbon dioxide. Chin J Chem, 2021, 39(2), 440–462.
[34] Joarder B, Lin JB, Romero Z, Shimizu GK. Single crystal proton conduction study of a metal organic framework of modest water stability. J Am Chem Soc, 2017, 139(21), 7176–7179.
[35] Ye Y, Gong L, Xiang S, Zhang Z, Chen B. Metal–organic frameworks as a versatile platform for proton conductors. Adv Mater, 2020, 32(21), 1907090.
[36] Huang Z, Yu H, Wang L, Liu X, Lin T, Haq F, . . . Lemenovskiy DA. Ferrocene-contained metal organic frameworks: From synthesis to applications. Coord Chem Rev, 2021, 430, 213737.
[37] Wong NE, Ramaswamy P, Lee AS, Gelfand BS, Bladek KJ, Taylor JM, . . . Shimizu GK. Tuning intrinsic and extrinsic proton conduction in metal–organic frameworks by the lanthanide contraction. J Am Chem Soc, 2017, 139(41), 14676–14683.
[38] Li B, Dong X, Wang H, Ma D, Tan K, Jensen S, . . . Li J. Capture of organic iodides from nuclear waste by metal-organic framework-based molecular traps. Nat Commun, 2017, 8(1), 485.
[39] Gai S, Fan R, Zhang J, Zhou X, Xing K, Zhu K, . . . Yang Y. Fabrication of highly stable metal–organic frameworks and corresponding hydrophobic foam through a reticular chemistry strategy for simultaneous organic micropollutant and insoluble oil removal from wastewater. J Mater Chem A, 2021, 9(6), 3369–3378.
[40] Karmakar A, Li J. The best of both worlds: An MOP/COF-based hybrid material for highly selective and very fast sequestration of toxic oxoanions from water. 2020.
[41] Yohannes A, Li J, Yao S. Various metal organic frameworks combined with imidazolium, quinolinum and benzothiazolium ionic liquids for removal of three antibiotics from water. J Mol Liq, 2020, 318, 114304.
[42] Zheng Y, Sun FZ, Han X, Xu J, Bu XH. Recent progress in 2D metal-organic frameworks for optical applications. Adv Opt Mater, 2020, 8(13), 2000110.
[43] Wang K, Bi R, Huang M, Lv B, Wang H, Li C, . . . Zhang Q. Porous cobalt metal–organic frameworks as active elements in battery–supercapacitor hybrid devices. Inorg Chem, 2020, 59(10), 6808–6814.
[44] Zhang Q, Zhou Z, Pan Z, Sun J, He B, Li Q, . . . Yao Y. All-metal-organic framework-derived battery materials on carbon nanotube fibers for wearable energy-storage device. Adv Sci, 2018, 5(12), 1801462.
[45] Zhong M, Kong L, Zhao K, Zhang YH, Li N, Bu XH. Recent progress of nanoscale metal-organic frameworks in synthesis and battery applications. Adv Sci, 2021, 8(4), 2001980.
[46] Liu J, Huang J, Zhang L, Lei J. Multifunctional metal–organic framework heterostructures for enhanced cancer therapy. Chem Soc Rev, 2021, 50(2), 1188–1218.
[47] Samanta SK, Isaacs L. Biomedical applications of metal organic polygons and polyhedra (MOPs). Coord Chem Rev, 2020, 410, 213181.
[48] Simon-Yarza T, Mielcarek A, Couvreur P, Serre C. Nanoparticles of metal-organic frameworks: On the road to in vivo efficacy in biomedicine. Adv Mater, 2018, 30(37), 1707365.
[49] Zhao SN, Zhang Y, Song SY, Zhang HJ. Design strategies and applications of charged metal organic frameworks. Coord Chem Rev, 2019, 398, 113007.
[50] Liu X, Xiao Y, Zhang Z, You Z, Li J, Ma D, Li B. Recent progress in metal-organic frameworks@ cellulose hybrids and their applications. Chin J Chem, 2021, 39(12), 3462–3480.
[51] Tung TT, Tran MT, Feller JF, Castro M, Van Ngo T, Hassan K, . . . Losic D. Graphene and metal organic frameworks (MOFs) hybridization for tunable chemoresistive sensors for detection of volatile organic compounds (VOCs) biomarkers. Carbon, 2020, 159, 333–344.
[52] Kumar R, Raut D, Ramamurty U, Rao CNR. Remarkable improvement in the mechanical properties and CO2 uptake of MOFs brought about by covalent linking to graphene. Angew Chem Int Ed, 2016, 55(27), 7857–7861.
[53] Liang J, Liang Z, Zou R, Zhao Y. Heterogeneous catalysis in zeolites, mesoporous silica, and metal–organic frameworks. Adv Mater, 2017, 29(30), 1701139.

[54] Koo J, Hwang IC, Yu X, Saha S, Kim Y, Kim K. Hollowing out MOFs: Hierarchical micro-and mesoporous MOFs with tailorable porosity via selective acid etching. Chem Sci, 2017, 8(10), 6799–6803.

[55] Kitao T, Zhang Y, Kitagawa S, Wang B, Uemura T. Hybridization of MOFs and polymers. Chem Soc Rev, 2017, 46(11), 3108–3133.

[56] Jin P, Tan W, Huo J, Liu T, Liang Y, Wang S, Bradshaw D. Hierarchically porous MOF/polymer composites via interfacial nanoassembly and emulsion polymerization. J Mater Chem A, 2018, 6(41), 20473–20479.

[57] Bian W, Chen J, Chen Y, Xu W, Jia J. ORIGINAL RESEARCH A novel waste paper cellulose-based Cu-MOF hybrid material threaded by PSS for lithium extraction with high adsorption capacity and selectivity. Cellulose, 2021, 28(5), 3041–3054. 10.1007/s10570-021-03707-y.

[58] Jabbar SQ. Cellulose-based hybrid composites enabled by metal organic frameworks for CO2 capture: The effect of cellulosic substrate. 2022, 0–31.

[59] Li Q. et al. Injectable and self-healing chitosan-based hydrogel with MOF-loaded α-lipoic acid promotes diabetic wound healing. Mater Sci Eng C, 2021, 131(October), 112519. 10.1016/j.msec.2021.112519.

[60] Liu L. et al. Multifunctional Ag@MOF-5@chitosan non-woven cloth composites for sulfur mustard decontamination and hemostasis. Dalt Trans, 2019, 48(20), 6951–6959. 10.1039/c9dt00503j.

[61] Yousefian M, Rafiee Z. Cu-metal-organic framework supported on chitosan for efficient condensation of aromatic aldehydes and malononitrile. Carbohydr Polym, 2020, 228(no. September 2019), 10.1016/j.carbpol.2019.115393.

[62] Omer AM, Abd El-Monaem EM, Abd El-Latif MM, El-Subruiti GM, Eltaweil AS. Facile fabrication of novel magnetic ZIF-67 MOF@aminated chitosan composite beads for the adsorptive removal of Cr(VI) from aqueous solutions. Carbohydr Polym, 2021, 265(April), 118084. 10.1016/j.carbpol.2021.118084.

[63] Ehsani A, Bigdeloo M, Assefi F, Kiamehr M, Alizadeh R. Ternary nanocomposite of conductive polymer/chitosan biopolymer/metal organic framework: Synthesis, characterization and electrochemical performance as effective electrode materials in pseudocapacitors. Inorg Chem Commun, 2020, 115(February), 107885. 10.1016/j.inoche.2020.107885.

[64] Zeyadi M, Almulaiky YQ. Chitosan-Based metal-organic framework for Stabilization of β-glucosidase: Reusability and storage stability. Heliyon, 2023, 9(11), e21169. 10.1016/j.heliyon.2023.e21169.

[65] Gouda M, Ibrahim HIM, Negm A. Chitosan containing nano zn-organic framework: Synthesis, characterization and biological activity. Polymers (Basel), 2022, 14(7), 10.3390/polym14071276.

[66] Valadi FM, Shahsavari S, Akbarzadeh E, Gholami MR. Preparation of new MOF-808/chitosan composite for Cr(VI) adsorption from aqueous solution: Experimental and DFT study. Carbohydr Polym, 2022, 288, 119383, https://doi.org/10.1016/j.carbpol.2022.119383.

[67] Chen J. et al. Fabrication and adsorption mechanism of chitosan/Zr-MOF (UiO-66) composite foams for efficient removal of ketoprofen from aqueous solution. Chem Eng J, 2022, 431, 134045, https://doi.org/10.1016/j.cej.2021.134045.

[68] Pooresmaeil M, Namazi H. Chitosan coated Fe3O4@Cd-MOF microspheres as an effective adsorbent for the removal of the amoxicillin from aqueous solution. Int J Biol Macromol, 2021, 191, 108–117, https://doi.org/10.1016/j.ijbiomac.2021.09.071.

[69] Zhang S. et al. Preparation of novel poly(sodium p-styrenesulfonate)/sodium alginate hydrogel incorporated with MOF-5 nanoparticles for the adsorption of Pb(II) and tetracycline. J Mol Struct, 2024, 1300, 137313, https://doi.org/10.1016/j.molstruc.2023.137313.

[70] Ehsani A, Aghdasinia H, Farshchi ME, Rostamnia S, Khataee A. Synthesis of sodium alginate/carboxy-methyl cellulose/Cu-based metal-organic framework composite for adsorption of tetracycline from aqueous solution: Isotherm, kinetic and thermodynamic approach. Surf Interfaces, 2023, 36, 102506, https://doi.org/10.1016/j.surfin.2022.102506.

[71] Bradai M, AitRadi M, Zeggai FZ, Karkachi N, Meghabar R. In-situ Synthesis of Highly Potent Antibacterial Copper-Based MOFs/Sodium Alginate Composite Beads. ChemistrySelect, 2024, 9(13), e202304855. https://doi.org/10.1002/slct.202304855.

[72] Mahmoud ME, Tharwat RM, Abdelfattah AM, Hassan SSM. U(VI) capture from water-based systems by decorated nanohybrid of Zn-BTC MOF with GQDs-rGO and alginate hydrogel. J Environ Chem Eng, 2023, 11(5), 110497. https://doi.org/10.1016/j.jece.2023.110497.

[73] Li M, Huang W, Tang B, Fang Q, Ling X, Lv A. Characterizations and n-Hexane vapor adsorption of a series of MOF/Alginates. Ind Eng Chem Res, 2020, 59(42), 18835–18843. 10.1021/acs.iecr.9b06744.

[74] Zhuang Y, Kong Y, Wang X, Shi B. Novel one step preparation of a 3D alginate based MOF hydrogel for water treatment. New J Chem, 2019, 43(19), 7202–7208. 10.1039/c8nj06031b.

[75] Eltaweil AS, Mamdouh IM, Abd El-Monaem EM, El-Subruiti GM. Highly efficient removal for methylene blue and Cu2+ onto UiO-66 metal-organic framework/carboxylated graphene oxide-incorporated sodium alginate beads. ACS Omega, 2021, 6(36), 23528–23541. 10.1021/acsomega.1c03479.

[76] Wu C, Dan Y, Tian D, Zheng Y, Wei S, Xiang D. Facile fabrication of MOF(Fe)@alginate aerogel and its application for a high-performance slow-release N-fertilizer. Int J Biol Macromol, 2020, 145, 1073–1079, 10.1016/j.ijbiomac.2019.09.200.

[77] Zirak Hassan Kiadeh S, Ghaee A, Farokhi M, Nourmohammadi J, Bahi A, Ko FK. Electrospun pectin/modified copper-based metal–organic framework (MOF) nanofibers as a drug delivery system. Int J Biol Macromol, 2021, 173, 351–365, 10.1016/j.ijbiomac.2021.01.058.

[78] Kadirov MK. et al. A nickel-based pectin metal-organic framework as a hydrogen oxidation reaction catalyst for proton-exchange-membrane fuel cells. ChemistrySelect, 2019, 4(16), 4731–4734. 10.1002/slct.201802955.

[79] Jasim Al-Khafaji HH. et al. Synthesis of a novel Cu/DPA-MOF/OP/CS hydrogel with high capability in antimicrobial studies. Front Chem, 2023, 11(August), 1–10. 10.3389/fchem.2023.1236580.

[80] Liang Y. et al. Pectin functionalized metal-organic frameworks as dual-stimuli-responsive carriers to improve the pesticide targeting and reduce environmental risks. Coll Surf B Biointerfac, 2022, 219, 112796, https://doi.org/10.1016/j.colsurfb.2022.112796.

[81] Dong Y. et al. Influence of microwave-assisted synthesis on the structural and textural properties of mesoporous MIL-101 (Fe) and NH2-MIL-101 (Fe) for enhanced tetracycline adsorption. Mater Chem Phys, 2020, 251, 123060.

[82] Wang R. et al. High-quality Al@ Fe-MOF prepared using Fe-MOF as a micro-reactor to improve adsorption performance for selenite. J Hazard Mater, 2019, 364, 272–280.

[83] Peter S. et al. Chitin and chitosan based composites for energy and environmental applications: A review. Waste Biomass Valori, 2021, 12, 4777–4804.

[84] Nikolov S. et al. Robustness and optimal use of design principles of arthropod exoskeletons studied by ab initio-based multiscale simulations. J Mech Behav Biomed Mater, 2011, 4(2), 129–145.

[85] Paulino AT. et al. Characterization of chitosan and chitin produced from silkworm crysalides. Carbohydr Polym, 2006, 64(1), 98–103.

[86] Anitha A. et al. Chitin and chitosan in selected biomedical applications. Prog Polym Sci, 2014, 39(9), 1644–1667.

[87] Singh P, Nagendran R. A comparative study of sorption of chromium (III) onto chitin and chitosan. Appl Water Sci, 2016, 6, 199–204.

[88] El Knidri H. et al. Extraction, chemical modification and characterization of chitin and chitosan. Int J Biol Macromol, 2018, 120, 1181–1189.

[89] Sadjadi S. et al. Composite of MOF and chitin as an efficient catalyst for photodegradation of organic dyes. Int J Biol Macromol, 2021, 182, 524–533.

[90] Tovar GI. et al. Chitin/metal-organic framework composites as wide-range adsorbent. ChemSusChem, 2021, 14(14), 2892–2901.

[91] Zhao R-N. et al. Cyclodextrin-based metal-organic framework materials: Classifications, synthesis strategies and applications in variegated delivery systems. Carbohydr Polym, 2023, 121198.

[92] Lee J-U. et al. Noncovalent complexes of cyclodextrin with small organic molecules: Applications and insights into host–guest interactions in the gas phase and condensed phase. Molecules, 2020, 25(18), 4048.

[93] Morin-Crini N. et al. 130 years of cyclodextrin discovery for health, food, agriculture, and the industry: A review. Environ Chem Lett, 2021, 19(3), 2581–2617.

[94] Hamedi A. et al. A γ-cyclodextrin-based metal–organic framework (γ-CD-MOF): A review of recent advances for drug delivery application. J Drug Target, 2022, 30(4), 381–393.

[95] Qin G. et al. Polyvinylpyrrolidone-assisted synthesis of highly water-stable cadmium-based metal–organic framework nanosheets for the detection of metronidazole. RSC Adv, 2021, 11(55), 34842–34848.

[96] Oulkhir A. et al. A critical review on natural surfactants and their potential for sustainable mineral flotation. Rev Environ Sci Bio/Technol, 2023, 22(1), 105–131.

[97] Mallakpour S, Nikkhoo E, Hussain CM. Application of MOF materials as drug delivery systems for cancer therapy and dermal treatment. Coord Chem Rev, 2022, 451, 214262.

[98] Zhou Y. et al. Solidification of volatile D-Limonene by cyclodextrin metal-organic framework for pulmonary delivery via dry powder inhalers: In vitro and in vivo evaluation. Int J Pharm, 2021, 606, 120825.

[99] Hu Z. et al. Enhanced stability and controlled release of menthol using a β-cyclodextrin metal-organic framework. Food Chem, 2022, 374, 131760.

[100] Liu Y. et al. Simultaneous solubilization and extended release of insoluble drug as payload in highly soluble particles of γ-cyclodextrin metal-organic frameworks. Int J Pharm, 2022, 619, 121685.

[101] Li S. et al. Synthesis of γ-cyclodextrin metal-organic framework as ethylene absorber for improving postharvest quality of kiwi fruit. Food Hydrocoll, 2023, 136, 108294.

[102] Dummert SV. et al. Cyclodextrin metal–organic frameworks and derivatives: Recent developments and applications. Chem Soc Rev, 2022, 51(12), 5175–5213.

Index

https://doi.org/10.1515/9783111447094-021

De Gruyter Series in Green Bioactive Heterocycles

Volume 1
György Keglevich and Bubun Banerjee (Eds.)
Non-Conventional Synthesis, 2023
ISBN 978-3-11-099226-7, e-ISBN 978-3-11-098018-9

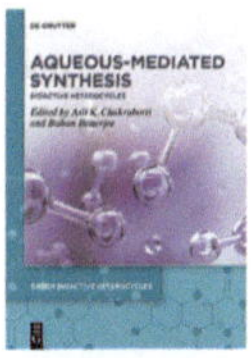

Volume 2
Asit K. Chakraborti, Bubun Banerjee (Eds.)
Aqueous-Mediated Synthesis, 2024
ISBN 978-3-11-099726-2, e-ISBN 978-3-11-098562-7

Volume 3
Yunfei Du, Bubun Banerjee (Eds.)
Non-Metal Catalyzed Synthesis, 2023
ISBN 978-3-11-099728-6, e-ISBN 978-3-11-098547-4

Volume 4
Sreekantha B. Jonnalagadda, Bubun Banerjee (Eds.)
Solvent-Free Synthesis, 2023
ISBN 978-3-11-099730-9, e-ISBN 978-3-11-098546-7

Volume 5
Basudeb Basu, Bubun Banerjee (Eds.)
Multicomponent Synthesis, 2023
ISBN 978-3-11-099733-0, e-ISBN 978-3-11-098531-3

Volume 6
Dilip Kumar Maiti, Bubun Banerjee (Eds.)
Bioactive Three-Membered Heterocycles, 2025
ISBN 978-3-11-138631-7, e-ISBN 978-3-11-138637-9

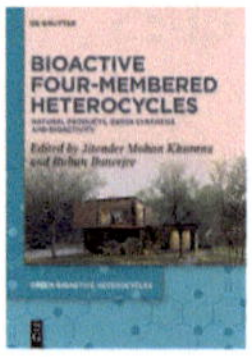

Volume 7
Jitender Mohan Khurana, Bubun Banerjee (Eds.)
Bioactive Four-Membered Heterocycles, 2025
ISBN 978-3-11-138655-3, e-ISBN 978-3-11-138657-7

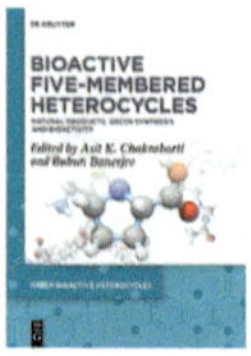

Volume 8
Asit K. Chakraborti, Bubun Banerjee (Eds.)
Bioactive Five-Membered Heterocycles, 2025
ISBN 978-3-11-138716-1, e-ISBN 978-3-11-138752-9

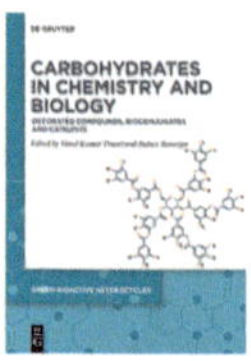

Volume 9
Vinod Kumar Tiwari, Bubun Banerjee (Eds.)
Carbohydrates in Chemistry and Biology, 2025
ISBN 978-3-11-144702-5, e-ISBN 978-3-11-144709-4